Advances in Experimental Medicine and Biology provides a platform for scientific contributions in the main disciplines of the biomedicine and the life sciences. This series publishes thematic volumes on contemporary research in the areas of microbiology, immunology, neurosciences, biochemistry, biomedical engineering, genetics, physiology, and cancer research. Covering emerging topics and techniques in basic and clinical science, it brings together clinicians and researchers from various fields.

Advances in Experimental Medicine and Biology has been publishing exceptional works in the field for over 40 years, and is indexed in SCOPUS, Medline (PubMed), Journal Citation Reports/Science Edition, Science Citation Index Expanded (SciSearch, Web of Science), EMBASE, BIOSIS, Reaxys, EMBiology, the Chemical Abstracts Service (CAS), and Pathway Studio.

2020 Impact Factor: 2.622

Advances in Experimental Medicine and Biology

Volume 1384

Editors
Thomas Penzel
Sleep Medicine Center
Charite Universitätsmedizin Berlin
Berlin, Berlin, Germany

Roberto Hornero
Biomedical Engineering Group
University of Valladolid
Valladolid, Spain

Centro de Investigación Biomédica en
Red Bioingeniería
Biomateriales y Nanomedicina
(CIBER-BBN)
Valladolid, Spain

ISSN 0065-2598 ISSN 2214-8019 (electronic)
Advances in Experimental Medicine and Biology
ISBN 978-3-031-06412-8 ISBN 978-3-031-06413-5 (eBook)
https://doi.org/10.1007/978-3-031-06413-5

This Springer imprint is published by the registered company Springer Nature Switzerland AG
The registered company address is: Gewerbestrasse 11, 6330 Cham, Switzerland

Thomas Penzel • Roberto Hornero
Editors

Advances in the Diagnosis and Treatment of Sleep Apnea

Filling the Gap Between Physicians and Engineers

Springer

Preface

Sleep apnea is a sleep disorder with a very high prevalence and many health consequences. As such it is a major health burden (Benjafield et al., 2019). Sleep apnea has been systematically explored only a little more than 40 years now (Guilleminault & Dement, 1978). Major impacts of sleep apnea are sleepiness and associated risks for accidents (Bonsignore et al., 2021). Major health impacts are cardiovascular risk and pathophysiological traits, even if this is currently much debated when focusing on the apnea-hypopnea index as the measure for sleep apnea severity (Arnaud et al., 2020). Sleep apnea is a disorder which is a chronic condition and can be treated successfully.

The disorders of sleep-disordered breathing have largely supported the growth of sleep medicine in general from a small specialty field to a major spectrum of disorders in the arena of medical specialties. This activity helped to convert the niche field of sleep research into sleep medicine, a clinical discipline with its own departments, its own center certification, physician certification, dedicated conferences, journals, and research activities. The recognition and importance have grown so much that the new International Classification of Disorders by WHO in its 11th version, being launched in 2022, has added a new section on sleep and wake disorders with its own range of codes. This worldwide recognition will enable the growth of medical education on sleep physiology, sleep pathology, and specific sleep disorders.

The diagnostic field for sleep disorders, and for sleep apnea specifically, is strongly linked to the development of new and recent methods, which allow long-term recording and analysis of physiological functions during sleep. Sleep and sleep apnea are not just identified by taking a single blood sample or by a single measurement by a physician at a visit, but sleep recording requires the continuous recording of biosignals. This is comparable to monitoring of vital functions during anesthesia or intensive care. Because of this methodological challenge, biomedical engineering as well as new sensor and analysis technologies are closely linked to the development of sleep apnea diagnosis. New technologies helped to a large extent develop new diagnostic and treatment modalities for sleep-disordered breathing. Sleep apnea diagnostic research is now linked to the development of new wearables, nearables, and smartphone apps, and profits much from the ubiquitous development of photoplethysmography recording everywhere.

Artificial intelligence is playing a very important role in analyzing sleep recordings and, particularly, in automatizing several of the stages of sleep apnea diagnosis. Since the generalization of computerized analysis in the 1990s, automated processing of cardiorespiratory and neuromuscular signals from polysomnographic studies provided a number of indices able to assist sleep experts in the characterization of the disease (Shokoueinejad et al., 2017). Parameterization of the influence of apneic events on biological system dynamics has relied on widely known techniques from the engineering field, such as spectral and nonlinear analysis. Currently, there is a demand for novel alternative metrics able to overcome the limitations of the standard apnea-hypopnea index concerning its low association with patient symptoms and outcomes (Malhotra et al., 2021). In this regard, signal processing and pattern recognition are going to play a key role. In addition, machine learning has also shown its usefulness in the last decades (Uddin et al., 2018) and, like many other areas in our society, sleep apnea diagnosis is rapidly entering the deep learning era (Mostafa et al., 2019) and big data. These new analytical techniques, along with the advances in health device development, are the main hope for reaching a reliable diagnostic paradigm shift. One that finally could cope with the disease prevalence, personalized interventions, and runaway spending.

Beyond the widespread application of machine learning methods to automate polysomnography scoring and to provide sleep experts with tools for automated diagnosis, artificial intelligence has also the potential to significantly improve the management of sleep apnea treatment. Recent advances in the framework of big data together with remote monitoring capability of novel treatment devices are able to promote conventional sleep medicine towards a real personalized medicine. Identification of refined clinical phenotypes of patients will allow the development of precision interventions, enabling the quick identification of the treatment option that best fits the particular characteristics of a patient (Watson & Fernández., 2021). Similarly, machine learning is able to accurately model patient's adherence from usage data (pressure setting, residual respiratory events, mask leaks) derived from portable treatment devices, improving the efficacy of available therapies (Goldstein et al., 2020). Thus, artificial intelligence is going to significantly change the management of sleep apnea treatment in the short term.

This volume gives a basis of current knowledge on sleep research, sleep medicine, and sleep apnea, with a strong focus on new challenges and new research directions in the diagnosis of sleep apnea and its treatment. The volume contains three sections: the first one is on physiology and pathophysiology, the second one is on diagnostic advances, and the third one is on treatment advances. Each chapter author was asked to not only describe the state of the art but also develop visions for future research as seen from their special angle and viewpoint.

As editors, we think that the volume can serve as an introduction to the field of sleep-disordered breathing, can serve as a basis for educating in sleep-disordered breathing, and can immediately stimulate and trigger new research in physiology, clinical trials, and biomedical engineering for sensors and analysis methodologies.

Berlin, Berlin, Germany — Thomas Penzel
Valladolid, Spain — Roberto Hornero

References

Arnaud, C., Bochaton, T., Pépin, J. L., & Belaidi, E. (2020). Obstructive sleep apnoea and cardiovascular consequences: Pathophysiological mechanisms. *Archives of Cardiovascular Disease, 113*:350—358

Benjafield, A. V., Ayas, N. T., Eastwood, P. R., Heinzer, R., Ip, M. S. M., Morrell, M. J., Nunez, C. M., Patel, S. R., Penzel, T., Pepin, J. L. D., Peppard, P. E., Sinha, S., Tufik, S., Valentine, K., & Malhotra, A. (2019). Estimating the global prevalence and burden of obstructive sleep apnoea: a literature-based analysis. *Lancet Respiratory Medicine, 7*:687–698. 10.1016/S2213-2600(19)30198-5

Bonsignore, M.R., Randerath, W., Schiza, S., Verbraecken, J., Elliott, M. W., Riha, R., Barbe, F., Bouloukaki, I., Castrogiovanni, A., Deleanu, O., Goncalves, M., Leger, D., Marrone, O., Penzel, T., Ryan, S., Smyth, D., Teran-Santos, J., Turino, C., McNicholas, W. T. (2021). European Respiratory Society statement on sleep apnoea, sleepiness and driving risk. *European Respiratory Journal, 57*: 2001272 doi: 10.1183/13993003.01272-2020

Goldstein, C. A., Berry, R. B., Kent, D. T., Kristo, D. A., Seixas, A. A., Redline, S., Westover, M. B., Abbasi-Feinberg, F., Aurora, R. N., Carden, K. A., Kirsch, D. B., Malhotra, R. K., Martin, J. L., Olson, E. J., Ramar, K., Rosen, C. L., Rowley, J. A., Shelgikar, A. V. (2020). Artificial intelligence in sleep medicine: An American Academy of Sleep Medicine position statement. *Journal of Clinical Sleep Medicine, 16*(4):605-607. 10.5664/jcsm.8288

Guilleminault, C., & Dement, W. C. (eds) (1978). *Sleep apnea syndromes*. New York: Alan R. Liss Inc.

Malhotra, A., Ayappa, I., Ayas, N., Collop, N., Kirsch, D., Mcardle, N., Mehra, R, Pack, A. I., Punjabi, N., White, D. P., & Gottlieb, D. J. (2021). Metrics of sleep apnea severity: beyond the apnea-hypopnea index. *Sleep, 44*(7):1-16. 10.1093/sleep/zsab030

Mostafa, S. S., et al. (2019). A systematic review of detecting sleep apnea using deep learning. *Sensors, 19.22*: 4934. 10.3390/s19224934

Shokoueinejad, M., Fernandez, C., Carroll, E., Wang, F., Levin, J., Rusk, S., Glattard, N., Mulchrone, A., Zhang, X., Xie, A., Teodorescu, M., Dempsey, J., & Webster, J. (2017). Sleep apnea: a review of diagnostic sensors, algorithms, and therapies. *Physiol Meas, 38*:R204–R252 doi: 10.1088/1361-6579/aa6ec6

Uddin, M. B., Chow, C. M., & S. W. Su. (2018). Classification methods to detect sleep apnea in adults based on respiratory and oximetry signals: A systematic review. *Physiological Measurement, 39.3*: 03TR01. 10.1088/1361-6579/aaafb8

Watson, N. F., & Fernandez, C. R. (2021). Artificial intelligence and sleep: Advancing sleep medicine. *Sleep Med Rev, 59*:101512. 10.1016/j.smrv.2021.101512

Contents

Part I

Physiology

1 An Overview on Sleep Medicine

Alex Iranzo

Abstract

Sleep plays an important role in homeostasis, brain plasticity, clearance of neurotoxins, cognition, memory, concentration, performance, and the regulation of the temperature, endocrine and immunological systems. Insufficient, disorganized, and poor-quality sleep impacts performance, cognition, and safety, carries social and economic consequences and predisposes to obesity, excessive daytime sleepiness, fatigue, arterial hypertension, diabetes, stroke, coronary arterial disease, Alzheimer's disease, depression, and anxiety. Consequently, the search of sleeping well and sufficiently aims to be happy, healthy, and being productive at work, social and family levels. Therefore, one of the fundamental pillars of health is sleeping an adequate number of hours, follow regular sleep-wake habits and identify sleep disorders. There is a wide variety of sleep disorders that may impact the patient quality of life such as obstructive sleep apnea, chronic insomnia, narcolepsy, delayed sleep-wake phase disorder and Kleine-Levin syndrome. The need to study sleep and its disturbances made the appearance of Sleep Medicine. This is a relatively new discipline that was born in the second half of the twentieth century and aims to promote good sleep hygiene and detect and treat those sleep disorders impairing the subject quality of life. Moreover, the field has expanded to fields such as the evaluation of pediatrics, women, aging, shift work, sports, forensic aspects, and its socioeconomic impact.

Keywords

sleep medicine · sleep · wakefulness · sleep habits · sleep habits · sleep-wake cycle · sleep hygiene · sleep disorders · obstructive sleep apnea

1.1 The Origin and Regulation of Sleep

General Considerations on Sleep and Its Biological Importance We spend about one-third of our lives asleep. Animals (e.g., mammals, birds, reptiles) also sleep. But sleeping seems to be a waste of time because while we are asleep, we do not enjoy, we do not work, we do not acquire new information, we do not love, and we do not relate to other people or perform social activities. However, it is obvious that if sleeping served no purpose, it would be a major evolution-

A. Iranzo (✉)
Neurology Service, Hospital Clínic de Barcelona, Barcelona, Spain

University of Barcelona, Idibaps, Ciberned, Barcelona, Spain
e-mail: airanzo@clinic.cat

T. Penzel, R. Hornero (eds.), *Advances in the Diagnosis and Treatment of Sleep Apnea*, Advances in Experimental Medicine and Biology 1384, https://doi.org/10.1007/978-3-031-06413-5_1

ary mistake. Then, why do we sleep and what is the point of sleeping? The answer seems obvious and is based on common sense; we sleep at night to rest, to recover, and to acquire maximum energy and vitality when we wake up each morning to a start a bright new day. We get tired mentally and physically throughout the day, and then we must go back to sleep to rest again. And this occurs every day, repeating itself throughout life in cycles that our brain programs about every 24 hours.

Sleep is a state characterized by the suspension of consciousness, loss of response to external stimuli, and decreased motor activity, adopting a typical posture in humans of lying down with eyes closed. Sleep is a function of the brain. It is a natural state that is necessary, periodic, cyclic, and reversible. It is essential for life, since its absolute deprivation, at least in mammals and rodents, leads to death in less than a few weeks. Although there is increasing information about its organization and functioning, we still do not know its ultimate biological meaning. However, the scientific study of sleep has allowed us to learn that sleep is not a passive state because when we are asleep, the cells, especially the neurons, carry out processes that they do not perform while the subject is awake (Sullivan et al., 2022; Bódizs & European Sleep Research Society, 2021). In other words, neurons perform functions and adopt different roles depending on whether the subject is asleep or awake. Although it may seem that sleep is a passive behavior, the truth is that this state is characterized by a complex network of neurological mechanisms that converge in a state where the energetic and metabolic needs of the brain do not disappear. On the contrary, the metabolic rate in some regions of the brain is even higher than those of wakefulness in some specific stages. Sleep, then, is a dynamic state where neurons remain active, performing functions different from those they perform during wakefulness. It has been suggested that during sleep it is possible to consolidate memory, control body temperature, regulate the immune and endocrine systems, and encode our emotions, temperamental and psychological stability. In addition, the sleep state promotes cortical plasticity, neuronal synapses, receptor replacement, and the clearance of some proteins such as amyloid (Sullivan et al., 2022; Bódizs & European Sleep Research Society, 2021).

Sleep Organization Scientific interest in the function and functioning of sleep has advanced significantly since around 1970 when the first sleep laboratories and medical centers specialized in sleep disorders and research were set up in the United States. We have been able to learn partially how sleep is organized and regulated thanks to information obtained through cellular records, neuroanatomy, laboratory experiments with animals and human healthy volunteers, genetic manipulation in animals, optogenetics, neuroimaging, the development of nocturnal polysomnography, and the study of certain human diseases. Researchers looked for the holy grail by trying to find the part of the brain and the substance that was the regulator of sleep onset and maintenance. They failed to identify a single brain region and a single neurotransmitter system. Scientists found that the situation was much more complex and that there was no single player. Thus, it has been identified that various structures of the nervous system located in the brainstem, thalamus, hypothalamus, basal forebrain, and limbic system are specifically involved in the origin and regulation of sleep. Numerous neurotransmitters are also involved in the origin and maintenance of sleep such as noradrenaline, acetylcholine, serotonin, histamine, dopamine, adenosine, GABA, glutamate, glycine, melatonin, and hypocretin (Adamantidis et al., 2021; Siegel, 2022).

The modulation of the sleep-wake cycle depends on homeostatic factors (the more time awake means the greater the need to sleep, and the more time asleep means the greater the need to be awake), circadian factors (the greatest pressure to fall asleep is during the day around 15:00 and during the night around 3:00, body temperature decreases throughout the night, and the presence of light and darkness), external factors (noise, change of schedule, boredom, anxiety), and individual differences (age, sex, ethnicity, animal species). Humans sleep at night and are awake during the day while rodents sleep during the day and are awake at night. Healthy women tend to have a more continuous and deeper sleep than men. Asians have an easier time falling asleep in common situations (e.g., riding the subway) or are less able to withstand jet lag. Dolphins and other aquatic mammals may have one hemisphere asleep and the other awake (Sullivan et al., 2022; Bódizs & European Sleep Research Society, 2021).

Most sleep data have been obtained by continuous nocturnal recording of three electrophysiological variables: cortical electrical activity (electroencephalogram, EEG), eye movements (electrooculogram, EOG), and muscle activity (electromyogram, EMG). During wakefulness, with eyes closed, the EEG records a mixture of relatively fast moderate-amplitude frequencies designated as beta (13–25 cycles/s or Hz) and alpha (8–12 Hz), the EMG of the mentalis muscle in the chin shows the greatest degree of tonic activity, and the EOG records rapid eye movements and eye blinking (Berry. et al., 2020).

Sleep Stages Sleep is divided into the non-REM stage (without rapid eye movements) and the REM stage (with rapid eye movements). In a young healthy adult, non-REM sleep occupies approximately 75% of the time and REM sleep the remaining 25%. Non-REM sleep is divided into the stages N1 (drowsiness, 10% of the total sleep time), N2 (light sleep, 45% of the time), and N3 (deep sleep, 20% of the time) (Berry. et al., 2020) (Fig. 1.1).

During stage N1, eye movements become slow, EMG activity decreases compared to wakefulness and the EEG shows slow frequencies (4–7 Hz theta) of low amplitude and isolated bursts of higher voltage waves appear over the center of the skull (vertex sharp waves). In stage N2, K-complexes (bursts of slow 1 Hz, high amplitude waves in the central and frontal areas) and sleep spindles (bursts of rhythmic activity at 12–15 Hz) appear. K-complexes have a cortical activation origin. Sleep spindles represent the block in the thalamus of the transmission from sensory impulses to the cortex. As the depth of sleep increases, delta activity (1–3 Hz) of high amplitude is recorded, which, if it occupies more than 20% of the time analyzed, is designated as stage N3. In this period of deep sleep N3, eye movements are undetectable and muscle tone is still present. In REM sleep there are rapid eye movements, disorganized EEG with muscle atonia (Berry. et al., 2020).

The Effect of Age on Sleep In a young healthy adult, sleep begins with several minutes of stage N1, followed by stages N2 and N3. At this point, the sequence is reversed, and after N3 and N2 are recorded again, the first REM sleep stage appears. This first period of REM sleep occurs about 60–100 min after sleep onset. REM sleep is characterized by rapid eye movements like those of wakefulness (although with eyes closed), minimal or absent muscle activity (the subject is practically paralyzed) and the cortical activity seen in the EEG is intermediate between that of wakefulness and stage N1. After this sequence of sleep stages (N1-N2-N3-N2-REM), the process is repeated, alternating 60–90 min of non-REM sleep (N1, N2, and N3 phases) with 15–30 min of REM sleep. N3 sleep predominates in the first third of the night and REM sleep in the last third. Overall, 75% of sleep is non-REM and 25% is REM in the healthy young adult (Berry. et al., 2020).

These proportions change throughout life. Newborns and infants experience longer total sleep time than any other age groups. In newborns, the total daily duration of sleep may be 14 to 16 hours. REM sleep in infants accounts for a higher percentage of total sleep, at the expense of N2 sleep. Newborns move directly from wakefulness to REM sleep until they are 3 to 4 months of age, after which they move directly to non-REM sleep, and then to non-REM sleep. In the elderly, there is an overall decrease in total sleep time compared to adults. In the elderly, the N3 stage decreases by 10% or more, and the N2 stage of sleep increases by 5%. Sleep latency increases, as well as the number and duration of

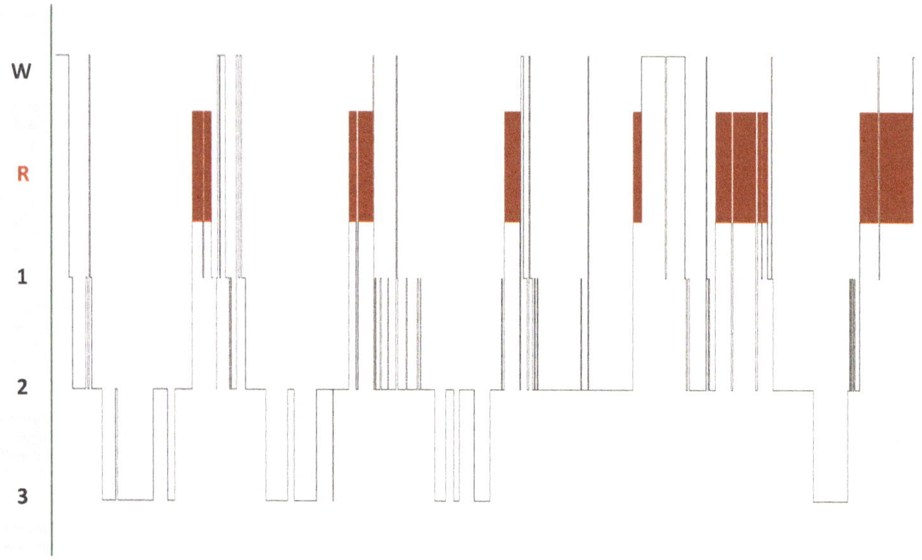

Fig. 1.1 The normal hypnogram of a young adult. W wakefulness, R REM sleep. 1: Stage N1; 2: Stage N2; 3: Stage N3

nocturnal brief awakening periods. There is also an increase in daytime naps, an earlier sleep and awakening phase, and greater intolerance to time changes (Sullivan et al., 2022; Bódizs & European Sleep Research Society, 2021).

The Effect of Medications and Diseases on Sleep Medications can influence sleep architecture; antidepressants decrease the proportion of REM sleep and increase REM sleep latency, while benzodiazepines increase the proportion of light N2 sleep by decreasing deep N3 sleep. It is important to note that in neurodegenerative diseases such as Alzheimer's disease and Parkinson's disease the neurodegenerative process affects the brain structures that regulate sleep. In these diseases the sleep architecture becomes abnormal and typical findings are reduction or absence of both K complexes and sleep spindles, REM sleep without muscular atonia, reduction in total sleep time and in stage N3. In severe cases, the sleep stages are impossible to recognize and a new system to score sleep needs to be defined because the classic (N1, N2, N3, REM stages) cannot be recognized. These are the cases of several dementias (e.g., dementia with Lewy bodies), parkinsonisms (e.g., multiple system atrophy, progressive supranuclear palsy), autoimmune diseases (e.g., anti-IgLON 5 parasomnia), and prion diseases (e.g., fatal familial insomnia). In the extreme form of this situation wakefulness and sleep are almost impossible to recognize, a state that has been termed *status dissociatus*. Besides, in healthy people medications can influence sleep architecture; benzodiazepines increase the proportion of light N2 sleep by decreasing deep N3 sleep, while antidepressants decrease the proportion of REM sleep and increase REM sleep latency. As million of people take antidepressants and subsequently, they have a chronic REM sleep reduction, it is striking to see that these people apparently do not present any (cognitive) deficits due to the loss of REM sleep.

Biological and Circadian Influences on Sleep In addition to changes in EEG, EOG, and EMG, there are an important number of physiological variables that are modified during normal sleep. For example, in the REM phase, temperature control becomes dependent on ambient temperature, unlike in non-REM sleep and wakefulness, where it is independent. In the REM phase, the ventilatory response to CO_2 decreases, blood pressure increases, and cardiac output, respiratory rate, cerebral oxygen consumption, and blood flow in the limbic system, visual associative cortex, and brainstem also increase. In this phase, above all, the dreams that we usually remember upon awakening are produced. Another of the most outstanding characteristics of the REM phase is the paralysis of the voluntary musculature to protect ourselves physically from dream-enacting. Only the diaphragm and the extrinsic ocular muscles partially maintain their usual activity (although brief distal movements of the limbs can be detected, and newborns grimace). This paralysis is due to an inhibition of the spinal motor neurons by the reticular formation nuclei of the brainstem (Sullivan et al., 2022; Bódizs & European Sleep Research Society, 2021).

The 24-hourly rhythmic oscillation of sleep-wake behavior does not depend only on the presence of light and darkness, but is imprinted in our organism, since after suppressing external influences, a similar biological rhythm persists. Many biological functions have rhythmic variations of about 24 hours, including hormone secretion, body temperature, urinary potassium excretion, gastric secretion, and renal acid secretion. Body temperature, for example, has a very stable circadian rhythm of about 24 h, with a minimum in the early morning hours and a maximum in the late afternoon. The period of lowest temperature is usually synchronized with the period of night-time sleep (Sullivan et al., 2022; Bódizs & European Sleep Research Society, 2021).

Neuromodulation of sleep The origin, modulation, and regulation of sleep are very complex, and we still have much to learn. Many neurotransmitters and brain structures are involved (Figs. 1.2 and 1.3). In a very simplified way, we can say that the areas and transmitters that are activated in wakefulness are the posterior and lateral region of the hypothalamus (containing hypocretin/orexin), the dorsal raphe in the pons (containing serotonin), the locus coeruleus in the pons (containing noradrenaline), the pedunculopontine nucleus in the pons (containing acetylcholine), the ventral tegmental area in the midbrain (containing dopamine), and the mammillary tubercles in the midbrain (containing histamine). In wakefulness, all these structures innervate with their neurotransmitters the cerebral cortex which makes it receptive to sensory stimuli and ready to perform motor and cognitive actions. The transition from wakefulness to sleep is made by a gradual flip-flop switch system where the forces of sleep overcome those of wakefulness, as if it were a balance that tipped on the onset of sleep. One of these factors that modulate the origin of sleep is the accumulation of adenosine during wakefulness (which comes from cell catabolism) and the secretion of melatonin released from the pineal gland (or epiphysis) during darkness. The onset of non-REM sleep is determined by influences reaching the preoptic nucleus of the hypothalamus (which contains GABA) after a decrease in the activity of the nuclei that maintained wakefulness (posterior hypothalamus, dorsal raphe, locus coeruleus, pedunculopontine nucleus, ventral tegmental area, and mammillary tubercles) and its transmitters (hypocretin/orexin, serotonin, noradrenaline, acetylcholine, dopamine, and histamine). In non-REM sleep there is a 40% decrease in metabolism with respect to wakefulness, with increased metabolism in the thalamus (sleep spindles) and cortex (K-complexes and delta waves). In REM sleep there is an important decrease of hypocretin, noradrenaline, serotonin, and histamine, with increased activity of acetylcholine, GABA, glutamate, and glycine. During REM sleep the cells of

	Wakefulness	No REM	REM
Hypocretin	⇑ ⇑	⇓	0
Ach	⇑	⇓	⇑ ⇑
NA, Ser, His	⇑ ⇑	⇓	0
GABA	0	⇑ ⇑	⇑ ⇑

Fig. 1.2 Main neurotransmitters and their activity during wakefulness, non-REM sleep, and REM sleep

Neurotransmittters	Neurological regions	Function promoted
Noradrenaline	Locus coeruleus in the pons	Wakefulness (N1 y N2)
Serotonin	Dorsal raphe in the pons	Wakefulness (N1 y N2)
Histamine	Tuberommamilary nucleus (hypothalamus)	Wakefulness (N1 y N2)
Acetylcholine	Pedunculopontine nucleus (pons), Meynert nucleus (Basal prosencephalon), amygdala	Wakefulness and REM
GABA	Ventrolateral preoptic nucleus (hypotahalamus), locus subcoeruleus (pons), lateral dorsal pontin area (pons) magnocelularis nucleus (bulb)	N1, N2, N3 and REM
Glutamate	Locus subcoeruleus (pons), amygdala, cortex	Wakefulness and REM
Hypocretin/orexin	Posterior hypothalamus	Wakefulness (N1, N2, N3)
Melatonin	Epiphysis, suprachiasmtic nucleus (Hypothalamus)	N1, N2, N3, REM

Fig. 1.3 Neurotransmitters and brain areas involved in wakefulness, non-REM sleep, and REM sleep

the nucleus subcoeruleus of the pons (containing GABA) stimulate those of the nucleus magnocellularis of the medulla (containing glycine) and these inhibit the motor neurons of the anterior horn of the spinal cord resulting in muscle paralysis (Fig. 1.4). Other cells of the nucleus subcoeruleus (containing glutamate) activate the occipital cortex during REM sleep. The amygdala and hippocampus (containing glutamate and acetylcholine) are also active in REM sleep and they regulate emotions, mood, and memory. In REM sleep there is hyperactivation of the brainstem, limbic system, and occipital cortex, while there is inhibition of the frontal cortex. This explains why in REM sleep, there is muscular paralysis, and experienced dreams have visual components and emotional content. Darkness stimulates the suprachiasmatic nucleus of the hypothalamus which sends a signal to the epiphysis to secrete melatonin. Melatonin will synchronize the biological rhythms of our organism including the endocrine, hormonal, immunological, and neuronal activity. Therefore, the suprachiasmatic nucleus is called the circadian clock. (Sullivan et al., 2022; Bódizs & European Sleep Research Society, 2021; Adamantidis et al., 2021; Siegel, 2022).

Still, there are many open issues to be covered. They include the ultimate role of sleep, the reason why the structure of sleep and its stages are so different between different animals and different species, the function of dreams and dream content, the role of REM sleep when many people take antidepressants which are drugs that reduce the quantity of REM sleep, to learn the most important brain regions and neurotransmitter systems that modulate sleep to implement therapeutic targets, and to know why among humans, there are people and groups prone to daytime sleepiness and others not.

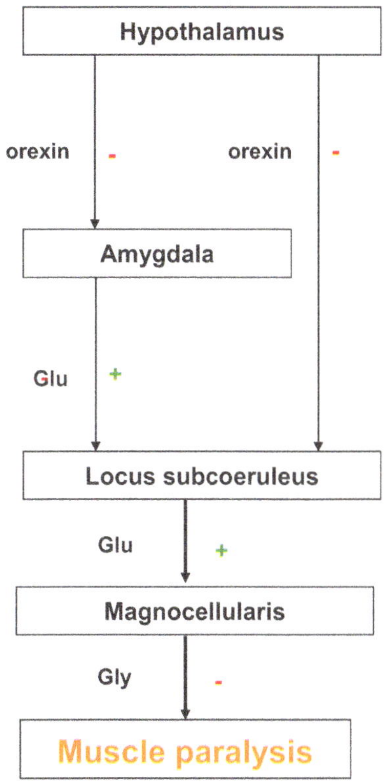

Fig. 1.4 Modulation of sleep atonia in REM sleep. Glu glutamate, Gly glycine. Green cross indicates activation. Red mark indicates inhibition

1.2 Sleep Impairment

The importance of sleep is reflected in two distinct situations which lead to impairment and negative consequences: sleep deprivation (when the amount of sleep is not sufficient) and sleep disorders (when the systems that regulate sleep cease to function properly).

Loss of Sleep Quantity In adults, a continuous sleep of 7–8 hours is restorative. In some cultures, total sleep may be divided into a 7-hour night period and a physiological mid-afternoon nap of half an hour. Sleep in the industrialized world is in chronic deficit, due in part to evening light exposure, which delays sleep onset and truncates sleep depending on morning work or school schedules. In societies with electricity that live modern lifestyles, sleep onset is delayed, and nocturnal sleep duration is reduced. The frenetic pace of social life and economic demands has changed our lifestyle habits (Banks et al., 2022a; Holst et al., 2021). In the United States, the percentage of individuals between 45 and 64 years of age who said they slept less than 7 hours at night was 23% in 1985 and 33% in 2004. The same was true for the younger and older age groups; in individuals between 18 and 29 years of age, the percentage of subjects who slept less than 7 hours in 1985 and 2004 increased from 22% to 26%, and in subjects between 65 and 74 years of age it increased from 15% to 21%. Thus, we are facing a reduction of total sleep time imposed by the lifestyle we have chosen. This was very different in the time of Sherlock Holmes, whose creator, Sir Arthur Conan Doyle, slept enough hours at night to lead an orderly life, and thus be able to be fresh in the morning and create authentic masterpieces of literature. However, other literary genius such as Kafka and Dickens chose to write at night, stealing hours of their sleep.

Accumulative data obtained from animal and human experiments have shown that sleep deprivation is associated with poorer cognitive performance, poorer working memory, poorer attention, poorer concentration, increased daytime sleepiness, increased anxiety and irritability, low self-esteem and emotional tone, fatigue, decreased libido and reduced creativity and productivity (Banks et al., 2022a). This situation leads to errors (wrong decisions, wrong planning), personal conflicts, traffic accidents, economic consequences, abuse of stimulants, sedatives, hypnotics, alcohol, drugs, and predisposition to certain diseases. Insufficient or poor-quality sleep at night has been associated with an imbalance of some substances that regulate mood, appetite, and weight such as orexin, leptin, ghrelin, insulin, and cortisol. This predisposes to diseases such as obesity, arterial hypertension, type 2 diabetes mellitus, myocardial infarction, stroke, depression, some types of cancer, Alzheimer's disease, and infections. In short, little, or fragmented sleep leads to poorer health and economic productivity (Banks et al., 2022a; Holst et al., 2021).

Loss of Sleep Quality Not only may be affected by reduction of total sleep time but also by poor sleep quality due to sleep disorders. Sleep disorders can be classified into four major groups: (1) disorders associated with difficulty in falling and maintaining sleep (e.g., chronic insomnia, restless legs syndrome); (2) disorders associated with excessive daytime sleepiness (e.g., insufficient nigh time sleep, narcolepsy, obstructive sleep apnea, idiopathic hypersomnia); (3) disorders characterized by abnormal sleep behaviors or parasomnias (e.g., sleepwalking, night terrors, REM sleep behavior disorder), and (4) sleep-wake rhythm disorders, e.g. (jet lag, shift work, delayed sleep phase syndrome, advanced sleep phase syndrome). The study of these disorders and the evaluation of the patients that suffer from them has shown the same consequences in health and social and economic life that I have described in the situation of sleep deprivation.

The causes of these disorders are multiple and consist of a genetic predisposition (e.g., polymorphisms in restless legs syndrome, HLA haplotypes related to sleepwalking and narcolepsy, gene mutations in some circadian disorders), an anatomical defect (e.g., narrowing of the upper airway in obstructive sleep apnea), damage of brain structures that regulate sleep (e.g., the suprachiasmatic nucleus in Alzheimer's disease, the nucleus subcoeruleus in REM sleep behavior disorder), deficiency of neurotransmitters (e.g., hypocretin in narcolepsy), deficiency of substances (e.g., iron in restless legs syndrome), a predisposing personality (e.g., insomnia), poorly acquired habits (e.g., insomnia, circadian rhythm disorder), psychiatric disorders such as anxiety and depression that induce insomnia or hypersomnia, other symptoms that may hinder the onset and maintenance of sleep (pain, muscle stiffness, cardiac and respiratory failure) and the effect of some drugs that cause insomnia, hypersomnia, sleep-related eating syndrome, and nightmares (Malow, 2022; American Academy of Sleep Medicine, 2014).

Sleep disorders are very common in the general population, affecting approximately 25% of individuals (Banks et al., 2022b; Malow, 2022; American Academy of Sleep Medicine, 2014). They affect children, adults, and the elderly. The most frequent disorder is insomnia, both acute and chronic forms. Snoring, obstructive apneas, sleepwalking, sleep paralysis and restless legs syndrome are also very frequent, each of them separately affecting up to 5% of the population in a transient or chronic manner. Other less frequent disorders such as idiopathic REM sleep behavior disorder, nocturnal eating syndrome, catathrenia, narcolepsy, anti-IgLON 5 disease and fatal familial insomnia are less frequent, but it is important to know about them because of their importance. From the point of view of gender, we should also know that pregnancy and menopause are associated with specific sleep disorders in women, and that in men nocturia due to prostate problems is a cause of sleep fragmentation. From the neurological point of view, we should know that some autoimmune diseases (anti-IgLON5 disease, multiple sclerosis), neurodegenerative diseases (Parkinson's disease, multiple system atrophy, Alzheimer's disease) and neoplastic diseases (tumors of the central nervous system, paraneoplastic syndromes) are frequently associated with sleep disorders and that these can be a frequent cause of consultation with the neurologist. Psychiatric conditions such as depression, anxiety, bipolar disorder, autism, attention deficit hyperactivity disorder, and posttraumatic stress disorder are usually associated with sleep disorders such as insomnia, changes in sleep, pattern, excessive daytime sleepiness, and nightmares.

1.3 Sleep Medicine

The knowledge that sleep deprivation and sleep disorders are harmful to the health of the individual led to the birth of sleep medicine in the second half of the twentieth century; the discipline that studies the origin, characteristics diagnosis, and therapy of sleep disturbances and disorders.

Sleep as a Medical Discipline The fact that people began to visit the doctor seeking medical advice for unrefreshed sleep, insomnia, excessive daytime sleepiness, or abnormal sleep behaviors led to the establishment of Sleep Medicine Centers in the United States, Japan, and some European countries by the 1970s (Pelayo & Dement, 2022). These centers are composed by medical staff, technicians, and beds to attend to patients and perform sleep studies at night. Physicians, psychologists, nurses, technicians, dentists, and biologists began to take interest in sleep and its disorders and to specialize in sleep medicine (Cirignotta et al., 2021; Dogas et al., 2021). Organizations devoted to sleep medicine such as the American Academy of Sleep Medicine, Sleep Research Society, World Association of Sleep Medicine, World Federation of Sleep Research, World Sleep Society, European Sleep Research Society, Asian Sleep Research Society, Australian Sleep Association, Canadian Sleep Society, and Federation of Latin American Sleep Societies were founded. Thus, the figure of the sleep expert or sleep doctor was born but training and official certification was needed not only for physicians but also for technicians and dentists (Cirignotta et al., 2021; Dogas et al., 2021). Sleep expert is now an official title accredited by societies such as the European Research Society or the Spanish Sleep Society. These sleep experts came from various medical specialties such as neurology, pulmonology, internal medicine, psychiatry, surgery, and otorhinolaryngology. They realized that the patient with a sleep problem could benefit from several of these experts from different medical specialties working together. This is how the concept of the multidisciplinary sleep center was born. In addition, several journals devoted to sleep exist including Sleep, Sleep Medicine, Journal of Sleep Research, Journal of Clinical Sleep Medicine and Sleep Breathing (Pelayo & Dement, 2022).

Sleep as an Important Factor of Public Health The correct quality and quantity of sleep has become a public health problem, besides having social and economic repercussions (Upender, 2022). This has led to a growing interest in obtaining the right quality of sleep (along with a healthy diet, exercise, and proper mental health). This is executed in social programs, documentaries, books for the general population, conferences, and debates. The idea is that institutions and health programs, advised by physicians and sleep experts, promote quality sleep as a therapeutic and wellness weapon. Herein, I show three different examples of how sleep medicine is recognized as an important factor in performance and quality of life.

One example is how good sleep quality and quantity may impact on a well performance in professional sports such as football (soccer). In addition to the coach, the soccer teams incorporated physical trainers, cooks, dieticians, and psychologists into their staff. Now important local teams and national teams have started to look for experts in sleep medicine to teach the players how to sleep well and sufficient time before playing a match or starting an important tournament such as the World Cup. Hotels or resorts where players stay before the game should be selected properly avoiding a noisy placement and distance from supporters. Achieving a good night's sleep, the night before the game is important, and it is said that this was one of the several factors why the Netherlands lost to Germany the 1974 World Cup final in Munich. In addition, when tournaments are held on another continent, it is advisable for teams to travel a few weeks or days before the first match to mitigate the impact of jet lag. Teams know that in these situations the advice of a sleep expert can help their players' performance. In these cases, behavioral sleep medicine and proper sleep habits are essential, especially when most medications used as sleep inducers are banned and are not recommended to be taken chronically. Institutions such as the Fútbol Club Barcelona have understood very well this situation and are investigating the quality and quantity of sleep among lower categories in male and female players between 7 and 25 years of age (Merayo et al., 2021).

Another example is the campaign that exits to incorporate a permanent winter time with more daylight in the mornings to promote health and higher performance, particularly in children

(Czeisler & Buxton, 2022). Researchers, institutions, and Sleep Societies, including the Spanish Sleep Society, advise that the most convenient for health is that there should be a stable timetable without changes during the year, and to permanently maintain winter time (GMT + 1). This provides greater exposure to sunlight during the most common work and school hours (from eight in the morning to five in the afternoon), especially in the early hours of the morning. This position is supported by scientific studies that show that winter time (1) promotes a more stable biological rhythm than summer time, (2) improves intellectual performance, and (3) helps to reduce the onset of diseases such as cardiovascular disease, obesity, insomnia, and depression. Winter time would be the most beneficial for the population, especially for the groups most sensitive to time changes (besides those suffering from sleep and health disorders) such as children and the elderly. However, there is a thorny, long, and winding road to achieve this change, which is full of difficulties and obstacles, such as the social and economic impact it would entail (Czeisler & Buxton, 2022).

Finally, the coronavirus disease pandemic (COVID-19) has also put sleep medicine in check. The lockdown situation has exposed the individual to social isolation, fear, anxiety, depressive symptoms, changes in sleep pattern, insomnia, and nightmares (Kryger & Goldstein, 2022). This emphasizes the importance of mental health and sleep health interventions in unexpected situations such as social isolation and posttraumatic stress disorder. The lockdown led to expansion of telemedicine, a tool that many sleep doctors have learnt to use since the pandemia while following patients with insomnia, obstructive sleep apnea, sleepwalking, and restless legs syndrome.

Sleep Management From the clinical point of view, the diagnosis approach is usually made by clinical history, specific sleep scales, and sleep studies such as polysomnography, the multiple sleep latency test, home sleep testing, and actigraphy (Grote et al., 2021; Mathis et al., 2021; Penzel, 2022). The new sleep medicine was born when the physician had to learn to inquire about sleep habits and to learn, understand, and recognize a wide variety of sleep complaints and sleep disorders (Grote et al., 2021). Doctors learned that a good medical history is the first step to approach an individual who consults for a sleep problem. It should always be done with the patient present and if possible, with an observer who is able to corroborate or identify alterations during sleep of which the patient himself/herself may not be aware while is asleep. The clinical sleep history consists first of asking about sleep habits: what he/she does before going to bed, what time he/she goes to bed, how long it takes to fall asleep, what he/she does before falling asleep, what he/she does to fall asleep, the occurrence of noises or movements of the bed partner, the use of a cell phone or television or radio or book in bed before falling asleep, the postures adopted when sleeping, how often the patient wakes up at night, what he/she does when waking up in the middle of the night, how long it takes to fall back to sleep, the presence of enuresis and nocturia, the presence of an alarm clock or a watch where the subject may look at in the idle of the night, the time of last awakening, if the last awakening is spontaneous or with alarm clock, work shifts, variations of schedules between working days and holidays, variations of habits between working and holiday periods. Physicians also have to identify sleep disorders asking about issues such as the difficulty encountered in falling asleep or staying asleep, whether on waking up in the morning the sleep has been restful, whether the patient wakes up with a dry mouth, whether he/she takes naps and their duration, whether there is excessive daytime sleepiness, snoring and witnessed apneas, the content of dreams and nightmares, whether there is discomfort in the legs, whether he/she has fallen out of bed and whether nocturia is a problem to be taken into account. It has also to pay attention at the type of medication the subject takes, if any. It is also important to ask partners, relatives, and caregivers if they agree with the patients' impression of their sleep quality. People close to the patient can tell us about situations that the patients do not recognize or are not aware such as hypersomnia, snoring, apneas, or abnormal behaviors during their sleep. Physicians also need to pay attention to family history and past medical history. It is important to note that the diagnoses of restless legs syndrome, insomnia, and snoring are made by clinical history and no sleep scales or sleep studies are needed to make their diagnoses (Grote et al., 2021).

Sleep scales and questionnaires were developed for screening purposes to identify sleep disorders and sleep symptoms, and to evaluate their magnitude and impact. For example, the Pittsburgh Sleep Quality Index is a widely used instrument that evaluates the subject quality of sleep, sleep habits, and sleep disturbances. The Insomnia Severity Index evaluates the severity of sleep initiation, sleep maintenance, and early awakening. Excessive Daytime Sleepiness can be evaluated by the Epworth Sleepiness Scale, the Stanford Sleepiness Scale, and the Karolinska Sleepiness Scale, that were designed for the public. As screening, obstructive sleep apnea may be suggested by instruments such as STOP-BANG and the Berlin Questionnaire. Formal diagnostic criteria for restless legs syndrome were developed by the International Restless Legs Syndrome Study Group. This group also developed a rating scale as a measurement instrument for assessing severity of restless legs syndrome's symptoms. This scale can be used to monitor the effect of a medication for the symptomatology of the syndrome. The diagnosis of REM sleep behavior disorder requires video-polysomnography. There are several screening questionnaires and single questions for the screening of this parasomnia. However, the specificity of these instruments is low, making video-polysomnography the gold standard for the diagnosis of REM sleep behavior disorder. Besides, the third edition of the book International Classification of Sleep Disorders, published in 2014 by the American Academy of Sleep Medicine, contains the diagnostic criteria of most of the sleep disorders.

The sleep expert has also investigated the sleep disorders, and the scientific advances in the field during the last 40 years have been very important. Advances have been in nearly all fields; clinical (e.g., the description of different disorders such as cathatrenia, sexsomnia, REM sleep behavior disorder), genetic (e.g., the association of periodic leg movements in sleep to polymorphisms in the gene MEIS1), immunologic (e.g., the association of narcolepsy linked to some adjuvants of vaccines), biological (e.g., the discovery that narcolepsy was linked to hypocretin deficit), imaging (e.g., thalamic abnormalities in Klein-Levine syndrome), technical (home sleep studies) and therapeutic (e.g., mandibular advancement devices). Here I give three examples of (1) how a new disease has been identified in patients who first sought medical advice because of sleep problems, (2) how a sleep disorder may be the first manifestation of a neurodegenerative disease, and (3) how a sleep disorder is recognized as a cardiovascular risk factor.

Anti-IgLON5 disease is a novel neurological disease initially described in 2014 that affects the adult (Sabater et al., 2014). As in fatal familial insomnia, the cases described in the seminal description of the disease initially sought medical advice in a sleep center complaining of restless and unrefreshed sleep, nocturia, abnormal behaviors during sleep and mild sleepiness. Polysomnography showed abnormal sleep architecture where conventional stages were difficult to recognize, obstructive sleep apnea, stridor, purposeful behaviors in non-REM sleep and REM sleep behavior disorder. This unique sleep pattern has been well characterized. Later, it has been shown that the first manifestation may also consist in abnormal gait, cognitive impairment, or bulbar symptoms such as dysphagia. The disease is identified when the serum and cerebrospinal fluid contains antibodies against the neuronal protein IgLON5. There is a strong HLA association, absence of coexistent autoimmune disorders, neoplasms, and neurodegenerative diseases. Neuropathology shows a unique pattern defined by tau deposits in the brainstem and hypothalamus impairing some nuclei that regulate sleep. The challenge is to find an effective therapeutic approach since conventional immunotherapies (e.g., steroids, rituximab, immunoglobulins) are usually ineffective and the disease seems to be disabling and progressive.

Isolated REM sleep behavior disorder is a condition that affects people over 50 years of age which is manifested by unpleasant dreams (e.g., being attacked or chased) and vigorous behaviors during sleep (e.g., punching, screaming) that may result in injuries (Iranzo et al., 2016). In this condition, polysomnography shows REM sleep with increased muscle activity linked to abnormal behaviors. The formal description in the medical literature of this parasomnia was made in 1986, but 10 years later the same group of investigators showed that patients with this REM sleep abnormality develop dementia and parkinsonism fulfilling the diagnostic criteria of dementia with Lewy bodies, Parkinson's disease, and multiple system atrophy. As these three neurodegenerative diseases are originated by abnormal deposits in the nervous system of the protein synuclein, sleep experts looked for this protein in subjects with IRBD and this was found in the cerebrospinal fluid and peripheral organs (colon, salivary glands, and skin). This indicates that IRBD is in most of the cases the first manifestation of a synucleinopathy. The challenge is to implement a neuroprotective strategy in these patients to stop the neurodegenerative process and avoid the appearance of dementia and motor abnormalities. Today, development of a neuroprotective therapy is an unmet need in IRBD.

Obstructive sleep apnea is recognized to be one of the most common sleep disorders affecting about 2–4% of the adult population. It is caused by repetitive episodes of upper way collapse during sleep which are associated with arousals that produce sleep fragmentation and oxyhemoglobin desaturation. From the clinical point of view it is usually linked to snoring, unrefreshed sleep upon awakening, and excessive daytime sleepiness. The first thing that attracted attention in this syndrome is that excessive daytime sleepiness could be severe and affect the social, family, and marital life of the patients. Moreover, it was highlighted that obstructive sleep apnea predisposed to road accidents as the patients had an important tendency to fall asleep at the wheel. However, follow-up of these patients and extensive research has shown that obstructive sleep apnea impacts not only performance and safety but is recognized as a condition that is associated with several cardiovascular factors including obesity, hypertension, arrhythmias, congestive heart failure, diabetes plus vascular diseases such as stroke and coronary artery disease (Randerath et al., 2018). Obstructive sleep apnea belongs to the metabolic syndrome. The link between obstructive sleep apnea and cancer is also a subject of recent investigations.

1.4 Future Directions

Sleep Medicine is a very wide and new field. Based on ongoing research we can expect several developments in the diagnosis and management of sleep disorders. They include the following:

- New technology is required to simplify the access to polysomnography and easier and faster ways to score sleep.
- The implementation of artificial intelligence, telemedicine, new devices and instruments, cell phone applications, development of new medications and neuroprotective strategies are promising issues under development in the field of sleep medicine.
- There is a need to reach the general population to teach them how important is sleep. Governmental social programs should include the implementation of good sleep quality and quantity to promote health, in association with other aspects such as adequate exercise and diet and bad habits such as smoking and alcoholism.
- To find an effective treatment to manage disorders such as insomnia without the need of medications. When medications are needed, they should target the physiopathological basis of the disorder (e.g., hypocretin agonists in narcolepsy).

References

Sullivan, S. S., Carskadon, M. A., Dement, W. C., & Jackson, C. (2022). Normal human sleep: an overview. In M. Kryger, T. Roth, C. A. Goldstein, & W. C. Dement (Eds.), *Principles and practice of sleep medicine* (pp. 16–26). Elsevier.

Bódizs R. Theories of the functions of sleep. In: Sleep medicine textbook. European Sleep Research Society. Eds: Bassetti C, McNicholas W, Paunio T, Peigneux P. European Sleep Research Society 2021:41–55.

Adamantidis A, Fort P, Luppi HV. The neurophysiology and neurobiology of wakefulness and non-Rapid eye movement (NREM) sleep. In: Sleep medicine textbook. European Sleep Research Society. Eds: Bassetti C, McNicholas W, Paunio T, Peigneux P. European Sleep Research Society 2021:7–14.

Siegel, J. (2022). Rapid eye movement sleep control and function. In M. Kryger, T. Roth, C. A. Goldstein, & W. C. Dement (Eds.), *Principles and practice of sleep medicine* (pp. 78–85). Elsevier.

Berry. R. B., Quan, S. F., & Abreu, A. R., et al. for the American Academy of Sleep Medicine. (2020). *The AASM Manual for the scoring of sleep and associated events: Rules, terminology, and technical specifications. Version 2.6.* Darien; American Academy of Sleep Medicine.

Banks, S., Dorrian, J., Basner, M., Kaizi-Lutu, M., & Dinges, D. F. (2022a). Sleep deprivation. In M. Kryger, T. Roth, C. A. Goldstein, & W. C. Dement (Eds.), *Principles and practice of sleep medicine* (pp. 47–51). Elsevier.

Holst SC, Landolt HP. The neurophysiology and neurobiology of wakefulness and non-Rapid eye movement (NREM) sleep. In: Sleep medicine textbook. European Sleep Research Society. Eds: Bassetti C, McNicholas W, Paunio T, Peigneux P. European Sleep Research Society 2021:67–80.

Banks, S., Dorrian, J., Basner, M., Kaizi-Lutu, M., & Dinges, D. F. (2022b). Sleep deprivation. In M. Kryger, T. Roth, C. A. Goldstein, & W. C. Dement (Eds.), *Principles and practice of sleep medicine* (pp. 47–51). Elsevier.

Malow, B. A. (2022). Approach to the patient with disordered sleep. In M. Kryger, T. Roth, C. A. Goldstein, & W. C. Dement (Eds.), *Principles and practice of sleep medicine* (pp. 622–624). Elsevier.

American Academy of Sleep Medicine (Ed.). (2014). *International Classification of Sleep Disorders* (3rd ed.). American Academy of Sleep Medicine.

Pelayo, R., & Dement, W. C. (2022). History of sleep physiology and medicine. In M. Kryger, T. Roth, C. A. Goldstein, & W. C. Dement (Eds.), *Principles and practice of sleep medicine* (pp. 3–15). Elsevier.

Cirignotta F, Moldini S, Dogas Z. Organization of Sleep medicine Centers. In: Sleep medicine textbook. European Sleep Research Society. Eds: Bassetti C, McNicholas W, Paunio T, Peigneux P. European Sleep Research Society 2021:1039–1048.

Dogas Z, Schiza SE. Training initiatives in sleep medicine. In: Sleep medicine textbook. European Sleep Research Society. Eds: Bassetti C, McNicholas W, Paunio T, Peigneux P. European Sleep Research Society 2021:1039–1048.

Upender, R. P. (2022). Sleep medicine, public policy and public health. In M. Kryger, T. Roth, C. A. Goldstein, & W. C. Dement (Eds.), *Principles and practice of sleep medicine* (pp. 699–707). Elsevier.

Merayo, A., Gallego, J. M., Sans, O., Iranzo, A., Sugimoto, D., & Rodas, G. (2021). Quantity and quality of sleep in young players of a professional football club. *Science and Medicine in Football*. https://doi.org/10.1080/24733938.2021.1962541

Czeisler, C. A., & Buxton, O. M. (2022). Human circadian timing system and sleep-wake regulation. In M. Kryger, T. Roth, C. A. Goldstein, & W. C. Dement (Eds.), *Principles and practice of sleep medicine* (pp. 375–389). Elsevier.

Kryger, M., & Goldstein, C. A. (2022). COVID-19 and sleep. In M. Kryger, T. Roth, C. A. Goldstein, & W. C. Dement (Eds.), *Principles and practice of sleep medicine* (pp. 1999–2009). Elsevier.

Grote L, Puertas FJ. The clinical interview and clinical examination In: Sleep medicine textbook. European Sleep Research Society. Eds: Bassetti C, McNicholas W, Paunio T, Peigneux P. European Sleep Research Society 2021:167–180.

Mathis J, de Lacy S, Roth C, Hill EA. Measuring: monitoring sleep and wakefulness. In: Sleep medicine textbook. European Sleep Research Society. Eds: Bassetti C, McNicholas W, Paunio T, Peigneux P. European Sleep Research Society 2021:181–200.

Penzel, T. (2022). Home sleep testing for sleep-related breathing disorders. In M. Kryger, T. Roth, C. A. Goldstein, & W. C. Dement (Eds.), *Principles and practice of sleep medicine* (pp. 1925–1929). Elsevier.

Sabater, L., Gaig, C., Gelpi, E., Bataller, L., Lewerenz, J., Torres-Vega, E., Contreras, A., Giometto, B., Compta, Y., Embid, C., Vilaseca, I., Iranzo, A., Santamaria, J., Dalmau, J., & Graus, F. (2014). A novel non-rapid-eye movement and rapid-eye-movement parasomnia with sleep breathing disorder associated with antibodies to Iglon5: An observational study: a case series, characterisation of the antigen, and postmortem study. *Lancet Neurology, 13*, 575–586.

Iranzo, A., Santamaria, J., & Tolosa, E. (2016). Idiopathic rapid eye movement sleep behaviour disorder: diagnosis, management, and the need for neuroprotective interventions. *Lancet Neurology, 15*, 405–419.

Randerath, W., Bassetti, C. L., Bonsignore, M. R., et al. (2018). Challenges and perspectives in obstructive sleep apnoea: Report by an ad hoc working group of the Sleep Disordered Breathing Group of the European Respiratory Society and the European Sleep Research Society. *The European Respiratory Journal, 52*(3), 1702616. https://doi.org/10.1183/13993003.02616-2017

2 Covering the Gap Between Sleep and Cognition – Mechanisms and Clinical Examples

Javier Gomez-Pilar, Gonzalo C. Gutiérrez-Tobal, and Roberto Hornero

Abstract

A growing number of studies have shown the strong relationship between sleep and different cognitive processes, especially those that involve memory consolidation. Traditionally, these processes were attributed to mechanisms related to the macroarchitecture of sleep, as sleep cycles or the duration of specific stages, such as the REM stage. More recently, the relationship between different cognitive traits and specific waves (sleep spindles or slow oscillations) has been studied. We here present the most important physiological processes induced by sleep, with particular focus on brain electrophysiology. In addition, recent and classical literature were reviewed to cover the gap between sleep and cognition, while illustrating this relationship by means of clinical examples. Finally, we propose that future studies may focus not only on analyzing specific waves, but also on the relationship between their characteristics as potential biomarkers for multiple diseases.

Keywords

Sleep · Cognition · Sleep spindles · Slow oscillations · Slow waves

2.1 Why We Need to Sleep?

Surprisingly, after decades of research, there is still no consensus or a clear answer to this question. This is probably not due to a lack of knowledge of the sleep functions, but to the number of functions it performs both for the brain and for the whole body. Finally, after multiple studies, we are now able to understand some of them.

J. Gomez-Pilar (✉) · G. C. Gutiérrez-Tobal
Biomedical Engineering Group, University of Valladolid, Valladolid, Spain

Centro de Investigación Biomédica en Red en Bioingeniería, Biomateriales y Nanomedicina (CIBER-BBN), Valladolid, Spain
e-mail: javier.gomez@gib.tel.uva.es

R. Hornero
Biomedical Engineering Group, University of Valladolid, Valladolid, Spain

Centro de Investigación Biomédica en Red en Bioingeniería, Biomateriales y Nanomedicina (CIBER-BBN), Valladolid, Spain

Mathematics Research Institute of the University of Valladolid (IMUVa), Valladolid, Spain

T. Penzel, R. Hornero (eds.), *Advances in the Diagnosis and Treatment of Sleep Apnea*, Advances in Experimental Medicine and Biology 1384, https://doi.org/10.1007/978-3-031-06413-5_2

More than 40 years ago, the famous researcher Allan Rechtschaffen, accepted that sleep functions should be of unquestionable utility, since "if sleep does not serve an absolutely vital function, then it is the biggest mistake the evolutionary process has ever made". Although Rechtschaffen's intuition was correct, he probably did not imagine the number of functions of sleep, which include the elimination of toxins (Xie et al., 2013), regulation of glucose level (Van Cauter et al., 2008) and endocrine functions (van Cauter et al., 2007), stimulation of immune function (Ganz, 2012), modulation of emotional brain processes (Walker, 2009), or reinforcement of learning and memory mechanisms (Antony et al., 2019; Fang et al., 2019; Fernandez & Lüthi, 2020; Schabus et al., 2004), among others (see the review from Assefa and colleagues (ZAssefa et al., 2015) for different theories of the sleep functions).

In this chapter, we are interested in addressing one of these sleep functions, in particular delving into the proven relationship between sleep and cognition across the lifespan (Murawski et al., 2018; Ohayon et al., 2004; Reynaud et al., 2018; Yaffe et al., 2014), as well as its link with a large number of diverse pathologies (Ferrarelli & Tononi, 2017; Gutiérrez-Tobal et al., 2021; Vgontzas & Pavlović, 2018; Weng et al., 2020). Accordingly, it is essential to mention sleep spindles as a mechanism that plays a central role in cognitive processes, such as memory consolidation (Fogel, Albouy, et al., 2017; Fogel & Smith, 2011; Fogel, Vien, et al., 2017). Therefore, our main aim is to provide a synthesis of the role of sleep, with special focus on sleep spindles, and the relationship between sleep abnormalities and diverse pathologies.

2.2 Sleep Electrophysiology

2.2.1 Acquisition of the Electroencephalogram

The usual way to acquire the neuronal electrical signal is the use of the electroencephalogram (EEG). The equipment usually used is between 8 and 64 channels, although there are already systems with more than 1000 electrodes (ref). The sampling frequency depends on the equipment but is usually not less than 128 Hz or more than 1000 Hz. Although these are the usual characteristics, the acquisition of the EEG during sleep is usually performed in specialized Sleep Units, where many other signs are usually acquired, such as those from a polysomnography (PSG) (Jafari & Mohsenin, 2010).

Given the great variability of acquisition characteristics, the American Academy of Sleep Medicine (AASM) suggests minimum characteristics for EEG acquisition during sleep (Iber et al., 2007). Among them, they recommend a desirable sampling rate of 500 Hz, establishing the minimum into 200 Hz. In this way, according to Nyquist's theorem, it is possible to analyze frequencies up to 100 Hz. However, for clinical utility a high-frequency filter of 35 Hz is also recommended. Additionally, electrode impedance must keep under 5 KΩ and the minimum resolution should be 12 bits per sample.

2.2.2 Sleep Stages and the Cyclical Sleep

Sleep is far from uniform. Conversely, it is essentially cyclical, with cycles lasting about 90 minutes on average. However, the duration of each cycle is highly variable, increasing its duration throughout the night (Březinová, 1974). During a typical 8-hour restful sleep, there are usually between four to six cycles chained in a row (Keenan, 1999). Within these cycles, there are different stages of sleep that, according to the latest version of the AASM guide (Iber et al., 2007), are divided into two main periods: rapid eye movement (REM) and non-rapid eye movement (NREM). While REM stage is not divided into other subphases, NREM, in turn, consists of three different stages: N1, N2, and N3.

It is known that the duration of these sleep stages is not constant with age. In particular, as we get older, there is an increasing percentage of sleep in N1 and N2 stages, while the percentage of time in N3 and REM is decreased, resulting in less restful sleep and, sometimes, increased age-related cognitive decline (Feinsilver, 2003; Ohayon et al., 2004). It seems, therefore, that each stage of sleep has a specific function and that small percentual alterations in their duration have a great influence in both the short and the long term.

If we take a closer look at what happens in each of the stages of sleep, we can see that each one has well-differentiated characteristics:

N1) Stage 1 is essentially a transition stage from "wake" to "sleep" states, and it usually lasts just one to five minutes (Březinová, 1974). During N1 sleep, the body starts to slow down, giving rise to periods of brief and sudden movements (hypnogogic jerks) (Vetrugno & Montagna, 2011). Brain activity slows down too, and the alpha frequencies (in adults) are no longer the most dominant (Iber et al., 2007). As sleep cycles occur, phase N1 serves as a reset to restart a new cycle, but an uninterrupted sleep may not spend much more time in N1 throughout the night.

N2) During N2, the body reduces its temperature, relaxes the muscles, and slows the heart and breathing rates. At the same time, eye movement stops, and brain waves lower their dominant frequency relative to N1 (Schönauer & Pöhlchen, 2018). At this time, brief bursts of activity, characteristic of this stage, begin to emerge: the *sleep spindles* (Schönauer & Pöhlchen, 2018). Among the various functions of spindles (some of them are addressed in the next subsection), it is known that they help resist being woken up by external stimuli (Walker, 2009). Although the N2 stage can last from 10 to 25 minutes during the first sleep cycle, it lengthens as the night progresses, reaching approximately half of the total sleep time (Březinová, 1974).

N3) Stage 3 is also known as deep sleep. During this stage it is more difficult to wake someone up. Muscle tone, pulse, and respiratory rate decrease further (Diekelmann & Born, 2010). Something similar occurs with brain activity: thalamocortical neurons fall into a hyperpolarized state, resulting in slow waves (SW) between 0.5 and 4.5 Hz (i.e., delta activity) (Bernardi et al., 2018). During the first few sleep cycles, the N3 stages typically last between 20 and 40 minutes. As one goes through the cycles, this stage gets shorter, and more time is spent in REM sleep instead.

REM) Paradoxically, during REM sleep, brain activity increases, reaching levels of complexity that resemble activity during wakefulness, or at least N1 (Zilio et al., 2021). The body experiences atony except for the eyes that move rapidly, reason why this stage receives its name. Although dreams can occur at any stage of sleep, they are more common and intense in REM sleep, which is believed to be related to certain cognitive functions such as memory, learning, and creativity (Cai et al., 2009). REM stages are lengthened, especially in the second half of the night, lasting up to an hour.

The cyclical repetitions of the sleep phases described above are chained in a repeating pattern, which is usually represented by a hypnogram (see Fig. 2.1). Although with certain limitations, there are various automatic methods to identify the sleep phases from the EEG signal (Boostani et al., 2017), so it is common in clinical practice that this identification is not carried out manually (Aboalayon et al., 2016).

2.2.3 The Nested Hierarchy of Electrophysiological Waves during Sleep

In each of the sleep stages, there is a dominant oscillation activity easily measurable by means of the EEG signal. This dominant signal is fundamentally slower than the EEG during wakefulness. Nonetheless, there is a complex microarchitecture, comprising both slow and fast non-stationary burst events (Gorgoni et al., 2020). Thus, while some networks such as visual, auditory, somatomotor, and the default mode remain almost unchanged during sleep relative to wakefulness (Larson-Prior et al., 2009), different brain waves, such as slow oscillations (SOs), spindles

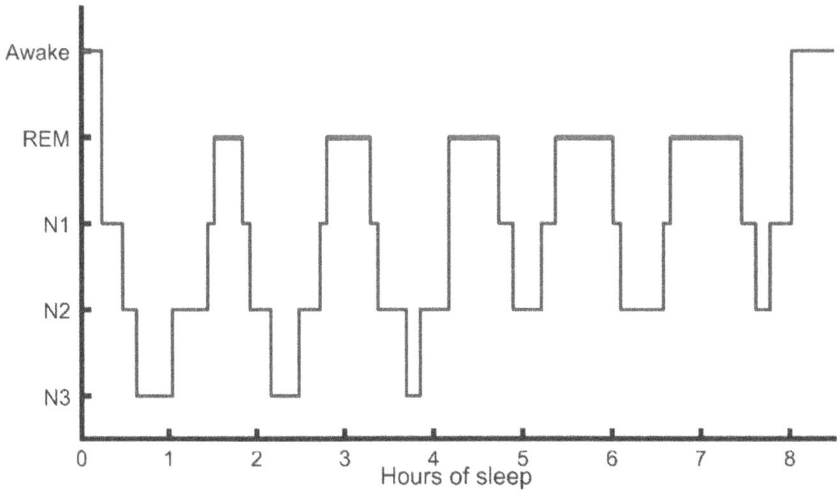

Fig. 2.1 Hypnogram. Representation of sleep stages as a function of time. This hypnogram shows typical sleep architecture with the majority of slow-wave sleep (N3) in the first half of the night, while REM sleep majority is in the last half, with progressive longer durations

and ripples, are generated through activation of rhythmic neuronal thalamocortical connections. These waves do not occur in isolation, but are elicited within a well-defined nested hierarchy, where SOs are thought to have a relevant role in their organization (Gomez-Pilar et al., 2021; Staresina et al., 2015).

SOs are oscillations around 0.75 Hz that, during their up-state, facilitate the production of spindles (Ngo et al., 2019; Staresina et al., 2015), which are easily recognized as burst between 11 and 16 Hz (Antony et al., 2019), i.e., signa band (see Fig. 2.2 for an example of the nesting between SOs and spindles). In turn, sleep spindles facilitate the firing of ripples in the hippocampus, high frequency bursts around 100 Hz (Axmacher et al., 2008; Staresina et al., 2015).

Although the function of these neuronal triggering chain reactions is still not fully understood, the dynamic interaction of these waves is believed to be closely related to the exchange of information between distributed cortical regions, promoting various cognitive functions (Axmacher et al., 2008; Ngo et al., 2019; Staresina et al., 2015).

2.3 Memory Consolidation – The Role of Sleep Spindles

Memory processes begin with the neural encoding of experiences, which results in storage "within" the brain (Harrison & Horne, 2000; Poh & Chee, 2017; Stickgold & Walker, 2005). However, without post-encoding memory processes, this initial encoding does not persist over time. Therefore, the so-called memory consolidation is necessary for long-term storage.

Thanks to sleep deprivation studies, it is known that sleep plays an important role in the encoding processes during wakefulness (Drummond et al., 2000). Even more interesting are some recent studies that have shown that sleep strongly influences memory consolidation (Fogel, Albouy, et al., 2017; Hahn et al., 2019). Although the precise underlying processes are still unknown, we have gained valuable clues about them.

Traditionally, a link between REM and memory has been stablished both in human (Siegel, 2001) and animal studies (Pearlman, 1979). More recently, NREM sleep has been associated with

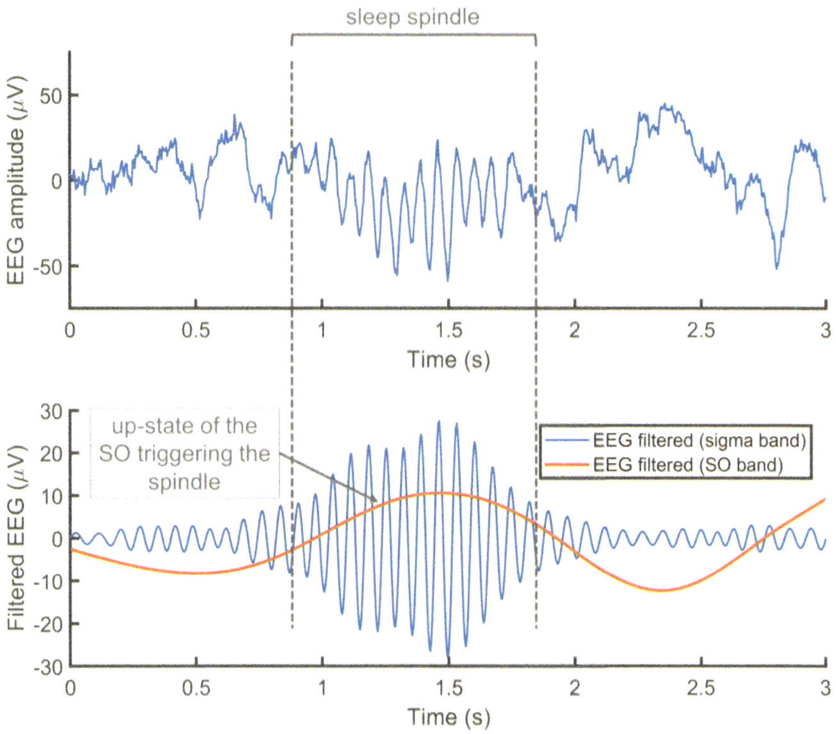

Fig. 2.2 Oscillatory hierarchy nested during sleep. The upper panel shows the EEG signal at the Cz electrode during a spindle event (during N2). The lower panel represents the signal filtered to show the slow oscillation (low pass filtering between 0 and 1 Hz) and show the sleep spindle (bandpass filtering between 11 and 16 Hz). It can be seen how, just after the up-state of the slow oscillation, the spindle is elicited

memory consolidation, especially the stages related to the appearance of sleep spindles (Cairney et al., 2018). These memory consolidation processes during NREM stages are based on the strengthening of particular memory pathways through the delivery of auditory cues (Cairney et al., 2017), a procedure known as targeted memory reactivation (TMR) (Cairney et al., 2018). Interestingly, the time window that coincides with spindle activity overlaps with the TMR process (Cairney et al., 2018). This fact, together with the positive correlation between spindle density and cognitive performance (Fogel & Smith, 2011), procedural memory (Fogel & Smith, 2006), or IQ (Fang et al., 2017), highlights the role of sleep spindles in the service of memory consolidation.

2.4 Is There Room for Slow Oscillations?

As previously stated, sleep spindles are not isolated events. Changes in electrophysiological activity are often mediated by an external stimulus, varying from ongoing activity to task-related activity elicited by external stimulus. However, this transition between states can also be mediated by "internal stimulus", eliciting what is known as internally evoked activity (i.e., internally-guided cognition) (Nakao et al., 2012). Sleep spindles could be considered an example of this internal evoked activity triggered by SOs and elicited during their up-state.

This relationship was evidenced in a recent study in which pre-spindle and spindle activity strong correlations were reported (Gomez-Pilar et al., 2021). Curiously, these correlations were stronger than wake-related evoked activity (Wolff et al., 2019). In other words, this suggests that the brain dynamics associated with SOs determine with great fidelity the characteristics of the following spindle. Whether SOs and the posterior spindle interact with the sleep spindles following an additive (Arieli et al., 1996) or non-additive (Huang et al., 2017) model remains unclear. Non-additive models are based on the assumption that there is a nonlinear superposition between the different waves of the brain activity, which is between SOs and sleep spindles. It would be associated with higher uniformity of the activity, which facilitates the information processing in the cortex (Monier et al., 2003; White et al., 2012). This increased stability would lead to a more structured dynamics enhancing the data predictability (Gershenson & Fernández, 2012). Being aware of the repetitive and uniform patterns in closed loop between the thalamus, reticular nucleus, and the neocortex during SOs and spindle generation (Schönauer & Pöhlchen, 2018), this stability would play a fundamental role for sending information units to distributed neocortical sites for long-term storage. Therefore, a non-additive model in which SOs have a fundamental role is, in principle, presented as a more likely model during sleep for memory consolidation. This is supported by a previous study focused on boosting SOs through transcranial stimulation (Marshall et al., 2006), instead of stimulating the generation of spindles (Berner et al., 2006; Ladenbauer et al., 2017). However, future work is required to support this hypothesis.

2.5 Consequences of Poor Sleep Quality – Illustrative Examples

At this point, we can be confident of the relevant role that sleep has not only in a number of cognitive processes, especially those related to encoding and memory consolidation processes, but also in metabolic processes (van Cauter et al., 2007; Van Cauter et al., 2008). Then, it is worth asking what effects may arise related to pathologies that cause a reduction in the quality of sleep. Or, in the opposite direction, a poor quality of sleep can increase the probability of developing (or worsening) certain diseases?

The number of diseases in which a close relationship with sleep has been found is far from negligible, and it seems to be constantly increasing, such as sleep apnea, migraine, Alzheimer's disease, schizophrenia (all the above are explained below in this section), schizoaffective disorders (Castelnovo et al., 2018), Parkinson (Latreille et al., 2015), or Asperger's syndrome (Godbout et al., 2000), among others. We here present some illustrative examples about the importance of sleep quality and health. Although in some cases the consequences of poor sleep quality that are not related to cognition are mentioned, the main focus is cognition from a neurophysiological point of view.

2.5.1 Non-pathological or Quasi-Pathological Consequences

The effects of a poor sleep quality on behavior and cognition have been fundamentally assessed by sleep deprivation studies. These cognitive – and metabolic – deficits are accentuated if the poor quality of sleep is prolonged in time, without the affected individual being fully aware of it (Goel et al., 2009).

The causes for sleep deprivation, or at least a reduction in its quality, that are not directly related to any pathology are very diverse and range from individual lifestyle to specific shifting in sleep period in relation to the circadian cycle (e.g., due to shift work) (Orzeł-Gryglewska, 2010). If sleep deprivation is total, the consequences depend largely on the number of sleepless nights (Orzeł-Gryglewska, 2010). However, there is great interindividual variability that suggests the influence of genetic alleles associated with differential cognitive vulnerability to sleep loss (Goel et al., 2009). The consequences range from tremor and increased muscle tone (when sleep deprivation is for a single night) to disturbances in reasoning and orientation, visual and tactile hallucinations, fatigue, irritability, and delusions, when sleep deprivation is for 4 or 5 days (Orzeł-Gryglewska, 2010).

Although sustained total sleep deprivation is not common in healthy individuals, sleep problems constitute a global epidemic that threatens the health and quality of life of around 40% of the adult population (Ohayon & Partinen, 2002). This prevalence is similar in children (Fricke-Oerkermann et al., 2007) and is even increased in the elderly (Foley et al., 1995). These problems often do not have a direct tangible effect, but the long-term consequences are of paramount importance, highlighting obesity, diabetes mellitus, hypertension, and decreased cognitive performance, among others (Calhoun & Harding, 2010; Van Cauter & Knutson, 2008).

2.5.2 Sleep Apnea and Cognitive Consequences

Obstructive sleep apnea (OSA) is probably one of the pathologies that most obviously affects healthy and restorative sleep. OSA is mainly characterized by repetitive pharyngeal collapse during sleep, leading to intermittent interruptions of breathing (apnea) (Malhotra & White, 2002). This usually leads to arousals that disrupt the cyclical architecture of sleep (Ferreira et al., 2020; Korkalainen et al., 2021).

Interestingly, recent studies have reported that OSA also has effects on specific oscillations, such as the progressive slowdown of SOs directly related to the severity of the disease (Gutiérrez-Tobal et al., 2021). It has been suggested that this deceleration could be due to an inhibitory effect on thalamus produced by OSA (Gutiérrez-Tobal et al., 2021). Previous studies in rats have shown that suppression of the role of the thalamus leads to a deceleration of the typical frequency of SO, leading to cortical attempts to substitute the role of the thalamus (David et al., 2013). Together, although speculative, we hypothesize that OSA directly influences the neural underpinning involved in the SOs generation (Gutiérrez-Tobal et al., 2021).

As previously stated, SOs are precursors and facilitators of the generation of spindles. Therefore, if SOs are affected, it seems reasonable to think that there would be alterations in the density of spindles beyond the interruptions of the sleep cycle. This is supported by previous studies that show alterations in the spindles in patients with OSA, both in the pediatric population (Brockmann et al., 2018; Weichard et al., 2016), as well as in adults (Ahuja et al., 2018).

The effects on different cognitive processes (especially those related to memory consolidation) that patients with OSA may develop due to hypoxia and sleep fragmentation are still not entirely understood. What is clear, however, is that the fastest intellectual changes happen during school-age (Fry & Hale, 2000), which explains the focus of the increasing number of OSA studies and its related changes in micro and macro sleep architecture in this population (Brockmann et al., 2018; Gruber et al., 2013; Gutiérrez-Tobal et al., 2021).

2.5.3 Migraine and Sleep – A Bidirectional Relationship?

The relationship between sleep and migraine can be interpreted as a bidirectional relationship. In fact, insomnia can be seen as both a cause and a consequence of migraine (Vgontzas & Pavlović, 2018). This leads researchers to think that migraine and sleep problems are "two sides of the same coin", that is, that they both have a common underlying pathophysiology (Vgontzas & Pavlović, 2018). In the outstanding review of Vgontzas and Pavlović (2018), the glymphatic system was proposed as a possible common mechanism. This system is responsible for macroscopic waste removal, primarily active during sleep (Iliff et al., 2012). On the other hand, cortical spreading depression – a wave of excitation followed by inhibition in cortical neurons that may be a direct cause of aura phase that precedes migraine headache – has been shown to cause impaired glymphatic flow (Schain et al., 2017). Therefore, a deterioration in this system could produce an accumulation of waste products that would contribute to later migraine attacks.

2.5.4 The Role of Glymphatic System and Sleep Spindles in Alzheimer's Disease

The accumulation of amyloid-β peptide in the brain appears to be the trigger for a series of events that lead to Alzheimer's disease (Ju et al., 2014). Given that sleep deprivation increases the amyloid-β peptide concentrations, glymphatic system – in charge of removing this toxic substance (Iliff et al., 2012) – seems to be the link between sleep Alzheimer's disease (AD).

Nonetheless, this does not appear to be the only link between AD and sleep. It is well-known that AD is a disease characterized by memory impairments. On the other hand, we have previously shown a number of studies that link sleep spindles functions and memory consolidation. With these precedents, previous studies have searched for a direct relationship between spindles and AD (see (Weng et al., 2020) for a recent review). As might be expected, it is observed that a higher density of spindles is inversely related to the evolution of AD (Gorgoni et al., 2016; Liu et al., 2019). Even more noticeable, a recent positron emission tomography (PET) study showed that the nesting hierarchy between SOs and spindles was altered and predicted accumulated tau levels in the medial frontal cortex (Winer et al., 2019), which is significantly more hyperphosphorylated in AD than in the normal adult brain (Iqbal et al., 2010). Therefore, albeit speculative, the alterations in SOs and spindles produced by OSA could be a potential underlying mechanism for the well-known relationship between OSA and AD (Kheirandish-Gozal et al., 2016). These findings in AD support our previous hypothesis about the importance of the relationship between SOs and spindles (and not spindles alone) for memory consolidation processes.

2.5.4.1 Sleep Spindles as Biomarker of Schizophrenia

Sleep disorders have been associated with the onset of psychosis (Benson, 2015; Zhang et al., 2020). These disorders are unrelated to pharmacological treatment since this association has been reproduced in patients with schizophrenia without antipsychotic medication (Chouinard et al., 2004). Sleep disturbances in schizophrenia patients do not only correspond to alterations in its macroarchitecture (Poulin et al., 2003; Yang & Winkelman, 2006) (i.e., the distribution of time spent in different sleep stages), but also in its microarchitecture (Ferrarelli et al., 2007; Göder et al., 2015) (i.e., characteristics of the waves associated with each stage of sleep). This concordance could have a genetic origin, since both sleep fingerprints, such as spindles (Goldschmied et al., 2021), and schizophrenia (Cao et al., 2019) appear to be highly heritable and share common aspects. For example, the risk gene in schizophrenia that encodes a calcium channel (Lubeiro et al., 2020), CACNA1I, plays a critical role in the generation of spindles in the thalamus (Steullet et al., 2018).

Among the abnormalities found in the sleep microarchitecture in schizophrenia, the reduction in the density of spindles stands out (Ferrarelli et al., 2007). The production of spindles begins with the inhibition of the thalamocortical neurons mediated by the gabaergic inhibition of the reticular nucleus (Berry et al., 2012; Steriade, 2003). This process is followed by glutamatergic rebound peaks that cause cortical neurons to oscillate at the typical spindle frequency (Contreras & Steriade, 1996). Therefore, spindle production depends entirely on the inhibitory onset of the reticular nucleus, which is known to show structural and biochemical abnormalities in schizophrenia (Court et al., 2002; Smith et al., 2001; Steullet et al., 2018). Furthermore, spindle production is governed by an orchestrated organization of inhibitory (gabaergic) and excitatory (glutamatergic) neurons. This excitatory-inhibitory balance is altered in schizophrenia (Kehrer, 2008; Northoff & Gomez-Pilar, 2021), especially affecting the thalamus, as has recently been discovered (Quiñones et al., 2021).

Together, these findings show that the spindle generation process in schizophrenia is disrupted, reducing the density of spindles and likely producing other changes in sleep architecture. Therefore, spindles are postulated as a noticeable biomarker of increasing importance in schizophrenia.

2.6 Conclusion

As new studies appear, the relationship between restful sleep and health is increasingly evident. In this relationship, the role of spindles has gained much relevance due to its proven importance with memory consolidation processes. However, recent studies have shown that SOs are at least equally important in many of these processes. Future studies should be directed to analyze whether the relationship between SOs and spindles is altered in different sleep-related pathologies – and not just the spindles themselves. If so, changes in their relationship could shed new light on the pathophysiological mechanisms involved.

Acknowledgments This research has been developed under the grants PID2020-115468RB-I00 and RTC-2017-6516-1 funded by "Ministerio de Ciencia e Innovación/Agencia Estatal de Investigación/10.13039/501100011033/" and ERDF A way of making Europe; under the R + D + i project "Análisis y correlación entre la epigenética y la actividad cerebral para evaluar el riesgo de migraña crónica y episódica en mujeres" ("Cooperation Programme Interreg V-A Spain-Portugal POCTEP 2014–2020") funded by 'European Commission' and ERDF; and by "CIBER en Bioingeniería, Biomateriales y Nanomedicina (CIBER-BBN)" through "Instituto de Salud Carlos III" co-funded with ERDF funds.

References

Aboalayon, K., Faezipour, M., Almuhammadi, W., & Moslehpour, S. (2016). Sleep stage classification using EEG signal analysis: A comprehensive survey and new investigation. *Entropy, 18*(9), 272. https://doi.org/10.3390/e18090272

Ahuja, S., Chen, R. K., Kam, K., Pettibone, W. D., Osorio, R. S., & Varga, A. W. (2018). Role of normal sleep and sleep apnea in human memory processing. *Nature and Science of Sleep, 10*, 255–269. https://doi.org/10.2147/NSS.S125299

Antony, J. W., Schönauer, M., Staresina, B. P., & Cairney, S. A. (2019). Sleep spindles and memory reprocessing. *Trends in Neurosciences, 42*(1), 1–3. https://doi.org/10.1016/j.tins.2018.09.012

Arieli, A., Sterkin, A., Grinvald, A., & Aertsen, A. (1996). Dynamics of ongoing activity: Explanation of the large variability in evoked cortical responses. *Science (New York, N.Y.), 273*(5283), 1868–1871. https://doi.org/10.1126/science.273.5283.1868

Axmacher, N., Elger, C. E., & Fell, J. (2008). Ripples in the medial temporal lobe are relevant for human memory consolidation. *Brain, 131*(7), 1806–1817. https://doi.org/10.1093/brain/awn103

Benson, K. L. (2015). Sleep in Schizophrenia. *Sleep Medicine Clinics, 10*(1), 49–55. https://doi.org/10.1016/j.jsmc.2014.11.001

Bernardi, G., Siclari, F., Handjaras, G., Riedner, B. A., & Tononi, G. (2018). Local and widespread slow waves in stable NREM sleep: Evidence for distinct regulation mechanisms. *Frontiers in Human Neuroscience, 12*, 248. https://doi.org/10.3389/fnhum.2018.00248

Berner, I., Schabus, M., Wienerroither, T., & Klimesch, W. (2006). The significance of sigma neurofeedback training on sleep spindles and aspects of declarative memory. *Applied Psychophysiology and Biofeedback, 31*(2), 97–114. https://doi.org/10.1007/s10484-006-9013-7

Berry, R. B., Budhiraja, R., Gottlieb, D. J., Gozal, D., Iber, C., Kapur, V. K., … Tangredi, M. M. (2012). Rules for scoring respiratory events in sleep: Update of the 2007 AASM manual for the scoring of sleep and associated events. *Journal of Clinical Sleep Medicine, 08*(05), 597–619. https://doi.org/10.5664/jcsm.2172

Boostani, R., Karimzadeh, F., & Nami, M. (2017). A comparative review on sleep stage classification methods in patients and healthy individuals. *Computer Methods and Programs in Biomedicine, 140*, 77–91. https://doi.org/10.1016/j.cmpb.2016.12.004

Březinová, V. (1974). Sleep cycle content and sleep cycle duration. *Electroencephalography and Clinical Neurophysiology, 36*, 275–282. https://doi.org/10.1016/0013-4694(74)90169-2

Brockmann, P. E., Damiani, F., Pincheira, E., Daiber, F., Ruiz, S., Aboitiz, F., … Bruni, O. (2018). Sleep spindle activity in children with obstructive sleep apnea as a marker of neurocognitive performance: A pilot study. *European Journal of Paediatric Neurology, 22*(3), 434–439. https://doi.org/10.1016/j.ejpn.2018.02.003

Cai, D. J., Mednick, S. A., Harrison, E. M., Kanady, J. C., & Mednick, S. C. (2009). REM, not incubation, improves creativity by priming associative networks. *Proceedings of the National Academy of Sciences, 106*(25), 10130–10134. https://doi.org/10.1073/pnas.0900271106

Cairney, S. A., Vali Guttesen, A. A., El Marj, N., & Staresina, B. P. (2018). Memory consolidation is linked to spindle-mediated information processing during Sleep. *Current Biology, 28*(6), 948–954.e4. https://doi.org/10.1016/j.cub.2018.01.087

Cairney, S. A., Sobczak, J. M., Lindsay, S., & Gaskell, M. G. (2017). Mechanisms of memory retrieval in slow-wave sleep. *Sleep, 40*(9), zsx114. https://doi.org/10.1093/sleep/zsx114

Calhoun, D. A., & Harding, S. M. (2010). Sleep and hypertension. *Chest, 138*(2), 434–443. https://doi.org/10.1378/chest.09-2954

Cao, H., Ingvar, M., Hultman, C. M., & Cannon, T. (2019). Evidence for cerebello-thalamo-cortical hyperconnectivity as a heritable trait for schizophrenia. *Translational Psychiatry, 9*(1), 192. https://doi.org/10.1038/s41398-019-0531-5

Castelnovo, A., Graziano, B., Ferrarelli, F., & D'Agostino, A. (2018). Sleep spindles and slow waves in schizophrenia and related disorders: main findings, challenges and future perspectives. *European Journal of Neuroscience, 48*(8), 2738–2758. https://doi.org/10.1111/ejn.13815

Chouinard, S., Poulin, J., Stip, E., & Godbout, R. (2004). Sleep in untreated patients with schizophrenia: A meta-analysis. *Schizophrenia Bulletin, 30*(4), 957–967. https://doi.org/10.1093/oxfordjournals.schbul.a007145

Contreras, D., & Steriade, M. (1996). Spindle oscillation in cats: The role of corticothalamic feedback in a thalamically generated rhythm. *The Journal of Physiology, 490*(1), 159–179. https://doi.org/10.1113/jphysiol.1996.sp021133

Court, J., Spurden, D., Lloyd, S., McKeith, I., Ballard, C., Cairns, N., ... Perry, E. (2002). Neuronal nicotinic receptors in dementia with Lewy bodies and schizophrenia. *Journal of Neurochemistry, 73*(4), 1590–1597. https://doi.org/10.1046/j.1471-4159.1999.0731590.x

David, F., Schmiedt, J. T., Taylor, H. L., Orban, G., Di Giovanni, G., Uebele, V. N., ... Crunelli, V. (2013). Essential thalamic contribution to slow waves of natural sleep. *Journal of Neuroscience, 33*(50), 19599–19610. https://doi.org/10.1523/JNEUROSCI.3169-13.2013

Diekelmann, S., & Born, J. (2010). The memory function of sleep. *Nature Reviews Neuroscience, 11*(2), 114–126. https://doi.org/10.1038/nrn2762

Drummond, S. P. A., Brown, G. G., Gillin, J. C., Stricker, J. L., Wong, E. C., & Buxton, R. B. (2000). Altered brain response to verbal learning following sleep deprivation. *Nature, 403*(6770), 655–657. https://doi.org/10.1038/35001068

Fang, Z., Ray, L. B., Owen, A. M., & Fogel, S. M. (2019). Brain activation time-locked to sleep spindles associated with human cognitive abilities. *Frontiers in Neuroscience, 13*, 46. https://doi.org/10.3389/fnins.2019.00046

Fang, Z., Sergeeva, V., Ray, L. B., Viczko, J., Owen, A. M., & Fogel, S. M. (2017). Sleep spindles and intellectual ability: Epiphenomenon or directly related? *Journal of Cognitive Neuroscience, 29*(1), 167–182. https://doi.org/10.1162/jocn_a_01034

Feinsilver, S. H. (2003). Sleep in the elderly. *Clinics in Geriatric Medicine, 19*(1), 177–188. https://doi.org/10.1016/S0749-0690(02)00064-2

Fernandez, L. M. J., & Lüthi, A. (2020). Sleep spindles: Mechanisms and functions. *Physiological Reviews, 100*(2), 805–868. https://doi.org/10.1152/physrev.00042.2018

Ferrarelli, F., Huber, R., Peterson, M. J., Massimini, M., Murphy, M., Riedner, B. A., ... Tononi, G. (2007). Reduced sleep spindle activity in schizophrenia patients. *American Journal of Psychiatry, 164*(3), 483–492. https://doi.org/10.1176/ajp.2007.164.3.483

Ferrarelli, F., & Tononi, G. (2017). Reduced sleep spindle activity point to a TRN-MD thalamus-PFC circuit dysfunction in schizophrenia. *Schizophrenia Research, 180*, 36–43. https://doi.org/10.1016/j.schres.2016.05.023

Ferreira, C. B., Schoorlemmer, G. H., Rocha, A. A., & Cravo, S. L. (2020). Increased sympathetic responses induced by chronic obstructive sleep apnea are caused by sleep fragmentation. *Journal of Applied Physiology, 129*(1), 163–172. https://doi.org/10.1152/japplphysiol.00811.2019

Fogel, S., Albouy, G., King, B. R., Lungu, O., Vien, C., Bore, A., ... Doyon, J. (2017). Reactivation or transformation? Motor memory consolidation associated with cerebral activation time-locked to sleep spindles. *PLoS One, 12*(4), e0174755. https://doi.org/10.1371/journal.pone.0174755

Fogel, S., & Smith, C. (2006). Learning-dependent changes in sleep spindles and stage 2 sleep. *Journal of Sleep Research, 15*(3), 250–255. https://doi.org/10.1111/j.1365-2869.2006.00522.x

Fogel, S., & Smith, C. (2011). The function of the sleep spindle: A physiological index of intelligence and a mechanism for sleep-dependent memory consolidation. *Neuroscience and Biobehavioral Reviews, 35*(5), 1154–1165. https://doi.org/10.1016/j.neubiorev.2010.12.003

Fogel, S., Vien, C., Karni, A., Benali, H., Carrier, J., & Doyon, J. (2017). Sleep spindles: A physiological marker of age-related changes in gray matter in brain regions supporting motor skill memory consolidation. *Neurobiology of Aging, 49*, 154–164. https://doi.org/10.1016/j.neurobiolaging.2016.10.009

Foley, D. J., Monjan, A. A., Brown, S. L., Simonsick, E. M., Wallace, R. B., & Blazer, D. G. (1995). Sleep complaints among elderly persons: An epidemiologic study of three communities. *Sleep, 18*(6), 425–432. https://doi.org/10.1093/sleep/18.6.425

Fricke-Oerkermann, L., Plück, J., Schredl, M., Heinz, K., Mitschke, A., Wiater, A., & Lehmkuhl, G. (2007). Prevalence and course of sleep problems in childhood. *Sleep, 30*(10), 1371–1377. https://doi.org/10.1093/sleep/30.10.1371

Fry, A. F., & Hale, S. (2000). Relationships among processing speed, working memory, and fluid intelligence in children. *Biological Psychology, 54*(1–3), 1–34. https://doi.org/10.1016/S0301-0511(00)00051-X

Ganz, F. D. (2012). Sleep and immune function. *Critical Care Nurse, 32*(2), e19–e25. https://doi.org/10.4037/ccn2012689

Gershenson, C., & Fernández, N. (2012). Complexity and information: Measuring emergence, self-organization, and homeostasis at multiple scales. *Complexity, 18*(2), 29–44. https://doi.org/10.1002/cplx.21424

Göder, R., Graf, A., Ballhausen, F., Weinhold, S., Baier, P. C., Junghanns, K., & Prehn-Kristensen, A. (2015). Impairment of sleep-related memory consolidation in schizophrenia: Relevance of sleep spindles? *Sleep Medicine, 16*(5), 564–569. https://doi.org/10.1016/j.sleep.2014.12.022

Godbout, R., Bergeron, C., Limoges, É., Stip, E., & Mottron, L. (2000). A laboratory study of sleep in Asperger's syndrome. *NeuroReport, 11*(1), 127–130. https://doi.org/10.1097/00001756-200001170-00025

Goel, N., Rao, H., Durmer, J., & Dinges, D. (2009). Neurocognitive consequences of sleep deprivation. *Seminars in Neurology, 29*(04), 320–339. https://doi.org/10.1055/s-0029-1237117

Goldschmied, J. R., Lacourse, K., Maislin, G., Delfrate, J., Gehrman, P., Pack, F. M., ... Warby, S. C. (2021). Spindles are highly heritable as identified by different spindle detectors. *Sleep, 44*(4), zsaa230. https://doi.org/10.1093/sleep/zsaa230

Gomez-Pilar, J., Gutiérrez-Tobal, G. C., Poza, J., Fogel, S., Doyon, J., Northoff, G., & Hornero, R. (2021). Spectral and temporal characterization of sleep spindles—Methodological implications. *Journal of Neural Engineering, 18*(3), 036014. https://doi.org/10.1088/1741-2552/abe8ad

Gorgoni, M., Lauri, G., Truglia, I., Cordone, S., Sarasso, S., Scarpelli, S., ... De Gennaro, L. (2016). Parietal fast sleep spindle density decrease in Alzheimer's disease and amnesic mild cognitive impairment. *Neural Plasticity, 2016*, 1–10. https://doi.org/10.1155/2016/8376108

Gorgoni, M., Scarpelli, S., Reda, F., & De Gennaro, L. (2020). Sleep EEG oscillations in neurodevelopmental disorders without intellectual disabilities. *Sleep Medicine Reviews, 49*, 101224. https://doi.org/10.1016/j.smrv.2019.101224

Gruber, R., Wise, M. S., Frenette, S., Knäauper, B., Boom, A., Fontil, L., & Carrier, J. (2013). The association between sleep spindles and IQ in healthy school-age children. *International Journal of Psychophysiology, 89*(2), 229–240. https://doi.org/10.1016/j.ijpsycho.2013.03.018

Gutiérrez-Tobal, G. C., Gomez-Pilar, J., Kheirandish-Gozal, L., Poza, J., Álvarez, D., Campo, F., ... Hornero, R. (2021). Pediatric sleep apnea: The overnight electroencephalogram as a phenotypic biomarker. *Forntiers in Neuroscience, 15*, 644697.

Hahn, M., Joechner, A., Roell, J., Schabus, M., Heib, D. P., Gruber, G., ... Hoedlmoser, K. (2019). Developmental changes of sleep spindles and their impact on sleep-dependent memory consolidation and general cognitive abilities: A longitudinal approach. *Developmental Science, 22*(1), e12706. https://doi.org/10.1111/desc.12706

Harrison, Y., & Horne, J. A. (2000). Sleep loss and temporal memory. *The Quarterly Journal of Experimental Psychology Section A, 53*(1), 271–279. https://doi.org/10.1080/713755870

Huang, Z., Zhang, J., Longtin, A., Dumont, G., Duncan, N. W., Pokorny, J., ... Northoff, G. (2017). Is there a nonadditive interaction between spontaneous and evoked activity? Phase-dependence and its relation to the temporal structure of scale-free brain activity. *Cerebral Cortex, 27*(2), 1037–1059. https://doi.org/10.1093/cercor/bhv288

Iber, C., Ancoli-Israel, S., Chesson, A., & Quan, S. (2007). The AASM manual for the scoring of sleep and associated events: Rules. *Terminology and Technical Specification*. Retrieved from http://ci.nii.ac.jp/naid/10024500923/en/

Iliff, J. J., Wang, M., Liao, Y., Plogg, B. A., Peng, W., Gundersen, G. A., ... Nedergaard, M. (2012). A Paravascular pathway facilitates CSF flow through the brain parenchyma and the clearance of interstitial solutes, including amyloid. *Science Translational Medicine, 4*(147), 147ra111–147ra111. https://doi.org/10.1126/scitranslmed.3003748

Iqbal, K., Liu, F., Gong, C.-X., & Grundke-Iqbal, I. (2010). Tau in Alzheimer disease and related tauopathies. *Current Alzheimer Research, 7*(8), 656–664. https://doi.org/10.2174/156720510793611592

Jafari, B., & Mohsenin, V. (2010). Polysomnography. *Clinics in Chest Medicine, 31*(2), 287–297. https://doi.org/10.1016/j.ccm.2010.02.005

Ju, Y.-E. S., Lucey, B. P., & Holtzman, D. M. (2014). Sleep and Alzheimer disease pathology—A bidirectional relationship. *Nature Reviews Neurology, 10*(2), 115–119. https://doi.org/10.1038/nrneurol.2013.269

Keenan, S. A. (1999). Normal human sleep. *Respiratory Care Clinics of North America, 5*(3), 319–331, vii. Retrieved from http://www.ncbi.nlm.nih.gov/pubmed/10419578.

Kehrer, C. (2008). Altered excitatory-inhibitory balance in the NMDA-hypofunction model of schizophrenia. *Frontiers in Molecular Neuroscience*, 1. https://doi.org/10.3389/neuro.02.006.2008

Kheirandish-Gozal, L., Philby, M. F., Alonso-Álvarez, M. L., Terán-Santos, J., & Gozal, D. (2016). Biomarkers of Alzheimer disease in children with obstructive sleep apnea: Effect of adenotonsillectomy. *Sleep, 39*(6), 1225–1232. https://doi.org/10.5665/sleep.5838

Korkalainen, H., Leppanen, T., Duce, B., Kainulainen, S., Aakko, J., Leino, A., ... Toyras, J. (2021). Detailed assessment of sleep architecture with deep learning and shorter epoch-to-epoch duration reveals sleep fragmentation of patients with obstructive sleep apnea. *IEEE Journal of Biomedical and Health Informatics, 25*(7), 2567–2574. https://doi.org/10.1109/JBHI.2020.3043507

Ladenbauer, J., Ladenbauer, J., Külzow, N., de Boor, R., Avramova, E., Grittner, U., & Flöel, A. (2017). Promoting sleep oscillations and their functional coupling by transcranial stimulation enhances memory consolidation in mild cognitive impairment. *The Journal of Neuroscience, 37*(30), 7111–7124. https://doi.org/10.1523/JNEUROSCI.0260-17.2017

Latreille, V., Carrier, J., Lafortune, M., Postuma, R. B., Bertrand, J.-A., Panisset, M., Chouinard, S., & Gagnon, J.-F. (2015). Sleep spindles in Parkinson's disease may predict the development of dementia. *Neurobiology of Aging, 36*(2), 1083–1090. https://doi.org/10.1016/j.neurobiolaging.2014.09.009

Larson-Prior, L. J., Zempel, J. M., Nolan, T. S., Prior, F. W., Snyder, A. Z., & Raichle, M. E. (2009). Cortical

network functional connectivity in the descent to sleep. *Proceedings of the National Academy of Sciences, 106*(11), 4489–4494. https://doi.org/10.1073/pnas.0900924106
Liu, S., Benson, T. A., & Reiter, M. K. (2019). Efficient and safe network updates with suffix causal consistency. In *Proceedings of the fourteenth EuroSys conference 2019* (pp. 1–15). ACM. https://doi.org/10.1145/3302424.3303965
Lubeiro, A., Fatjó-Vilas, M., Guardiola, M., Almodóvar, C., Gomez-Pilar, J., Cea-Cañas, B., ... Molina, V. (2020). Analysis of KCNH2 and CACNA1C schizophrenia risk genes on EEG functional network modulation during an auditory oddball task. *European Archives of Psychiatry and Clinical Neuroscience, 270*(4), 433–442. https://doi.org/10.1007/s00406-018-0977-0
Malhotra, A., & White, D. P. (2002). Obstructive sleep apnoea. *The Lancet, 360*(9328), 237–245. https://doi.org/10.1016/S0140-6736(02)09464-3
Marshall, L., Helgadóttir, H., Mölle, M., & Born, J. (2006). Boosting slow oscillations during sleep potentiates memory. *Nature, 444*(7119), 610–613. https://doi.org/10.1038/nature05278
Monier, C., Chavane, F., Baudot, P., Graham, L. J., & Frégnac, Y. (2003). Orientation and direction selectivity of synaptic inputs in visual cortical neurons. *Neuron, 37*(4), 663–680. https://doi.org/10.1016/S0896-6273(03)00064-3
Murawski, B., Wade, L., Plotnikoff, R. C., Lubans, D. R., & Duncan, M. J. (2018). A systematic review and meta-analysis of cognitive and behavioral interventions to improve sleep health in adults without sleep disorders. *Sleep Medicine Reviews, 40*, 160–169. https://doi.org/10.1016/j.smrv.2017.12.003
Nakao, T., Ohira, H., & Northoff, G. (2012). Distinction between externally vs. internally guided decision-making: Operational differences, meta-analytical comparisons and their theoretical implications. *Frontiers in Neuroscience, 6*, 31. https://doi.org/10.3389/fnins.2012.00031
Ngo, H.-V. V., Seibold, M., Boche, D. C., Mölle, M., & Born, J. (2019). Insights on auditory closed-loop stimulation targeting sleep spindles in slow oscillation up-states. *Journal of Neuroscience Methods, 316*, 117–124. https://doi.org/10.1016/j.jneumeth.2018.09.006
Northoff, G., & Gomez-Pilar, J. (2021). Overcoming rest–task divide—Abnormal temporospatial dynamics and its cognition in schizophrenia. *Schizophrenia Bulletin, 47*(3), 751–765. https://doi.org/10.1093/schbul/sbaa178
Ohayon, M. M., Carskadon, M. A., Guilleminault, C., & Vitiello, M. V. (2004). Meta-analysis of quantitative sleep parameters from childhood to old age in healthy individuals: Developing normative sleep values across the human lifespan. *Sleep, 27*(7), 1255–1273. https://doi.org/10.1093/sleep/27.7.1255
Ohayon, M. M., & Partinen, M. (2002). Insomnia and global sleep dissatisfaction in Finland. *Journal of Sleep Research, 11*(4), 339–346. https://doi.org/10.1046/j.1365-2869.2002.00317.x
Orzeł-Gryglewska, J. (2010). Consequences of sleep deprivation. *International Journal of Occupational Medicine and Environmental Health, 23*(1). https://doi.org/10.2478/v10001-010-0004-9
Pearlman, C. A. (1979). REM sleep and information processing: Evidence from animal studies. *Neuroscience & Biobehavioral Reviews, 3*(2), 57–68. https://doi.org/10.1016/0149-7634(79)90034-4
Poh, J.-H., & Chee, M. W. L. (2017). Degradation of cortical representations during encoding following sleep deprivation. *NeuroImage, 153*, 131–138. https://doi.org/10.1016/j.neuroimage.2017.01.080
Poulin, J., Daoust, A.-M., Forest, G., Stip, E., & Godbout, R. (2003). Sleep architecture and its clinical correlates in first episode and neuroleptic-naive patients with schizophrenia. *Schizophrenia Research, 62*(1–2), 147–153. https://doi.org/10.1016/S0920-9964(02)00346-8
Quiñones, G. M., Mayeli, A., Yushmanov, V. E., Hetherington, H. P., & Ferrarelli, F. (2021). Reduced GABA/glutamate in the thalamus of individuals at clinical high risk for psychosis. *Neuropsychopharmacology, 46*(6), 1133–1139. https://doi.org/10.1038/s41386-020-00920-4
Reynaud, E., Vecchierini, M.-F., Heude, B., Charles, M.-A., & Plancoulaine, S. (2018). Sleep and its relation to cognition and behaviour in preschool-aged children of the general population: A systematic review. *Journal of Sleep Research, 27*(3), e12636. https://doi.org/10.1111/jsr.12636
Schabus, M., Gruber, G., Parapatics, S., Sauter, C., Klösch, G., Anderer, P., ... Zeitlhofer, J. (2004). Sleep spindles and their significance for declarative memory consolidation. *Sleep, 27*(8), 1479–1485. https://doi.org/10.1093/sleep/27.7.1479
Schain, A. J., Melo-Carrillo, A., Strassman, A. M., & Burstein, R. (2017). Cortical spreading depression closes paravascular space and impairs glymphatic flow: Implications for migraine headache. *The Journal of Neuroscience, 37*(11), 2904–2915. https://doi.org/10.1523/JNEUROSCI.3390-16.2017
Schönauer, M., & Pöhlchen, D. (2018). Sleep spindles. *Current Biology, 28*(19), R1129–R1130. https://doi.org/10.1016/j.cub.2018.07.035
Siegel, J. M. (2001). The REM sleep-memory consolidation hypothesis. *Science, 294*(5544), 1058–1063. https://doi.org/10.1126/science.1063049
Smith, R. E., Haroutunian, V., Davis, K. L., & Meador-Woodruff, J. H. (2001). Expression of excitatory amino acid transporter transcripts in the thalamus of subjects with schizophrenia. *American Journal of Psychiatry, 158*(9), 1393–1399. https://doi.org/10.1176/appi.ajp.158.9.1393
Staresina, B. P., Bergmann, T. O., Bonnefond, M., Van Der Meij, R., Jensen, O., Deuker, L., ... Fell, J. (2015). Hierarchical nesting of slow oscillations, spindles and ripples in the human hippocampus during sleep. *Nature Neuroscience, 18*(11), 1679–1686. https://doi.org/10.1038/nn.4119

Steriade, M. (2003). The corticothalamic system in sleep. *Frontiers in Bioscience, 8*(4), 1043. https://doi.org/10.2741/1043

Steullet, P., Cabungcal, J.-H., Bukhari, S. A., Ardelt, M. I., Pantazopoulos, H., Hamati, F., … Berretta, S. (2018). The thalamic reticular nucleus in schizophrenia and bipolar disorder: Role of parvalbumin-expressing neuron networks and oxidative stress. *Molecular Psychiatry, 23*(10), 2057–2065. https://doi.org/10.1038/mp.2017.230

Stickgold, R., & Walker, M. (2005). Memory consolidation and reconsolidation: What is the role of sleep? *Trends in Neurosciences, 28*(8), 408–415. https://doi.org/10.1016/j.tins.2005.06.004

van Cauter, E., Holmbäck, U., Knutson, K., Leproult, R., Miller, A., Nedeltcheva, A., … Spiegel, K. (2007). Impact of sleep and sleep loss on neuroendocrine and metabolic function. *Hormone Research in Pædiatrics, 67*(1), 2–9. https://doi.org/10.1159/000097543

Van Cauter, E., & Knutson, K. L. (2008). Sleep and the epidemic of obesity in children and adults. *European journal of endocrinology, 159*(suppl_1), S59–S66. https://doi.org/10.1530/EJE-08-0298

Van Cauter, E., Spiegel, K., Tasali, E., & Leproult, R. (2008). Metabolic consequences of sleep and sleep loss. *Sleep Medicine, 9*, S23–S28. https://doi.org/10.1016/S1389-9457(08)70013-3

Vetrugno, R., & Montagna, P. (2011). Sleep-to-wake transition movement disorders. *Sleep Medicine, 12*, S11–S16. https://doi.org/10.1016/j.sleep.2011.10.005

Vgontzas, A., & Pavlović, J. M. (2018). Sleep disorders and migraine: Review of literature and potential pathophysiology mechanisms. *Headache: The Journal of Head and Face Pain, 58*(7), 1030–1039. https://doi.org/10.1111/head.13358

Walker, M. P. (2009). The role of sleep in cognition and emotion. *Annals of the New York Academy of Sciences, 1156*(1), 168–197. https://doi.org/10.1111/j.1749-6632.2009.04416.x

Weichard, A. J., Walter, L. M., Hollis, S. L., Nixon, G. M., Davey, M. J., Horne, R. S. C., & Biggs, S. N. (2016). Association between slow-wave activity, cognition and behaviour in children with sleep-disordered breathing. *Sleep Medicine, 25*, 49–55. https://doi.org/10.1016/j.sleep.2016.06.004

Weng, Y.-Y., Lei, X., & Yu, J. (2020). Sleep spindle abnormalities related to Alzheimer's disease: A systematic mini-review. *Sleep Medicine, 75*, 37–44. https://doi.org/10.1016/j.sleep.2020.07.044

White, B., Abbott, L. F., & Fiser, J. (2012). Suppression of cortical neural variability is stimulus- and state-dependent. *Journal of Neurophysiology, 108*(9), 2383–2392. https://doi.org/10.1152/jn.00723.2011

Winer, J. R., Mander, B. A., Helfrich, R. F., Maass, A., Harrison, T. M., Baker, S. L., … Walker, M. P. (2019). Sleep as a potential biomarker of tau and β-amyloid burden in the human brain. *The Journal of Neuroscience, 39*(32), 6315–6324. https://doi.org/10.1523/JNEUROSCI.0503-19.2019

Wolff, A., Yao, L., Gomez-Pilar, J., Shoaran, M., Jiang, N., & Northoff, G. (2019). Neural variability quenching during decision-making: Neural individuality and its prestimulus complexity. *NeuroImage, 192*. https://doi.org/10.1016/j.neuroimage.2019.02.070

Xie, L., Kang, H., Xu, Q., Chen, M. J., Liao, Y., Thiyagarajan, M., … Nedergaard, M. (2013). Sleep drives metabolite clearance from the adult brain. *Science, 342*(6156), 373–377. https://doi.org/10.1126/science.1241224

Yaffe, K., Falvey, C. M., & Hoang, T. (2014). Connections between sleep and cognition in older adults. *The Lancet Neurology, 13*(10), 1017–1028. https://doi.org/10.1016/S1474-4422(14)70172-3

Yang, C., & Winkelman, J. W. (2006). Clinical significance of sleep EEG abnormalities in chronic schizophrenia. *Schizophrenia Research, 82*(2–3), 251–260. https://doi.org/10.1016/j.schres.2005.10.021

ZAssefa, S., Diaz-Abad, M., Wickwire, E. M., & Scharf, S. M. (2015). The functions of sleep. *AIMS Neuroscience, 2*(3), 155–171. https://doi.org/10.3934/Neuroscience.2015.3.155

Zhang, Y., Quiñones, G. M., & Ferrarelli, F. (2020). Sleep spindle and slow wave abnormalities in schizophrenia and other psychotic disorders: Recent findings and future directions. *Schizophrenia Research, 221*, 29–36. https://doi.org/10.1016/j.schres.2019.11.002

Zilio, F., Gomez-Pilar, J., Cao, S., Zhang, J., Zang, D., Qi, Z., … Northoff, G. (2021). Are intrinsic neural timescales related to sensory processing? Evidence from abnormal behavioral states. *NeuroImage, 226*, 117579. https://doi.org/10.1016/j.neuroimage.2020.117579

3 Obstructive Sleep Apnoea: Focus on Pathophysiology

Walter T. McNicholas

Abstract

Obstructive sleep apnoea (OSA) is characterized by recurring episodes of upper airway obstruction during sleep and the fundamental abnormality reflects the inability of the upper airway dilating muscles to withstand the negative forces generated within the upper airway during inspiration. Factors that result in narrowing of the oropharynx such as abnormal craniofacial anatomy, soft tissue accumulation in the neck, and rostral fluid shift in the recumbent position increase the collapsing forces within the airway. The counteracting forces of upper airway dilating muscles, especially the genioglossus, are negatively influenced by sleep onset, inadequacy of the genioglossus responsiveness, ventilatory instability, especially post arousal, and loop gain. Recent reports indicate that multiple endotypes reflecting OSA pathophysiology are present in individual patients. A detailed understanding of the complex pathophysiology of OSA encourages the development of therapies targeted at these pathophysiological endotypes and facilitates a move towards precision medicine as a potential alternative to continuous positive airway pressure therapy in selected patients.

Keywords

Obstructive sleep apnoea · Pathophysiology · Upper airway anatomy · Ventilatory control · Arousal · Endotype · Treatment

3.1 Introduction

Obstructive sleep apnoea (OSA) is characterized by recurring episodes of upper airway obstruction during sleep, leading to markedly reduced (hypopnoea) or absent (apnoea) airflow at the nose/mouth. The condition is usually associated with loud snoring and intermittent hypoxaemia, and apnoeas are typically terminated by brief micro-arousals, which result in sleep fragmentation and diminished amounts of slow wave sleep (SWS) and rapid-eye-movement (REM) sleep (Deegan & Mcnicholas, 1995). Patients with OSA are usually unaware of this sleep disturbance, but the changes in sleep architecture contribute significantly to the prominent symptoms of unrefreshing sleep and excessive daytime sleepiness (EDS) typically reported by many of these patients (Lévy et al., 2015). Furthermore, the intermittent hypoxaemia and sleep fragmentation associated with OSA generate cell and molecular responses that generate systemic

W. T. McNicholas (✉)
School of Medicine, University College Dublin, Department of Respiratory and Sleep Medicine, St. Vincent's Hospital Group, Dublin, Ireland
e-mail: walter.mcnicholas@ucd.ie

T. Penzel, R. Hornero (eds.), *Advances in the Diagnosis and Treatment of Sleep Apnea*, Advances in Experimental Medicine and Biology 1384, https://doi.org/10.1007/978-3-031-06413-5_3

inflammation, sympathetic excitation, and other responses that predispose to comorbidities, especially cardiometabolic and neuropsychiatric (McNicholas, 2019).

While the detailed pathophysiology of OSA is complex, the fundamental abnormality reflects the inability of the upper airway dilating muscles to withstand the negative forces generated within the upper airway during inspiration. In the normal setting, upper airway dilating muscles contract in a coordinated manner that is timed with each inspiration, thus counteracting the negative pressures that are generated within the upper airway during inspiration. Factors that increase these negative pressures or diminish the efficacy of dilating muscle contraction upset this balance and thus predispose to upper airway obstruction (Deegan & Mcnicholas, 1995). Any factor that results in narrowing of the upper airway will increase upper airway negative pressures during inspiration, thus promoting collapse (Fig. 3.1).

The present review explores the various factors contributing to an imbalance of forces within the upper airway that predispose to obstruction, discusses the mechanisms by which obstruction occurs, reviews the more recent evidence regarding pathophysiological endotypes and phenotypes that may help predict the development of OSA, and, finally, reviews the emerging role of targeted therapy for OSA based on individual pathophysiological mechanisms.

3.2 Pharyngeal Pressure

The most important factor contributing to increased negative pharyngeal pressure during inspiration is narrowing of the oropharyngeal airway, which results in increased upper airway resistance during inspiration (Lévy et al., 2015). There are many potential causes of such narrowing, which include structural narrowing because of craniofacial bony morphology, soft tissue accumulation in and around the oropharynx because of factors such as obesity or adenotonsillar hypertrophy, and transient factors such as fluid accumulation that gravitates towards the neck in the recumbent position.

3.2.1 Craniofacial Morphology

Most patients with OSA demonstrate a narrowed oropharyngeal airway that can be clinically assessed by the Mallampati score, which is graded 1–4 depending on the degree of narrowing (McNicholas, 2008a). The typical patient with OSA has a score of 3 or 4 (Yu & Rosen, 2020). There is increasing evidence that genetic

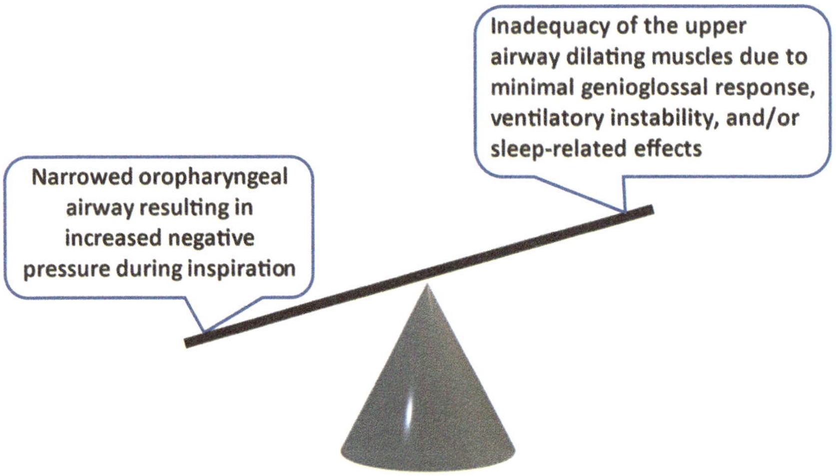

Fig. 3.1 Balance of forces affecting the patency of the upper airway with factors resulting in increased negative intrapharyngeal pressure and factors that reduce dilating muscle contraction promoting airway collapse

factors play a major role in this anatomical narrowing and, thus, are present from birth (Chen et al., 2018; Chi et al., 2014). Cephalometric and computed tomographic (CT) studies of the head and neck have demonstrated bony dimensions in the lower face and neck that result in narrowing of the upper airway (Neelapu et al., 2017; Sakat et al., 2016) (Abramson et al., 2010), and clinical assessment demonstrates micro/retrognathia in many of these patients (McNicholas, 2008a). On lateral cephalometry studies, OSA patients have a variety of anatomical abnormalities, including an abnormally small airway below the base of the tongue, a long bulky soft palate, an inferiorly placed hyoid bone and retrognathia (Rivlin et al., 1984).

Children with the Robin sequence (Bravo et al., 2005) or Treacher-Collins syndrome (Moraleda-Cibrián et al., 2014) are especially prone to OSA because of bony changes to the lower face and/or mandible that result in structural narrowing of the oropharyngeal airway (Tan et al., 2016). Micrognathia, which is the central feature of the Robin sequence, is particularly associated with OSA, as a small and/or retropositioned mandible places the base of the tongue closer to the posterior pharyngeal wall and interferes with the efficiency of the genioglossus muscle in keeping the tongue out of the narrowed pharynx (Sher, 1992). Indeed, the important role of such factors in this context is demonstrated by a case report from our department of a young girl with the Robin sequence who presented at the age of 12 with severe OSA complicated by right heart dysfunction and was successfully treated with nasal continuous airway pressure (CPAP) but resolved the OSA following growth of the mandible during puberty to the extent that CPAP therapy was no longer necessary (Kiely et al., 1998).

3.2.2 Soft Tissue Accumulation

Soft tissue accumulation in and around the upper airway can predispose to OSA by narrowing the oropharyngeal lumen. The two major entities in this context are obesity and adenotonsillar hypertrophy.

Obesity is closely linked to OSA and the role of central obesity in the pathophysiology of OSA occurs at different levels. The accumulation of fat in the neck results in oropharyngeal narrowing, which increases the collapsibility of the upper airway, and abdominal obesity reduces traction on the upper airway, which further predisposes to increased collapsibility (Deegan & Mcnicholas, 1995). 70% of patients with OSA are obese (Tuomilehto et al., 2013), and conversely, 50% of patients with a body mass index over 40 have an AHI over 10 (Resta et al., 2001). A higher body mass index (BMI) typically results in more severe OSA, especially in males and in younger subjects.

Adenotonsillar hypertrophy is associated with increased soft tissue within the oropharyngeal airway, which reduces the cross-sectional area and increases oropharyngeal collapsibility. Adenotonsillar hypertrophy is an important contributing factor in paediatric OSA, often in association with obesity (Dayyat et al., 2009). This form of OSA is amenable to surgical treatment by tonsillectomy (Stradling et al., 1990), although surgery may not be curative if there is co-existing obesity and/or an otherwise narrowed upper airway (Dayyat et al., 2009).

Infiltration of the upper airway soft tissues can reduce the upper airway lumen, as occurs in myxoedema, acromegaly, involvement by neoplastic processes, and mucopolysaccharidoses, and all such disorders have been associated with a predisposition to OSA (Grunstein et al., 1991; Orr et al., 1981).

3.2.3 Fluid Accumulation

Fluid accumulation, as occurs in patients with congestive heart failure, predisposes to OSA by the gravitational behaviour of oedema. Nocturnal redistribution of fluid in the recumbent position to dependent areas of the body such as the parapharyngeal soft tissues increases upper airway resistance and collapsibility (White & Bradley, 2013). Dietary sodium intake has been reported to be closely correlated with the severity of OSA in patients with heart failure, likely as a consequence of fluid retention and redistribution (Kasai et al., 2011). Furthermore, non-obese male subjects with venous insufficiency who wore compression stockings during the day to limit fluid accumulation had a reduction in AHI of 36% when compared to those not wearing stockings (Redolfi et al., 2011). While these observations imply that diuretic therapy to remove excess fluid should benefit OSA, a randomized controlled trial of patients with severe OSA reported that sodium restriction and diuretic therapy resulted in only a modest improvement in AHI, implying that fluid accumulation only partially explains the aetiology of OSA in patients with heart failure (Fiori et al., 2018).

In patients with end-stage renal failure, fluid accumulation with associated nocturnal redistribution in the recumbent position results in oropharyngeal narrowing, like heart failure, which predisposes to OSA. In a group of 40 patients with end-stage renal failure on haemodialysis, 70% had an AHI >15, and these patients had a greater total body extracellular fluid volume, including neck, thorax and leg volumes despite no difference in BMI compared to those with an AHI <5 (Lyons et al., 2017). Furthermore, excess fluid removal by dialysis has been demonstrated to reduce the severity of OSA. One report indicated that removal of 2.2 L of fluid during a single ultrafiltration session resulted in a 36% fall in AHI, which also correlated with the volume of fluid removed (Lyons et al., 2015).

Patients with chronic obstructive pulmonary disease (COPD) may also develop OSA and the chronic bronchitis phenotype appears to be most susceptible (McNicholas, 2017). This phenotype is more prone to right heart failure, and the associated peripheral oedema may be an important factor in predisposing to OSA.

3.2.4 Nasal Obstruction

The primary route of breathing, especially while asleep, is through the nose, and there has been considerable interest in the role of nasal obstruction in the pathophysiology of OSA. The upper airway can be viewed as a Starling Resistor (Fig. 3.2) with the nose as the fixed inlet for breathing and the oropharyngeal airway as the collapsible segment leading to the fixed downstream segment of the lower respiratory tract (McNicholas, 2008b). This model supports a role for nasal obstruction in increasing upper airway collapsibility. Nasal obstruction can be fixed, such as occurs with a deviated nasal septum, or variable, as may be seen in seasonal rhinitis. The available evidence supports a role for variable nasal obstruction in the pathophysiology of OSA (McNicholas, 2008b; McNicholas et al., 1982), and active therapy of rhinitis with intranasal corticosteroids has been reported to reduce AHI in patients with mild to moderate OSA (Kiely et al., 2004). On the other hand, a randomized control trial of surgery for fixed nasal obstruction in patients with OSA reported little benefit in terms of AHI reduction (Koutsourelakis et al., 2008), supporting the view, somewhat surprisingly, that fixed nasal obstruction is not a significant factor in the pathophysiology of OSA.

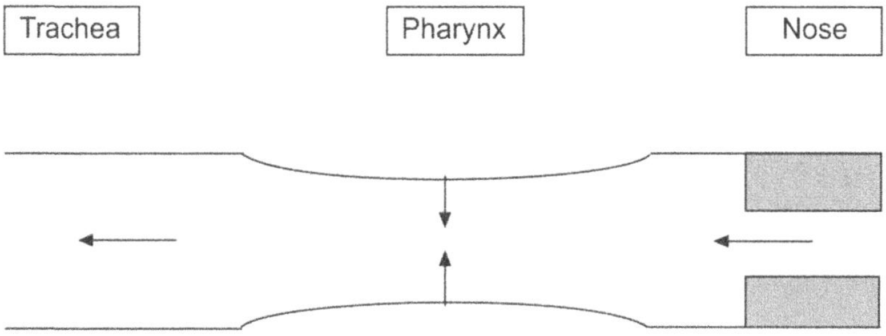

Fig. 3.2 Starling resistor model of the upper airway

3.2.5 Other Factors Influencing Upper Airway Calibre

The position of the head and neck has a significant influence on pharyngeal patency and varying head position between flexion and extension can cause significant variations in size of the retroglossal space and hyoid position on lateral cephalometry (Davies & Stradling, 1990). Neck flexion makes the upper airway more collapsible, whereas neck extension makes it more resistant to collapse, irrespective of changes in general body posture (Wilson et al., 1980). The supine posture also has an adverse effect on upper airway patency. Pharyngeal cross-sectional area is reduced from the upright to the supine position in both apnoeic and nonapnoeic snorers (Yildirim et al., 1991), and the supine posture effect appears to be due to gravitational forces acting to narrow the upper airway (Fouke & Strohl, 1987).

3.3 Upper Airway Dilator Muscle Function

The oropharyngeal airway is not a rigid structure and patency of this segment of the upper airway is dependent on the contraction of pharyngeal dilator muscles, especially the genioglossus, which act to stiffen the collapsible segment during inspiration (Deegan & Mcnicholas, 1995). These muscles contract in a phasic manner that is coordinated with inspiration and contraction of these muscles precedes the contraction of the diaphragm by milliseconds. Activity of these upper airway muscles is modulated by chemical stimuli, vagal input, changes in upper airway pressure, and baroreceptor activity (Brouillette & Thach, 1980).

In the setting of OSA, breathing through a narrowed upper airway generates a greater suction pressure and, thus, greater collapsing force, and pharyngeal dilator muscles must, therefore, contract more forcefully to prevent upper airway obstruction. This situation results in higher dilating muscle activity being evident during wakefulness, which diminishes to a greater extent than normal subjects during sleep, thus predisposing to upper airway obstruction (Mezzanotte et al., 1996). Progressive hypercapnia, hypoxia, asphyxia and negative pressure application all produce an augmenting drive to upper airway dilator muscles (Brouillette & Thach, 1980). Furthermore, genioglossus muscle activity varies with sleep stage, and is lowest in rapid-eye-movement (REM) sleep (Carberry et al., 2016), thus making the upper airway most collapsible in this sleep stage. This reduction across sleep stages is similar in patients with OSA to that seen in normal subjects and is similar in males and females (Eckert et al., 2009). Increased genioglossus muscle tone is associated with spontaneous periods of stable flow limited breathing in OSA and reductions in genioglossus activity during REM may explain the higher severity of OSA in that stage (Jordan et al., 2009).

The complexity of the upper airway musculature makes it unlikely that dysfunction of a single muscle group is responsible for OSA, but the genioglossus appears to be the most important, which pulls the tongue forward and opposes pharyngeal collapse. Muscles causing forward movement of the hyoid bone (geniohyoid, sternohyoid, and thyrohyoid) result in enlargement and stabilization of the pharyngeal airway, and the supine posture is associated with forward movement of the hyoid bone, which acts to limit the collapsibility of the airway in this position (Yildirim et al., 1991).

The degree of upper airway muscle preactivation prior to diaphragmatic contraction varies with respiratory drive (Strohl et al., 1980), and this could represent a compensatory attempt to open the airway before airway pressure is lowered by contraction of the diaphragm. Overall, the role of upper airway muscles in the pathophysiology of OSA appears to be more relating to inadequate compensation in the face of increasingly negative pressure during inspiration in patients with OSA, rather than a primary deficiency in the function of these muscles. This inadequacy is compounded by the observation that upper airway dilating muscles, as skeletal muscles, demonstrate a greater decrease in activity during sleep than the diaphragm as a normal physiological response to sleep, especially during REM (Mezzanotte et al., 1996).

3.4 Respiratory Control

Contraction of the upper airway muscles and diaphragm respond in a similar manner to hypercapnia, hypoxia and airway occlusion (Brouillette & Thach, 1980), which suggests that central control mechanisms of upper airway and respiratory pump muscles in humans are closely related. However, there appear to be quantitative differences in the response to different stimuli. For example, oxygen breathing decreases genioglossal more than diaphragmatic electromyographic (EMG) activity, whereas hypercapnia and prolonged occlusion produce greater increase in genioglossal compared to diaphragmatic EMG (Brouillette & Thach, 1980).

The pattern of recurring apnoea frequently observed in OSA supports an instability of ventilatory control similar to periodic breathing and upper airway obstruction is most likely when diaphragmatic and genioglossal inspiratory EMG activity are at the lowest point of the cycle (Deegan & Mcnicholas, 1995). EMG activity progressively increases through the later stages of apnoea until the upper airway reopens, at which time the increase in genioglossal EMG is typically greater than that of the diaphragm (Dempsey et al., 2010). The period immediately following resolution of the apnoea is usually characterized by hyperventilation for several breaths, following which both EMGs then decrease in activity, which predisposes to further obstruction.

3.4.1 Apnoea Threshold

Normal subjects demonstrate fluctuations in ventilation associated with the transition from wakefulness to non-REM sleep, which is due to a reduction in the carbon dioxide (CO2) drive to breathe and the exposing of a sensitive apnoeic threshold that is critically CO2 dependent (Phillipson, 1978). The pivotal role of hypocapnia in this apnoea threshold is demonstrated by the observation that adding even small amounts of CO2 to the inspired air of patients with Cheyne Stokes Breathing can be sufficient to resolve the associated central apnoeas (Dempsey et al., 2010).

In OSA, the apnoea threshold is further amplified by the ventilatory overshoot that occurs after the termination of obstructive apnoea resulting in CO_2 reduction and thus predisposing to further apnoea. Such predisposition is initially towards central apnoea but the associated reduction in upper airway muscle activity contributes to upper airway collapse and associated obstructive apnoea. The CO_2-responsive apnoea threshold is particularly evident in non-REM sleep and there appears to be no evident threshold during phasic REM sleep (Skatrud & Dempsey, 1983). Furthermore, the periodic breathing associated with heart failure is rarely present in REM sleep. Additional factors that may contribute to further apnea post hyperventilation include lung stretch receptor and baroreceptor stimulation (Deegan & Mcnicholas, 1995).

3.5 Sleep Effects

During wakefulness, patients with OSA typically breathe normally, which is a consequence of the waking stimulus to breathe and associated tonic stimulation of the upper airway dilating muscles. However, with sleep onset, upper airway muscle tone diminishes, resulting in a more collapsible upper airway. EMG activity of the diaphragm and upper airway dilating muscles in healthy humans show reductions in amplitude associated with the transition from wakefulness to non-REM sleep, typically accompanied by a mild hypoventilation and a significant increase in upper airway resistance (Dempsey et al., 2010). Sleep is associated with a bigger reduction in upper airway EMG activity compared to that of the respiratory pump muscles, and this effect is greatest in REM sleep. This differential effect further compromises upper airway patency during inspiration.

The relative timing of phasic inspiratory EMG activity of the upper airway to diaphragmatic and ribcage muscle activity varies during sleep in OSA (Hudgel & Harasick, 1990). Around the onset of obstruction, upper airway muscle EMG activity may fall behind the ribcage EMG, which facilitates airway collapse, but the normal pattern is restored as the apnoea progresses, thus facilitating airway reopening (Hudgel & Harasick, 1990). A clinical model of a disturbed timing relationship between upper airway and diaphragmatic contraction predisposing to OSA is seen in patients with diaphragmatic palsy treated with an electrophrenic pacemaker. Such patients are susceptible to OSA because the pacemaker results in diaphragmatic contraction at times other than when upper airway muscles contract.

3.5.1 Loop Gain

The predisposition to apnoea associated with recurring cycles of hyper/hypoventilation during sleep varies considerably relating to the respiratory control system gain and sleep state stability. Ventilatory instability depends on the loop gain of the respiratory control system. In general terms, loop gain refers to the stability of a system controlled by a feedback loop. In the context of respiratory control, loop gain refers to the gain of the negative-feedback loop that regulates ventilation in response to a ventilatory disturbance. Variations in loop gain may constitute an important potential contributing factor to obstructive apnoea. A high loop gain occurs where the magnitude of the increase in ventilation following apnoea is high, thus increasing ventilatory system instability and increasing the likelihood of subsequent apnoea.

Two types of respiratory control system gain are evident, namely plant gain and controller gain, which are both determinants of loop gain and consequent ventilatory stability. Plant gain relates to the background drive to breathe. A higher ventilatory drive protects against apnoea by requiring a larger additional transient hyperventilation and hypocapnia to reach the apnoeic threshold (low plant gain). Conversely, a reduced ventilatory drive and associated hypoventilation increases susceptibility to apnoea, by requiring only small transient ventilatory overshoots to reach the apnoeic threshold (high plant gain). Controller gain relates to chemoreponsiveness, especially the hypercapnic ventilatory response, and quantitively describes the slope of the change in ventilation in response to CO_2. An increased slope results in an increased susceptibility to apnoea even in the setting of background hyperventilation and low plant gain. However, loop gain can be difficult to measure, and there are few clinical studies that have explored this variable in the setting of OSA. Thus, the importance of loop gain as an inherent contributor to OSA pathophysiology remains uncertain.

3.5.2 Arousal

Termination of apnoea is usually associated with brain arousal, and thus, the arousal response may be an important protective mechanism (Eckert & Malhotra, 2008). However, the physiological events associated with arousal may have deleterious consequences that contribute to the pathophysiology of OSA, both by contributing to daytime sleepiness because of sleep disturbance, but more importantly, by predisposing to further upper airway collapse, thus predisposing to repetitive apnoeas (McNicholas, 1998). Studies of transient upper airway occlusion in normal sleeping subjects demonstrate that, if the occlusion is associated with arousal, hyperventilation and associated hypocapnia occurs following apnoea termination, whereas if no arousal occurs, hyperventilation is limited, and CO_2 may rise. OSA patients appear to more reliant on arousal at the termination of apnoea than normal subjects (Jordan et al., 2007). The post-apnoeic hyperventilation and fall in PCO_2 associated with arousal may reduce respiratory drive, and the resulting reduced drive to the UA muscles may predispose to further obstructive apnoea, and a repetitive cycle of recurring apnoeas may ensue (McNicholas, 1998). It has been proposed that arousal is not necessary for the upper airways to reopen and that the consequences of arousal at apnea termination are largely deleterious by promoting ventilatory instability (Younes, 2004). However, a recent report indicated that the intensity of respiratory cortical arousals is a distinct pathophysiologic feature and is associated with disease severity in patients with OSA (Bahr et al., 2021).

Factors relating to obstructive apnoea that contribute to the arousal response include inspiratory efforts against an occluded airway, hypoxia, and hypercapnia. In humans, hypercapnia is a more important stimulus to arousal than hypoxia. Increasing ventilatory effort is an important factor in the arousal response, possibly mediated by mechanoreceptor feedback from respiratory muscles and/or from pressure-sensitive mechanoreceptors in the upper airway. Overall, increasing ventilatory effort appears to be the most important stimulus to arousal, and the stimulus to arousal from hypoxia and hypercapnia may be mediated principally through stimulating an increased ventilatory effort (Deegan & Mcnicholas, 1995).

The arousal response varies in patients with OSA and can be quantified by the arousal threshold. Experimentally, this threshold is measured as the minimum oesophageal pressure generated on the breath preceding arousal during a respiratory load or occlusion, and can be quantified non-invasively by polysomnography (Sands et al., 2017). As a group, OSA patients tend to have a higher arousal threshold than normal subjects, although there is considerable inter-subject variability in both groups. A low arousal threshold is an important potential contributing factor to OSA pathophysiology and may represent a therapeutic target in selected patients (Eckert et al., 2011).

3.6 Pathophysiological Endotypes and Phenotypes

The relevance of physiological, non-anatomic factors in the pathophysiology of OSA has been generating major interest in recent years (Randerath et al., 2018). These factors can be related to the underlying aetiology, referred to as endotype, and/or clinical manifestation, referred to as phenotype (Edwards et al., 2019), and may be viewed as a continuum from the genotype to personalized treatment options based on the individual endotype. Inadequate responsiveness of the genioglossus muscle, the arousal threshold, the critical closing pressure of the upper airway, and the stability of the respiratory control system defined by factors such as loop gain, define distinct endotypes of OSA that may be amenable to specific treatment approaches (Randerath et al., 2018). In one report of subjects with and without OSA, similar proportions of subjects, roughly one third each, had the endotypic traits of a minimal genioglossus muscle responsiveness during sleep, a low arousal threshold, or a high loop gain, and 28% of subjects had more than one of these traits (Eckert et al., 2013).

Phenotypes of pharyngeal dysfunction in OSA, such as collapsibility and pharyngeal muscle compensation, are evident from spontaneous changes in ventilation and ventilatory drive during sleep, which may be noninvasively assessed by polysomnography (Sands et al., 2018). There appear to be gender differences in OSA endotypes, with one report indicating that women demonstrate lower loop gain, less airway collapsibility, and lower arousal threshold in NREM sleep (Won et al., 2019), and endotypes explained 30% of the relative sex differences in NREM.

3.7 Integrated Pathophysiology

While the fundamental deficit in the pathophysiology of OSA relates to the inability of the upper airway dilating muscles to maintain a patent airway, the foregoing discussion indicates that many factors contribute to this basic pathophysiology. These factors vary in importance in different patients and in different sleep stages. For example, ventilatory drive withdrawal has recently been reported to be a more important mechanism of OSA than reduced genioglossus muscle compensation in REM sleep (Messineo et al., 2022). Overall, an insufficiency in drive to the upper airway dilating muscles for whatever reason, be it due to sleep-related factors such as the arousal threshold, respiratory control factors such as loop gain, or inadequate dilating muscle compensation, these factors interact to varying and overlapping degrees to result in the increased negative intrapharyngeal pressure that is a consequence of airway narrowing being sufficient to collapse the oropharyngeal airway (Fig. 3.1).

3.8 Implications for Treatment

While the basic deficit of increased upper airway collapsibility in OSA can be readily reversed by CPAP therapy, a detailed understanding of the pathophysiology opens the potential for other management options and has the subject of extensive research, especially in recent years (Schütz et al., 2021). Inadequate upper airway dilating muscle compensation may be improved by targeted pharmacotherapy. Desipramine, which is a tricyclic antidepressant (TCA) that inhibits the norepinephrine reuptake receptor in the central nervous system, reduces the sleep-related loss of genioglossus activity and improves pharyngeal collapsibility in healthy humans (Taranto-Montemurro, Edwards, et al., 2016), and has been reported to reduce the AHI in OSA patients who demonstrate minimal genioglossus muscle compensation (Taranto-Montemurro, Sands, et al., 2016). Another norepinephrine reuptake inhibitor (atomoxetine) combined with an antimuscarinic (oxybutynin) have also been reported to substantially reduce AHI in patients with OSA (Taranto-Montemurro et al., 2019).

Sleep-induced reduction in respiratory motor neurone output can be reversed by electrical stimulation of the hypogossal nerve and this therapeutic approach is gaining support as a potential alternative therapy to CPAP (Heiser et al., 2021; Strollo et al., 2014). Acetazolamide may benefit OSA in selected patients with a high loop gain and has the added potential benefit of reducing blood pressure (Edwards et al., 2012; Eskandari et al., 2018). Diuretic therapy may also benefit OSA, especially in patients with fluid overload, by reducing nocturnal rostral fluid shift (Revol et al., 2020). Zolpidem increases sleep efficiency and the respiratory arousal threshold without changing sleep apnoea severity and pharyngeal muscle activity (Messineo et al., 2020).

Soft tissue accumulation in and around the oropharynx that contributes to airway narrowing can be treated medically or surgically, as appropriate. Children with adenotonsillar hypertrophy and OSA benefit from surgical removal (Stradling et al., 1990) and adults with OSA and central obesity benefit from weight reduction, induced by bariatric surgery (Currie et al., 2021) or medically by intensive dietary measures and/or pharmacological therapy (Chirinos et al., 2014). Liraglutide, which is a long-acting glucagon-like peptide one receptor agonist, has been reported to induce weight loss and lead to a significant reduction in AHI in patients with OSA (Blackman et al., 2016).

The role of oxygen therapy in the management of OSA is uncertain, although a recent report suggests that oxygen supplementation may benefit OSA acutely, possibly by reducing the arousal response (Joosten et al., 2021).

3.9 Conclusion

The complex pathophysiology of OSA offers opportunities to develop targeted therapy based on an understanding of the multiple interacting factors that contribute to upper airway collapse in individual patients. These measures offer the opportunity for precision therapy as an alternative to the established uniform therapy of CPAP.

References

Abramson, Z., Susarla, S., August, M., Troulis, M., & Kaban, L. (2010). Three-dimensional computed tomographic analysis of airway anatomy in patients with obstructive sleep apnea. *Journal of Oral and Maxillofacial Surgery, 68*, 354–362.

Bahr, K., Geisler, V., Huppertz, T., Groppa, S., Matthias, C., Gouveris, H., & Muthuraman, M. (2021). Intensity of respiratory cortical arousals is a distinct pathophysiologic feature and is associated with disease severity in obstructive sleep apnea patients. *Brain Sciences, 11*, 282.

Blackman, A., Foster, G. D., Zammit, G., Rosenberg, R., Aronne, L., Wadden, T., Claudius, B., Jensen, C. B., Mignot, E., & on behalf of the, S. S. G. (2016). Effect of liraglutide 3.0 mg in individuals with obesity and moderate or severe obstructive sleep apnea:

The SCALE sleep apnea randomized clinical trial. *International Journal of Obesity, 40*, 1310–1319.

Bravo, G., Ysunza, A., Arrieta, J., & Pamplona, M. C. (2005). Videonasopharyngoscopy is useful for identifying children with Pierre Robin sequence and severe obstructive sleep apnea. *International Journal of Pediatric Otorhinolaryngology, 69*, 27–33.

Brouillette, R. T., & Thach, B. T. (1980). Control of genioglossus muscle inspiratory activity. *Journal of Applied Physiology: Respiratory, Environmental and Exercise Physiology, 49*, 801–808.

Carberry, J. C., Jordan, A. S., White, D. P., Wellman, A., & Eckert, D. J. (2016). Upper airway collapsibility (Pcrit) and pharyngeal dilator muscle activity are sleep stage dependent. *Sleep, 39*, 511–521.

Chen, H., Cade, B. E., Gleason, K. J., Bjonnes, A. C., Stilp, A. M., Sofer, T., Conomos, M. P., Ancoli-Israel, S., Arens, R., Azarbarzin, A., Bell, G. I., Below, J. E., Chun, S., Evans, D. S., Ewert, R., Frazier-Wood, A. C., Gharib, S. A., Haba-Rubio, J., Hagen, E. W., … Redline, S. (2018). Multiethnic meta-analysis identifies RAI1 as a possible obstructive sleep apnea-related quantitative trait locus in men. *American Journal of Respiratory Cell and Molecular Biology, 58*, 391–401.

Chi, L., Comyn, F. L., Keenan, B. T., Cater, J., Maislin, G., Pack, A. I., & Schwab, R. J. (2014). Heritability of craniofacial structures in normal subjects and patients with sleep apnea. *Sleep, 37*, 1689–1698.

Chirinos, J. A., Gurubhagavatula, I., Teff, K., Rader, D. J., Wadden, T. A., Townsend, R., Foster, G. D., Maislin, G., Saif, H., Broderick, P., Chittams, J., Hanlon, A. L., & Pack, A. I. (2014). CPAP, weight loss, or both for obstructive sleep apnea. *The New England Journal of Medicine, 370*, 2265–2275.

Currie, A. C., Kaur, V., Carey, I., Al-Rubaye, H., Mahawar, K., Madhok, B., Small, P., Mcglone, E. R., & Khan, O. A. (2021). Obstructive sleep apnea remission following bariatric surgery: A national registry cohort study. *Surgery for Obesity and Related Diseases, 17*, 1576–1582.

Davies, R. J., & Stradling, J. R. (1990). The relationship between neck circumference, radiographic pharyngeal anatomy, and the obstructive sleep apnoea syndrome. *The European Respiratory Journal, 3*, 509–514.

Dayyat, E., Kheirandish-Gozal, L., Sans Capdevila, O., Maarafeya, M. M. A., & Gozal, D. (2009). Obstructive sleep apnea in children: Relative contributions of body mass index and Adenotonsillar hypertrophy. *Chest, 136*, 137–144.

Deegan, P. C., & Mcnicholas, W. T. (1995). Pathophysiology of obstructive sleep apnoea. *The European Respiratory Journal, 8*, 1161–1178.

Dempsey, J. A., Veasey, S. C., Morgan, B. J., & O'Donnell, C. P. (2010). Pathophysiology of sleep apnea. *Physiological Reviews, 90*, 47–112.

Eckert, D. J., & Malhotra, A. (2008). Pathophysiology of adult obstructive sleep apnea. *Proceedings of the American Thoracic Society, 5*, 144–153.

Eckert, D. J., Malhotra, A., Lo, Y. L., White, D. P., & Jordan, A. S. (2009). The influence of obstructive sleep apnea and gender on genioglossus activity during rapid eye movement sleep. *Chest, 135*, 957–964.

Eckert, D. J., Owens, R. L., Kehlmann, G. B., Wellman, A., Rahangdale, S., Yim-Yeh, S., White, D. P., & Malhotra, A. (2011). Eszopiclone increases the respiratory arousal threshold and lowers the apnoea/hypopnoea index in obstructive sleep apnoea patients with a low arousal threshold. *Clinical Science (London, England), 120*, 505–514.

Eckert, D. J., White, D. P., Jordan, A. S., Malhotra, A., & Wellman, A. (2013). Defining phenotypic causes of obstructive sleep apnea. Identification of novel therapeutic targets. *American Journal of Respiratory and Critical Care Medicine, 188*, 996–1004.

Edwards, B. A., Redline, S., Sands, S. A., & Owens, R. L. (2019). More than the sum of the respiratory events: Personalized medicine approaches for obstructive sleep apnea. *American Journal of Respiratory and Critical Care Medicine, 200*, 691–703.

Edwards, B. A., Sands, S. A., Eckert, D. J., White, D. P., Butler, J. P., Owens, R. L., Malhotra, A., & Wellman, A. (2012). Acetazolamide improves loop gain but not the other physiological traits causing obstructive sleep apnoea. *The Journal of Physiology, 590*, 1199–1211.

Eskandari, D., Zou, D., Grote, L., Hoff, E., & Hedner, J. (2018). Acetazolamide reduces blood pressure and sleep-disordered breathing in patients with hypertension and obstructive sleep apnea: A randomized controlled trial. *Journal of Clinical Sleep Medicine, 14*, 309–317.

Fiori, C. Z., Martinez, D., Montanari, C. C., Lopez, P., Camargo, R., Sezerá, L., Gonçalves, S. C., & Fuchs, F. D. (2018). Diuretic or sodium-restricted diet for obstructive sleep apnea-a randomized trial. *Sleep, 41*, zsy016.

Fouke, J. M., & Strohl, K. P. (1987). Effect of position and lung volume on upper airway geometry. *Journal of Applied Physiology (1985), 63*, 375–380.

Grunstein, R. R., Ho, K. Y., & Sullivan, C. E. (1991). Sleep apnea in acromegaly. *Annals of Internal Medicine, 115*, 527–532.

Heiser, C., Steffen, A., Hofauer, B., Mehra, R., Strollo, P. J., Jr., Vanderveken, O. M., & Maurer, J. T. (2021). Effect of Upper Airway Stimulation in Patients with Obstructive Sleep Apnea (EFFECT): A randomized controlled crossover trial. *Journal of Clinical Medicine, 10*, 2880.

Hudgel, D. W. W., & Harasick, T. (1990). Fluctuation in timing of upper airway and chest wall inspiratory muscle activity in obstructive sleep apnea. *Journal of Applied Physiology (1985), 69*, 443–450.

Joosten, S. A., Tan, M., Wong, A.-M., Landry, S. A., Leong, P., Sand, S. A., Beatty, C., Thomson, L., Stonehouse, J., Turton, A., Hamilton, G. S., & Ba, E. (2021). A randomized controlled trial of oxygen therapy for patients who do not respond to upper airway surgery for obstructive sleep apnea. *Journal of Clinical Sleep Medicine, 17*, 445–452.

Jordan, A. S., Wellman, A., Heinzer, R. C., Lo, Y. L., Schory, K., Dover, L., Gautam, S., Malhotra, A., &

White, D. P. (2007). Mechanisms used to restore ventilation after partial upper airway collapse during sleep in humans. *Thorax, 62*, 861–867.

Jordan, A. S., White, D. P., Lo, Y.-L., Wellman, A., Eckert, D. J., Yim-Yeh, S., Eikermann, M., Smith, S. A., Stevenson, K. E., & Malhotra, A. (2009). Airway dilator muscle activity and lung volume during stable breathing in obstructive sleep apnea. *Sleep, 32*, 361–368.

Kasai, T., Arcand, J., Allard, J. P., Mak, S., Azevedo, E. R., Newton, G. E., & Bradley, T. D. (2011). Relationship between sodium intake and sleep apnea in patients with heart failure. *Journal of the American College of Cardiology, 58*, 1970–1974.

Kiely, J. L., Deegan, P. C., & Mcnicholas, W. T. (1998). Resolution of obstructive sleep apnoea with growth in the Robin sequence. *The European Respiratory Journal, 12*, 499–501.

Kiely, J. L., Nolan, P., & Mcnicholas, W. T. (2004). Intranasal corticosteroid therapy for obstructive sleep apnoea in patients with co-existing rhinitis. *Thorax, 59*, 50–55.

Koutsourelakis, I., Georgoulopoulos, G., Perraki, E., Vagiakis, E., Roussos, C., & Zakynthinos, S. G. (2008). Randomised trial of nasal surgery for fixed nasal obstruction in obstructive sleep apnoea. *The European Respiratory Journal, 31*, 110–117.

Lévy, P., Kohler, M., Mcnicholas, W. T., Barbé, F., Mcevoy, R. D., Somers, V. K., Lavie, L., & Pépin, J. L. (2015). Obstructive sleep apnoea syndrome. *Nature Reviews. Disease Primers, 1*, 1–20.

Lyons, O. D., Chan, C. T., Yadollahi, A., & Bradley, T. D. (2015). Effect of ultrafiltration on sleep apnea and sleep structure in patients with end-stage renal disease. *American Journal of Respiratory and Critical Care Medicine, 191*, 1287–1294.

Lyons, O. D., Inami, T., Perger, E., Yadollahi, A., Chan, C. T., & Bradley, T. D. (2017). The effect of fluid overload on sleep apnoea severity in haemodialysis patients. *The European Respiratory Journal, 49*, 1601789.

Mcnicholas, W. T. (1998). Arousal in the sleep apnoea syndrome: A mixed blessing? *The European Respiratory Journal, 12*, 1239–1241.

Mcnicholas, W. T. (2008a). Diagnosis of obstructive sleep apnea in adults. *Proceedings of the American Thoracic Society, 5*, 154–160.

Mcnicholas, W. T. (2008b). The nose and OSA: Variable nasal obstruction may be more important in pathophysiology than fixed obstruction. *The European Respiratory Journal, 32*, 3–8.

Mcnicholas, W. T. (2017). COPD-OSA overlap syndrome: Evolving evidence regarding epidemiology, clinical consequences, and management. *Chest, 152*, 1318–1326.

Mcnicholas, W. T. (2019). Obstructive sleep apnoea and comorbidity - an overview of the association and impact of continuous positive airway pressure therapy. *Expert Review of Respiratory Medicine, 13*, 251–261.

Mcnicholas, W. T., Tarlo, S., Cole, P., ZAMEL, N., Rutherford, R., Griffin, D., & Phillipson, E. A. (1982). Obstructive apneas during sleep in patients with seasonal allergic rhinitis. *American Review of Respiratory Disease, 126*, 625–628.

Messineo, L., Eckert, D. J., Lim, R., Chiang, A., Azarbarzin, A., Carter, S. G., & Carberry, J. C. (2020). Zolpidem increases sleep efficiency and the respiratory arousal threshold without changing sleep apnoea severity and pharyngeal muscle activity. *The Journal of Physiology, 598*, 4681–4692.

Messineo, L., Eckert, D. J., Taranto-Montemurro, L., Vena, D., Azarbarzin, A., Hess, L. B., Calianese, N., white, D. P., Wellman, A., Gell, L., & Sands, S. A. (2022). Ventilatory drive withdrawal rather than reduced genioglossus compensation as a mechanism of obstructive sleep apnea in REM sleep. *American Journal of Respiratory and Critical Care Medicine, 205*, 219–232.

Mezzanotte, W. S., Tangel, D. J., & White, D. P. (1996). Influence of sleep onset on upper-airway muscle activity in apnea patients versus normal controls. *American Journal of Respiratory and Critical Care Medicine, 153*, 1880–1887.

Moraleda-Cibrián, M., Edwards, S. P., Kasten, S. J., Berger, M., Buchman, S. R., & O'Brien, L. M. (2014). Symptoms of sleep disordered breathing in children with craniofacial malformations. *Journal of Clinical Sleep Medicine, 10*, 307–312.

Neelapu, B. C., Kharbanda, O. P., Sardana, H. K., Balachandran, R., Sardana, V., Kapoor, P., Gupta, A., & Vasamsetti, S. (2017). Craniofacial and upper airway morphology in adult obstructive sleep apnea patients: A systematic review and meta-analysis of cephalometric studies. *Sleep Medicine Reviews, 31*, 79–90.

Orr, W. C., Males, J. L., & Imes, N. K. (1981). Myxedema and obstructive sleep apnea. *The American Journal of Medicine, 70*, 1061–1066.

Phillipson, E. (1978). Control of breathing during sleep. *The American Review of Respiratory Disease, 118*, 909–939.

Randerath, W., Bassetti, C. L., Bonsignore, M. R., Farre, R., Ferini-Strambi, L., Grote, L., Hedner, J., Kohler, M., Martinez-Garcia, M., Mihaicuta, S., Montserrat, J., Pepin, J.-L., Pevernagie, D., Pizza, F., Polo, O., Riha, R., Ryan, S., Verbraecken, J. & Mcnicholas, W. T. 2018. Challenges and perspectives in obstructive sleep Apnoea: Report by an ad hoc working group of the sleep disordered breathing Group of the European Respiratory Society and the European Sleep Research Society. *European Respiratory Journal*, 1702616. https://doi.org/10.1183/13993003.02616-2017. [Epub ahead of print].

Redolfi, S., Arnulf, I., Pottier, M., Lajou, J., Koskas, I., Bradley, T. D., & Similowski, T. (2011). Attenuation of obstructive sleep apnea by compression stockings in subjects with venous insufficiency. *American Journal of Respiratory and Critical Care Medicine, 184*, 1062–1066.

Resta, O., Foschino-Barbaro, M. P., Legari, G., Talamo, S., Bonfitto, P., Palumbo, A., Minenna, A., Giorgino, R., & De Pergola, G. (2001). Sleep-related breathing disorders, loud snoring and excessive daytime sleepiness in obese subjects. *International Journal of Obesity and Related Metabolic Disorders, 25*, 669–675.

Revol, B., Jullian-Desayes, I., Bailly, S., Tamisier, R., Grillet, Y., Sapène, M., Joyeux-Faure, M., Pépin, J.-L., Grillet, Y., Sapène, M., & Pépin, J.-L. (2020). Who may benefit from diuretics in OSA?: A propensity score-match observational study. *Chest, 158*, 359–364.

Rivlin, J., Hoffstein, V., Kalbfleisch, J., Mcnicholas, W., Zamel, N., & Bryan, A. C. (1984). Upper airway morphology in patients with idiopathic obstructive sleep-apnea. *American Review of Respiratory Disease, 129*, 355–360.

Sakat, M. S., Sütbeyaz, Y., Yüceler, Z., Kantarci, M., Kilic, K., & Kurt, S. (2016). Cephalometric measurements with multislice computed tomography in patients with obstructive sleep apnea syndrome. *The Journal of Craniofacial Surgery, 27*, 82–86.

Sands, S. A., Edwards, B. A., Terrill, P. I., Taranto-Montemurro, L., Azarbarzin, A., Marques, M., Hess, L. B., White, D. P., & Wellman, A. (2018). Phenotyping pharyngeal pathophysiology using polysomnography in patients with obstructive sleep apnea. *American Journal of Respiratory and Critical Care Medicine, 197*, 1187–1197.

Sands, S. A., Terrill, P. I., Edwards, B. A., Taranto Montemurro, L., Azarbarzin, A., Marques, M., De Melo, C. M., Loring, S. H., Butler, J. P., White, D. P., & Wellman, A. (2017). Quantifying the arousal threshold using polysomnography in obstructive sleep apnea. *Sleep, 41*.

Schütz, S. G., Dunn, A., Braley, T. J., Pitt, B., & Shelgikar, A. V. (2021). New frontiers in pharmacologic obstructive sleep apnea treatment: A narrative review. *Sleep Medicine Reviews, 57*, 101473.

Sher, A. E. (1992). Mechanisms of airway obstruction in Robin sequence: Implications for treatment. *The Cleft Palate-Craniofacial Journal, 29*, 224–231.

Skatrud, J. B., & Dempsey, J. A. (1983). Interaction of sleep state and chemical stimuli in sustaining rhythmic ventilation. *Journal of Applied Physiology: Respiratory, Environmental and Exercise Physiology, 55*, 813–822.

Stradling, J. R., Thomas, G., Warley, A. R., Williams, P., & Freeland, A. (1990). Effect of adenotonsillectomy on nocturnal hypoxaemia, sleep disturbance, and symptoms in snoring children. *Lancet, 335*, 249–253.

Strohl, K. P., Hensley, M. J., Hallett, M., Saunders, N. A., & Ingram, R. H., Jr. (1980). Activation of upper airway muscles before onset of inspiration in normal humans. *Journal of Applied Physiology: Respiratory, Environmental and Exercise Physiology, 49*, 638–642.

Strollo, P. J., Soose, R. J., Maurer, J. T., DE Vries, N., Cornelius, J., Froymovich, O., Hanson, R. D., Padhya, T. A., Steward, D. L., Gillespie, M. B., Woodson, B. T., Van De Heyning, P. H., Goetting, M. G., Vanderveken, O. M., Feldman, N., Knaack, L., & Strohl, K. P. (2014). Upper-airway stimulation for obstructive sleep apnea. *The New England Journal of Medicine, 370*, 139–149.

Tan, H.-L., Kheirandish-Gozal, L., Abel, F., & Gozal, D. (2016). Craniofacial syndromes and sleep-related breathing disorders. *Sleep Medicine Reviews, 27*, 74–88.

Taranto-Montemurro, L., Edwards, B. A., Sands, S. A., Marques, M., Eckert, D. J., White, D. P., & Wellman, A. (2016). Desipramine increases genioglossus activity and reduces upper airway collapsibility during non-REM sleep in healthy subjects. *American Journal of Respiratory and Critical Care Medicine, 194*, 878–885.

Taranto-Montemurro, L., Messineo, L., Sands, S. A., Azarbarzin, A., Marques, M., Edwards, B. A., Eckert, D. J., White, D. P., & Wellman, A. (2019). The combination of atomoxetine and oxybutynin greatly reduces obstructive sleep apnea severity. A randomized, placebo-controlled, double-blind crossover trial. *American Journal of Respiratory and Critical Care Medicine, 199*, 1267–1276.

Taranto-Montemurro, L., Sands, S. A., EDWARDS, B. A., Azarbarzin, A., Marques, M., De Melo, C., Eckert, D. J., White, D. P., & Wellman, A. (2016). Desipramine improves upper airway collapsibility and reduces OSA severity in patients with minimal muscle compensation. *European Respiratory Journal, 48*, 1340–1350.

Tuomilehto, H., Seppä, J., & Uusitupa, M. (2013). Obesity and obstructive sleep apnea – Clinical significance of weight loss. *Sleep Medicine Reviews, 17*, 321–329.

White, L. H., & Bradley, T. D. (2013). Role of nocturnal rostral fluid shift in the pathogenesis of obstructive and central sleep apnoea. *The Journal of Physiology, 591*, 1179–1193.

Wilson, S. L., Thach, B. T., Brouillette, R. T., & Abu-Osba, Y. K. (1980). Upper airway patency in the human infant: Influence of airway pressure and posture. *Journal of Applied Physiology: Respiratory, Environmental and Exercise Physiology, 48*, 500–504.

Won, C. H. J., Reid, M., Sofer, T., Azarbarzin, A., Purcell, S., White, D., Wellman, A., Sands, S., & Redline, S. (2019). Sex differences in obstructive sleep apnea phenotypes, the multi-ethnic study of atherosclerosis. *Sleep, 43*, zsz274.

Yildirim, N., Fitzpatrick, M. F., Whyte, K. F., Jalleh, R., Wightman, A. J., & Douglas, N. J. (1991). The effect of posture on upper airway dimensions in normal subjects and in patients with the sleep apnea/hypopnea syndrome. *The American Review of Respiratory Disease, 144*, 845–847.

Younes, M. (2004). Role of arousals in the pathogenesis of obstructive sleep apnea. *American Journal of Respiratory and Critical Care Medicine, 169*, 623–633.

Yu, J. L., & Rosen, I. (2020). Utility of the modified Mallampati grade and Friedman tongue position in the assessment of obstructive sleep apnea. *Journal of Clinical Sleep Medicine, 16*, 303–308.

4 Diagnosis of Obstructive Sleep Apnea in Patients with Associated Comorbidity

Félix del Campo, C. Ainhoa Arroyo, Carlos Zamarrón, and Daniel Álvarez

Abstract

Obstructive sleep apnea (OSA) is a heterogeneous disease with many physiological implications. OSA is associated with a great diversity of diseases, with which it shares common and very often bidirectional pathophysiological mechanisms, leading to significantly negative implications on morbidity and mortality. In these patients, underdiagnosis of OSA is high. Concerning cardiorespiratory comorbidities, several studies have assessed the usefulness of simplified screening tests for OSA in patients with hypertension, COPD, heart failure, atrial fibrillation, stroke, morbid obesity, and in hospitalized elders.

The key question is whether there is any benefit in the screening for the existence of OSA in patients with comorbidities. In this regard, there are few studies evaluating the performance of the various diagnostic procedures in patients at high risk for OSA. The purpose of this chapter is to review the existing literature about diagnosis in those diseases with a high risk for OSA, with special reference to artificial intelligence-related methods.

Keywords

Obstructive sleep apnea · Comorbidities · Diagnosis · Polysomnography · Respiratory event · Sleep staging · Home sleep apnea testing · Screening · Decision support system · Artificial intelligence · Machine learning

4.1 Introduction

Obstructive sleep apnea (OSA) is a chronic, complex, and heterogeneous respiratory disease of high prevalence in the general population, with important health consequences. OSA is associated with a great diversity of diseases. In the clinical evaluation of these patients cardiorespiratory, cerebrovascular, and metabolic comorbidities potentially linked with OSA should be investigated (Bonsignore et al., 2019; Marin-Oto et al., 2019). It shares common and very often

F. del Campo · D. Álvarez (✉)
Pneumology Department, Río Hortega University Hospital, Valladolid, Spain

Biomedical Engineering Group (GIB), University of Valladolid, Valladolid, Spain

Centro de Investigación Biomédica en Red en Bioingeniería, Biomateriales y Nanomedicina (CIBER-BBN), Instituto de Salud Carlos III, Madrid, Spain
e-mail: daniel.alvarez.gonzalez@uva.es

C. A. Arroyo
Pneumology Department, Río Hortega University Hospital, Valladolid, Spain

C. Zamarrón
Division of Respiratory Medicine, Hospital Clínico Universitario, Santiago de Compostela, Spain

T. Penzel, R. Hornero (eds.), *Advances in the Diagnosis and Treatment of Sleep Apnea*, Advances in Experimental Medicine and Biology 1384, https://doi.org/10.1007/978-3-031-06413-5_4

bidirectional pathophysiological mechanisms, which have significant implications on morbidity and mortality. The most frequent comorbidities are found in the vascular field, respiratory or metabolic among others. Associated diseases vary according to sex, being more frequent in patients with a higher severity of OSA.

Table 4.1 shows the most frequent comorbidities in which screening studies have been performed due to the high possibility of presenting OSA. In these patients, underdiagnosis of OSA is high. There are many reasons for this, including the fact that these patients often do not present with daytime sleepiness or the usual symptoms of OSA. Moreover, the symptoms of the disease themselves often mask the presence of OSA. Added to this is the fact that there is little diagnostic suspicion on the part of the physician (Costa et al., 2015). Hence the importance of early diagnosis in order to initiate treatment as soon as possible. In most of these comorbidities, the treatment of OSA is a therapeutic objective in itself, as it acts as a risk factor.

The key question is whether there is any benefit in the screening for the existence of OSA in patients with comorbidities. Currently, there are not enough studies to establish the existence of a benefit in the general asymptomatic population (Jonas et al., 2017; Rosen et al., 2017). Given the high frequency of OSA in these diseases with the possible benefit of treatment, the need for diagnostic studies in these patients can be assumed. However, some authors advocate the need to confirm the benefits of treatment through randomized studies, especially in relation to CPAP treatment, as a step prior to the need for screening (Sanchez-de-la-Torre et al., 2021, Kee et al., 2018). Moreover, there are few studies evaluating the performance of the various diagnostic procedures in patients at high risk for OSA (Treptow et al., 2015).

Table 4.2 summarizes the main approaches to the abbreviated diagnosis of OSA in the presence of comorbidities using simplified tools. Among the level IV procedures, pulse oximetry has been one of the most exhaustively studied biological signals for screening. Table 4.3 shows the characteristics of the main approaches to OSA diagnosis in patient with comorbidities based on the analysis of pulse oximetry. Biomedical signal processing techniques and artificial intelligence-based tools have hardly been applied to evaluate their usefulness in the group of diseases where there is a high risk of associated OSA.

The purpose of this chapter is to review the existing knowledge regarding diagnosis in those diseases with a high risk for OSA, with special reference to artificial intelligence-related methods.

Table 4.1 Main conditions commonly related to OSA where abbreviated screening tests have been assessed

High-risk patients	Obesity (BMI >35) Chronic obstructive pulmonary diseases Asthma Congestive heart failure Atrial fibrillation Refractory hypertension Type 2 diabetes Stroke TIA Pulmonary hypertension High-risk driving populations Preoperative for bariatric surgery Chronic renal failure Retinal vein occlusion Pregnancy Down syndrome

4.2 Chronic Obstructive Pulmonary Disease (COPD)

Both COPD and OSA are two very prevalent diseases in the general population, which are associated with high morbidity, especially in the area of cardiovascular disease. Their association has been widely described in the literature. Both diseases are characterized by low-grade inflammation (Zamarrón et al., 2008). Their association increases morbimortality and the costs associated with them, which makes it necessary to maintain an integral vision of the patient, being able to identify both diseases early and optimize their control (Jelic, 2008). The use of CPAP in these patients has been shown to reduce mortality (Marin et al., 2010).

Table 4.2 Evidences on the effectiveness of portable monitoring for OSA detection in patients with comorbidities

Author	Population	Method and setting	Goal	Se (%)	Sp (%)	ICC
Oliveira et al., (2012)	26 COPD patients showing symptoms of suffering from OSA	Method: Manual AHI Setting. SpO_2 from RP in-lab and at home	Agreement:			
			PSG vs. RP_{LAB}	–	–	0.61
			PSG vs. RP_{HOME}	–	–	0.47
			RP_{LAB} vs. RP_{HOME}			0.47
To et al., (2021)	74 patients Psychiatric illnesses Stroke Ischemic heart diseases Chronic kidney diseases Others	Method: Manual analyses Setting: In-hospital PSG and NOX-T3	AHI > 5.	97	100	0,85
			AHI >10.	89	88	
			AHI >15.	86	94	
Jen et al., (2020)	36 COPD patients	Setting: Validation Wach-Pat Automatic analyses	Agreement:			
			AHI >5.	95.8	55.6	
			AHI >15.	92.3	65	
			AHI >30.	88.9	95.8	
Chang et al., (2019)	90 COPD patients	Setting: Validation NoX-T3	AHI >5	95%	78%	
Quintana-Gallego et al., (2004)	90 chronic heart failure patients	Setting: In lab PSG. Polygraphy at home. Manual analyses	OSA and CSA detection			
			AHI >5.	82.5	88.6	
			AHI >10.	79.3	97.8	
			AHI >15.	68.4	94.6	
Tauman et al., (2020)	111 atrial fibrillation patients	Setting: Polygraphy and PSG in lab	AHI >15	88	63	
De Vries et al., (2015)	90 heart failure patients	Home PSG-apnea link	AHI >15	92.9	91.9	0,85
Araujo et al., (2018)	35 heart failure patients	PSG Apnea link	AHI >15	83.3	91.3	
Aurora et al., (2018)	53 heart failure patients	PSG type 2 Type 3 sleep study	AHI >5	95.8	80	0.94
Li et al., (2021)	84 heart failure patients	PSG in lab Polygraphy type 3	AHI >5	86.7	76.5	
Kepplingeret al., (2013)	61 stroke patients	Cardiorespiratory polygraph	AHI >15 AHI >30.	60 77.8		

AHI apnea-hypopnea index, *COPD* chronic obstructive pulmonary disease, *CSA*: central sleep apnea, *GOLD* Global Initiative for Chronic Obstructive Lung Disease, $MLP_{OX\text{-}LAB}$ multilayer perceptron artificial neural network trained with oximetric recordings from portable oximetry in the hospital, $MLP_{OX\text{-}HOME}$ multilayer perceptron artificial neural network trained with oximetric recordings from portable oximetry at home, *ODI4* oxygen desaturation index of 4%, *OSA* obstructive sleep apnea, *PSG* polysomnography, RP_{LAB} in-hospital respiratory polygraphy, RP_{HOME} respiratory polygraphy at home, *RP* respiratory polygraphy, SpO_2 blood oxygen saturation

Table 4.3 Summary of the studies using oximetry to assist in OSA diagnosis in patients with comorbidities and especial risk groups

Author (year)	Dataset (n)	Gold standard	Aim	Variables from oximetry	Classification technique	Se (%)	Sp (%)	Acc (%)
Ward et al., (2012)	173 patients with congestive heart failure regardless suspicion of OSA	Unattended PSG (either in-lab or at-home)	Binary classification (AHI ≥15 events/h)	ODI3 from portable oximetry	ODI3 > 7.5 desaturations/h	97	32	–
Aaronson et al., (2012)	56 stroke patients admitted to rehabilitation regardless suspicion of OSA	In-hospital attended RP	Binary classification (AHI ≥15 events/h)	ODI4 from RP	ODI4 > 15 desaturations/h	77	100	–
Scott et al., (2014)	59 COPD admitted for pulmonary rehabilitation regardless of suspicion of OSA	In-hospital PSG	Binary classification (AHI ≥15 events/h)	Visual inspection and ODI4 from in-lab portable oximetry	Manual visual inspection Automated ODI4	59 60	60 63	– –
Andrés-Blanco et al., (2017)	407 patients suspected of OSA with and without COPD	In-hospital PSG	Regression of AHI, common cut-offs	Statistical, spectral, and nonlinear	MLP ANN: $AHI_{LAB} \geq 15$			
					Non-COPD	97.5	58.6	87.3
					COPD	96.2	56.3	86.8
					$AHI_{HOME} \geq 15$			
					Non-COPD	97.5	24.1	78.2
					COPD	86.5	37.5	75.0
Lajoie et al., (2020)	674 COPD patients	In hospital PSG Home nocturnal oximetry	Binary Classification	Visual inspection	Cyclical changes			
Mohammadieh et al., (2021)	98 patients atrial fibrillation	In hospital PSG Home HSAT III	Severity categories AHI	AHI ODI_3	ODI 4,95 AHI=5.15	84.4 79.7	79.4 88.2	0.87 0.89
Sharma et al., (2017)	105 patients hospitalized heart failures	In-hospital Apnea link High resolution oximetry	Binary Classification AHI >5	ODI_3	ODI>5	89.8	50	83.8
Siarnick et al., (2021)	49 patients hospitalized stroke	In hospital PSG Oximetry before 7 days	Binary classification (AHI ≥15 events/h)	Variability index ODI3	ODI 15.3	90.5	75	
Lin et al., (2018)	Stroke	Home HSAT	Binary classification	Variability index ODI3	ODI>5	88.4	91.7	89.3

Acc accuracy, *AHI* apnea-hypopnea index, AHI_{HOME} estimated apnea-hypopnea index from at-home oximetry, AHI_{LAB} apnea-hypopnea index from PSG, *ANN* artificial neural network, *COPD* chronic obstructive pulmonary disease, *HSAT* home sleep apnea test, *MLP* Multilayer perceptron, *ODI3* oxygen desaturation index ≥3%, *ODI4* oxygen desaturation index ≥4%, *OSA* obstructive sleep apnea, *PSG* polysomnography, *RP* respiratory polygraphy, *Se* sensitivity; *Sp* specificity, *Var. ind.* variability index

Despite the high prevalence of its association, nocturnal polysomnography is not routinely recommended in COPD patients. In the same regard, spirometry is not routinely performed in clinical practice in patients with OSA.

As in the rest of the comorbidities, the diagnosis of OSA is often underestimated even though these patients often present symptomatology in relation to sleep quality. Gunduz found in his study that 58% of patients with COPD and no symptoms had OSA (Gunduz et al., 2018).

Nocturnal polysomnography would be indicated in patients with COPD in the presence of excessive daytime sleepiness, observed nocturnal apneas, morning headache as well as if *cor pulmonale* or polycythemia is present (McNicholas, 2017). Similarly, the presence of clinical deterioration disproportionate to pulmonary function, with the presence of excessive daytime sleepiness, polycythemia, or pulmonary hypertension with a baseline PaO_2 greater than 60 mmHg point to the diagnosis of OSA. Recently, the American Thoracic Society in its clinical guidelines on non-invasive ventilation in patients with COPD and chronic hypercapnic respiratory failure recommends that before starting ventilation it is necessary to perform an OSA screening using the STOP-BANG questionnaire (Macrea et al., 2020).

The most optimal method of diagnosing OSA in these patients is not determined (Malhotra et al., 2018). Very different clinical questionnaires have been used for the prediction of OSA in COPD patients: Epworth, STOP-BANG, Berlin Questionnaire, and Sleep Apnea Clinical Score. Most of them include small population sizes, presenting poor performance with high sensitivity and moderate specificity.

Thus, in a study carried out in a Chinese population using the Berlin and STOP-BANG questionnaires, the sensitivity and specificity achieved was similar to that obtained in patients without COPD (Wu et al., 2020), although greater diagnostic accuracy was obtained in patients with less pulmonary involvement measured in terms of forced expiratory volume in 1 second (FEV_1) or forced vital capacity (FVC). Xiong et al., compared the diagnostic performance of various questionnaires in 335 patients with COPD and OSA, finding that the questionnaire with the best performance was the Berlin questionnaire (Xiong et al., 2019). Particularly, for an AHI >15 events/h, they obtained a sensitivity of 77.6%, and a specificity of 55% and an area under the receiver operating characteristics curve (AUC) equal to 0.737. The specificity was higher in patients with severe OSA.

Arsian et al., in a comparative study of the usefulness of various questionnaires (Berlin, STOP-BANG and Epworth) in patients with suspected OSA, evaluated the impact of various comorbidities in 1003 patients, finding that the STOP-BANG showed the highest sensitivity and the highest PPV (97% and 91.4%, respectively) (Arsian et al., 2020). The STOP-BANG showed high sensitivity in the group of patients with comorbidities (hypertension, diabetes *mellitus*, coronary artery disease, COPD, asthma), while notably lower values of specificity were reached with respect to the group without comorbidity (Arsian et al., 2020).

Other authors have evaluated the symptoms of OSA themselves in the development of other predictive models. Thus, the study by Soler et al. does not show that clinical features such as male sex, body mass index (BMI), or neck circumference are relevant in the prediction of OSA in these patients (Soler et al., 2017) Other authors, such as Faria et al., were interested in a new predictive model called Sleep Apnea Clinical Score and randomly applied it to 24 patients with COPD referred to PSG, in order to assess both the BQ and the Epworth sleepiness scale. They reported that their predictive model had a better diagnostic performance, with an AUC of 0.82, higher than that obtained by the other questionnaires. In addition, a sensitivity close to 60% was obtained, although in this study the STOP-BANG was not included (Faria et al., 2015).

There are very few studies that evaluated the usefulness of respiratory polygraphy in patients with COPD. Oliveira et al., in one of the first studies in this regard, evaluated the usefulness of a respiratory polygraphy (Stardust) in patients with COPD (Oliveira et al., 2012). The study was initially performed in 72 patients, in GOLD stages II and III, but due to difficulties during enrollment, finally only 26 patients were included, which implied the presence of a high failure rate. The intra-class correlation coefficient (ICC) between the AHI derived from the respiratory polygraphy and that from standard PSG was 0.61 (0.28–0.8) in the hospital setting and 0.47 (0.11–0.72) at the patient's home. Graphical analysis showed a tendency to overestimate the AHI in mild cases and underestimate it in the more severe ones. The authors conclude that there is insufficient significant evidence to support the use of this diagnostic procedure in these subgroups of patients.

For the same aim, Chang et al. evaluated 90 patients diagnosed with COPD with a high probability of OSA who underwent home polygraphy (NoxT3) followed by a one-week in-hospital polysomnographic and polygraphic recording (Chang et al., 2019). The home study showed good agreement with the AHI and the rest of the studies, especially in the most severe cases. The authors reported 95% sensitivity, 78% specificity, 88% positive predictive value (PPV), and 89% negative predictive value (NPV) compared to PSG, with a kappa coefficient of 0.746 for an AHI >5 events/h. The failure rate was 5.6%. The CT90 obtained was higher in polygraphic recordings, both at home and in the hospital. The authors highlight the usefulness of these systems in patients with COPD, whose results are similar to those obtained in patients without comorbidity. Additionally, they reported similar results between manual and automatic analyses. The same equipment has been used for this purpose in the presence of various comorbidities: psychiatric diseases, stroke, ischemic heart disease, chronic kidney disease, and others, obtaining an underestimation of severity in each of the groups of diseases, finding a greater dispersion in relation to the concordance of NOX-T3 and polysomnography (To et al., 2021).

The usefulness of peripheral arterial tonometry in patients with COPD versus polysomnography was evaluated by Holmedahl et al. in 16 patients with COPD (Holmedahl et al., 2019). Concerning sleep staging, they obtained an accuracy of 63% and an agreement of 0.418 kappa, while an ICC of 0.957 (CI95% 0.878–0.985) was reached for the AHI estimation task. It is noticeable that the concordance was lower than that previously obtained in control patients and OSA (lower specificity). However, the accuracy for AHI was adequate. One of the limitations of the study was the small sample size, as well as its inability to differentiate between central and obstructive apneas.

Jen et al. evaluated the usefulness of the WatchPAT system with respect to polysomnography in 33 patients diagnosed with COPD (Jen et al., 2020). WatchPAT is a new device that records the peripheral arterial tone (PAT), heart rate, oximetry, actigraphy, position, snoring, and chest movements. It shows good agreement with the AHI, unaffected by the severity of lung function. The WatchPAT system overestimated total sleep and REM sleep time. The agreement with polysomnography was 78.8%, with an overestimation of the AHI in 18.2% of the cases, concluding that the WatchPAT is a good alternative test in patients with COPD for a severe-to-moderate degree of OSA. In patients with an AHI >15 events/h, they obtained a sensitivity of 92.3% and a specificity of 65%. A cut-off point of 20 events/h allowed for the same degree of severity, a sensitivity of 76.9%, and a specificity of 90%. Its main difficulty lies when the events are very close in time.

The role of nocturnal oximetry as an aid in the diagnosis of OSA in patients with COPD presents important limitations, because of the desaturations linked to COPD that these patients present during sleep. However, it has the advantage of being easy to access, being one of the tools most widely used as a screening test in patients with suspected of OSA. (Del Campo et al., 2018). Therefore, the design and validation of automated techniques for OSA detection based on unsupervised oximetry at home is justified in the context of COPD patients, which can benefit from an early therapy by means of CPAP.

One of the first approaches to the diagnostic utility of nocturnal oximetry in the diagnosis of OSA in patients with respiratory diseases was performed by Pépin et al. (1991). These authors attempted to evaluate the diagnostic behavior of the delta index in different respiratory diseases. The number of COPD patients included in the study was notably small (only eight subjects), which significantly limits the generalizability of their results. The value of the delta index obtained in patients with COPD was very low compared to other respiratory diseases also analyzed in the study, although they obtained a high and more balanced sensitivity-specificity pair.

Scott et al. sought to develop a strategy to interpret nocturnal pulse oximetry and evaluate its ability to detect OSA in patients with stage 3 and 4 COPD (Scott et al., 2014). Consecutive COPD patients referred for simultaneous oximetry and polysomnography were studied. Patients were diagnosed with OSA if the polysomnographic AHI was >15 events/h. These criteria consisted of visually identifying oximetry "events" (sustained desaturation ≥4% in 1 hour time scale), "patterns" (≥ 3 similar desaturation cycles/15 minutes time scale), and the automated oxygen desaturation index (ODI). AUC, sensitivity, specificity, and accuracy were computed. Of the 59 patients (27 males), 31 had OSA (53%). Among these 59 patients, 35 were correctly identified as having OSA corresponding to an accuracy of 59%, with a sensitivity and specificity of 59% and 60%, respectively. The AUC was 0.57 (CI95%: 0.55 to 0.59). Using a computerized software for scoring desaturation events (hypoxemia ≥4% for ≥10 s) and using a cutoff of ≥15 events/h (of sleep time) for diagnostic criteria, the sensitivity was 60%, the specificity 63%, and the AUC was 0.64 (CI95% 0.62–0.66) (Scott et al., 2014). Interpretation of pulse oximetry tracing was of modest diagnostic value in identifying OSA in patients with moderate to severe COPD.

Lajoie et al., within the INOX clinical trial (multicenter, randomized, double-blind, placebo-controlled trial of nocturnal oxygen therapy in patients with COPD and nocturnal oxygen desaturation), performed a polysomnography on those desaturating patients who presented a cyclic desaturation pattern suggestive of OSA, confirming the existence of OSA in 50% of the patients and concluding that the oximetry tracing is not useful (Lajoie et al., 2020). However, the study was performed in a small sample population and with particular constraints.

In one of the few existing studies applying machine learning techniques, Andrés et al. evaluated the usefulness of an automated diagnostic algorithm for OSA diagnosis in COPD patients based on nocturnal oximetry. They extracted statistical, spectral, and nonlinear characteristics from the oximetry signal, which fed a regression multilayer perceptron (MLP) artificial neural network aimed at estimating the AHI, both in the hospital and at home (Andrés-Blanco et al., 2017). The algorithm was validated in patients with and without COPD. A high ICC was obtained both in the hospital (0.937 vs. 0.936) and at the patient's home (0.731 vs. 0.788). For an AHI >15 events/h, the algorithm reached 87.3% and 86.8% accuracy in patients with and without COPD in the supervised hospital setting, respectively, while it reached 78.2% and 75% at home. It is concluded that an algorithm based on a MLP neural network model can be a good, simplified test in patients with moderate-to-severe OSA regardless of the presence of associated COPD.

Another area of interest, given the implications between OSA and the different comorbidities, is the diagnosis of these conditions in patients referred for OSA (Bar et al., 2021). This is the case of Levy et al., who tried to identify the presence of COPD using nocturnal oximetry in patients undergoing diagnostic PSG due to clinical suspicion of OSA on the basis that very often these patients are underdiagnosed (Levy et al., 2021). The study was performed in 350 patients, 70 of whom had COPD. Clinical and oximetric characteristics were used as input to the automated algorithm. Both logistic regression and random forest were assessed for this task. The random forest model obtained an AUC of 0.94 and a F1 score of 0.89.

COPD, like other respiratory comorbidities, may need continuous monitoring of CO_2. Often the presence of hypoventilation is not easily evidenced in these patients. The measurement of CO_2 in exhaled air (end-tidal CO_2) has become a reliable diagnostic method to assess the presence of hypoventilation during sleep in these patients (Mayer et al., 2017).

4.3 Cardiovascular Diseases

Several studies have shown the existence of a high prevalence of OSA in patients with cardiovascular disease, estimated at 40–60%, although the prevalence varies depending on the type of cardiovascular disease. Despite this, as in the rest of the comorbidities, OSA is very often underdiagnosed (Costa et al., 2015), mainly because these patients do not present the usual symptoms. Given the high prevalence of OSA in cardiovascular diseases together with the benefits of treatment, it is useful to design and assess abbreviated tests for these patients (McEvoy et al., 2016).

A recent scientific statement by the American Heart Association regarding OSA and cardiovascular disease (Yeghiazarians et al., 2021) recommends the following indications for screening for OSA: resistant/poorly controlled hypertension, pulmonary hypertension, and recurrent atrial fibrillation after cardioversion or ablation. Screening for the presence of symptoms was recommended as well in the following situations: heart failure, tachy-brady syndrome, sick sinus syndrome, ventricular tachycardia, survivors of sudden cardiac death, and stroke. Likewise, as future research directions and areas of research, it should be highlighted the use of artificial intelligence and machine learning for the processing and identification of actionable data in OSA patients and the development of personalized therapies (Gutierrez-Tobal et al., 2019).

4.3.1 Atrial Fibrillation

Atrial fibrillation is one of the most prevalent arrhythmias in the general population. It is highly frequent in patients with OSA (Traaen et al., 2020), both in men and women, with a prevalence ranging from 49% to 62%. Both entities share common pathophysiological mechanisms of a complex nature. It is accepted that treatment with CPAP can reduce recurrences of these episodes, especially in patients with episodes of paroxysmal atrial fibrillation, although most of these studies are observational (Youssef et al., 2018). It is important to know that atrial fibrillation is currently a common reason for consultation in sleep units, which demands further analysis.

OSA is considered a modifiable risk factor by most clinical guidelines, recommending its screening in a broad sense (Calkins et al., 2017), although it is not clearly specified how and when the sleep study should be indicated. The European Cardiology Society recommends screening for OSA in patients with asymptomatic AF before initiating rhythm control treatment such as catheter ablation (Hindricks et al., 2021), although its implementation in clinical practice is not established. For other authors, screening would be justified in patients with AF if the patient had an episode of stroke or suffer from recurrent arrhythmias (Marulanda-Londoño & Chaturvedi, 2017). In this regard, there is great interest in determining the most optimal option for diagnosing these patients (Kadhim et al., 2020), as well as optimizing their diagnosis and treatment, given that very often there is a lack of coordination between cardiologists and sleep units (Desteghe et al., 2021).

OSA is frequently underdiagnosed in patients with AF. As with other comorbidities, it is always necessary to ask patients about symptoms related to sleep-disordered breathing. Several questionnaires have been used for screening (BQ, STOP-BANG, Non-OSA), although these questionnaires have not been validated in this population (Genta et al., 2017; Mohammadieh et al., 2021), being considered of little value as a screening method as they have a low negative predictive value and a low specificity (Ranjan, 2020). A main limitation is the absence of somnolence in these patients, so that the application of the Epworth test will provide a low sensitivity. The Berlin questionnaire shows high sensitivity (86–100%) but lower specificity (30–89%). In 579 patients with AF, Traaen et al. reported a sensitivity of 84% and specificity of 45% using the STOP-BANG, with respiratory polygraphy as a diagnostic method (Traaen, 2020). The authors attributed the lower performance to the lack of drowsiness reported by these patients.

In one of the few comparative studies, May et al. evaluated the efficacy of these questionnaires in patients with atrial fibrillation with respect to a control group, with polysomnography being the reference method (May et al., 2020). They included 150 patients in each of the groups. The authors assessed the Epworth, STOP-BANG, BQ, and NoSAS questionnaires, as well as a new model based on snoring, age, neck circumference, and BMI. In both groups, the clinical questionnaires showed worse performance in patients with AF, except for the STOP-BANG. Thus, in the presence of AF they obtained an AUC of 0.75 (CI95% 0.66–0.86) and 0.79 using the NoSAS, for an AHI> 15 events/h as cutoff for clinical diagnosis. The inclusion in the model of clinical variables such as neck circumference, BMI, snoring, and age improves the results obtained by the STOP-BANG. The model reached a sensitivity of 45% and a specificity of 97% for an AHI> 15 events/h. The inclusion of physiological variables such as heart rate or left atrial volume did not improve the performance of model.

Starkey et al. tried to evaluate the usefulness of the Berlin and NoSAS questionnaires, as well as a technique called acoustic pharyngometry, in 188 patients using the ApneaLink as a diagnostic method (Starkey et al., 2021). They concluded that the questionnaires were not useful to predict OSA in these patients. With the same purpose, Delesie et al. evaluated the usefulness of these questionnaires in 100 patients referred to study for atrial fibrillation, to whom a polysomnographic study was performed (Delesie et al., 2021). None of them showed sufficient discriminative ability (OSA50, BQ, STOP-BANG, MOODS, SACS, and Epworth), with an AUC <0.70 in the detection of severe OSA. In these patients, Abumuamar et al. also found that these questionnaires present low specificity (Abumuamar et al., 2018).

With respect to the use of respiratory polygraphy, its diagnostic accuracy is not as clearly established as in studies performed with oximetry, which makes it necessary to search for accurate and validated techniques (Hendricks, 2020). Thus, *Linz* et al. performed in-hospital polysomnography independent of clinical suspicion in 439 patients, subsequently obtaining the oximetry signal from the PSG (Linz et al., 2018). The prevalence of severe-to-moderate OSA was 33.9%. The authors evaluated the performance of the desaturation index using a new automatic algorithm that takes into account resaturation after desaturation in order to increase specificity. For AHI >15 events/h, they found an AUC of 0.951(0.929–0.972), while 0.932 was reached for an AHI >30 events/h. With a desaturation index cutoff point of 4.1, they obtained a sensitivity of 91% and a specificity of 83% for an AHI >15 events/h, thus being useful to rule out the disease, showing a negative predictive value of 95%.

Mohammadieh et al. evaluated the usefulness of various clinical questionnaires and the value of the oximetry tracing extracted from a respiratory polygraph (apnea-link) performed at the patient's home in a series of 98 patients referred for AF (Mohammadieh et al., 2021). In this study, the ODI showed excellent diagnostic accuracy for an AHI >5 events/h, with an AUC of 0.874. Similarly, using the automated scoring tool, the ApneaLink reached 0.925 AUC for moderate and 0.925 AUC for severe OSA.

In a multicenter study, Tauman et al. evaluated the usefulness of automatic analysis with WatchPAT versus PSG in 101 patients with AF (Tauman et al., 2020). He obtained a good correlation, as well as 88% sensitivity and 63% specificity, with 0.89 PPV and an AUC of 0.85 for a cutoff of AHI >15 events/h. A kappa agreement of 0.42 was obtained with respect to sleep phases, being higher in the absence of AF episodes during the night. There were no significant differences neither in relation to the persistence or not of episodes of AF during the night nor concerning medication.

In these patients, the use of new generation implanted pacemakers has been used to assess the presence of OSA by incorporating a respiratory monitoring algorithm, although one of its drawbacks is the inability to assess the duration of apneas. A recent meta-analysis evaluated 5 cohort studies using the measurement derived from transthoracic impedance provided by various electronic devices and Holters, in order to assess its usefulness in screening for OSA (Wyckmans et al., 2021), being of particular interest in patients with severe OSA. In the same regard, Gonçalves et al. achieved a diagnosis of 62% in 81 patients who underwent pacemaker implantation, reporting an AUC 0.76 and a sensitivity of 78% (Gonçalves et al., 2019). Algorithms implemented in implanted defibrillators (apnea scan system) have also been used for this aim. Thus, in 25 patients with AF, Defaye et al. obtained an ICC of 0.67 (CI 95% 0.39–0.84) with respect to polysomnography. For a cut-off point of 30 events/h, they obtained a sensitivity of 100% and a specificity of 80% (Defaye et al., 2019).

This type of device has also proven usefulness in monitoring AF patients, especially those for whom OSA is not evident in the first study. An example of monitoring is the non-contact biomotion radar sensor (SleepMinder™; ResMed) that allows monitoring over long periods of time and has been used in the evaluation of patients with atrial fibrillation or in patients with heart failure.

4.3.2 Chronic Ischemic Heart Disease

In chronic ischemic heart disease, clinical questionnaires do not accurately predict the presence of OSA in the patients. Szymanski et al. used a model for the identification of risk factors in the development of OSA based on clinical parameters. In their model, they use logistic regression based on clinical and echocardiographic data from patients who have suffered an acute myocardial infarction (Szymanski et al., 2015). Their model takes into account left ventricular diastolic diameter, interventricular septal thickness, diagnosis of hypertension, BMI, and diastolic pressure, all of which are independent risk factors for a high risk of OSA, reaching 0.87 AUC.

4.3.3 Chronic Heart Failure

OSA is highly prevalent in patients with heart failure, estimated at 47–76%. It is accepted that the association between OSA and heart failure has implications in the prognosis of the disease (Valika & Costanzo 2017), being frequent in the presence of both central and obstructive apneas. Various societies such as the American College of Cardiology (ACC), the American Heart Association (AHA), and the Heart Failure Society of America have pointed out in their respective guidelines the importance of diagnosing the existence of a sleep-disordered breathing and initiating the correct treatment in these patients (Yancy et al., 2017). Screening in these types of patients has been performed in two contexts: in the stable phase or during hospital admission because of an exacerbation, the latter being considered a good opportunity, although the performance of screening methods in these patients is widely debated (Series, 2015).

The Epworth sleepiness scale, Stop-Bang, and the Berlin questionnaire have been used as the main screening questionnaires. Parisot et al. proposed a clinical scale in which age, BMI, the New York Heart Association (NYHA) scale, and sex are combined to help identify OSA in patients with heart failure using multivariate logistic regression (Parisot et al., 2015). For an AHI $\geq$ 5 events/h, they obtained 78.9% sensitivity, 61.5% specificity, and an AUC of 0.73.

Type III polygraphs have been shown to be useful in these subjects, both in patients hospitalized for exacerbation and in chronic forms. In one of the first studies, Quintana-Gallego et al. evaluated the usefulness of home polygraphy in 75 patients with heart failure who underwent hospital polysomnography and respiratory polygraphy in a randomized way over a period of 30 days (Quintana-Gallego et al., 2004). For an AHI cut-off point >5 events/h, the diagnostic accuracy was 78.6%, while for an AHI>15 events/h the sensitivity was 68.4% and the specificity 94.6%. On the other hand, respiratory polygraphy was able to detect the presence of both central and obstructive apneas (Quintana-Gallego et al., 2004).

De Vries et al. used the ApneaLink device in 90 patients with stable chronic heart failure, using home polysomnography as a reference (De Vries et al., 2015). These authors obtained an ICC of 0.85 (0.78–0.90) with a kappa coefficient of 0.59 for classification into the common degrees of OSA severity using automated scoring with

ApneLink. For an AHI >15 events/h, the sensitivity obtained was 92.9% and the specificity 91.9%. In this study, the AHI was overestimated in more than two-thirds by the portable device, considering its usefulness just to rule out the disease (De Vries et al., 2015).

Araujo et al. used the ApneaLink in 35 patients with heart failure, simultaneously performing a PSG as reference (Araujo et al., 2018). Using a cutoff of AHI >15 events/h, they obtained a sensitivity of 83.3%, specificity of 91.3%, accuracy of 88.6%, and 0.93 AUC. For values above AHI >20 events/h, automated scoring with the ApneaLink showed a trend towards underestimate. The authors found greater efficacy in patients with more severe OSA (Araujo et al., 2018).

Similarly, Aurora et al. evaluated the usefulness of this portable sleep monitoring in 57 patients admitted for heart failure. They reported a significant performance, with 95.8% sensitivity and 80% specificity, obtaining better results in relation to central apneas (Aurora et al., 2018).

Li et al. used a type 3 equipment to identify the different respiratory events in 84 patients admitted for exacerbation of heart failure. For an AHI >5 events/h, they reported a sensitivity of 86.7%, specificity of 76.5%, and a PPV of 92.9%. The equipment used was able to identify both obstructive and central apneas, as well as Cheyne-Stokes respiration (Li et al., 2021).

Sharma et al. performed a prospective study in a population composed of 105 patients admitted for heart failure who underwent simultaneous high-resolution oximetry and respiratory polygraphy (ApneaLink) as a reference method (Sharma et al., 2017). The presence of OSA was confirmed in 87% of the sample. For an ODI of 5 events/h, the sensitivity was 89.8%, specificity 50%, and accuracy 83%. In patients with an AHI >30 events/h, sensitivity remained high and specificity increased to 87.6%. However, saturation values differed between methods, probably due to the use of different oximeters (Sharma et al., 2017).

Central sleep apneas are frequent in patients with heart failure, and there are hardly any studies that evaluate the usefulness of simplified diagnostic approaches in this type of apneas. Thus, polysomnography continues to be the reference diagnostic method in this context. However, it is advisable for this type of equipment to use inductive plethysmography bands for the detection of respiratory effort. Within the simplified procedures, such as ApneaLink, an algorithm for Cheyne-Stokes breathing detection using the flow cannula is available. Using this algorithm, the recognition of this respiratory pattern achieved a sensitivity of 87% and a specificity of 94% (Weinreich et al., 2009). These same authors used spectral entropy to automatically detect this pattern (Weinreich et al., 2008). Similarly, the usefulness of detecting the presence of Cheyne-Stokes respiration by means of neural networks based on the spectral analysis of oximetry has been described, obtaining also a high performance. Using respiratory polygraphy, Li et al. obtained a sensitivity of 94.6% for the detection of Cheyne-Stokes respiration (Li et al., 2021).

In its initial design, the WatchPAT system did not have the possibility of identifying the presence of central apneas, an aspect of particular importance in diseases such as heart failure, where patients may present central sleep apnea. This device currently has a specific module for the identification of central apneas. Accordingly, in a recent multicenter study performed in 11 centers that included 84 patients with heart failure and/or atrial fibrillation, correlations higher than 0.8 were reached for both AHI and central AHI, obtaining in the latter case 67% sensitivity, 100% specificity, 100% PPV, and 94.7% NPV for an AHI >15 events/h, with a kappa index of 0.77 (Pillar et al., 2020).

Concerning the use of machine learning approaches, artificial neural networks have been applied to identify certain respiratory patterns, such as the presence of Cheyne-Stokes respiration, using a probabilistic neural network based on spectral analysis, oximetric indices, and the delta index (El-Solh et al., 2003; Weinreich et al., 2008).

4.4 Cerebrovascular Diseases

In a systematic review by Dong et al., an overall prevalence of OSA in patients with cerebrovascular disease was found to be 61.9% (Dong et al., 2018). In this framework, OSA is considered an independent risk factor for stroke. Treatment with CPAP reduces the risk of suffering a stroke episode, as well as having a beneficial effect on sleepiness, quality of life, and blood pressure control. However, its efficacy on the occurrence of new events is in doubt, as the researchers of the SAVE study found no evidence of a reduction in events including stroke episodes (McEvoy et al., 2016).

The high prevalence and possible treatment implications in these patients (Seiler et al., 2019) have led various scientific societies, such as the American Heart Association-American Stroke Association, to publish secondary stroke prevention guidelines in order to recommend that patients with ischemic stroke or transient ischemic attack (TIA) should consider an OSA assessment for diagnosis (Kleindorfer et al., 2021). However, these recommendations have hardly been implemented in clinical practice or accepted by all societies (Warner et al., 2019). In a study involving a total of 1000 patients, only 17% were offered a sleep test (Brown et al., 2020) and were hardly asked about symptoms related to OSA within the first 3 months. Unfortunately, the guidelines on this subject have not changed significantly. This emphasizes the need for randomized studies to ascertain the benefits of CPAP in these types of patients.

The study of the association between sleep apnea and stroke has been carried out both at the hospital level in the acute phase (Huhtakangas et al., 2019) and in the follow-up of these patients, although in the latter case the prevalence of OSA may be overestimated.

Clinical questionnaires have not been shown to be useful as screening methods in patients who have had a stroke (Sico et al., 2017; Takala et al., 2018), since they have moderate sensitivity and low specificity. Other authors even question the need for a pretest questionnaire given the high pretest probability of OSA in patients with cerebrovascular disease.

Several questionnaires have been used in these types of patients, mainly the Berlin and the Stop-Bang questionnaires (Boulos et al., 2016; Senaratna et al., 2017). Some authors have proposed modifications to the Stop-Bang questionnaire (Boulos et al., 2019) to increase its diagnostic performance, removing the neck circumference item due to its low impact in these patients, and incorporating oxygen saturation values, either ODI4% or presenting an oxyhemoglobin saturation < 88%, which added one point to the questionnaire. For their study, Boulos et al. used either polysomnography or the ApneaLink at home in 231 patients. For an AHI >15 events/h and a cutoff point of 3 desaturations, they obtained a sensitivity of 98.5%, but a very low specificity of 23%, although the diagnostic performance was superior to that of the questionnaire. This is a consequence of the absence of somnolence and obesity in these patients.

Katzan et al. retrospectively assessed a modification of the Stop-Bang questionnaire in patients who had previously undergone the questionnaire and polysomnography (Katzan et al., 2016). In 208 patients, they created six logistic regression-based predictive models, obtaining better results with the use of continuous variables than with dichotomous variables as in the STOP-BANG. The authors found high sensitivity in all the automated models, while low specificity. The proposed model was able to detect 14% more patients with OSA.

Similarly, the development of predictive models based on logistic regression has been attempted by other authors (Siarnick et al., 2021). Thus, Siarnik et al. included clinical and echocardiographic characteristics, such as BMI, diastolic dysfunction, and history of wake-up stroke onset, as input variables to a model, which was applied to 120 stroke patients. The proposed model reached a sensitivity of 82.9%, a specificity of 71.9%, and an AUC of 0.81 in patients with severe-to-moderate OSA. The performance was lower for central apneas (Siarnick et al., 2021).

Sico et al. developed a new model (sleep inventory) based on symptoms and anthropometric measurements, using home polysomnography as a reference (Sico et al., 2017). This model achieved an AUC of 0.73, failing to classify 25% of patients. It reached high sensitivity and very low specificity both in development and validation stages.

Respiratory polygraphy is a good alternative to polysomnography in these patients (Boulos et al., 2021; Bravata et al., 2017; Saletu et al., 2018), although it is necessary to select the appropriate diagnostic procedure. Using the SOMNOcheck polygraph for detecting moderate-to-severe OSA, Kepplinger et al. found a sensitivity of 94.7% in 61 patients with mild ischemic accidents (Kepplinger et al., 2013). Similarly, in the context of a rehabilitation unit, Saletu et al. only studied those patients who presented a positive result in the polygraphy, so he eventually assessed 33 patients, reporting a good concordance in the Bland-Altman plot (Saletu et al., 2018).

Boulos et al. conducted a comparative study in 250 post-stroke patients randomized to home study versus PSG (Boulos et al., 2021), although 94 patients were included in the group of home sleep apnea test and 71 in the polysomnography group. They found a higher prevalence of OSA in the group of patients who underwent a home study, with a higher proportion of patients with CPAP being more cost-effective.

Huhtakangas et al. assessed the feasibility of OSA screening in the acute phase of ischemic stroke using automatically and manually scored cardiorespiratory polygraphy (Huhtakangas et al., 2019). A diagnosis of OSA was confirmed in 111 (59.3%) out of 187 subjects. Automated scoring properly identified respiratory events. A high agreement was obtained (ICC = 0.869), being inferior for central and mixed apneas.

In a multicenter, prospective study conducted in 1330 patients who presented an ischemic stroke, Brown et al. found a prevalence of 67%. They used machine learning algorithms to build different models for automated diagnosis (Random Forests, Boosted Regression Models, XGBoost, Deep Learning and Stacked Ensembles) (Brown et al., 2019). One of the limitations of the study is that the reference test is the ApneaLink. For an AHI >10 events/h and by means of a Random Forest approach, they reached an AUC of 0.75, correctly classifying 72.5% of the validation samples. Superior performance was achieved compared to that obtained with a logistic regression and the rest of the algorithms assessed, but the gain was small, showing AUC ranging 0.68–0.73. The most important variables in the model were neck circumference, BMI, waist circumference, age, NIHSS, and pre-stroke daytime sleepiness (Brown et al., 2019).

Oximetry has been widely used in these patients. In patients recovering from stroke, ODI4% provided a sensitivity of 77% and a specificity of 100% in patients with moderate-to-severe OSA (Aaronson et al., 2012). In the same regard, Lin et al. studied 254 patients undergoing an ApneaLink study, showing that an ODI <5 ruled out the disease and an ODI >5 confirmed moderate-to-severe OSA for an RDI >15 events/h, with a specificity of 96.4% (Lin et al., 2018). However, they did not perform polysomnography as reference standard.

Siarnick et al. evaluated the usefulness of pulse oximetry in 420 patients with acute stroke (Siarnik et al., 2020). A control polysomnography was conducted, although the proportion of patients performing both tests was low. With an ODI-based cutoff point of 15.3, the authors found a sensitivity of 90.5%, specificity of 75% for moderate-to-severe OSA, correctly classifying 81.6% with an AUC of 0.86 (CI95% 0.76–0.97) (Siarnik et al., 2020).

Boulos et al. extracted the oximetry from either the PSG or the ApneaLink in 231 patients who had a stroke in the previous year (Boulos et al., 2016). The STOP-BANG was performed as abbreviated test as well. A score < 3 achieved the highest sensitivity, while 4 led to the highest specificity. They included in the STOP-BANG questionnaire certain oximetric values, improving the performance of this tool, being capable of identifying both high- and low-risk patients. It is important to note that the authors removed the neck circumference item from the questionnaire (Boulos et al., 2016).

Although deep learning techniques have been applied in the field of sleep-disordered breathing

(Vaquerizo-Villar et al., 2021), they have been scarcely used in patients with additional comorbidities. Bernardini et al. proposed an algorithm based on ECG and saturation monitoring obtained from unselected patients, to which they applied a convolutional-based deep-learning framework to detect apneas events (Bernardini et al., 2021). The authors validated their algorithm in 30 patients using in-laboratory polysomnograpy as reference.

Leino et al. used the oximetric recording as input to an algorithm also based on a convolutional neural network and they assessed its usefulness as a screening test for OSA in patients with cerebrovascular disease (Leino et al., 2021). The algorithm was previously developed in patients without cerebrovascular disease, while the authors proposed to assess its generalizability in these types of patients. The design group was composed of 1379 oximetry recordings obtained by means of a home polygraph (Embletta) and validated in 77 patients admitted for ischemic stroke or TIA who underwent a polygraph study, as well as in 394 patients with suspected OSA. A 4% drop in hypopneas was used instead of the common 3% decrease. The deep-learning model was trained to estimate the respiratory event index (REI). The agreement was close to 80% in the classification by degree of severity, although it was higher in the suspected OSA group. Errors in REI estimation appeared in apneas without desaturation. The main inconvenience is that central apneas, which are frequent in these patients, are not estimated. The ICC was 0.982 in patients with OSA and 0.972 in cerebrovascular disease patients, being the sensitivity and specificity high in both groups and in all degrees of severity. A correct classification of the categories was obtained in 88.3% and 77.9%, although the accuracy was better in the OSA suspicion group for a cutoff point of AHI >15 events/h. The sensitivity was 97.3% and the specificity 98.6% in the first group, while 92.3% sensitivity and 96.1% specificity were achieved in the second group.

Capnography monitoring has been used as a screening method for OSA in stroke patients. Dziewas et al. (2005) found a significant correlation between the AHI estimated from capnography and that derived from respiratory polygraphy. Assessing a population composed of patients with an AHI >15 events/h and using a cutoff point of 5 events/h for the estimated AHI from capnography, they achieved 100% positive predictive value, 86% negative predictive value, 87% sensitivity, and 100% specificity. Nevertheless, a trend to overestimation was observed.

All these studies have great heterogeneity in terms of design and timing of the disease, while the post-stroke data underestimate the true prevalence.

4.5 Diabetes

OSA is frequently associated with type 2 diabetes *mellitus*. It is estimated that 55%–85% of patients with this type of diabetes have also concomitant OSA (Tahrani et al., 2015). Several studies have shown that OSA contributes to the presence of glucose intolerance and the development of insulin resistance, hindering its control and leading to the appearance of vascular complications (Lindberg et al., 2012). The influence of CPAP treatment on glucose metabolism is not well known. The studies found in the literature report contradictory findings, although better results are linked with long-term therapy. Currently, there is a lack of evidence concerning the potential benefit of screening for OSA in these patients. Nevertheless, it is accepted to perform a diagnostic test in those patients showing symptoms (Donovan et al., 2017), although the most appropriate diagnosis method is not clearly established. In one of the few studies in this context, Chen et al. analyze the diagnostic ability of nocturnal oximetry derived from standard PSG along with other clinical variables in 440 patients with diabetes. The authors report a high diagnostic accuracy (AUC 0.94) for an ODI >5 events/h, with a sensitivity of 92% and a specificity of 73%, while for an ODI >25 events/h the sensitivity was 93% and the specificity 85% (Chen et al., 2021). Kurinami et al. (2018) analyze body composition data obtained via electrical bioimpedance of 186 patients with decompensated type 2 diabetes *mellitus* who required admission. They obtained an AUC of 0.70, with a great imbalance in the sensitivity-specificity pair (27.1% vs. 90.5%). In addition, the presence of OSAS was confirmed using a conservative diagnostic threshold (RDI >19 events/h). The use of clinical questionnaires (STOP-BANG, Berlin) shows no difference among them in terms of performance, being their overall diagnostic capacity suboptimal.

Acknowledgments This work was supported in part by "Ministerio de Ciencia e Innovación-Agencia Estatal de Investigación" and "European Regional Development Fund (FEDER)" under Projects PID2020-115468RB-I00 and RTC-2017-6516-1, in part by "Sociedad Española de Neumología y Cirugía Torácica (SEPAR)" under project 649/2018, by "Sociedad Española de Sueño (SES)" under project "Beca de Investigación SES 2019", and in part by "Centro de Investigación Biomédica en Red en Bioingeniería, Biomateriales y Nanomedicina, (CIBER-BBN), Spain" through "Instituto de Salud Carlos III" co-funded with FEDER funds. The work of Daniel Álvarez was supported by a "Ramón y Cajal" grant (RYC2019-028566-I) from the "Ministerio de Ciencia e Innovación – Agencia Estatal de Inves-tigación" co-funded by the European Social Fund.

References

Aaronson, J. A., van Bezeij, T., van den Aardweg, J. G., van Bennekom, C. A. M., & Hofman, W. F. (2012). Diagnostic accuracy of nocturnal oximetry for detection of sleep apnea syndrome in stroke rehabilitation. *Stroke, 43*, 2491–2493.

Abumuamar, A. M., Dorian, P., Newman, D., & Shapiro, C. M. (2018). The Stop-Bang questionnaire shows an insufficient specificity for detecting obstructive sleep apnea in patients with atrial fibrillation. *Journal of Sleep Research, 27*, 1–7.

Andrés-Blanco, A. M., Alvarez, D., Crespo, A., Arroyo, A., Cerezo, A., Gutierrez Tobal, G., et al. (2017). Assessment of automated analysis of portable oximetry as a screening test for moderate-to-severe sleep apnea in patients with chronic obstructive pulmonary disease. *PLoS One, 12*(11), e0188094.

Araujo, I., Marques, F., Andre, S., Araujo, M., Marques, S., Ferreira, R., et al. (2018). Diagnosis of sleep apnea in patients with stable chronic heart failure using a portable sleep test diagnostic device. *Sleep & Breathing, 22*(3), 749–755.

Arsian, B. O., Hosgor, Z. Z., & Orman, M. N. (2020). Which screening questionnaire is best for predicting obstructive sleep apnea in the sleep clinic population considering age, gender, and comorbidities? *Turkish Thoracic Journal, 21*(6), 383–389.

Aurora, R. N., Patil, S. P., & Punjabi, N. M. (2018). Portable sleep monitoring for diagnosing sleep apnea in hospitalized patients with heart failure. *Chest, 154*(1), 91–98.

Bar, N., Sobel, J. A., Penzel, T., Shamay, Y., & Bejar, J. (2021). From sleep medicine to medicine during sleep. A clinical perspective. *Physiological Measurement, 42*, 044006.

Bernardini, A., Brunello, A., Gigli, G. L., Montanari, A., & Saccomanno, N. (2021). AIOSA: An approach to the automatic identification of obstructive sleep apnea events based on deep learning. *Artificial Intelligence in Medicine, 118*, 102133.

Boulos, M. I., Colelli, D. R., Vaccarino, S. R., Kamra, E., Murray, B. J., & Swartz, R. H. (2019). Using a modified version of the "STOP-BANG" questionnaire and nocturnal oxygen desaturation to predict obstructive sleep apnea after stroke or TIA. *Sleep Medicine, 56*, 177–183.

Boulos, M. I., Kamra, M., Colelli, D. R., Kirolos, N., Gladstone, D., Boyle, K., et al. (2021). SLEEP SMART (Sleep Apnea Screening Using Mobile Ambulatory Recorders After TIA/Stroke): A randomized controlled trial. *Stroke, 53*, 710.

Boulos, M. I., Wan, A., Im, J., Elias, S., Frankul, F., & Atalla, M. (2016). Identifying obstructive sleep apnea after stroke/TIA: Evaluating four simple screening tools. *Sleep Medicine, 21*, 133–139.

Bonsignore, M. R., Baiamonte, P., Mazzuca, E., Castrogiobavvi, A., & Marrone, O. (2019). Obstructive sleep apnea and comorbidities: A dangerous liaison. *Multidisciplinary Respiratory Medicine, 14*, 8.

Bravata, D. M., McClain, V., Austin, C., Ferguson, J., Burrus, N., Miech, E. J., et al. (2017). Diagnosing and managing sleep apnea in patients with chronic cerebrovascular disease: A randomized trial of a home-based strategy. *Sleep & Breathing, 21*, 713–725.

Brown, D. L., He, K., Kim, S., Hsu, C.-W., Case, E., Chervin, R. D., et al. (2020). Prediction of sleep-disordered breathing after stroke. *Sleep Medicine, 75*, 1–6.

Brown, D. L., Jiang, X., Li, C., Case, E., Sozener, C. B., Chervin, R. D., et al. (2019). Sleep apnea screening is uncommon after stroke. *Sleep Medicine, 59*, 90.

Calkins, H., Hindricks, G., Cappato, R., Kim, Y. H., Saad, E. S., Aguinaga, L., et al. (2017). 2017 HRS/EHRA/ECAS/ APHRS/SOLAECE expert consensus statement on catheter and sugical ablation of atrial fibrillation. *Heart Rhythm, 14*(10), e275–e244.

Costa, L. E., Uchoa, C. H. G., Harmon, R. R., Bortolotto, L. A., LorenFiho, G., & Drager, L. F. (2015). Potential underdiagnosis of obstructive sleep apnea in the cardiology outpatient setting. *Heart, 101*(16), 1288–1292.

Chang, Y., Xu, L., Han, F., Keenan, B. T., Kneeland-Szanto, E., Zhang, R., et al. (2019). Validation of the Nox-T3 portable monitor for diagnosis of obstructive sleep apnea in patients with chronic obstructive pulmonary disease. *Journal of Clinical Sleep Medicine, 15*(4), 587–596.

Chen, L., Tang, W., Wang, C., Chen, D., Gao, Y., Ma, W., et al. (2021). Diagnostic accuracy of oxygen desaturation index for sleep-disordered breathing in patients with diabetes. *Frontiers in Endocrinology, 12*, 598470.

Del Campo, F., Crespo, A., Cerezo-Hernández, A., Gutiérrez-Tobal, G. C., Hornero, R., & Álvarez, D. (2018). Oximetry use in obstructive sleep apnea. *Expert Review of Respiratory Medicine, 12*(8), 665–681.

Defaye, P., Mendelson, M., Tamisier, R., Jacon, P., Venier, S., Arnold, N., et al. (2019). Validation of an apnea and hypopnea detection algorithm implemented in implantable cardioverter defibrillators. The AIRLESS study. *Scientific Reports, 9*, 9597.

Delesie, M., Knaepen, L., Hendrikcs, B., Huygen, L., Verbraeken, J., Weytjens, K., et al. (2021). The value of screening questionnaires/scoring scales for obstructive sleep apnoea in patients with atrial fibrillation. *Archives of Cardiovascular Diseases*. https://doi.org/10.1016/j.acvd.2021.08.002

Desteghe, L., Hendricks, J. M. L., Heidbuchel, H., Potpara, T. S., Lee, G. A., & Linz, D. (2021). Obstructive sleep apnoea testing and management in atrial fibrillation patients: A joint survey by the European heart rhythm association (EHRA) and the Association of Cardiovascular Nurses and Allied Professions (ACNAP). *Europace, 23*(10), 1667–1674.

De Vries, G. E., Van der Wal, H. H., Kerstjens, H. A. M., Van Deursen, V. M., Stegenga, B., Van Veldhuisen, D. J., et al. (2015). Validity and predictive value of portable two channel sleep screening tool in the identification of sleep apnea in patients with heart failure. *Journal of Cardiac Failure, 21*, 848–855.

Dong, R., Dong, Z., Liu, H., Shi, F., & Du, J. (2018). Prevalence, risk factors, outcomes, and treatment of obstructive sleep apnea in patients with cerebrovascular disease: A systematic review. *Journal of Stroke and Cerebrovascular Diseases, 27*(6), 1471–1480.

Donovan, L. M., Rueschman, M., Weng, M., Basu, N., Dudley, K. A., Bakker, J. P., et al. (2017). Effectiveness of an obstructive sleep apnea screening and treatment program in patients with type 2 diabetes. *Diabetes Research and Clinical Practice, 134*, 145–152.

Dziewas, R., Hopmann, B., Humpert, M., Böntert, M., Dittrich, R., Lüdemann, P., et al. (2005). Capnography screening for sleep apnea in patients with acute stroke. *Neurological Research, 27*, 83–87.

El-Solh, A. A., Magalang, U. J., Mador, M. J., Dmochowski, J., Veeramachaneni, S., Saberi, A., et al. (2003). The utility of neural network in the diagnosis of Cheyne-stokes respiration. *Journal of Medical Engineering & Technology, 27*, 54–58.

Faria, A. C., da Costa, C. H., & Rufino, R. (2015). Sleep apnea clinical score, Berlin questionnaire, or Epworth sleepiness scale: Which is the best obstructive sleep apnea predictor in patients with COPD? *International Journal of General Medicine, 8*, 275–281.

Genta, P. R., Drager, L. F., & Lorenzi-Filho, G. (2017). Screening for obstructivesleep apnea in patients with atrial fibrillation. *Sleep Medicine Clinics, 12*, 99–105.

Gonçalves, I. S., Agostinho, J. R., Silva, G., Guimarães, T., Bernardes, A., Santos, I., et al. (2019). Accuracy and utility of a pacemaker respiratory monitoring algorithm for the detection of obstructive sleep apnea in patients with atrial fibrillation. *Sleep Medicine, 61*, 88–94.

Gunduz, C., Basoglu, O. K., & Tasbakan, M. S. (2018). Prevalence of overlap syndrome in chronic obstructive pulmonary disease patients without sleep apnea symptoms. *The Clinical Respiratory Journal, 12*(1), 105–112.

Gutierrez-Tobal, G. C., Alvarez, D., Crespo, A., Del Campo, F., & Hornero, R. (2019). Evaluation of machine-learning approaches to estimate sleep apnea severity from at-home oximetry recordings. *IEEE Journal of Biomedical and Health Informatics, 23*(2), 882–892.

Hendricks, J. M. (2020). Screening for obstructive sleep apnoea in patients with atrial fibrillation: Much more than a screening tool alone. *IJC Heart Vasc, 31*, 100655.

Hindricks, G., Potpara, T., Dagres, N., Arbelo, E., Bax, J. J., Blomström, C., et al. (2021). 2020 ESC guidelines for the diagnosis and management of atrial fibrillation developed in collaboration with the European Association of Cardio-Thoracic Surgery (EACTS). *European Heart Journal, 42*, 373–498.

Holmedahl, N. H., Fjeldstad, O. M., Engan, H., Saxvig, I. W., & Gronli, J. (2019). Validation of peripheral arterial tonometry as tool for sleep assessment in chronic obstructive pulmonary disease. *Scientific Reports, 9*, 19392.

Huhtakangas, J. K., Huhtakangas, J., Bloigu, R., & Saaresranta, T. (2019). Unattended sleep study in screening for sleep apnea in the acute phase of ischemic stroke. *Sleep Medicine, 65*, 121–126.

Jelic, S. (2008). Diagnostic and therapeutic approach to coexistent chronic obstructive pulmonary disease and obstructive sleep apnea. *International Journal of COPD, 3*, 269–275.

Jen, R., Orr, J. E., DeYoung, P., Smales, E., Malhotra, A., & Owens, R. L. (2020). Accuracy of WatchPAT for the diagnosis of obstructive sleep apnea in patients with chronic obstructive pulmonary disease. *COPD, 17*(1), 34–39.

Jonas, D. E., Amick, H. R., Feltner, C., Weber, R. P., Arvanitis, M., Stine, A., et al. (2017). *Screening for obstructive sleep apnea in adults: An evidence review for the US preventive services task force: Evidence synthesis no. 146*. Agency for Healthcare Research and Quality. AHRQ publication 14–05216-EF-1.

Kadhim, K., Lau, D. H., & Sandres, P. (2020). Sleep apnea in atrial fibrillation highly prevalent, highly relevant, but most patients are not somnolent. *Int J Cardiol Heart Vasc, 26*, 100463.

Katzan, I. L., Thompson, N. R., Uchino, K., & Foldvary-Schaefer, N. (2016). A screening tool for obstructive sleep apnea in cerebrovascular patients. *Sleep Medicine, 21*, 70–76.

Kleindorfer, D. O., Towfighi, A., Chaturvedi, S., Cockroft, K. M., Gutierrez, J., Lombardi-Hill, D., et al. (2021). 2021 guideline for the prevention of stroke in patients with stroke and transient ischemic attack: A guideline from the American Heart Association/American Stroke Association. *Stroke, 52*(7), e364–e467.

Kee, K, Dixon, J, Shaw, J, Vulikh, E, Schlaich, M, Kaye, D. M., et al. (2018). Comparison of commonly used questionnaires to identify obstructive sleep apnea in a high-risk population. *J Clin Sleep Med. 14(12):2057–2064.*

Kepplinger, J., Barlinn, K., & Albright, K. (2013). Early sleep apnea screening on a stroke unit is feasible in patients with acute cerebral ischemia. *Journal of Neurology, 260*, 1343–1350.

Kurinami, N., Sugiyamaa, S., Ijima, H., Yoshida, A., Hieshima, K., Miyamoto, F., et al. (2018). Clinical usefulness of the body muscle-to-fat ratio for screening obstructive sleep apnea syndrome in patients with inadequately controlled type 2 diabetes mellitus. *Diab Res Clin Pract, 143*, 134–139.

Leino, A., Nikkonen, S., Kainulainen, S., Korkalainen, H., Töyräs, J., Myllymaa, S., et al. (2021). Neural network analysis of nocturnal SpO2 signal enables easy screening of sleep apnea in patients with acute cerebrovascular disease. *Sleep Medicine, 79*, 71–78.

Lajoie, A., Series, F., Bernard, S., Bernard, E., Egea-SantaOlalla, C. J., Abad Fernandez, A., et al. (2020). Reliability of home nocturnal oximetry in the diagnosis of overlap syndrome in COPD. *Respiration, 99*(2), 132–139.

Levy, J., Alvarez, D., Rosemberg, A. A., Alexandrovich, A., del Campo, F., & Behar, J. (2021). Digital oximetry biomarkers for assessing respiratory functions: Standards of measurement, physiological interpretation, and clinical use. *Npg Digital Medicine, 4*, 1–14.

Li, S., Xu, L., Dong, X., Zhang, X., Keenan, B. T., Hang, F., et al. (2021). Home sleep apnea testing of adults with chronic heart failure. *Journal of Clinical Sleep Medicine, 17*(7), 1453–1463.

Lin, S. H., Branson, C., Park, L., Leung, J., Doshi, N., & Auerbach, S. H. (2018). Oximetry as an accurate tool for identifying moderate to severe sleep apnea in patients with acute stroke. *Journal of Clinical Sleep Medicine, 14*(12), 2065–2073.

Lindberg, E., Theorell-Haglow, J., Svensson, M., Gislason, T., Berne, C., & Janson, C. (2012). Sleep apnea and glucose metabolism: A long-term follow-up in a community-based sample. *Chest, 142*, 935–942.

Linz, D., Kadhim, K., Brooks, A. G., Elliot, A. D., Hendricks, J. M., Lau, D. H., et al. (2018). Diagnostic accuracy of overnight oximetry for the diagnosis of sleep-disordered breathing in atrial fibrillation patients.Int. *Journal of Cardiology, 272*, 155–161.

Macrea, M., Oczkowski, S., Rochwerg, B., Branson, R. D., Celli, B., Coleman, M., et al. (2020). Long-term noninvasive ventilation in chronic stable hypercapnic chronic obstructive pulmonary disease. An official American Thoracic Society clinical practice guideline. *American Journal of Respiratory and Critical Care Medicine, 202*(4), e74–e87.

Marin-Oto, M., Vicente, E. E., & Marin, J. M. (2019). Lon term management of obstructive sleep apnea and its comorbidities. *Multidisciplinary Respiratory Medicine, 14*, 21.

Marin, J. M., Soriano, J. B., Carrizo, S. J., Boldova, A., & Celli, B. R. (2010). Outcomes in patients with chronic obstructive pulmonary disease and obstructive sleep apnea. The overlap syndrome. *American Journal of Respiratory and Critical Care Medicine, 182*, 325–331.

Marulanda-Londoño, E., & Chaturvedi, S. (2017). The interplay between obstructive sleep apnea and atrial fibrillation. *Frontiers in Neurology, 8*, 668.

Mayer, G., Schwalmstadt-Treysa, M. A., Regensburg, B. B., et al. (2017). S3-guideline non-restorative sleep / sleep disorders – Chapter "sleep related breathing disorders". *Somnologie, 20*(Suppl 2), S97–S180.

McEvoy, R. D., Antic, N. A., Heeley, E., Luo, Y., Ou, Q., Zhang, X., et al. (2016). CPAP for prevention of cardiovascular events in obstructive sleep apnea. *The New England Journal of Medicine, 375*, 919–931.

McNicholas, W. T. (2017). COPD-OSA overlap syndrome: Evolving evidence regarding epidemiology, clinical consequences, and management. *Chest, 152*(6), 1318–1326.

Malhotra, A., Schwartz, A. R., Schneider, H., Owens, R. L., DeYoung, P., Han, M. K., et al. (2018). ATS assembly on sleep and respiratory neurobiology. Research priorities in pathophysiology for sleep-disordered breathing in patients with chronic obstructive pulmonary dis-ease. An Official American Thoracic Society Research Statement. *American Journal of Respiratory and Critical Care Medicine, 197*(3), 289–299.

May, A. M., Wang, L., Kwon, D. H., Van Wagoner, D. R., Chung, M. K., Dalton, J. E., & Mehra, R. (2020). Sleep apnea screening instrument evaluation and novel model development and validation in the paroxysmal atrial fibrillation population. *IJC Heart & Vasculature, 31*, 100624.

Mohammadieh, A. M., Sutherland, K., Kanagaratnam, L. B., Whalley, D. W., Gillett, M. J., & Cistulli, P. A. (2021). Clinical screening tools for obstructive sleep apnea in a population with atrial fibrillation: A diagnostic accuracy trial. *Journal of Clinical Sleep Medicine, 17*(5), 1015–1024.

Oliveira, M. G., Nery, L. E., Santos-Silva, R., Sartori, D. E., Alonso, F. F., Togeiro, S. M., et al. (2012). Is portable monitoring accurate in the diagnosis of obstructive sleep apnea syndrome in chronic pulmonary obstructive disease? *Sleep Medicine, 13*, 1033–1038.

Parisot, J., Damy, T., Gellen, B., Covali-Noroc, A., Bodez, D., Rappeneau, S., et al. (2015). Sleep-disordered breathing in chronic heart failure: Development and validation of a clinical screening score. *Sleep Medicine, 16*(9), 1094–1101.

Pépin, J. L., Lévy, P., Lepaulle, B., & Guilleminault, C. (1991). Does oximetry contribute to the detection of

apneic events? Mathematical processing of the SaO2 signal. *Chest, 99*, 1151–1157.
Pillar, G., Berall, M., Berry, R., Etzioni, T., Shrater, N., & Hwang, D. (2020). Detecting central sleep apnea in adult patients using WatchPAT— A multicenter validation study. *Sleep & Breathing, 24*, 387–398.
Quintana-Gallego, E., Villa-Gil, M., Carmona-Bernal, C., Botebol-Benhamou, G., Martínez-Martínez, A., Sánchez-Armengol, A., et al. (2004). Home respiratory polygraphy for diagnosis of sleep-disordered breathing in heart failure. *The European Respiratory Journal, 24*(3), 443–448.
Ranjan, R. (2020). Is it time to screen all patients for sleep apnea before atrial fibrillation ablation? *Journal of the American College of Cardiology, 6*(12), 1507–1508.
Rosen, I. M., Kirsch, D. B., Chervin, R. D., Carden, K. A., Ramar, K., Aurora, R. N., et al. (2017). American Academy of sleep medicine Board of Directors. Clinical use of a home sleep apnea test: An American Academy of sleep medicine position statement. *Journal of Clinical Sleep Medicine, 13*(10), 1205–1207.
Saletu, M., Kotzian, S., Schwarzinger, A., Haider, S., Spatt, J., & Saletu, B. (2018). Home sleep apnea testing is a feasible and accurate method to diagnose obstructive sleep apnea in stroke patients during in-hospital rehabilitation. *Journal of Clinical Sleep Medicine, 14*, 1495–1501.
Sanchez-de-la-Torre, M., Lee, C.-H., & Barbé, F. (2021). Obstructive sleep apnea and atrial fibrillation: We need to go step by step. *Journal of Clinical Sleep Medicine, 17*(5), 869–870.
Scott, A. S., Baltzan, M. A., & Wolkove, N. (2014). Examination of pulse oximetry tracings to detect obstructive sleep apnea in patients with advanced chronic obstructive pulmonary disease. *Canadian Respiratory Journal, 21*, 171–175.
Seiler, A., Camilo, M., Korostovtseva, L., Haynes, A. G., Brill, A. K., Horvath, T., et al. (2019). Prevalence of sleep disordered breathing after stroke and TIA. A meta-analysis. *Neurology, 92*, e648–e654.
Series, F. (2015). Should all congestive heart failure patients have a routine sleep apnea screening? Pro. *Canadian Journal of Cardiology, 31*, 935–939.
Senaratna, C. V., Perret, J. L., Matheson, M. C., Lodge, C. J., Lowe, A. J., Cassim, R., et al. (2017). Validity of the Berlin questionnaire in detecting obstructive sleep apnea: A systematic review and meta-analysis. *Sleep Medicine Reviews, 36*, 116–124.
Sharma, S., Mather, P. J., Chowdhury, A., Gupta, S., Mukhtar, U., & Willes, L. (2017). Sleep overnight monitoring for apnea in patients hospitalized with heart failure (SOMA-HF study). *Journal of Clinical Sleep Medicine, 13*(10), 1185–1190.
Siarnick, P., Jurik, M., Klobučníkováa, K., Kollár, B., Pirošováb, M., Malik, M., et al. (2021). Sleep apnea prediction in acute ischemic stroke (SLAPS score): A derivation study. *Sleep Medicine, 77*, 23–28.
Siarnik, P., Jurík, M., Veverka, J., Klobučníkováa, K., Kollár, B., Turcani, P., et al. (2020). Pulse oximetric routine examination of sleep apnea in acute stroke (PRESS). *Sleep Medicine, 73*, 208–212.
Sico, J., Yaggi, H., Ofner, S., Cocato, J., Austin, C. H., Fergurson, J., et al. (2017). Development, validation, and assessment of an ischemic stroke or transient ischemic attack-specific prediction tool for obstructive sleep apnea. *Journal of Stroke and Cerebrovascular Diseases, 26*, 1745–1754.
Soler, X., Liao, S. Y., Marin, J. M., Lorenzi-Filho, G., Jen, R., DeYoung, P., et al. (2017). Age, gender, neck circumference, and Epworth sleepiness scale do not predict obstructive sleep apnea (OSA) in moderate to severe chronic obstructive pulmonary disease (COPD): The challenge to predict OSA in advanced COPD. *PLoS One, 12*(5), e0177289.
Starkey, S. Y., Jonasson, D. R., Alexis, S., Su, S., Johal, R., Sweeney, P., et al. (2021). Screening for obstructive sleep apnea in an atrial fibrillation population: What's the best test? *CJC Open, 3*, 442–449.
Szymanski, F. M., Platek, A. E., & Filipiak, K. J. (2015). Is obstructive sleep apnea associated with the risk of ischemic stroke in patients with atrial fibrillation? *International Journal of Cardiology, 2015*(184), 481–482.
Tahrani, A. A. (2015). Obstructive sleep Apnoea and vascular disease in patients with type 2 diabetes. *Eur Endocrinol, 11*(2), 81–89.
Takala, M., Puustinen, J., Rauhala, E., et al. (2018). Prescreening of sleep-disordered breathing after stroke: A systematic review. *Brain and Behavior: A Cognitive Neuroscience Perspective, 8*, e01146.
Tauman, R., Berall, M., Berry, R., Etzioni, T., Shrater, N., Hwang, D., et al. (2020). Watch-PAT is useful in the diagnosis of sleep apnea in patients with atrial fibrillation. *Nat Sci Sleep, 12*, 1115–1121.
To KW, Chan TO, Chan WC, Choo KL, Hui DSC. (2021) Using a portable monitoring device for diagnosing obstructive sleep apnea in patients with multiple coexisting medical illnesses. Clin Respir J. 15(10):1104–1112.
Traaen, G. M., Overland, B., Aakeroy, L., Hunt, T. E., Bendz, C., Sande, L., et al. (2020). Prevalence, risk factors, and type of sleep apnea in patients with paroxysmal atrial fibrillation. *Int J Cardiol Heart Vasc, 26*, 100463.
Treptow, E., Oliveira, M. G., Moreira, G. A., Togeiro, S., Nery, L. E., Tufyk, S., et al. (2015). Update on the use of portable monitoring system for the diagnosis of sleep apnea in specific population. *World J Respirol., 5*(1), 17–27.
Valika, A., Costanzo, M. R. (2017). Sleep-disordered breathing during congestive heart failure: To intervene or not to intervene? *Cardiac Failure Review, 3*(2), 134–139.
Vaquerizo-Villar, F., Álvarez, D., Kheirandish-Gozal, L., Gutiérrez-Tobal, G. C., Barroso-García, V., Santamaría-Vázquez, E., del Campo, F., Gozal, D., & Hornero, R. (2021). A convolutional neural network architecture to enhance oximetry ability to diagnose pediatric obstructive sleep apnea. *IEEE*

Journal of Biomedical and Health Informatics, 25(8), 2906–2916.

Warner, J. J., Harrington, R. A., Sacco, R. L., et al. (2019). Guidelines for the early management of patients with acute ischemic stroke: 2019 update to the 2018 guidelines for the early management of acute ischemic stroke. *Stroke, 50*, 3331–3332.

Weinreich, G., Armitstead, J., Töpfer, V., Wang, Y. M., Wang, Y., & Teschler, H. (2009). Validation of apnea link as screening device for Cheyne-stokes respiration. *Sleep, 32*(4), 553–557.

Weinreich, G., Armitstead, J., & Teschler, H. (2008). Pattern recognition of obstructive sleep apnoea and Cheyne-stokes respiration. *Physiological Measurement, 29*, 869–878.

Wyckmans, M., Tukanov, S., Winters, R., Stinissen, R., Vermeulen, H., Dendale, P., et al. (2021). Pacemaker guide screening for severe sleep apnea, a possible option for patients with atrial fibrillation. A systematic review and meta-analysis. *Pacing and Clinical Electrophysiology, 44*(8), 1421–1431.

Wu, Q., Xie, L., Li, W., Xiang, G., Hu, W., Jiang, H., et al. (2020). Pulmonary function influences the performance of Berlin Questionnaire, modified Berlin Questionnaire, and STOP-Bang score for screening obstructive sleep apnea in subjects with chronic obstructive pulmonary disease. *International Journal of COPD, 15*, 1207–1216.

Xiong, M., Hu, W., Dong, M., Wang, M., Chen, J., Xiong, H., et al. (2019). The screening value of ESS, SACS, BQ, and SBQ on obstructive sleep apnea in patients with chronic obstructive pulmonary disease. *International Journal of COPD, 14*, 2497–2505.

Yancy, C. W., Jessup, M., Bozkurt, B., Butler, B., Casey, D. E., Jr., Colvin, M. M., et al. (2017). 2017 ACC/AHA/HFSA focused update of the 2013 ACCF/AHA guideline for the management of heart failure: A report of the American College of Cardiology/American Heart Association task force on clinical practice guidelines and the Heart Failure Society of America. *Coll Cardiol, 70*(6), 776–803.

Yeghiazarians, J. H., Tietjens, J. R., Redline, S., Brown, D. L., El-Sherif, N., et al. (2021). Obstructive sleep apnea and cardiovascular disease: A scientific statement from the American Heart Association. *Circulation, 144*(3), e56–e67.

Youssef, I., Kamran, H., Yacoub, M., Patel, N., Goulbourne, C., Kumar, S., et al. (2018). Obstructive sleep apnea as a risk factor for atrial fibrillation: A meta-analysis. *J Sleep Disord Ther., 7*(1), 1000282.

Zamarrón, C., García, V., Morete, E., & del Campo, F. (2008). Association of chronic obstructive pulmonary diseases and obstructive sleep apnea consequences. *International Journal of COPD, 3*(4), 1–12.

5 Pediatric Obstructive Sleep Apnea: What's in a Name?

Allan Damian and David Gozal

Abstract

Obstructive sleep apnea is a highly prevalent disease across the lifespan and imposes substantial morbidities, some of which may become irreversible if the condition is not diagnosed and treated in a timely fashion. Here, we focus on the clinical and epidemiological characteristics of pediatric obstructive sleep apnea, describe some of the elements that by virtue of their presence facilitate the emergence of disrupted sleep and breathing and its downstream consequences, and also discuss the potential approaches to diagnosis in at-risk children.

Keywords

Adenotonsillar hypertrophy · Obesity · Sleep fragmentation · Arousal · Respiratory loading · Upper airway · Craniofacial · Neuromuscular tone · Upper airway reflexes · Polysomnography · Polygraphy · Intermittent hypoxia · Hypercapnia · Alveolar hypoventilation

Obstructive sleep apnea (OSA) is defined by the American Academy of Sleep Medicine (AASM) as a sleep-related breathing disorder that involves the presence of recurrent decrease or complete cessation of airflow despite ongoing efforts to breathe (American Academy of Sleep Medicine (AASM), 2021). In turn, these events may cause progressive reductions in the blood oxygen concentration along with elevations in carbon dioxide, and ultimately may be terminated arousals (either electroencephalographic or autonomic) corresponding to the occurrence of sleep fragmentation and leading among multiple other potential consequences to excessive daytime sleepiness (EDS). Unlike in adults however, the explicit manifestations of EDS in children may vary with such daytime symptoms often including behavioral problems like hyperactivity, sometimes leading to a (mis)diagnosis of attention deficit hyperactivity

A. Damian
Departments of Neurology, University of Missouri School of Medicine, Columbia, MO, USA

Comprehensive Sleep Medicine Program, University of Missouri School of Medicine, Columbia, MO, USA

D. Gozal (✉)
Comprehensive Sleep Medicine Program, University of Missouri School of Medicine, Columbia, MO, USA

Department of Child Health and the Child Health Research Institute, University of Missouri School of Medicine, Columbia, MO, USA
e-mail: gozald@health.missouri.edu

T. Penzel, R. Hornero (eds.), *Advances in the Diagnosis and Treatment of Sleep Apnea*, Advances in Experimental Medicine and Biology 1384, https://doi.org/10.1007/978-3-031-06413-5_5

disorder (ADHD) (Chung et al., 2016; Chervin et al., 2002; Chervin & Archbold, 2001; Smith et al., 2017; Spruyt & Gozal, 2011; O'Brien et al., 2004a). Of note, the typical presentation of EDS consists in the inordinate tendency to fall asleep in common situations such as watching tv, during school class, etc., in other words, reflecting the high propensity for sleep that is induced by the lack of sleep continuity and the changes in sleep states provoked by the respiratory events. Regardless of whether the patient has predominantly EDS or presents with EDS-induced hyperactive symptoms, the presence of such morbidities interferes with the quality of life, and especially in children, with their cognitive and academic performance (Gozal & Pope, 2001; Gozal, 1998; Beebe & Byars, 2011; Wu et al., 2021; Harding et al., 2021). Since the reversibility of such consequences is still unclear, particularly when OSA is left untreated or treated late in the course of the disease, early recognition and treatment are imperative. The disturbances in airflow in the upper respiratory tract may be due to specific physiologic and anatomic factors, including congenital malformations, although it is usually a combination of such elements that accounts for the majority of the cases. The gold standard in diagnosis remains the in-lab attended polysomnogram (PSG), despite the emerging popularity of home sleep apnea tests (HSAT) and arterial tonometry (e.g., WatchPat™) in adult Sleep Medicine practice. The latter two are not yet approved for routine use in Pediatric Sleep Medicine below a certain age, and as such access to testing may be somewhat problematic in underserved areas. Treatment of OSA in children initially consists of optimization of the therapy of concurrent issues such as weight management, orthodontics, allergic rhinitis, and asthma, but for the most part, children suffering from OSA will traditionally undergo adenotonsillectomy (T&A) as their first line of therapy, and eventually positive airway pressure (PAP) treatment when residual OSA occurs after T&A.

This chapter aims to provide an overview of pediatric OSA. We aim to present the pathophysiology and presentation of OSA in children. We also emphasize key differences in the pediatric clinical phenotypes versus the phenotypes of OSA in adults.

5.1 Historical Perspective and Epidemiology

Pediatric OSA was first described by Guilleminault, and colleagues in 1976 as a distinct entity, different from the one in adults, whereas the same author emphasized in 1981 that compared to adult OSA, pediatric OSA presented with more behavioral problems, particularly school problems, hyperactivity, nocturnal enuresis, sleep terrors, depression, insomnia, and psychiatric issues (Huang & Guilleminault, 2017).

A 2012 technical report by the American Academy of Pediatrics (AAP) estimated the prevalence of pediatric OSA as ranging from 1.1 to 5.7% (Marcus et al., 2012). However, this estimate may be lower than the actual frequency of the condition, owing to the under-diagnosis of sleep problems, including sleep-disordered breathing (SDB) in children, and the increasing trends in obesity affecting the pediatric population. Other factors contributing to underestimation of the actual true prevalence may include the low, albeit increasing awareness of parents to snoring and OSA in general, the under-documentation of the diagnosis in patients' charts, and the overall lack of screening for sleep-related issues during well-child visits (Meltzer et al., 2010). For example, the prevalence of habitual snoring, i.e., snoring three or more times per week and loud enough to gain recognition by the caretakers, was 11.7% in the pediatric community (O'Brien et al., 2003), while a more recent study in Japan estimated possible OSA in school-aged children at around 9.5% (Tsukada et al., 2018).

5.2 Risk Factors

Disturbances in upper airway functioning, mainly a reduction in upper airway functional dimensions during sleep, contribute to OSA.

Table 5.1 Common pediatric disorders affecting upper airway characteristics and associated with obstructive sleep apnea syndrome

I. Craniofacial anomalies
Apert syndrome
Crouzon syndrome
Pfeiffer syndrome
Treacher-Collins syndrome
Pierre Robin sequence
Stickler syndrome
Nager syndrome
Hallerman-Streif syndrome
Goldenhar syndrome
Rubenstein-Taybi syndrome
Down syndrome (trisomy 21)
Beckwith-Wiedemann syndrome
Achondroplasia
Klippel-Feil syndrome
Marfan syndrome
Choanal stenosis
Mucopolysaccharidoses (e.g., Hunter syndrome, hurler syndrome)
Trisomies 13 and 18
II. Neuromuscular disorders
Cerebral palsy
Spina bifida
Syringobulbia
Syringomyelia
Myasthenia gravis
Moebius syndrome
Arnold-Chiari malformation
Poliomyelitis
III. Miscellaneous disorders
Adenotonsillar hypertrophy[a]
Obesity[a]
Allergic rhinitis
Sickle cell disease
Asthma
Tonsillar tumors
Glossomegaly (macroglossia)
Thyroid tumors

[a]Constitute the majority of cases in otherwise healthy children

Risk factors can be divided into anatomic, mostly congenital or developmental; physiologic, comprising neurologic and inflammatory conditions; and miscellaneous conditions that encompass complications of certain other conditions or disorders. A summary is presented in Table 5.1.

5.3 Anatomic Considerations

The upper respiratory tract comprises the region from the external nares, all the way down to the larynx and trachea. Its main purpose is to warm and conduct the air to the lower respiratory tract. Several areas in this region can cause narrowing, and therefore partial or complete obstruction, which in turn may lead to OSA (Katz & D'Ambrosio, 2008).

5.4 Upper Airway Anatomy

5.4.1 Nasal Passages

Infants are thought to be obligate nose breathers. Therefore, any obstruction in the nasal passages, such as caused by congenital malformations (e.g., choanal stenosis, craniofacial syndromes), or inflammation (e.g., upper respiratory infection), or poor tone (e.g., hypotonia, decreased neuromotor tone) may result in OSA. Apart from frank obstruction (e.g., bilateral choanal atresia), obstruction may be caused by narrowing of passages or crowding of structures. Congenital malformations involving the facial, especially the maxillary and mandibular bones, such as presented in Table 5.1 can lead to crowding of the tonsils, adenoids, and other soft tissues. Children with all these conditions may present with OSA soon after birth (Arens et al., 2021). Macroglossia by itself may not necessarily lead to OSA (Follmar et al., 2014), but rather be the risk determinant of glossoptosis (Schaaf Jr et al., 2010). Decreased neuromotor tone may further reduce the airways size by facilitating the occurrence of glossoptosis, and hypopharyngeal collapse during sleep (Arens et al., 2021). Children with Down syndrome may be more prone to OSA owing to hypotonia and upper airway crowding (Arens et al., 2021; Goffinsky et al., 2015). Obstruction can also occur with irritation of the mucosa that leads to edema and laryngospasm, such as in the case of gastroesophageal reflux disease (GERD). GERD appears to be the most common co-morbidity in younger children, even if the temporal association between gastroesophageal episodes and upper airway obstruction is not consistently present (Goffinsky et al., 2015; Qubty et al., 2014; Nobile et al., 2019; Quitadamo et al., 2020).

5.4.2 Pharynx

There are key differences in the upper airways of infants versus those of older children and adults (Otteson et al., 2021; Arens et al., 2021; Chun & Arvedson, 2021). The infant larynx is higher in the neck, at the level of the second and fourth cervical vertebrae (Otteson et al., 2021). This brings the epiglottis closer to the uvula, which allows for temporally independent suckling and breathing alternations (Arens et al., 2021). As the infant matures, the larynx migrates down the level of the fifth vertebra at around 18 months of age (Arens et al., 2021), and eventually at the level of the seventh cervical vertebra in adulthood (Otteson et al., 2021).

The pharynx is generally divided into three regions (Arens et al., 2021):

1. The nasopharynx is located superior to the soft palate and is continuous with the nasal passages.
2. The oropharynx is below the soft palate, and above the larynx – it is continuous with the oral cavity and is bounded by the posterior third of the tongue, anteriorly. It can be further divided into two regions: the retropalatal region, between the hard and soft palates; and the retroglossal region, between the tip of the soft palate up to the tip of the epiglottis. Of note that because of the more superior location of the larynx, infants and young children are more likely to sustain obstructive episodes within the retropalatal region (Katz et al., 2012).
3. The hypopharynx from the tip of the epiglottis communicates with the cavity of the larynx.

The minimum cross-sectional area of the upper airway is usually situated at the level of the adenoids and soft palate (Arens et al., 2021; Isono et al., 1998). This "overlap region" in the retropalatal region was further defined as being the area where the adenoids overlap with the tonsils and soft palate (Arens et al., 2003; Fregosi et al., 2003). Various studies have shown that the dynamic fluctuations in this overlap region are sixfold higher in pediatric OSA than in controls (Arens et al., 2005), attesting to the increased instability and collapsibility of the upper airway (Gozal & Burnside, 2004; Huang et al., 2012).

5.4.3 Soft Tissues: Tonsils and Adenoids

Arens et al. studied the somatic growth relationships between the soft and bony tissues surrounding the upper airways in children ranging from 1 to 11 years of age (Arens et al., 2002). They reported that in healthy children there was proportional growth of the tongue, soft palate, and adenoids with the nasopharyngeal airway. There was also proportional growth of the mandible. Lastly fat pads appear to grow proportionally as well (Arens et al., 2002). However, in children with OSA, disproportional overgrowth of the tonsils and adenoids was apparent (Arens et al., 2021).

In most cases, adenotonsillectomy leads to improvement of breathing symptoms in OSA (Suen et al., 1995). One meta-analysis showed improvement in sleep outcomes, compared with no surgery, but sustainability of such findings beyond 12 months follow-up was not studied (Chinnadurai et al., 2017). It is thought that in about 10–15% of otherwise healthy children with OSA, adenotonsillectomy will not result in normalization of the breathing patterns (Arens et al., 2021; Tal et al., 2003; Martinot et al., 2018). Moreover, the success rates of adenotonsillectomy in obese children are remarkably low (See Table 5.2) (Bhattacharjee et al., 2010), such that presumed "cure" is estimated by some as yielding only 25–35% (Tauman et al., 2006a). Several risk factors have been identified for persistence of OSAS after adenotonsillectomy (Alonso-Álvarez et al. 2015; Lee et al. 2020; Huang & Guilleminault, 2017; Marcus et al., 2013; Bhattacharjee et al., 2010; Tauman et al., 2006a;

Boudewyns et al., 2017; Suri et al., 2015), as presented in Table 5.2:

5.4.4 Functional Considerations Underlying OSA in Children

It is clear that simply having marked adenotonsillar hypertrophy (sometimes referred to as "kissing tonsils") or a small retrognathic mandible are not sufficient to manifest as OSA in children (Fig. 5.1). Therefore, other elements contribute to the pathophysiology of OSA. Functional considerations that provide important determinants to the generation of OSA in at-risk children include ventilatory drive, response to resistive loading, and neuromotor tone.

5.4.5 Ventilatory Drive

Children with associated central nervous system disorders, such as spina bifida, may also have associated disordered breathing that is particularly manifest during sleep. This may be due to an altered central chemosensitivity, dysfunctional or insensitive carotid peripheral chemoreceptors, or a combination of both, as well as intervening reductions in muscle tone.

Table 5.2 Factors associated with persistence of OSA after surgical adenotonsillectomy

High pre-operative AHI
Obesity
High-arched palate
Mallampati score III–IV
Male gender
Age > 7 years
African-American ethnicity
Allergic rhinitis
Asthma

Central apnea or the occurrence of pauses in airflow without any respiratory effort is an important form of sleep-disordered breathing but is clearly outside the scope of this chapter and will not be the main subject of this discussion. In brief, these events occur in healthy infants and also in older children, are particularly prominent in premature infants, and are related to ongoing maturation of the respiratory control system. It appears, however that central apneas can occur more frequently in children with OSA (Del-Rio Camacho et al., 2019). Del-Rio Camacho et al. also showed that CA worsens with increasing OSA severity, and that adenotonsillectomy may improve both. A high central apnea index is thought to be related to a hypersensitive, rather than depressed peripheral chemoreceptors (Harman et al., 2020). Of note, in otherwise normal children the relative contribution of central chemosensitivity is probably small except in specific circumstances, such as when accompanied by long-term adaptations to elevated carbon dioxide related to increased upper airway resistance loads (Marcus et al., 1994; Gozal et al., 2013; Nava-Guerra et al., 2016).

5.4.6 Inspiratory Resistive Loading

In the context of Respiratory Muscle Training (RMT), Inspiratory Resistive Loading (IRL)

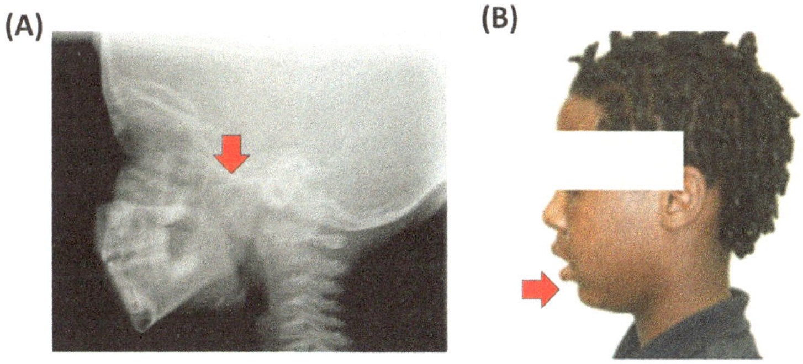

Fig. 5.1 Examples of enlarged adenoids (panel **a**) and retrognathia (panel **b**)

refers to the imposition of increasing respiratory resistances during the inspiratory phase, whereby the subject breathes through a tubing with a one-way valve. Exhalation is unimpeded, whereas inspiration can be loaded to varying resistances. In children with OSA, there appears to be a blunted arousal response to the increased upper airway resistive loads that occur with the narrowing or collapse of the airway during sleep (Marcus et al., 2013; Kohyama & Hasegawa, 2002; Marcus et al., 1999). The adaptive mechanisms of arousal in the context of the balance between sleep pressure generated from sleep fragmentation and the arousal response to inspiratory and expiratory loads are important determinants of apnea duration and overall thresholds at which the autonomic or cortical arousals will occur (Marcus et al., 1998b; Saito et al., 2002).

5.4.7 Arousals from Sleep

In general, arousals occur as a result of an obstructive event. Arousals during sleep are thought to be a protective mechanism since transition to wakefulness immediately recruits neural mechanisms governing the drive to breathe and restore many of the functional elements that due to their relative compromise were determinants of the upper airway increased resistance during sleep states. Accordingly, arousals coincide with increased dilator muscle activity, reduced upper airway resistance, and restoration of normal ventilation (Arens et al., 2021). In fact, during the initial few breaths immediately following an arousal, increased ventilation (hyperventilation) and often sighs (augmented breaths) will occur. This phenomenon is the result of the wakefulness drive to breathe at the same time that the eupneic carbon dioxide threshold is lowered by the transition from sleep to wakefulness. However, since the duration of such arousals is usually short-lived and sleep resumes, the eupneic threshold will increase again, and central apnea may emerge and will last until the carbon dioxide levels increase till reaching the eupneic threshold. Such events, which may occur in situations where the respiratory mechanics are not optimal and functional residual capacity of the lungs is still reduced, may lead to hypoxemia, the latter potentially resulting in periodic breathing. The increased number of arousals during sleep will lead to fragmentation (discontinuity of sleep), which even if it does not reduce the total duration of sleep will alter the microarchitecture and homeostasis of sleep, promoting the emergence of EDS. Indeed, in a murine model of sleep fragmentation mimicking the recurrent arousals observed in OSA patients, many of the consequences of OSA can be recapitulated and suggest potent induction of oxidative stress and inflammation (Ramesh et al., 2012; Nair et al., 2011; Gozal et al., 2010; Kim et al., 2011; Wang et al., 2014; Carreras et al., 2014).

There are unique aspects to the relationship between obstructive events and arousals in children, compared to adults. First, arousals in children occur more during REM sleep (Nino et al., 2021; Seppä-Moilanen et al., 2021; Bathory & Tomopoulos, 2017). Second, arousals associated with obstructive events are less frequent compared to adults, most likely because of the increased sleep pressure and the elevated arousal thresholds in children. Indeed, healthy children in general, and more so those with OSA, appear to have a higher threshold for arousal compared to adults (Moreira et al., 2005; Marcus et al., 1998a; Arens & Marcus, 2004). Nevertheless, in both adults and children there are reciprocal interactions between the occurrence of respiratory-related arousals and spontaneous arousals, attesting to the overall initial ability to compensate and maintain homeostasis but upon reaching a respiratory arousal index threshold, sleep pressure (an equivalent of EDS) will start becoming manifest and increase with increasing severity of OSA (O'Brien et al., 2004b; Tauman et al., 2004, 2006b).

5.4.8 Neuromotor Tone

Changes in upper airway neuromotor tone also play an important role in the etiology of OSA. Upper airway neuromotor tone and reflexes are increased in normal infants and children compared to adults, perhaps as a compensatory response for the relatively narrow airway (Follmar et al., 2014). However, subtle reductions in neuromotor recruitment or diseases that are characterized by reduced neuromotor tone can markedly aggravate underlying anatomical predisposition to increased upper airway resistance and thereby facilitate the emergence of OSA. Clinical examples include conditions such as neuromuscular disorders, cerebral palsy, but also genetic syndrome with hypotonia (e.g., Down syndrome, Prader-Willi syndrome, etc.).

5.4.9 Special Population: Childhood Obesity

Even though we have already alluded to the contribution of obesity to the risk and severity of OSA in children, it seems important to further emphasize this issue. In a study of approximately 250 prospectively community recruited, otherwise healthy obese children, we found a prevalence of OSA ranging from 21.5% to 39.5%, depending on the specific cut-off being selected for diagnosis (Alonso-Álvarez et al., 2014). Thus, there is an overall increased prevalence of OSA among children with obesity compared to those with normal weight, which is likely contributed by the crowdedness of the upper airway structures (as reflected for example by the Mallampati score) (Dayyat et al., 2009). A more recent cross-sectional study estimated that the OSA prevalence was about 44.6% in children with overweight/obesity compared with 9.1% in the normal-weight group (Andersen et al., 2019), thereby confirming many of the previous studies. Likewise, OSA was associated with a significantly increased risk of obesity (Bachrach et al., 2021). The exact mechanism of this relationship is not known. However, it is known that both OSA and obesity share a common inflammatory pathway (Hakim et al., 2015; Kheirandish-Gozal & Gozal, 2019), and that they may therefore facilitate the increased risk of each other.

5.5 Clinical Presentation

5.5.1 History

It is quite uncommon for parents to mention any concerns with their child's sleep unless explicitly asked. Snoring is frequently viewed by caretakers as a normal "trait" and not a worrisome symptom of a disease. Clinicians should screen for snoring and other concerning symptoms like frank apnea, as well as daytime symptoms such as excessive daytime sleepiness (EDS), behavioral problems (e.g., hyperactivity, inattention), or academic underachievement. Older children are able to verbalize EDS when asked. Screening tools are available such as the Epworth Sleepiness Scale for Children (Boudewyns et al., 2017; Janssen et al., 2017).

Other clues to the presence of OSA and other less severe forms of sleep-disordered breathing include a history of prematurity, neuromuscular disorders, or congenital anomalies involving the structures of the face and neck. Other conditions commonly associated with OSA include obesity, which may present with associated co-morbidities encompassing the metabolic syndrome. Bedtime enuresis, bruxism, sleepwalking and other parasomnias including night terrors and nightmares, morning headaches, and mood disorders are all frequent manifestations of underlying OSA in children and therefore should be not only screened for but also evaluated for underlying OSA (Tan & Kaditis, 2021; Kaditis et al., 2017; Joosten et al., 2017).

5.5.2 Physical Examination

Oropharyngeal examination should include an assessment of the Mallampati score and tonsillar size, as both are associated with the severity of OSA (Gipson et al., 2019). Nasal passages should be evaluated for patency and if possible endoscopic assessment of the nasal turbinates and adenoids would be desirable. The configuration of the jaw in relation to the maxilla (i.e., overbite or overjet) should be noted. Micrognathia or retrognathia can contribute to a reduced oropharyngeal space, even for a normal-sized tongue (Fig. 5.1). Scalloping on the edges of the tongue may indicate this reduced space. Children with obesity, especially if severe, may present with physical signs like excess fat around the neck and acanthosis nigricans.

5.5.3 Differential Diagnosis

The differential diagnosis for OSA is given in Table 5.2.

Primary snoring
Central sleep apnea
Periodic breathing
GERD
Obesity hypoventilation syndrome
Narcolepsy
Idiopathic hypersomnia
Insufficient sleep syndrome
Sleep-related movement disorder
Sleep-related epilepsy

5.6 Diagnosis

The gold standard in the diagnosis of OSA is an in-lab attended overnight polysomnographic study (PSG). It is recommended that the child be evaluated by a duly accredited sleep center (Kirk et al., 2017a). During this study, multiple leads will be placed to monitor parameters such as electroencephalogram (EEG), electromyogram (EMG), and several breathing parameters including nasal airflow, carbon dioxide and oximetry, and chest and abdominal movements. There are several handouts available on how to help desensitize younger children to this procedure such as to minimize what has been termed a "first-night effect", i.e., reduction in sleep efficiency and in sleep state representation that may yield misleading conclusions. Many sleep centers also offer acclimatization walk-throughs to improve the reliability of the single-night PSG.

5.6.1 AASM Scoring Guidelines

The AASM has set forth guidelines consisting of recommended parameters to be reported and scoring rules in its manual (Berry et al., 2020). These parameters can be grouped into several categories. General parameters include derivations for electrooculogram (EOG), EEG, chin EMG, airflow signals, respiratory signals, oxygen saturation, body position, and electrocardiogram, among others (Figs. 5.2 and 5.3). Other categories listed in the manual include sleep scoring data (e.g., recording time, total sleep time, etc.), arousal events, cardiac events, movement events, and respiratory events. The rules for children apply from two months post-term and older (<18 year of age). A summary of the different types of respiratory events scored in children, and their corresponding AASM rules are summarized in Table 5.3.

The apneas and hypopneas are averaged per hour and leads to the Apnea-Hypopnea Index (AHI). On the other hand, when the RERAs are averaged together with the former two, this leads to the Respiratory Disturbance Index (RDI). The cutoff pediatric values are given in Table 5.4.

5.7 Alternatives to PSG

The PSG remains as the gold standard in the diagnosis of pediatric OSA (Dehlink & Tan, 2016). This test, however, may be technically difficult to perform or impractical in certain settings. Younger children may need a high degree of acclimatization or desensitization due to the large number of leads being placed on them, or simply because they are sleeping in an unfamiliar place (Gipson et al., 2019). In resource-limited settings, PSGs simply may be too expensive if at all available due to economic circumstances. As alternatives, several modalities have been proposed (Dehlink & Tan, 2016), and they are given in the following subsections.

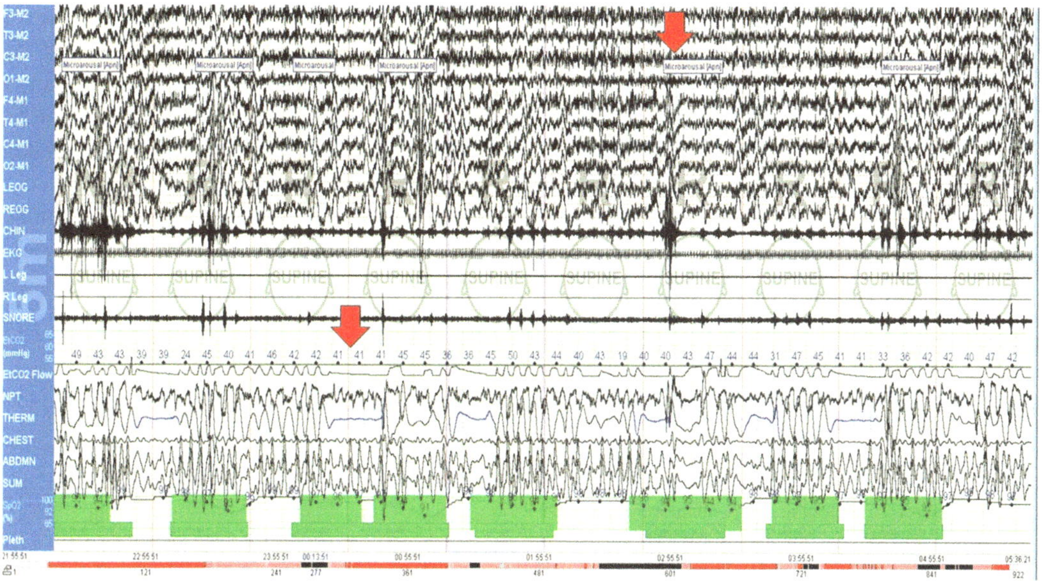

Fig. 5.2 Multiple obstructive apneic events (left arrow and blue highlights) followed by oxyhemoglobin desaturations (green highlights) and arousals (right arrow)

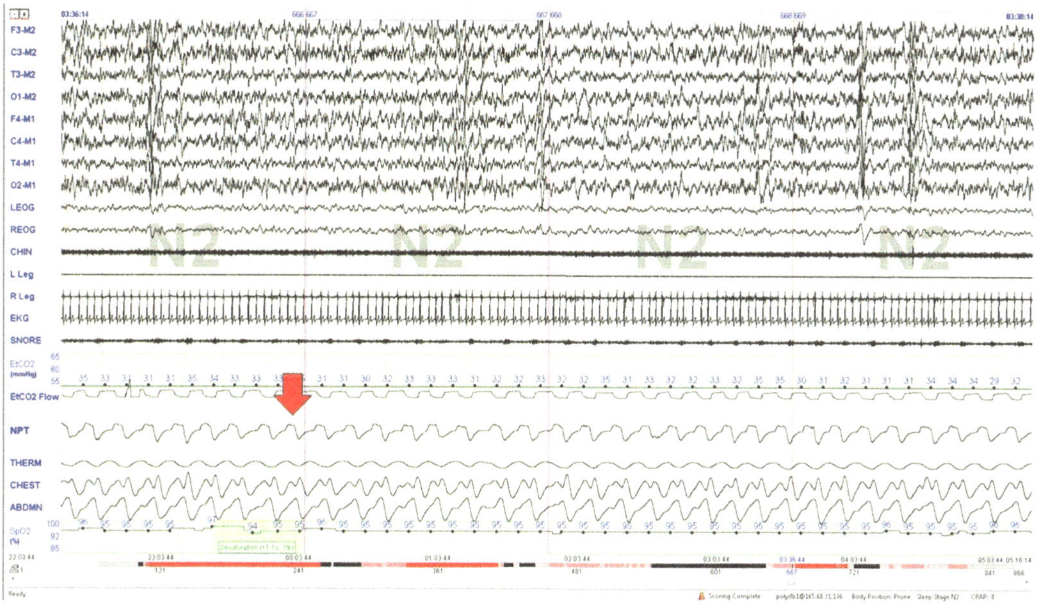

Fig. 5.3 Snoring with flow limitation (arrow) but no evidence of gas exchange abnormalities or sleep disruption

5.7.1 Sleep Clinical Record (SCR)

Villa et al. proposed a scoring system (Sleep Clinical Score or SCS) that takes into account items from the patients' history and physical examination (Villa et al., 2013). The first item considered in the SCR consists of physical examination findings like nasal septum deviation, nasal obstruction, oral breathing, orthodontic examination findings, and tonsillar grading, among others. The second item considered is based on a questionnaire to calculate the

Table 5.3 Respiratory events and a summary of the corresponding summary description (Tauman et al., 2004)

Apnea	Obstructive apneas are scored when there is at least a 90% decrease in the signal from the oronasal thermal airflow sensor.
Hypopnea	This is sometimes referred to as "partial apneas," because they represent a decrease in peak signal excursion from at least 30% to below 90%. The duration must last at least 2 breaths, with a decrease in oxygen saturation of at least 3%, or is associated with an arousal.
RERA	This stands for respiratory effort-related arousal. These are respiratory events that do not meet the above criteria for apneas and hypopneas but are related to arousals. It is also characterized by one or more of the following: Increasing respiratory effort, flattening of the inspiratory portion of the nasal pressure transducer, snoring, or an elevation in EtCo2 above the pre-event baseline.
Snoring	Snoring is not scored and its reporting is optional.

Table 5.4 Pediatric AHI cutoff values (per hour) for the diagnosis of OSA (Dehlink & Tan, 2016)

Normal	<1
Mild	1 to </= 5
Moderate	>5 to </=10
Severe	>10

Table 5.5 McGill oximetry scoring system and levels of severity of nocturnal hypoxemia (Kaditis et al., 2016; Nixon et al., 2004)

1 (normal or inconclusive)	Baseline SpO2 > 95% with fewer than 3 clusters of events
2 (mildly abnormal)	3 or more clusters of desaturations are present with at least 3 SpO2 drops <90% but not less than 85%
3 (moderately abnormal)	3 or more clusters of desaturations are present with at least 3 SpO2 drops <85% but not less than 80%
4 (severely abnormal)	3 or more clusters of desaturations are present with at least 3 SpO2 drops <80%

Brouillette score (Brouillette et al., 1984). The third item is based on a questionnaire about ADHD symptoms (Villa et al., 2013; DuPaul et al., 2001). The SCS was defined as positive when the final score was >/=6.5. Their study showed that this value had a positive correlation with AHI: a positive score had an increased probability of OSA while a negative score (<6.5) had a lowered probability of OSA. Their study concluded that their scoring may accurately exclude OSA, with no further testing or treatment necessary.

5.7.2 Nocturnal Oximetry

Many authors have proposed overnight oximetry as an alternative to PSG for the diagnosis of OSA (Kaditis et al., 2016; Brouillette et al., 2000; Nixon et al., 2004; Hornero et al., 2017). The main argument for it is the practicality of only having a single channel and therefore, a much lower technical and financial cost of its performance. Several studies have looked at clusters of desaturation events. It is assumed that clusters of desaturations probably occur with respiratory events during REM and N2 sleep (Kaditis et al., 2016). The study is considered diagnostic for OSA if three or more clusters of desaturations are present with at least three SpO2 drops <90%. Brouillette et al. proposed that the above trend in SpO2 had an increased probability of OSA, but it did not rule it out (Brouillette et al., 2000). On the other hand, Nixon et al. proposed that overnight oximetry can also be used to estimate the severity of OSA (Nixon et al., 2004). Table 5.5 shows this scoring system.

Villa et al proposed a combination algorithm of using both the McGill oximetry score, and SCR in resource-limited settings (Villa et al., 2015). In their study, children with positive SCR scores (>/=6.5) were further classified as having a MOS score of either 1 or > 1. They showed that a positive SCR and a MOS score > 1 had a high positive predictive value, but low negative predictive value for identifying an AHI greater than 5 in children. They concluded that this algorithm was applicable in about two-thirds of children – accurately classifying their severity of OSA.

5.7.3 Polygraphy

Respiratory polygraphy is similar to PSG but without the EOG, EEG, and EMG channels (Dehlink & Tan, 2016). This method of diagnosis is used in Europe and is proposed by some as being a valid alternative to PSG (Alonso Alvarez et al., 2008). However, the official AASM position is that this may underestimate the total AHI (Tan et al., 2014).

5.7.4 Portable Studies

Home Sleep Apnea Tests (HSAT) that could be done at home are an accepted way of diagnosing OSA in adult Sleep Medicine practice, as per the AASM guideline (Caples et al., 2021). Several vendors are available with multiple channel options which usually include nasal airflow, thermistor, chest effort, abdominal movement, pulse oximetry, and body position. Another modality is the WatchPAT™ which uses peripheral arterial tonometry (PAT) technology. WatchPAT™ combines actigraphy with a PAT signal probe that measures changes in arterial volume, which corresponds with activation of the sympathetic nervous system (Tanphachitr et al., 2018). An algorithm was developed to characterize the association between this sympathetic activation and REM (Tanphachitr et al., 2018; Herscovici et al., 2007).

There have been many suggestions of performing some of these home sleep studies in children (Ross & Redline, 2020; Bhattacharjee, 2019; Gozal et al., 2015) owing to reasons mentioned at the top of this section.

The official position of the AASM is that HSATs are not recommended for use in diagnosing OSA below 18 years of age (Kirk et al., 2017b). At the time of publication (2017) of their official position statement, the AASM identified several issues with the use of HSATs in children. First, the success rate in having an adequate study may be diminished when a caregiver places the sensors instead of a technician. Second, there was paucity of data in children at the time of publication of the position statement. Last, HSATs lacked EEG and CO_2 channels that may lead to an underestimation of AHI (Kirk et al., 2017b). However, this is an evolving area, and in light of substantial advances in technology and other considerations, it is likely that such recommendations will be modified soon, and home-based diagnostic studies will become the routine approach to the snoring child (Gozal et al., 2015).

Acknowledgments This work was partially supported by the Leda Sears Foundation. D.G. is supported by NIH grants AG061824 and HL140548, and by Tier 2 and TRIUMPH grants from the University of Missouri.

References

Alonso Alvarez, M. L., Teran Santos, J., Cordero Guevara, J. A., Navazo Eguia, A. I., Ordax Carbajo, E., Masa Jimenez, J. F., & Pelay, R. (2008). Reliability of respiratory polygraphy for the diagnosis of sleep-apnea hypopnea syndrome in children. *Archivos de Bronconeumología, 44*(6), 318–323. Spanish. PMID:18559221.

Alonso-Álvarez, M. L., Cordero-Guevara, J. A., Terán-Santos, J., Gonzalez-Martinez, M., Jurado-Luque, M. J., Corral-Peñafiel, J., Duran-Cantolla, J., Kheirandish-Gozal, L., & Gozal, D. (2014). Obstructive sleep apnea in obese community-dwelling children: The NANOS study. *Sleep, 37*(5), 943–949. https://doi.org/10.5665/sleep.3666. PMID: 24790273; PMCID: PMC3985101.

Alonso-Álvarez, M. L., Terán-Santos, J., Navazo-Egüia, A. I., Martinez, M. G., Jurado-Luque, M. J., Corral-Peñafiel, J., Duran-Cantolla, J., Cordero-Guevara, J. A., & Kheirandish-Gozal, L. (2015). Gozal D; Spanish sleep network. Treatment outcomes of obstructive sleep apnoea in obese community-dwelling children: The NANOS study. *The European Respiratory Journal, 46*(3), 717–727. https://doi.org/10.1183/09031936.00013815. Epub 2015 Jun 11.

American Academy of Sleep Medicine (AASM) | Sleep | Medical. Obstructive sleep apnea. www.aasm.org/. Accessed 10 Oct 2021.

Andersen, I. G., Holm, J. C., & Homoe, P. (2019). Obstructive sleep apnea in children and adolescents with and without obesity. *European Archives of Oto-Rhino-Laryngology, 276*(3), 871–878. https://doi.org/10.1007/s00405-019-05290-2

Arens, R., & Marcus, C. L. (2004). Pathophysiology of upper airway obstruction: At developmental perspective. *Sleep, 27*(5), 997–1019. https://doi.org/10.1093/sleep/27.5.997

Arens, R., McDonough, J. M., Corbin, A. M., Hernandez, M. E., Maislin, G., Schwab, R. J., & Pack, A. I. (2002). Linear dimensions of the upper airways structure during development: Assessment by magnetic resonance imaging. *American Journal of Respiratory and Critical Care Medicine, 165*(1), 117–122. https://doi.org/10.1164/ajrccm.165.1.2107140

Arens, R., McDonough, J. M., Corbin, A. M., Rubin, N. K., Carroll, M. E., Pack, A. I., Liu, J., & Udupa, J. K. (2003). Upper airway size analysis by magnetic resonance imaging of children with obstructive sleep apnea syndrome. *American Journal of Respiratory and Critical Care Medicine, 167*(1), 65–70. https://doi.org/10.1164/rccm.200206-613OC

Arens, R., Sin, S., McDonough, J. M., Palmer, J. M., Dominguez, T., Meyer, H., Wootton, D. M., & Pack, A. I. (2005). Changes in upper airway size during tidal breathing in children with obstructive sleep apnea syndrome. *American Journal of Respiratory and Critical Care Medicine, 171*(11), 1298–1304. https://doi.org/10.1164/rccm.200411-1597OC

Arens, R., Sin, S., & Wootton, D. (2021). Pathophysiology of obstructive sleep apnea in childhood. In *Pediatric sleep medicine* (pp. 437–452). Springer.

Bachrach, K., Danis, D. O., 3rd, Cohen, M. B., & Levi, J. R. (2021). The relationship between obstructive sleep apnea and pediatric obesity: a nationwide analysis. *Annals of Otology, Rhinology and Laryngology, 131*, 520. https://doi.org/10.1177/00034894211028489

Bathory, E., & Tomopoulos, S. (2017). Sleep regulation, physiology and development, sleep duration and patterns, and sleep hygiene in infants, toddlers, and preschool-age children. *Current Problems in Pediatric and Adolescent Health Care, 47*(2), 29–42. https://doi.org/10.1016/j.cppeds.2016.12.001. Epub 2017 Jan 20. PMID: 28117135.

Beebe, D. W., & Byars, K. C. (2011). Adolescents with obstructive sleep apnea adhere poorly to positive airway pressure (PAP), but PAP users show improved attention and school performance. *PLoS One, 6*(3), e16924. https://doi.org/10.1371/journal.pone.0016924. PMID: 21437285; PMCID: PMC3060091.

Berry, R. B., Quan, S. F., Abreu, A. R., Bibbs, M. L., DelRosso, L., Harding, S. M., Mao, M. M., Plante, D. T., Pressman, M. R., Troester, M. M., & Vaugh, B. V. The AASM manual for the scoring of sleep and associated events: Rules, terminology and technical specifications v2.6. 2020.

Bhattacharjee, R. (2019). Ready for primetime? Home sleep apnea tests for children. *Journal of Clinical Sleep Medicine, 15*(5), 685–686. https://doi.org/10.5664/jcsm.7748. Published 2019 May15.

Bhattacharjee, R., Kheirandish-Gozal, L., Spruyt, K., Mitchell, R. B., Promchiarak, J., Simakajornboon, N., Kaditis, A. G., Splaingard, D., Splaingard, M., Brooks, L. J., Marcus, C. L., Sin, S., Arens, R., Verhulst, S. L., & Gozal, D. (2010). Adenotonsillectomy outcomes in treatment of obstructive sleep apnea in children: A multicenter retrospective study. *American Journal of Respiratory and Critical Care Medicine, 182*(5), 676–683. https://doi.org/10.1164/rccm.200912-1930OC. Epub 2010 May 6.

Boudewyns, A., Abel, F., Alexopoulos, E., Evangelisti, M., Kaditis, A., Miano, S., Villa, M. P., & Verhulst, S. L. (2017). Adenotonsillectomy to treat obstructive sleep apnea: Is it enough? *Pediatric Pulmonology, 52*(5), 699–709. https://doi.org/10.1002/ppul.23641

Brouillette, R., Hanson, D., & David, R. (1984). A diagnostic approach to suspected obstructive sleep apnea in children. *The Journal of Pediatrics, 105*, 10–14.

Brouillette, R. T., Morielli, A., Limanis, A., Waters, K. A., Luciano, R., & Ducharme, F. M. (2000). Nocturnal pulse oximetry as an abbreviated testing modality for pediatric obstructive sleep apnea. *Pediatrics, 105*, 405–412.

Caples, S. M., Anderson, W. M., Calero, K., Howell, M., & Hashmi, S. D. (2021). Use of polysomnography and home sleep apnea tests for the longitudinal management of obstructive sleep apnea in adults: An American Academy of Sleep Medicine clinical guidance statement. *Journal of Clinical Sleep Medicine, 17*(6), 1287–1293. https://doi.org/10.5664/jcsm.9240. PMID: 33704050; PMCID: PMC8314660.

Carreras, A., Zhang, S. X., Peris, E., Qiao, Z., Gileles-Hillel, A., Li, R. C., Wang, Y., & Gozal, D. (2014). Chronic sleep fragmentation induces endothelial dysfunction and structural vascular changes in mice. *Sleep, 37*(11), 1817–1824. https://doi.org/10.5665/sleep.4178. PMID: 25364077; PMCID: PMC4196065.

Chervin, R. D., & Archbold, K. H. (2001). Hyperactivity and polysomnographic findings in children evaluated for sleep-disordered breathing. *Sleep, 24*(3), 313–320. https://doi.org/10.1093/sleep/24.3.313

Chervin, R. D., Archbold, K. H., Dillon, J. E., Panahi, P., Pituch, K. J., Dahl, R. E., & Guilleminault, C. (2002). Inattention, hyperactivity, and symptoms of sleep-disordered breathing. *Pediatrics, 109*(3), 449–456. https://doi.org/10.1542/peds.109.3.449

Chinnadurai, S., Jordan, A. K., Sathe, N. A., Fonnesbeck, C., McPheeters, M. L., & Francis, D. O. (2017). Tonsillectomy for obstructive sleep-disordered breathing: A meta-analysis. *Pediatrics, 139*(2), e20163491. https://doi.org/10.1542/peds.2016-3491

Chun, R., & Arvedson, J. (2021). Structure and development of the upper respiratory system. In *Fuhrman and Zimmerman's pediatric critical care* (pp. 453–461). Philadelphia.

Chung, S., Hodges, E. K., Ruzicka, D. L., Hoban, T. F., Garetz, S. L., Guire, K. E., Felt, B. T., Dillon, J. E., Chervin, R. D., & Giordani, B. (2016). Improved behavior after adenotonsillectomy in children with higher and lower IQ. *International Journal of Pediatric Otorhinolaryngology, 80*, 21–25. https://doi.org/10.1016/j.ijporl.2015.11.005. Epub 2015 Nov 10. PMID: 26746606; PMCID: PMC4706995.

Dayyat, E., Kheirandish-Gozal, L., Sans Capdevila, O., Maarafeya, M. M. A., & Gozal, D. (2009). Obstructive sleep apnea in children: relative contributions of body mass index and adenotonsillar hypertrophy. *Chest, 136*(1), 137–144. https://doi.org/10.1378/chest.08-2568. Epub 2009 Feb 18. PMID: 19225059; PMCID: PMC2716713.

Dehlink, E., & Tan, H. L. (2016). Update on paediatric obstructive sleep apnea. *Journal of Thoracic Disease, 8*(2), 224–235. https://doi.org/10.3978/j.issn.2072-1439.2015.12.04

Del-Rio Camacho, G., Medina Castillo, L., Rodriguez-Catalan, J., Soto Insuga, V., & Gomez, G. T. (2019). Central sleep apnea in children with obstructive sleep apnea syndrome an improvement following addendum tonsillectomy. *Pediatric Pulmonology, 54*(11), 1670–1675. https://doi.org/10.1002/ppul.24469

DuPaul, G. J., McGoey, K. E., & Eckert, T. L. (2001). VanBrakeJ. Preschool children with attention-deficit/ hyperatcitivy disorder: Impairments in behavioral, social, and school functioning. *Journal of the American Academy of Child and Adolescent Psychiatry, 40*(5), 508–515. https://doi.org/10.1097/00004585-200105000-00009

Follmar, A., Dentino, K., Abramowicz, S., & Padwa, B. L. (2014). Prevalance of sleep-disordered breathing in patients with Beckwith-Wideman syndrome. *The Journal of Craniofacial Surgery, 25*(5), 1814–1817. https://doi.org/10.1097/SCS0000000000000991

Fregosi, R. F., Quan, S. F., Kaemingk, K. L., Morgan, W. J., Goodwin, J. L., Cabrera, R., & Gmitro, A. (2003). Sleep-disordered breathing, pharyngeal size and soft tissue anatomy in children. *Journal of Applied Physiology (1985), 95*(5), 2030–2038. https://doi.org/10.1152/japplphysiol.00293.2003

Gipson, K., Lu, M., & Kinane, T. B. (2019). Sleep disordered breathing in children. *Pediatrics in Review, 40*(1), 3–13. https://doi.org/10.1542/pir.2018-0142

Goffinsky, A., Stanley, M. A., Shpherd, M., Duvall, N., Jenkinson, S. B., Davis, C., Bull, M. J., & Roper, R. (2015). Obstructive sleep apnea in young infants with Down syndrome evaluated in a Down syndrome specialty clinic. *American Journal of Medical Genetics. Part A, 167A*(2), 324–330. https://doi.org/10.1002/ajmg.a.36903

Gozal, D. (1998). Sleep-disordered breathing and school performance in children. *Pediatrics, 102*(3 Pt 1), 616–620. https://doi.org/10.1542/peds.102.3.616

Gozal, D., & Burnside, M. M. (2004). Increased upper airway collapsibility in children with obstructive sleep apnea during wakefulness. *American Journal of Respiratory and Critical Care Medicine, 169*(2), 163–167. https://doi.org/10.1164/rccm.200304-590OC. Epub 2003 Oct 2.

Gozal, D., Hakim, F., & Kheirandish-Gozal, L. (2013). Chemoreceptors, baroreceptors, and autonomic deregulation in children with obstructive sleep apnea. *Respiratory Physiology & Neurobiology, 185*(1), 177–185. https://doi.org/10.1016/j.resp.2012.08.019. Epub 2012 Aug 29. PMID: 22954503; PMCID: PMC3513509.

Gozal, D., Kheirandish-Gozal, L., & Kaditis, A. G. (2015). Home sleep testing for the diagnosis of pediatric obstructive sleep apnea: The times they are a changing...! *Current Opinion in Pulmonary Medicine, 21*(6), 563–568. https://doi.org/10.1097/MCP.0000000000000205. PMID: 26390329.

Gozal, D., & Pope, D. W., Jr. (2001). Snoring during early childhood and academic performance at ages thirteen to fourteen years. *Pediatrics, 107*(6), 1394–1399. https://doi.org/10.1542/peds.107.6.1394

Gozal, D., Serpero, L. D., Kheirandish-Gozal, L., Capdevila, O. S., Khalyfa, A., & Tauman, R. (2010). Sleep measures and morning plasma TNF-alpha levels in children with sleep-disordered breathing. *Sleep, 33*(3), 319–325. https://doi.org/10.1093/sleep/33.3.319. PMID: 20337189; PMCID: PMC2831425.

Hakim, F., Kheirandish-Gozal, L., & Gozal, D. (2015). Obesity an altered sleep: A pathway to metabolic derangements in children? *Seminars in Pediatric Neurology, 22*(2), 77–85. https://doi.org/10.1016/j.spen.2015.04.006

Harding, R., Schaughency, E., Haszard, J. J., Gill, A. I., Luo, R., Lobb, C., Dawes, P., & Galland, B. (2021). Sleep-related breathing problem trajectories across early childhood and academic achievement-related performance at age eight. *Frontiers in Psychology, 12*, 661156. https://doi.org/10.3389/fpsyg.2021.661156. PMID: 34267700; PMCID: PMC8276242.

Harman, K., Weichard, A. J., Davey, M. J., Horne, R. S. C., Nixon, G. M., & Edwards, B. A. (2020). Assessing ventilatory control stability in children with and without an elevated central apnea index. *Respirology, 25*(2), 214–220. https://doi.org/10.1111/resp.13606

Herscovici, S., Pe'er, A., Papya, S., & Lavie, P. (2007). Detecting REM sleep from the finger: An automatic REM sleep algorithm based on peripheral arterial tone (PAT) and actigraphy. *Physiological Measurement, 28*(2), 129–140. https://doi.org/10.1088/0967-3334/28/2/002. Epub 2006 Dec 12. PMID: 17237585.

Hornero, R., Kheirandish-Gozal, L., Gutiérrez-Tobal, G. C., Philby, M. F., Alonso-Álvarez, M. L., Álvarez, D., Dayyat, E. A., Xu, Z., Huang, Y. S., Tamae Kakazu, M., Li, A. M., Van Eyck, A., Brockmann, P. E., Ehsan, Z., Simakajornboon, N., Kaditis, A. G., Vaquerizo-Villar, F., Crespo Sedano, A., Sans Capdevila, O., … Gozal, D. (2017). Nocturnal oximetry-based evaluation of habitually snoring children. *American Journal of Respiratory and Critical Care Medicine, 196*(12), 1591–1598. https://doi.org/10.1164/rccm.201705-0930OC. PMID: 28759260; PMCID: PMC5754445.

Huang, J., Pinto, S. J., Yuan, H., Katz, E. S., Karamessinis, L. R., Bradford, R. M., Gallagher, P. R., Hannigan, J. T., Nixon, T., Ward, M. B., Lee, Y. N., & Marcus, C. L. (2012). Upper airway collapsibility and genioglossus activity in adolescents during sleep. *Sleep, 35*(10), 1345–1352. https://doi.org/10.5665/sleep.2110. PMID: 23024432; PMCID: PMC3443760.

Huang, Y.-S., & Guilleminault, C. (2017). Pediatric obstructive sleep apnea: Where do we stand? *Sleep-Related Breathing Disorders.*, 136–144. https://doi.org/10.1159/000470885

Isono, S., Shimada, A., Utsugi, M., Konno, A., & Nishino, T. (1998). Comparison of static mechanical properties of the passive pharynx between normal children and children with sleep-disordered breathing. *American*

Journal of Respiratory and Critical Care Medicine, 157(4 Pt 1), 1204–1212. https://doi.org/10.1164/ajrccm.157.4.9702042

Janssen, K. C., Phillipson, S., O'Connor, J., & Johns, M. W. (2017). Validation of the Epworth sleepiness scale for children and adolescents using rasch analysis. *Sleep Medicine, 33*, 30–35. https://doi.org/10.1016/j.sleep.2017.01.014. Epub 2017 Feb 12. PMID: 28449902.

Joosten, K. F., Larramona, H., Miano, S., Van Waardenburg, D., Kaditis, A. G., Vandenbussche, N., & Ersu, R. (2017). How do we recognize the child with OSAS? *Pediatric Pulmonology, 52*(2), 260–271. https://doi.org/10.1002/ppul.23639. Epub 2016 Nov 1. PMID: 27865065.6.

Kaditis, A., Kheirandish-Gozal, L., & Gozal, D. (2016). Pediatric OSAS: Oximetry can provide answers when polysomnography is not available. *Sleep Medicine Reviews, 27*, 96–105. https://doi.org/10.1016/j.smrv.2015.05.008. Epub 2015 Jun 4. PMID: 26146027.

Kaditis, A. G., Alonso Alvarez, M. L., Boudewyns, A., Abel, F., Alexopoulos, E. I., Ersu, R., Joosten, K., Larramona, H., Miano, S., Narang, I., Tan, H. L., Trang, H., Tsaoussoglou, M., Vandenbussche, N., Villa, M. P., Van Waardenburg, D., Weber, S., & Verhulst, S. (2017). ERS statement on obstructive sleep disordered breathing in 1- to 23-month-old children. *The European Respiratory Journal, 50*(6), 1700985. https://doi.org/10.1183/13993003.00985-2017. PMID: 29217599.

Katz, E. S., & D'Ambrosio, C. M. (2008). Pathophysiology of pediatric obstructive sleep apnea. *Proceedings of the American Thoracic Society, 5*(2), 253–262. https://doi.org/10.1513/pats.200707-111MG. PMID: 18250219; PMCID: PMC2645256.

Katz, E. S., Mitchell, R. B., & D'Ambrosio, C. M. (2012). Obstructive sleep apnea in infants. *American Journal of Respiratory and Critical Care Medicine, 185*(8), 805–816. https://doi.org/10.1164/rccm.201108-1455CI. Epub 2011 Dec 1. PMID: 22135346; PMCID: PMC5448577.

Kheirandish-Gozal, L., & Gozal, D. (2019). Obstructive sleep apnea and inflammation: Proof of concept based on two illustrative cytokines. *International Journal of Molecular Sciences, 3*, 459.

Kim, J., Hakim, F., Kheirandish-Gozal, L., & Gozal, D. (2011). Inflammatory pathways in children with insufficient or disordered sleep. *Respiratory Physiology & Neurobiology, 178*(3), 465–474. https://doi.org/10.1016/j.resp.2011.04.024. Epub 2011 May 5. PMID:21569868; PMCID: PMC3168951.

Kirk, V., Baughn, J., D'Andrea, L., Friedman, N., Galion, A., Garetz, S., Hassan, F., Wrede, J., Harrod, C. G., & Malhotra, R. K. (2017a). American academy of sleep medicine position paper for the use of a home sleep apnea test for the diagnosis of OSA in children. *Journal of Clinical Sleep Medicine, 13*(10), 1199–1203. https://doi.org/10.5664/jcsm.6772. PMID: 28877820; PMCID: PMC5612636.

Kirk, V., Baughn, J., D'Andrea, L., Friedman, N., Galion, A., Garetz, S., Hassan, F., Wrede, J., Harrod, C. G., & Malhotra, R. K. (2017b). American Academy of Sleep Medicine Position Paper for the use of a home sleep apnea test for the diagnosis of OSA in children. *Journal of Clinical Sleep Medicine, 13*(10), 1199–1203. https://doi.org/10.5664/jcsm.6772. PMID: 28877820; PMCID: PMC5612636.

Kohyama, J., & Hasegawa, T. (2002). Subcortical arousal response in child patients with obstructive sleep apnea. *Sleep Medicine, 3*(Suppl 2), S33–S36. https://doi.org/10.1016/s1389-9457(02)00161-2. PMID: 14592376.

Lee, T., Wulfovich, S., Kettler, E., & Nation, J. (2020). Incidence of cure and residual obstructive sleep apnea in obese children after tonsillectomy and adenoidectomy stratified by age group. *International Journal of Pediatric Otorhinolaryngology, 139*, 110394. https://doi.org/10.1016/j.ijporl.2020.110394. Epub 2020 Sep 28.

Marcus, C. L., Brooks, L. J., Ward, S. D., Draper, K. A., Gozal, D., Halbower, A. C., Jones, J., Lehmann, C., Schechter, M. S., Sheldon, S., Shiffman, R. N., & Spruyt, K. (2012). Diagnosis and management of childhood obstructive sleep apnea syndrome. *Pediatrics, 130*(3), 714–714. https://doi.org/10.1542/peds.2012-1672

Marcus, C. L., Gozal, D., Arens, R., Basinski, D. J., Omlin, K. J., Keens, T. G., & Ward, S. L. (1994). Ventilatory responses during wakefulness in children with obstructive sleep apnea. *American Journal of Respiratory and Critical Care Medicine, 149*(3 Pt 1), 715–721. https://doi.org/10.1164/ajrccm.149.3.8118641. PMID: 8118641.

Marcus, C. L., Lutz, J., Carroll, J. L., & Bamford, O. (1998a). Arousal and ventilatory responses during sleep in children obstructive sleep apnea. *Journal of Applied Physiology (Bethesda, MD: 1985), 84*(6), 1926–1936. https://doi.org/10.1152/jappl.1998.84.6.1926

Marcus, C. L., Lutz, J., Carroll, J. L., & Bamford, O. (1998b). Arousal and ventilatory responses during sleep in children with obstructive sleep apnea. *Journal of Applied Physiology (Bethesda, MD: 1985), 84*(6), 1926–1936. https://doi.org/10.1152/jappl.1998.84.6.1926. PMID: 9609786.

Marcus, C. L., Moore, R. H., Rosen, C. L., Giordani, B., Garetz, S. L., Taylor, H. G., Mitchell, R. B., Amin, R., Katz, E. S., Arens, R., Paruthi, S., Muzumdar, H., Gozal, D., Thomas, N. H., Ware, J., Beebe, D., Snyder, K., Elden, L., Sprecher, R. C., … Childhood Adenotonsillectomy Trial (CHAT). (2013). A randomized trial of adenotonsillectomy for childhood sleep apnea. *The New England Journal of Medicine, 368*(25), 2366–2376. https://doi.org/10.1056/NEJMoa1215881. Epub 2013 May 21. PMID: 23692173; PMCID: PMC3756808.

Marcus, C. L., Moreira, G. A., Bamford, O., & Lutz, J. (1999). Response to inspiratory resistive loading during sleep in normal children and children with obstructive apnea. *Journal of Applied Physiology*

(Bethesda, MD: 1985), 87(4), 1448–1454. https://doi.org/10.1152/jappl.1999.87.4.1448. PMID: 10517777.

Martinot, J. B., Le-Dong, N. N., Denison, S., Guénard, H. J., Borel, J. C., Silkoff, P. E., Pepin, J. L., & Gozal, D. (2018). Persistent respiratory effort after adenotonsillectomy in children with sleep-disordered breathing. *The Laryngoscope, 128*(5), 1230–1237. https://doi.org/10.1002/lary.26830. Epub 2017 Aug 22.

Meltzer, L. J., Johnson, C., Crosette, J., Ramos, M., & Mindell, J. A. (2010). Prevalence of diagnosed sleep disorders in pediatric primary care practices. *Pediatrics, 125*(6), 1416–1418. https://doi.org/10.1542/peds.2009-2725

Moreira, G. A., Tufik, S., Nery, L. E., Lutz, J., Verfaille, K., Luan, X., & Marcus, C. L. (2005). Acoustic arousal responses in children with obstructive sleep apnea. *Pediatric Pulmonology, 40*(4), 300–305. https://doi.org/10.1002/ppul.20219. PMID: 15880403.

Nair, D., Zhang, S. X., Ramesh, V., Hakim, F., Kaushal, N., Wang, Y., & Gozal, D. (2011). Sleep fragmentation induces cognitive deficits via nicotinamide adenine dinucleotide phosphate oxidase-dependent pathways in mouse. *American Journal of Respiratory and Critical Care Medicine, 184*(11), 1305–1312. https://doi.org/10.1164/rccm.201107-1173OC. Epub 2011 Aug 25. PMID: 21868506; PMCID: PMC3262045.

Nava-Guerra, L., Tran, W. H., Chalacheva, P., Loloyan, S., Joshi, B., Keens, T. G., Nayak, K. S., Davidson Ward, S. L., & Khoo, M. C. (2016). Model-based stability assessment of ventilatory control in overweight adolescents with obstructive sleep apnea during NREM sleep. *Journal of Applied Physiology (Bethesda, MD: 1985), 121*(1), 185–197. https://doi.org/10.1152/japplphysiol.01081.2015. Epub 2016 May 12. PMID: 27174926; PMCID: PMC4967255.

Nino, G., Aziz, J., Weiss, M., Allen, M., Lew, J., Manrique, M., Mantilla-Rivas, E., McGrath, J. L., Rogers, G. F., & Oh, A. K. (2021). Defining age-related OSA features in Robin sequence using polysomnographic-based analyses of respiratory arousal responses and gas-exchange parameters. *The Cleft Palate-Craniofacial Journal, 17*, 10556656211055017. https://doi.org/10.1177/10556656211055017. Epub ahead of print. PMID: 34787016.

Nixon, G. M., Kermack, A. S., Davis, G. M., Manoukian, J. J., Brown, K. A., & Brouillette, R. T. (2004). Planning adenotonsillectomy in children with obstructive sleep apnea: The role of overnight oximetry. *Pediatrics, 113*, e19–e25.

Nobile, S., Marchionni, P., Noviello, C., & Carnielli, V. P. (2019). Correlation between cardiorespiratory events and gastro-esophageal reflux in preterm and term infants: Analysis of predisposing factors. *Early Human Development, 134*, 14–18. https://doi.org/10.1016/j.earlhumdev.2019.05.003. Epub 2019 May 18.

O'Brien, L. M., Holbrook, C. R., Mervis, C. B., Klaus, C. J., Bruner, J. L., Raffield, T. J., Rutherford, J., Mehl, R. C., Wang, M., Tuell, A., Hume, B. C., & Gozal, D. (2003). Sleep and neurobehavioral characteristics of 5- to 7-year-old children with parentally reported symptoms of attention-deficit/hyperactivity disorder. *Pediatrics, 111*(3), 554–563. https://doi.org/10.1542/peds.111.3.554

O'Brien, L. M., Mervis, C. B., Holbrook, C. R., Bruner, J. L., Klaus, C. J., Rutherford, J., Raffield, T. J., & Gozal, D. (2004a). Neurobehavioral implications of habitual snoring in children. *Pediatrics, 114*(1), 44–49. https://doi.org/10.1542/peds.114.1.44

O'Brien, L. M., Tauman, R., & Gozal, D. (2004b). Sleep pressure correlates of cognitive and behavioral morbidity in snoring children. *Sleep, 27*(2), 279–282. https://doi.org/10.1093/sleep/27.2.279. PMID: 15124723.

Otteson, T., Richardson, C., & Shah, J. (2021). Diseases of the upper respiratory tract. In *Fuhrman and Zimmerman's pediatric critical care* (pp. 524–535). Elsevier, Philadelphia.

Qubty, W. F., Mrelashvili, A., Kotagal, S., & Lloyd, R. (2014). Comorbidities in infants with obstructive sleep apnea. *Journal of Clinical Sleep Medicine, 10*(11), 1213–1216. https://doi.org/10.5664/jcsm.4204

Quitadamo, P., Giorgio, V., Zenzeri, L., Baldassarre, M., Cresi, F., Borrelli, O., & Salvatore, S. (2020). Apnea in preterm neonates: what's the role of gastroesophageal reflux? A systematic review. *Digestive and Liver Disease, 52*(7), 723–729. https://doi.org/10.1016/j.dld.2020.03.032. Epub 2020 May 15.

Ramesh, V., Nair, D., Zhang, S. X., Hakim, F., Kaushal, N., Kayali, F., Wang, Y., Li, R. C., Carreras, A., & Gozal, D. (2012). Disrupted sleep without sleep curtailment induces sleepiness and cognitive dysfunction via the tumor necrosis factor-α pathway. *Journal of Neuroinflammation, 9*, 91. https://doi.org/10.1186/1742-2094-9-91. PMID:22578011; PMCID: PMC3411474.

Ross, K. R., & Redline, S. (2020). Is it time to head home for the night? Home sleep testing in young children. *Annals of the American Thoracic Society, 17*(10), 1207–1209. https://doi.org/10.1513/AnnalsATS.202008-970ED

Saito, Y., Ezure, K., Kobayashi, M., Ito, M., Saito, K., & Osawa, M. (2002). A review of functional and structural components of the respiratory center involved in the arousal response. *Sleep Medicine, 3*(Suppl 2), S71–S74. https://doi.org/10.1016/s1389-9457(02)00170-3. PMID: 14592385.

Schaaf, W. E., Jr., Wootten, C. T., Donnelly, L. F., Ying, J., & Shott, S. R. (2010). Findings on MR sleep studies as biomarkers to predict outcome of genioglossus advancement in the treatment of obstructive sleep apnea in children and young adults. *American Journal of Roentgenology, 194*(5), 1204–1209. https://doi.org/10.2214/AJR.09.3254

Seppä-Moilanen, M., Andersson, S., & Kirjavainen, T. (2021). Spontaneous and apnea arousals from sleep in preterm infants. *Pediatric Research, 89*(5), 1261–1267. https://doi.org/10.1038/s41390-020-1068-2. Epub 2020 Jul 18. PMID: 32682326.

Smith, D. L., Gozal, D., Hunter, S. J., & Kheirandish-Gozal, L. (2017). Parent-reported behavioral and psychiatric problems mediate the relationship between sleep-disordered breathing and cognitive deficits in school-aged children. *Frontiers in Neurology, 8*, 410. https://doi.org/10.3389/fneur.2017.00410. PMID: 28848496; PMCID: PMC5554505.

Spruyt, K., & Gozal, D. (2011). Sleep disturbances in children with attention- deficit/hyperactivity disorder. *Expert Review of Neurotherapeutics, 11*(4), 565–577. https://doi.org/10.1586/ern.11.7. PMID: 21469929; PMCID: PMC3129712.

Suen, J. S., Arnold, J. E., & Brooks, L. J. (1995). Adenotosillectomy for the treatment of obstructive sleep apnea in children. *Archives of Otolaryngology – Head & Neck Surgery, 121*(5), 525–530. https://doi.org/10.1001/archotol.1995.01890050023005

Suri, J. C., Sen, M. K., Venkatachalam, V. P., Bhool, S., Charma, R., Elias, M., & Adhikari, T. (2015). Outcome of adenotonsillectomy for children with sleep apnea. *Sleep Medicine, 16*, 1181–1186.

Tal, A., Lieberman, A., & Tarasiuk, A. (2003). Sleep characteristics following adenotonsillectomy in children with obstructive sleep apnea syndrome. *Chest, 124*(3), 948–953. https://doi.org/10.1378/chest.124.3.948

Tan, H. L., Gozal, D., Ramirez, H. M., Bandla, H. P. R., & Kheirandish-Gozal, L. (2014). Overnight polysomnography versus respiratory polygraphy in the diagnosis of pediatric obstructive sleep apnea. *Sleep, 37*, 255–260.

Tan, H. L., & Kaditis, A. G. (2021). Phenotypic variance in pediatric obstructive sleep apnea. *Pediatric Pulmonology, 56*(6), 1754–1762. https://doi.org/10.1002/ppul.25309. Epub 2021 Feb 16. PMID: 33543838.

Tanphachitr, A., Thianboonsong, A., Banhiran, W., Vathanophas, V., & Ungkanont, K. (2018). Watch peripheral arterial tonometry in the diagnosis of pediatric obstructive sleep apnea. *Otolaryngology and Head and Neck Surgery, 159*(1), 166–172. https://doi.org/10.1077/0194599818768215. Epub2018Apr10. PMID:29631515.

Tauman, R., Gulliver, T. E., Krishna, J., Montgomery-Downs, H. E., O'Brien, L. M., Ivanenko, A., & Gozal, D. (2006a). Persistence of obstructive sleep apnea syndrome in children after adenotonsillectomy. *The Journal of Pediatrics, 149*(6), 803–808. https://doi.org/10.1016/j.jpeds.2006.08.067

Tauman, R., O'Brien, L. M., Barbé, F., Iyer, V. G., & Gozal, D. (2006b). Reciprocal interactions between spontaneous and respiratory arousals in adults with suspected sleep-disordered breathing. *Sleep Medicine, 7*(3), 229–234. https://doi.org/10.1016/j.sleep.2005.09.009. Epub 2006 Mar 24. PMID: 16564216.

Tauman, R., O'Brien, L. M., Holbrook, C. R., & Gozal, D. (2004). Sleep pressure score: A new index of sleep disruption in snoring children. *Sleep, 27*(2), 274–278. https://doi.org/10.1093/sleep/27.2.274. PMID: 15124722.

Tsukada, E., Kitamura, S., Enomoto, M., Moriwaki, A., Kamio, Y., Asada, T., Arai, T., & Mishima, K. (2018). Prevalence of childhood obstructive sleep apnea syndrome and its role in daytime sleepiness. *PLoS One, 13*(10), 6–9. https://doi.org/10.1371/journal.pone.0204409

Villa, M. P., Paolino, M. C., & Castaldo, R. (2013). Sleep clinical record: An aid to rapid and accurate diagnosis of paediatric sleep disordered breathing. *The European Respiratory Journal, 41*(6), 1355–1361. https://doi.org/10.1183/09031936.00215411

Villa, M. P., Pietropaoli, N., Supino, M. C., Vitelli, O., Rabasco, J., Evangelisti, M., Del Pozzo, M., & Kaditis, A. G. (2015). Diagnosis of pediatric obstructive sleep apnea syndrome in settings with limited resources. *JAMA Otolaryngology. Head & Neck Surgery, 141*(11), 990–996. https://doi.org/10.1001/jamaoto.2015.2354. PMID: 26540025.

Wang, Y., Carreras, A., Lee, S., Hakim, F., Zhang, S. X., Nair, D., Ye, H., & Gozal, D. (2014). Chronic sleep fragmentation promotes obesity in young adult mice. *Obesity (Silver Spring), 22*(3), 758–762. https://doi.org/10.1002/oby.20616. Epub 2013 Oct 16. PMID: 24039209; PMCID: PMC3947647.

Wu, D., Au, V. H., Yang, B., Horne, S. J., Weedon, J., Bernstein, M. J., & Goldstein, N. A. (2021). Impact of Adenotonsillectomy on homework performance in children with obstructive sleep-disordered breathing. *The Annals of Otology, Rhinology, and Laryngology, 7*, 34894211062543. https://doi.org/10.1177/00034894211062543. Epub ahead of print.

6 Treatment of Cheyne-Stokes Respiration in Heart Failure with Adaptive Servo-Ventilation: An Integrative Model

Wen-Hsin Hu and Michael C. K. Khoo

Abstract

The SERVE-HF (Treatment of Predominant Central Sleep Apnea by Adaptive Servo Ventilation in Patients with Heart Failure) multicenter trial found a small but significant increase in all-cause and cardiovascular mortality in patients assigned to adaptive servo-ventilation (ASV) versus guideline-based medical treatment. To better understand the physiological underpinnings of this clinical outcome, we employ an integrative computer model to simulate congestive heart failure with Cheyne-Stokes respiration (CHF-CSR) in subjects with a broad spectrum of underlying pathogenetic mechanisms, as well as to determine the in silico changes in cardiopulmonary and autonomic physiology resulting from ASV. Our simulation results demonstrate that while the elimination of CSR through ASV can partially restore cardiorespiratory and autonomic physiology toward normality in the vast majority of CHF phenotypes, the degree of restoration can be highly variable, depending on the combination of CHF mechanisms in play. The group with the lowest left ventricular ejection fraction (LVEF) appears to be most vulnerable to the potentially adverse effects of ASV, but the level of pulmonary capillary wedge pressure (PCWP) plays an important role in determining the nature of these effects.

Keywords

Heart Failure · Cheyne-Stokes Respiration · Sleep Apnea · Adaptive Servo-Ventilation · Mathematical Model · Computer Simulation · Autonomic Regulation · Integrative Physiology

6.1 Introduction

Congestive heart failure (CHF) is recognized to be a clinical syndrome associated with a constellation of symptoms and signs that occur in the presence of structural and functional cardiac abnormalities (Cowie & Poole-Wilson, 2013). In the course of CHF progression, diminished cardiac output resulting from a dysfunctional heart and maladaptive peripheral vasculature, abnormally high chemoreceptor sensitivity, impaired baroreflex function, hypervolemia, and other factors may act in combination to promote ventilatory instability in the form of Cheyne-Stokes respiration (CSR) (Cowie & Poole-Wilson, 2013; Emdin et al., 2017; Dempsey & Smith, 2014; Naughton, 1998). The presence of CSR generally signals poor prognosis in the advanced stages of

W.-H. Hu · M. C. K. Khoo (✉)
Department of Biomedical Engineering, University of Southern California, Los Angeles, CA, USA
e-mail: khoo@usc.edu

T. Penzel, R. Hornero (eds.), *Advances in the Diagnosis and Treatment of Sleep Apnea*, Advances in Experimental Medicine and Biology 1384, https://doi.org/10.1007/978-3-031-06413-5_6

CHF, and statistical analyses have shown CSR to be an independent risk factor for higher mortality in patients with this syndrome (Linz et al., 2018; Lorenzi-Filho et al., 2005; Lanfranchi et al., 1999; Naughton, 2016; Javaheri et al., 2007).

Beyond pharmacological treatment with acetazolamide and theophylline, the therapeutic options for CSR include supplemental O_2 administration, phrenic nerve stimulation, and ventilatory assistance of various modalities – continuous positive airway pressure (CPAP), bilevel positive airway pressure (BiPAP), and adaptive servo-ventilation (ASV). For two decades, ASV has been reported to consistently and effectively eliminate CSR, ameliorate sleep quality, restore blood oxygen level, and improve cardiac function in terms of left ventricular ejection fraction (LVEF) (Teschler et al., 2001). However, an international multicenter, large-scale randomized control trial (Treatment of Predominant Central Sleep Apnea by Adaptive Servo Ventilation in Patients With Heart Failure, or "SERVE-HF") reported the somewhat surprising finding of higher cardiovascular mortality among ASV-treated subjects (Cowie et al., 2015). Since then, both the American Academy of Sleep Medicine and a joint task force of the American College of Cardiology Foundation and the American Heart Association have updated their guidelines for treating central sleep apnea in heart failure with reduced ejection fraction (HFrEF) to specify ASV as a contraindication (Aurora et al., 2016; Yancy et al., 2017). Following this large clinical trial, several smaller follow-up studies have been undertaken to better understand the unexpected results. A sub-study based on cardiac imaging and biomarkers ruled out adverse cardiac remodeling or worsening of the CHF syndrome as likely mechanisms contributing to increased cardiovascular mortality (Cowie et al., 2018). Moreover, a post-hoc multi-state analysis revealed that the increased mortality was derived from cardiovascular death without previous hospital admission, presumably sudden cardiac death, and occurred at higher rate in subjects with low left ventricular ejection fraction (LVEF < 30%) (Eulenburg et al., 2016). Based on other studies, there was no dose dependence of ASV usage on cardiovascular mortality in the on-treatment analysis (Woehrle et al., 2017) and the stratification of biomarkers for CSR percentage and mortality showed no overlap suggesting CSR severity itself was likely not responsible for the higher mortality (Ferreira et al., 2020a,b).

To better understand the physiological underpinnings of the SERVE-HF outcomes, we adopt an alternative approach by employing an integrative computer model of cardiorespiratory control to determine how the autonomic nervous system, cardiovascular system, and respiratory system respond to ASV in "simulated patients" with a diversity of conditions in CHF that are accompanied by CSR. The rationale for this "in silico" approach is that the complexities of the outcomes that arise from multiple interacting physiological mechanisms make it difficult to distinguish cause from effect, and thus it becomes useful to turn to the rigorous, though simplified, framework inherent in a mathematical model to help extricate the various influences from one another.

6.2 Methods

6.2.1 "In Silico Subjects"

To take into account the broad range of physiological conditions that have been measured in CHF patients with CSR as well as the various stages in CHF progression, we vary the parameters/initial conditions in three major domains—left ventricular function, total blood volume, and chemoreflex gain to simulate a large subject population with different combinations of pathophysiologic changes and a broad range of severities (Table 6.1). Left ventricular systolic function characterized by the elastance (determined by both the baseline value and the sympathetic-modulated gain) has been demonstrated to correlate well with LVEF (Mirsky et al., 1987) and is varied in our model to simulate the HFrEF population. Hypervolemia, in the form of intravascular volume overload, is frequently detected even in non-edematous CHF

Table 6.1 Simulated subject populations defined by different combinations of model parameters or initial conditions representing a range of levels of left ventricular dysfunction (LVD), intravascular hypervolemia (IVH), and chemoreflex gain upregulation (CGU)

CHF factor	Description	
LVD level	**LV systolic function[a] (normalized)**	**LV diastolic function[b] (normalized)**
0[c]	1.00	1.0
1	0.50	0.7
2	0.40	0.7
3	0.35	0.7
4	0.30	0.7
5	0.25	0.7
6	0.20	0.7
7	0.15	0.7
IVH level	**Total blood volume[d] (mL)**	
0[c]	5300	
1	6300	
2	7300	
CGU level	**Chemoreflex gain factor[e] (normalized)**	
0[c]	1	
1	3	
2	6	

[a]Factor multiplying maximum LV elastance and maximum septum elastance in normal subjects
[b]Factor multiplying LV diastolic compliance in normal subjects
[c]Baseline case: parameters assume values for healthy adult
[d]Initial volume of each vascular compartment is proportionally scaled to total blood volume
[e]Factor by which both central and peripheral chemoreflex gains are increased in heart failure; these gain changes are assumed to be the same for both respiratory and cardiovascular parameters influenced by the chemoreflexes

(Androne et al., 2004). These two factors contribute to prolonged circulation time. The ventilatory response to hypercapnia has been found to be increased in CHF (Solin et al., 2000; Wilcox et al., 1998), and thus in the model we introduced factors consistent with chemoreflex gain upregulation. It is well-established that prolonged circulatory delay and augmented ventilatory chemoreflex gain work together to promote respiratory instability (Khoo, 1991).

6.2.2 Computer Model

The model we employed in this work used as its foundation "PNEUMA", an existing integrative computational model, originally developed to integrate the cardiovascular, respiratory, and sleep/wake control systems in the context of a variety of sleep-related breathing disorders (Cheng et al., 2010). The main systems are constructed in a hierarchical manner with interactions within each level and across hierarchical levels. These multiple sets of interactions underscore the complexities inherent in real-life physiological control. Figure 6.1 depicts the scope of the model and the interactions among the systems. The "controller" portion of the model includes local vascular autoregulation at the level of the brain and elsewhere, along with the three major groups of reflexes: (a) the baroreflex, (b) pulmonary slow adaptive stretch reflex, and (c) peripheral and central chemoreflexes. The parameters representing all these reflexes are modified by slow wave activity regulated by the sleep mechanism. In the original version of PNEUMA, the baroreflex consisted of only the reflex arc with input from the arterial baroreceptors. However, in the current model, we also incorporate input from the cardiopulmonary baroreceptors from changes in left atrial pressure, with the net effect of blunting overall baroreflex gain and exaggerating sympathovagal balance at high filling pressures, as suggested by studies in human heart failure (Millar et al., 2015; Floras, 2009; Floras & Ponikowski, 2015).

The single compartment that was used to represent CO_2 and O_2 exchange in the body tissues in PNEUMA is now divided into five parallel peripheral body compartments (brain, coronary circulation, muscle, splanchnic, extra-splanchnic) receiving arterial blood. The mixing of blood gases during transport of blood from the lungs to the chemoreceptors was previously characterized by a volume-invariant time delay and fixed time-constants in PNEUMA, but in the current model, these processes are modeled with time delays and mixing and convection compartments that can vary dynamically.

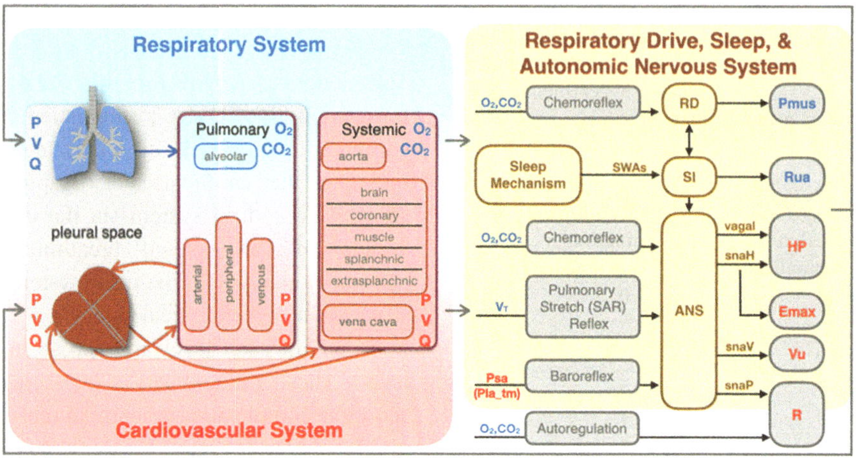

Fig. 6.1 A schematic diagram of the comprehensive physiology-based model (based on PNEUMA) that includes the "plant" portions of the respiratory (gas exchange & pulmonary mechanics) and cardiovascular (circulatory mechanics and hemodynamics) systems, and the "controller" portion that consists of cerebral and local vascular autoregulation, respiratory drive generation (RD), sleep mechanism, and autonomic nervous system (ANS). The respiratory and cardiovascular systems are coupled through the physiological processes occurring at multiple levels in the model, including mechanical coupling through the common pleural space. Pressures (*P*), volumes (*V*), and flows (*Q*) in either media of gas (blue) or blood (red) are interdependent. Gas exchange and mixing occurring within each body compartment are modeled to yield local oxygen (O_2) and carbon dioxide (CO_2) levels at all air-blood interfaces and blood-interstitium interfaces. The large array of cardiovascular, respiratory, and neural variables generated by the plant portions are fed back to the controller that in turn adjusts the parameters of and inputs to the plant. Key parameters and inputs include: Pmus pressure generated by inspiratory muscle movement, Rua upper airway resistance, HP heart period, Emax maximum ventricular elastance, Vu unstressed volume of vascular bed, R vascular resistance

In the coronary circulation, the metabolic rates (O_2 consumption and CO_2 production) are now adjusted in proportion to the rate stress product (RSP) (Braunwald, 1971; Strauer, 1979; Devereux et al., 2015). The dynamic myocardial metabolic demand and supply relationship enables the current model to predict various scenarios of burden imposed on the myocardium by CHF, as well as the potential effects resulting from ASV therapy. In the current model of the heart, the effect of elastance in the septum is incorporated to simulate interventricular volume redistribution or effectively the dynamic change of both ventricular elastance or compliance beyond pacemaker-activated ventricular contraction. This introduces the capability of addressing different conditions in which blood is returned to the heart under CHF as well as during ASV therapy.

The model is implemented using the Simulink® (Mathworks, Inc., Natick, MA) programming environment that is platform-independent. Simulink programs take the form of interconnected graphical objects, similar to the block diagrams of classical control theory.

6.2.3 ASV

The intervention of ASV is incorporated in the model by simulating the settings in SERVE-HF study using the algorithm employed in the AutoSet CS system (Resmed, Sydney, Australia). There is positive end-expiratory pressure (PEEP) set to a fixed 5 cmH_2O; along with pressure support varying between a minimum of 3 cmH_2O and a maximum of 10 cmH_2O. The servomechanism takes the feedback from the subject's minute ventilation and provides pressure support through a high-gain integral controller (0.3 cmH_2O per L/min per sec) to match 90% of a moving average of the minute ventilation in the past 3 minutes (Teschler et al., 2001; Cowie et al., 2015).

6.2.4 "Protocol" for Model Simulations and Subsequent Analyses

Each simulation starts 1 hour before sleep onset. The duration of sleep in any given simulation run is determined by the in-built sleep mechanism (Cheng et al., 2010), depending on how much the sleep process is fragmented by arousals or reversion to the waking state during the night; thus predicted sleep duration can vary from simulation run to simulation run. The running sleep index (SI) is set to 0 in wakefulness and 1 in sleep; thus, the average SI value that is close to unity indicates high sleep continuity in the nighttime, but values that are substantially lower than unity indicate substantial sleep fragmentation. The total simulation time for each run is 12 hours, including the hour of wakefulness ("daytime") before sleep, a variable duration of sleep ("nighttime"), and the subsequent period of wakefulness ("daytime"). In our analyses of the simulated "data", computations are made on the nighttime sleep segment or daytime segment after allowance of 1000 seconds transition time from the start of each state. To quantify the degree of ventilatory instability, we used the apnea-hypopnea index (AHI) and selected the threshold of AHI ≥15 events per hour, consistent with the criteria used in SERVE-HF, to screen for cases representing CSR (Cowie et al., 2015). The feature of upper airway collapsibility is bypassed in this model, although the original version of PNEUMA was capable of simulating obstructive sleep apnea. Thus, the apneas simulated by this model are purely central in nature. Table 6.2 lists the pertinent indices and model variables/parameters analyzed in the current study. In particular, we highlight PCWP (pulmonary capillary wedge pressure) which we deduce from the model simulations by capturing the value of end-diastolic left ventricular pressure at the end of expiration; at this point in the cardiac and breathing cycles, left ventricular pressure is equilibrated with left atrial pressure. We focused only on simulated CHF "subjects" with nighttime LVEF ≤45%, representing HFrEF.

6.3 Results

6.3.1 Illustration of ASV Effect on Respiratory, Cardiovascular, and Autonomic Variables

A typical response of CHF-CSR "subject" treated with ASV, as simulated by the model, is shown in Fig. 6.2. Before ASV intervention, the characteristic CSR breathing pattern of waxing and waning is not only observed in the respiratory variables but also entrained across cardiovascular and neural variables through cardiorespiratory coupling and interaction with other regulatory subsystems. ASV works to provide out-of-phase positive airway pressure to complement the subject's insufficient spontaneous ventilatory drive. Once it is applied to the subject, the output of the ventilator is adjusted according to the input from the subject, so that more ventilatory assistance is provided when the subject is breathing less and minimal assistance is given when there is adequate drive, thus stabilizing the ventilatory pattern. The stabilized pattern lasts only during ASV application. After ASV is disconnected, the subject resumes spontaneous breathing and the previous CSR patterns recur.

6.3.2 Comprehensive Summary of ASV Effect on Ventricular Function and Respiratory Stability

Figure 6.3 provides a comprehensive "map" of the relationships among key cardiac, vascular, and respiratory indices derived from the model simulations in the untreated state and during ASV application, covering the large range of parameter combinations that represent the span of CHF conditions explored in this study. Out of 72 simulations, 55 cases show CHF with night-time LVEF ≤45%, among which 48 cases exhibit CSR with AHI ≥15 events per hour (eph). The predicted AHI in each untreated case (closed circles) is coded on a color scale, ranging from dark blue

Table 6.2 Clinical indices investigated in the study

Index	Description	Definition/calculation	References
Respiratory			
AHI	Apnea-hypopnea index (eph)	Measured in episodes per hour Apnea event: >95% reduction in tidal volume lasting for ≥10 seconds Hypopnea event: one of the following Recommended: >30% reduction in tidal volume with ≥4% O_2 desaturation and lasting for ≥10 seconds Alternative: >50% reduction in tidal volume with either ≥3% O_2 desaturation or an arousal, and lasting for ≥10 seconds	Berry et al. (2012)
CL	Cycle length (s)	Time span between two consecutive apnea/hypopnea; CL = AHL + hyperpnea length	
AHL	Apnea-hypopnea length (s)	Time span across the beginning and end of apnea/hypopnea	
CSR	Cheyne stokes respiration (%)	Proportion of time spent on CSR – defined by ≥3 consecutive central apnea/hypopnea cycles	Berry et al. (2012)
VT	Tidal volume (mL)	Lung volume change during each breath	
BF	Breathing frequency (bpm)	Measured in breath per minute	
VE	Ventilation (L/min)	VT × BF	
WOB	Work of breathing (J/min)	BF × $\int_{breath} P_{mus,\,in}\, dV_{lung}$, expressed in power $P_{mus,in}$: pressure exerted by inspiratory muscle and assumes passive expiration	
Peripheral tissue health			
SaO_2	Arterial oxygen saturation (%)		
$PaCO_2$	Arterial carbon dioxide partial pressure (mmHg)		
O_2ERt	Peripheral tissue oxygen extraction rate (%)	The model assumes fixed metabolic rate and no extraction limitation of the peripheral tissue	
Cardiovascular			
Cardiac (left ventricular) performance			
CO	Cardiac output (L/min)	HR × SV	
HR	Heart rate (bpm)	Measured in beat per minute	
SV	Stroke volume (mL)	Difference in left ventricular end-diastolic and end-systolic volume	
CW	Cardiac work (J/min)	HR × $\int_{beat} P_{tm,\,lv}\, dV_{lv}$, expressed in power $P_{tm,lv}$: left ventricular transmural pressure	
SW	Stroke work (mmHg*L)	$(ESP_{tm} - EDP_{tm}) \times SV$ ESPtm/ EDPtm: left ventricular transmural end-systolic/diastolic pressure	
CE	Mechanical cardiac efficiency (%)	$\dfrac{HR \times SW}{m\dot{V}O_2 \times \dfrac{LVS}{LVS + RVS} \times f}$ *f*: factor converting O_2 utilization to energy production, which assumes myocardium oxidizes free fatty acid and glucose equally, $f = 20.4$ J/mL of O_2 The middle term in the denominator reflects the factor by which myocardial oxygen consumption of the left heart is adjusted in proportion to SW (calculated for the left heart only) in the numerator	Schipke (1994), Steendijk and Brinke (2008), Westerhof (2000), Wong et al. (2011)

(continued)

Table 6.2 (continued)

Index	Description	Definition/calculation	References
LVEF	Left ventricular ejection fraction (%)	$\frac{SV}{EDV}$	
Myocardial health			
CcO_2	Coronary oxygen concentration (mL/mL)		
$CcCO_2$	Coronary carbon dioxide concentration (mL/mL)		
O_2ERc	Myocardial oxygen extraction rate (%)	The model assumes variable myocardial metabolic rate and no extraction limitation	
LVS/ RVS	Left/ right ventricular wall stress (mmHg)	$LVS = \frac{P_{tm,lv} \times \sqrt[3]{V_{lv}}}{2h}$ $RVS = \frac{P_{tm,rv} \times \sqrt[3]{V_{rv}}}{2h}$ where h = ventricular wall thickness. Assumption in calculations: isotropic volume change	Aurigemma (2017)
RSP	Rate stress product (bpm*mmHg)	HR × (LVS + RVS) Adopted in the model to dynamically modify myocardial metabolic rate	Devereux et al. (2015), Hoffman and Buckberg (2014)
mVO_2	Myocardial oxygen consumption rate (mL/s)	$Q_c \times (CaO_2 - C_cO_2)$	
Q_c	Coronary flow (mL/s)	=myocardial perfusion	
MOB	Myocardial oxygen balance ((mL/s)/(mL/s))	$\frac{Q_c \times C_aO_2}{mVO_2}$	
Other			
EDV	Left ventricular end-diastolic volume (mL)		
EDPtm	Left ventricular transmural end-diastolic pressure (mmHg)	Left ventricular filling pressure referenced to pleural pressure	
PCWP	Pulmonary capillary wedge pressure (mmHg)	Clinical index of left ventricular filling pressure =EDP measured at end expiration	Ryan et al. (2012)
TPR	Total peripheral vascular resistance (mmHg*s/mL)	$\frac{MAP - CVP}{CO}$, MAP: mean arterial pressure, CVP: central venous pressure	
Neural/sleep regulation			
snaP/ snaH	Efferent sympathetic neural activities to the peripheral vasculature/ the heart (spikes/s)	snaP: regulates peripheral vascular resistance snaH: regulates heart rate and contractility of both ventricles	
Vagal	Efferent parasympathetic neural activities to the heart (spikes/s)	Regulates heart rate	
SI	Sleep index (unitless)	Determined by interaction of modeled sleep mechanism and respiratory drive. SI = 0: wakefulness, SI = 1: deep sleep	

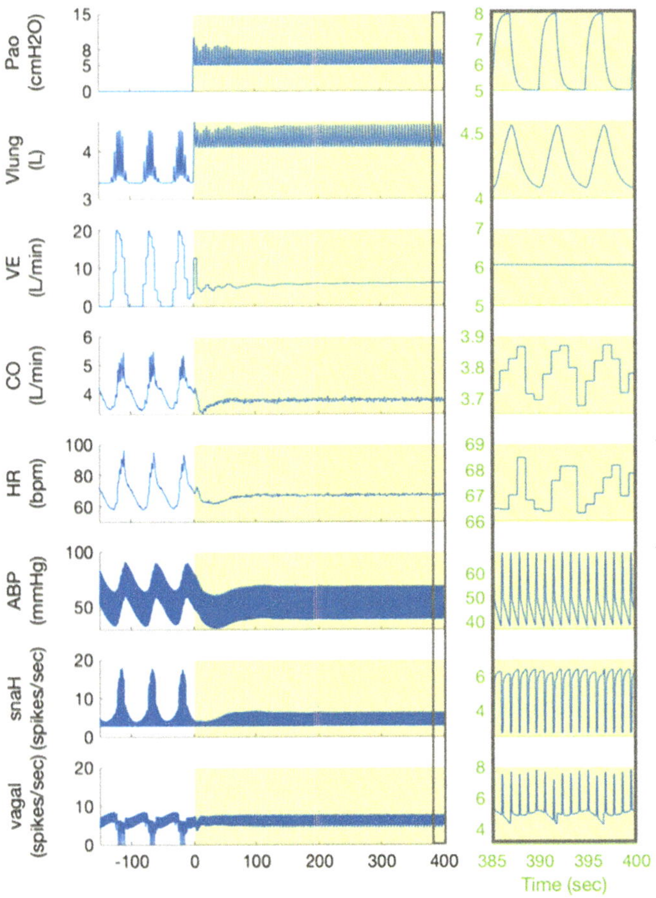

Fig. 6.2 The response of a "subject" with CHF-CSR settings (LVD = 2, IVH = 0, CGU = 1, see Table 6.2) to ASV applied during 0 ~ 500 seconds (shaded in yellow). Vlung lung volume, VE ventilation, CO cardiac output, HR heart rate, ABP arterial blood pressure, snaH sympathetic neural activity to regulate the heart, Vagal parasympathetic neural activity to regulate the heart. The middle column shows an expanded view of the time-window of 385 ~ 400 seconds, displayed within the thin vertical box (black borders) in the left column

representing AHI~0 eph to bright yellow representing AHI~120 eph (Fig. 6.3).

The colored dotted curves in Fig. 6.3 represent the equivalents of the Frank-Starling relationship between left ventricular performance and preload over the range of CHF conditions simulated in the study. Left ventricular performance is quantified using stroke work (SW), since SW takes into account both left ventricular pressure and stroke volume. Preload is quantified using transmural end-diastolic pressure (EDPtm), in which end-diastolic pressure is referenced to pleural pressure.

Starting with the blue dotted curves and accompanying symbols that are closest to the top left of the graph, these represent the cases where left ventricular function is normal (LVD = 0). As total blood volume increases with intravascular hypovolemia (IVH > 0), both SW and EDPtm also increase (following the trajectories represented by the blue dotted curves). At normal levels of chemoreflex gain (small dark blue circles), these cases exhibit no periodic breathing (dark blue color representing AHI~0 eph). But with substantially increased chemoreflex gain (large yellow circles), periodic breathing with high values of AHI (~100 eph) occurs. SW and EDPtm are little changed by the increases in chemoreflex gain. Treating these periodic breathing (including CSR) cases with ASV leads to the elimination of

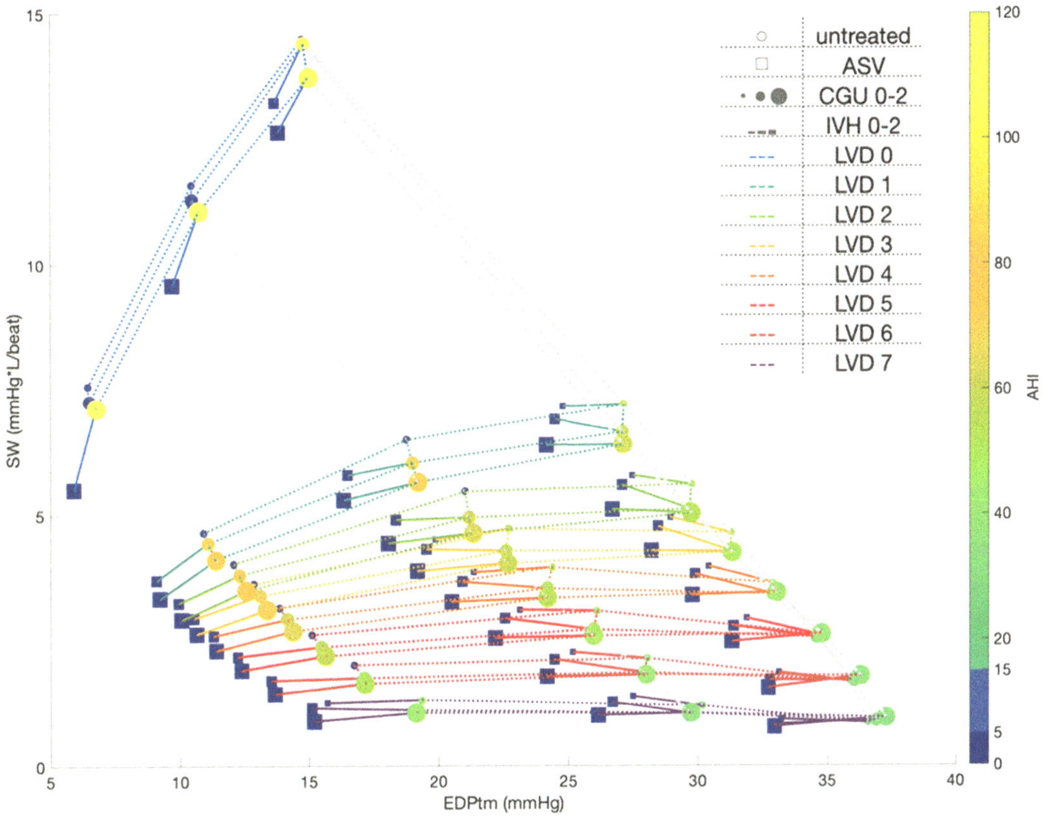

Fig. 6.3 Ventricular function-respiratory stability diagram. Ventricular function is shown with stroke work (SW)-transmural end-diastolic pressure (EDPtm). The degree of ventilatory instability is quantified using the apnea-hypopnea index (AHI, color-mapped marker). There are 72 cases uniquely defined by left ventricular dysfunction (LVD, connected by color-coded line), intravascular hypervolemia (IVH, connected by thickness-coded grey line), and chemoreflex gain upregulation (CGU, coded by marker size). Those with AHI ≥15 eph (59 cases of circle beyond blue color on the left) are further treated with ASV (results with ASV displayed as squares each connected by line to its untreated baseline (circles))

periodic breathing (represented by the dark blue squares). ASV also produces a significant reduction in both SW and EDPtm.

LVD is increased in integer steps from zero to simulate progressive impairment in left ventricular systolic function. In Fig. 6.3, the relationships between SW and EDPtm (Frank-Starling curves) become progressively flatter until they almost assume a slightly negative slope in the most severe case studied (LVD = 7, displayed in purple), consistent with the depressed ventricular function conditions first explored in humans by Braunwald and colleagues in the 1960s (Braunwald et al., 1967). The factor that most influences the extent to which these SW-EDPtm curves are depressed is the degree of LVD, whereas intravascular hypervolemia works to stretch these curves toward the right, i.e., the effect is largely the elevation of EDPtm. Chemoreflex gain, on the other hand, primarily influences ventilatory instability – in general, with higher gain contributing to higher AHI. For example, in mild CHF (LVD = 1 & IVH = 0), CSR with AHI ≥15 eph appears when chemoreflex gains are larger than in normal subjects. Nevertheless, it should be noted that all three factors (LVD, IVH, and CGU) contribute in varying degrees to the occurrence and severity of CSR. For instance, with increasing LVD, more cases show AHI beyond the threshold for CSR (marked with color beyond dark blue); however, AHI paradoxically displays a clear tendency to decrease. Similarly, as IVH increases, more AHI surpasses the threshold while its value becomes lower but the effect from IVH is noticeably weaker than the effect from LVD. While these results may appear odd, closer inspection of the simulations shows that increasing LVD and IVH act to increase the cycle duration of the CSR periodicities. When quantified in terms of the number of "events per hour", the longer cycle durations translate to *smaller* values of AHI. With progressively deteriorating left ventricular systolic function (e.g., LVD > 2), application of ASV continues to eliminate CSR, but unlike the cases with mild CHF, there is now a tendency for little reduction or even increase in SW whereas EDPtm is substantially decreased (lower parts of Fig. 6.3).

Based on our finding from all our model simulations of a strong monotonic relationship between left ventricular systolic function and left ventricular ejection fraction (LVEF), we stratify the large number of simulation cases by their LVD levels into three categories which correspond to a spectrum of LVEF below 45%. As displayed in Table 6.3, these categories are (a) "L" representing low LVEF (LVEF $\leq$30%) in the group of LVD = 6 ~ 7, (b) "M" representing medium LVEF (30% < LVEF $\leq$ 36%) in the group of LVD = 4 ~ 5, and (c) "H" representing high LVEF (36% < LVEF $\leq$ 45%) in the group of LVD = 2 ~ 3. It should be noted that the range of LVEF values explored in each group is consistent with that reported in the sub-study of SERVE-HF (Eulenburg et al., 2016); "high LVEF" merely refers to the highest tertile of the CHF range, which is still substantially lower than the levels (~60%) found in healthy subjects.

Figure 6.4 displays box-plots (displaying 25th, median and 75th percentiles) of several key physiologic parameters at baseline (untreated state during sleep, left panel) for each of the 3 LVEF levels, along with their corresponding values or changes following ASV therapy (right panel). For comparison, the corresponding simulation values for the normal (healthy) subject are displayed as purple horizontal lines and the quartiles for the CHF only subjects (LVEF $\leq$45% and AHI <15 eph) are displayed as blue lines and bands. To the right of the simulation results, the corresponding experimental values of these parameters, as reported in SERVE-HF for all the subjects studied, are shown for comparison, displayed as burgundy horizontal lines (means or medians) with the-same-tone shaded regions ((interquartile) ranges or 95% confidence intervals). The top left plot shows that increasing LVEF level (progressively improved left ventricular function) is associated with increased AHI. This echoes what was presented in Fig. 6.3: as LVD increases (and LVEF decreases), the accompanying reduction in cardiac output occurring with or without hypervolemia leads to prolonged circulatory delay, which in turn increases CSR cycle duration, thereby reducing AHI. ASV acts to eliminate CSR in all these cases (right panel).

The simulation results show a small improvement on the order of ~2% in LVEF for all LVEF levels with ASV application, remarkably similar to the data reported in SERVE-HF (mean change ~ +1.83%). ASV is predicted to reduce left ventricular stress (LVS) by ~13% on average for all LVEF levels. In SERVE-HF, LVS was not measured directly, but measurements of NT-proBNP, a biomarker for LVS, showed a roughly 10% reduction (Braunwald, 2008; Krittayaphong et al., 2008).

Table 6.3 (first line of each row) provides a comparison of these LVEF groups in terms of the various key clinical indices associated with their untreated baseline states. The values showing significant difference among groups include clinical indices associated with severity and pattern of the CSR periodicities, nadir arterial O_2 saturation, cardiac performance in terms of cardiac output, cardiac efficiency, and LVEF, as well as sympathetic neural drives. There are common features among the untreated baseline states of the 3 LVEF groups: (a) all exhibit hypocapnia (average ~ 35.9 mmHg vs 43.8 mmHg in the normal subject and 43.4 mmHg in the CHF-only group); (b) increased left ventricular stress at the expense of the augmented ventricular volume and end-diastolic pressure; and (c) reduced vagal tone. Upon ASV application (second line of each row in Table 6.3), all three groups show:

Table 6.3 ASV treatment effect stratified by LVEF group

Group (LVEF)	L (<30%)		M (30–36%)		H (36–45%)	
	n = 17		n = 16		n = 15	
AHI (eph)	37.89	(±5.65)	48.52	(±8.39)	58.45	(±9.76)
	0.00	(±0.00)	0.00	(±0.00)	0.00	(±0.00)
CL (s)	96.67	(±13.38)	76.11	(±12.70)	63.08	(±10.14)
	0.00	(±0.00)	0.00	(±0.00)	0.00	(±0.00)
AHL (s)	62.76	(±14.01)	48.72	(±11.73)	38.54	(±7.12)
	0.00	(±0.00)	0.00	(±0.00)	0.00	(±0.00)
VT (mL)	320.31	(±51.88)	321.62	(±45.88)	335.05	(±43.19)
	512.36	(±17.74)	508.11	(±13.95)	508.22	(±12.89)
VE (L/min)	7.28	(±0.80)	6.69	(±0.58)	6.49	(±0.42)
	6.48	(±0.33)	6.39	(±0.24)	6.37	(±0.21)
WOB (J/min)	8.03	(±1.31)	7.19	(±1.34)	6.72	(±1.30)
	0.99	(±0.23)	0.93	(±0.16)	0.91	(±0.12)
SaO_2min (%)	92.63	(±1.73)	93.42	(±1.20)	94.14	(±1.01)
	97.13	(±0.31)	97.11	(±0.26)	97.12	(±0.25)
$PaCO_2$ (mmHg)	35.03	(±2.43)	36.00	(±2.41)	36.76	(±2.30)
	38.18	(±2.38)	38.87	(±2.10)	39.07	(±1.87)
O_2ERt (%)	39.49	(±2.56)	33.07	(±1.69)	29.74	(±1.56)
	38.88	(±3.34)	32.75	(±2.46)	29.68	(±2.49)
TPR (mmHg*s/mL)	0.65	(±0.04)	0.70	(±0.04)	0.74	(±0.04)
	0.58	(±0.07)	0.66	(±0.08)	0.70	(±0.08)
SBP (mmHg)	53.71	(±6.83)	71.66	(±8.39)	84.12	(±10.06)
	51.69	(±8.50)	70.24	(±11.09)	82.28	(±13.95)
CO (L/min)	3.26	(±0.30)	3.98	(±0.28)	4.43	(±0.35)
	3.21	(±0.37)	3.94	(±0.38)	4.37	(±0.48)
HR (bpm)	67.98	(±6.26)	65.56	(±4.14)	65.00	(±3.19)
	64.58	(±5.48)	64.31	(±3.14)	64.02	(±2.26)
SV (mL)	48.88	(±6.38)	61.27	(±6.46)	68.63	(±7.59)
	50.04	(±6.75)	61.55	(±7.40)	68.52	(±8.82)
EDV (mL)	189.57	(±15.20)	182.52	(±17.57)	174.57	(±19.44)
	182.00	(±18.23)	174.66	(±20.90)	166.48	(±22.80)
EDPtm (mmHg)	28.52	(±7.54)	25.83	(±7.61)	23.00	(±7.37)
	25.00	(±7.61)	22.67	(±7.65)	20.20	(±7.33)
LVEF (%)	25.85	(±2.63)	33.64	(±1.69)	39.40	(±1.53)
	27.53	(±2.74)	35.30	(±1.97)	41.25	(±1.97)
CW (J/min)	11.93	(±3.71)	24.78	(±4.73)	36.26	(±5.46)
	11.06	(±4.18)	24.03	(±5.47)	34.95	(±7.75)
LVS (mmHg)	98.21	(±20.38)	100.16	(±21.60)	99.14	(±22.90)
	85.11	(±21.50)	88.71	(±23.77)	88.35	(±25.61)
RSP (bpm*mHg)	12.30	(±1.94)	11.98	(±2.28)	11.79	(±2.56)
	9.97	(±2.25)	10.32	(±2.60)	10.28	(±2.86)
mVO_2 (mL/s)	0.53	(±0.09)	0.54	(±0.11)	0.55	(±0.13)
	0.43	(±0.10)	0.46	(±0.12)	0.48	(±0.14)
Q_c (mL/sec)	5.91	(±0.98)	6.10	(±1.21)	6.23	(±1.43)
	4.68	(±1.08)	5.16	(±1.34)	5.36	(±1.56)
O_2ERc (%)	45.92	(±0.35)	45.26	(±0.26)	44.99	(±0.23)
	46.43	(±0.58)	45.63	(±0.46)	45.24	(±0.36)
MOB ((mL/s)/(mL/s))	2.17	(±0.02)	2.21	(±0.01)	2.22	(±0.01)
	2.15	(±0.03)	2.19	(±0.02)	2.21	(±0.02)

(continued)

Table 6.3 (continued)

Group (LVEF)	L (<30%)		M (30–36%)		H (36–45%)	
	n = 17		n = 16		n = 15	
CE (%)	4.60	(±1.74)	9.19	(±2.07)	13.25	(±2.01)
	5.01	(±1.86)	9.74	(±1.90)	13.64	(±1.60)
snaP (spikes/s)	8.37	(±3.36)	5.58	(±2.04)	4.30	(±0.55)
	2.98	(±0.43)	3.01	(±0.38)	3.01	(±0.33)
snaH (spikes/s)	6.09	(±1.64)	5.13	(±1.22)	4.86	(±0.97)
	4.42	(±1.68)	4.62	(±1.24)	4.71	(±0.94)
vagal (spikes/s)	4.89	(±0.48)	4.96	(±0.25)	4.99	(±0.16)
	5.01	(±0.28)	5.16	(±0.22)	5.26	(±0.20)
SI (unitless)	0.68	(±0.04)	0.71	(±0.03)	0.73	(±0.02)
	0.98	(±0.04)	0.99	(±0.01)	1.00	(±0.01)

Data in each cell: mean (±SD) of the untreated nighttime cases in the 1st row (shaded in grey) and following application of ASV in the 2nd row

AHI apnea-hypopnea index, *CL* cycle length, *AHL* apnea-hypopnea length, *VT* tidal volume, *VE* ventilation, *WOB* work of breathing, SaO_2min nadir arterial O_2 saturation, $PaCO_2$ arterial CO_2 partial pressure, O_2ERt tissue oxygen extraction rate, *TPR* total peripheral resistance, *SBP* arterial systolic blood pressure, *CO* cardiac output, *HR* heart rate, *SV* stroke volume, *EDV* left ventricular end-diastolic volume, *EDPtm* left ventricular end-diastolic transmural pressure, *LVEF* left ventricular ejection fraction, *CW* cardiac work, *LVS* left ventricular stress, *RSP* rate stress product, mVO_2 myocardial O_2 consumption, Q_c coronary flow, O_2ERc myocardial O_2 extraction rate, *MOB* myocardial oxygen balance, *CE* cardiac efficiency, *snaP* sympathetic neural activity to peripheral vasculature, *snaH* sympathetic neural activity to the heart, *vagal* parasympathetic neural activity to the heart, *SI* sleep index

(a) consistent alleviation of the burden of hypoxemia, (b) reduced work of breathing, (c) reduced ventricular stress, (d) reduction in sympathetic cardiac and peripheral vascular neural tone, and (e) reduced peripheral vascular resistance.

The model predicts on average a roughly 20% decrease in coronary perfusion during ASV and thus about the same reduction in oxygen supply to the heart (Table 6.3 and Fig. 6.5). At the same time, with ASV application, left ventricular stress is reduced, along with systolic blood pressure and heart rate. As such, myocardial oxygen demand is also reduced. Thus, myocardial oxygen balance remains largely unchanged between the untreated nighttime state and nighttime ASV. Model simulations of daytime wakefulness, where it is assumed that the "subject" would be breathing spontaneously without ASV, generally display no CSR. Comparison between the untreated nighttime (sleep) and the untreated daytime (wake) conditions suggests that elimination of CSR without the application of ASV increases coronary blood flow. On the other hand, elimination of CSR through ASV reduces coronary blood flow below its untreated nighttime (sleep) baseline. This is due largely to the application of the positive end-expiratory pressure (PEEP) that accompanies ASV. Similar effects can be observed for cardiac output (Table 6.3 and Fig. 6.5, lower panel), except that the changes are substantially smaller (on the order of ~5% on average across all cases). However, the coronary blood flow levels for all 3 LVEF groups in the untreated nighttime (sleep) state are roughly the same, whereas there are significant differences in baseline (untreated nighttime/sleep) cardiac output across LVEF groups. In the low LVEF ("L") group, the baseline cardiac output is 16% lower than the group with mild CHF without CSR (blue line and band) and 35% lower than the normal subject (purple line). The very large variability in coronary blood flow and cardiac output between the various cases within the L group should also be noted. In particular, the effect of ASV on cardiac output in the cases within the L group can range between ~ −10% (i.e., reduction) and ~ +10% (i.e., increase).

The model predicts higher peripheral sympathetic activity (snaP) and lower vagal tone in the majority of all LVEF groups in the untreated

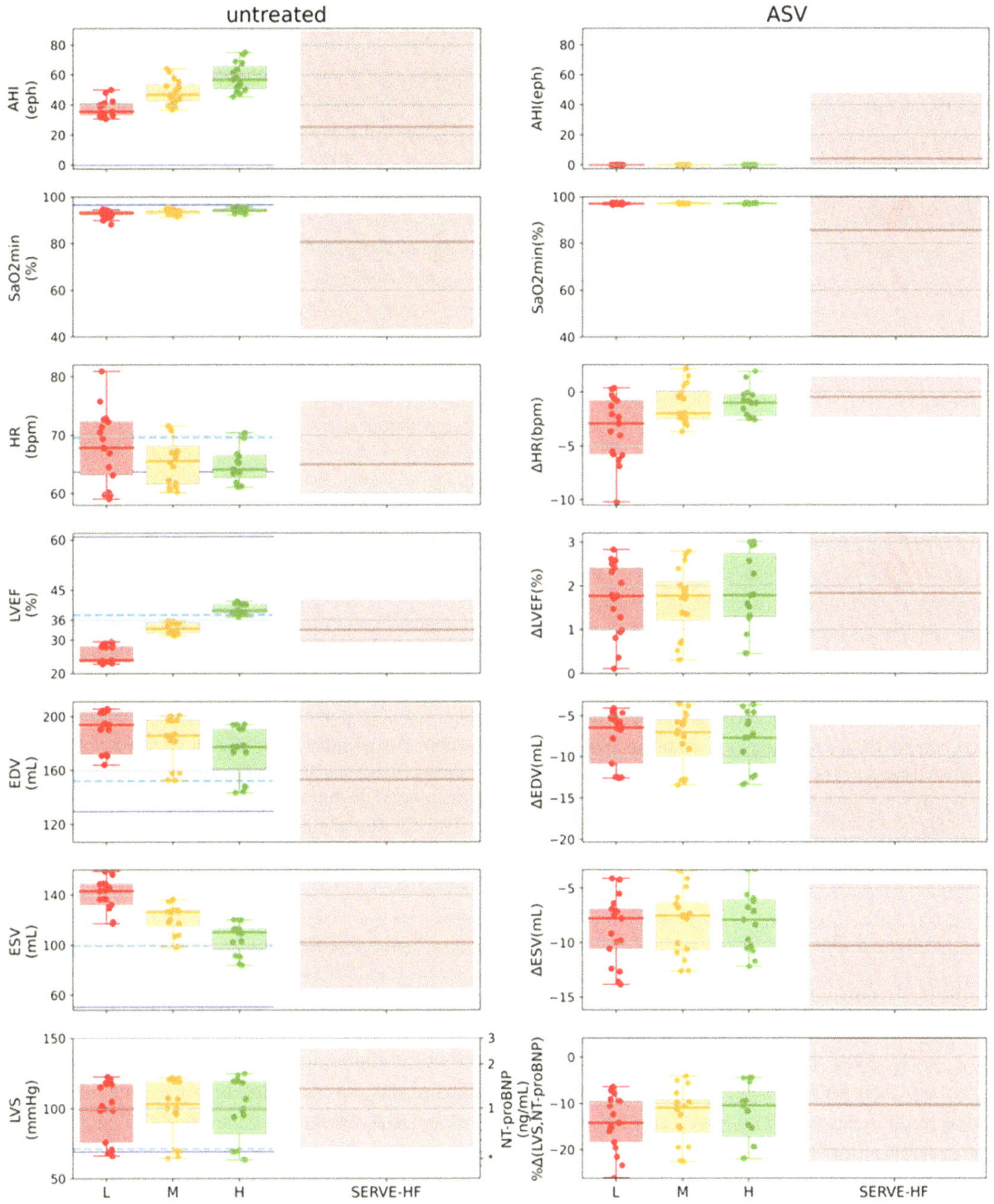

Fig. 6.4 ASV effects predicted by model vs corresponding results reported in SERVE-HF study. Left: untreated CSR cases of untreated nighttime values, with the background of the normal "subject" (LVD = 0, IVH = 0, CGU = 0) in purple and the CHF-only group in blue. Along with the simulation data are the baseline values reported in SERVE-HF shown in burgundy. Right: overnight-ASV data from PNEUMA vs post-12-month-ASV data from SERVE-HF. The simulation CSR cases are grouped by their left ventricular dysfunction resulting in low (L), medium (M), and high (H) left ventricular ejection fraction. The box plots of our simulation data highlight the 25th, 50th, and 75th percentile whereas those of SERVE-HF are presented with their mean & range in AHI and SaO_2min ($n = 666$, (Cowie et al., 2015)), and in the rest of the metrics with quartile values of the untreated while mean & 95% confidence interval of the change by ASV ($n = 159$, (Cowie et al., 2018)). AHI apnea-hypopnea index, SaO_2min minimum O_2 saturation, HR heart rate, LVEF left ventricular ejection fraction, EDV end-diastolic volume, ESV end-systolic volume, LVS left ventricular stress, NT-proBNP N-terminal prohormone of B-type natriuretic peptide. The bottom left panel shows our simulation data in LVS compared to SERVE-HF subgroup study data in NT-proBNP (* referenced to the normal as 0.45 ng/mL, (Krittayaphong et al., 2008; Januzzi et al., 2006)), while the bottom right panel shows both data sets in terms of percentage change from untreated (nighttime) levels

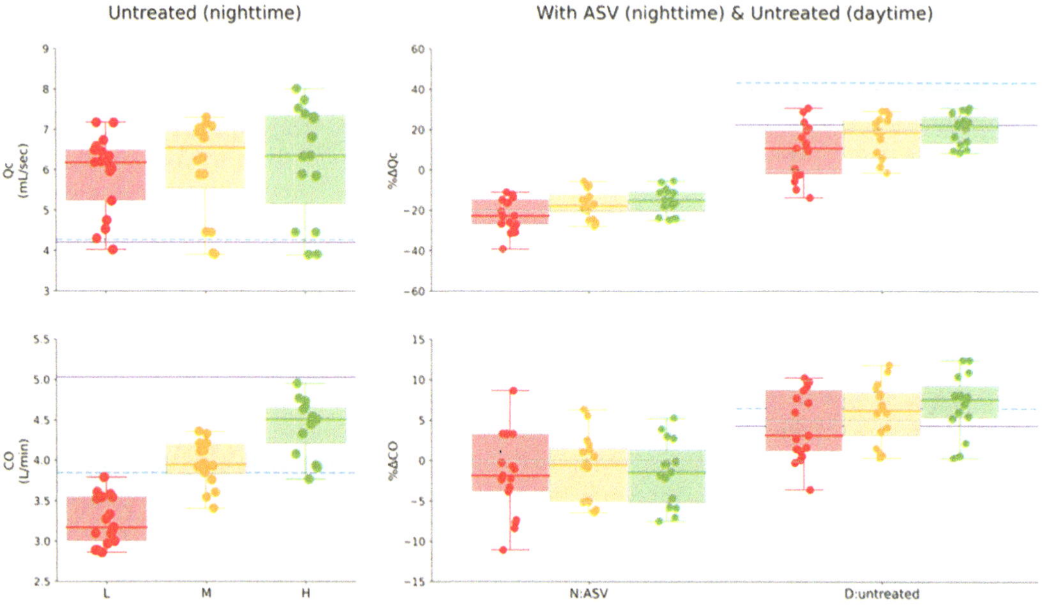

Fig. 6.5 Myocardial perfusion and cardiac output under various conditions. Top: coronary flow (Q_c, part of the determinant of myocardial oxygen supply). Bottom: cardiac output (CO). Left: nighttime untreated CSR cases characterized by low (L), medium (M), and high (H) LVEF groups shown with quartile values in box plots and individual case values in scatter plots. They are referenced to the normal subject (purple line) and CHF-only group (blue line and band) in the background. Right: percentage change of the corresponding indices of various conditions with respect to their untreated nighttime selves. The conditions include nighttime treated with ASV and untreated daytime. Also displayed for reference are the changes due to changes in state from sleep (nighttime, untreated) to wakefulness (daytime but without ASV treatment) in the normal subject (purple line) and CHF-only group (blue line and band)

nighttime state versus the corresponding autonomic activity levels in the normal (purple line) and CHF only group (blue line and band) (Fig. 6.6, left panel). Interestingly, sympathetic drive to the heart (snaH) is predicted to be below the corresponding levels in CHF only. Application of ASV during sleep reduces snaP substantially and snaH to a smaller extent, while increasing vagal drive in the majority of cases (Fig. 6.6, right panel). The daytime (wakefulness) state without ASV leads to similar reductions in snaP and snaH, but there is a substantial withdrawal of vagal drive.

The results presented in Fig. 6.7 enable further exploration of the large degree of variability observed in some of the key indices within and across LVEF groups, particularly within the lowest LVEF group. For simplicity, the sub-groups are categorized based on untreated nighttime LVEF level ("L" or ≤30% vs "M+H" or >30%) and pulmonary capillary wedge pressure (PCWP: ≤25 mmHg vs >25 mmHg). The thin lines that connect the untreated cases (shown in pink) to their ASV counterparts (displayed in turquoise) highlight the changes produced by ASV application. The key results that stand out are: (a) for the "L" group with low PCWP, ASV leads to significant drops in both coronary blood flow and cardiac output to levels that are much lower (in absolute terms) than their corresponding values in the normal and CHF-only groups; (b) ASV produces a reduction in sympathetic activity and increase in vagal tone in all cases, except for a subgroup of "L" cases with high PCWP where there is instead a *reduction* in vagal tone along with very large decrease in sympathetic drive.

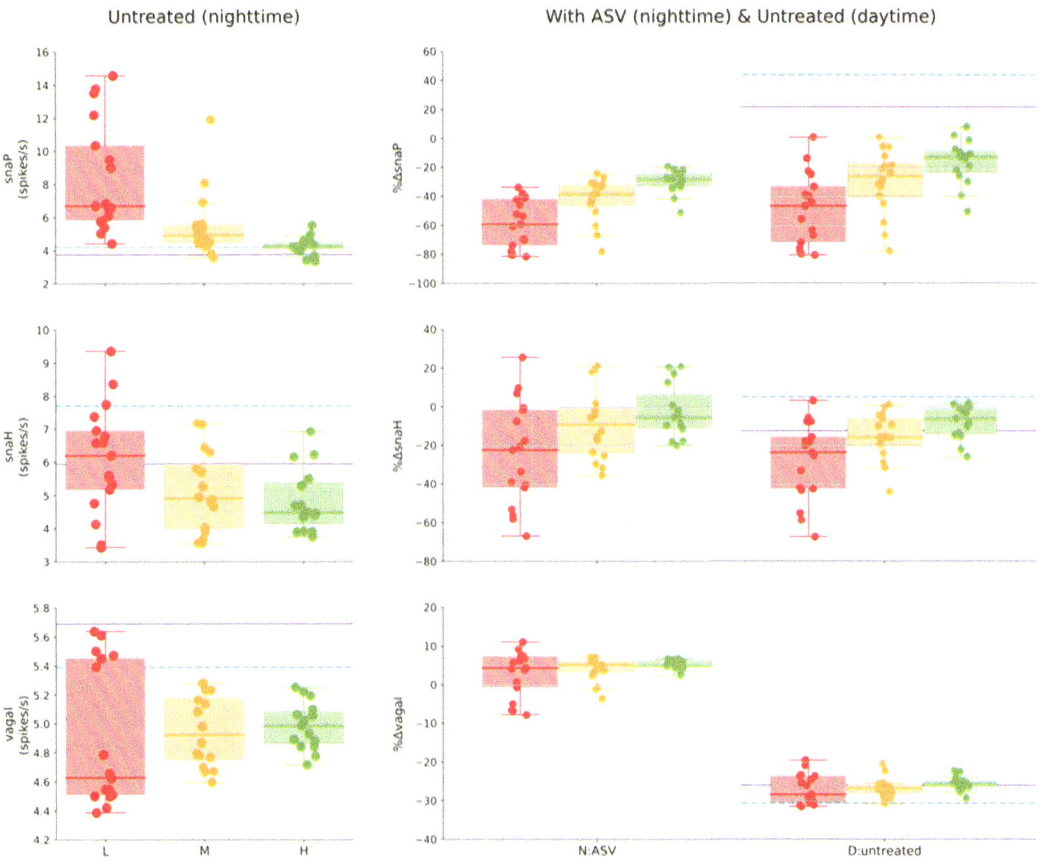

Fig. 6.6 Autonomic efferent activities under various conditions. Top: sympathetic neural activities to peripheral vasculature (snaP). Middle: sympathetic neural activities to the heart (snaH). Bottom: parasympathetic neural activities to the heart (vagal). Left: nighttime untreated CSR cases characterized by low (L), medium (M), and high (H) LVEF groups shown with quartile values in box plots and individual case values in scatter plots. They are referenced to the normal subject (purple line) and CHF-only group (blue line and band) in the background. Right: percentage change of the corresponding indices of various conditions with respect to their untreated nighttime selves. The conditions include nighttime treated with ASV and untreated daytime. Also displayed for reference are the changes due to changes in state from sleep (nighttime, untreated) to wakefulness (daytime but without ASV treatment) in the normal subject (purple line) and CHF-only group (blue line and band)

6.4 Discussion

6.4.1 Is CSR the Consequence of or Compensatory Mechanism to CHF?

The simulations conducted with the current model demonstrate that CSR in CHF can be produced by a combination of different levels of ventricular dysfunction, intravascular hypervolemia, and chemoreflex gain upregulation. It is well known that CSR frequently accompanies CHF at various stages of disease progression and the effects of this ventilatory instability are often viewed as an indicator of poor prognosis. A countervailing perspective is that the entrained ventilation, perfusion, and neural activities in CSR, though occurring on a temporally non-uniform fashion, can be viewed as an exaggerated form of respiratory sinus arrhythmia that works, on average, to optimize the efficacy of pulmonary gas exchange (Yasuma & Hayano, 2017). Moreover, the associated respiratory alkalosis resulting from the tendency for CSR to result in hypocapnia is thought to shield cardiac myocytes from

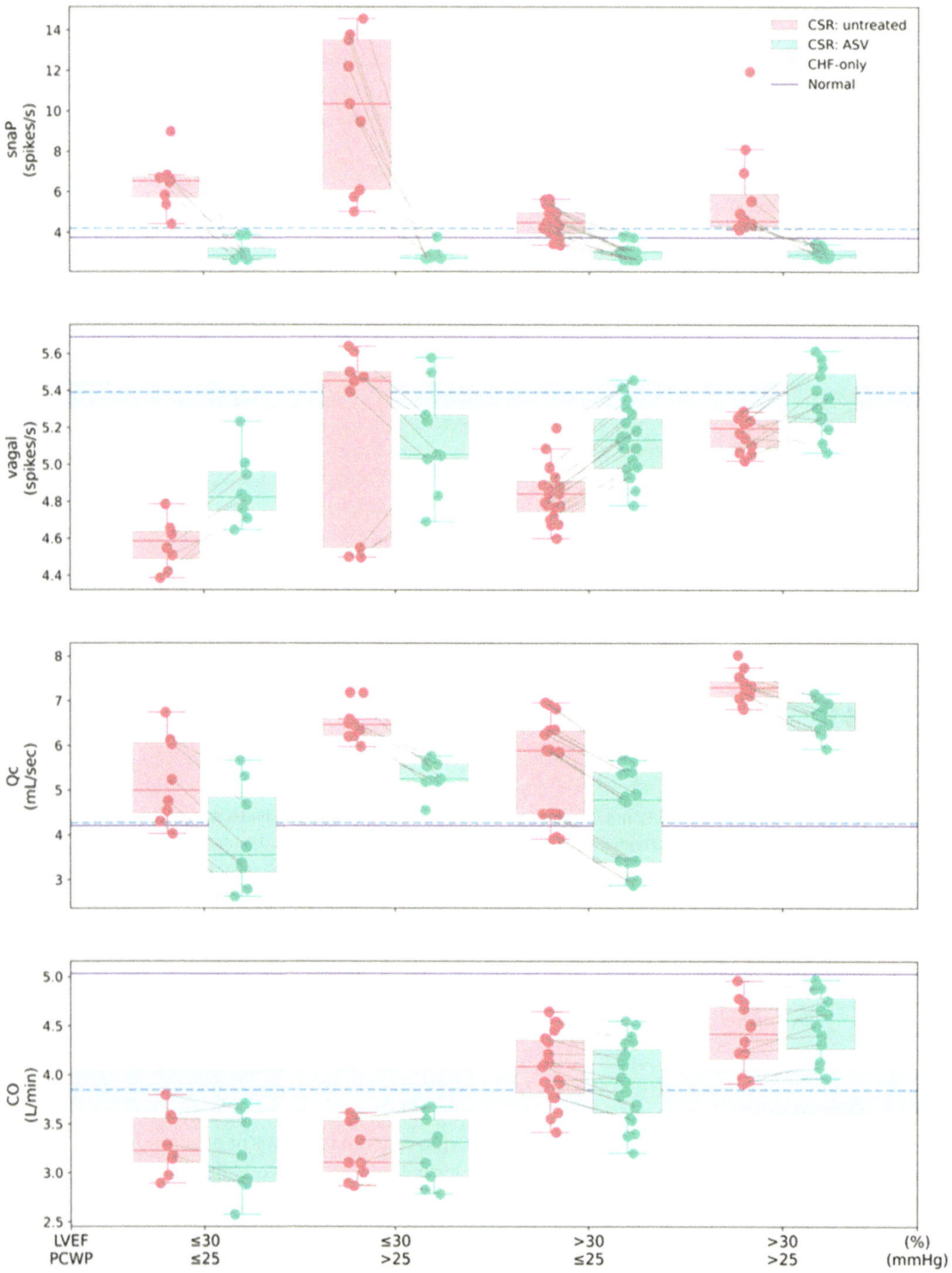

Fig. 6.7 ASV effect stratified by left ventricular ejection fraction (LVEF) and pulmonary capillary wedge pressure (PCWP). From top to bottom, there are sympathetic neural activities to peripheral vasculature (snaP), vagal tone, coronary flow (Q_c), and cardiac output (CO). CSR groups of the four combinations of high and low in LVEF and PCWP are displayed from left to right. Each of the four groups is further shown with within-subject comparison between the untreated nighttime (pink) and ASV-treated nighttime (turquoise). The corresponding predictions for the normal subject (purple line) and CHF-only group (blue line and band) are displayed for reference

cell death promoted by hypoxia-acidosis (Graham et al., 2004), as well as prevent cardiac arrhythmia or reduced contractility caused by acidosis (Crampin et al., 1842). Thus, CSR could represent a compensatory strategy to improve cardiopulmonary function in CHF. On the other hand, the periodic surges of sympathetic activity (which can lead to increased probability of life-threatening arrhythmia), fragmented sleep due to intermittent arousals, with the accompanying elevated risk of mortality, provide the key rationale for treatment to suppress CSR (Naughton, 2012).

An unexpected finding from this simulation study is that, while the apnea-hypopnea index (AHI) is considered the "gold standard" for assessing severity of CSR and other forms of periodic respiration, ventilatory patterns with the same degree of modulation can have lower AHI simply due to an increase in cycle duration, since there will be fewer "respiratory events" within a fixed time-frame. Increased cycle duration will generally occur with prolonged circulation delay as a consequence of hypervolemia and/or significant reduction in cardiac output.

6.4.2 The Impact of ASV on CHF-CSR Includes Restoring Stable Breathing and Elevating Intrathoracic Pressure

The beneficial effects of ASV have been reported extensively when considered from a variety of perspectives: overnight impact (Teschler et al., 2001; Oldenburg et al., 2015; Haruki et al., 2011), steady improvement over months/years (Cowie et al., 2015; Philippe et al., 2006; O'Connor et al., 2017; Oldenburg et al., 2018; Miyata et al., 2012; Koyama et al., 2013), and in responding patient populations, such as subjects with heart failure with preserved ejection fraction (HFpEF) (Bitter et al., 2010; Yoshihisa et al., 2013) or reduced ejection fraction (HFrEF) (Bitter et al., 2013; Arzt et al., 2013), and CSR with or without OSA (Arzt et al., 2013; Kasai et al., 2010; Lyons et al., 2017; O'Connor et al., 2017). Compared to other types of respiration-related therapy – such as nasal oxygen, continuous positive airway pressure (CPAP), or bilevel positive airway pressure (BiPAP), a number of previous studies have reported ASV to be more efficacious (Kasai et al., 2010, 2013; Teschler et al., 2001; Randerath et al., 2012; Philippe et al., 2006; Fietze et al., 2008); on the other hand, some meta-analyses have been more equivocal about its relative benefits (Sharma et al., 2012; Yang & Sawyer, 2016; Schwarz et al., 2019; Wang et al., 2021). SERVE-HF was the first large-scale, randomized multicenter trial to compare the effects of ASV therapy to conventional medical treatment in CHF patients with HFrEF (LVEF $\leq$ 45%) and predominantly central sleep apnea with AHI $\geq$15. Unexpectedly, the study found significantly higher all-cause mortality and cardiovascular mortality in the patients who received ASV. A follow-up study (Woehrle et al., 2017) showed similar findings to SERVE-HF in that patients treated with ASV had higher risk of cardiovascular death, but this risk was not related to the duration of nighttime usage of ASV.

In general, our simulation results confirm the efficacy of ASV in eliminating CSR, improving arterial O_2, and alleviating breathing workload, as has been demonstrated in experimental studies. As for the cardiovascular and neurohormonal variables, our results are compatible with published clinical findings that ASV facilitates stress release of the heart chambers, improves LVEF and cardiac function, reduces sympathetic neural activity, and improves sleep continuity by reducing arousals from sleep. However, it is important to note that all the effects predicted from the model are based on simulations reflecting overnight ASV application and these effects are temporary and reversible following the discontinuation of ASV. The current model codifies only what is known quantitatively about the physiological mechanisms operating in CHF and does not include any mechanisms for a "carry-over effect". In contrast, some clinical reports have suggested potential sustained improvement of the underlying disease (i.e., cardiac reverse remodeling) (Haruki et al., 2011; Miyata et al., 2012; Yoshihisa et al., 2011), but these remain speculative at this time.

6.4.3 ASV Significantly Reduces Coronary Flow

During ASV intervention, the positive airway pressure reduces ventricular stress thus curtailing myocardial oxygen demand. The increase in pleural pressure in ASV also mechanically impedes coronary flow (Fessler et al., 1990). Accompanying the metabolic change, there is a reduction in coronary flow and therefore a reduction in myocardial oxygen supply. Indeed, the model predicts a very small, and likely insignificant, reduction in myocardial oxygen balance. However, the model calculations are relatively simple and do not take into account coronary microvasculopathy (i.e., capillary rarefaction and impaired endothelium-dependent vasodilation or chronic hypoxia-induced systemic vasoconstriction (van de Wouw et al., 2020; Tsagalou et al., 2008)) or altered myocardial energetics (i.e., shifted myocardial metabolism with limited cardiac metabolic reserve and flexibility under mitochondrial dysfunction (De Marco et al., 1988; Neglia et al., 2007; Snyder et al., 2020)). These additional factors may contribute toward undermining myocardial oxygen reserve in these patients and put the heart tissue at risk of ischemia, which the current model does not predict.

The model predicts small reductions (~<5%) in cardiac output and somewhat larger (~20%) decrease in coronary blood flow during ASV for most of the conditions examined in our study (Fig. 6.5). These changes are most heterogenous and variable in the lowest LVEF group. A secondary factor that can account for some of this variability is the baseline PCWP level. If the low LVEF group is partitioned into a sub-group that has low PCWP (≤25 mmHg) and higher PCWP (>25 mmHg), the cases with low LVEF and low PCWP have the most severe reductions in cardiac output (>10%) and coronary blood flow (~40%). As well, application of ASV during the night reduces coronary blood flow by ~20% from the untreated (nighttime) state. But assuming that ASV is not used in daytime wakefulness, the simulations show that coronary blood flow can become 20% higher than the untreated nighttime level. Thus, a patient who is on ASV during sleep at nighttime but is off ASV during the day could experience diurnal swings in coronary blood flow of 40% amplitude (Fig. 6.5). This raises the question of whether the repetitive reduction and reperfusion of cardiac tissue over the 24-hour cycle in the long term might have adverse effects akin to small-scale ischemia-reperfusion injury, and thus promote the occurrence of arrhythmias (Yellon & Hausenloy, 2007; Manning & Hearse, 1984).

6.4.4 ASV Further Alters Sympathovagal Balance That Is Already Abnormal in CHF-CSR

In heart failure progression, sympathetic hyperactivity and reduced vagal tone play a pivotal role in regulating not only cardiac but also renal, systemic vasculature, and metabolic function. It is involved in the beginning of cardiac dysfunction as a compensatory mechanism to restore systemic perfusion; however, over time this compensatory measure exhausts the heart, further deteriorating cardiac function, triggering maladaptive remodeling as part of the vicious cycle (Cowie & Poole-Wilson, 2013; van Bilsen et al., 2017). There have been debates in the literature as to whether CSR causes further sympathetic excitation under parasympathetic withdrawal due to the intermittent hypercapnic hypoxia and arousals that accompany the periodic ventilation and apneas (Lanfranchi et al., 1999) or if it is the consequence of the deteriorating CHF associated with upregulated circulating catecholamines (Mansfield et al., 2003). It is recognized that the autonomic imbalance toward sympathetic dominance can lead to increased risk of lethal arrhythmias in CHF (Schwartz & De Ferrari, 2011; Sanchez et al., 2020). Nevertheless, our model predicts that CSR leads to a substantial elevation of peripheral sympathetic drive (especially in the low EF group) accompanied paradoxically with diminished central sympathetic drive and significant withdrawal of vagal tone. These combinations of changes in autonomic activities bring about peripheral vasoconstriction and ineffectiveness in cardiac pumping leading to fluid redistribution that promotes central congestion. However, it should be cautioned that the model equations relating autonomic activity to other physiological variables are based largely on empirical relationships derived from animal studies (Cheng et al., 2010). Thus, the aforementioned predictions of this model require validation in humans. Future studies of CHF-CSR would benefit from surrogate measures of cardiac sympathetic activity, such as heart rate variability, or peripheral sympathetic activity derived from noninvasively measured peripheral arterial tonometry.

The model also predicts differential effects of ASV on autonomic activity in the lowest LVEF versus the other 2 LVEF groups (Fig. 6.6). The lowest LVEF group appears to be most impacted by ASV, but the effects are also very heterogeneous within the group. Considering sub-groups with low versus high PCWP provides a better accounting of this variability. As displayed in Figs. 6.6 and 6.7 (upper 2 panels), although there is variability in the magnitude of the effects, ASV generally reduces sympathetic tone and increases vagal tone across LVEF levels. This partial restoration toward more normal levels of sympathovagal balance is likely a beneficial change. However, in the case of the sub-group that has the lowest LVEF and high PCWP, sympathetic drive is decreased from very high levels, but this occurs in conjunction with a concomitant depression of vagal tone toward levels even further from normality (Fig. 6.7). It remains unclear what the clinical implications are for the anomalous changes in autonomic drives in this sub-group, but one would suspect that interventions that reduce rather than increase vagal tone in the face of large decreases in sympathetic activity are more likely to favor susceptibility to arrhythmias. The significant effect of PCWP suggests that left ventricular filling pressure or central congestion plays a pivotal role in the model. Not only is it the index that reflects left ventricular function and demonstrates the correlation with intravascular hypervolemia (Androne et al., 2004), but it also is the source that blunts baroreflex sensitivity via the activation of the paradoxical cardiopulmonary reflex (Mortara et al., 1997).

6.4.5 What Could Explain the Higher Mortality Among CHF-CSR with Low EF Treated with ASV?

The clinical evidence so far revealed that the unexpected higher cardiovascular mortality comes from the ASV arm in the low ejection fraction group (LVEF < 30%) without previous hospital admission or life-saving intervention (Eulenburg et al., 2016); moreover, a subset analysis rules out adverse remodeling and worsening heart failure as the cause (Cowie et al., 2018); thus leaving sudden cardiac death (arrhythmia) as the potential culprit. There are two main mechanisms we found from our simulation results that could contribute to arrhythmogenicity: (1) the large diurnal swing of coronary flow upon nocturnal ASV treatment especially in low EF group that is most vulnerable to ischemic reperfusion injury, and (2) autonomic imbalance with higher sympathetic and lower vagal tone. Our model predicts the differential responses of CHF-CSR to ASV among various EF groups. On top of that, our model highlights PCWP as another potent factor in determining the ASV effects. While CSR is consistently eliminated by ASV in all groups, removing CSR appears to result in different outcomes in the various groups of CHF-CSR. This may be in part due to the varying degrees of elevation in intrathoracic pressure resulting from ASV in the various CHF-CSR endotypes. CSR appears to be more than just the consequence of severe CHF during disease progression. Perhaps, instead of directly targeting CSR elimination, the treatment regimen should focus more on addressing myocardial reverse remodeling or closely monitoring the pertinent parameters to finetune the treatment regime (e.g., fluid intake) in the various stages of CHF.

6.5 Limitations

The current simulation model omits the inclusion of some factors pertinent to CHF that might render the impact of either ASV or CSR less appreciated in the results. For example, the provision of intrinsic PEEP or elevated lung volume for oxygen stores during the hyperventilation phase in CSR would benefit those with atelectasis (Naughton, 2012; Linz et al., 2016). As well as those with obstructive sleep apnea, pulmonary edema, or mitral regurgitation could benefit from the PEEP setting or mean positive airway pressure of ASV (Piper, 2020; Chadda et al., 2002; Kinoshita et al., 2017). Additionally, altered skeletal muscle/myocardial energetics and endothelial dysfunction whether caused by chronic hypoxia, which lead to ventricular dysfunction and further hypoxia, could contribute to the vicious cycle in CHF progression (Ventura-Clapier et al., 2004; Mettauer et al., 2006). Not incorporating these complications in the model might explain our predictions of relatively higher baseline oxygen level than what was reported in the SERVE-HF study.

An important limitation in this model is the focus on only central apnea and hypopnea and the exclusion of any obstructive component, even though in many people with CHF there is significant prevalence of both central and obstructive sleep apnea (Javaheri, 2006; Schulz et al., 2007). The original version of this model (PNEUMA) was in fact developed to study the pathogenesis and effects of obstructive sleep apnea (Cheng et al., 2010), and thus the inclusion of upper airway obstruction can be included in the simulations relatively easily. This represents a direction we would pursue in future work, the results of which could be compared with the outcomes of the ongoing ADVENT-HF trial (Lyons et al., 2017), which includes patients with both central and obstructive apneas.

There are many other details in the model that are missing, since the physiological underpinnings remain largely unexplored. The quantitative characterization of some of the neural reflexes is inadequate. For instance, we adopted Ursino's model of autonomic regulation (Magosso & Ursino, 2001; Albanese et al., 2016) and factored in the pathological cardiopulmonary excitatory reflex by expanding the submodel that was originally meant to represent the arterial (carotid) baroreflex. However, using this extended combination of the arterial and cardiopulmonary baroreflexes leads to a simplification of the scope and dynamics of the control of responses to independent changes in arterial and atrial pressure. Moreover, the cardiopulmonary reflex not only takes both excitatory and inhibitory pathways that result in nonuniform and nonlinear behavior in efferent fibers (Millar et al., 2015) but also exhibits interaction with the chemoreflexes (Du & Chen, 2007). Also not included in our model are other pathological reflexes such as pulmonary C fiber reflex that may become important with pulmonary edema or influences from the renin-angiotensin aldosterone system. The former causes a rapid and shallow breathing pattern that undermines baseline pulmonary function and may modulate the influence of PCWP on triggering central apnea (Solin et al., 1999). The renin-angiotensin aldosterone system likely plays an important role in neurohumoral activation in response to hypervolemia (Packer, 1988) but this system is missing from the current model.

The current model also assumed that ASV would not be applied during daytime and CSR would not occur spontaneously by introducing a "wakefulness" neural drive that substantially reduces the $PaCO_2$ apneic threshold. However, in cases with severe CHF-CSR, periodic breathing could persist during daytime wakefulness if ASV is not administered. Further investigation with the model could also be directed at determining whether continuing ASV through daytime is necessary to mitigate the adverse consequences that contribute to cardiovascular mortality.

6.6 Conclusion

The present in silico model of CHF-CSR incorporates the key mechanisms of ventricular dysfunction, intravascular hypervolemia, and chemoreflex hypersensitivity. By exploring a range of parameter combinations across these mechanisms, the model is able to simulate a broad spectrum of phenotypes in levels of LVEF, CSR cycle length, and chemoreflex gain. The impact of ASV includes both positive airway pressure in the lungs and the effective regularization of CSR. But the model simulations clearly demonstrate that while the elimination of CSR through ASV can partially restore cardiorespiratory and autonomic physiology back toward more normal conditions in most cases, the degree of restoration can be highly variable, depending on the combination of mechanisms in play. The lowest LVEF group appears to be most vulnerable to the potentially adverse effects of ASV. The subgroup with lowest LVEF and low PCWP tends to have the greatest reductions in cardiac output and coronary blood flow during ASV at nighttime but significant reversals in the daytime when ASV is not applied. The subgroup with lowest LVEF and high PCWP appears to be susceptible to excessive sympathetic reduction along with concomitant vagal withdrawal during the administration of ASV, which works against the restoration of normal sympathovagal balance. Based on the findings generated by the model, we speculate that the increased risk of myocardial ischemia/reperfusion injury following long-term exposure to repetitive large diurnal fluctuations in coronary blood and/or reduced vagal protection from arrhythmia might be potential factors that could explain the small but significant increase in cardiovascular mortality reported in the SERVE-HF study.

References

Albanese, A., Cheng, L., Ursino, M., & Chbat, N. W. (2016). An integrated mathematical model of the human cardiopulmonary system: Model development. *American Journal of Physiology-Heart and Circulatory Physiology, 310*, H899–H921.

Androne, A. S., Hryniewicz, K., Hudaihed, A., Mancini, D., Lamanca, J., & Katz, S. D. (2004). Relation of unrecognized hypervolemia in chronic heart failure to clinical status, hemodynamics, and patient outcomes. *The American Journal of Cardiology, 93*(10), 1254–1259.

Arzt, M., Schroll, S., Series, F., Lewis, K., Benjamin, A., Escourrou, P., et al. (2013). Auto-servoventilation in heart failure with sleep apnoea: A randomised controlled trial. *The European Respiratory Journal, 42*(5), 1244–1254.

Aurigemma, G. P. (2017). Quantitative evaluation of left ventricular structure, wall stress and systolic function. In C. M. Otto (Ed.), *Practice of clinical echocardiography* (Fifth ed., pp. 107–127). Elsevier Inc.

Aurora, R. N. R., Bista, S. S. R., Casey, K. K. R., Chowdhuri, S., Kristo, D. A., Mallea, J. M., et al. (2016). Updated adaptive servo-ventilation recommendations for the 2012 AASM guideline: "The treatment of central sleep apnea syndromes in adults: Practice parameters with an evidence-based literature review and meta-analyses". *Journal of Clinical Sleep Medicine, 12*(5), 757–761.

Berry, R. B., Budhiraja, R., Gottlieb, D. J., Gozal, D., Iber, C., Kapur, V. K., et al. (2012). Rules for scoring respiratory events in sleep: Update of the 2007 AASM manual for the scoring of sleep and associated events. *Journal of Clinical Sleep Medicine, 8*(5), 597–619.

Bitter, T., Westerheide, N., Faber, L., Hering, D., Prinz, C., Langer, C., et al. (2010). Adaptive servoventilation in diastolic heart failure and Cheyne-Stokes respiration. *The European Respiratory Journal, 36*(2), 385–392.

Bitter, T., Gutleben, K. J., Nölker, G., Westerheide, N., Prinz, C., Dimitriadis, Z., et al. (2013). Treatment of Cheyne-Stokes respiration reduces arrhythmic events in chronic heart failure. *Journal of Cardiovascular Electrophysiology, 24*(10), 1132–1140.

Braunwald, E. (1971). Control of myocardial oxygen consumption. *The American Journal of Cardiology, 27*(4), 416–432.

Braunwald, E. (2008). Biomarkers in heart failure management. *The New England Journal of Medicine, 358*(20), 2148–2159.

Braunwald, E., Ross, J., & Sonnenblick, E. H. (1967). Mechanisms of contraction of the normal and failing heart. *The New England Journal of Medicine, 277*(19), 1012–1022.

Chadda, K., Chadda, K., Annane, D., Annane, D., Hart, N., Hart, N., et al. (2002). Cardiac and respiratory effects of continuous positive airway pressure and noninvasive ventilation in acute cardiac pulmonary edema. *Critical Care Medicine, 30*(11), 2457–2461.

Cheng, L., Ivanova, O., Fan, H.-H. H., & Khoo, M. C. K. (2010). An integrative model of respiratory and cardiovascular control in sleep-disordered breathing. *Respiratory Physiology & Neurobiology, 174*(1–2), 4–28.

Cowie, M. R., & Poole-Wilson, P. A. (2013). Pathophysiology of heart failure. In C. Rosendorff (Ed.), *Essential cardiology: Principles and practice* (pp. 327–345). Springer Science+Business.

Cowie, M. R., Woehrle, H., Wegscheider, K., Angermann, C., d'Ortho, M.-P., Erdmann, E., et al. (2015). Adaptive servo-ventilation for central sleep apnea in systolic heart failure. *The New England Journal of Medicine, 373*(12), 1095–1105.

Cowie, M. R., Woehrle, H., Wegscheider, K., Vettorazzi, E., Lezius, S., Koenig, W., et al. (2018). Adaptive servo-ventilation for central sleep apnoea in systolic heart failure: Results of the major substudy of SERVE-HF. *European Journal of Heart Failure, 20*(3), 536–544.

Crampin, E. J., Smith, N. P., Langham, A. E., Clayton, R. H., & Orchard, C. H. (1842). Acidosis in models of cardiac ventricular myocytes. *Philosophical Transactions of the Royal Society A - Mathematical*

Physical and Engineering Sciences, 2006(364), 1171–1186.

De Marco, T., Chatterjee, K., Rouleau, J. L., & Parmley, W. W. (1988). Abnormal coronary hemodynamics and myocardial energetics in patients with chronic heart failure caused by ischemic heart disease and dilated cardiomyopathy. *American Heart Journal, 115*(4), 809–815.

Dempsey, J. A., & Smith, C. A. (2014). Pathophysiology of human ventilatory control. *The European Respiratory Journal, 44*(2), 495–512.

Devereux, R. B., Bang, C. N., Roman, M. J., Palmieri, V., Boman, K., Gerdts, E., et al. (2015). Left ventricular wall stress-mass-heart rate product and cardiovascular events in treated hypertensive patients: Life study. *Hypertension, 66*(5), 945–953.

Du, Y. H., & Chen, A. F. (2007). A "love triangle" elicited by electrochemistry: Complex interactions among cardiac sympathetic afferent, chemo-, and baroreflexes. *Journal of Applied Physiology, 102*(1), 9–10.

Emdin, M., Giannoni, A., & Passino, C. (Eds.). (2017). *The breathless heart*. Springer International Publishing.

Eulenburg, C., Wegscheider, K., Woehrle, H., Angermann, C., D'Ortho, M. P., Erdmann, E., et al. (2016). Mechanisms underlying increased mortality risk in patients with heart failure and reduced ejection fraction randomly assigned to adaptive servoventilation in the SERVE-HF study: Results of a secondary multistate modelling analysis. *The Lancet Respiratory Medicine, 4*(11), 873–881.

Ferreira, J. P., Duarte, K., Woehrle, H., Cowie, M. R., Wegscheider, K., Angermann, C., et al. (2020a). Biomarkers in patients with heart failure and central sleep apnoea: Findings from the SERVE-HF trial. *ESC Heart Failure, 7*(2), 503–511.

Ferreira, J. P., Duarte, K., Woehrle, H., Cowie, M. R., Angermann, C., D'Ortho, M.-P., et al. (2020b). Bioprofiles and mechanistic pathways associated with Cheyne-Stokes respiration: Insights from the SERVE-HF trial. *Clinical Research in Cardiology, 109*(7), 881–891.

Fessler, H. E., Brower, R. G., Wise, R., & Permutt, S. (1990). Positive pleural pressure decreases coronary perfusion. *American Journal of Physiology-Heart and Circulatory Physiology, 258*(3 Pt 2):H814–H820.

Fietze, I., Blau, A., Glos, M., Theres, H., Baumann, G., & Penzel, T. (2008). Bi-level positive pressure ventilation and adaptive servo ventilation in patients with heart failure and Cheyne-Stokes respiration. *Sleep Medicine, 9*(6), 652–659.

Floras, J. S. (2009). Sympathetic nervous system activation in human heart failure. Clinical implications of an updated model. *Journal of the American College of Cardiology, 54*(5), 375–385.

Floras, J. S., & Ponikowski, P. (2015). The sympathetic/parasympathetic imbalance in heart failure with reduced ejection fraction. *European Heart Journal, 36*(30), 1974–1982.

Graham, R. M., Frazier, D. P., Thompson, J. W., Haliko, S., Li, H., Wasserlauf, B. J., et al. (2004). A unique pathway of cardiac myocyte death caused by hypoxia-acidosis. *The Journal of Experimental Biology, 207*(18), 3189–3200.

Haruki, N., Takeuchi, M., Kaku, K., Yoshitani, H., Kuwaki, H., Tamura, M., et al. (2011). Comparison of acute and chronic impact of adaptive servo-ventilation on left chamber geometry and function in patients with chronic heart failure. *European Journal of Heart Failure, 13*(10), 1140–1146.

Hoffman, J. I. E., & Buckberg, G. D. (2014). The myocardial oxygen supply: Demand index revisited. *Journal of the American Heart Association, 3*(1), 1–10.

Januzzi, J. L., Van Kimmenade, R., Lainchbury, J., Bayes-Genis, A., Ordonez-Llanos, J., Santalo-Bel, M., et al. (2006). NT-proBNP testing for diagnosis and short-term prognosis in acute destabilized heart failure: An international pooled analysis of 1256 patients: The international collaborative of NT-proBNP study. *European Heart Journal, 27*(3), 330–337.

Javaheri, S. (2006). Sleep disorders in systolic heart failure: A prospective study of 100 male patients. The final report. *International Journal of Cardiology, 106*(1), 21–28.

Javaheri, S., Shukla, R., Zeigler, H., & Wexler, L. (2007). Central sleep apnea, right ventricular dysfunction, and low diastolic blood pressure are predictors of mortality in systolic heart failure. *Journal of the American College of Cardiology, 49*(20), 2028–2034.

Kasai, T., Usui, Y., Yoshioka, T., Yanagisawa, N., Takata, Y., Narui, K., et al. (2010). Effect of flow-triggered adaptive servo-ventilation compared with continuous positive airway pressure in patients with chronic heart failure with coexisting obstructive sleep apnea and Cheyne-Stokes respiration. *Circulation Heart Failure, 3*(1), 140–148.

Kasai, T., Kasagi, S., Maeno, K.-I., Dohi, T., Kawana, F., Kato, M., et al. (2013). Adaptive servo-ventilation in cardiac function and neurohormonal status in patients with heart failure and central sleep apnea nonresponsive to continuous positive airway pressure. *JACC Heart Failure, 1*(1), 58–63.

Khoo, M. C. K. (1991). Periodic breathing. In R. Crystal, J. West, P. Barnes, N. Cherniack, & E. Weibel (Eds.), *The lung: Scientific foundations* (pp. 1419–1431). Raven Press, Ltd.

Kinoshita, M., Okayama, H., Kawamura, G., Shigematsu, T., Takahashi, T., Kawata, Y., et al. (2017). Beneficial effects of rapid introduction of adaptive servo-ventilation in the emergency room in patients with acute cardiogenic pulmonary edema. *Journal of Cardiology, 69*(1), 308–313.

Koyama, T., Watanabe, H., Tamura, Y., Oguma, Y., Kosaka, T., & Ito, H. (2013). Adaptive servo-ventilation therapy improves cardiac sympathetic nerve activity in patients with heart failure. *European Journal of Heart Failure, 15*(8), 902–909.

Krittayaphong, R., Boonyasirinant, T., Saiviroonporn, P., Thanapiboonpol, P., Nakyen, S., & Udompunturak, S. (2008). Correlation between NT-pro BNP levels and left ventricular wall stress, sphericity index and extent of myocardial damage: A magnetic resonance imaging study. *Journal of Cardiac Failure, 14*(8), 687–694.

Lanfranchi, P. A., Braghiroli, A., Bosimini, E., Mazzuero, G., Colombo, R., Donner, C. F., et al. (1999). Prognostic value of nocturnal Cheyne-Stokes respiration in chronic heart failure. *Handbook of Clinical Nutrition and Aging, 99*, 1435–1440.

Linz, D., Fox, H., Bitter, T., Spießhöfer, J., Schöbel, C., Skobel, E., et al. (2016). Impact of SERVE-HF on management of sleep disordered breathing in heart failure: A call for further studies. *Clinical Research in Cardiology, 105*(7), 563–570.

Linz, D., Baumert, M., Catcheside, P., Floras, J., Sanders, P., Lévy, P., et al. (2018). Assessment and interpretation of sleep disordered breathing severity in cardiology: Clinical implications and perspectives. *International Journal of Cardiology, 271*, 281–288.

Lorenzi-Filho, G., Genta, P. R., Figueiredo, A. C., & Inoue, D. (2005). Cheyne-Stokes respiration in patients with congestive heart failure: Causes and consequences. *Clinics, 60*(4), 333–344.

Lyons, O. D., Floras, J. S., Logan, A. G., Beanlands, R., Cantolla, J. D., Fitzpatrick, M., et al. (2017). Design of the effect of adaptive servo-ventilation on survival and cardiovascular hospital admissions in patients with heart failure and sleep apnoea: The ADVENT-HF trial. *European Journal of Heart Failure, 19*(4), 579–587.

Magosso, E., & Ursino, M. (2001). A mathematical model of CO2 effect on cardiovascular regulation. *American Journal of Physiology-Heart and Circulatory Physiology, 281*(5), H2036–H2052.

Manning, A. S., & Hearse, D. J. (1984). Reperfusion-induced arrhythmias: Mechanisms and prevention. *Journal of Molecular and Cellular Cardiology, 16*(6), 497–518.

Mansfield, D., Kaye, D. M., La Rocca, H. B., Solin, P., Esler, M. D., & Naughton, M. T. (2003). Raised sympathetic nerve activity in heart failure and central sleep apnea is due to heart failure severity. *Circulation, 107*(10), 1396–1400.

Mettauer, B., Zoll, J., Garnier, A., & Ventura-Clapier, R. (2006). Heart failure: A model of cardiac and skeletal muscle energetic failure. *Pflügers Archiv: European Journal of Physiology, 452*(6), 653–666.

Millar, P. J., Murai, H., & Floras, J. S. (2015). Paradoxical muscle sympathetic reflex activation in human heart failure. *Circulation, 131*(5), 459–468.

Mirsky, I., Tajimi, T., & Peterson, K. L. (1987). The development of the entire end-systolic pressure-volume and ejection fraction-afterload relations: A new concept of systolic myocardial stiffness. *Circulation, 76*(2), 343–356.

Miyata, M., Yoshihisa, A., Suzuki, S., Yamada, S., Kamioka, M., Kamiyama, Y., et al. (2012). Adaptive servo ventilation improves Cheyne-Stokes respiration, cardiac function, and prognosis in chronic heart failure patients with cardiac resynchronization therapy. *Journal of Cardiology, 60*(3), 222–227.

Mortara, A., La Rovere, M. T., Pinna, G. D., Parziale, P., Maestri, R., Capomolla, S., et al. (1997). Depressed arterial baroreflex sensitivity and not reduced heart rate variability identifies patients with chronic heart failure and nonsustained ventricular tachycardia: The effect of high ventricular filling pressure. *American Heart Journal, 134*(5 I), 879–888.

Naughton, M. T. (1998). Pathophysiology and treatment of Cheyne-Stokes respiration. *Thorax, 53*(6), 514–518.

Naughton, M. T. (2012). Cheyne–Stokes respiration: Friend or foe? *Thorax, 67*(4), 357–360.

Naughton, M. T. (2016). Epidemiology of central sleep apnoea in heart failure. *International Journal of Cardiology, 206*, S4–S7.

Neglia, D., De Caterina, A., Marraccini, P., Natali, A., Ciardetti, M., Vecoli, C., et al. (2007). Impaired myocardial metabolic reserve and substrate selection flexibility during stress in patients with idiopathic dilated cardiomyopathy. *American Journal of Physiology-Heart and Circulatory Physiology, 293*(6), 3270–3278.

O'Connor, C. M., Whellan, D. J., Fiuzat, M., Punjabi, N. M., Tasissa, G., Anstrom, K. J., et al. (2017). Cardiovascular outcomes with minute ventilation–targeted adaptive servo-ventilation therapy in heart failure: The CAT-HF trial. *Journal of the American College of Cardiology, 69*(12), 1577–1587.

Oldenburg, O., Spießhöfer, J., Fox, H., Prib, N., & Horstkotte, D. (2015). Performance of conventional and enhanced adaptive servoventilation (ASV) in heart failure patients with central sleep apnea who have adapted to conventional ASV. *Sleep & Breathing, 19*(3), 795–800.

Oldenburg, O., Wellmann, B., Bitter, T., Fox, H., Buchholz, A., Freiwald, E., et al. (2018). Adaptive servo-ventilation to treat central sleep apnea in heart failure with reduced ejection fraction: The Bad Oeynhausen prospective ASV registry. *Clinical Research in Cardiology, 107*(8), 719–728.

Packer, M. (1988). Neurohormonal interactions and adaptations in congestive heart failure. *Circulation, 77*(4), 721–730.

Philippe, C., Stoïca-Herman, M., Drouot, X., Raffestin, B., Escourrou, P., Hittinger, L., et al. (2006). Compliance with and effectiveness of adaptive servoventilation versus continuous positive airway pressure in the treatment of Cheyne-Stokes respiration in heart failure over a six month period. *Heart, 92*(3), 337–342.

Piper, A. J. (2020). Advances in non-invasive positive airway pressure technology. *Respirology, 25*(4), 372–382.

Randerath, W. J., Nothofer, G., Priegnitz, C., Anduleit, N., Treml, M., Kehl, V., et al. (2012). Long-term auto-servoventilation or constant positive pressure in heart failure and coexisting central with obstructive sleep apnea. *Chest, 142*(2), 440–447.

Ryan, J. J., Rich, J. D., Thiruvoipati, T., Swamy, R., Kim, G. H., & Rich, S. (2012). Current practice for determining pulmonary capillary wedge pressure predisposes to serious errors in the classification of patients with pulmonary hypertension. *American Heart Journal, 163*(4), 589–594.

Sanchez, A. M., Germany, R., Lozier, M. R., Schweitzer, M. D., Kosseifi, S., & Anand, R. (2020). Central sleep apnea and atrial fibrillation: A review on pathophysiological mechanisms and therapeutic implications. *International Journal of Cardiology. Heart & Vasculature, 30*, 100527.

Schipke, J. D. (1994). Cardiac efficiency. *Basic Research in Cardiology, 89*(3), 207–240.

Schulz, R., Blau, A., Börgel, J., Duchna, H. W., Fietze, I., Koper, I., et al. (2007). Sleep apnoea in heart failure. *The European Respiratory Journal, 29*(6), 1201–1205.

Schwartz, P. J., & De Ferrari, G. M. (2011). Sympathetic-parasympathetic interaction in health and disease: Abnormalities and relevance in heart failure. *Heart Failure Reviews, 16*(2), 101–107.

Schwarz, E. I., Scherff, F., Haile, S. R., Steier, J., & Kohler, M. (2019). Effect of treatment of central sleep apnea/Cheyne-Stokes respiration on left ventricular ejection fraction in heart failure: A network meta-analysis is not the answer. *Journal of Clinical Sleep Medicine, 15*(12), 1817–1825.

Sharma, B. K., Bakker, J. P., McSharry, D. G., Desai, A. S., Javaheri, S., & Malhotra, A. (2012). Adaptive servoventilation for treatment of sleep-disordered breathing in heart failure: A systematic review and meta-analysis. *Chest, 142*(5), 1211–1221.

Snyder, J., Zhai, R., Lackey, A. I., & Sato, P. Y. (2020). Changes in myocardial metabolism preceding sudden cardiac death. *Frontiers in Physiology, 11*(June), 1–18.

Solin, P., Bergin, P., Richardson, M., Kaye, D. M., Walters, E. H., & Naughton, M. T. (1999). Influence of pulmonary capillary wedge pressure on central apnea in heart failure. *Circulation, 99*(12), 1574–1579.

Solin, P., Roebuck, T., Johns, D. P., Walters, E. H., & Naughton, M. T. (2000). Peripheral and central ventilatory responses in central sleep apnea with and without congestive heart failure. *American Journal of Respiratory and Critical Care Medicine, 162*(6), 2194–2200.

Steendijk, P., & Brinke, E. A. (2008). Invasive assessment of cardiac efficiency. *Hear Metab., 39*, 33–36.

Strauer, B. E. (1979). Myocardial oxygen consumption in chronic heart disease: Role of wall stress, hypertrophy and coronary reserve. *The American Journal of Cardiology, 44*(4), 730–740.

Teschler, H., Döhring, J., Wang, Y. M., & Berthon-Jones, M. (2001). Adaptive pressure support servo-ventilation: A novel treatment for Cheyne-Stokes respiration in heart failure. *American Journal of Respiratory and Critical Care Medicine, 164*(4), 614–619.

Tsagalou, E. P., Anastasiou-Nana, M., Agapitos, E., Gika, A., Drakos, S. G., Terrovitis, J. V., et al. (2008). Depressed coronary flow reserve is associated with decreased myocardial capillary density in patients with heart failure due to idiopathic dilated cardiomyopathy. *Journal of the American College of Cardiology, 52*(17), 1391–1398.

van Bilsen, M., Patel, H. C., Bauersachs, J., Böhm, M., Borggrefe, M., Brutsaert, D., et al. (2017). The autonomic nervous system as a therapeutic target in heart failure: A scientific position statement from the Translational Research Committee of the Heart Failure Association of the European Society of Cardiology. *European Journal of Heart Failure, 19*(11), 1361–1378.

van de Wouw, J., Sorop, O., van Drie, R. W. A., van Duin, R. W. B., Nguyen, I. T. N., Joles, J. A., et al. (2020). Perturbations in myocardial perfusion and oxygen balance in swine with multiple risk factors: A novel model of ischemia and no obstructive coronary artery disease. *Basic Research in Cardiology, 115*(2), 1–18.

Ventura-Clapier, R., Garnier, A., & Veksler, V. (2004). Energy metabolism in heart failure. *The Journal of Physiology, 555*(1), 1–13.

Wang, J., Covassin, N., Dai, T., Fan, Z., Yisilamu, P., Sun, D., et al. (2021). Therapeutic value of treating central sleep apnea by adaptive servo-ventilation in patients with heart failure: A systematic review and meta-analysis. *Heart & Lung: The Journal of Critical Care, 50*(2), 344–351.

Westerhof, N. (2000). Cardiac work and efficiency. *Cardiovascular Research, 48*(1), 4–7.

Wilcox, I., McNamara, S., Dodd, M., & Sullivan, C. (1998). Ventilatory control in patients with sleep apnoea and left ventricular dysfunction: Comparison of obstructive and central sleep apnoea. *The European Respiratory Journal, 11*, 7–13.

Woehrle, H., Cowie, M. R., Eulenburg, C., Suling, A., Angermann, C., D'Ortho, M.-P., et al. (2017). Adaptive servo ventilation for central sleep apnoea in heart failure: SERVE-HF on-treatment analysis. *The European Respiratory Journal, 50*(2), 1601692.

Wong, Y. Y., Ruiter, G., Lubberink, M., Raijmakers, P. G., Knaapen, P., Marcus, J. T., et al. (2011). Right ventricular failure in idiopathic pulmonary arterial hypertension is associated with inefficient myocardial oxygen utilization. *Circulation. Heart Failure, 4*(6), 700–706.

Yancy, C. W., Jessup, M., Bozkurt, B., Butler, J., Casey, D. E., Colvin, M. M., et al. (2017). 2017 ACC/AHA/HFSA focused update of the 2013 ACCF/AHA guideline for the management of heart failure: A report of the American College of Cardiology/American Heart Association task force on clinical practice guidelines and the Heart Failure Society of Amer. *Journal of the American College of Cardiology, 70*(6), 776–803.

Yang, H., & Sawyer, A. M. (2016). The effect of adaptive servo ventilation (ASV) on objective and subjective outcomes in Cheyne-Stokes respiration (CSR) with central sleep apnea (CSA) in heart failure (HF): A systematic review. *Heart & Lung: The Journal of Critical Care, 45*(3), 199–211.

Yasuma, F., & Hayano, J. (2017). Respiratory sinus arrhythmia and entraining heartbeats with Cheyne-Stokes respiration: Cardiopulmonary works to be minimal by synchronizing heartbeats with breathing.

In *Clinical assessment of the autonomic nervous system* (pp. 129–146). Springer Japan.

Yellon, D. M., & Hausenloy, D. J. (2007). Myocardial reperfusion injury. *The New England Journal of Medicine, 357*(11), 1121–1135.

Yoshihisa, A., Shimizu, T., Owada, T., Nakamura, Y., Iwaya, S., Yamauchi, H., et al. (2011). Adaptive servo ventilation improves cardiac dysfunction and prognosis in chronic heart failure patients with Cheyne-Stokes respiration. *International Heart Journal, 52*(4), 218–223.

Yoshihisa, A., Suzuki, S., Yamaki, T., Sugimoto, K., Kunii, H., Nakazato, K., et al. (2013). Impact of adaptive servo-ventilation on cardiovascular function and prognosis in heart failure patients with preserved left ventricular ejection fraction and sleep-disordered breathing. *European Journal of Heart Failure, 15*(5), 543–550.

Part II

Diagnostic Innovations

Automated Scoring of Sleep and Associated Events

7

Peter Anderer, Marco Ross, Andreas Cerny, and Edmund Shaw

Abstract

Conventionally, sleep and associated events are scored visually by trained technologists according to the rules summarized in the American Academy of Sleep Medicine Manual. Since its first publication in 2007, the manual was continuously updated; the most recent version as of this writing was published in 2020. Human expert scoring is considered as gold standard, even though there is increasing evidence of limited interrater reliability between human scorers. Significant advances in machine learning have resulted in powerful methods for addressing complex classification problems such as automated scoring of sleep and associated events. Evidence is increasing that these autoscoring systems deliver performance comparable to manual scoring and offer several advantages to visual scoring: (1) avoidance of the rather expensive, time-consuming, and difficult visual scoring task that can be performed only by well-trained and experienced human scorers, (2) attainment of consistent scoring results, and (3) proposition of added value such as scoring in real time, sleep stage probabilities per epoch (hypnodensity), estimates of signal quality and sleep/wake-related features, identifications of periods with clinically relevant ambiguities (confidence trends), configurable sensitivity and rule settings, as well as cardiorespiratory sleep staging for home sleep apnea testing. This chapter describes the development of autoscoring systems since the first attempts in the 1970s up to the most recent solutions based on deep neural network approaches which achieve an accuracy that allows to use the autoscoring results directly for review and interpretation by a physician.

Keywords

Feature extraction · Classifier architecture · Deep learning · Sleep stage probabilities · Hypnodensity graph · Physician-ready autoscoring

Disclosure Statement: Peter Anderer, Marco Ross, and Andreas Cerny are part-time employees of Philips. Edmund Shaw is full-time employee of Philips. Peter Anderer is in addition freelancer and shareholder in The Siesta Group.

P. Anderer (✉)
Philips Sleep and Respiratory Care, Vienna, Austria

The Siesta Group Schlafanalyse GmbH, Vienna, Austria
e-mail: peter.anderer@philips.com

M. Ross · A. Cerny
Philips Sleep and Respiratory Care, Vienna, Austria

E. Shaw
Philips Sleep and Respiratory Care, Pittsburgh, PA, USA

T. Penzel, R. Hornero (eds.), *Advances in the Diagnosis and Treatment of Sleep Apnea*, Advances in Experimental Medicine and Biology 1384, https://doi.org/10.1007/978-3-031-06413-5_7

7.1 Development of Autoscoring Systems: From Simple Decision Trees to Deep Neural Network Classifiers

7.1.1 Problem Statement

The first generally accepted rules for visual sleep scoring were provided by the recommendations of Rechtschaffen and Kales (R&K) published in 1968. According to this manual, wake, rapid eye movement (REM) sleep, non-rapid eye movement (NREM) sleep stages S1 to S4, and movement time were distinguished. In 2007, the Manual for the Scoring of Sleep and Associated Events was published by the American Academy of Sleep Medicine (AASM, Iber et al., 2007). Concerning the visual classification of sleep stages, these rules were intended to replace the rules by Rechtschaffen and Kales (1968). An important advance of the AASM manual was the inclusion of technical and digital specifications for polysomnographic (PSG) recordings, a section on parameters to be reported for PSGs, and rules for visual scoring of cortical arousals as well as cardiac, respiratory, and movement events. Since publication, the AASM manual has been updated several times. As of this writing, the most recent update, version 2.6, was published in January 2020 (Berry et al., 2020). While the rules for the visual classification of sleep stages remained more or less unchanged from version to version, other sections, most notably the scoring of respiratory events, have had major updates. Nevertheless, even in the most recent version, two definitions for the scoring of hypopneas coexist, a recommended and an acceptable rule. Which rule applied must be specified in the PSG report because the resulting apnea-hypopnea index (AHI, i.e., number of apneas + hypopneas per hour sleep) can vary significantly depending on the applied rule. A full PSG scoring includes the scoring of sleep stages, arousals, cardiac events, periodic limb movements, and respiratory events such as obstructive apneas, mixed apneas, central apneas, (obstructive and central) hypopneas, oxygen desaturations, and snoring events. Consequently, visual scoring of a PSG not only is time-consuming but also requires well-trained and experienced human scorers. Nevertheless, visual scoring of PSGs is always subjective, and studies investigating interrater reliability revealed that the consistency between scorers is rather limited (Danker-Hopfe et al., 2004, 2009; Penzel et al., 2013; Rosenberg & Van Hout, 2013; Younes et al., 2016, 2018; Cesari et al., 2021).

7.1.2 Autoscoring According to Rechtschaffen and Kales

Given the shortcomings of visual scoring, it is not surprising that right after the R&K manual for visual sleep scoring was presented in 1968, numerous attempts at a computer-assisted identification of sleep stages have been published (Itil, 1969; Larsen & Walter, 1970; Smith & Karacan, 1971; Martin et al., 1972; Gaillard & Tissot, 1973; Hoffmann et al., 1984; Kubicki et al., 1989; Schwaibold et al., 2002). Martin et al. (1972) applied a simple decision tree using electroencephalographic (EEG) and electrooculographic (EOG) signals. Stanus et al. (1987) developed and compared two methods for automatic sleep scoring: one based on an autoregressive model and another one based on spectral bands and Bayesian decision theory using one EEG, two EOG, and an electromyographic (EMG) channel. The EOG was used to detect eye movements and the EMG to assess the muscle tone. Schaltenbrand et al. (1993) applied artificial neural networks for sleep stage classification using 17 features extracted from PSG signals, and Pardey et al. (1996) combined artificial neural networks with fuzzy logic. Fell et al. (1996) examined automatic sleep scoring using in addition to spectral features nonlinear features (correlation dimension, Kolmogorov entropy, Lyapunov exponent) and concluded that such measures carry additional information not captured with spectral features. Park et al. (2000) built a hybrid rule- and case-based system. A decision tree-like algorithm was used by Louis et al. (2004). Many of these methods were developed only based on a limited set of data and/or had never been validated in an independent sample. To produce a robust and valid automatic sleep stager suitable for clinical and pharmacological studies, the method must be validated in a large sample of subjects of both sexes covering the respective age range, including both healthy controls and patients with sleep disturbances. Since the visual scorings of the training set serve as the "gold standard" for the automatic classifier, their quality is crucial as well.

Our first version of the Somnolyzer autoscoring system was developed and validated based on a dataset that fulfilled all these requirements (Anderer et al., 2005). The used SIESTA dataset consists of 588 PSGs, 394 PSGs from healthy subjects (2 PSGs each in 98 males and 99 females aged 20–95 years with approximately the same number of males and females per decade) and 194 PSGs from patients with sleep-disordered breathing, insomnia related to generalized anxiety disorder, as well as mood disorders, periodic limb movement disorders, and Parkinson's disease (Klösch et al., 2001). One random half of this dataset was used for training and the other half for validation. Each recording was first scored by an expert from the recording lab and thereafter independently by a randomly assigned second scorer from one of the other seven participating sleep labs. Finally, the two scorings were reviewed by a scorer in a third lab, who took the final (consensus) decision. In total, 30 human experts from 8 European sleep labs participated in this study, and thus there was no bias due to a preference of 1 scorer or sleep lab (school). For determining the thresholds and optimization of the total procedure, the consensus scorings were considered as "gold standard." In addition, 9 types of sleep/wake-related patterns were visually identified and marked by 12 experienced scorers from 7 sleep labs, including episodes with alpha waves, vertex sharp waves, sleep spindles, k-complexes, delta waves, saw-tooth waves, slow eye movements (SEMs), rapid eye movements, and artifacts. The final scoring system (Somnolyzer) consisted of a raw data quality check, a feature extraction algorithm (density and intensity of sleep/wake-related patterns such as sleep spindles, delta waves, SEMs, and REMs), a classifier designed as an expert system, and a rule-based smoothing procedure for the start and the end of stage REM. The validation in 286 PSGs revealed an overall epoch-by-epoch agreement of 80% (Cohen's kappa: 0.72) between the Somnolyzer and the human expert scoring, as compared with an interrater reliability of 77% (Cohen's kappa: 0.68) between 2 human experts scoring the same 286 PSGs. Comparing two Somnolyzer-assisted analyses after a structured quality control by two human experts revealed an interrater reliability close to 1 (Cohen's kappa: 0.991), which confirmed that the variability induced by the quality control procedure can be neglected. Thus, the validation study proved the high reliability and validity of the Somnolyzer-assisted scoring system. Indeed, this study was acknowledged as one of only two evidence level 1 studies for sleep classification according to R&K by the Digital Task Force Committee of the American Academy of Sleep Medicine (AASM) (Penzel et al., 2007). The applicability in clinical routine and sleep studies was demonstrated in a sleep laboratory study on single and repeated dose effects of paroxetine, alprazolam, and their combination in healthy young volunteers (Barbanoj et al., 2005) and in a sleep laboratory studies in insomnia in somatoform pain disorder on differences to controls and acute effects of trazodone (Saletu et al., 2005). This validity was further confirmed by Svetnik et al. (2007) in 164 PSGs of 82 subjects in a clinical trial using zolpidem in a phase advance model of transient insomnia.

7.1.3 Autoscoring According to AASM

After the AASM Manual for the Scoring of Sleep and Associated Events was published in 2007 (Iber et al., 2007), the scoring algorithm had to be adapted to the new rules and had to be tested again for validity as compared with visual scoring. In the AASM classification, sleep stages S1–S4 are referred to as N1, N2 and N3, with N3 comprising the slow-wave sleep stages S3 and S4. The stage rapid eye movement (REM) is referred to as stage R and wake as stage W, and the stage "movement time" does not exist anymore. According to the AASM manual, a minimum of three EEG derivations sampling activity from the frontal, central, and occipital regions have to be recorded. One major change in the rules defining the start and end of the different sleep stages concerns the termination of stage N2 due to a cortical arousal. According to the AASM manual, cortical arousals, whether or not associated with an increase in EMG muscle tone, may reflect a change from N2 to N1. Changing back to N2 thereafter requires the reappearance of a K-complex unassociated with an arousal or a sleep spindle. For updating the Somnolyzer algorithm, 72 PSGs from the SIESTA database were visually rescored according to the AASM rules by 2 independent scorers out of a pool of 7 sleep experts from 3 European sleep labs. Interrater reliability testing showed that the overall agreement for human scorings according to the AASM standard was slightly higher than that for scorings according to the R&K standard. Cohen's kappa was 0.76 for the AASM standard and 0.74 for the R&K standard (Danker-Hopfe et al., 2009). The authors concluded that the integration of frontal, central, and occipital leads improved interrater reliability, but that this advantage was counteracted by the rather low interrater reliability for scoring cortical arousals, which may define the end of N2. The agreement between Somnolyzer-assisted and the visual scorings revealed a kappa value of 0.75–0.76 and was thus comparable to that of the human experts (Anderer et al., 2010). In 2015, Punjabi et al. (2015) validated the Somnolyzer autoscoring system in PSGs from 97 patients, which had been scored by 4 technologists for sleep staging, arousals, and respiratory events. The authors reported a high degree of agreement between manual and automated scoring of the apnea-hypopnea index and substantial concordance in the arousal index, total sleep time, and sleep efficiency. They concluded that automated analysis of PSGs using the Somnolyzer system provides results that are comparable to manual scoring for commonly used metrics in sleep medicine.

7.1.4 Machine Learning Approaches

In recent years, several papers have been published that applied deep neural network approaches for sleep staging based on large datasets for training and validation. For a comprehensive review, see Fiorillo et al. (2019). Sun et al. (2017) reported a kappa of 0.68 in an independent validation sample from a clinical dataset (n = 1000). Biswal et al. (2018) used a recurrent neural network (RNN) approach and reported a kappa value of 0.73 in a large independent validation set from the Sleep Heart Health Study (SHHS, n = 5804). Also, in 2018, Patanaik et al. (2018) applied a deep learning approach and reported a kappa of 0.74 in an independent validation dataset from the Sleep Disorder Unit in Singapore (n = 210). Stephansen et al. (2018) presented a neural network analysis for diagnosing narcolepsy and reported kappa values between 0.72 and 0.77 versus unbiased consensus scorings derived from six expert scorings in a validation set from a study examining the impact of sleep-disordered breathing in women aged 40–57 years (n = 70). Malafeev et al. (2018) compared machine learning algorithms for sleep classification based on random forests or artificial neural networks using features or raw data as inputs and reported high performance for all four combinations for all stages except for stage N1. They reported, however, no kappa values for the five-class comparison. Zhang et al. (2019) trained again a recurrent neural network with long short-term memory (LSTM) units and reported a kappa up to 0.70 in an independent validation set from the Study of Osteoporotic Fractures (SOF, n = 461) and the Osteoporotic Fractures in Men study (MrOS, n = 2907). More recently, Abou Jaoude et al. (2020) used once more a similar RNN approach and reported kappa values up to 0.69 in another large independent validation set study from the Home Positive Airway Pressure (homePAP) study (n = 243) and Apnea, Bariatric Surgery, and CPAP (ABC) study (n = 129). In 2021, Cesari et al. (2021) applied the Stanford-STAGES algorithm as described in Stephansen et al. (2018) to 1066 records from the Study of Health in Pomerania and reported kappa values of 0.55 and 0.68 versus 2 independent expert scorings, respectively, with a kappa value of 0.66 between the 2 expert scorings.

In a further development step of the Somnolyzer algorithm, we kept the artifact processing and feature extraction module unchanged, but replaced the expert system based on a decision tree by a bidirectional RNN with LSTM units. The supervised deep learning algorithm was trained to obtain R&K sleep stage probabilities in 472 out of the 588 PSGs from the SIESTA database, and the remaining 116 PSGs were used for early stopping to prevent the model from overfitting. Note that the large number of training samples is an essential requirement for using machine learning approaches; Sun et al. (2017) showed that epoch-by-epoch Cohen's kappa improved with increasing training PSG recordings until saturation was reached when at least 300 recordings have been included in the training phase. RNNs differ from feed-forward neural networks by redirecting outputs back to inputs. This enables bidirectional models to consider context from the past and the future, which makes it especially suitable for modeling temporal data. Thus, for the assignment of a sleep stage to a given 30-s epoch, not only the features derived from this 30-s epoch but rather the feature distribution throughout the entire recording is considered. In a further step, onset and duration of arousals, sleep spindles, and k-complexes were added to the feature set, and a convolutional neural network (CNN) followed by another bidirectional LSTM layer was trained using the data from the 72 SIESTA PSGs that had been scored according to AASM criteria, to sub-classify NREM sleep stages accordingly. The resulted network was integrated in the Somnolyzer sleep scoring system version 4.0. The first validation was based on the dataset that had been used in the Punjabi et al. (2015) validation ($n = 97$). Epoch-by-epoch Cohen's kappa agreement was determined as compared to a consensus scoring without the assessed scorer included in the consensus. The kappa coefficients between the four manual scorers and the manual consensus scoring ranged between 0.70 and 0.79. The kappa coefficient between auto and consensus scoring was 0.79, showing that the autoscoring based on the RNN classifier was equal to the best human expert scorer (Anderer et al., 2018).

Thus, there is now convincing evidence that sleep staging using deep neural network approaches can achieve agreements to the "gold standard" of manual scoring with an accuracy comparable to the interrater reliability between manual scorers, suggesting that these artificial intelligence systems are valid alternatives to manual scoring while having the advantage of consistency.

7.2 Validation of an Artificial Intelligence-Based Autoscoring System for PSGs

PSG autoscoring systems must score sleep stages and the associated events, such as cortical arousals, apneas and hypopneas, oxygen desaturation, and periodic leg movements. Thus, in addition to Cohen's kappa reflecting the epoch-by-epoch agreement for categorical sleep stages, measures reflecting the agreement in event scoring must be presented. Many of the metrics used in clinical practice are event indices, indicating the number of events per hour of sleep. A common measure for comparing the reliability between two measurement methods for such continuous parameters is the intraclass correlation coefficient (ICC) for absolute agreement (ICC (2,1) according to Shrout & Fleiss, 1979 and Koo & Li, 2016). Here, we are presenting a clinical validation study for the Somnolyzer autoscoring system for scoring sleep and associated events using Cohen's kappa and ICC statistics. The purpose of this validation study was to demonstrate the robustness of algorithm performance when applied to data drawn from the National Sleep Research Resource (NSRR) (Dean 2nd et al., 2016; Zhang et al., 2018), a publicly available dataset of sleep studies, each scored by a single technologist.

7.2.1 Methods

This study was approved by the Western Institutional Review Board (20192296). All PSG data were de-identified, and therefore the requirement for informed consent was waived.

7.2.1.1 PSG Identification and Scoring

The PSGs used in this study were originally collected in the Apnea, Bariatric Surgery, and CPAP trial (ABC; n = 24) (Bakker et al., 2018), the HomePAP trial (n = 180) (Rosen et al., 2012), and the Multi-Ethnic Study of Atherosclerosis study (MESA; n = 224) (Chen et al., 2015). After applying parameters to ensure a wide range of disease severity and to ensure a recording of ≥4 hours per study, PSGs were selected from each dataset at random. The original sleep staging and event identification were left unchanged from the manual scoring originally performed for each study following American Academy of Sleep Medicine (AASM) recommendations. For the MESA and ABC studies, hypopneas were identified when associated with a ≥3% SpO_2 desaturation and/or an arousal. In the HomePAP study, hypopneas were identified when associated with a ≥4% SpO_2 desaturation. Limb movements data based on leg-EMG were available only in the HomePAP study. Each PSG was analyzed in Sleepware G3 software containing Somnolyzer 4.1 (Philips, Monroeville, PA, USA) following the original scoring criteria after applying the same lights off/on times.

Somnolyzer scored sleep stages and arousals based on all available EEG, EOG, and chin EMG channels (ABC and HomePAP, F4-M1, C4-M1, O2-M1, F3-M2, C3-M2, O1-M2; MESA, F4-M1, C4-M1, O2-M1; left and right EOG and chin EMG for all three studies). In all three studies, Somnolyzer scored respiratory events based on oronasal thermal airflow, airflow recorded with a nasal pressure transducer, respiratory effort with thoracic and abdominal respiratory inductance plethysmography (RIP) belts, and the oxygen saturation signal based on pulse oximetry. In the HomePAP study, Somnolyzer scored leg movements based on EMG from left and right anterior tibialis muscle. No leg EMG channels were available in the ABC and MESA study.

Based on published interrater reliability studies that present epoch-by-epoch comparisons of sleep stages (Danker-Hopfe et al., 2009; Magalang et al., 2013; Cesari et al., 2021), Cohen's kappa for all sleep stages, as well as for wake and REM, typically show substantial agreement (>0.60), while for N1, N2, and N3, Cohen's kappa values typically show only moderate agreement (>0.4). Published studies that have included the ICC for assessing continuous sleep variables (Punjabi et al., 2015; Magalang et al., 2013; Malhotra et al., 2013) typically show good inter-reliability (>0.75) for apnea-hypopnea index (AHI), total sleep time (TST), sleep efficiency (SE), oxygen desaturation index (ODI), and periodic leg movements in sleep index (PLMSI), while only moderate inter-reliability (>0.50) has been demonstrated for arousal index (ArI) and respiratory sleep events (total apneas, obstructive apneas, central apneas, mixed apneas, and hypopneas). Thus, our predefined thresholds for the lower 95% confidence intervals (CIs) were 0.60 for Cohen's kappa based on epoch-by-epoch comparisons of all sleep stages, 0.75 for the ICC between the Somnolyzer-AHI and the manual-AHI, 0.5 for the ICC between the Somnolyzer-ArI and the manual-ArI, and 0.75 for the ICC between the Somnolyzer-PLMSI and the manual-PLMSI.

7.2.1.2 Statistical Power

An a priori power calculation was undertaken for all four endpoints described in our hypotheses above. The study was powered based on the weakest effect size (ICC between the Somnolyzer-ArI and manual-ArI).

7.2.1.3 Statistical Analyses

All analyses were performed using MATLAB R2019b, validated against IBM SPSS (version 19.0.0.2). To assess sleep staging performance, we undertook an epoch-by-epoch comparison of manual staging and Somnolyzer staging and calculated a kappa statistic across all sleep stages (W/N1/N2/N3/R), as well as each individual stage, along with 95% CIs. We also calculated accuracy for each sleep stage discrimination, that is, the percentage of all epochs that were correctly identified by Somnolyzer. To assess the performance of respiratory event, arousal, and limb movement identification, we computed the ICC for absolute agreement between Somnolyzer and manual scoring of the AHI, ArI, and PLMSI, along with a 95% CI. In addition to the performance targets adopted in our hypotheses, we compared the lower limit of the kappa 95% CI values to the thresholds defined by Landis and Koch (1977) as follows: 0.0–0.2 slight agreement; 0.21–0.40 fair agreement; 0.41–0.60 moderate agreement; 0.61–0.80 substantial agreement; and 0.81–1.0 almost perfect or perfect agreement. The lower limits of the ICC 95% CIs were compared against the thresholds defined by Koo and Li (2016) as follows: <0.5 poor reliability; 0.5 to <0.75 moderate reliability, 0.75 to <0.90 good reliability; and ≥0.90 excellent reliability.

7.2.2 Results

A total of 428 PSGs were randomly selected from within disease severity categories (AHI<5, 5 to <15, 15 to <30 and ≥30 events/hour). A small number were removed due to invalid signals (complete or partial) or missing manual scoring. The final sample sizes for each hypothesis were therefore $n = 426$ (sleep staging and AHI), $n = 425$ (ArI), and $n = 174$ (PLMSI; based on HomePAP data only). Descriptive information is provided in Table 7.1. As anticipated, with the sampling strategy and the nature of each study, each sample contained participants with a wide range of disease severity (AHIs in MESA 0–88 events/hour; HomePAP 0–113 events/hour; ABC 15–115 events/hour).

7.2.2.1 Sleep Staging

Cohen's kappa based on an epoch-by-epoch comparison of all sleep stages between Somnolyzer and manual scoring was 0.739 (95% CI 0.737–0.741); see Table 7.2. The lower bound of the CI (0.737) exceeded the prespecified threshold of 0.60, supporting this hypothesis. Sleep staging accuracy was 80.7% across all sleep stages (W/N1/N2/N3/R), 94.2% for wake, 87.5% for N1, 86.6% for N2, 95.9% for N3, and 97% for REM. The results of the epoch-by-epoch comparison of sleep stage scoring are presented as confusion matrix in Table 7.3.

7.2.2.2 Respiratory Events

The ICC between the Somnolyzer-AHI and the manual-AHI was 0.969 (95% CI 0.957–0.976); see upper part of Table 7.4. The lower bound of the CI (0.957) was higher than the prespecified threshold of 0.75, supporting this hypothesis. Figure 7.1 presents the Bland-Altman plot of the Somnolyzer-AHI against the manual-AHI.

7.2.2.3 Arousals

The ICC between the Somnolyzer-ArI and the manual-ArI was 0.794 (95% CI 0.668–0.864); see upper part of Table 7.4. The lower bound of the CI (0.668) was higher than the prespecified threshold of 0.50, supporting this hypothesis.

7.2.2.4 Periodic Limb Movements

In a sample of 174 PSGs, the ICC between the Somnolyzer-PLMSI and the manual-PLMSI was 0.907 (95% CI 0.877–0.930); see upper part of Table 7.4. The lower bound of the CI (0.877) was higher than the prespecified threshold of 0.75, supporting this hypothesis.

7.2.3 Discussion

In this study, an automated review of key sleep staging and event detection parameters met prespecified performance thresholds, demonstrating that the Somnolyzer scoring solution generates results that agree with human scoring. Our performance targets were set at the lower margin of the agreement across expert human scorers in the literature. In these studies, Cohen's kappa for discrimination of all sleep stages (W/N1/N2/N3/R), as well as for wake and REM individually, typically show substantial agreement (>0.60), while for N1, N2, and N3, Cohen's kappa values typically show only moderate agreement (>0.4) (Danker-Hopfe et al., 2009; Magalang et al.,

Table 7.1 Descriptive demographic and clinical information (n = 426 PSGs)

Variable	MESA (n = 224)	HomePAP (n = 178)	ABC (n = 24)
Age (years)	69.8 ± 8.8	46.1 ± 12.1	50.5 ± 9.1
Gender (number; %)			
• Female	110; 49.1%	89; 50.0%	14; 58.3%
• Male	114; 50.9%	89; 50.0%	10; 41.7%
Ethnicity (number; %)	–		
• Hispanic		14; 7.9%	2; 8.3%
• Non-Hispanic		163; 91.6%	22; 91.7%
• Unknown		1; 0.6%	–
Race (number; %)	–		
• White		124; 69.7%	16; 66.7%
• African American		44; 2.7%	4; 16.7%
• Other		10; 5.6%	16.7%
Ethnicity/race composite (number)		–	–
• White/Caucasian	84; 37.5%		
• Chinese American	36; 16.1%		
• African American	53; 23.7%		
• Hispanic	51; 22.8%		
• Other	–		
Body mass index (kg/m²)	–	40.0 ± 9.2	38.6 ± 2.9
Neck circumference (cm)			
Systolic blood pressure (mmHg)	–	125.4 ± 12.9	–
Diastolic blood pressure (mmHg)	–	79.8 ± 9.0	–
AHI_{PSG} (events/hour)	28.0 ± 17.6 (range 1–88)	13.0 ± 16.7 (range 0–113)	50.5 ± 9.1 (range 15–115)
SDB severity (number; %)			
• None	17; 7.6%	36; 38.8%	0; 0.0%
• Mild	17; 7.6%	59; 33.1%	0; 0.0%
• Moderate	107; 47.8%	31; 17.4%	4; 16.7%
• Severe	83; 37.1%	19; 10.7%	20; 83.3%
Epworth Sleepiness Scale score (/24)	5.8 ± 3.9	14.2 ± 3.7	–
Total sleep time per PSG (hours)	5.9 ± 1.4	5.7 ± 1.1	7.0 ± 1.5
Recording time (hours)	10.6 ± 1.2	7.8 ± 1.1	8.4 ± 0.2

2013). Good interrater reliability (ICC >0.75) has been demonstrated for the AHI, TST, sleep efficiency, ODI, and PLMSI, while only moderate interrater reliability (ICC >0.50) has been demonstrated for the ArI (Punjabi et al., 2015; Magalang et al., 2013; Malhotra et al., 2013). By exceeding our performance targets, we can conclude that the variability between Somnolyzer and human scoring in the current study is no more than the variability observed across expert scorers and, therefore, Somnolyzer scoring provides output that is ready for review and interpretation by a physician.

Key strengths of the current study include the large, ethnically/racially diverse sample, which supports the generalizability of algorithm performance, as well as the fact that the PSGs were collected in a range of clinical and research settings using various data collection platforms and montages. Our selection criteria ensured that the performance of Somnolyzer was assessed across the full range of disease severity. The most important limitation of our study was the fact that each PSG was scored by a single technologist. Given the aforementioned variability across scorers (Rosenberg & Van Hout, 2013, 2014), our comparator may not represent the true "gold standard" of manual scoring (Penzel et al., 2013). Note that, however, numerous different technologists scored the 426 PSGs and thus the aforemen-

Table 7.2 Comparison of Somnolyzer- and manually scored sleep staging

Sleep stage	Kappa (95% CI)	Performance threshold	Accuracy (%)
W/N1/N2/N3/R	0.739 (0.737–0.741)	0.60	80.7
Wake	0.853 (0.851–0.855)		94.2
N1	0.457 (0.452–0.461)		87.5
N2	0.721 (0.719–0.723)		86.8
N3	0.731 (0.727–0.735)		95.9
REM	0.868 (0.865–0.870)		97.0

All analyses $n = 426$

The lower limits of each CI can be compared against thresholds defined by Landis and Koch (1977) as follows: 0.0–0.2 slight agreement; 0.21–0.40 fair agreement; 0.41–0.60 moderate agreement; 0.61–0.80 substantial agreement; and 0.81–1.0 almost perfect or perfect agreement

The figures relevant to predefined hypothesis are underlined

tioned variability across scorers is reflected in the manual scorings used as the comparator in this study. In that respect, the comparator reflects standard practice in most clinical settings in which a single technologist is responsible for scoring each sleep study. The Somnolyzer 4.1 automatic scoring solution is a validated tool that offers operational efficiencies and can be easily implemented into existing workflows. The automated scoring results have been shown to agree with human scoring in a variety of settings and with the use of multiple data acquisition technologies. The consistency of an automated scoring process can be beneficial in both clinical and research settings by minimizing inter- and intra-scoring variability. The implementation of Somnolyzer 4.1 has the potential to free up clinical staff to perform other duties and improve the end-to-end sleep care experience; confirmation of these outcomes would be beneficial.

7.3 Added Value of Autoscoring Systems: From the Hypnodensity to Confidence Trends

Besides the high reproducibility of autoscoring systems, they offer several additional benefits that could help in interpreting a sleep study faster and more accurate.

7.3.1 Scoring in Real Time

Autoscoring systems can be integrated in acquisition systems to perform scoring in real time. The Somnolyzer autoscoring system was integrated in Sleepware G3. Due to the real-time analysis, the results are available during the acquisition (with a delay of 7 min to ensure the necessary context information), and the final scoring, ready for review and interpretation by the physician, is available immediately after the end of the recording. Scoring results during the acquisition are, for instance, valuable in split-night studies where a patient is diagnosed and, if applicable, receives treatment titration in a single night. According to the AASM guidelines, a split-night study may be performed if an AHI ≥40 events/hour is documented during 2 h of a diagnostic study or may be considered for an AHI of 20–40 events/hour based on clinical judgment (Epstein et al., 2009). For the example of a split-night study shown in Fig. 7.2, the autoscoring revealed in real time that the total sleep time in the diagnostic part was 115 min with both slow wave sleep and REM sleep periods and the 94 obstructive events (predominantly obstructive hypopneas) resulted in an AHI of 49.1 events/hour.

7.3.2 Scoring According to Different Rules

Specifically, for the scoring of hypopneas, various criteria have been used. In the first version of the AASM manual, for instance, the recommended rule for scoring hypopneas required a

Table 7.3 Confusion matrix for epoch-by-epoch sleep staging (n = 426)

		Manual staging				
		Wake	N1	N2	N3	REM
Somnolyzer staging	Wake	99,828	5,705	1,880	101	1,439
		87.1%	12.2%	1.2%	0.3%	2.6%
	N1	10,405	28,843	19,761	120	3,545
		9.1%	61.7%	12.3%	0.3%	6.4%
	N2	2,207	10,359	130,683	9,823	2,357
		1.9%	22.2%	81.5%	27.2%	4.2%
	N3	112	47	6,826	26,100	14
		0.1%	0.1%	4.3%	72.2%	0.0%
	REM	2,093	1,802	1,281	2	48,271
		1.8%	3.9%	0.8%	0.0%	86.8%

All analysis n = 426
Gray cells indicate the raw count and percentage of epochs of each manually scored sleep stage that were correctly identified by Somnolyzer

≥30% drop in flow amplitudes for ≥10 s and a ≥4% oxygen desaturation from pre-event baseline. The alternative rule required a ≥50% drop in flow amplitudes for ≥10 s and a ≥3% oxygen desaturation and/or an associated arousal for confirmation. In version 2.6 of the AASM manual, the recommended rule requires a ≥30% drop in flow amplitudes for ≥10 s and a ≥3% oxygen desaturation and/or an associated arousal for confirming a hypopnea. The recommended rule from 2007 is labeled as "acceptable" in version 2.6. Several studies have described the large impact of the various criteria for scoring hypopneas on the apnea hypopnea index (Redline et al., 2000; Ruehland et al., 2009; Duce et al., 2015; Kapur et al., 2017); see Berry et al. (2012) for a comprehensive review. Indeed, autoscoring systems can offer the ability to select and, if required, to change the criteria settings which allows direct comparisons between the different rule settings for individual patients.

In 2012, we investigated the effects of changing the scoring criteria for hypopneas on the AHI and the resulting sleep-disordered breathing (SDB) severity classification in a study based on 15 PSGs in patients with suspected OSAS (8 females, 7 males, aged 28–56 years) (Anderer et al., 2012). An AHI <5 indicates no SDB, 5 ≤ AHI < 15 indicates mild SDB, 15 ≤ AHI 30 indicates moderate, and an AHI ≥ 30 indicates severe SDB. Table 7.5 summarizes the AHIs for the 15 patients using 5 different scoring criteria: 4.A (recommended 2007 rule, acceptable 2020 rule), 4.B (alternative 2007 rule), 4.A* (recommended 2020 rule for HSAT), 4.AB (recommended 2020 rule for PSG), and Chicago (the American Academy of Sleep Medicine Task Force consensus paper rule (1999)). Changing the desaturation criteria from 4% to 3% affected the classification in just one patient (compare 4.A with 4.A* in Table 7.5). If, however, an arousal could confirm a hypopnea as well, the classification in 10 out of the 15 patients is affected (com-

Table 7.4 Comparison of Somnolyzer- and manually scored sleep and event metrics

Metric	ICC (95% CI)	Performance threshold
Respiratory, limb movement, and desaturation events:		
AHI (events/hour)	0.969 (0.957–0.976)	0.75
ArI (events/hour)	0.794 (0.668–0.864)	0.50
PLMSI (events/hour)	0.907 (0.877–0.930)	0.75
Total apneas (number)	0.848 (0.733–0.904)	
Total hypopneas (number)	0.898 (0.757–0.946)	
ODI (events/hour)	0.990 (0.987–0.992)	
Sleep staging:		
Sleep efficiency (%)	0.927 (0.904–0.943)	
Total sleep time (minutes)	0.938 (0.920–0.951)	
Time in N1 (minutes)	0.690 (0.372–0.826)	
Time in N2 (minutes)	0.815 (0.778–0.846)	
Time in N3 (minutes)	0.772 (0.728–0.809)	
Time in NREM (minutes)	0.919 (0.875–0.944)	
Time in REM (minutes)	0.908 (0.888–0.925)	

All analyses $n = 426$ except for ArI which was based on $n = 425$ and PLMSI which was based on $n = 174$ (HomePAP data only)
The lower limits of each CI can be compared against thresholds defined by Koo and Li (2016), as follows: <0.5 poor agreement; 0.5 to <0.75 moderate agreement, 0.75 to <0.90 good agreement; and ≥0.90 excellent agreement
The figures relevant to the predefined hypotheses are underlined

pare 4.A with 4.AB). Thus, while the higher sensitivity for detecting desaturations increased the AHIs only minorly (in patient a90000173, for instance, from 0.8 to 2.6 events/hour), the inclusion of arousals significantly increased the AHIs (in the same patient from 2.6 to 24.6 events/hour). All four patients classified as normal if arousals are not considered for hypopnea confirmation (4.A, acceptable 2020 rule) showed a mild to moderate abnormal AHI when hypopneas could be confirmed by arousals as well (4.AB, recommended 2020 rule). Thus, autoscoring enables to explore the impact of the different criteria for confirming hypopneas and to report the AHI based both on the acceptable and on the recommended rule for individual patients, offering the interpreting physician further insights in patient's disease characteristics.

7.3.3 Scoring with Different Sensitivity Settings

While there are the well-established AASM rules for scoring PSGs, the interpretation of the rules may vary substantially between scorers, specifically for epochs or events with equivocal features (Rosenberg & Van Hout, 2013, 2014; Younes et al., 2016, 2018). Experts, when scoring these epochs, may be biased toward sensitivity or specificity. Younes et al. (2018) showed, for instance, that some technologists scored stage N3 sleep when delta wave duration was well below 6 s, whereas for others, much greater durations were required. Thus, by varying sensitivity settings, an autoscoring can mimic these different interpretations, assuming that each scorer interprets the rules consistently. The autoscoring system Somnolyzer has the option to select different sensitivities for arousal, spindle/k-complex, slow wave, apnea, and hypopnea event detection. We investigated the effects of changing these sensitivity settings in a study based on ten PSGs in ten apnea patients (five diagnostic, two titration, and three split nights; Anderer et al., 2016). All PSGs were manually scored independently by eight experts and by

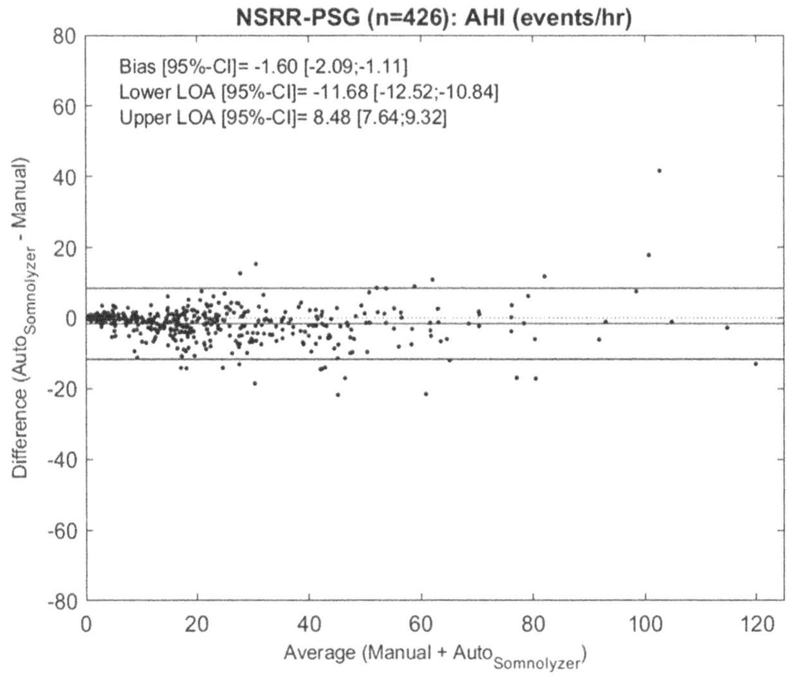

Fig. 7.1 Bland-Altman plot of the Somnolyzer-AHI against the manual-AHI ($n = 426$) Mean difference of −1.60 events/hour; lower and upper limits of agreement −11.68 and 8.48, respectively

Somnolyzer with various sensitivity settings according to the recommended AASM version 2.2 rules (Berry et al., 2015). As can be seen in Fig. 7.3, sleep parameters derived from the manual scorings varied considerably between the eight scorers (time in N1, 29–127 min; time in N2, 125–209 min; time in N3, 19–56 min; time in R, 42–63 min; number of arousals, 86–193 events; number of apneas+hypopneas, 173–255 events). With the default (= balanced) setting, Somnolyzer autoscoring was close to the mean of the eight manual scorings (time in N1, 82 and 85 min; time in N2, 184 and 176 min; time in N3, 41 and 42 min; time in R, 59 and 56 min; number of arousals, 160 and 142 events; number of apneas+hypopneas, 246 and 219 events, for the Somnolyzer scoring and the mean of the eight manual scorings, respectively; Fig. 7.3). As can be seen in Fig. 7.4, showing precision (positive predictive value)-recall (sensitivity) plots for the manual scorings (crosses) and the autoscoring with five different sensitivity settings (circles), the autoscoring perfectly mimics the variability observed in the eight manual scorers, by varying the sensitivity settings from maximal precision to maximal sensitivity. Note that the manual arousal scoring with a sensitivity of 33% as shown in the left upper precision-recall plot in Fig. 7.4 is considered as outlier and thus it is not covered with any of the autoscoring sensitivity settings. Most importantly, Somnolyzer autoscoring with default settings (green circles in Fig. 7.4) is perfectly balanced between sensitivity and precision, and by merely varying the sensitivity settings, the variability observed in manual scorings can be explained.

7.3.4 Estimating Sleep Stage Probabilities per Epoch (Hypnodensity)

The primary output of an RNN used for scoring sleep stages are sleep stage probabilities per 30-s epoch. They can be plotted in pseudo-color graphs (see Figs. 7.2 and 7.5, 7.6, 7.7 for examples) and have been referred to as hypnodensity by Stephansen et al. (2018). Interestingly, if multiple human scorings are available, a hypnodensity graph can be derived also based on the human scorings, visualizing ambiguous epochs (MAN-Hypnodensity in Fig. 7.5). Consequently, the sleep stage probabilities derived from multiple manual scorers and from

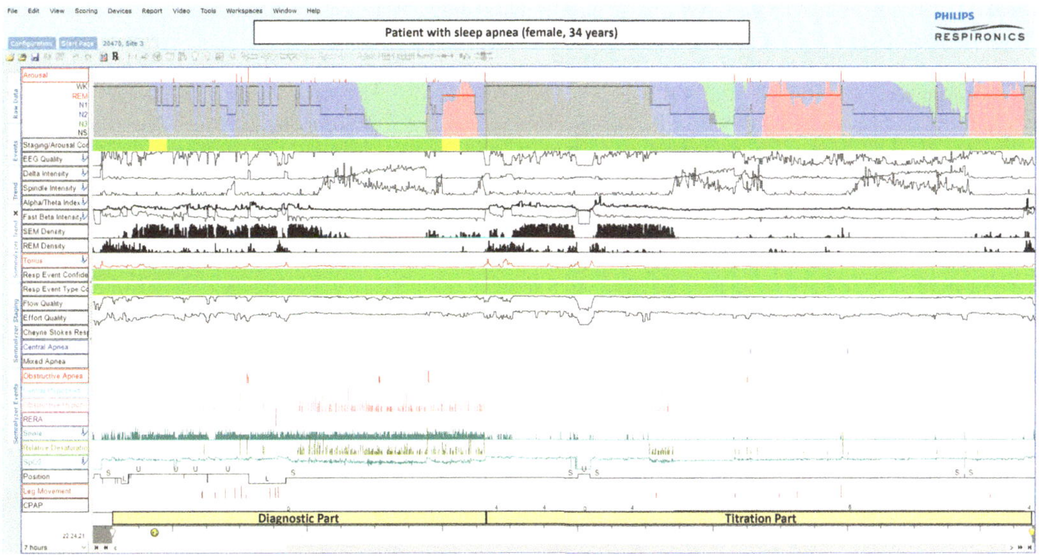

Fig. 7.2 A representative example for autoscoring trends as visualized in Sleepware G3: Split-night study in a patient with sleep apnea (female, 34 years)
The trends from top to bottom are as follows: "Arousal," scored arousal events; "WK, REM, N1, N2, N3," the hypnogram superimposed on the hypnodensity graph with the color codes W, gray; REM, red; N1, cyan; N2, blue; and N3, green; "Staging/Arousal Confidence," confidence trend for guiding the reviewer by a traffic light system to areas that need more careful review of sleep and arousal scoring; "EEG Quality," estimation of the signal quality of the EEG derivation most relevant for the staging; "Delta Intensity," intensity (based on duration and amplitude) of delta waves per 30-s epoch; "Spindle Intensity," intensity (based on duration and amplitude) of sleep spindle activity per 30-s epoch; "Alpha/Theta Index," quotient of the activity in the alpha band to the activity in the theta band per 30-s epoch; "Fast Beta Intensity," intensity (based on duration and amplitude) of the activity in the fast-beta band per 30-s epoch; "SEM Density," density (based on duration) of slow eye movements per 30-s epoch; "REM Density," density (based on duration) of rapid eye movements per 30-s epoch; "Tonus," chin EMG tonus; "Resp Event Confidence," confidence trend for guiding the reviewer by a traffic light system to areas that need more careful review of respiratory event scoring; "Resp Event Type Confidence," confidence trend for guiding the reviewer by a traffic light system to areas that need more careful review of scoring the type of a respiratory event; "Flow Quality," estimation of the signal quality of the flow channel most relevant for the scoring respiratory events; "Effort Quality," estimation of the signal quality of the effort channel most relevant for scoring the type of a respiratory event; "Cheyne Stokes Res," indicates the presence/absence of Cheyne-Stokes breathing; "Central Apnea," scored central apnea events; "Mixed Apnea," scored mixed apnea events; "Obstructive Apnea," scored obstructive apnea events; "Central Hypopnea," scored central hypopnea events; "Obstructive Hypopnea," scored obstructive hypopnea events; "RERA," scored respiratory effort-related arousal events; "Snore," intensity of snoring; "Relative Desaturations," scored oxygen desaturations ≥3%; "SpO_2," arterial oxygen saturation level; "Position," body position; "Leg Movement," scored leg movements; "CPAP," CPAP pressure

RNN classifiers can be directly compared. In Fig. 7.6, the manually derived and the autoscoring-derived probability curves per sleep stage are compared for the same study shown in Figs. 7.2 and 7.5. The respective ICCs for absolute agreement (ICC (2,1) according to Shrout & Fleiss, 1979) between these probabilities are as follows: for stage W, 0.97; for N1, 0.79; for N2, 0.89; for N3, 0.96; for R, 0.95; and over all five stages, 0.93. Table 7.6 summarizes the ICCs for the studies used in the Punjabi et al. (2015) paper. Thus, the ICC between manually derived probabilities and autoscoring-derived probabilities was 0.862 over all five stages indicating good agreement (>0.75) according to Koo and Li (2016). For stages W and R, the agreement was even excellent (>0.90) and as expected from the rather low interrater reliability for NREM sub-classification (e.g., Younes et al., 2018) only moderate for N1 and N3 (>0.50) but good for N2 (>0.75). These results demonstrate that the sleep stage probabilities derived from

Table 7.5 Classification of apnea severity based on AHI in patients with suspected OSAS for various hypopnea scoring criteria (*n*:15)

Acq-Number	4.A ≥30% drop & ≥4% des	4.B ≥50% drop & ≥3% des or arousal	4.A* ≥30% drop & ≥3% des	4.AB ≥30% drop & ≥3% des or arousal	Chicago ≥50% drop or ≥30% drop & ≥3% des or arousal
a9000173	.8	8.7	2.6	24.6	32.6
a9000172	1.3	10.6	4.6	26.2	31.1
a9000159	1.9	5.6	3.7	9.8	17.5
a9000164	1.7	6.7	4.0	15.0	20.9
a9000161	5.3	13.0	9.5	23.5	28.6
a9010175	6.8	19.1	11.3	27.8	37.8
a9020221	7.3	7.2	9.4	12.0	15.2
a9000174	8.4	16.3	14.1	25.0	32.0
a9020222	10.4	16.8	13.1	27.6	35.6
a9020337	14.3	30.0	21.4	36.0	49.3
a9020339	15.1	17.1	16.5	17.7	51.9
a9000165	26.1	28.2	29.2	34.5	42.9
a9020331	67.9	78.2	74.3	79.6	86.4
a9000157	83.4	94.2	88.5	99.4	106.6
a9000171	107.8	111.1	110.8	113.0	115.4

Normal (green), AHI < 5; mild (yellow), 5 ≤AHI<15; moderate (blue), 15 ≤AHI<30; severe (red), AHI ≥ 30
Note: 4.A (recommended AASM 2007 rule, acceptable AASM 2020 rule), 4.B (alternative AASM 2007 rule), 4.A* (recommended AASM 2020 rule for HSAT), 4.AB (recommended AASM 2020 rule for PSG), and Chicago (the 1999 Chicago consensus paper)

Table 7.6 Intraclass correlation coefficients (ICC) as well as the upper and lower 95% confidence intervals (CI) comparing sleep stage probabilities based on four human scorings with Somnolyzer autoscoring ($n = 97$)

Sleep stage	Lower 95% CI	ICC	Upper 95% CI
W/N1/N2/N3/R	0.861	0.862	0.862
W	0.938	0.939	0.940
N1	0.562	0.598	0.630
N2	0.727	0.801	0.849
N3	0.657	0.727	0.778
R	0.951	0.953	0.955

autoscoring are in good agreement with the sleep stage probabilities derived from multiple human scorers. Moreover, all sleep scoring parameters to be reported for PSG studies according to the AASM manual (total sleep time, time and percentage in each sleep stage, sleep and REM latency, and wake after sleep onset) can be derived directly from the hypnodensity. This provides the possibility to obtain robust estimates of the sleep scoring parameters that mimic averaging the results from a large pool of human experts (Bakker et al., 2022).

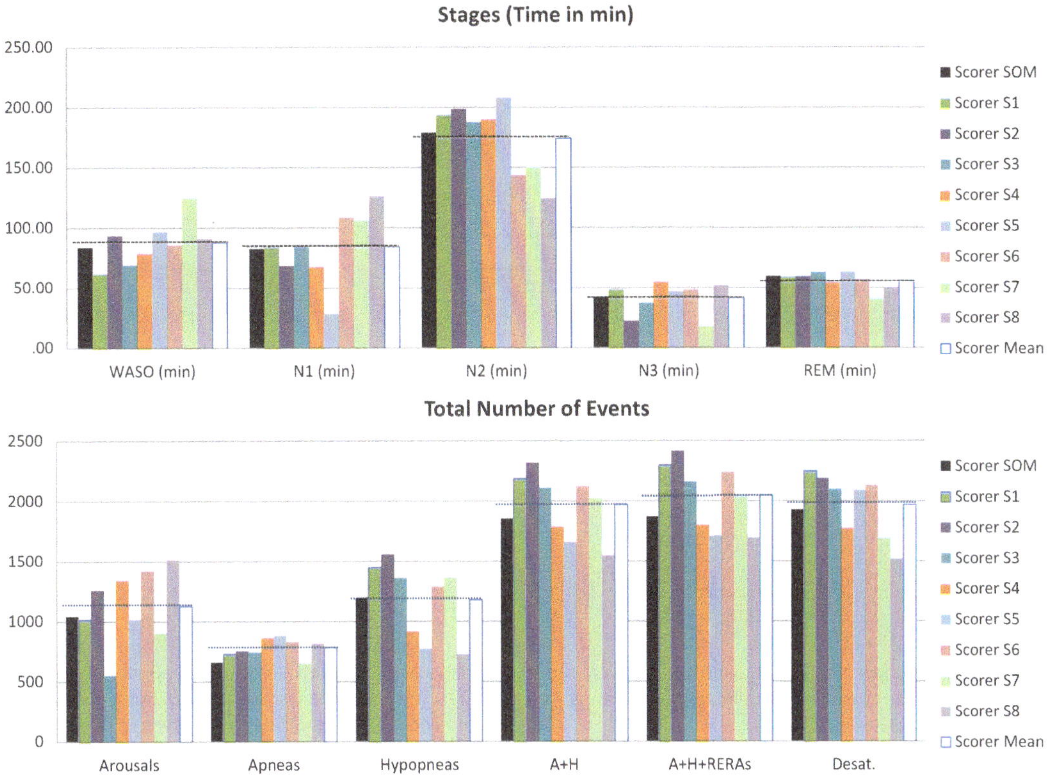

Fig. 7.3 Comparison of parameters derived from autoscoring and multiple manual scoring in ten apnea patients (five diagnostic studies, two titration studies, and three split-night studies)
Autoscoring with balanced (= default) sensitivity settings (Scorer SOM, black) as compared to eight independent manual scorers (Scorer S1–S8). In addition, the mean of all eight manual scorers (Scorer Mean, white with thin horizontal lines to facilitate comparisons) is shown for the parameters derived from sleep staging (upper part) and for the number of scored events (lower part)

7.3.5 Estimating Signal Quality

Artifact processing is a critical step in scoring PSG signals (see Anderer et al., 1999 for a review). As part of the Somnolyzer feature extraction for the electrophysiological signals, we minimize artifacts due to line and ECG interference and identify continuous and transient low-frequency artifacts (such as sweat and electrode pop artifacts), high-frequency artifacts (such as muscle bursts or movement artifacts), as well as ocular artifacts (such as eye movements and eye blinks). Concerning respiratory signals, artifacts due to body movements or sensor displacements are identified. Quality trends (EEG Quality, Flow Quality, and Effort Quality in Fig. 7.2) based on the detected artifacts allow both, an overview of the respective signal quality in the total recording and references to periods with poor signal quality.

7.3.6 Identification of Periods with Clinically Relevant Ambiguities (Confidence Trends)

Low interrater reliability is a well-known limitation with manual scoring of sleep stages (specifically N1) and cortical arousals as well as with manual scoring of disordered breathing events (Whitney et al., 1998; Bliwise et al., 1984; Loredo et al., 1999; Collop, 2002; Redline et al., 2007; Bonnet et al., 2007; Rosenberg &

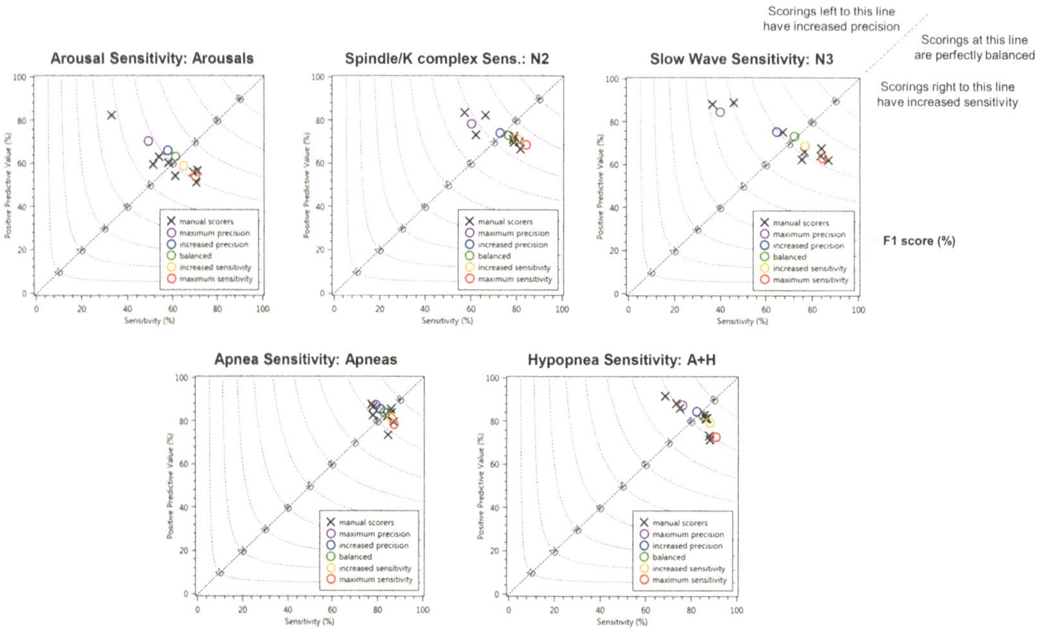

Fig. 7.4 Somnolyzer autoscoring with five different sensitivity settings as compared to eight manual scorings for the same dataset shown in Fig. 7.3
The plots depict sensitivities (x-axis) versus positive predictive values (y-axis) based on epoch-by-epoch comparison for sleep stages N3 and N2 and event-by-event comparison for arousals, apneas, and apneas+hypopneas (A + H). The contour lines (magenta) indicate constant F1 scores. The scale for the F1 score is given in the diagonal line. The five different sensitivity settings of the Somnolyzer autoscoring are plotted as colored circles; the eight manual scorings are plotted as black crosses

Van Hout, 2013, 2014; Younes et al., 2016, 2018). Luckily, not all ambiguities are clinically relevant, i.e., affect the parameters that influence clinical diagnosis. In respect to sleep staging and arousal detection, the RNN-derived sleep stage probabilities can be used to identify clinically relevant periods with ambiguities possibly affecting sleep onset (sleep latency), the first REM epoch (REM latency), and the sleep cycles (REM periods). Visualizing such periods in a confidence trend might help to concentrate the reviewer to these clinically relevant periods. The trend "Staging/Arousal Confidence" in Fig. 7.2 gives an example of such periods at sleep onset (first marked yellow period) and REM onset (second marked yellow period). As shown in Fig. 7.5, these periods of ambiguity based on the hypnodensity from autoscoring (from 10% to 90% probability of the respective sleep stage) correspond perfectly with the ambiguities seen for sleep onset and REM onset between the four human scorers. Averaged over the 97 studies that had been used in the Punjabi et al. (2015) paper, only 2.7% of the recording time was labeled as periods with clinically relevant ambiguities for scoring sleep stages, 1.8% for scoring respiratory events, and 2.2% for scoring the types of the respiratory events. Consequently, the confidence trends are designed to guide the reviewer by a traffic light system to the few areas of a study (in the average 2–3% of the recording time) that need a more careful review.

7.3.7 Visualization of Sleep/Wake-Related Features

The sleep staging rules in the AASM manual define the frequency bands for EEG background activity as well as various sleep/wake-related features such as slow waves, sleep spindles, k-complexes, or slow and rapid eye movements (Berry et al., 2020). The Somnolyzer autoscoring

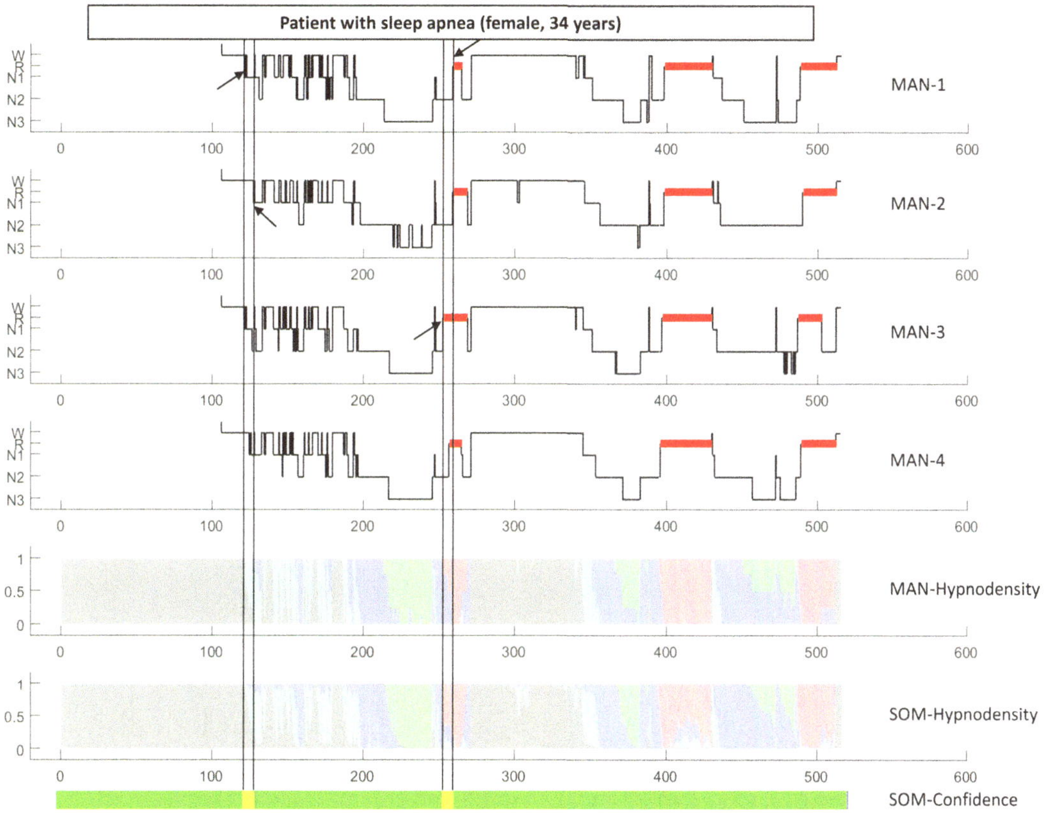

Fig. 7.5 A representative example of four individual-scored hypnograms, the hypnodensity charts derived from manual and autoscoring and the confidence trend derived from the autoscored hypnodensity for the same PSG shown in Fig. 7.2
The figure displays the hypnograms (MAN-1 to MAN-4), the hypnodensity graphs derived from the four manual scorings (MAN-Hypnodensity) and from autoscoring (SOM-Hypnodensity), as well as the confidence trend (Staging/Arousal Confidence) derived from the autoscored hypnodensity. Color codes for the hypnodensity: W, gray; REM, red; N1, cyan; N2, blue; and N3, green. The color-coded confidence trend guides the reviewer by a traffic light system to areas of a study that need a more careful review. In case of little ambiguity in the data, the confidence trend is green. The areas marked in yellow indicate ambiguity for the first epoch scored as sleep (sleep latency) and the first epoch scored as REM (REM latency). The yellow areas start with the first epoch with at least 10% probability for stage N1 or R and end with the first epoch with at least 90% probability of the respective sleep stage. Note that these areas based on the autoscored hypnodensity perfectly reflect the sleep onset and REM onset variability between the four manual scorers

system detects most of these patterns in the feature extraction module. An overview of the main electrophysiological patterns and the methods used for their automatic detection can be found in Anderer et al. (2005). In Fig. 7.2, a representative subset of these features is visualized as trends together with the hypnogram superimposed on the hypnodensity: delta intensity (a measure of sleep depth, indicating N3), spindle intensity (an indicator specifically for N2), alpha/theta index (a measure of vigilance in wake with eyes closed), fast beta intensity (indicating excitatory activity in wake with eyes open), SEM and REM densities (indicating eye movements in wake or REM periods), and chin EMG tonus (for differentiating wake and REM periods). As can be seen in Fig. 7.2, these features, detected from different sensors (EEG, EOG, and EMG), speak in most epochs for a certain sleep stage and therefore confirm each other, both in the diagnostic and in the titration part. Consequently, in this example, the confidence trend for staging and arousals is green (high confidence) almost all over the recording.

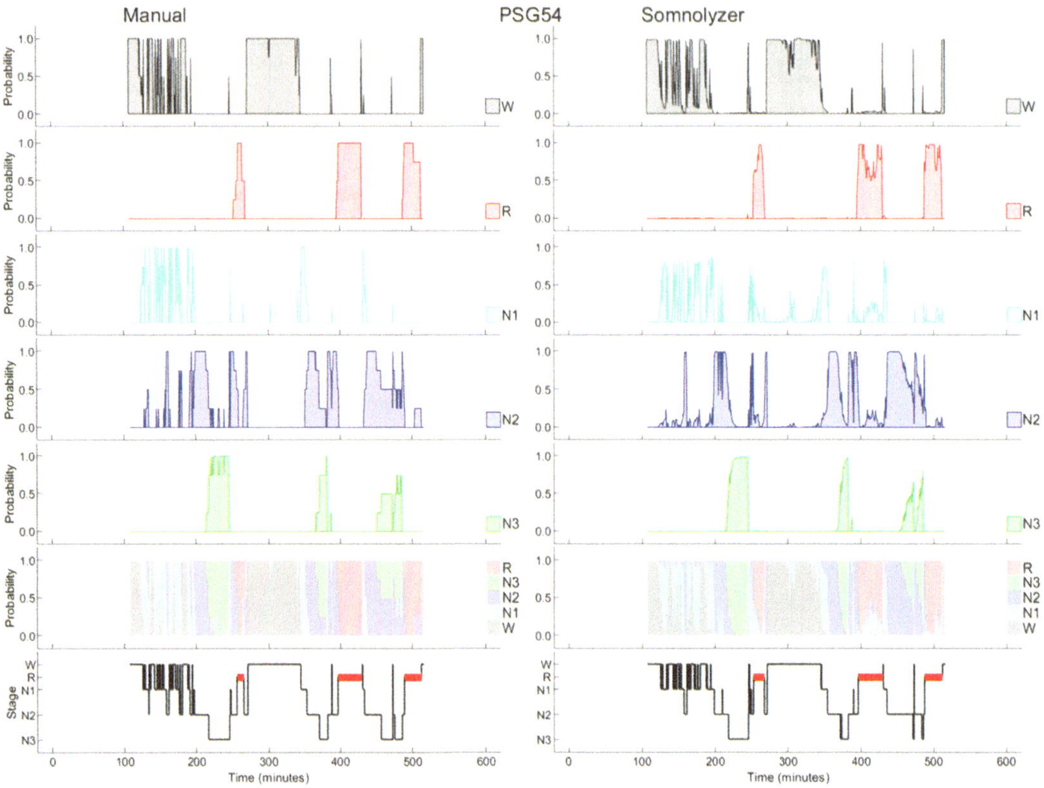

Fig. 7.6 A representative example comparing the probabilities per sleep stage as derived from multiple manual scorings (left) and from autoscoring (right) for the same PSG shown in Fig. 7.2
The figure displays the probabilities for stages W, N1, N2, N3, and R as well as the hypnodensity chart which combines the individual sleep stage probabilities into a stacked area graph. The bottom graph shows the hypnogram. The plots on the left side are from the four manual scorings, and the hypnogram is based on the majority vote (in the case of ties, scorers demonstrating a higher consensus with the group were considered more reliable, and thus their assessments were weighted heavier than the other scorers); the plots on the right side are the Somnolyzer autoscoring results. Color codes for the hypnodensity: W, gray; REM, red; N1, cyan; N2, blue; and N3, green. The ICC between the manually derived and autoscored probabilities are as follows: for W, 0.97; N1, 0.79; N2, 0.89; N3, 0.96; R, 0.95; and over all five stages, 0.93. Cohen's kappa between the human majority hypnogram (bottom left) and the autoscored hypnogram (bottom right) is 0.82

7.3.8 Cardiorespiratory Sleep Staging for Home Sleep Apnea Testing (HSAT)

Typically, no neurological signals are recorded in HSAT, and thus standard sleep scoring is not applicable. Consequently, HSATs are less sensitive than PSGs in the detection of sleep-disordered breathing (SDB) since recording time (RT) rather than total sleep time (TST) is used to define the denominator of the respiratory event index (REI). A false negative test based on the REI may lead to harm to the patient resulting from denial of beneficial therapy (Bianchi & Goparaju 2017). Thus, estimating sleep reliably based on cardiorespiratory signals in HSATs could compensate for this shortcoming. In 2015, Fonseca et al. (2015) presented their first attempts to estimate sleep based on cardiorespiratory signals using manually engineered features and a linear discriminant classifier and reported a Cohen's kappa of 0.49 for the four-stage comparison (wake, light sleep, deep sleep, and REM sleep) validated in 48 healthy subjects. By incorporating time information and replacing the classifier by a conditional random field, Cohen's kappa increased to 0.53 in 100 healthy subjects (Fonseca et al., 2018). In a further development, we trained a deep neural network (CNN +

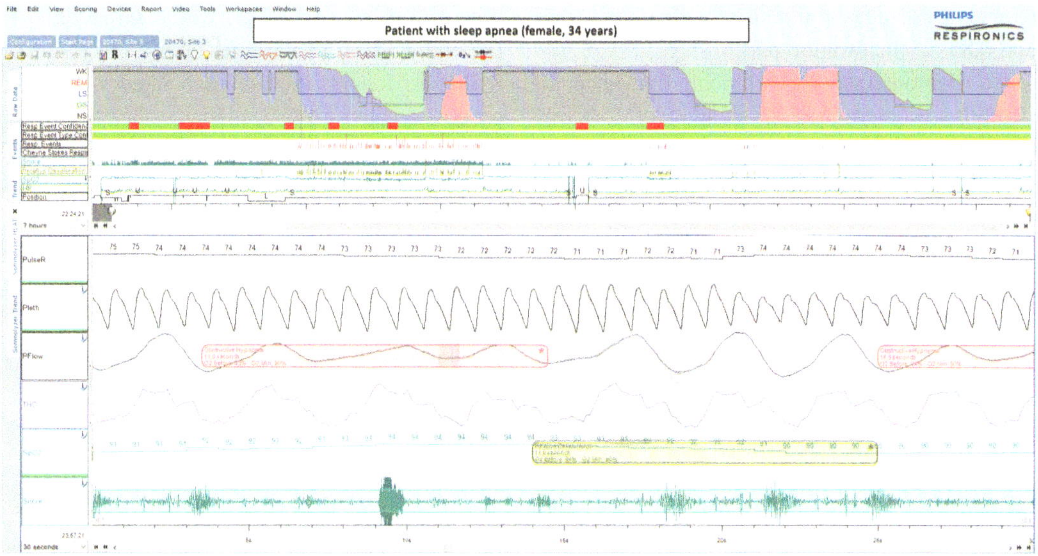

Fig. 7.7 A representative example for cardiorespiratory sleep staging (CReSS) as visualized in Sleepware G3 for the same study shown in Fig. 7.2
Upper part (entire night): The trends from top to bottom are as follows: "WK, REM, LS, DS," the hypnogram superimposed on the hypnodensity graph with the color codes W, gray; REM, red; light sleep (LS), blue; and deep sleep (DS), green; "Resp Event Confidence," confidence trend for guiding the reviewer by a traffic light system to areas that need more careful review of respiratory event scoring; "Resp Event Type Confidence," confidence trend for guiding the reviewer by a traffic light system to areas that need more careful review of scoring the type of a respiratory event; "Resp. Events," respiratory events; "Cheyne Stokes Res," indicates the presence/absence of Cheyne-Stokes breathing; "Snore," intensity of snoring; "Relative Desaturations," scored oxygen desaturations ≥3%; "SpO_2," arterial oxygen saturation level; and "Position," body position
Lower part (30-s window): The signals from top to bottom are as follows: "PulseR," pulse rate in beats/min; "Pleth," photoplethysmography signal; "PFlow," nasal pressure airflow; "THO," thoracic effort; "SpO_2," arterial oxygen saturation level; and "Snore," snoring sound signal
Note that the cardiorespiratory sleep staging shown in the upper panel is based on the "Pleth," "PFlow," and "THO" signals, as shown for the 30-s window in the lower panel, with an autoscored obstructive hypopnea with associated oxygen desaturation

LSTM) and further increased kappa to 0.61 in 195 healthy subjects and 97 patients (Radha et al., 2019) as well as to 0.60 in 389 patients (Fonseca et al., 2020). In a recent development step, we used a deep learning approach not only for the classifier but also for determining high-level features and thereby increased kappa to 0.68 in 296 studies from the MESA dataset with photoplethysmography (PPG), nasal pressure airflow, and respiratory inductance plethysmography (RIP) as input signals (Bakker et al., 2021). Most importantly, the kappa of 0.68 reflecting substantial agreement between sleep staging determined by the algorithm and the gold standard of manual PSG-based sleep staging was consistent across the full spectrum of sleep-disordered breathing severity (for further details, see Bakker et al., 2021).

To demonstrate the performance of the cardiorespiratory sleep staging (CReSS) for the study shown in Fig. 7.2, we deleted all electrophysiological channels from the PSG recording and reanalyzed the study with CReSS based on the HSAT channels (PPG, airflow, thoracic RIP belt). The resulting hypnogram superimposed on the hypnodensity is shown in Fig. 7.7. Cohen's kappa for the four-stage comparison between the CReSS-derived and the PSG-derived hypnograms (W, LS=N1 + N2, DS=N3, REM) is 0.70 for the diagnostic part and 0.74 for the titration part. The substantial agreement between the CReSS-derived and the PSG-derived sleep stage probabilities (compare the hypnodensities in Figs. 7.2 and 7.7) is confirmed by the ICCs between the two probability curves aggregated over all four stages of 0.83 for the diagnostic part and 0.85 for the titration part.

To demonstrate the clinical relevance of CReSS, we determined the number of correctly diagnosed patients by HSAT as compared to the gold standard AHI based on PSG-derived total sleep time (TST) in the 296 studies from the MESA dataset for a threshold of 15 events per hour. As can be seen in Table 7.7, using the CReSS-derived TST instead of the recording time as denominator for the calculation of the indices reduced the false negative diagnosis from 33 patients (11.1%) to 5 patients (1.7%). The ICC between the apnea-hypopnea index based on CReSS-determined total sleep time (AHI_{CReSS}) and the manual AHI_{PSG} was 0.971 (95% CI 0.955 to 0.980) indicating excellent agreement (0.955 > 0.9) according to Koo and Li (2016). In contrast, the ICC between the respiratory event index based on recording time (REI_{RT}) and the manual AHI_{PSG} was only 0.853 (95% CI 0.512–0.934) indicating only moderate agreement (0.512 > 0.5). Note that the lower bound of the 95% CI of the ICC between AHI_{CReSS} and AHI_{PSG} is higher than the upper bound of the 95% CI of the ICC between REI_{RT} and AHI_{PSG} (0.955 > 0.934) and thus the AHI_{CReSS} is a more accurate estimate of the AHI_{PSG} than the REI_{RT}.

The autoscoring system may offer the option of using autonomic responses (respiratory event-related increases in heart rate) as surrogates of cortical arousals to increase the sensitivity for detecting SDB in HSAT even further by approximating the hypopnea scoring rules as recommended in the AASM scoring manual version 2.6 for PSG recordings. As shown in the paragraph "Scoring according to different rules," considering the arousals for hypopnea confirmation increases the SDB severity classification for several patients (compare values in columns 4.A*, corresponding to recommended 2020 rule for HSAT, and 4.AB, corresponding to recommended 2020 rule for PSG, in Table 7.5). Moreover, REM-related OSA according to the definition by Mokhlesi and Punjabi (2012) could be detected based on CReSS-determined REM sleep with a clinically relevant accuracy (sensitivity, 91%, and specificity, 98%). Note that patients with a relatively low overall AHI may be experiencing severe OSA during REM, which is particularly important given that events taking place during REM are longer and are associated with more pronounced hypoxemia, higher sympathetic activation, and greater surges in blood pressure and heart rate (Findley et al., 1985; Somers et al., 1995; Peppard et al., 2009). Consequently, REM-related OSA is associated with adverse cardiovascular, metabolic, and neurocognitive outcomes. For a comprehensive review on risks for adverse health outcomes and on novel treatments of REM-related OSA, see Varga and Mokhlesi (2019).

7.4 Future Directions

Further studies should investigate whether the traditional assignment of a sleep stage per epoch (hypnogram) might be replaced by sleep stage probabilities (hypnodensity) since the latter reflects the ambiguity observed between human scorers while providing all the information contained in a hypnogram. Moreover, all sleep stage parameters required for reporting per AASM guidance can be determined directly from sleep stage probabilities.

Further head-to-head analyses should be undertaken to evaluate the various publicly or commercially available autoscoring systems in large datasets with multiple scorers, such as the

Table 7.7 Confusion matrices comparing the diagnostic performance of HSAT with and without CReSS to gold standard PSG (*n* = 296)

		AHI_{PSG}	
		<15 events/hour	≥15 events/hour
REI_{RT}	<15 events/hour	**139 (47.0%)**	33 (11.1%)
	≥15 events/hour	0 (0.0%)	**124 (41.9%)**
AHI_{CReSS}	<15 events/hour	**139 (47.0%)**	5 (1.7%)
	≥15 events/hour	0 (0.0%)	**152 (51.4%)**

AHI_{PSG}: Apnea-hypopnea index based on PSG-determined total sleep time

REI_{RT}: Respiratory event index based on recording time (RT)

AHI_{CReSS}: Apnea-hypopnea index based on CReSS-determined total sleep time dataset generated via the AASM inter-scorer reliability program (Rosenberg & Van Hout, 2013, 2014). For systems offering full PSG analysis including arousals, respiratory events, and periodic leg movement detection in addition to sleep staging, evaluation of the scoring performance of all scored events and calculated indices would have to be included in the analysis. Access of autoscoring systems to quality assurance programs such as those utilized for accreditation of technologists would allow for more thorough and systematic assessments of autoscoring performance, enabling clinicians and researchers to determine the best fit-for-purpose approach for their needs. Depending on the results of this endeavor, the AASM Facility Standards for Accreditation concerning autoscoring will need to be reconsidered. As of this writing, these guidelines require that automated scoring must be reviewed epoch-by-epoch and edited by staff for accuracy (https://j2vjt3dnbra3ps7ll1clb4q2--wpengine.netdna-ssl.com/wp-content/uploads/2019/05/AASM-Facility-Standards-for-Accreditation-8.2020.pdf).

Moreover, alternative metrics beyond the traditional AHI for determining the severity of sleep apnea could be implemented in autoscoring systems, such as the hypoxic burden as suggested by Azarbarzin et al. (2019). The hypoxic burden is determined by measuring the respiratory event-associated area under the desaturation curve from pre-event baseline. The authors showed in a large sample from the Sleep Disorder in Older Men (MrOS) and the SHHS that the hypoxic burden strongly predicted cardiovascular disease-related mortality, indicating that not only the frequency (as measured by the AHI) but the depth and duration of sleep-related upper airway obstructions (as measured by the hypoxic burden) are important disease-characterizing features. For the split-night study, presented in Fig. 7.2, Sleepware G3 reported for the diagnostic part a hypoxic burden of 67.5 (%min)/hour and for the therapeutic part 6.7 (%min)/hour indicating a 90% reduction of the hypoxic burden due to treatment. A hypoxic burden of 67.5 (%min)/hour is equivalent to approximately 23 min of 3% desaturation per hour, while a hypoxic burden of 6.7 (%min)/hour is equivalent to approximately 2 min of 3% desaturation per hour. Malhotra et al. (2021) listed further alternative metrics of sleep apnea severity: arousal intensity (as a distinct pathophysiological trait); odds ratio product (as a metric that quantifies sleep depth); and cardiorespiratory coupling (measuring cardiopulmonary interactions dynamically during sleep using a single-lead ECG signal).

Finally, the aforementioned large number of successfully applied machine learning approaches for scoring sleep confirms impressively the potential of artificial intelligence applications for sleep diagnostics. These methods can identify complex patterns in empirical data and can take (spatio)-temporal context into account, which allows a direct application of deep learning algorithms to raw data. Deep learning algorithms can be trained on targets such as manually scored sleep stages (supervised deep learning), but also hypothesis-free without targets (unsupervised deep learning) based on large datasets (big data), which preferably contain also clinical outcome data. Such datasets could also be used to identify physiological biomarkers based on the recorded PSG signals to complement the usual biomarkers taken from clinical chemistry as suggested by Penzel et al. (2017). The authors suggest that such physiological biomarkers might be more appropriate to characterize functional characteristics, as seen in the variety of sleep disorders. An example for a successful construction and evaluation of a narcolepsy biomarker can be found in Stephansen et al. (2018). They created a biomarker for narcolepsy that achieved with a specificity of 96% and a sensitivity of 91% similar performance to the current clinical gold standard, the multiple sleep latency test (MSLT), but only requires a single sleep study. Moreover, they showed that addition of human leukocyte antigen (HLA) or other genetic typing in the model can increase the specificity above 99% without loss of sensitivity. Steiger et al. (2015) presented a review on sleep EEG biomarkers for the diagnosis, treatment, and prognosis of depression. The authors concluded that sleep EEG variables are among the biomarkers that should be included in the classification of mood disorders. Even healthy volunteers with a risk gene of depression show subtle sleep EEG changes. In healthy subjects at high risk for affective disorders, elevated REM density was found. In patients with depression disturbed sleep continuity, REM sleep disinhibition and impaired non-REM sleep were the characteristic findings.

References

Abou Jaoude, M., Sun, H., Pellerin, K. R., et al. (2020). Expert-level automated sleep staging of long-term scalp electroencephalography recordings using deep learning. *Sleep, 43*(11), 1–12.

American Academy of Sleep Medicine Task Force. (1999). Sleep-related breathing disorders in adults: Recommendations for syndrome definition and measurement techniques in clinical research. The Report of an American Academy of Sleep Medicine Task Force. *Sleep, 22*, 667–689.

Anderer, P., Roberts, S. J., Schlögl, A., et al. (1999). Artifact processing in computerized analysis of sleep EEG – A review. *Neuropsychobiology, 3*, 150–157.

Anderer, P., Gruber, G., Parapatics, S., et al. (2005). An E-health solution for automatic sleep classification according to Rechtschaffen and Kales: Validation study of the Somnolyzer 24 x 7 utilizing the Siesta database. *Neuropsychobiology, 51*, 115–133.

Anderer, P., Moreau, A., Woertz, M., et al. (2010). Computer-assisted sleep classification according to the standard of the American Academy of Sleep Medicine: Validation study of the AASM version of the Somnolyzer 24 x 7. *Neuropsychobiology, 62*, 250–264.

Anderer, P., Ross, M., Moreau, A., et al. (2012). Effects of inter-rule as opposed to inter-rate variability on the severity of the obstructive sleep apnoea syndrome. *Journal of Sleep Research, 21*(Suppl 1), 228.

Anderer, P., Ross, M., Cerny, A., et al. (2016). Scoring of sleep and associated events: Modelling interrater variability by adjusting sensitivity settings of an auto scoring system. *Journal of Sleep Research, 25*(Suppl 1), 160.

Anderer, P., Ross, M., Cerny, A., & Moreau, A. (2018). Scoring sleep with artificial intelligence enables quantification of sleep stage ambiguity. *Journal of Sleep Research, 27*(Suppl 1), 13.

Azarbarzin, A., Sands, S. A., Stone, K. L., et al. (2019). The hypoxic burden of sleep apnoea predicts cardiovascular disease-related mortality: The Osteoporotic fractures in Men Study and the Sleep Heart Health Study. *European Heart Journal, 40*, 1149–1157.

Bakker, J. P., Tavakkoli, A., Rueschman, M., et al. (2018). Gastric banding surgery versus continuous positive airway pressure for obstructive sleep apnea: A randomized controlled trial. *American Journal of Respiratory and Critical Care Medicine, 197*(8), 1080–1083.

Bakker, J. P., Ross, M., Vasko, R., et al. (2021). Estimating sleep stages using cardiorespiratory signals: Validation of a novel algorithm across a wide range of sleep-disordered breathing severity. *Journal of Clinical Sleep Medicine, 17*(7), 1343–1354.

Bakker, J. P., Ross, M., Cerny, A., et al. (2022). Scoring sleep with artificial intelligence enables quantification of sleep stage ambiguity: Hypnodensity based on multiple expert scorers and auto-scoring. *Sleep*. https://doi.org/10.1093/sleep/zsac154

Barbanoj, M. J., Clos, S., Romero, S., et al. (2005). Sleep laboratory study on single and repeated dose effects of paroxetine, alprazolam and their combination in healthy young volunteers. *Neuropsychobiology, 51*, 134–147.

Berry, R. B., Budhiraja, R., Gottlieb, D. J., et al. (2012). Rules for scoring respiratory events in sleep: update of the 2007 AASM Manual for the Scoring of Sleep and AssociatedEvents. *Journal of Clinical Sleep Medicine 8*(5), 597–619.

Berry, R. B., Brooks, R., Gamaldo, C. E., et al; for the American Academy of Sleep Medicine. (2015). *The AASM manual for the scoring of sleep and associated events: Rules, terminology and technical specifications, version 2.2*. American Academy of Sleep Medicine.

Berry, R. B., Albertario, C. L., Harding, S. M., et al. (2020). For the American Academy of sleep medicine. In *The AASM manual for the scoring of sleep and associated events: Rules, terminology and technical specifications. Version 2.6*. American Academy of Sleep Medicine.

Bianchi, M. T., & Goparaju, B. (2017). Potential underestimation of sleep apnea severity by at-home kits: rescoring in-laboratory polysomnography without sleep staging. *Journal of Clinical Sleep Medicine 13*(4), 551–555.

Biswal, S., Sun, H., Goparaju, B., et al. (2018). Expert-level sleep scoring with deep neural networks. *Journal of the American Medical Informatics Association, 25*, 1643–1650.

Bliwise, D., Bliwise, N. G., Kraemer, H. C., & Dement, W. (1984). Measurement error in visually scored electrophysiological data: Respiration during sleep. *Journal of Neuroscience Methods, 12*, 49–56.

Bonnet, M. H., Doghramji, K., Roehrs, T., et al. (2007). The scoring of arousal in sleep: Reliability, validity, and alternatives. *Journal of Clinical Sleep Medicine, 3*, 133–145.

Cesari, M., Stefani, A., Penzel, T., et al. (2021). Interrater sleep stage scoring reliability between manual scoring from two European sleep centers and automatic scoring performed by the artificial intelligence–based Stanford-STAGES algorithm. *Journal of Clinical Sleep Medicine, 17*(6), 1237–1247.

Chen, X., Wang, R., Zee, P., et al. (2015). Racial/ethnic differences in sleep disturbances: The Multi-Ethnic Study of Atherosclerosis (MESA). *Sleep, 38*(6), 877–888.

Collop, N. A. (2002). Scoring variability between polysomnography technologists in different sleep laboratories. *Sleep Medicine, 3*, 43–47.

Danker-Hopfe, H., Kunz, D., Gruber, G., et al. (2004). Interrater reliability between scorers from eight European sleep laboratories in subjects with different sleep disorders. *Journal of Sleep Research, 13*, 63–69.

Danker-Hopfe, H., Anderer, P., Zeitlhofer, J., et al. (2009). Interrater reliability for sleep scoring according to the Rechtschaffen & Kales and the new AASM standard. *Journal of Sleep Research, 18*(1), 74–84.

Dean, D. A., 2nd, Goldberger, A. L., Mueller, R., et al. (2016). Scaling up scientific discovery in sleep medicine: The National Sleep Research Resource. *Sleep, 39*(5), 1151–1164.

Duce, B., Milosavljevic, J., & Hukins, C. (2015). The 2012 AASM respiratory event criteria increase the incidence of hypopneas in an adult sleep center population. *Journal of Clinical Sleep Medicine, 11*(12), 1425–1431.

Epstein, L. J., Kristo, D., Strollo, P. J., et al. (2009). Clinical guideline for the evaluation, management and long-term care of obstructive sleep apnea in adults. *Journal of Clinical Sleep Medicine, 5*(3), 263–276.

Fell, J., Röschke, J., Mann, K., & Schäffner, C. (1996). Discrimination of sleep stages: A comparison between spectral and nonlinear EEG measures. *Electroencephalography and Clinical Neurophysiology, 98*, 401–410.

Findley, L. J., Wilhoit, S. C., & Suratt, P. M. (1985). Apnea duration and hypoxemia during REM sleep in patients with obstructive sleep apnea. *Chest, 87*, 432–436.

Fiorillo, L., Puiatti, A., Papandrea, M., et al. (2019). Automated sleep scoring: A review of the latest approaches. *Sleep Medicine Reviews, 48*, 1–12.

Fonseca, P., Long, X., Radha, M., et al. (2015). Sleep stage classification with ECG and respiratory effort. *Physiological Measurement, 36*(10), 2027–2040.

Fonseca, P., den Teuling, N., & Long, X. (2018). A comparison of probabilistic classifiers for sleep stage classification. *Physiological Measurement, 39*(5), 055001.

Fonseca, P., van Gilst, M. M., Radha, M., et al. (2020). Automatic sleep staging using heart rate variability, body movements, and recurrent neural networks in a sleep disordered population. *Sleep, 43*(9), zsaa048.

Gaillard, J., & Tissot, R. (1973). Principles of automatic analysis of sleep records with a hybrid system. *Computers and Biomedical Research, 6*, 1–13.

Hoffmann, R., Moffitt, A., Wells, R., et al. (1984). Quantitative description of sleep stage electrophysiology using digital period analytic techniques. *Sleep, 7*, 356–364.

Iber, C., Ancoli-Israel, S., Chesson, A., Quan, S. F., & for the American Academy of Sleep Medicine. (2007). *The AASM manual for the scoring of sleep and associated events: Rules, terminology and technical specifications* (1st ed.). American Academy of Sleep Medicine.

Itil, T. M. (1969). Automatic classification of sleep stages and the discrimination of vigilance changes using digital computer methods. *Agressologie, 10, Suppl*, 603–610.

Kapur, V. K., Auckley, D. H., Chowdhuri, S., et al. (2017). Clinical practice guideline for diagnostic testing for adult obstructive sleep apnea: An American Academy of Sleep Medicine clinical practice guideline. *Journal of Clinical Sleep Medicine, 13*(3), 479–504.

Klösch, G., Kemp, B., Penzel, T., et al. (2001). The SIESTA project polygraphic and clinical database. *IEEE Engineering in Medicine and Biology Magazine, 20*, 51–57.

Koo, T. K., & Li, M. Y. (2016). A guideline of selecting and reporting intraclass correlation coefficients for reliability research. *Journal of Chiropractic Medicine, 15*(2), 155–163.

Kubicki, S., Holler, L., Berg, I., et al. (1989). Sleep EEG evaluation: A comparison of results obtained by visual scoring and automatic analysis with the Oxford sleep stager. *Sleep, 12*, 140–149.

Landis, J. R., & Koch, G. G. (1977). The measurement of observer agreement for categorical data. *Biometrics, 33*(1), 159–174.

Larsen, L., & Walter, D. (1970). On automatic methods of sleep staging by EEG spectra. *Electroencephalography and Clinical Neurophysiology, 28*, 459–467.

Loredo, J. S., Clausen, J. L., Ancoli-Israel, S., & Dimsdale, J. E. (1999). Night-to-night arousal variability and interscorer reliability of arousal measurements. *Sleep, 22*, 916–920.

Louis, R. P., Lee, J., & Stephenson, R. (2004). Design and validation of a computer-based sleep-scoring algorithm. *Journal of Neuroscience Methods, 133*, 71–80.

Magalang, U. J., Chen, N. H., Cistulli, P. A., et al. (2013). Agreement in the scoring of respiratory events and sleep among international sleep centers. *Sleep, 36*(4), 591–596.

Malafeev, A., Laptev, D., & Bauer, S. (2018). Automatic human sleep stage scoring using deep neural networks. *Frontiers in Neuroscience, 12*, Article 781.

Malhotra, A., Younes, M., Kuna, S. T., et al. (2013). Performance of an automated polysomnography scoring system versus computer-assisted manual scoring. *Sleep, 36*(4), 573–582.

Malhotra, A., Ayappa, I., Ayas, N., et al. (2021). Metrics of sleep apnea severity: Beyond the apnea-hypopnea index. *Sleep, 44*(7), zsab030.

Martin, W., Johnson, L., Viglione, S., et al. (1972). Pattern recognition of EEG-EOG as a technique for all-night sleep stage scoring. *Electroencephalography and Clinical Neurophysiology, 32*, 417–427.

Mokhlesi, B., & Punjabi, N. M. (2012). "REM-related" obstructive sleep apnea: An epiphenomenon or a clinically important entity? *Sleep, 35*, 5–7.

Pardey, J., Roberts, S., Tarassenko, L., & Stradling, J. (1996). A new approach to the analysis of the human sleep/wakefulness continuum. *Journal of Sleep Research, 5*, 201–210.

Park, H.-J., Oh, J.-S., Jeong, D.-U., & Park, K.-S. (2000). Automated sleep stage scoring using hybrid rule-and case-based reasoning. *Computers and Biomedical Research, 33*, 330–349.

Patanaik, A., Ong, J. L., Gooley, J. J., et al. (2018). An end-to-end framework for real-time automatic sleep stage classification. *Sleep, 41*(5), zsy041.

Penzel, T., Hirshkowitz, M., Harsh, J., et al. (2007). Digital analysis and technical specifications. *Journal of Clinical Sleep Medicine, 3*, 109–120.

Penzel, T., Zhang, X., & Fietze, I. (2013). Inter-scorer reliability between sleep centers can teach us what to improve in the scoring rules. *Journal of Clinical Sleep Medicine, 9*(1), 89–91.

Penzel, T., Fietze, I., & Veauthier, C. (2017). The need for a reliable sleep EEG biomarker. *Journal of Clinical Sleep Medicine, 13*(6), 771–772.

Peppard, P. E., Ward, N. R., & Morrell, M. J. (2009). The impact of obesity on oxygen desaturation during sleep-disordered breathing. *American Journal of Respiratory and Critical Care Medicine, 180*, 788–793.

Punjabi, N. M., Shifa, N., Dorffner, G., et al. (2015). Computer-assisted automated scoring of polysomnograms using the Somnolyzer system. *Sleep, 38*, 1555–1566.

Radha, M., Fonseca, P., Moreau, A., et al. (2019). Sleep stage classification from heart-rate variability using long short-term memory neural networks. *Scientific Reports, 9*(1), 14149.

Rechtschaffen, A., & Kales, A. (Eds.). (1968). *A manual of standardized terminology, techniques, and scoring systems for sleep stages of human subjects*. UCLA Brain Information Service/Brain Research Institute.

Redline, S., Kapur, V. K., Sanders, M. H., et al. (2000). Effects of varying approaches for identifying respiratory disturbances on sleep apnea assessment. *American Journal of Respiratory and Critical Care Medicine, 161*, 369–374.

Redline, S., Budhiraja, R., Kapur, V., et al. (2007). The scoring of respiratory events in sleep: Reliability and validity. *Journal of Clinical Sleep Medicine, 3*, 169–200.

Rosen, C. L., Auckley, D., Benca, R., et al. (2012). A multisite randomized trial of portable sleep studies and positive airway pressure autotitration versus laboratory-based polysomnography for the diagnosis and treatment of obstructive sleep apnea: The HomePAP study. *Sleep, 35*(6), 757–767.

Rosenberg, R. S., & Van Hout, S. (2013). The American Academy of Sleep Medicine inter-scorer reliability program: Sleep stage scoring. *Journal of Clinical Sleep Medicine, 9*(1), 81–87.

Rosenberg, R. S., & Van Hout, S. (2014). The American Academy of Sleep Medicine Inter-scorer Reliability program: Respiratory events. *Journal of Clinical Sleep Medicine, 10*(4), 447–454.

Ruehland, W. R., Rochford, P. D., O'Donoghue, F. J., et al. (2009). The new AASM criteria for scoring hypopneas: Impact on the apnea hypopnea index. *Sleep, 32*(2), 150–157.

Saletu, B., Prause, W., Anderer, P., et al. (2005). Insomnia in somatoform pain disorder: Sleep laboratory studies on differences to controls and acute effects of trazodone, evaluated by the Somnolyzer 24 x 7 and the Siesta database. *Neuropsychobiology, 51*, 148–163.

Schaltenbrand, N., Lengelle, R., & Macher, J.-P. (1993). Neural network model: Application to automatic analysis of human sleep. *Computers and Biomedical Research, 26*, 157–171.

Schwaibold, M., Schochlin, J., & Bolz, A. (2002). Automated sleep stage detection with a classical and a neural learning algorithm – Methodological aspects. *Biomedizinische Technik. Biomedical Engineering, 47 Suppl 1*, 318–320.

Shrout, P. E., & Fleiss, J. L. (1979). Intraclass correlations: Uses in assessing rater reliability. *Psychological Bulletin, 86*, 420–428.

Smith, J. R., & Karacan, I. (1971). EEG sleep stage scoring by an automatic hybrid system. *Electroencephalography and Clinical Neurophysiology, 31*, 231–237.

Somers, V. K., Dyken, M. E., Clary, M. P., & Abboud, F. M. (1995). Sympathetic neural mechanisms in obstructive sleep apnea. *The Journal of Clinical Investigation, 96*, 1897–1904.

Stanus, E., Lacroix, B., Kerkhofs, M., & Mendlewicz, J. (1987). Automated sleep scoring: A comparative reliability study of two algorithms. *Electroencephalography and Clinical Neurophysiology, 66*, 448–456.

Steiger, A., Pawlowski, M., & Kimura, M. (2015). Sleep electroencephalography as a biomarker in depression. *ChronoPhysiology and Therapy, 5*, 15–25.

Stephansen, J. B., Olesen, A. N., Olsen, M., et al. (2018). Neural network analysis of sleep stages enables efficient diagnosis of narcolepsy. *Nature Communications, 9*(1), 5229.

Sun, H., Jia, J., Goparaju, B., et al. (2017). Large-scale automated sleep staging. *Sleep, 40*(10), zsx139.

Svetnik, V., Ma, J., Soper, K., et al. (2007). Evaluation of automated and semi-automated scoring of polysomnographic recordings from a clinical trial using zolpidem in the treatment of insomnia. *Sleep, 20*, 1562–1574.

Varga, A. W., & Mokhlesi, B. (2019). REM obstructive sleep apnea: Risk for adverse health outcomes and novel treatments. *Sleep & Breathing, 23*(2), 413–423.

Whitney, C. W., Gottlieb, D. J., Redline, S., et al. (1998). Reliability of scoring respiratory disturbance indices and sleep staging. *Sleep, 21*, 749–757.

Younes, M., Raneri, J., & Hanly, P. (2016). Staging sleep in polysomnograms: Analysis of inter-scorer variability. *Journal of Clinical Sleep Medicine, 12*(6), 885–898.

Younes, M., Kuna, S. T., Pack, A. I., et al. (2018). Reliability of the American Academy of Sleep Medicine rules for assessing sleep depth in clinical practice. *Journal of Clinical Sleep Medicine, 14*, 205–213.

Zhang, G. Q., Cui, L., Mueller, R., et al. (2018). The National Sleep Research Resource: Towards a sleep data commons. *Journal of the American Medical Informatics Association, 25*(10), 1351–1358.

Zhang, L. D., Fabbri, D., Upender, R., & Kent, D. (2019). Automated sleep stage scoring of the Sleep Heart Health Study using deep neural networks. *Sleep, 42*(11), zsz159.

8 Conventional Machine Learning Methods Applied to the Automatic Diagnosis of Sleep Apnea

Gonzalo C. Gutiérrez-Tobal, Daniel Álvarez, Fernando Vaquerizo-Villar, Verónica Barroso-García, Javier Gómez-Pilar, Félix del Campo, and Roberto Hornero

Abstract

The overnight polysomnography shows a range of drawbacks to diagnose obstructive sleep apnea (OSA) that have led to the search for artificial intelligence-based alternatives. Many classic machine learning methods have been already evaluated for this purpose. In this chapter, we show the main approaches found in the scientific literature along with the most used data to develop the models, useful and large easily available databases, and suitable methods to assess performances. In addition, a range of results from selected studies are presented as examples of these methods. Very high diagnostic performances are reported in these results regardless of the approaches taken. This leads us to conclude that conventional machine learning methods are useful techniques to develop new OSA diagnosis simplification proposals and to act as benchmark for other more recent methods such as deep learning.

Keywords

Sleep apnea · Machine learning · Sleep Heart Health Study · Childhood Adenotonsillectomy Trial · Classification · Regression · Biomedical signal processing · Airflow · Blood oxygen saturation · Electrocardiogram

8.1 Introduction

The technical complexity, costs, and logistic-associated problems in the diagnosis of obstructive sleep apnea (OSA) have driven the scientific community to search for new simpler and automatic alternatives to standard polysomnography (PSG) (Ghegan et al., 2006). One of the most common and ambitious approximations to achieve this goal has been the implementation of systems or algorithms based on the study of a reduced set of information from the PSG. Usually,

G. C. Gutiérrez-Tobal (✉) · D. Álvarez · F. Vaquerizo-Villar · V. Barroso-García · J. Gómez-Pilar · R. Hornero
Centro de Investigación Biomédica en Red, Bioingeniería, Biomateriales, Nanomedicina, Madrid, Spain

Biomedical Engineering Group, University of Valladolid, Valladolid, Spain
e-mail: gonzalo.gutierrez@gib.tel.uva.es

F. del Campo
Centro de Investigación Biomédica en Red, Bioingeniería, Biomateriales, Nanomedicina, Madrid, Spain

Biomedical Engineering Group, University of Valladolid, Valladolid, Spain

Sleep Unit, Pneumology Service, Hospital Universitario Rio Hortega, Valladolid, Spain

T. Penzel, R. Hornero (eds.), *Advances in the Diagnosis and Treatment of Sleep Apnea*, Advances in Experimental Medicine and Biology 1384, https://doi.org/10.1007/978-3-031-06413-5_8

the automatic analysis of only a very small set of signals – out of a maximum of 32 recorded during PSG – has been conducted, with the investigation on a single one being a very frequent approach (Uddin et al., 2018; Mendonça et al., 2019; Gonzalo C Gutiérrez-Tobal et al., 2021c). In this regard, overnight blood oxygen saturation (SpO_2), airflow (AF), and electrocardiogram (ECG) are among the most analyzed signals (Uddin et al., 2018; Mendonça et al., 2019; Gutiérrez-Tobal et al., 2021c).

Since the beginning of the twenty-first century, machine learning techniques have gained an increasing role when improving the performance of automatic health-related diagnostic tools, and OSA has not been an exception. A three-step methodology, the so-called feature engineering approach, has been traditionally applied to the problem (Vaquerizo-Villar et al., 2021). This strategy begins with the "feature extraction" stage, in which the data – usually an overnight signal – are analyzed following one or several complementary analytical techniques, such as spectral, non-linear, or time-frequency methods. The purpose of this step is to characterize the signal or signals under study, so that the original raw or pre-processed data becomes information of interest for the problem. Then, an optional but useful automatic "feature selection" stage is conducted to ensure that all the information extracted in the previous step is as relevant for your problem and as complementary to each other as possible (Guyon & Elisseeff, 2003). The third stage is what involves machine learning. It could be termed simply "machine learning" stage or, depending on the context, "classification" stage, "regression" stage, or, in a more general way, "pattern recognition" stage (Bishop, 2006).

Certainly, the latest deep learning methods are able to avoid the two first stages in the above-described feature engineering approach (Ian et al., 2016). However, many traditional methods are still used nowadays in the context of OSA diagnosis and constitute a valid and very useful benchmark to compare the results obtained with any new approximation to the problem. Accordingly, this chapter aims at exposing the interested readers to a set of conventional machine learning tools that have proven their usefulness to help in the automatic OSA diagnosis. As shown in the next sections, it is not a minor challenge to outperform some of these methods, so any new algorithm must demonstrate that demanding an extra effort, on either data or computation, is justified.

This chapter continues with a data section, which is dedicated to briefly present the information traditionally analyzed in the OSA diagnosis simplification. Then, a methods section explains the two main machine learning approaches (classification and regression) that have shown usefulness in this problem. It includes introducing some specific examples used in OSA context along with a brief explanation on their rationale, as well as appropriate references where the readers will be able to gain insight into these methods. Next, a results section shows some of the highest performances achieved using them. Finally, the "Discussion and Conclusions" section analyzes the most important information included in this chapter.

8.2 Data Analyzed in the Simplification of Sleep Apnea Diagnosis

Several chapters of this book are specifically devoted to describing useful sources of information in the context of sleep apnea. Therefore, this section is only a short introduction on those that have been more frequently used along with machine learning approaches. These include some overnight biomedical signals recorded during PSG and other clinical and demographic data. In addition, we present some popular public databases that have been used in dozens of different studies to gain insight into sleep apnea in both adults and children.

8.2.1 Typical Overnight Biomedical Signals

AF, SpO_2, and ECG (including the ECG-derived heart rate variability or HRV) have been extensively analyzed in the context of simplifying OSA diagnosis in the last decades (Uddin et al., 2018; Mendonça et al., 2019; Gutiérrez-Tobal et al., 2021c). Usually, the recordings are acquired during the night with the same equipment used in the PSG, but there exist a substantial number of scientific studies using devices specifically dedicated to acquiring each signal alone. In addition, the most common approach has focused on the analysis of single-channel signals, but some studies also analyzed the usefulness of automatically combining the information from two or more of them.

8.2.1.1 Airflow (AF)

As explained in dedicated chapters of this book, one of the most important indicators of the presence and severity of OSA is the apnea-hypopnea index (AHI) (Iber et al., 2007; Berry et al., 2012, 2017). AHI accounts for the number of apnea – complete cessation of the respiratory cycle – and hypopnea events, significant reduction of the respiratory amplitude, per hour of sleep (Berry et al., 2012). These qualitative definitions of apneas and hypopneas are detailed in the rules for scoring respiratory events published and updated by the American Academy of Sleep Medicine (Iber et al., 2007; Berry et al., 2012, 2017). A reduction of 90% in AF is mandatory to annotate an apnea event (Berry et al., 2012), showing a duration of a minimum of two respiratory cycles in pediatric patients and 10 seconds in adults. In the case of hypopneas, a 30% drop in AF suffices, but the event needs to be accompanied by either a 3% drop in the SpO_2 signal or an arousal (Berry et al., 2012). The minimum duration requirement of the AF drop is also two respiratory cycles for children and 10 seconds for adults. According to these definitions, in which AF plays a key role, the study of this signal is a natural choice to search for simpler OSA diagnostic alternatives.

When scoring these respiratory events, it is necessary to consider that apneas must be counted using an oronasal thermal sensor, whereas hypopneas are annotated using a nasal pressure sensor (Berry et al., 2012). This is because of the complementary performances of these two kinds of probes when detecting each of the event types (Bahammam, 2004). This also needs to be considered when using machine learning techniques that only focus on detecting apneas and hypopneas. However, in machine learning approaches not conducting event detection, but full characterization of the overnight AF signal, recent studies have shown similar performances using single-channel AF approaches regardless if thermal or nasal pressure sensors were used (Gutiérrez-Tobal et al., 2013; Gutierrez-Tobal et al., 2016).

8.2.1.2 Blood Oxygen Saturation (SpO_2)

Blood oxygen drops – or desaturations – are typical effects caused by apneic events (Iber et al., 2007; Berry et al., 2012, 2017). Actually, we have already shown that 3% desaturations are directly involved in the hypopnea definition. Additional important advantages need to be considered that have led SpO_2 to be probably the most analyzed and successful signal when simplifying OSA diagnosis, in both adults and children. The first one is that it is easily acquired using a single-channel pulse oximetry placed on a finger (or a toe in babies). This is very comfortable compared to all the channels required to conduct a full PSG. As a result, the associated portable technology is highly developed, which facilitates to move the diagnostic test to patients' homes. A second advantage is that the overnight blood oxygen saturation gathers not only the information regarding the apneic events but also the health prognosis associated with the condition. In this regard, 3% and 4% oxygen desaturation indices (ODI3 and ODI4), cumulative time under 90% of saturation (CT90), or, more recently, hypoxic burden have been linked to different negative health consequences in OSA presence (Azarbarzin et al., 2019; Karhu et al., 2021). Finally, as shown in the next sections, the results reached when applying machine learning methods to SpO_2 are among the highest in the related scientific literature.

8.2.1.3 Electrocardiogram and Heart Rate Variability (ECG/HRV)

The natural cardiorespiratory coupling is one of the main reasons behind the study of ECG to help simplify OSA diagnosis. This coordination has been found to increase during the night in the presence of sleep apnea (Riedl et al., 2014), being one of its expressions the occurrence of a clear bradycardia/tachycardia pattern following the apneic events (Penzel et al., 2003). Moreover, the ECG was one of the first biomedical signals studied, and it is still one of the most analyzed in different health contexts, which very often provides a comfortable scientific knowledge background on which to justify the interpretations of eventual results (Acharya et al., 2006). Similarly, OSA in adults is known to be significantly associated with cardiovascular morbidity (Newman et al., 2001). Together, these aspects have led to an intensive scientific activity regarding the simplification of OSA diagnosis based on ECG information (Penzel et al., 2002). Particularly common has been the investigations on HRV, which offers a nexus between OSA and the autonomic nervous system (Acharya et al., 2006). An additional advantage of the HRV information is that it can be surrogated in some contexts by the pulse rate variability signal (PRV) (Gil et al., 2010), which can be easily obtained from a pulse oximeter.

8.2.2 Other Sources of Information

The clinical analysis of PSG is the result of the examination of a range of up to 32 biomedical channels. Consequently, it is not surprising that several approaches explored the combination of the information from two or three of the above-mentioned biomedical signals along with the use of machine learning techniques (Garde et al., 2014; Álvarez et al., 2020; Jiménez-García et al., 2020). In addition, other single- or combined-channel approaches have been evaluated. In this regard, the use of overnight snoring sounds (Solà-Soler et al., 2012), thoracic and/or abdominal movements (Lin et al., 2017), photoplethysmography (Gil et al., 2010; Lázaro et al., 2014), or the electroencephalography (Gonzalo C. Gutiérrez-Tobal, Gomez-Pilar, et al., 2021b), among others, have been also explored with promising results.

Moreover, machine learning has been also used with data other than those from PSG. Demographic, social, clinical, and anthropometric variables have been also used as source of information to train machine learning models with ability to diagnose OSA (El-Solh et al., 1999; Skotko et al., 2017; Gonzalo C Gutiérrez-Tobal et al., 2021c). These have been used most often combined within them and with the information obtained from the PSG, such as overnight biomedical signals.

8.2.3 Important Databases

Large and commonly used databases are very useful both to properly train and validate the machine learning models and to share a reference to which compare the performance from different methods. Unfortunately, freely available large databases are very uncommon in OSA context, if there exist. However, the National Research Sleep Resource offers several very large sleep-related databases with only minor requirements to be accomplished. Here, we briefly introduce two of them that have been used in dozens of OSA-related studies from adults and children, namely, the Sleep Heart Health Study (SHHS) database and the Childhood Adenotonsillectomy Trial (CHAT) database, respectively.

8.2.3.1 Sleep Heart Health Study (SHHS)

The SHHS was originally designed to evaluate whether OSA is an independent risk factor for the development of cardiovascular morbidity in adults (Newman et al., 2001). The database comprises at-home conducted PSGs from 5804 individuals older than 40 years who were recruited from several previous cohorts aimed at evaluating cardiovascular risks (Quan et al., 1997). It is divided into SHHS1, with a first round of sleep data and recordings from all the participants, and SHHS2, with a follow-up at-home PSG conducted on 2647 participants 5 years later. Accordingly, longitudinal studies are possible when using this database. In total, 8451 full PSGs are available to use it as source of information in machine learning-based studies, including annotations such as respiratory events or sleep stages, along with a wide range of clinical, social, and anthropometric variables (Quan et al., 1997; Newman et al., 2001).

8.2.3.2 Childhood Adenotonsillectomy Trial (CHAT)

The aim of the CHAT randomized study was to analyze the effects of a treatment based on the removing of tonsils and adenoids in a cohort of OSA-affected children (Marcus et al., 2013). To assess these effects, PSGs from 1447 children between 5 and 9 years were conducted, from 464 who were randomized to adenotonsillectomy treatment (206 children) or the alternative watchful waiting with supportive care (198 children) (Marcus et al., 2013). Accordingly, these participants underwent a baseline PSG and a follow-up PSG 7 months later, once completing the treatment or the alternative. A wide range of clinical, sociodemographic, cognitive, and anthropometric variables is also available (Marcus et al., 2013). As in the case of SHHS, the follow-up conducted on the children allows for longitudinal studies taking into account that there is a therapeutic intervention between the two PSGs. In addition to the randomized children, the PSGs from the non-randomizing are also available to develop the machine learning approaches. However, the set of additional variables is dramatically reduced compared to the randomized set.

8.3 Methods: Classic Machine Learning Approaches in Sleep Apnea Diagnosis

In accordance with the purpose of automatically diagnosing OSA, supervised learning is the most common strategy followed in the scientific literature. Particularly, both classification and regression approaches have been frequently implemented. OSA presence and severity are routinely categorized by using AHI thresholds in clinical practice, which leads to classification methods. Moreover, AHI can be also directly estimated, thus leading to regression approaches. In this section, we also introduce the ways in which the performance of the OSA-related machine learning methods should be assessed for both classification and regression.

8.3.1 Classification

There are two typical ways to implement classification approaches in OSA diagnosis context: binary classification and multiclass classification. In addition, these may have different purposes. On the one hand, classification may focus on directly assigning subjects into two (presence vs. absence of OSA) or more (presence and severity of OSA) categories. This should be the final goal of any automatic diagnostic approach. On the other hand, however, classification may also focus on detecting apneic events, and this can be also implemented as binary classification (apneic vs. normal signal segments) or multiclass classification (apneas/hypopneas/normal or obstructive apneas/central apneas/normal, etc.).

8.3.1.1 Binary Classification

Over the years, the clinicians have focused on AHI thresholds to assess whether a person suffers from OSA. Ten and 15 events per hour (e/h) have been commonly used in adults, and 1 e/h, 3 e/h, and 5 e/h in children, the exact cut-off evolving as the corresponding medical associations proposed new rules (Iber et al., 2007; Berry et al., 2012, 2017). In accordance with these thresholds, one of the machine learning approaches has focused on automatically detecting the presence of the illness, that is, classifying subjects into OSA positive (above or equal the AHI cut-off) or OSA negative (below the AHI cut-off). Different classic machine learning methods have been used to implement this approach. Linear discriminant analysis (LDA) is one of the most typical classification procedures (Bishop, 2006) and has been evaluated in both adults and children in OSA context. LDA assumes a linear relationship between the predictors (variables used as the data to predict OSA) and the target (the variable containing the OSA-positive and OSA-negative labels). Despite its relatively simplicity, LDA has reached promising results when discriminating OSA-positive and OSA-negative patients using information from SpO_2 (Marcos et al., 2009), SpO_2 + PRV (Garde et al., 2014), and HRV (Martín-Montero et al., 2021). Logistic regression (LR) (Hosmer & Lemeshow, 1989) is a standard in binary classification and has been also evaluated with SpO_2 (Marcos et al., 2009; Álvarez et al., 2010, 2013) and AF (Barroso-García et al., 2017), in both adults and children. LR uses the logistic formulae to transform the output resulting from a linear regression into a non-linear posterior probability (Hosmer & Lemeshow, 1989), that is, given the predictors, the probability of belonging to the OSA-positive class – as defined by the AHI cut-off used. Accordingly, LR avoids the limitation of the linear relationship assumption.

This limitation can be also minimized with more complex and modern methods such as artificial neural networks (ANNs) and support vector machines (SVMs) (Bishop, 2006). SVMs are machine learning algorithms that transform the data into a higher-dimensional space so that the distance between data points with different labels – in this case, OSA positive and negative – is maximized (Bishop, 2006). This is equivalent to choosing a decision boundary between classes for which the distance to the closest data point, the so-called margin, is maximized (Bishop, 2006). Accordingly, the decision boundary is defined by several of these data points termed support vectors. Some examples of SVM binary classification in OSA context can be found applied to SpO_2 (Álvarez et al., 2013) and ECG (Khandoker et al., 2009; Chen et al., 2015). On the other hand, several ANNs have been evaluated in OSA binary classification approaches (Marcos et al., 2008; Morillo & Gross, 2013), being multi-layer perceptron (MLP) one common approach that has become one of the most successful machine learning methods in any problem. ANNs are algorithms inspired in the biological neural networks, such as the human brain. Accordingly, MLP arrange computing units, also known as perceptrons or neurons, in several massively connected layers: input, hidden, and output (Bishop, 2006). The input layer is composed of one neuron for each feature or variable used as predictor. These input neurons are connected through weights with all the neurons in the next layer, which is part of the hidden layers. There can be as many hidden layers as the designers may consider appropriate. However, one single hidden layer is known to be able to provide universal approximations (Bishop, 2006). This means that, provided that your data gather information enough for your problem, one single hidden layer should suffice to model the function that transform your predictors into your desired target. In any case, both the number of hidden layers and the number of neurons per hidden layer are hyperparameters of the model to be tuned during the training process. Finally, each neuron of the last hidden layer – if there is more than one – is connected to all the neurons in the output layer, which in the case of the binary classification approach is a single neuron that offers the posterior probability of belonging to the OSA-positive class. During the training process of the MLP (and other ANNs), all the weights connecting all the neurons of the network are optimized using the well-known backpropagation algorithm (Bishop, 2006), which is one of the most remarkable milestones of machine learning. Another feature of ANNs is that each neuron has an associated activation function that combines the outputs – including weights – from previous layers into a single output, being logistic or softmax functions typically used in classification and linear functions in regression problems (Bishop, 2006).

8.3.1.2 Multiclass Classification

In recent years, as more sleep data has been available for scientific purposes, the focus of OSA diagnosis simplification has gone from binary classification to the determination of both OSA presence and severity, which naturally fits multiclass classification. There exist AHI thresholds for the definition of OSA severity categories in both adults and children, being the latter much more restrictive. Nowadays, the most clinically used ones are probably as follows (Flemons et al., 1999; Tan et al., 2014, 2017):

- *Adults*: no OSA if AHI < 5 e/h; mild OSA if 5 e/h ≤ AHI < 15 e/h; moderate OSA if 15 e/h ≤ AHI < 30 e/h; and severe OSA if 30 e/h ≤ AHI
- *Children*: no OSA if AHI < 1 e/h; mild OSA if 1 e/h ≤ AHI < 5 e/h; moderate OSA if 5 e/h ≤ AHI < 10 e/h; and severe OSA if 10 e/h ≤ AHI

As in the case of binary classification, several multiclass approaches have been already evaluated in OSA context. LDA and LR models were also used in the multiclass problem along with SpO_2 data (Gutiérrez-Tobal et al., 2019), the latter needing an additional "one-vs.-all" strategy to upgrade the binary approach. MLP and other ANNs have been also developed with both SpO_2 data (Gutiérrez-Tobal et al., 2019), SpO_2 + AF data (Barroso-García et al., 2021), and clinical, anthropometric, and demographic variables (Skotko et al., 2017). In this regard, from an implementation point of view, only minor changes in the architecture are needed to develop multiclass ANNs instead of binary ones, such as equaling the number of output neurons to the number of classes. The interested readers should notice, however, that data requirements usually increase as more classes are targeted and that multiclass overall performance tends to be lower than the binary one.

Ensemble learning methods have been also used to address the multiclass problem. As deduced from its name, this family of machine learning methods conduct the classification task as the result of the combination of the classification of several single models, typically termed "base classifiers." These can be any of the abovementioned methods, but simpler ones are preferred to increase the generalization ability of the final classification (Witten et al., 2011). Bagging ensemble learning algorithms have been tested, including the remarkable random forest (RF) method used with SpO_2 data (Deviaene et al., 2019). Bagging is the acronym for "bootstrap aggregating," which indicates the basic methodology behind this method. In essence, the original data is subsampled with replacement to form, typically, a high number of bootstrap replicates of these data (Kuncheva, 2014). A different classifier is trained for each of these replicates, and its decision is only one vote for the final classification task, which is conducted based on the decisions from all classifiers. RF follows this elementary scheme using decision trees as base classifiers. In addition, RF includes more sources of variability in the training of its classifiers by randomly varying the features and the decision trees hyperparameters involved within each bootstrap iteration (Kuncheva, 2014). Boosting ensemble learning methods have been also applied in OSA-related multiclass tasks, as is the case of the well-known AdaBoost (for "adaptive boosting"), used with SpO_2 (Gutiérrez-Tobal et al., 2019), AF (Gutierrez-Tobal et al., 2016), and SpO_2 + AF (Jiménez-García et al., 2020; Barroso-García et al., 2021). In contrast to bagging, boosting methods are iterative algorithms in which each new classifier is trained using the same data, but accounting for the errors made by previous classifiers. In this regard, misclassified data points in previous iterations are weighted to give them more importance, thus increasing the chances to be rightly classified in the current and next iterations (Witten et al., 2011). Another difference with bagging is that the vote of each classifier is dependent on its error so that the ones with higher performance contribute more to the final decision (Witten et al., 2011).

8.3.2 Regression

The automatic AHI estimation is another popular approach when simplifying OSA diagnosis. Instead of training machine learning methods to directly assign subjects (or epochs) into different OSA severity categories (or events), this strategy looks for assigning an AHI to each subject. As the clinical use of AHI thresholds has evolved over the years, and there are still some limitations regarding the OSA severity categories and the actual health state of the patients (Penzel et al., 2015; Korkalainen et al., 2019), the AHI estimation has the advantage of being relatively transparent to future changes in thresholding criteria.

There exists an extensive literature focused on regression methods and OSA diagnosis simplification. They focus on both simpler algorithms, such as multiple linear regression applied to clinical data (Wu et al., 2017) and more complex methods already mentioned such as MLP or SVM applied to clinical (El-Solh et al., 1999), SpO_2 (Marcos et al., 2012; Hornero et al., 2017; Rolón et al., 2017; Xu et al., 2019; Rolon et al., 2020), AF (Álvarez et al., 2020; Barroso-García et al., 2021), and SpO_2 + AF data (Álvarez et al., 2020; Barroso-García et al., 2021). Moreover, boosting ensemble learning methods have been also evaluated, such as least-square boosting (LSBoost) using SpO_2 information (Gonzalo C. Gutiérrez-Tobal, Álvarez, et al., 2021a). In this regression task, rather than focusing on previously misclassified data points, the boosting algorithm LSBoost looks for computing the remaining residual error (between the actual and the estimated AHI) that was not able to be estimated in previous iterations (Bühlmann & Hothorn, 2007).

8.3.3 Machine Learning Performance Assessment and Validation

8.3.3.1 Underfitting and Overfitting

Machine learning faces two main issues regardless of the problem and the approach considered. The first one, underfitting, relates to the inability of the method to learn the function it is intended for. Two aspects are often behind underfitting, unsuitable learning algorithm or unsuitable input information, the latter caused by either data scarcity or low quality. However, provided that a proper study design has been conducted, underfitting is not the main drawback that machine learning may confront. Very often, overfitting is a much more challenging aspect. Overfitting refers to an excessive fitting of the machine learning algorithm to the training sample, thus resulting in poor generalization ability when evaluated in new (test) data (Bishop, 2006). Most of machine learning methods can be affected by overfitting. Accordingly, several strategies can be followed to minimize this effect. Increasing the size of the training set is a common option but it is not always possible. More usual are the methods based on "regularization." Under this name, there are a wide range of strategies based on adding a penalty term to the learning of the method during the training process, so that it can model a more general function instead of adjusting to the particularities of the training data (Bishop, 2006). The interested reader should be aware that this is not an issue that can be obviated, especially when using relatively complex methods such as ANNs or SVM. Even the ensemble learning methods, which have a natural well-known robustness against overfitting (Witten et al., 2011), can benefit from using regularization techniques (Bühlmann & Hothorn, 2007).

8.3.3.2 Validation Strategy

There exists an intimate relationship between overfitting and the way in which the machine learning models should be validated. As the risk for overfitting exists, evaluating the models using training data would most probably lead to overoptimistic performance results (Bishop, 2006; Witten et al., 2011). For the same reason, the hyperparameters needed for some of the abovementioned methods – including the regularization term – should be chosen based on the results from an independent dataset. Finally, a reliable performance should be derived from a third previously unseen (test) dataset. This would be a classic and robust validation strategy, which would include a training group for model parameter estimation, a validation group for hyperparameter tuning, and a test group for assessing the performance of the final model.

Ideally, there should be a validation group for each freedom degree of the machine learning method used, including hyperparameters and possible previous feature selection stages. However, data scarcity is very common in healthcare problems, and the use of only three groups (training/validation/test) is usually accepted. Several cautions need to be considered when distributing the data among these three groups. First, the more the data in your training set, the better the chances for developing a more accurate model. Second, the data distribution of the validation and test groups should be as similar as possible. This means that if your test group has 50% OSA-positive and 50% OSA-negative subjects, your validation group should have similar proportions. Third, machine learning methods tend to favor the correct classification of the majority classes in classification problems, as well as the range of values more represented in regression problems. Consequently, provided that the classification of all your classes (or the estimation within all range of values) is equally important, it is also advisable that your training data is well-balanced. Finally, commonly used proportions in data distribution include 60–80% for training and 10–20% for each of the validation and test groups. However, in the rare cases in which data availability is not a problem, these proportions can vary if the other advice is considered.

A final consideration is needed regarding data scarcity. As mentioned above, healthcare-related machine learning problems tend to lack data, and OSA diagnosis simplification is not an exception (Gonzalo C Gutiérrez-Tobal et al., 2021c). Accordingly, split data in three independent groups is not often possible. A usual solution is to emulate one of the groups (typically the validation or the test group) using statistical methodologies such as bootstrapping, jackknife, leave-one-out cross-validation, or k-fold cross-validation (Bishop, 2006; Witten et al., 2011).

8.3.3.3 Performance Statistics

Performance assessment in binary problems is based on different combinations of the number of true positive (TP), false negative (FN), true negative (TN), and false positive (FP) subjects or events. In this sense, sensitivity (Se, also known as recall), specificity (Sp), and accuracy (Acc) are important metrics to evaluate the percentage of positive, negative, and total number of subjects/events rightly classified, respectively:

$$\mathrm{Se} = \frac{\mathrm{TP}}{\mathrm{TP}+\mathrm{FN}} * 100 \qquad (8.1)$$

$$\mathrm{Sp} = \frac{\mathrm{TN}}{\mathrm{TN}+\mathrm{FP}} * 100 \qquad (8.2)$$

$$\mathrm{Acc} = \frac{\mathrm{TP}+\mathrm{TN}}{\mathrm{TP}+\mathrm{TN}+\mathrm{FN}+\mathrm{FP}} * 100. \qquad (8.3)$$

Useful statistics are also positive and negative predictive values (PPV, also known as precision, and NPV), which account for the percentage of success when assigning a data point within one class (e.g., positive) or the another (e.g., negative):

$$\mathrm{PPV} = \frac{\mathrm{TP}}{\mathrm{TP}+\mathrm{FP}} * 100 \qquad (8.4)$$

$$\mathrm{NPV} = \frac{\mathrm{TN}}{\mathrm{TN}+\mathrm{FN}} * 100. \qquad (8.5)$$

Moreover, positive and negative likelihood ratios (LR+ and LR−) account for the ratios of the true positive rate to the false positive rate and the false negative rate to the true negative rate, respectively. In the next definitions, Se and Sp are also taken as rates instead of percentages:

$$\mathrm{LR+} = \frac{\mathrm{Se}}{1-\mathrm{Sp}} \qquad (8.6)$$

$$\mathrm{LR-} = \frac{1-\mathrm{Se}}{\mathrm{Sp}}. \qquad (8.7)$$

These metrics, however, are affected by class imbalance to some extent. Therefore, they are often complemented with the receiver-operating characteristics (ROC) analysis (Zweig & Campbell, 1993). ROC is based on a plot representing Se vs. 1-Sp (in unit proportion) computed for a range of possible decision thresholds from the same output, which in the case of binary machine learning could be the posterior probability of belonging to the class of interest. One possible application of this analysis is the estimation of a suitable threshold to act as a trade-off between Se and Sp (Zweig & Campbell, 1993), i.e., the threshold that minimizes biases due to class imbalance. Other possible uses include to measure the overall performance of a model and, in turn, the comparison of the performance of different models. In this sense, the perfect performance would be achieved by a machine learning model that reaches the point of the plot Se = 1 and 1-Se = 0.

To properly quantify the overall performance, however, it is common to estimate the area under the ROC curve (AROC), which may range between 0 and 1, showing AROC = 0.5 the less discriminative power (Zweig & Campbell, 1993).

All the abovementioned metrics can be also used for evaluating the performance of a multi-class classification approach in each of the thresholds used to determine OSA severity categories. In addition, specific statistics can be used to assess the overall performance in the multi-class problem. Cohen's kappa, which can be also used in binary classification, is one of the most helpful as it measures the agreement between the actual and the estimated class by correcting it by the agreement occurred by chance (Witten et al., 2011). Values closer to 1 (or 100%) mean higher agreement, whereas values closer to 0 indicate lower agreement. Acc adapted to the number of classes is another useful metric to evaluate multi-class performance.

As the definition of OSA severity classes, either binary or multiclass, is conducted based on AHI, the corresponding assessing metrics can be also used to evaluate regression approaches provided that the estimated AHI is properly transformed into the OSA-related classes. Moreover, there exist specific analytical tools to evaluate the similarity between the estimated and the actual AHI. One typical example is the intraclass correlation coefficient (ICC) (Chen & Barnhart, 2008), which measures the agreement between continuous variables. Accordingly, values closer to 1 indicate higher degree of agreement, whereas values closer to 0 mean lower degree of agreement. However, contrary to other statistics such as Pearson's correlation, ICC accounts for systematic errors to estimate agreement (Chen & Barnhart, 2008). Finally, a typical and very useful method to graphically assess the agreement in AHI estimations is the Bland-Altman plot (Bland & Altman, 1986). This method shows the difference between the estimated and the actual continuous variable against the mean of the two values (Bland & Altman, 1986). In addition, it provides possible bias for the estimation (the mean of the differences of all data points) and the limits of the agreement (mean ± 1.96*standard deviation of the differences of all data points). These limits are useful to evaluate whether the estimation can be used as a surrogate for the actual continuous variable (Giavarina, 2015).

8.4 Selected Results from the Literature

Table 8.1 displays some results selected from the literature regarding classic machine learning performance in OSA diagnosis simplification. Showing as many approaches as possible has been one important objective when selecting the results to be included in the table. Accordingly, there are studies focused on adults and children, using different overnight signals (alone and combined) and clinical data, and up to eight different machine learning methods. Validation strategies also vary among studies. In addition, these works are divided into the three main approaches explained above: binary classification, multiclass classification, and regression. The metrics included in the table have been chosen as a trade-off between those reported in the studies and those highlighted as important in the previous sections. An interesting point is the range of methods that can be used to evaluate performance in each machine learning approach. Whereas binary classification is limited to very specific statistics, multiclass classification and, specially, regression approaches can be assessed with an increasing number of methods, thus providing a more complete picture of their performance. Unfortunately, not all the studies provided data to show or estimate all the statistics. Moreover, in some of the studies, the machine learning task focuses on the subjects, whereas in others it focuses on the apneic events. However, as we think that the most valuable approach implies to provide a final diagnosis, we only show those results that end up assigning subjects into one OSA class, regardless the specific purpose of the machine learning method.

As observed, very high diagnostic performance can be achieved using classic machine learning methods. Similarly, all the data involved

Table 8.1 Reported diagnostic performance in selected previous studies following binary classification, multiclass classification, and regression approaches

Study	Population (*N*)	Data source	ML method	Validation	AHI (e/h)	Se (%)	Sp (%)	Acc (%)	ICC	Kappa[a]
Binary classification										
Khandoker et al. (2009)	Adults (125)	ECG	SVM	loo-cv + test	10	92.3	93.8	92.9	–	–
Álvarez et al. (2013)	Adults (316)	SpO_2	SVM	Training + validation + test	10	95.2	80.0	84.5	–	–
Garde et al. (2014)	Children (146)	SpO_2 + PRV	LR	loo-cv + -fold cv	5	88.4	83.6	84.9	–	–
Martín-Montero et al. (2021)	Children (1738)	HRV	LDA	Training + test	5	63.8	84.7	82.8	–	–
Multiclass classification										
Gutierrez-Tobal et al. (2016)	Adults (317)	AF	AdaBoost	Training + Bootstrap + test	5	87.1	80.0	86.5	–	0.432
					15	85.9	72.9	81.0		
					30	74.2	90.6	82.5		
Skotko et al. (2017)	Children (102)	Clinical	LLM	Cross-validation	1	75.6	50.9	61.8	–	–
					5	72.2	54.8	57.8		
Deviaene et al. (2019)	Adults (5793)	SpO_2	RF	Training + 10-fold cv + test	5	83.5	88.0	84.3	–	0.547
					15	75.6	95.8	87.0		
					30	77.3	97.7	94.3		
Jiménez-García et al. (2020)	Children (974)	SpO_2 + AF	AdaBoost	Training + bootstrap + test	1	89.2	37.3	79.2	–	0.398
					5	76.0	85.7	82.1		
					10	62.7	97.7	90.3		
Regression										
El-Solh et al. (1999)	Adults (269)	Clinical	MLP	10-fold cv	10	94.9	64.7	–	0.850[b]	–
					15	95.3	60.0	–		
					20	95.5	73.4	–		
Hornero et al. (2017)	Children (4191)	SpO_2	MLP	Training + loo-cv + test	1	84.0	53.2	75.2	0.785	0.348
					5	68.2	87.2	81.7		
					10	68.7	94.1	90.2		
Álvarez et al. (2020)	Adults (239)	SpO_2 + AF	SVM	Training + loo-cv + test	5	95.6	83.3	94.8	0.930	0.610
					15	96.0	72.7	90.6		
					30	93.6	98.0	95.8		

(continued)

Table 8.1 (continued)

Study	Population (*N*)	Data source	ML method	Validation	AHI (e/h)	Se (%)	Sp (%)	Acc (%)	ICC	Kappa[a]
Gutiérrez-Tobal, Álvarez, et al. (2021a)	Adults (5793)	SpO_2	LSBoost	Training + validation + test	5	93.8	56.3	89.2	0.900	0.561
					15	87.0	84.1	85.3		
					30	82.2	96.3	94.6		

ML machine learning, *AHI* apnea-hypopnea index, *Se* sensitivity, *Sp* specificity, *Acc* 2 classes accuracy, *ICC* intraclass correlation coefficient, *ECG* electrocardiogram, *SVM* support vector machines, *loo-cv* leave-one-out cross-validation, SpO_2 oxygen saturation, *PRV* pulse rate variability, *LR* logistic regression, *HRV* heart rate variability, LDA linear discriminant analysis, *AF* airflow, *LLM*, logic learning machine, *RF* random forest, *MLP* multi-layer perceptron

[a]4 class Cohen's kappa

[b]Pearson's correlation

in the studies can reach high statistics. The reader should notice that the results from those works with higher number of participants – and more independent groups in the validation strategy – should be initially considered as more robust. We kindly invite them to examine the original studies to evaluate whether this assumption is true. It is also observed that those studies focused on children achieved lower diagnostic performances. The reader can also check in the literature that this is not an effect due to the non-systematic selection of the studies, but a general tendency. Traditionally, the study of pediatric OSA has gathered much less attention than adult OSA. Consequently, efforts, resources, and data have been scarce, thus resulting in lower knowledge compared to adult OSA. In addition, the AHI rules for establishing pediatric OSA are tighter. All these limitations have favored that there is still a gap between the performances reached in adults and children.

Among the studies, Martin-Montero et al. (Martín-Montero et al., 2021) involve the non-randomized group of the CHAT database along with a private database from the University of Chicago, USA. Similarly, Deviaene et al. and Gutiérrez-Tobal et al. (Deviaene et al., 2019; Gonzalo C. Gutiérrez-Tobal, Álvarez, et al., 2021a) involved the SHHS database. For the sake of simplicity, only results from 5793 subjects (the SHHS1 subgroup) are shown. However, the studies also reported diagnostic results from the follow-up subgroup (SHHS2) with 2647 recordings and, in the case of Gutiérrez-Tobal et al., a high pre-test probability subgroup with 322 recordings from Hospital Universitario Rio Hortega from Valladolid, Spain.

8.5 Discussion and Conclusions

In this chapter, we focused on the most typical classic machine learning approaches involved in OSA diagnosis simplification, thus setting aside deep learning techniques. We have shown specific machine learning methods, the data regularly used with them, as well as large and easily available adult's and children's databases. We have also exposed useful strategies to measure and validate the performance of the machine learning methods, and we have shown a variety of studies in which this performance is high.

One first take-away idea to be highlighted is that there exists a wide range of successful machine learning methods applied to OSA diagnosis simplification. They covered both classification (either binary or multiclass) and regression approaches, the latter being more easily evaluated in depth. Other interesting key point is that many of the data from the PSG (SpO_2, AF, ECG/HRV, etc.) gather information enough to obtain accurate machine learning methods, as reflected by the high diagnostic performances shown in the studies involved in Table 8.1, and in many others referenced within this chapter. This implies that those methods to be evaluated in the future would need to not only justify a possible increase in the performance but also the eventual rise in data requirements and computational costs.

The studies we chose as examples also reflect a lack of homogeneity in the validation strategy. This is an issue that is also present in the scientific literature (Gonzalo C Gutiérrez-Tobal et al., 2021c) and hinders the comparison between the different methods. As mentioned in the past sections, the ideal training/validation/test strategy is greatly influenced by data scarcity. Accordingly, the problem is closely related to the different sample sizes of the studies, which involve a number of subjects ranging from moderate (102) to high (5793). This underlines the need to make available for the scientific community large databases such as CHAT and SHHS.

The previous idea is particularly important in the case of pediatric OSA. The gap between the machine learning performance in adults and children can be partially attributed to the more restrictive diagnosis rules for children. However, large samples such as CHAT can be very useful to increase the knowledge of pediatric OSA and develop more accurate machine learning models.

Finally, despite the high performance shown in several of the studies referenced in this chapter, it is very difficult to find machine learning-based systems implemented in real clinical environments. One possible reason for this issue is that clinicians and healthcare providers and managers perceive these methods as a black box, thus preventing them from completely relying on their predictions (Gonzalo C Gutiérrez-Tobal et al., 2021c). Accordingly, the machine learning designers who expect their work to be finally implemented will need to put extra efforts in explaining the decisions taken by their automatic models (Adadi & Berrada, 2018).

To sum up, traditional machine learning methods have proven to be very useful in the automatic OSA diagnosis simplification. Accordingly, they are still valid options both to develop new proposals and to act as benchmark for future methods.

Acknowledgments This work was supported in part by "Ministerio de Ciencia e Innovación-Agencia Estatal de Investigación" and the "European Regional Development Fund (FEDER)" under projects PID2020-115468RB-I00 and RTC-2017-6516-1, in part by the "European Commission" and "FEDER" under project "Análisis y correlación entre la epigenética y la actividad cerebral para evaluar el riesgo de migraña crónica y episódica en mujeres" ("Cooperation Programme Interreg V-A Spain-Portugal POCTEP 2014–2020"), in part by Sociedad Española de Neumología y Cirugía Torácica (SEPAR) under project 649/2018, in part by Sociedad Española de Sueño (SES) under project "Beca de Investigación SES 2019," and in part by "Centro de Investigación Biomédica en Red en Bioingeniería, Biomateriales y Nanomedicina (CIBER-BBN), Spain," through "Instituto de Salud Carlos III" co-funded with FEDER funds. The work of Fernando Vaquerizo-Villar was supported by a "Ayuda para contratos predoctorales para la Formación de Profesorado Universitario (FPU)" grant from the Ministerio de Educación, Cultura y Deporte (FPU16/02938). The work of Daniel Álvarez was supported by a "Ramón y Cajal" grant (RYC2019-028566-I) from the "Ministerio de Ciencia e Innovación-Agencia Estatal de Investigación" co-funded by the European Social Fund.

References

Acharya, U. R., et al. (2006). Heart rate variability: A review. *Medical and Biological Engineering and Computing, 44*(12), 1031–1051. https://doi.org/10.1007/s11517-006-0119-0

Adadi, A., & Berrada, M. (2018). Peeking inside the black-box: A survey on explainable artificial intelligence (XAI). *IEEE Access* [Preprint]. https://doi.org/10.1109/ACCESS.2018.2870052

Álvarez, D., et al. (2010). Multivariate analysis of blood oxygen saturation recordings in obstructive sleep apnea diagnosis. *IEEE Transactions on Biomedical Engineering, 57*(12), 2816–2824. https://doi.org/10.1109/TBME.2010.2056924

Álvarez, D., et al. (2013). Assessment of feature selection and classification approaches to enhance information from overnight oximetry in the context of sleep apnea diagnosis, *23*(5). https://doi.org/10.1142/S0129065713500202

Álvarez, D., et al. (2020). A machine learning-based test for adult sleep apnoea screening at home using oximetry and airflow. *Scientific Reports* [Preprint]. https://doi.org/10.1038/s41598-020-62223-4

Azarbarzin, A., et al. (2019). The hypoxic burden of sleep apnoea predicts cardiovascular disease-related mortality: The Osteoporotic Fractures in Men Study and the sleep heart health study. *European Heart Journal, 40*(14), 1149–1157. https://doi.org/10.1093/EURHEARTJ/EHY624

Bahammam, A. (2004). Comparison of nasal prong pressure and thermistor measurements for detecting respiratory events during sleep. *Respiration, 71*(4), 385–390. https://doi.org/10.1159/000079644

Barroso-García, V., et al. (2017). Irregularity and variability analysis of airflow recordings to facilitate the diagnosis of paediatric sleep apnoea-hypopnoea syndrome. *Entropy, 19*(9), 447. https://doi.org/10.3390/E19090447

Barroso-García, V. et al. (2021). Wavelet analysis of overnight airflow to detect obstructive sleep apnea in children. *Sensors, 21*(4). https://doi.org/10.3390/s21041491

Berry, R. B., et al. (2012). Rules for scoring respiratory events in sleep: Update of the 2007 AASM manual for the scoring of sleep and associated events. *Journal of Clinical Sleep Medicine, 8*(05), 597–619.

Berry, R. B., et al. (2017). AASM scoring manual updates for 2017 (version 2.4). *Journal of Clinical Sleep Medicine, 13*(05), 665–666.

Bishop, C. M. (2006). *Pattern recognition and machine learning*. Springer.

Bland, J. M., & Altman, D. G. (1986). Statistical methods for assessing agreement between two methods of clinical measurement. *The Lancet, 327*(8476), 307–310. https://doi.org/10.1016/S0140-6736(86)90837-8

Bühlmann, P., & Hothorn, T. (2007). Boosting algorithms: Regularization, prediction and model fitting. *Statistical Science* [Preprint]. https://doi.org/10.1214/07-STS242

Chen, C. C., & Barnhart, H. X. (2008). Comparison of ICC and CCC for assessing agreement for data without and with replications. *Computational Statistics and Data Analysis, 53*(2), 554–564. https://doi.org/10.1016/j.csda.2008.09.026

Chen, L., Zhang, X., & Song, C. (2015). An automatic screening approach for obstructive sleep apnea diagnosis based on single-lead electrocardiogram. *IEEE*

Transactions on Automation Science and Engineering, 12(1). https://doi.org/10.1109/TASE.2014.2345667

Deviaene, M., et al. (2019). Automatic screening of sleep apnea patients based on the SpO 2 signal. *IEEE Journal of Biomedical and Health Informatics* [Preprint]. https://doi.org/10.1109/JBHI.2018.2817368

El-Solh, A. A., et al. (1999). Validity of neural network in sleep apnea. *Sleep, 22*(1). https://doi.org/10.1093/sleep/22.1.105

Flemons, W. W., et al. (1999). Sleep-related breathing disorders in adults: Recommendations for syndrome definition and measurement techniques in clinical research. *Sleep*. https://doi.org/10.1093/sleep/22.5.667

Garde, A., et al. (2014). Development of a screening tool for sleep disordered breathing in children using the phone oximeterTM," *PLoS One, 9*(11). https://doi.org/10.1371/journal.pone.0112959

Ghegan, M. D., et al. (2006). Laboratory versus portable sleep studies: A meta-analysis. *The Laryngoscope, 116*(6), 859–864. https://doi.org/10.1097/01.mlg.0000214866.32050.2e

Giavarina, D. (2015). Understanding bland altman analysis. *Biochemia Medica, 25*(2). https://doi.org/10.11613/BM.2015.015

Gil, E., et al. (2010). "Photoplethysmography pulse rate variability as a surrogate measurement of heart rate variability during non-stationary conditions," *Physiological Measurement, 31*(9). https://doi.org/10.1088/0967-3334/31/9/015

Gutiérrez-Tobal, G. C., et al. (2013). Pattern recognition in airflow recordings to assist in the sleep apnoea-hypopnoea syndrome diagnosis. *Medical and Biological Engineering and Computing, 51*(12), 1367–1380. https://doi.org/10.1007/s11517-013-1109-7

Gutierrez-Tobal, G. C., et al. (2016). Utility of AdaBoost to detect sleep apnea-hypopnea syndrome from single-channel airflow. *IEEE Transactions on Biomedical Engineering, 63*(3), 636–646. https://doi.org/10.1109/TBME.2015.2467188

Gutiérrez-Tobal, G. C., et al. (2019). Evaluation of machine-learning approaches to estimate sleep apnea severity from at-home oximetry recordings. *IEEE Journal of Biomedical and Health Informatics, 23*(2), 882–892. https://doi.org/10.1109/JBHI.2018.2823384

Gutiérrez-Tobal, G. C., Álvarez, D., et al. (2021a). Ensemble-learning regression to estimate sleep apnea severity using at-home oximetry in adults. *Applied Soft Computing, 111*. https://doi.org/10.1016/j.asoc.2021.107827

Gutiérrez-Tobal, G. C., Gomez-Pilar, J., et al. (2021b). Pediatric sleep apnea: The overnight electroencephalogram as a phenotypic biomarker. *Frontiers in Neuroscience*, 1448. https://doi.org/10.3389/FNINS.2021.644697

Gutiérrez-Tobal, G. C., et al. (2021c). Reliability of machine learning to diagnose pediatric obstructive sleep apnea: Systematic review and meta-analysis. *Pediatric Pulmonology* [Preprint].

Guyon, I., & Elisseeff, A. (2003). An introduction to variable and feature selection. *Journal of Machine Learning Research, 3*, 1157–1182. https://doi.org/10.1162/153244303322753616

Hornero, R., et al. (2017). Nocturnal oximetry-based evaluation of habitually snoring children. *American Journal of Respiratory and Critical Care Medicine, 196*(12), 1591–1598. https://doi.org/10.1164/rccm.201705-0930OC

Hosmer, D., & Lemeshow, S. (1989). *Applied logistic regression*. Available at: http://ecsocman.hse.ru/text/19164818/. Accessed: 2 Nov 2021.

Ian, G., Yoshua, B., & Aaron, C. (2016). *Deep learning* - Ian Goodfellow, Yoshua Bengio, Aaron Courville - Google Books. *MIT Press* [Preprint].

Iber, C., et al. (2007). The AASM manual for the scoring of sleep and associated events: Rules, terminology and technical specification. *Journal of Clinical Sleep Medicine* [Preprint].

Jiménez-García, J., et al. (2020). Assessment of airflow and oximetry signals to detect pediatric sleep apnea-hypopnea syndrome using AdaBoost. *Entropy* [Preprint]. https://doi.org/10.3390/E22060670

Karhu, T., et al. (2021). Longer and deeper desaturations are associated with the worsening of mild sleep apnea: The sleep heart health study," *Frontiers in Neuroscience* [Preprint]. https://doi.org/10.3389/fnins.2021.657126

Khandoker, A. H., Palaniswami, M., & Karmakar, C. K. (2009). Support vector machines for automated recognition of obstructive sleep apnea syndrome from ECG recordings. *IEEE Transactions on Information Technology in Biomedicine, 13*(1). https://doi.org/10.1109/TITB.2008.2004495

Korkalainen, H., et al. (2019). Mortality-risk-based apnea–hypopnea index thresholds for diagnostics of obstructive sleep apnea. *Journal of Sleep Research* [Preprint]. https://doi.org/10.1111/jsr.12855

Kuncheva, L. I. (2014). *Combining pattern classifiers: Methods and algorithms*. John Wiley & Sons.

Lázaro, J., et al. (2014). Pulse rate variability analysis for discrimination of sleep-apnea-related decreases in the amplitude fluctuations of pulse photoplethysmographic signal in children. *IEEE Journal of Biomedical and Health Informatics* [Preprint]. https://doi.org/10.1109/JBHI.2013.2267096

Lin, Y. Y., et al. (2017). Sleep apnea detection based on thoracic and abdominal movement signals of wearable piezoelectric bands. *IEEE Journal of Biomedical and Health Informatics, 21*(6). https://doi.org/10.1109/JBHI.2016.2636778

Marcos, J. V., et al. (2008) Utility of multilayer perceptron neural network classifiers in the diagnosis of the obstructive sleep apnoea syndrome from nocturnal oximetry. *Computer Methods and Programs in Biomedicine, 92*(1). https://doi.org/10.1016/j.cmpb.2008.05.006

Marcos, J. V., et al. (2009). Assessment of four statistical pattern recognition techniques to assist in obstructive sleep apnoea diagnosis from nocturnal oximetry. *Medical Engineering & Physics, 31*(8), 971–978. https://doi.org/10.1016/J.MEDENGPHY.2009.05.010

Marcos, J. V., et al. (2012). Automated prediction of the apnea-hypopnea index from nocturnal oximetry recordings. *IEEE Transactions on Biomedical Engineering, 59*(1), 141–149. https://doi.org/10.1109/TBME.2011.2167971

Marcus, C. L., et al. (2013). A randomized trial of adenotonsillectomy for childhood sleep apnea. *New England Journal of Medicine* [Preprint]. https://doi.org/10.1056/nejmoa1215881

Martín-Montero, A., et al. (2021). Heart rate variability spectrum characteristics in children with sleep apnea. *Pediatric Research, 89*(7), 1771. https://doi.org/10.1038/S41390-020-01138-2

Mendonça, F., et al. (2019). A review of obstructive sleep apnea detection approaches. *IEEE Journal of Biomedical and Health Informatics, 23*(2), 825–837. https://doi.org/10.1109/JBHI.2018.2823265

Morillo, D. S., & Gross, N. (2013). Probabilistic neural network approach for the detection of SAHS from overnight pulse oximetry. *Medical and Biological Engineering and Computing, 51*(3), 305–315. https://doi.org/10.1007/s11517-012-0995-4

Newman, A. B., et al. (2001). Relation of sleep-disordered breathing to cardiovascular disease risk factors: The sleep heart health study. *American Journal of Epidemiology, 154*(1). https://doi.org/10.1093/aje/154.1.50

Penzel, T., et al. (2002). Systematic comparison of different algorithms for apnoea detection based on electrocardiogram recordings. *Medical & Biological Engineering & Computing, 40*(4), 402–407. https://doi.org/10.1007/BF02345072

Penzel, T., et al. (2003). Comparison of detrended fluctuation analysis and spectral analysis for heart rate variability in sleep and sleep apnea. *IEEE Transactions on Biomedical Engineering, 50*(10), 1143–1151. https://doi.org/10.1109/TBME.2003.817636

Penzel, T., Schöbel, C., & Fietze, I. (2015). Revise respiratory event criteria or revise severity thresholds for sleep apnea definition? *Journal of Clinical Sleep Medicine* [Preprint]. https://doi.org/10.5664/jcsm.5262

Quan, S. F., et al. (1997). The sleep heart health study: Design, rationale, and methods. *Sleep, 20*(12), 1077–1085.

Riedl, M., et al. (2014). Cardio-respiratory coordination increases during sleep apnea. *PLoS One, 9*(4). https://doi.org/10.1371/journal.pone.0093866

Rolón, R. E., et al. (2017). Discriminative methods based on sparse representations of pulse oximetry signals for sleep apnea–hypopnea detection. *Biomedical Signal Processing and Control* [Preprint]. https://doi.org/10.1016/j.bspc.2016.12.013

Rolon, R. E., et al. (2020). Automatic scoring of apnea and hypopnea events using blood oxygen saturation signals. *Biomedical Signal Processing and Control* [Preprint]. https://doi.org/10.1016/j.bspc.2020.102062

Skotko, B. G., et al. (2017). A predictive model for obstructive sleep apnea and Down syndrome. *American Journal of Medical Genetics, Part A* [Preprint]. https://doi.org/10.1002/ajmg.a.38137

Solà-Soler, J., et al. (2012). Multiclass classification of subjects with sleep apnoea-hypopnoea syndrome through snoring analysis. *Medical Engineering and Physics, 34*(9). https://doi.org/10.1016/j.medengphy.2011.12.008

Tan, H. L., et al. (2014). Overnight polysomnography versus respiratory polygraphy in the diagnosis of pediatric obstructive sleep apnea. *Sleep* [Preprint]. https://doi.org/10.5665/sleep.3392

Tan, H. L. et al. (2017). When and why to treat the child who snores? *Pediatric Pulmonology, 52*(3), 399–412. https://doi.org/10.1002/ppul.23658

Uddin, M., et al. (2018). Classification methods to detect sleep apnea in adults based on respiratory and oximetry signals: A systematic review. *iopscience.iop.org* [Preprint]. https://doi.org/10.1088/1361-6579/aaafb8

Vaquerizo-Villar, F., et al. (2021). A convolutional neural network architecture to enhance oximetry ability to diagnose pediatric obstructive sleep apnea. *IEEE Journal of Biomedical and Health Informatics* [Preprint]. https://doi.org/10.1109/JBHI.2020.3048901

Witten, I. H., Frank, E., & Hall, M. A. (2011). *Data mining: Practical machine learning tools and techniques*. Morgan Kaufmann.

Wu, D., et al. (2017). A simple diagnostic scale based on the analysis and screening of clinical parameters in paediatric obstructive sleep apnoea hypopnea syndrome. *Journal of Laryngology and Otology* [Preprint]. https://doi.org/10.1017/S0022215117000238

Xu, Z., et al. (2019). Cloud algorithm-driven oximetry-based diagnosis of obstructive sleep apnoea in symptomatic habitually snoring children. *The European Respiratory Journal* [Preprint]. https://doi.org/10.1183/13993003.01788-2018

Zweig, M. H., & Campbell, G. (1993) Receiver-operating characteristic (ROC) plots: A fundamental evaluation tool in clinical medicine. *Clinical Chemistry*. https://doi.org/10.1093/clinchem/39.4.561

Home Sleep Testing of Sleep Apnea

9

Martin Glos and Dora Triché

Abstract

Measurement methods with graded complexity for use in the lab as well as for home sleep testing (HST) are available for the diagnosis of sleep apnea, and there are different classification systems in existence. Simplified HST measurements, which record fewer parameters than traditional four- to six-channel devices, can indicate sleep apnea and can be used as screening tool in high-prevalence patient groups. Peripheral arterial tonometry (PAT) is a technique which can be suitable for the diagnosis of sleep apnea in certain cases. Different measurement methods are used, which has an influence on the significance of the results. New minimal-contact and non-contact technologies of recording and analysis of surrogate parameters are under development. If they are validated by clinical studies, it will be possible to detect sleep apnea in need of treatment more effectively. In addition, this could become a solution to monitor the effectiveness of such treatment.

Keywords

Obstructive sleep apnea · Sleep-related breathing disorders · Simplified sleep apnea diagnostics · Home sleep apnea testing · SCOPER criteria

9.1 Introduction

The diagnostic procedure for sleep apnea depends on the occurrence of symptoms, comorbidities, and the medical history of the patient. The reference method is the lab-based polysomnography (PSG), which allows comprehensive diagnostics for all kind of sleep-related breathing disorders (SRBD) by recording and characterization of different types of breathing pattern, cardiac activity, sleep structure, arousal, and behavior under controlled conditions. For home sleep apnea testing (HSAT), portable four- to six-channel systems and alternatively peripheral arterial tonometry (PAT) are used as out-of-center (OOC) procedures allowing diagnosis of obstructive sleep apnea (OSA) in a number of subjects (Stuck et al., 2020).

In addition, HSAT devices with a reduced number of channels, typically one to three, are available for initial screening for sleep apnea.

M. Glos (✉)
Interdisciplinary Center for Sleep Medicine, Charité – Universitätsmedizin Berlin, Berlin, Germany
e-mail: martin.glos@charite.de

D. Triché
Department of Respiratory Medicine, Allergology, Sleep Medicine, Paracelsus Medical University Nuremberg, Nuremberg General Hospital, Nuremberg, Germany

T. Penzel, R. Hornero (eds.), *Advances in the Diagnosis and Treatment of Sleep Apnea*, Advances in Experimental Medicine and Biology 1384, https://doi.org/10.1007/978-3-031-06413-5_9

Recently, smart wearable devices with new, minimal-contact, and non-contact measurement techniques have been developed, and they are currently under investigation for clinical use.

This chapter will give an overview of HSAT techniques, focusing on technologies using a limited number of channels, non-obtrusive sensors, and promising forward-looking technologies. The potential and limitations of diagnostic tools will be explained, and their diagnostic value will be outlined.

9.2 Classification of Methods for the Monitoring of Sleep Apnea at Home and in the Lab

An initial classification of methods of sleep measurement was introduced in 2003 by the American Association of Sleep Medicine (AASM) (Chesson et al., 2003), suggesting four categories ("levels") of recording systems:

- Type 1: supervised PSG, performed in sleep lab
- Type 2: non-supervised PSG, performed at home
- Type 3: portable sleep apnea monitoring with at least four channels (two respiratory variables, oxygen saturation, and pulse or heart rate)
- Type 4: the continuous measurement of one to two signals during sleep

The current AASM manual, version 2.6 (Berry et al., 2020), defines two different kinds of HSAT technologies for sleep apnea:

- Devices utilizing flow and/or effort measurement
- Devices utilizing PAT

This takes into account more recent device developments and new sensor technologies. In this context, the introduction of a new system of categorization, based on the so-called "SCOPER" criteria, has come to prominence. This system does not classify devices according to the number of recorded channels, but instead makes it possible to qualitatively graduate which of the functions affected by sleep apnea are recorded (Collop et al., 2011). The following functions are assessed:

- Of the sleep (Sleep – **S**)
- Of the cardiovascular system (Cardiovascular – **C**)
- Of the oxygen saturation (Oximetry – **O**)
- Of the body position (Position – **P**)
- Of the respiratory effort (Effort – **E**)
- Of the respiratory flow (Respiratory – **R**)

The complete diagnostic procedure for sleep apnea is country-specific. The AASM published a clinical practice guideline for diagnostic testing for OSA in adults (Kapur et al., 2017), whereas the German Sleep Society (DGSM) issued a S3 guideline initially in 2017 (Mayer et al., 2017), followed by a "Partial Update of the German S3 Guidelines on Sleep-Related Breathing Disorders in Adults" in 2020 (Stuck et al., 2020). According to these guidelines, polysomnography (PSG) is still the reference technique for diagnosing OSA.

The procedure, scoring, and documentation of the PSG technique are described in a manual of the AASM (Berry et al., 2020).

For the OOC diagnosis of OSA, the mentioned HSAT technologies are used, which are sufficient for the final diagnosis of OSA in cases of high pretest probability, i.e., with signs and symptoms that indicate an increased risk of moderate to severe OSA. However, this is true only for uncomplicated cases and does not apply if comorbid pulmonary, psychiatric, or neurological/neuromuscular diseases are present or if other forms of SRBD or other sleep disorders are suspected (Stuck et al., 2020; Kapur et al., 2017).

HSAT is usually performed unsupervised in the home environment with no objective recording of sleep structure. This may lead to an increased amount of measurement artifacts compared to PSG and subsequently to lower accuracy in the assessment of severity of nocturnal breathing disorders (Escourrou et al., 2015).

In addition, there are reduced systems recording typically one to three channels used for ambulatory diagnostics in OSA. At present, these simplified devices can only be used as an initial screening tool, e.g., to determine the pretest probability. They are currently not recommended for the final diagnosis of OSA (Stuck et al., 2020; Kapur et al., 2017).

9.3 Home Sleep Apnea Testing (HSAT) with Type 3 Portable Monitors

9.3.1 HSAT Utilizing Flow and/or Effort Parameters

The most established HSAT method is portable monitoring by using four to six signals. As a minimum, this must record airflow, chest and abdominal respiratory inductance plethysmography (RIP), and oximetry (Berry et al., 2020; Kapur et al., 2017). The standard HSAT procedures for airflow measurement are the use of either a nasal pressure transducer (nasal cannula) or an oronasal thermal airflow sensor. Nasal pressure sensors allow semi-quantitative measurements of airflow related to tidal volume and therefore are superior for hypopnea detection, although measuring the changes of temperature in front of the nose and mouth in inspiration and expiration by means of an oronasal thermal airflow sensor allows valid detection of respiratory events, especially hypopneas, in cases of mouth breathing (Berry et al., 2012). If HSAT is used for the monitoring of positive airway pressure therapy, the flow is derived from the device flow. To distinguish between obstructive and central apneas and hypopneas, movements of the thorax and abdomen are recorded. The change of circumference should preferably be recorded using RIP technology; alternative sensors are listed in the AASM manual (Berry et al., 2020). Changes in an electrical wire arranged as a coil lead to changes in a magnetic field, which is converted into differences of circumference. This measurement method is more exact but also more vulnerable and more expensive, as a strong pull on the wire leads to a defect. The snoring noise is either calculated from the flow curve or recorded by a snoring microphone placed at the neck. For cardiac rhythm monitoring, the standard method for HSAT is pulse recording performed by pulse oximetry. The more exact possibility of heart rate detection is to derive a single-channel ECG (Caples et al., 2007). The subject's body position is determined by using a 3D accelerometer sensor. Figure 9.1 illustrates a typical HSAT recording snapshot of 5 minutes duration during occurrence of episodes with obstructive apneas Fig. 1.

Some HSAT systems provide in addition information about the sleep-wake state by evaluating data from an inbuilt accelerometer sensor. Depending on the subject of investigation, it is possible to integrate EEG channels as well or to measure leg movements using EMG electrodes (Collop et al., 2011).

9.3.2 HSAT Utilizing Peripheral Arterial Tonometry (PAT)

As stated, PAT has been accepted as an additional OOC measurement method for the diagnosis of OSA and is included in the AASM manual (Berry et al., 2020). The measuring principle of PAT is based on recording the peripheral arterial vascular tone and thus the vascular volume. Activation of the sympathetic tone caused by apneas, hypopneas, respiratory effort-related arousals (RERAs), or other types of events accompanied by central nervous activation/arousal leads to vasoconstriction, controlled by α-adrenergic receptors in the smooth vascular muscles. A finger cuff tracks changes in light transmission through the fingertip, this factor being related to changes in the vessel diameter due to sympathetic and vagal modulation. In addition, the signal varies rhythmically with the frequency of heartbeats due to the pulsatile character of the vessel volume. For signal amplification and enhancement of signal-to-noise ratio, a subdiastolic pressure is applied in the cuff while recording (Penzel et al., 2004). A typical measuring device worn around the wrist records the following signals: PAT signal, pulse, oxygen saturation, and actimetry. As an optional addition, the body position and snoring sounds are recorded by an external sensor worn on the chest. An auto-

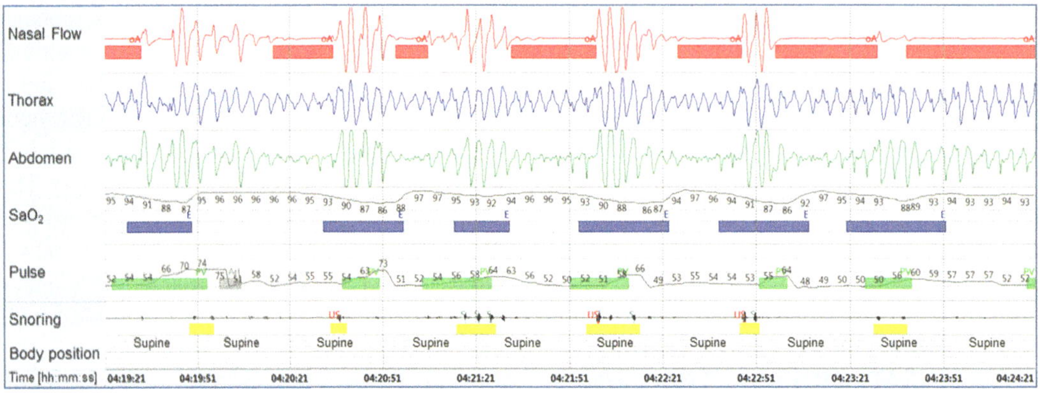

Fig. 9.1 Five-minute sample of a HSAT type 3 recording with channels of airflow (nasal cannula), thoracic and abdominal respiratory effort, oxygen saturation, pulse, snoring, and body position. The occurrence of obstructive apneas led to cessation of airflow (red marks), oxygen desaturations (purple marks), cyclic variation of heart rate (CVHR, green marks), as well as intermittent snoring episodes (yellow marks)

matic algorithm performs a preliminary analysis of the recorded data; a manual evaluation is possible and recommended. An advantage of this method compared to HSAT utilizing nasal flow and effort signals is the possibility of detecting arousals by changing the peripheral arterial vascular tone by activating the sympathetic nervous system. A differentiation between wakefulness and sleep and even individual sleep stages seems to be possible in this way, so that the method allows the calculation of an AHI based on total sleep time (TST) (Schöbel et al., 2018). This was demonstrated in a meta-analysis of Yalamanchali et al. (2013) which showed a very good correlation to PSG with regard to the sleep stage analysis. When it comes to the positive predictive value for OSA, this method is likewise considered to be well suited (Collop et al., 2011). In Germany, PAT technology became part of the 2020 update of the S3 guidelines on sleep-related breathing disorders in adults (Stuck et al., 2020).

9.4 Motivation and Indication for Use of Simplified HSAT Devices for Sleep Apnea

The clinical OOC and in-lab resources available for the abovementioned diagnostic procedures for sleep apnea are limited in terms of personnel, infrastructure, and technical equipment. The consequences are, on the one hand, long waiting lists for the diagnosis and for the start of appropriate therapy and, on the other, the difficulty of reducing the high number of previously undiagnosed affected persons with the existing structures. The ultimate aim in the future will be to identify more quickly those patients affected who require therapy. So broader screening and more effective therapy management are called for. This is particularly true of patients with certain underlying diseases that have a high prevalence of sleep apnea, especially for patients with cardiovascular risk diseases (arterial hypertension, pulmonary hypertension, CHD, stroke, arrhythmias, chronic heart failure) without the presence of all typical symptoms (Oldenburg et al., 2015).

From a clinical perspective, this puts a priority on the need to assess the value of the available simplified systems for sleep apnea and, at the same time, to test new digital technologies and low-contact/no-contact measurement methods for their clinical diagnostic functionality and suitability (Penzel et al., 2020).

Furthermore, these developments can also contribute to an increase in comfort by having the sensors on the body, a better sleep quality, and the improved accuracy of measurement systems. This also opens up opportunities of improving therapy monitoring – and thus the therapy outcome – of patients with sleep apnea in conjunction with new telemedical technologies.

9.5 Measurement Techniques Used for Simplified HSAT

9.5.1 Oximetry and Pulse Wave Analysis

Apnea and hypopnea events as markers of disordered breathing typically cause oxygen desaturations. Event-associated oxygen desaturation of ≥3% or ≥ 4% and respiratory flow reduction ≥30% for ≥10 seconds are criteria for detecting hypopnea events (Berry et al., 2020). The severity of oxygen desaturation depends on the duration of the event and the baseline oxygen saturation level. The exact drop in oxygen saturation is further determined by the oxygen binding curve and the placement of the sensor. Regardless of the complexity of the signal gathered by signal processing of red and infrared photoplethysmographic (PPG) signals from the small vessels in the fingertip, this itself is very intuitive and is therefore used without further measurement parameters as a diagnosis tool for sleep apnea. Typical scoring criteria of an ambulatory measurement are the number of oxygen desaturations (3% or 4%) per hour of recording time (Oxygen Desaturation Index – ODI); the minimum, mean, as well as maximum oxygen saturation during the recording time; and the percentage of the recording time with values below certain thresholds of oxygen saturation, e.g., ≤88% (Berry et al., 2020). Studies with home sleep testing investigating the ODI and the AHI in OSA patients have demonstrated a high measure of agreement (Dawson et al., 2015) and have shown a similar night-to-night variability (Fietze et al., 2004), although it should be noted that there are some sleep apnea patients who do not exhibit nocturnal desaturations to a large extent and may go undetected when this procedure is used alone (Fietze et al., 2004). The temporal dynamics of oxygen desaturation rates appear to be related to the degree of apnea-associated increases in blood pressure, as reported in a study by Wang et al. (Wang et al., 2020). Thus, a new marker of cardiovascular impairment due to sleep apnea may be available in the future. The degree of the tendency to fall asleep during the day also seems to correlate with apnea-associated oxygen desaturations during sleep (Zhang et al., 2020).

Devices from various manufacturers are available, which usually include a wrist-worn recording unit with display function in addition to the finger clip sensor as seen in Fig. 9.2.

In addition to oxygen saturation, the pulse rate is recorded typically. Some devices perform elaborated analysis of the PPG-based pulse wave morphology, for example, to estimate measures of cardiovascular risk (Cardiac Risk Indicator – CRI; Fig. 9.2, left panel) (Sommermeyer et al., 2016) or, in combination with a single-channel ECG, to evaluate pulse transit time (PTT). Considering the individual distance from the heart to the finger, one can subsequently calculate the pulse wave velocity (PWV) from beat-to-beat PTT values (Pielmus et al., 2021). PWV values are a marker of the vessel's stiffness, but temporal changes also reflect the blood pressure dynamics (Pielmus et al., 2021). By calibration considering intra-individual vessel properties, one gets an estimate of nocturnal BP dynamics during sleep, which allows, e.g., recognition of BP increases due to sleep apnea (Gehring et al., 2018).

The recorded data can be downloaded to a computer and visualized and evaluated offline. Some devices also enable data exchange with apps on mobile devices via a Bluetooth interface.

More recent developments from the consumer sector integrate LED-based photoplethysmography into the bottom of smartwatches or as shown in the example in Fig. 9.3 into finger ring-like devices. It enables the measurement of heart rate as well as of nocturnal oxygen saturation and subsequent estimation of sleep apnea severity "for everyone" by means of coupled apps, independent of a medical indication.

For the evaluation of these measurements with regard to SRBD, in addition to observing the abovementioned boundary conditions for the behavior of oxygen saturation, the fact remains that these devices have so far generally not been approved as medical devices and thus cannot replace medical diagnostics. In addition, data security issues for the cloud-based data storage

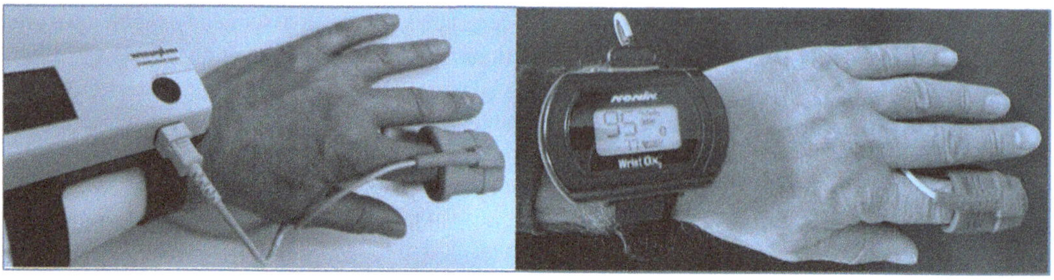

Fig. 9.2 Example of simplified HSAT devices (left panel, *SOMNOcheck micro CARDIO*, Weinmann/Löwenstein Medical SE & Co. KG; right panel, *WristOx$_2$ Model 3150*, Nonin Medical, Inc.) using photoplethysmographic signals from the finger for analysis of SpO_2 and pulse rate. In addition, the analysis of pulse wave morphology and concomitant cardiac risk is performed by the *SOMNOcheck micro CARDIO* (Sommermeyer et al., 2016)

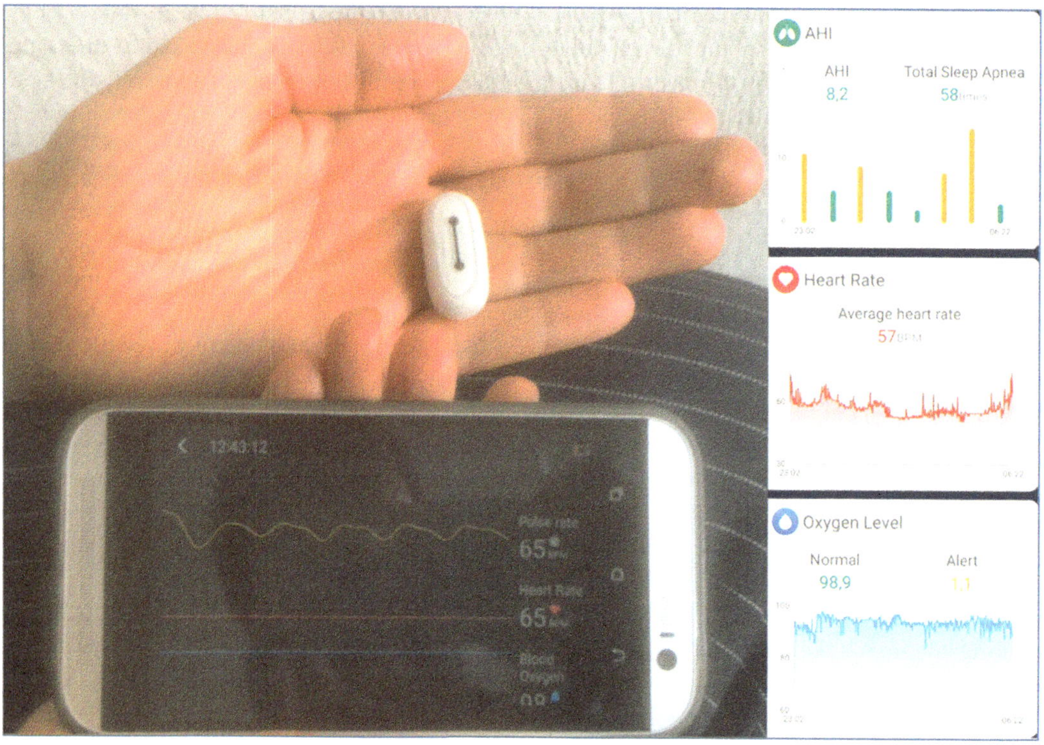

Fig. 9.3 Example of a mobile device out of the consumer sector ("wearable") for the estimation of sleep apnea severity by analysis of heart rate and oxygen saturation using finger photoplethysmography. Left panel shows the finger ring device which is connected to an app on the mobile phone for online data display and data transfer. Right panel shows results for sleep apnea severity based on cloud data analysis

and data processing remain unclear in many cases. Nevertheless, these and similar devices ("wearables") are becoming increasingly widespread. Some manufacturers are also conducting clinical-scientific studies and are striving to obtain approval for these measurements and evaluations for medical diagnostics in the future.

9.5.2 Nasal Flow

Changes of respiratory flow are key features of patterns of disturbed breathing during sleep in SRBD. These are characterized as either complete cessation (apneas) or partial reduction (hypopneas, RERAs) of flow for at least 10 seconds (Berry et al., 2020). For home testing as well as for in-lab recordings, thermistor/thermocouple and nasal pressure sensors (nasal cannula) are established qualitative measurement methods. They are a proven approach as an alternative to the gold standard technique – the pneumotachograph. In comparison to the thermistor/thermocouple, the nasal pressure sensors have the advantage of being able to detect respiratory flow reductions more sensitively and, in some cases, to a semi-quantitative degree, which is particularly important for hypopnea scoring. Reliable detection of events, especially apneas, with the nasal pressure sensor may however be limited in cases of mouth breathing.

In the field of simplified home sleep apnea testing using typically one to three channels, a number of devices are available which use nasal pressure signals (Crowley et al., 2013; Keshavarzi et al., 2018). Those may be supplemented by finger oximetry/plethysmography and, if necessary, by measurement of respiratory effort. By applying signal processing, the parameters respiratory flow and snoring can be determined from the "raw" signals of the nasal pressure sensor, while the additional parameters oxygen saturation and pulse rate are obtained from finger oximetry/plethysmography. Thus, apneas and hypopneas can be detected by these devices on an out-of-center basis. Furthermore, with additional respiratory effort measurement, these can in principle also be characterized as obstructive, mixed, or central apneas as well as patterns of periodic breathing, although reliability is limited due to the out-of-center setting. Another limitation is that the arousal criterion for hypopnea detection cannot be applied due to the absence of EEG recording.

9.5.3 ECG Measures

In SRBD, changes in autonomic tone occur due to sympathetic activation during apnea and hypopnea events. As a result, there is typically a temporal cyclic alternation of bradycardiac and tachycardiac episodes in the rhythm of each pattern of disordered breathing, which is called cyclic variation of heart rate (CVHR) (Penzel et al., 2016). CVHR is obtained from the ECG by beat-to-beat detection of the QRS complex and subsequently calculation of the resulting time series of heart rate.

Because CVHR is so characteristic of apneas and hypopneas, it is a suitable tool to elicit evidence for the presence of SRBD. In addition, the ECG exhibits respiration-dependent amplitude fluctuations of the QRS complex. This can also be used for the detection of SRBD and is called ECG-derived respiration (EDR) (De Chazal et al., 2009).

The combination of CVHR and EDR, as well as spectral analysis of heart rate variability (HRV), has been implemented in commercial analysis software of 24-hour ECG recorders ("Holter monitor"). It can give cardiologists an indication of the need for further investigations for SRBD. Figure 9.4 shows a sample of CVHR, EDR, and HRV from a Holter recording during the occurrence of apneas.

Furthermore, using mostly pulse rate instead of heart rate, the methods of analyzing CVHR and HRV along with other recorded information have already been implemented in wearable/consumer devices in order to provide the user with hints of the presence of SRBD (Fontana et al., 2019).

9.5.4 Transthoracic Impedance (TTI)

Measurement of TTI is a measurement technique mainly used in cardiology to obtain indications for SRBD. The measurement principle of TTI is that respiratory-associated volume changes of the thorax are recorded by measurement of electrical impedance. The SRBD pattern modulates the TTI signal accordingly, and amplitude-based detection algorithms and timing criteria are used to detect events. According to the pathophysiology, the sensitivity for central events is higher than for obstructive ones. Essentially, this method has been integrated into two classes of devices, a Holter ECG system (Mueller et al., 2006) and cardiac implantable electronic devices (CIED), such as pacemakers, cardioverter/defibrillators,

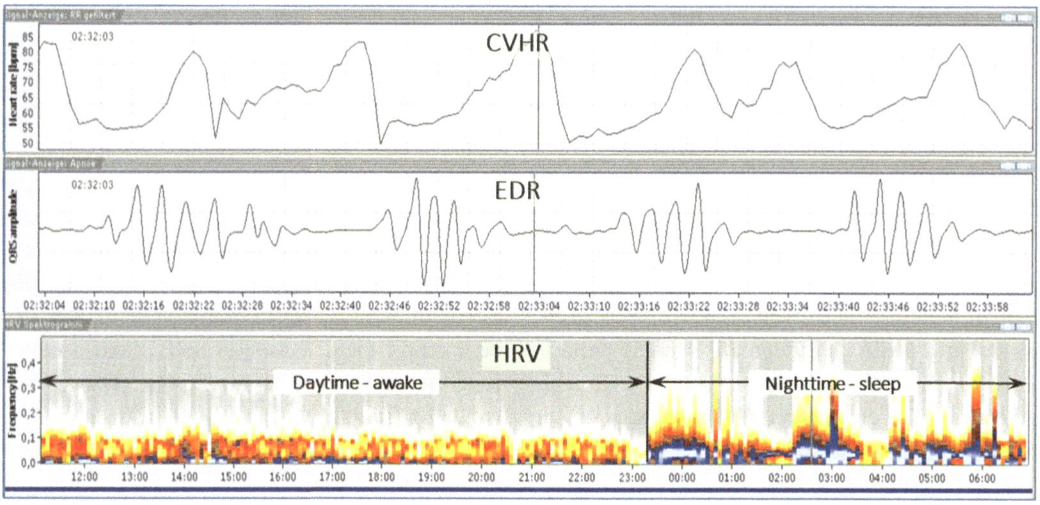

Fig. 9.4 Analysis for sleep apnea out of a 24-hour ECG recording ("Holter monitor"): 3-minute sample with cyclic variation of heart rate (CVHR, top) and electrocardiographic-derived respiration (EDR, middle) during apneas, as well as 24-hour time-frequency analysis of heart rate variability (spectrogram, bottom) showing increased low frequency (LF; 0.04 .. 015 Hz) indicating sympathetic activity at night due to the occurrence of sleep apnea. (Modified from (Glos & Schöbel, 2021))

and CRT defibrillators (Chen et al., 2019; Defaye et al., 2019; Dias et al., 2017). The advantage of both applications is that no additional device or electrodes are needed to perform the TTI measurement. Of course, the detected events usually cannot be characterized as apneas or hypopneas, nor can a distinction be made between obstructive, mixed, and central genesis. Nevertheless, a number of studies comparing this method with PSG have provided evidence that it is possible to reliably screen patients with cardiovascular disease for the risk of SRBD (Mueller et al., 2006; Chen et al., 2019; Defaye et al., 2019). Such patients can then be referred for further guideline-based SRBD diagnosis.

9.6 Surrogates of Respiration Gained by Minimal-Contact and Contactless Techniques

9.6.1 Sound Analyses

The recording and analysis of respiratory sounds (breathing, snoring, choking, gasping) can be used to estimate occurrence of SRBD as well. Measurement is performed in a low-contact fashion either by using an appropriate sensor applied to the skin or by using non-contact techniques such as microphones built into external devices.

An innovative approach is the combination of microphone and pressure transducer in one sensor. Use of such a combined sensor applied on the suprasternal notch makes it possible to calculate the respiratory flow, the respiratory effort, and the snoring by means of signal analysis (Fig. 9.5). Studies in comparison with PSG (partly also with esophageal pressure measurement) have shown that with high sensitivity and specificity, apneas and hypopneas can be detected and obstructive and central events can also be differentiated (Glos et al., 2019; Sabil et al., 2019).

In addition, there are numerous developments from the consumer sector which either use sound analyses of built-in mobile phone microphones or else work as stand-alone devices in order to obtain hints for the occurrence of SRBD. Scientifically substantiated clinical studies with these devices/smartphone apps have indeed been conducted and published (Narayan et al., 2019; Tiron et al., 2020), but they are gen-

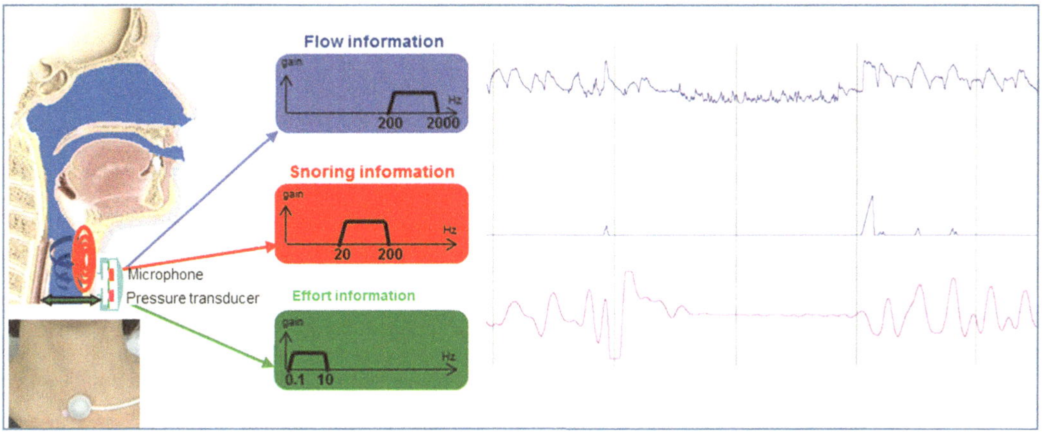

Fig. 9.5 Illustration showing a combination sensor in which respiratory flow and snoring are calculated from the microphone signal and respiratory effort is derived from the pressure sensor signal. The application of the sensor at a suprasternal position is shown on the bottom left. (Modified from (Glos & Schöbel, 2021))

erally not yet certified as medical devices for diagnostic purposes.

9.6.2 Movement Analyses

The movements of the thorax and abdomen caused by respiratory effort, the mechanical vibrations during snoring, and the mechanical cardiac action can be used as additional surrogate parameters for the detection of SRBD. As with sound analyses, data can also be acquired in a minimal-contact or non-contact fashion. With regard to this, these methods are particularly suitable for long-term recording, e.g., in connection with the monitoring of therapeutic outcome. A technological approach is the detection of micro-gravitations via pressure-sensitive/charge-sensitive sensor mattresses/strips (Tenhunen et al., 2013). Since heart movements are simultaneously present with respiratory mechanics (ballistocardiography), respiratory activity (Tenhunen et al., 2013), snoring (Perez-Macias et al., 2018), and heart rate (Paalasmaa et al., 2015) can also be recorded and analyzed using advanced signal processing algorithms. A number of such applications, usually coupled with dedicated smartphone apps on wearable devices, are already available commercially. A number of clinical studies suggest the suitability in principle of these methods for SRBD screening. However, further validation is needed to establish this method as a clinical tool (Fino & Mazzetti, 2019).

Additional approaches to screen for SRBD using movement analysis are technologies that either measure the reflection of high-frequency electrical impulses (radar) (Weinreich et al., 2014), of auditory waves (sonar) (Tiron et al., 2020), or of infrared light reflections (3D depth camera) (Coronel et al., 2019; Veauthier et al., 2019). The advantage of all these applications is that they operate "from the bedside table," completely without contact. Studies in comparison with PSG concluded the suitability of these methods for the calculation of pretest probability of SRBD, e.g., in patients with hypertension or heart failure (Crinion et al., 2020; Savage et al., 2016). As with other techniques analyzing surrogate parameters, approval procedures for authorizing this application as a medical device have not yet been completed.

9.7 Conclusion

Well-researched methods of examination for the diagnosis of sleep-related breathing disorders, such as four- to six-channel HSAT devices, have been established for many years. New tools are being developed and are increasingly coming to be validated. Some methods that are currently only used as wearables will be approved as medical devices in the future. All this will lead to an enrichment of diagnostic capabilities and an increase in resources to reduce the number of undiagnosed patients, as well as contributing to improvement in the treatment management of sleep apnea.

Literature

Stuck, B. A., Arzt, M., Fietze, I., et al. (2020). Partial update of the German S3 guideline sleep-related breathing disorders in adults. *Somnologie, 24*, 176–208.

Chesson, A. L., Berry, R. B., & Pack, A. (2003). Practice parameters for the use of portable monitoring devices in the investigation of suspected obstructive sleep apnea in adults. *Sleep, 26*, 907–913.

Berry, R. B., Quan, S. F., Abreu, A. R., et al. (2020). *The AASM manual for the scoring of sleep and associated events: Rules, terminology, and technical specification. Version 2.6*. American Academy of Sleep Medicine.

Collop, N. A., Tracy, S. L., Kapur, V., et al. (2011). Obstructive sleep apnea devices for out-of-center (OOC) testing: Technology evaluation. *Journal of Clinical Sleep Medicine, 7*, 531–548.

Kapur, V. K., Auckley, D. H., Chowdhuri, S., et al. (2017). Clinical practice guideline for diagnostic testing for adult obstructive sleep apnea: An American Academy of Sleep Medicine Clinical Practice Guideline. *Journal of Clinical Sleep Medicine, 13*, 479–504.

Mayer, G., Arzt, M., Braumann, B., et al. (2017). German S3 guideline nonrestorative sleep/sleep disorders, chapter "Sleep-Related Breathing Disorders in Adults," short version. *Somnologie, 21*, 290–301.

Escourrou, P., Grote, L., Penzel, T., et al. (2015). The diagnostic method has a strong influence on classification of obstructive sleep apnea. *Journal of Sleep Research, 24*, 730–738.

Berry, R. B., Budhiraja, R., Gottlieb, D. J., et al. (2012). Rules for scoring respiratory events in sleep: update of the 2007 AASM Manual for the Scoring of Sleep and Associated Events. Deliberations of the Sleep Apnea Definitions Task Force of the American Academy of Sleep Medicine. *Journal of Clinical Sleep Medicine, 8*, 597–619.

Caples, S. M., Rosen, C. L., Shen, W. K., et al. (2007). The scoring of cardiac events during sleep. *Journal of Clinical Sleep Medicine, 3*, 147–154.

Penzel, T., Kesper, K., Pinnow, I., et al. (2004). Peripheral arterial tonometry, oximetry and actigraphy for ambulatory recording of sleep apnea. *Physiological Measurement, 25*, 1025–1036.

Schöbel, C., Knorre, S., Glos, M., et al. (2018). Improved follow-up by peripheral arterial tonometry in CPAP-treated patients with obstructive sleep apnea and persistent excessive daytime sleepiness. *Sleep & Breathing, 22*, 1153–1160.

Yalamanchali, S., Farajian, V., Hamilton, C., et al. (2013). Diagnosis of obstructive sleep apnea by peripheral arterial tonometry: meta-analysis. *JAMA Otolaryngology–Head and Neck Surgery, 139*, 1343–1350.

Oldenburg, O., Arzt, M., Bitter, T., et al. (2015). Position paper "Sleep medicine in cardiology". *Der Kardiologe, 9*, 140–158.

Penzel, T., Fietze, I., & Glos, M. (2020). Alternative algorithms and devices in sleep apnoea diagnosis: what we know and what we expect. *Current Opinion in Pulmonary Medicine, 26*, 650–656.

Dawson, A., Loving, R. T., Gordon, R. M., et al. (2015). Type III home sleep testing versus pulse oximetry: is the respiratory disturbance index better than the oxygen desaturation index to predict the apnoea-hypopnoea index measured during laboratory polysomnography? *BMJ Open, 5*, e007956.

Fietze, I., Dingli, K., Diefenbach, K., et al. (2004). Night-to-night variation of the oxygen desaturation index in sleep apnoea syndrome. *The European Respiratory Journal, 24*, 987–993.

Wang, N., Meng, Z., Ding, N., et al. (2020). Oxygen desaturation rate as a novel intermittent hypoxemia parameter in severe obstructive sleep apnea is strongly associated with hypertension. *Journal of Clinical Sleep Medicine, 16*, 1055–1062.

Zhang, S., Meng, Z., Zhang, X., et al. (2020). The rate of decrease in oxygen desaturation during severe obstructive sleep apnea syndrome is correlated with subjective excessive daytime sleepiness. *Sleep & Breathing*.

Sommermeyer, D., Zou, D., Ficker, J. H., et al. (2016). Detection of cardiovascular risk from a photoplethysmographic signal using a matching pursuit algorithm. *Medical & Biological Engineering & Computing, 54*, 1111–1121.

Pielmus, A. G., Mühlstef, J., Bresch, E., et al. (2021). Surrogate based continuous noninvasive blood pressure measurement. *Biomedizinische Technik. Biomedical Engineering, 66*, 231–245.

Gehring, J., Gesche, H., Drewniok, G., et al. (2018). Nocturnal blood pressure fluctuations measured by using pulse transit time in patients with severe obstructive sleep apnea syndrome. *Sleep & Breathing, 22*, 337–343.

Crowley, K. E., Rajaratnam, S. M., Shea, S. A., et al. (2013). Evaluation of a single-channel nasal pressure device to assess obstructive sleep apnea risk in laboratory and home environments. *Journal of Clinical Sleep Medicine, 9*, 109–116.

Keshavarzi, F., Mehdizadeh, S., Khazaie, H., et al. (2018). Objective assessment of obstructive sleep apnea in normal pregnant and preeclamptic women. *Hypertension in Pregnancy, 37*, 154–159.

Penzel, T., Kantelhardt, J. W., Bartsch, R. P., et al. (2016). Modulations of heart rate, ECG, and cardio-respiratory

coupling observed in polysomnography. *Frontiers in Physiology, 7*, 460.
De Chazal, P., Heneghan, C., & Mcnicholas, W. T. (2009). Multimodal detection of sleep apnoea using electrocardiogram and oximetry signals. *Philosophical Transactions. Series A, Mathematical, Physical, and Engineering Sciences, 367*, 369–389.
Fontana, P., Martins, N. R. A., Camenzind, M., et al. (2019). Applicability of a textile ECG-belt for unattended sleep apnoea monitoring in a home setting. *Sensors (Basel), 19*, 3367.
Mueller, A., Fietze, I., Voelker, R., et al. (2006). Screening for sleep-related breathing disorders by transthoracic impedance recording integrated into a Holter ECG system. *Journal of Sleep Research, 15*, 455–462.
Chen, R., Chen, K., Dai, Y., et al. (2019). Utility of transthoracic impedance and novel algorithm for sleep apnea screening in pacemaker patient. *Sleep & Breathing, 23*, 741–746.
Defaye, P., Mendelson, M., Tamisier, R., et al. (2019). Validation of an apnea and hypopnea detection algorithm implemented in implantable cardioverter defibrillators. The AIRLESS study. *Scientific Reports, 9*, 9597.
Dias, M., Gonçalves, I., Amann, B., et al. (2017). Utility of new-generation pacemakers in sleep apnea screening. *Sleep Medicine, 37*, 27–31.
Glos, M., Sabil, A., Jelavic, K. S., et al. (2019). Tracheal sound analysis for detection of sleep disordered breathing. *Somnologie, 23*, 80–85.
Sabil, A., Schobel, C., Glos, M., et al. (2019). Apnea and hypopnea characterization using esophageal pressure, respiratory inductance plethysmography, and suprasternal pressure: a comparative study. *Sleep & Breathing, 23*, 1169–1176.
Narayan, S., Shivdare, P., Niranjan, T., et al. (2019). Noncontact identification of sleep-disturbed breathing from smartphone-recorded sounds validated by polysomnography. *Sleep & Breathing, 23*, 269–279.
Tiron, R., Lyon, G., Kilroy, H., et al. (2020). Screening for obstructive sleep apnea with novel hybrid acoustic smartphone app technology. *Journal of Thoracic Disease, 12*, 4476–4495.
Tenhunen, M., Elomaa, E., Sistonen, H., et al. (2013). Emfit movement sensor in evaluating nocturnal breathing. *Respiratory Physiology & Neurobiology, 187*, 183–189.
Perez-Macias, J. M., Tenhunen, M., Varri, A., et al. (2018). Detection of snores using source separation on an emfit signal. *IEEE Journal of Biomedical and Health Informatics, 22*, 1157–1167.
Paalasmaa, J., Toivonen, H., & Partinen, M. (2015). Adaptive heartbeat modeling for beat-to-beat heart rate measurement in ballistocardiograms. *IEEE Journal of Biomedical and Health Informatics, 19*, 1945–1952.
Fino, E., & Mazzetti, M. (2019). Monitoring healthy and disturbed sleep through smartphone applications: a review of experimental evidence. *Sleep & Breathing, 23*, 13–24.
Weinreich, G., Terjung, S., Wang, Y., et al. (2014). Validierung von SleepMinder® als Screeninggerät für die obstruktive Schlafapnoe. *Somnologie, 18*, 238–242.
Coronel, C., Wiesmeyr, C., Garn, H., et al. (2019). Measurement of respiratory effort in sleep by 3D camera and respiratory inductance plethysmography. *Somnologie, 23*, 86–92.
Veauthier, C., Ryczewski, J., Mansow-Model, S., et al. (2019). Contactless recording of sleep apnea and periodic leg movements by nocturnal 3-D-video and subsequent visual perceptive computing. *Scientific Reports, 9*, 16812.
Crinion, S. J., Tiron, R., Lyon, G., et al. (2020). Ambulatory detection of sleep apnea using a non-contact biomotion sensor. *Journal of Sleep Research, 29*, e12889.
Savage, H. O., Khushaba, R. N., Zaffaroni, A., et al. (2016). Development and validation of a novel non-contact monitor of nocturnal respiration for identifying sleep-disordered breathing in patients with heart failure. *ESC Heart Fail, 3*, 212–219.
Glos, M. T. D., & Schöbel, C. (2021). Sleep apnea screening using simplified systems. *Somnologie, 25*, 155–165.

ECG and Heart Rate Variability in Sleep-Related Breathing Disorders

10

Hua Qin, Fernando Vaquerizo-Villar, Nicolas Steenbergen, Jan F. Kraemer, and Thomas Penzel

Abstract

Here we discuss the current perspectives of comprehensive heart rate variability (HRV) analysis in electrocardiogram (ECG) signals as a non-invasive and reliable measure to assess autonomic function in sleep-related breathing disorders (SDB). It is a tool of increasing interest as different facets of HRV can be implemented to screen and diagnose SDB, monitor treatment efficacy, and prognose adverse cardiovascular outcomes in patients with sleep apnea. In this context, the technical aspects, pathophysiological features, and clinical applications of HRV are discussed to explore its usefulness in better understanding SDB.

Keywords

Heart rate variability · Sleep-related breathing disorders · Sleep apnea · Autonomic nervous system activity · Autonomic dysfunction · Time-domain analysis · Frequency-domain analysis · Nonlinear analysis · Detrended fluctuation analysis · Poincaré plot · Correlation dimension · Entropy · Symbolic dynamics · Recurrence plots · Chaotic invariant analysis

H. Qin
Interdisciplinary Center of Sleep Medicine, Charité-Universitätsmedizin Berlin, Berlin, Germany

National Clinical Research Center for Respiratory Disease, State Key Laboratory of Respiratory Disease, Guangzhou Institute of Respiratory Health, The First Affiliated Hospital of Guangzhou Medical University, Guangzhou Medical University, Guangzhou, China

F. Vaquerizo-Villar
Biomedical Engineering Group, Universidad de Valladolid, Valladolid, Spain

Centro de Investigación Biomédica en Red en Bioingeniería, Biomateriales y Nanomedicina, (CIBER-BBN), Valladolid, Spain

N. Steenbergen
Imperial College London School of Medicine, London, UK

J. F. Kraemer
Department of Physics, Humboldt Universität zu Berlin, Berlin, Germany

T. Penzel (✉)
Sleep Medicine Center, Charite Universitätsmedizin Berlin, Berlin, Berlin, Germany
e-mail: thomas.penzel@charite.de

T. Penzel, R. Hornero (eds.), *Advances in the Diagnosis and Treatment of Sleep Apnea*, Advances in Experimental Medicine and Biology 1384, https://doi.org/10.1007/978-3-031-06413-5_10

10.1 Introduction

Sleep-disordered breathing (SDB) has a detrimental impact on the autonomic nervous system (Dissanayake et al., 2021; Dempsey et al., 2010). The International Classification of Sleep Disorders (ICSD-3) classifies sleep-related breathing disorders into four categories: obstructive sleep apnea (OSA) disorders, central sleep apnea (CSA) syndromes, sleep-related hypoventilation disorders, and sleep-related hypoxemia disorder (Sateia, 2014). OSA and CSA are the most common disorders. Neither of these two main types would lead to intrinsic autonomic modulation as they have distinctive practical hemodynamic implications (Leung, 2009). Electrocardiogram (ECG) records the electrical activity generated by the polarization and depolarization of myocardial cells (Israel et al., 2005). ECG is an integrated signal manifested by morphology and cardiac rhythms. Specific waves (P wave, QRS complex, T wave) are formed in ECG morphology. The rate and regularity of heartbeats are regulated by the sinus node. Not limited to the identification of abnormal morphologies, numerical analysis techniques are also applied to the etiology of ECG, such as heart rate variability (HRV). HRV measures the variation between successive heartbeats in a sinus rhythm time series (Task Force of the European Society of Cardiology and the North American Society of Pacing and Electrophysiology, 1996). In HRV analysis, the R peak is the most frequently used to detect the normal sinus to normal sinus (NN) intervals. Both animal and human studies show that HRV may contribute to assessing the degree of central autonomic network to autonomic nervous system integration and central-peripheral neural feedback in a complex environment (Task Force of the European Society of Cardiology and the North American Society of Pacing and Electrophysiology, 1996; Pagani et al., 1986). Based on that, HRV provides information about the capacity of the heart for adaptive regulation to internal and external environmental challenges.

Research on HRV in sleep disorders has progressively increased. To date, HRV has been widely accepted as a common non-invasive test to evaluate autonomic nervous system function. In sleep studies, a single lead modified II ECG is recommended in polysomnography (PSG) according to the American Academy of Sleep Medicine (AASM) guidelines (Iber et al., 2007). Current apnea-hypopnea index (AHI) severity definitions only capture the respiratory aspects of disease heterogeneity among patients with SDB. Combining AHI with additional measures, such as HRV, may help to refine the characterization of disease severity and cardiovascular implications.

Given the easy access to ECG data extracted from PSG, HRV is a promising tool for the detection and prediction of respiratory events, identification of different sleep stages, monitoring and prognosis of cardiovascular outcomes, and assessment of treatment efficacy in sleep-disordered breathing (SDB) populations (Roche et al., 1999; Penzel et al., 2002; Le & Bukkapatnam, 2017; Penzel et al., 2003; Qin et al., 2021a; Zhang et al., 2020; Khoo et al., 2001; Glos et al., 2016). HRV during sleep serves as simple objective method for the evaluation of excessive daytime sleepiness, cognitive impairment, and mental disorders (e.g., depression and bipolar disorder) (Taranto Montemurro et al., 2014; Kong et al., 2021; Pawlowski et al., 2017; Idiaquez et al., 2014; Migliorini et al., 2012). Importantly, reduced HRV is strongly associated with cardiovascular mortality in a diverse spectrum of clinical populations (Hillebrand et al., 2013). Previous studies have shown that HRV during both daytime and night-time potentially facilitates the detection of abnormal cardiac autonomic modulation, suggesting the predictive and prognostic value of HRV in cardiovascular outcomes (Singh et al., 2018).

Another possible ECG morphology feature is R-wave amplitude. However, the use of this is limited, and there is little available literature in sleep apnea. A previous study investigated the use of a combinatorial R-wave amplitude-based respiratory sinus arrhythmia (RSA) and ECG-derived respiration (EDR) approach as a surrogate for respiratory inductance plethysmography to detect OSA and CSA, showing its ability to detect pre-recorded OSA and CSA diagnoses with 92.5% and 95.0% accuracy, respectively

(Khandoker & Palaniswami, 2011). These results indicated that an ECG-based neural network classification method could act as a potential surrogate to detect sleep apnea.

In this chapter, we address the technical aspects of HRV and its clinical applications in sleep study. Advanced methods and tools for ECG analysis and physiological interpretation for statistical, geometrical, spectral, and nonlinear HRV metrics derived from time-domain analysis, frequency-domain analysis, detrended fluctuation analysis, Poincaré plot, correlation dimension, entropies analysis, and symbolic dynamics will be presented.

10.2 Rationale and Scientific Basis of HRV in SDB

Fluctuations in heart rate during sleep apnea are obvious phenomena (Guilleminault et al., 1984). Heart rate is determined by sinoatrial node input from the central nervous system and medullary cardiovascular center autonomic outflow. Heart rate increases are mediated by the sympathetic nervous system, while decreases are mediated by the parasympathetic nervous system. The circadian pattern of HRV is a manifestation of alternations of sympathetic and parasympathetic activities during the day and night. The electrocardiographic changes observed in a 24-hour Holter recording show a periodic cosine HRV circadian pattern in healthy subjects (Bilan et al., 2005; Li et al., 2011). The sympathetic predominance assessed by LF (low frequency) possibly peaks in two periods (5:00–9:00 am and 4:00–6:00 pm), and the ratio of low frequency and high frequency (LF/HF) peaks in the afternoon (2:00–4:00 pm) and then keeps attenuating until the early hours of the night (Bilan et al., 2005; Li et al., 2011). The vagal control measured by high frequency (HF) and square root of the mean squared differences of consecutive NN intervals (RMSSD) remains increased during sleep hours at night, reaching its acrophase (3:00–5:00 am), and reaches its bottom in the afternoon (3:00–6:00 pm).

Previous findings demonstrated that OSA may have an abnormal circadian variation of autonomic activity, which may help explain the timing of the occurrence of cardiovascular events (Aydin et al., 2004; Noda et al., 1998). It appeared that the severity of OSA has a determined influence on the circadian rhythms of spectral HRV as LF, HF, and LF/HF differentiated significantly between mild and severe OSA (Noda et al., 1998). Significantly lower HF and higher LF/HF were found from 4:00 am to 12:00 pm in severe OSA compared to mild OSA and the non-OSA control (Noda et al., 1998). Compared to normal sleep, a dip in HF and a rise in LF/ HF from the early morning to daytime in severe OSA implied suppressed parasympathetic tone and vibrant sympathetic tone. The disturbed circadian patterns of autonomic modulation may contribute to daytime hypertension and an increased incidence of cardiovascular events such as myocardial infarction, stroke, and sudden cardiac death in early mornings.

In particular, the control of heart rate is critically dependent on sleep stages (Tobaldini et al., 2013). The time courses of sleep are classified to different sleep stages. There are three phases of non-rapid eye movement (NREM) sleep, including stage 1 (N1), stage 2 (N2), and stage 3 (N3) (Iber et al., 2007). It is shown that autonomic balance shifts from higher vagal activity during NREM sleep to higher sympathetic activity during rapid eye movement (REM) sleep in healthy subjects (Tobaldini et al., 2013; Vanoli et al., 1995). Autonomic regulation during wakefulness appears to be similar to that during REM.

It is known that OSA is characterized by partial and complete cessation of breathing, arousal, hypoxia, and sleep fragmentation (Dempsey et al., 2010). Given the genesis of HRV, those occurrences of OSA-related perturbations would result in various HRV patterns (Dissanayake et al., 2021; Sequeira et al., 2019). A cyclic bradycardia and tachycardia pattern is observed during sleep apneas (Guilleminault et al., 1984). This cardiac phenomenon, connected to central and autonomic reflexes, is featured by a decrease in heart rate during apnea episodes and an increase in heart rate during respiratory restoration. Parasympathetic tone predominates the time course of apnea, and sympathetic tone rises progressively and reaches the peak of outflow traffic at the termination of sleep apnea. Moreover, abnormal blood pressure variability and absolute heart rate occur later than abnormal cardiovascular variability. SDB is thought to exert deleterious effects on alterations in the autonomic nervous system through multiple mechanisms. Current findings suggest that autonomic dysfunction in OSA may be induced by vagal withdrawal, sympathetic hyperactivity, impaired baroreflex sensitivity, or a combination of these responses mediated by the baroreflex, chemoreflex, and mechanical reflex due to hypoxia, arousal, and pleural pressure swings (Narkiewicz et al., 1998; Somers et al., 1991, 1995). In addition, OSA may enhance the autonomic, hemodynamic, and ventilatory response to peripheral chemoreceptor activation stimulated by hypoxia due to elevated chemoreflex sensitivity. The duration of OSA and the degree of oxygen saturation also contribute to the sympathetic activation during OSA episodes (Narkiewicz & Somers, 2001; Jiang et al., 2017).

Recently, a systematic review attempted to summarize the evidence of associations between autonomic function and OSA (Dissanayake et al., 2021). It has generally shown that OSA is related to increased risk of cardiovascular diseases and worsened health outcomes as indicated by reduced HRV. Spatiotemporal HRV patterns in OSA population have been extensively investigated. The association between OSA and time-domain HRV was found to be clearer, while the relationship between OSA and frequency-domain HRV was less consistent.

Some studies find no significant differences in HRV between OSA subjects and healthy control. The reasons for confusing and inconsistent findings regarding temporal, spectral, and nonlinear HRV measures in OSA remain unclear. There are several important determinants of HRV. Age-related decline, gender, and racial differences in autonomic nervous system functions would be the important demographic confounders (Liao et al., 1995; Sloan et al., 2008). Breathing patterns, comorbidities, and other lifestyle behaviors like smoking, alcohol, caffeine intake, and physical activity are also known to have a recognizable influence on HRV analysis (Ucak et al., 2021; Gerritsen et al., 2001; Hayano et al., 1990; Malpas et al., 1991; Sondermeijer et al., 2002; Soares-Miranda et al., 2014). In addition, studies sometimes do not control for the aforementioned potential covariates in adjusted statistical analysis, which may result in misleading interpretations of data. Methodologically, the differences in HRV measurement protocol, recording length, resolution of raw data and analysis unit, as well as processing of the raw data (transformation methods and filters) also contribute to the mixed results. Nevertheless, time-domain measures demonstrate generally better reproducibility than frequency-domain measures under various experimental protocols.

HRV is associated with the severity of OSA (Aydin et al., 2004; Narkiewicz et al., 1998; Qin et al., 2021b). There is a trend with gradually decreased standard deviation of NN intervals (SDNN), increased LF, and decreased HF from mild to severe OSA. Furthermore, previous studies indicated long-lasting alterations in autonomic function in snoring subjects and OSA patients even during without the presence of respiratory events (Qin et al., 2021b; Gates et al., 2004; Ferini-Strambi et al., 1992). However, OSA screening in the mild cases is challenging as autonomic dysfunction may not always appear in patients with mild OSA (Blomster et al., 2015). Balachandran et al. found significantly decreased LF, decreased HF, and increased LF/HF between moderate to severe OSA without any symptoms and non-OSA subjects during daytime waking state using 5-minute ECG data (Balachandran et al., 2012). To determine which HRV feature is more sensitive to pre-clinical or subtle autonomic alternation in patients with OSA needs further large-scale studies.

OSA is a recognized risk factor for multiple adverse cardiovascular events such as hypertension, heart failure, arrhythmia, stroke, coronary heart disease, and sudden cardiac death (McNicholas & Bonsigore, 2007; Gami et al., 2005). Risk of cardiovascular morbidity and mortality varies among different OSA phenotypes such as symptom-based clusters. Particularly, OSA patients with excessive sleepiness have a significantly increased risk for the development of cardiovascular incidence and prevalence of adverse cardiovascular diseases than non-sleepy patients, which suggests excessively sleepy OSA patients are more likely to express certain underlying physiopathological traits that puts them at higher risk. Montemurro et al. found that sleepy patients with severe OSA have a lower very low frequency (VLF) in HRV than those who are non-sleepy, suggesting depressed sympathetic control (Taranto Montemurro et al., 2014). The possible mechanisms of high prevalence and risk of unpleasant health outcomes are poorly understood. Exploration of the role of HRV in OSA phenotypes and associated cardiovascular outcomes may contribute to discovering the explanation to those research questions.

However, electrocardiography can also give additional information. In particular, EDR can derive respiration rate from several sources, such as R-wave area, principal component analysis, and amplitude demodulation (Moody et al., 1985). Cardiopulmonary coupling (CPC) observes the phase difference between EDR and RR intervals to view the degree of influence respiration has on the cardiovascular system (Sadr & de Chazal, 2019). Several studies have shown that CPC can be used as an additional tool in the measure of sleep quality (Thomas et al., 2014). Additionally, CPC has shown promise as an accurate diagnostic tool in OSA, being able to diagnose OSA comparably to PSG-derived AHI and produce an automated AHI with a strong correlation with conventional PSG-derived AHI (Khandoker et al., 2009a). Overall, subjects with severe OSA show significantly less phase coupling than control subjects. Moreover CPC analyses the interactions between respiratory dysregulation and upper airway anatomical obstruction and different respiratory events (Thomas et al., 2005; Thomas et al., 2007). Due to the vast amounts of information that is recorded during sleep studies, the analysis of additional markers may prove vital to improving the diagnostic and monitoring efficiency in OSA as shown by the promise of HRV, EDR, and CPC.

10.3 HRV Measurements

Practice guidelines and guidance statements on HRV, a task force report of the European Society of Cardiology and the North American Society of Pacing and Electrophysiology, were published in 1996 (Task Force of the European Society of Cardiology and the North American Society of Pacing and Electrophysiology, 1996). In this respect, Fig. 10.1 shows an overview of the main steps for HRV extraction and analysis. This process starts with the ECG acquisition. As aforementioned, a single modified ECG lead II is recommended in PSG sleep studies. The guidelines of the AASM recommend ECG recordings with a minimum sample rate of 200 Hz and 500 Hz as recommended (Iber et al., 2007).

After acquisition, algorithms include a preprocessing to adapt ECG signals to the processing of HRV signal. ECG waveform is commonly distorted by noise and artifacts, such as baseline wander, electrode motion, and electromyographic interference. For this reason, filtering-based methods are applied to the ECG signals, which include bandpass linear filters to remove noise, as well as nonlinear transformations that correct inverted QRS complexes (Bansal et al., 2009).

The next step consists of R peak detection. Taking as input the preprocessed ECG signal, many automatic approaches have been proposed for the detection of QRS complexes, which includes the R peak position and amplitude. For this, many algorithms based on signal derivatives, digital filters, the wavelet transform, neural networks, the Hilbert transform, adaptive filters, and morphological transformations exist (Kohler et al., 2002). A popular one is the Pan and Tompkins method, which includes a two-stage bandpass filter, a derivative operation, a nonlinear transformation, an integration, and a thresholding operation (Pan & Tompkins, 1985).

Then the RR interval time series is derived. However, false R peak detection due to non-

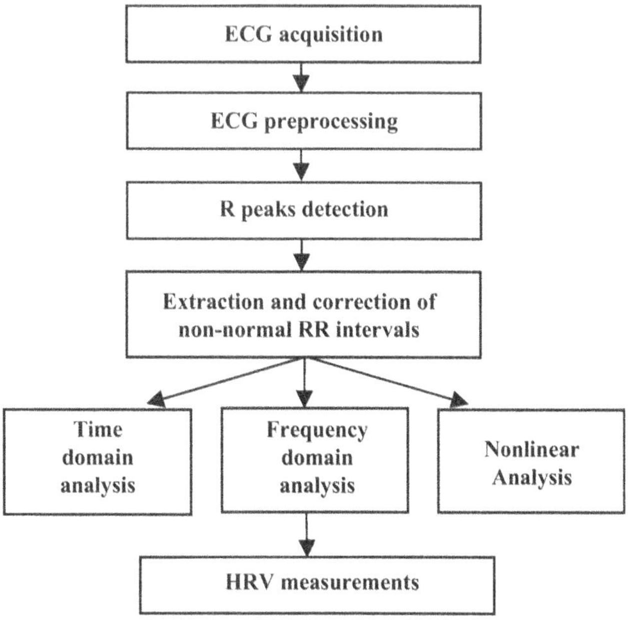

Fig. 10.1 Flowchart of heart rate variability analysis

removed noise can result in abnormal RR intervals. Ectopic beats can introduce irregularities in the HRV signal that affect HRV measurements. Therefore, a detection and correction of non-normal RR intervals follow using methods based on the empirical mode decomposition, neural network approaches, and wavelet-based methods. Conversely, the correction of non-normal beats can be performed using the deletion method, interpolation-based approaches, and adaptive methods.

Finally, HRV is analyzed using time-domain, frequency-domain, and nonlinear methods. Time-domain and frequency-domain analysis are utilized to measure the temporal and spectral properties of the RR tachogram (Figs. 10.2 and 10.3). In the case of frequency-domain analysis, the irregularly time-sampled HRV signal is interpolated and sampled before the application of Fourier transform-based methods These two classic approaches limited in nonstationary signal analysis as ECG signals are highly nonlinear and nonstationary. Sophisticated mathematical nonlinear approached toward RR variability to evaluate the nonlinearity of HRV. Furthermore, increasing data shows that nonlinear HRV measures may be superior to conventional HRV parameters in cardiovascular risk stratification (Voss et al., 1996).

Regarding HRV analysis, a chosen time window is critical for comparable HRV measurements as some variables are associated with time duration of data (Li et al., 2019a). With regard to this, there are ultra-short (<5 min), short-term (typically 5 min), and long-term (24 h) HRV analysis according to the selected length of ECG segments. Ultra-short HRV favors the HRV assessment involved in associations with events such as arousal and oxygen desaturation. Ultra-low frequency (ULF) and VLF computed from at least 20–30 min windows would be more robust and reliable. However, the reliability of ultra-short HRV features is an open question, partially due to the lack of well-established algorithms guiding investigators to systematically assess ultra-short HRV reliability. It is controversial that ultra-short time period HRV provides insufficient resolution to the feature of some HRV parameters. Long-term HRV could reflect the effects of metabolism, circadian rhythm, and daily activity on cardiovascular system. HRV from 24-h ECG recordings also has been proven to have high reproducibility. However, Dissanayake et al. found that it is more common to collect ECG data

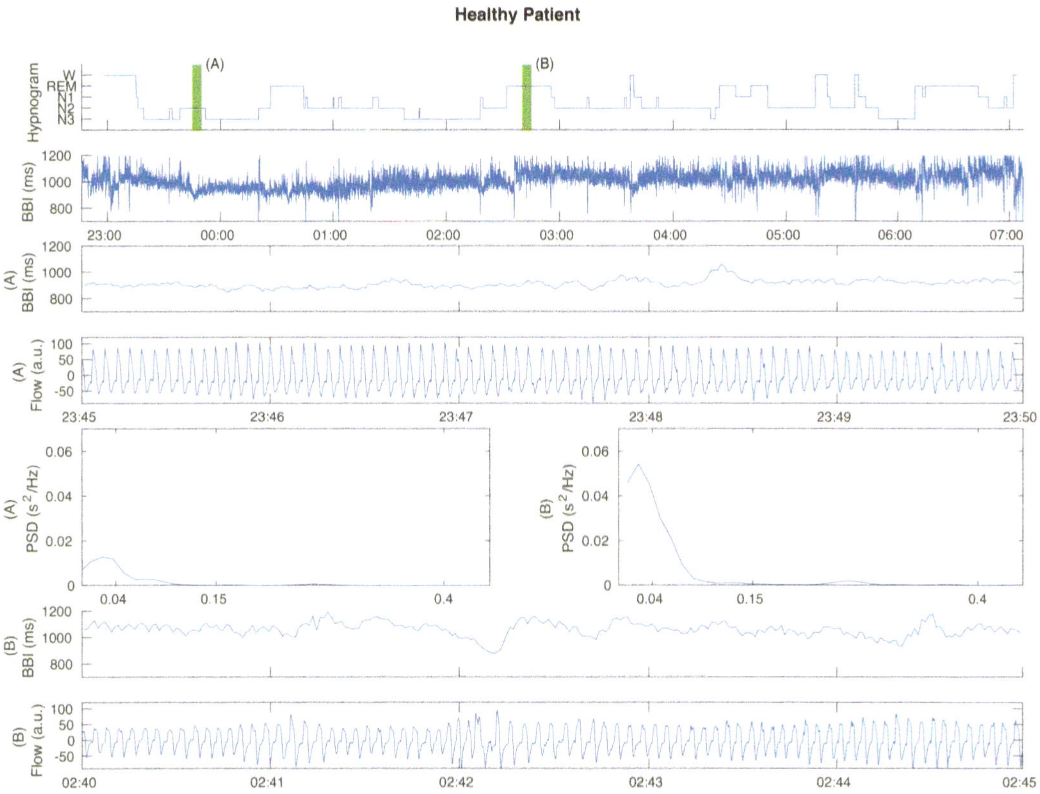

Fig. 10.2 Schematic representation of changes in beat-to-beat intervals (BBI) and power spectral density (PSD) from a patient without obstructive sleep apnea during 5-minute (A) non-rapid eye movement sleep stage 2 and (B) rapid eye movement sleep

from overnight PSG for HRV analysis than from 24-hour Holter recordings (Dissanayake et al., 2021). The strength of short-term HRV is easier to apply in the clinical practice and is affected by fewer factors. How to choose the time window for HRV analysis depends on the context of analysis.

The distributions of time-domain and frequency-domain measures are skewed in statistical analysis; the natural log transform of those values is frequently used. However, the absolute values of the HRV results (e.g., mean, standard deviations, median values, and interquartile ranges) would better facilitate comparisons to other studies, and mostly in clinical setting, thus, they should be reported. In terms of the elucidation for the heterogeneous findings, some methodological concerns such as insufficient resolution in ECG data and low sample size (e.g., underpowered samples or selective samples) should be taken into consideration.

10.3.1 Time-Domain Heart Rate Variability Analysis

Time-domain HRV analysis quantifies the amount of variation of heart rate in a given time series based on statistical methods. A majority of temporal HRV parameters measure the dispersion around their mean value in a recording period, including the mean value of normal-to-normal interval time series (meanNN), SDNN, the standard deviation of the 5-minute average of NN intervals (SDANN), the mean of the standard deviations of all the 5-minute NN intervals of a 24-h ECG recording (SDNNI), the ratio of SDNN divided by meanNN (CVNN), the square root of

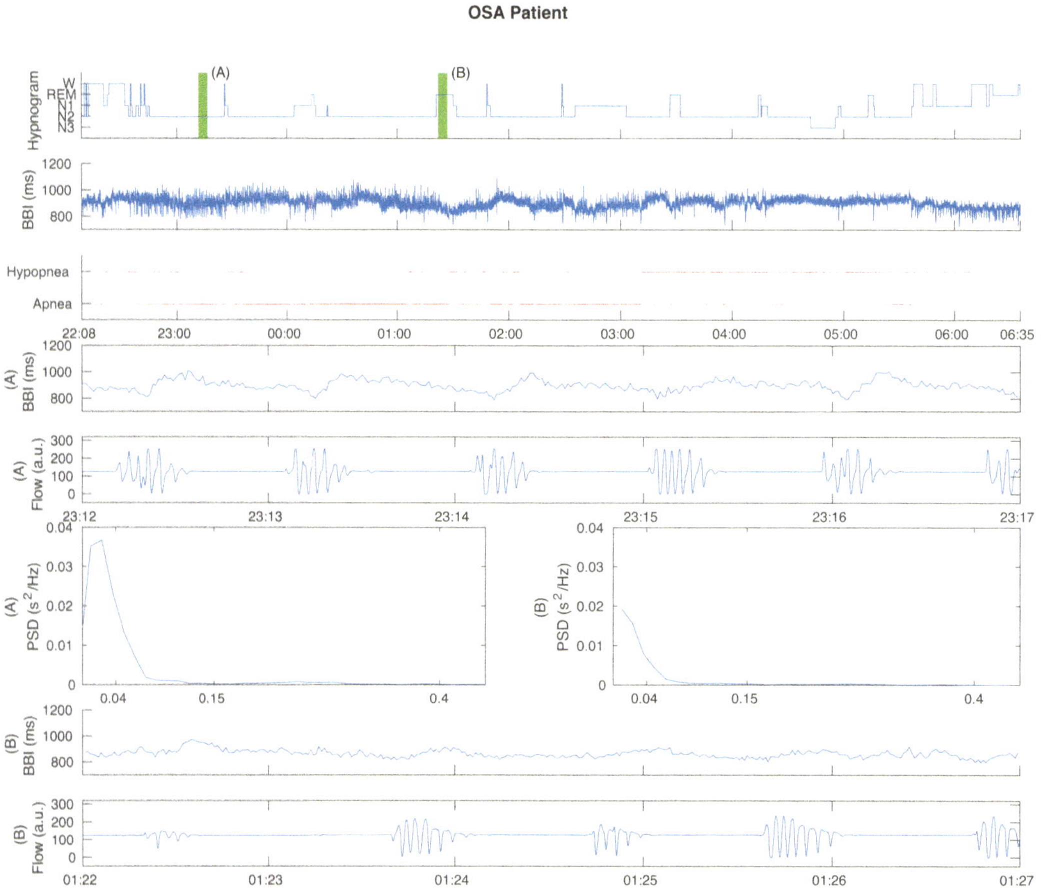

Fig. 10.3 Schematic representation of changes in beat-to-beat intervals (BBI) and power spectral density (PSD) from a patient with severe obstructive sleep apnea during 5-minute (A) non-rapid eye movement sleep stage 2 and (B) rapid eye movement sleep

the mean squared differences of consecutive NN intervals (RMSSD), and the percentage of NN > 50 ms counts divided by the total number of all NN intervals (pNN50).

Abnormalities in cardiovascular variability are correlated with an increased risk of cardiovascular events and mortality (Gerritsen et al., 2001). In myocardial patients with impaired autonomic function, increased risk can be attributed to life-threatening arrhythmias (La Rovere et al., 2001). Additionally, low HRV or low baroreflex sensitivity is also associated with increased risk of non-fatal cardiovascular events and hypertension in the general population. Lower HRV demonstrates 32–45% higher risk for the first emerging cardiovascular events in patients with no known history of cardiovascular diseases (Hillebrand et al., 2013). Kikuya et al. indicated that cardiovascular complications increased with reduced daytime HRV in the general population (Kikuya et al., 2000). Huikuri et al. found that nonlinear power-law relationship was the best predictor of all-cause mortality with significant 7.9 relative risk in population who is over 65 years old (Huikuri et al., 1998).

Notably, previous studies presented cut-off points of statistical HRV metrics for increased mortality risk. For example, reduced SDNN (e.g., <40 ms) or RMSSD (e.g., <25 ms) could be an indicator for cardiovascular risk stratification in certain population (Bigger Jr. et al., 1992). SDNN is the most common used time-domain index, which measures the total variance of HRV in the selected ECG length. SDNN plays an important role in predicting cardiovascular risk and the prognosis of cardiovascular outcomes in various pathophysiological conditions, such as in patients after myocardial infarction and the elderly (Hillebrand et al., 2013; Singh et al., 2018). Reduced SDNN is related to increased risk and incidence of adverse cardiovascular outcomes. Hillebrand et al. demonstrated a 1% decrease in SDNN is correlated with approximately a 1% increase in fatal and non-fatal cardiovascular events in individuals without recognized cardiovascular diseases in a dose-response meta-regression analysis (Hillebrand et al., 2013). Many cut-off points of SDNN are reported to have increased risk of all-cause and cardiovascular mortality in various populations (Singh et al., 2018). Findings on 24-hour SDNN calculated from Holter ECG data demonstrated up to 5.3 of risk ratio for cardiovascular death using SDNN<70 ms and 1.62–5.3 of risk ratio for all-cause mortality using SDNN<50 ms, SDNN<70 ms, and SDNN<93 ms (La Rovere et al., 1998; Kleiger et al., 1987; Zuanetti et al., 1996; Nolan et al., 1998). For mortality risk assessment using ultra-short-term resting SDNN, SDNN <20 ms is related to increased risk of 5-year CHD mortality (risk ratio 2.1, 95% CI [1.1–4.1]) and all-cause mortality (risk ratio 2.1, 95% CI [1.4–3.0]) (Dekker et al., 1997).

In SDB populations, Sankari et al. used total and sleep RR interval dip index (RRDI), computed by the number of RRI dips divided by total recording time and sleep time, respectively, to investigate the role of overnight alternation in heartbeats in cardiovascular risk in OSA patients from the Wisconsin Sleep Cohort (Sankari et al., 2019). Their findings suggested that the increased dips in sleep RRI are related to cardiovascular disease onset with a hazard ratio of 1.21 per 10-unit increment in RRDI. Patients with greater total RRDI are at higher risk of increased incidence of cardiovascular diseases and mortality with a 7.4 hazard ratio. However, there are no generally accepted cut-off points of SDNN to identify the risk of mortality in SDB population. Whether HRV measures are a good indicator of different cardiovascular risk phenotypes in OSA requires more investigation.

Geometric Measures

RR time series can also be converted in geometric patterns. From these patterns, the following geometric measures have been commonly used to characterize HRV dynamics (Task Force of the European Society of Cardiology and the North American Society of Pacing and Electrophysiology, 1996):

- HRV triangular index (HRVi), which is the integral of the histogram (total number of RR intervals) divided by the height of the histogram. It measures overall variability of RR intervals.
- Triangular interpolation of the NN interval histogram (TINN). TINN is the baseline width of the distribution measured as a base of a triangle, approximating the distribution of NN intervals using the mean square difference. TINN also expresses overall HRV.

Geometrical methods are highly insensitive to the analytical quality of RR time series, thus not being affected by artifacts or ectopic beats. These parameters are recommended in RR intervals of at least 20 minutes, being preferred 24 h recordings (Task Force of the European Society of Cardiology and the North American Society of Pacing and Electrophysiology, 1996; Rajendra Acharya et al., 2006). Thus, these methods are not appropriate to measure short-term changes in HRV, such as heart rate response to individual apneic events. Kim et al. extracted HRVi and TINN from HRV during sleep (Kim et al., 2015). TINN did not show statistical relevance among OSA groups, whereas HRVi was significantly higher in OSA patients (AHI≥15) than in controls (AHI<5), agreeing with other time-domain parameters that indicated that OSA patients have a diminished HRV and vagal predominance of heart control (Sequeira et al., 2019).

10.3.2 Frequency-Domain Heart Rate Variability Analysis

10.3.2.1 Conventional Frequency-Domain Analysis

Conventional frequency-domain analysis, also known as spectral analysis, identifies oscillatory components and quantifies the sympathetic and parasympathetic activities (Akselrod et al., 1981). It divides the power spectrum of HRV into individual frequency bands, including the ULF range (0–0.01 Hz), the VLF range (0.01–0.04 Hz), the LF range (0.04–0.15 Hz), and the HF range (0.15–0.4 Hz). Many transformation methods are used to decompose HRV spectra such as fast Fourier transform (FFT), autoregressive models, and wavelet transform. Limitations of FFT may include assumptions about the linearity and the stationary nature of the ECG data.

Based on these frequency ranges in power spectral density, their relative strengths are quantified as power. Relative spectral metrics are calculated, including low frequency expressed in normalized units (LF nu), high frequency expressed in normalized units (HF nu), and low-frequency to high-frequency power ratio (LF/HF). LF nu and HF nu are calculated as relative proportions, which equal to LF/(LF + HF) × 100 and HF/(LF + HF) × 100, respectively. The relative distribution of power in each frequency component is calculated as a percentage of TP minus ULF to give percentages of VLF (%VLF), LF (%LF), and HF (%HF). The HF component mainly reflects parasympathetic activity, corresponding to RSA. The reflection of LF is criticized as the LF component may present pure sympathetic tone or both sympathetic and vagal activity. However, LF is affected by other cardiac mechanisms such as baroreflex sensitivity. The physiological reflection of VLF is not clear, but it is believed that VLF is associated with regulatory mechanisms such as the renin-angiotensin system, thermoregulation, circadian oscillations, body temperature, and metabolism activity of the renin-angiotensin system (Taylor et al., 1998).

It is shown that there are strong correlations between time-domain and frequency-domain parameters ($r = 0.85$) (Kleiger et al., 1991). TP is approximately equal to the square of SDNN. ULF is closely associated with SDNN and SDANN. LF and VLF are linked to SDNNI. HF is related with RMSSD and PNN50. These correlations further help the verification of data quality. Increased sympathetic or decreased parasympathetic activity or both are considered as decreased HRV and vice versa.

Many of the aforementioned parameters have been investigated in sleep staging. Zemaityte et al. found that HR, mean, and SDNN decreased in sleep stages N1, N2, and N3, but increased during REM sleep (Zemaityte et al., 1984). Bonnet et al. and Otzenberger et al. confirmed these findings (Bonnet & Arand, 1997; Otzenberger et al., 1998). Scholz et al. showed that synchronized sleep was associated with a decreased LF/HF ratio and REM sleep with an increased LF/HF ratio (Scholz et al., 1997). The effect of age on nighttime cardiac vagal activity has also been explored. Crasset et al. compared young and elderly healthy subjects during wakefulness and sleep and found that elderly subjects had a lower HF during non-REM sleep than young subjects in the Sleep Heart Health Study (Crasset et al., 2001). The elderly patients showed HRV time- and frequency-domain parameters were at their lowest in slow-wave sleep. However, they also showed higher parameters in REM sleep than wakefulness, and the LF/HF ratio was not at its highest in REM sleep (Crasset et al., 2001).

Sleep stage estimation in diseased patients has also been explored. Post-MI patients showed a significantly higher LF/HF ratio than normal subjects in non-REM and REM sleep and then wakefulness (Vanoli et al., 1995). Contrastingly, normal subjects typically have a decrease in LF/HF ratio from wake to non-REM sleep. As mentioned before, SDB groups have had varied results in HRV studies compared to normal subjects (Dissanayake et al., 2021; Ucak et al., 2021). Shinar et al. found that there is no difference in VLF, LF, HF, LF/HF ratio, mean, and SDNN between control, OSA, and various sleep disorder groups (Shinar et al., 2006). It is evident that use of HRV in sleep stage estimation is possible, but that it needs to be adjusted for a host of confounding factors, including age and diseases.

Previous findings applying frequency-domain analysis demonstrated significantly diminished vagal activity and enhanced sympathetic activity with decreased HF, increased LF, and discordant LF/HF in patients with OSA compared to those without during sleep and daytime (Ucak et al., 2021). Numerous publications reported that HF gradually decreases and LF progressively increases from mild to severe OSA during wakefulness and sleep (Dissanayake et al., 2021; Ucak et al., 2021). Interestingly, Gula et al. suggested that changes in autonomic activity are disproportional to OSA severity, demonstrating elevated sympathetic predominance and discordant autonomic imbalance from mild to moderate OSA, compared to blunted responses in severe OSA (Gula et al., 2003). Qin et al. found similar patterns of cardiac autonomic control during wakefulness prior to sleep onset in only obese patients with different levels of OSA severity (Qin et al., 2021b). Their results suggested that among obese OSA patients, there was evidence of a shift to sympathetic hyperactivity in more severe OSA, as shown by a higher LF/HF ratio during wakefulness compared to obese non-OSA patients. Physiological and clinical values of these differences in autonomic response toward levels of OSA severity are unknown.

Low vagal tone is strongly associated with higher risk of cardiovascular diseases. Zhang et al. suggested sleep HRV metrics, particularly HF nu during all three sleep stages, HF during REM sleep, and LF during N3 sleep as precursors of potential cardiovascular disease outcomes using the subset data from the Sleep Heart Health Study (Zhang et al., 2020). Liao et al. also found that reduced HF is a 1.72 times greater hazard for incident coronary heart disease (95% confidence interval (CI) 1.17–2.51) in the Atherosclerosis Risk in Communities cohort (Liao et al., 1997). In contrast, Tsuji et al. implicated that lnLF is linked with all-cause mortality after adjusting for other risk factors in the early population recruited from the Framingham Heart Study (Tsuji et al., 1994).

10.3.2.2 Bispectral Analysis

Conventional frequency-domain analysis based on the Fourier transform cannot characterize nonlinear behaviors and non-Gaussian events, as the phase information is lost (Task Force of the European Society of Cardiology and the North American Society of Pacing and Electrophysiology, 1996). In contrast, bispectral analysis preserves both phase and amplitude information of the spectral components of a time series, which allows to reflect deviations from linearity and Gaussianity in the HRV, which typically occur during sleep in OSA patients (Martín-Montero et al., 2021). The bispectrum of a RR time series is obtained as follows:

$$B(f_1, f_2) = X(f_1) \cdot X(f_2) \cdot X^*(f_1 + f_2), f_1, f_2, = 0, \ldots, N$$

where $X(f)$ is the Fourier transform of the RR time series and f_1 and f_2 are the frequency indices of the 2-D representation of the bispectrum.

Atri et al. showed that the bispectrum plot of a normal HRV contains numerous quadratic (nonlinear) phase couplings (QPCs) between very low and respiratory-range frequencies, whereas the bispectrum of HRV in an OSA event has QPCs more concentrated on intermediate frequencies, which may be due to a loss in the synchronized rhythm of the heart. They also obtained 95% accuracy for apneic episodes detection using spectral and bispectral features (Atri & Mohebbi, 2015). Similarly, Martin-Montero et al. found that the nonlinear coupling observed in the bispectrum around VLF and respiratory peaks in severe OSA children are redistributed to frequency ranges related to apneic events. Their results also indicated that bispectral features computed in classical (VLF, LF, and HF) and OSA-specific bands (BW1, BW2, and BWres) can detect pediatric OSA more accurately than conventional spectral ones (Martín-Montero et al., 2021).

10.3.2.3 Wavelet Analysis

HRV dynamics is composed of slowly varying components (e.g., heart rate during sleep stages) and rapidly changing transient events (e.g., heart rate response to apneic events) (Qin et al., 2021a). Conventional spectral analysis is based on the short-time Fourier transform (STFT), which applies a fixed length window to analyze each segment of the time series, assuming stationarity. This fixed time-frequency resolution limits its capability to analyze HRV content. This limitation is overcome by the wavelet transform (WT), which employs short windows at high frequencies and long windows at low frequencies (Rioul & Vetterli, 1991), which makes it more suitable to analyze the nonstationary properties of HRV during sleep. Given a RR time series, the continuous wavelet transform (CWT) decomposes it onto a set of base functions, called wavelets, using the following expression (Rioul & Vetterli, 1991):

$$W(\tau,a) = \frac{1}{\sqrt{a}} \int RR(t) \cdot g^{*}\left(\frac{t-\tau}{a}\right) dt$$

where $g(t)$ is the basic wavelet prototype, known as mother wavelet, and τ and a are the time translation and scale factor, respectively. In the discrete wavelet transform (DWT), wavelet coefficients are computed only for power of 2 scales (Rioul & Vetterli, 1991).

Khandoker et al. reported that the variability of the DWT coefficients of the RR time series in the frequency range (0.019–0.071 Hz) allows better quantification of the frequency components associated with OSA events. Their results indicated that wavelet analysis of HRV and EDR signals provides useful information regarding the effect of sleep-related breathing disorder on cardiac rhythms (Khandoker et al., 2009b). Similarly, Mendez et al. observed that dynamic of the cardiorespiratory system is contained in the decomposition levels of the DWT that represent the recurrence of apneic events and the respiratory frequency, suggesting that the respiratory arrhythmia in the heart rate and the frequency of apneic episodes are the most important HRV-related information for OSA detection (Mendez et al., 2010).

10.3.3 Nonlinear Analysis

Conventional methods for HRV analysis in the time and frequency domains are often not enough to characterize the dynamics of HRV, as the mechanisms involved in the generation of the heartbeat also interact in a nonstationary and nonlinear way (Sunkaria, 2011). In this regard, nonlinear methods derived from the chaos theory have demonstrated its usefulness to further characterize the HRV. The main nonlinear methods used for HRV analysis are described in the following subsections.

10.3.3.1 Detrended Fluctuation Analysis

Detrended fluctuation analysis (DFA) is a nonlinear analysis technique widely used to quantify the fractal correlation properties of RR intervals (Peng et al., 1995). DFA provides a modified root-mean square analysis that allows to detect short-range and long-range correlations in nonstationary signals (Penzel et al., 2003; Rajendra Acharya et al., 2006). Given a RR time series of length N, the signal is first integrated (Peng et al., 1994):

$$y(j) = \sum_{i=1}^{k} RR(i) - RR_{avg}, k = 1,\ldots,N$$

where $RR(i)$ is the i-th RR interval and RR_{avg} is the average of the RR time series. Then, the integrated signal is divided into B non-overlapping time windows of equal length n. For each window, a least square line fit, $y_n(j)$, is obtained. Next, the integrated signal $y(j)$ is detrended by subtracting the local trend, $y_n(j)$, in each window. The root-mean square fluctuation of the integrated and detrended RR time series is calculated for all windows using the following expression:

$$F(n)=\sqrt{\frac{1}{B}\sum_{j=1}^{B}\left[y(j)-y_n(j)\right]^2}$$

This process is repeated over all window sizes (time scales) to obtain the relationship between the fluctuation function, $F(n)$, and the scale (n). The value of $F(n)$ typically increases with the scale of the RR time series (Peng et al., 1994). The fluctuations of HRV can be characterized by the scaling exponent, α, which measures the slope of the log-log plot representing $F(n)$ vs. n.

The slope of DFA describes the roughness of a time series, and the time series becomes smoother with larger values of α. In DFA, white Gaussian noise (totally random signal) leads to a scaling exponent value of 0.5, while Brownian noise generates a scaling exponent value of 1.5. The slope of DFA is close to 1 in a healthy young population. It decreases varying degrees according to different conditions. For example, the fractal scaling is low in patients with highly variable ECG signals (e.g., pre-ventricular contraction, atrial fibrillation, and ventricular fibrillation) and is high in patients with slowly variable ECG data (e.g., sick sinus syndrome, complete heart block, left bundle branch block, and ischemic/dilated cardiomyopathy) compared to $\alpha = 1$.

In general, the DFA plot ($F(n)$ vs. n) yields two scaling exponents: α_1, for short time scales, and α_2, for long time scales. In the analysis of HRV, Francis et al. compared the values of fractal α computed from DFA with the one from frequency-weighted spectral analysis to improve the understanding of the clinical implications of the scaling exponent. They suggested that a low α1 is related with low LF/HF and low %LF, while a low $\alpha 2$ is associated with high VLF/LF and high VLF/LF. These results establish not only a mathematical relationship between DFA and conventional frequency-domain analysis but also a clinical explanation for fractal parameters (Francis et al., 2002). In the OSA context, Penzel et al. identified two scaling regions of HRV: a short-time scale region ($10 \leq n < 40$ beats), related to the effect of breathing on the heart rate, and a long-time scale region ($70 \leq n < 300$ beats), related to the effects of sleep stages and the circadian rhythm. Penzel et al. suggested that the accuracy for scoring sleep apnea severity and sleep stages using scaling analysis increased to 74.4% and 85%, respectively, compared to frequency-domain analysis, 69.7% and 54.6% (Penzel et al., 2003). Compared to linear HRV metrics, Da Silva et al. demonstrated that DFA α2 derived from the full length of sleep ECG could be a potential indicator of identification of OSA severity (da Silva et al., 2015). Their results showed $\alpha 2 > 0.32$ referred to an 80% accuracy of prediction of moderate OSA (15 < AHI<30) and $\alpha 2 > 0.47$ referred to a 76% accuracy of prediction of moderate OSA (AHI>30). The advantage of DFA is that it addresses the methodological limitations of spectral analysis for nonlinear and nonstationary physiological data and easily removes noise and trends. The DFA method was commonly used to analyze long-term ECG data, which is limited in short-term HRV analysis.

10.3.3.2 Entropy Analysis

The nonlinear features of heartbeats evaluated by entropy have been reported as a sensitive indicator for the abnormal autonomic regulation in OSA. Entropy measures show the probability that similar patterns observed in time series sequence do not repeat. Many entropy-based approaches have been established to provide the amount of information in heart rate complexity such as approximate entropy (ApEn), sample entropy (SampEn), Shannon entropy, wavelet entropy, compression entropy, and multiscale entropy based on different algorithms (Henriques et al., 2020). In information theory, ApEn is introduced to evaluate the regularity and the unpredictability of oscillation presented in the time series data (Pincus, 1991). SampEn is a refinement of ApEn, condensing short and noisy time series data without assessing self-similar patterns in the equation (Richman & Moorman, 2000). The advantages of SampEn over ApEn are data length independence and relatively trouble-free implementation. Multiscale entropy (MSE) is an extension of SampEn to multiple time scales or signal resolutions (Costa et al., 2002). MSE can conduct the calculation even when the time scale of relevance in the time series is not known.

In terms of interpretation, a low value of ApEn reflects a regular pattern in the signal, while a high value of ApEn refers to a random pattern. For example, a reduction in ApEn is related with the high possibility of the presence of cardiac disease. Shannon entropy is also used to measure the complex dynamics of heartbeats. A higher Shannon entropy is reflective of more irregularity in HRV. Here, we briefly discuss some entropy measures used in sleep apnea.

- *Shannon entropy:* Shannon entropy of the HRV histogram allows to quantify the distribution of NN intervals:

$$\text{Shannon entropy} = -\sum_{j\in\Omega} q_j \log\left(q_j\right)$$

where q_j is the histogram of the RR time series. A higher value of the Shannon entropy indicates that there is more irregularity in HRV. Zhang et al. reported a positive correlation of the Shannon entropy of the degree distribution with OSA severity (Zhang et al., 2019). Qin et al. suggested that severe OSA patients (AHI>30) have a lower Shannon entropy during wakefulness, thus suggesting that OSA reduces HRV irregularity during wakefulness (Qin et al., 2021b).

- *Approximate entropy (ApEn):* In information theory, ApEn is introduced by Pincus et al. to evaluate the regularity and the unpredictability of oscillation presented in short and noisy time series data. ApEn allows to discriminate time series for which clear feature recognition is difficult by the evaluation of both dominant and subdominant patterns (Pincus, 1991). Given a RR time series of length N, RR(n) = {RR(n), RR(2), …, RR(N)}, ApEn is computed using the following expression:

$$\text{ApEn} = \frac{1}{N-m+1}\sum_{i=1}^{N-m+1}\log\left(C_i^m(r)\right) - \frac{1}{N-m}\sum_{i=1}^{N-m}\log\left(C_i^{m+1}(r)\right)$$

where $C_i^m(r)$ is the correlation integral, defined as:

$$C_i^m(r) = \frac{1}{N-m+1}\sum_{j=1}^{N-m+1}\theta\left(r - RR(i) - RR(j)\right) -$$

where θ is the Heaviside function and m, the length, and r, the tolerance window, are user-specified parameters. In terms of interpretation, a low value of ApEn reflects a regular pattern in the signal, while a high value of ApEn refers to a random pattern. For example, a reduction in ApEn is related with the high possibility of presence of cardiac disease (Rajendra Acharya et al., 2006). ApEn has an intrinsic bias caused by self-matching during its calculation, which makes it relatively inconsistent. This has led to the definition of modified versions of it. Li et al. proposed the sliding trend fuzzy approximate entropy (SlTr-fApEn) as a novel parameters to analyze HRV in OSA. They found that SlTr-fApEn had 85% accuracy for OSA screening, higher than the obtained with the LF/HF ratio (80%) (Li et al., 2019b). More recently, the same authors proposed the variance delay fuzzy approximate entropy (VD_fApEn), which reported 90% accuracy for OSA screening (Li et al., 2019b).

- *Sample entropy (SampEn):* Richman and Moorman introduced the SampEn to reduce the inherent bias caused by self-matching, in the ApEn, as well as to provide a result more independent on the time series length (Richman & Moorman, 2000). Given a RR time series of length N, RR(n) = {RR(1), RR(2), …, RR(N)}, $N - m + 1$ template vectors of length m are formed as $RR_m(i) = \{RR_i, RR_{i+1}, RR_{i+m-1}\}$. SampEn is defined as the negative natural logarithm of the conditional probability that two templates similar for m points remain similar (distance lower than the tolerance r) it if their lengths are increased in one sample (length $m + 1$):

$$\text{SampEn}\left(m,r,N\right) = -\ln\frac{A^m(r)}{B^m(r)}$$

where $A^m(r)$ and $B^m(r)$ are the total number of similar template vectors of lengths $m + 1$ and m that meet the distance criterion for each combination of $RR_m(i)$ and $RR_m(j)$, given $i \neq j$, respectively. SampEn has been widely employed to analyze HRV in many clinical domains, including the OSA context. Al-Angari et al. found statistically significant higher values of the SampEn in no-OSA than in OSA subjects, suggesting that complexity of HRV was significantly different between normal and OSA patients (Al-Angari & Sahakian, 2007). Similarly, Chang et al. also found a lower SampEn in the HRV of OSA patients than in controls. They also observed an increased SampEn which increased after continuous positive airway pressure (CPAP) treatment, suggesting that CPAP treatment normalized both cardiorespiratory decoupling and sympathovagal imbalance (Chang et al., 2013). Liang et al. proposed the nonparametric sample entropy (NPSampEn) as a novel index for short-term HRV analysis in the case of OSA. They reported a higher OSA screening accuracy (83.3%) than the LF/HF ratio (73.3%) and the SampEn (68.3%) (Liang et al., 2021a). Liang et al. reported a decreased SampEn during sleep-to-wake transitions in OSA patients, suggesting that a threshold-based SampEn could non-invasively detect them (Liang et al., 2021b).

- *Multiscale entropy (MSE):* On the basis of ApEn and SampEn, Costa et al. proposed the MSE. In contrast to conventional entropy measures, MSE computes entropy for different time scales of the time series, thus providing a measure of signal complexity (Costa et al., 2005). Given a RR time series of length N, RR(n) = {RR(1), RR(2), …, RR(N)}, the coarse-grained versions for each time scale τ is computed as follows:

$$y^{\tau}(j)=\frac{1}{\tau}\sum_{i=(j-1\tau+1)}^{j\tau} RR(i), 1 \le j < \frac{N}{\tau}$$

Hence, y^{τ} is is obtained by averaging the RR time series every τ samples without overlapping, being y^1 the original RR time series. Then, a single-scale entropy measure is computed for each of the coarse-grained versions. ApEn and SampEn are typically used as single-scale entropy measures (Costa et al., 2005). The curve representing the value of the single-scale entropy measure versus the scale allows to analyze the signal complexity. Pan et al. performed a MSE analysis of RR intervals during 10 minutes of N2 stage. They demonstrated that there is a high correlation between MSE and the AHI (Pan et al., 2015). Gutierrez-Tobal et al. also reported a higher value of MSE in OSA-positive subjects, as well as a high discrimination ability, with 85% accuracy for OSA detection. They also observed higher differences in MSE between OSA-positive (AHI≥10) and OSA-negative (AHI<10) in women than in men (Gutiérrez-Tobal et al., 2015).

10.3.3.3 Symbolic Dynamics

Symbolic dynamics is employed to measure randomness and predictability of cardiac rhythm. Symbolic dynamics transform RR intervals into a symbol sequence in order to analyze dynamic behavior (Kurths et al., 1995). Given a RR time series of length N, RR(i) = {RR(1), RR(2), …, RR(N)}, the RR intervals are first transformed into a symbolic dynamic representation using three-symbol ($w = 3$) words and an alphabet composed of four symbols.

$$S_i(X_i)=\begin{cases} 0: & \mu < RR(i) \le (1+\alpha)\mu \\ 1: & (1+\alpha)\mu < RR(i) < \infty \\ 2: & (1-\alpha)\mu < RR(i) \le \mu \\ 3: & 0 < RR(i) \le (1-\alpha)\mu \end{cases}$$

$S_1, S_2, S_3, \ldots S_N$. $S_i \in$ A, being A the alphabet of symbols = {0, 1, 2, 3, μ refers to the mean beat-to-beat interval, and α is a special parameter, with a typical value of 0.05. Then, the dynamics of the symbol sequence is analyzed by different parameters (Voss et al., 1996):

- Shannon entropy from the word distribution of the symbol sequence (Fwshannon). The Fwshannon measures complexity of the corresponding tachograms, with larger values meaning higher complexity.
- Renyi entropy from the word distribution of the symbol sequence (Fwrenyiq). Fwrenyiq also measures the complexity of the tachogram.

- Forbidden words (Forbword). Forbword is the number of words in the distribution of three-symbol words that never or seldom occur. Larger values of Forbword indicate a higher stability, since the number of forbidden words will be low when the time series is highly irregular.
- Standard deviation of the word sequence (wsdvar).
- Portion of low-variability (plvarxx) and high-variability patterns (phvarxx) in the NN interval time series. To compute these parameters, a simplified alphabet consisting of symbols "0" (difference between successive beats lower than xx) and "1" (difference between successive beats exceeds xx) is obtained. Observing six successive symbols of this alphabet, plvarxx (xx = 5, 10, 20, 50, and 100 ms) is computed as the probability of the word "000000," whereas phvarxx (xx = 5, 10, 20, 50, and 100 ms) is computed as the probability of the word "111111." plvarxx and phvarxx allow to detect intermittent decreased and intermittent increased variability, respectively.
- Percentage of probability of words consisting only of the symbols "0" and "2" (wpsum02) and the symbols "1" and "3" (wpsum13). Wpsum02 is a measure quantifying low variation in the mean of heartbeat intervals for decreased HRV, while wpsum13 is for increased HRV by quantifying high variation in the mean of heartbeat intervals.

Kabir et al. analyzed cardiorespiratory coordination in OSA patients using joint symbolic dynamics. They found that coupling between HRV and respiratory phase is reduced by OSA (Kabir et al., 2012) while being higher during slow wave sleep (Kabir et al., 2011). Ravelo-Garcia et al. developed a novel model using the combination of a symbolic dynamic maker (e.g., WPSUM13) and clinical variables (e.g., age, neck circumference, ESS, and intensity of snoring) for OSA screening with increased accuracy of 94% (sensitivity 88.71% and specificity 82.86%) than only using clinical information with the accuracy of 90% (sensitivity 87.10% and specificity 80%) (Ravelo-García et al., 2014). Recently, Qin et al. analyzed HRV in OSA patients recruited from the Sleep Apnea Global Interdisciplinary Consortium cohort during wakefulness using Fwshannon and Forbword from symbolic dynamics. Significantly lower values of Fwshannon and higher values of Forbword were obtained in patients with severe OSA, which displays the patterning of NN interval sequences becoming more monotonous as AHI increases. This indicates that heart rate in severe OSA patients does not adequately respond and adapt to endogenous and exogenous changes, due to blunted cardiac autonomic modulation (Qin et al., 2021b).

Information-based similarity indices proposed by Cui et al. used binary sequences to develop symbolic sequences derived from RR intervals (Cui et al., 2017). Wu et al. found that mild to moderate OSA tends to be overlooked when using HF, LF, and LF/HF calculated from frequency-domain analysis (Wu et al., 2021). Information-based similarity can differentiate mild to moderate OSA from severe OSA. However, it is disputable that they used 1-minute ECG instead of the 5-minute time window that is the standard recommended interval for spectral and dynamic HRV analysis. Preliminary evidence showed that minute-by-minute nonlinear HRV analysis has a good ability of real-time monitoring for OSA detection (Al-Angari & Sahakian, 2007; Li et al., 2018). The validity of those ultra-short-term HRV results still needs to be proven.

10.3.3.4 Poincaré Plots

The Poincaré plot analysis of HRV measures the nonlinearity of beat-to-beat dynamics based on a geometrical method, which portrays a scatter graph of RR intervals (RR_n) plotted against next RR intervals (RR_{n+1}) (Brennan et al., 2001). The following descriptors of the Poincaré plot are computed to display the geometrical and nonlinear features of the HRV dynamics.

- Standard deviation of the projection of the Poincaré plot in the perpendicular line to the line of identity, which is defined as:

$$SD1 = \frac{SD(RR_n - RR_{n+1})}{\sqrt{2}}$$

where SD denotes the standard deviation. $SD1$ measures the short-term NN interval variability (Brennan et al., 2001). SD1 is thought to reflect instantaneous beat-to-beat variability.

- Standard deviation of the projection of the Poincaré plot in the line of identity, which is defined as:

$$SD2 = \frac{SD(RR_n + RR_{n+1})}{\sqrt{2}}$$

where SD denotes the standard deviation. SD2 measures the long-term NN interval variability. SD2 is considered to be an index of global cardiac autonomic variability.

- Area of the ellipse characterized by SD1 and SD2 (A), which quantifies the total variability and is computed as:

$$A = 2\pi \cdot \cdot SD1 \cdot SD2$$

- SD1/SD2 ratio, which measures the influence of short-term vs. long-term NN variability.

Aljadeff et al. observed a greater dispersion in the Poincaré plots of the RR time series in children with OSA than in controls. They also found that beat-to-beat variation at slow rates was significantly increased in children with OSA, while variation at fast and intermediate heart rates was significantly reduced in these pediatric subjects (Aljadeff et al., 1997). Similarly, Rahman et al. reported significantly higher values in SD1 and SDRR of the RR intervals in adults with severe OSA than in adults with non-severe OSA, as well as an increase in the dispersion of the Poincaré plot with the increase of the hypopnea index (Rahman et al., 2018). In addition, an ensemble classifier fed with time-domain, frequency-domain, and Poincaré features obtained 87.5% for the detection of severe OSA. Chaidas et al. obtained that the width of the Poincaré plot of the RR intervals is reduced and morning urine norepinephrine concentration is increased in children with OSA (AHI>1) and moderate-to-severe nocturnal hypoxemia (Chaidas et al., 2014). This reflects that subjects with OSA have enhanced sympathetic activity and reduced parasympathetic drive. Limitations of Poincaré plot analysis include assumptions about normal distribution of RRI time series data and the elliptical shape of the plot.

10.3.3.5 Recurrence Plots

Recurrence plots (RP) were proposed by Eckmann et al. to visualize the recurrences of a dynamical system (Eckmann et al., 1987). RP analysis of HRV allows to capture nonlinear dynamics of a complex cardiorespiratory system during SDB. Nguyen et al. founded the cardiorespiratory change that occurs during apneic episodes leads to the appearance of diagonal and vertical patterns in the RP matrix. A soft decision fusion classifier using RQA features of HRV achieved 85.3% accuracy in the context of OSA diagnosis (Nguyen et al., 2014). Similarly, Martin-Gonzalez et al. demonstrated that RQA features from HRV contribute valuable information for apnea minutes discrimination (Martín-González et al., 2018). Rolink et al. suggested that RQA features from ECG, heart rate, and respiratory effort can discriminate between wakefulness and sleep stages (Rolink et al., 2015).

10.3.3.6 Chaotic Invariant Analysis

Beat-to-beat interval time series exhibit heterogeneous correlations in different disease states (Iyengar et al., 1996). Fractal behavior of HRV is highly correlated with abnormal respiratory patterns in patients with OSA (D'Addio et al., 2013). The correlation dimension (CD) is a commonly used indicator of the fractal dimension (Grassberger & Procaccia, 1983). CD is applied to capture correlations and feedback nonlinearities between the dynamic respiratory and cardiovascular systems by using a set of physiological signals from PSG to create a prediction model in sleep apnea. Nevertheless, limited articles on its significance and application in sleep medicine are available. From available literature, it has been shown that the value of D2 will be high for the chaotic RRI series and low for the less rhythmic RRI series. Previous studies indicated that D2 in respiratory movement obtained from inductive plethysmography could be a marker for OSA detection during daytime wakefulness and sleep. When applying correlation dimension to ECG signals, Naghsh et al. found that correlation dimension could differentiate between OSA and healthy subjects based on non-linear behavior of heartbeat (Naghsh et al., 2020).

10.4 Future Research Direction

The causal relationship between sleep apnea and autonomic abnormalities is not fully understood. Whether impaired autonomic modulation contributes to the development or worsening of sleep apnea via alterations in cardiorespiratory control or vascular stability of the upper airway requires longitudinal investigation. Additionally, increasing evidence supports that autonomic dysfunction may precede the development of functional cardiovascular impairment. Whether abnormal HRV is involved in the underlying etiology in SDB population-associated cardiovascular morbidity and mortality beyond as a risk factor is still a matter to discuss.

Advancements in big data management and predictive tools (e.g., machine learning) encourage the extensive study of HRV features. More research studies are encouraged to investigate the continuous physiological processes changing from normal breathing to apneic/hypopneic episodes during sleep. The prediction algorithms and approaches facilitate not only the potential mechanisms on how SDB predisposes the development of autonomic abnormalities and cardiovascular diseases but also the prediction of the onset of sleep apnea, which is in favor of proactive treatment.

Future research could take advantage of sleep and big data (https://sleepdata.org/) as well as existing datasets (e.g., the Sleep Heart Health Study, the Wisconsin Sleep Cohort, and the Sleep Apnea cardioVascular Endpoints (SAVE) Trial) to examine and validate the HRV as a cardiovascular risk stratification indicator to predict cardiovascular morbidity and mortality. However, there is a lack of evidence regarding the predictive and prognostic implication of altered HRV for cardiovascular events among different OSA subgroups (e.g., clinical symptom clusters, PSG-based phenotypes, cardiac phenotypes, and genomic phenotypes). These findings would improve decision-making capacity and lead to more precise treatment pathways.

In terms of the selection of HRV metrics, numerous HRV features help further the understanding of underlying associations between autonomic modulation and sleep apnea-related physiological changes. However, those that need to be reported depend on the research question. An evidence-based approach of HRV feature choice would reduce high computational load, particularly in machine learning models. Of particular note is that statistical properties in the time series of physiologic signals change over time, which make it complex and nonstationary. The combination of linear and nonlinear HRV analysis may provide more comprehensive information. The combination of EDR and CPC may also provide more information, as they have shown positive correlations. It is thus imperative to continue to extract more features of ECG morphology and apply them to PSG uses to create the most accurate diagnostic and monitoring tool possible.

The field is thus trending toward machine learning model tools with combinatorial analysis approaches, but this requires more investigation, particularly for explaining the results obtained with these models. However, it would inevitably be the most implementable and affordable clinical tool that does not require constant expert supervision and analysis.

Furthermore, it is important to investigate other sleep-disturbing diseases that may have

HRV changes during sleep. Hyperarousal processes are central to the pathophysiology of primary insomnia, and short sleep duration is known to be a risk factor for cardiovascular mortality. The autonomic effects of short sleep on mortality, however, have not convincingly been shown yet. There are a small number of studies comparing HRV between insomniacs and control subjects with mixed results, requiring more investigation to elucidate a definitive relationship. In depression, autonomic changes are correlated with altered mood states, sleep disturbance, and physical dysfunction. Some studies have shown reduced HRV indices due to altered autonomic activity in depression, signaling its potential for use in the diagnosis and monitoring of depression. However, more research is needed in animal models to provide supporting evidence for the investigation of HRV in such a capacity. Overall, further investigation of the HRV changes in other conditions with sleep disturbances could be helpful in exploring the detection of the condition and its underlying pathophysiology, as well as predicting the cardiovascular mortality risk in each of these conditions.

References

Akselrod, S., Gordon, D., Ubel, F. A., Shannon, D. C., Berger, A. C., & Cohen, R. J. (1981). Power spectrum analysis of heart rate fluctuation: A quantitative probe of beat-to-beat cardiovascular control. *Science (New York, N.Y.), 213*(4504), 220–222. https://doi.org/10.1126/science.6166045

Al-Angari, H. M., & Sahakian, A. V. (2007). Use of sample entropy approach to study heart rate variability in obstructive sleep apnea syndrome. *IEEE Transactions on Bio-medical Engineering, 54*(10), 1900–1904. https://doi.org/10.1109/tbme.2006.889772

Aljadeff, G., Gozal, D., Schechtman, V. L., Burrell, B., Harper, R. M., & Ward, S. L. (1997). Heart rate variability in children with obstructive sleep apnea. *Sleep, 20*(2), 151–157. https://doi.org/10.1093/sleep/20.2.151

Atri, R., & Mohebbi, M. (2015). Obstructive sleep apnea detection using spectrum and bispectrum analysis of single-lead ECG signal. *Physiological Measurement, 36*(9), 1963–1980. https://doi.org/10.1088/0967-3334/36/9/1963

Aydin, M., Altin, R., Ozeren, A., Kart, L., Bilge, M., & Unalacak, M. (2004). Cardiac autonomic activity in obstructive sleep apnea: Time-dependent and spectral analysis of heart rate variability using 24-hour Holter electrocardiograms. *Texas Heart Institute Journal, 31*(2), 132–136.

Balachandran, J. S., Bakker, J. P., Rahangdale, S., Yim-Yeh, S., Mietus, J. E., Goldberger, A. L., & Malhotra, A. (2012). Effect of mild, asymptomatic obstructive sleep apnea on daytime heart rate variability and impedance cardiography measurements. *The American Journal of Cardiology, 109*(1), 140–145. https://doi.org/10.1016/j.amjcard.2011.07.071

Bansal, D., Khan, M., & Salhan, A. K. A review of measurement and analysis of heart rate variability. In *2009 International Conference on Computer and Automation Engineering*, 8–10 March 2009. pp 243–246. https://doi.org/10.1109/ICCAE.2009.70.

Bigger, J. T., Jr., Fleiss, J. L., Steinman, R. C., Rolnitzky, L. M., Kleiger, R. E., & Rottman, J. N. (1992). Correlations among time and frequency domain measures of heart period variability two weeks after acute myocardial infarction. *The American Journal of Cardiology, 69*(9), 891–898. https://doi.org/10.1016/0002-9149(92)90788-z

Bilan, A., Witczak, A., Palusiński, R., Myśliński, W., & Hanzlik, J. (2005). Circadian rhythm of spectral indices of heart rate variability in healthy subjects. *Journal of Electrocardiology, 38*(3), 239–243. https://doi.org/10.1016/j.jelectrocard.2005.01.012

Blomster, H., Laitinen, T. P., Hartikainen, J. E., Laitinen, T. M., Vanninen, E., Gylling, H., Sahlman, J., Kokkarinen, J., Randell, J., Seppä, J., & Tuomilehto, H. (2015). Mild obstructive sleep apnea does not modulate baroreflex sensitivity in adult patients. *Nature and Science of Sleep, 7*, 73–80. https://doi.org/10.2147/nss.S82443

Bonnet, M. H., & Arand, D. L. (1997). Heart rate variability: Sleep stage, time of night, and arousal influences. *Electroencephalography and Clinical Neurophysiology, 102*(5), 390–396. https://doi.org/10.1016/s0921-884x(96)96070-1

Brennan, M., Palaniswami, M., & Kamen, P. (2001). Do existing measures of Poincaré plot geometry reflect nonlinear features of heart rate variability? *IEEE Transactions on Bio-medical Engineering, 48*(11), 1342–1347. https://doi.org/10.1109/10.959330

Chaidas, K., Tsaoussoglou, M., Theodorou, E., Lianou, L., Chrousos, G., & Kaditis, A. G. (2014). Poincaré plot width, morning urine norepinephrine levels, and autonomic imbalance in children with obstructive sleep apnea. *Pediatric Neurology, 51*(2), 246–251. https://doi.org/10.1016/j.pediatrneurol.2014.05.003

Chang, J. S., Lee, S. D., Ju, G., Kim, J. W., Ha, K., & Yoon, I. Y. (2013). Enhanced cardiorespiratory coupling in patients with obstructive sleep apnea following continuous positive airway pressure treatment. *Sleep Medicine, 14*(11), 1132–1138. https://doi.org/10.1016/j.sleep.2013.04.024

Costa, M., Goldberger, A. L., & Peng, C. K. (2002). Multiscale entropy analysis of complex physiologic time series. *Physical Review Letters, 89*(6), 068102. https://doi.org/10.1103/PhysRevLett.89.068102

Costa, M., Goldberger, A. L., & Peng, C. K. (2005). Multiscale entropy analysis of biological signals. *Physical Review. E, Statistical, Nonlinear, and Soft Matter Physics, 71*(2 Pt 1), 021906. https://doi.org/10.1103/PhysRevE.71.021906

Crasset, V., Mezzetti, S., Antoine, M., Linkowski, P., Degaute, J. P., & van de Borne, P. (2001). Effects of aging and cardiac denervation on heart rate variability during sleep. *Circulation, 103*(1), 84–88. https://doi.org/10.1161/01.cir.103.1.84

Cui, X., Chang, E., Yang, W.-H., Jiang, B., Yang, A. C., & Peng, C.-K. (2017). Automated detection of paroxysmal atrial fibrillation using an information-based similarity approach. *Entropy, 19*, 677. https://doi.org/10.3390/e19120677

D'Addio, G., Accardo, A., Fornasa, E., Cesarelli, M., & Felice, A. (2013). Fractal behaviour of heart rate variability reflects abnormal respiration patterns in OSAS patients, *40*.

da Silva, E. L. P., Pereira, R., Reis, L. N., Pereira, V. L., Jr., Campos, L. A., Wessel, N., & Baltatu, O. C. (2015). Heart rate detrended fluctuation indexes as estimate of obstructive sleep apnea severity. *Medicine, 94*(4), e516. https://doi.org/10.1097/md.0000000000000516

Dekker, J. M., Schouten, E. G., Klootwijk, P., Pool, J., Swenne, C. A., & Kromhout, D. (1997). Heart rate variability from short electrocardiographic recordings predicts mortality from all causes in middle-aged and elderly men. The Zutphen study. *American Journal of Epidemiology, 145*(10), 899–908. https://doi.org/10.1093/oxfordjournals.aje.a009049

Dempsey, J. A., Veasey, S. C., Morgan, B. J., & O'Donnell, C. P. (2010). Pathophysiology of sleep apnea. *Physiological Reviews, 90*(1), 47–112. https://doi.org/10.1152/physrev.00043.2008

Dissanayake, H. U., Bin, Y. S., Ucak, S., de Chazal, P., Sutherland, K., & Cistulli, P. A. (2021). Association between autonomic function and obstructive sleep apnea: A systematic review. *Sleep Medicine Reviews, 57*, 101470. https://doi.org/10.1016/j.smrv.2021.101470

Eckmann, J.-P., Kamphorst, S., & Ruelle, D. (1987). Recurrence plots of dynamical systems. *Europhysics Letters (epl), 4*, 973–977. https://doi.org/10.1209/0295-5075/4/9/004

Ferini-Strambi, L., Zucconi, M., Oldani, A., & Smirne, S. (1992). Heart rate variability during sleep in snorers with and without obstructive sleep apnea. *Chest, 102*(4), 1023–1027. https://doi.org/10.1378/chest.102.4.1023

Francis, D. P., Willson, K., Georgiadou, P., Wensel, R., Davies, L. C., Coats, A., & Piepoli, M. (2002). Physiological basis of fractal complexity properties of heart rate variability in man. *The Journal of Physiology, 542*(Pt 2), 619–629. https://doi.org/10.1113/jphysiol.2001.013389

Gami, A. S., Howard, D. E., Olson, E. J., & Somers, V. K. (2005). Day-night pattern of sudden death in obstructive sleep apnea. *The New England Journal of Medicine, 352*(12), 1206–1214. https://doi.org/10.1056/NEJMoa041832

Gates, G. J., Mateika, S. E., Basner, R. C., & Mateika, J. H. (2004). Baroreflex sensitivity in nonapneic snorers and control subjects before and after nasal continuous positive airway pressure. *Chest, 126*(3), 801–807. https://doi.org/10.1378/chest.126.3.801

Gerritsen, J., Dekker, J. M., TenVoorde, B. J., Kostense, P. J., Heine, R. J., Bouter, L. M., Heethaar, R. M., & Stehouwer, C. D. (2001). Impaired autonomic function is associated with increased mortality, especially in subjects with diabetes, hypertension, or a history of cardiovascular disease: The Hoorn Study. *Diabetes Care, 24*(10), 1793–1798. https://doi.org/10.2337/diacare.24.10.1793

Glos, M., Penzel, T., Schoebel, C., Nitzsche, G.-R., Zimmermann, S., Rudolph, C., Blau, A., Baumann, G., Jost-Brinkmann, P.-G., Rautengarten, S., Meier, J. C., Peroz, I., & Fietze, I. (2016). Comparison of effects of OSA treatment by MAD and by CPAP on cardiac autonomic function during daytime. *Sleep & Breathing, 20*(2), 635–646. https://doi.org/10.1007/s11325-015-1265-0

Grassberger, P., & Procaccia, I. (1983). Characterization of strange attractors. *Physical Review Letters, 50*(5), 346–344. https://doi.org/10.1103/PhysRevLett.50.346

Guilleminault, C., Connolly, S., Winkle, R., Melvin, K., & Tilkian, A. (1984). Cyclical variation of the heart rate in sleep apnoea syndrome. Mechanisms, and usefulness of 24 h electrocardiography as a screening technique. *Lancet, 1*(8369), 126–131. https://doi.org/10.1016/s0140-6736(84)90062-x

Gula, L. J., Krahn, A. D., Skanes, A., Ferguson, K. A., George, C., Yee, R., & Klein, G. J. (2003). Heart rate variability in obstructive sleep apnea: A prospective study and frequency domain analysis. *Annals of Noninvasive Electrocardiology: The Official Journal of the International Society for Holter and Noninvasive Electrocardiology, Inc, 8*(2), 144–149. https://doi.org/10.1046/j.1542-474x.2003.08209.x

Gutiérrez-Tobal, G., Alvarez, D., Gomez-Pilar, J., del Campo, F., & Hornero, R. (2015). Assessment of time and frequency domain entropies to detect sleep apnoea in heart rate variability recordings from men and women. *Entropy, 17*, 123–141. https://doi.org/10.3390/e17010123

Hayano, J., Yamada, M., Sakakibara, Y., Fujinami, T., Yokoyama, K., Watanabe, Y., & Takata, K. (1990). Short- and long-term effects of cigarette smoking on heart rate variability. *The American Journal of Cardiology, 65*(1), 84–88. https://doi.org/10.1016/0002-9149(90)90030-5

Henriques T, Ribeiro M, Teixeira A, Castro L, Antunes L, Costa-Santos C (2020) Nonlinear methods most applied to heart-rate time series: A review. 22 (3). doi:https://doi.org/10.3390/e22030309

Hillebrand, S., Gast, K. B., de Mutsert, R., Swenne, C. A., Jukema, J. W., Middeldorp, S., Rosendaal, F. R., & Dekkers, O. M. (2013). Heart rate variability and first cardiovascular event in populations without known cardiovascular disease: Meta-analysis and dose-response meta-regression. *Europace: European Pacing, Arrhythmias, and Cardiac Electrophysiology: Journal of the Working Groups on Cardiac Pacing, Arrhythmias, and Cardiac Cellular Electrophysiology of the European Society of Cardiology, 15*(5), 742–749. https://doi.org/10.1093/europace/eus341

Huikuri, H. V., Mäkikallio, T. H., Airaksinen, K. E., Seppänen, T., Puukka, P., Räihä, I. J., & Sourander, L. B. (1998). Power-law relationship of heart rate variability as a predictor of mortality in the elderly. *Circulation, 97*(20), 2031–2036. https://doi.org/10.1161/01.cir.97.20.2031

Iber, C., Ancoli-Israel, S., Chesson, A. L., & Quan, S. (2007). *The AASM manual for the scoring of sleep and associated events: Rules, terminology and technical specifications*. American Academy of Sleep Medicine.

Idiaquez, J., Santos, I., Santin, J., Del Rio, R., & Iturriaga, R. (2014). Neurobehavioral and autonomic alterations in adults with obstructive sleep apnea. *Sleep Medicine, 15*(11), 1319–1323. https://doi.org/10.1016/j.sleep.2014.05.030

Israel, S. A., Irvine, J. M., Cheng, A., Wiederhold, M. D., & Wiederhold, B. K. (2005). ECG to identify individuals. *Pattern Recognition, 38*(1), 133–142. https://doi.org/10.1016/j.patcog.2004.05.014

Iyengar, N., Peng, C. K., Morin, R., Goldberger, A. L., & Lipsitz, L. A. (1996). Age-related alterations in the fractal scaling of cardiac interbeat interval dynamics. *The American Journal of Physiology, 271*(4 Pt 2), R1078–R1084. https://doi.org/10.1152/ajpregu.1996.271.4.R1078

Jiang, J., Chen, X., Zhang, C., Wang, G., Fang, J., Ma, J., & Zhang, J. (2017). Heart rate acceleration runs and deceleration runs in patients with obstructive sleep apnea syndrome. *Sleep & Breathing = Schlaf & Atmung, 21*(2), 443–451. https://doi.org/10.1007/s11325-016-1437-6

Kabir, M. M., Dimitri, H., Sanders, P., Antic, R., Abbott, D., & Baumert, M. Quantification of cardio-respiratory interactions in patients with mild obstructive sleep apnea syndrome using joint symbolic dynamics. In *2011 Computing in Cardiology*, 18–21 Sept. 2011. pp 41–44.

Kabir, M. M., Dimitri, H., Sanders, P., Abbott, D., & Baumert, M. Joint symbolic dynamics as an effective approach for quantification of cardio-respiratory interaction in patients with obstructive sleep apnea syndrome. In *2012 7th International Conference on Electrical and Computer Engineering*, 20–22 Dec 2012. pp 133–136. https://doi.org/10.1109/ICECE.2012.6471503

Khandoker, A. H., & Palaniswami, M. (2011). Modeling respiratory movement signals during central and obstructive sleep apnea events using electrocardiogram. *Annals of Biomedical Engineering, 39*(2), 801–811. https://doi.org/10.1007/s10439-010-0189-x

Khandoker, A. H., Gubbi, J., & Palaniswami, M. (2009a). Automated scoring of obstructive sleep apnea and hypopnea events using short-term electrocardiogram recordings. *IEEE Transactions on Information Technology in Biomedicine: A Publication of the IEEE Engineering in Medicine and Biology Society, 13*(6), 1057–1067. https://doi.org/10.1109/titb.2009.2031639

Khandoker, A. H., Palaniswami, M., & Karmakar, C. K. (2009b). Support vector machines for automated recognition of obstructive sleep apnea syndrome from ECG recordings. *IEEE Transactions on Information Technology in Biomedicine: A Publication of the IEEE Engineering in Medicine and Biology Society, 13*(1), 37–48. https://doi.org/10.1109/titb.2008.2004495

Khoo, M. C., Belozeroff, V., Berry, R. B., & Sassoon, C. S. (2001). Cardiac autonomic control in obstructive sleep apnea: Effects of long-term CPAP therapy. *American Journal of Respiratory and Critical Care Medicine, 164*(5), 807–812. https://doi.org/10.1164/ajrccm.164.5.2010124

Kikuya, M., Hozawa, A., Ohokubo, T., Tsuji, I., Michimata, M., Matsubara, M., Ota, M., Nagai, K., Araki, T., Satoh, H., Ito, S., Hisamichi, S., & Imai, Y. (2000). Prognostic significance of blood pressure and heart rate variabilities: The Ohasama study. *Hypertension (Dallas, Tex : 1979), 36*(5), 901–906. https://doi.org/10.1161/01.hyp.36.5.901

Kim, Y. S., Kim, S. Y., Park, D. Y., Wu, H. W., Hwang, G. S., & Kim, H. J. (2015). Clinical implication of heart rate variability in obstructive sleep apnea syndrome patients. *The Journal of Craniofacial Surgery, 26*(5), 1592–1595. https://doi.org/10.1097/scs.0000000000001782

Kleiger, R. E., Miller, J. P., Bigger, J. T., Jr., & Moss, A. J. (1987). Decreased heart rate variability and its association with increased mortality after acute myocardial infarction. *The American Journal of Cardiology, 59*(4), 256–262. https://doi.org/10.1016/0002-9149(87)90795-8

Kleiger, R. E., Bigger, J. T., Bosner, M. S., Chung, M. K., Cook, J. R., Rolnitzky, L. M., Steinman, R., & Fleiss, J. L. (1991). Stability over time of variables measuring heart rate variability in normal subjects. *The American Journal of Cardiology, 68*(6), 626–630. https://doi.org/10.1016/0002-9149(91)90355-o

Kohler, B., Hennig, C., & Orglmeister, R. (2002). The principles of software QRS detection. *IEEE Engineering in Medicine and Biology Magazine, 21*(1), 42–57. https://doi.org/10.1109/51.993193

Kong, S. D. X., Hoyos, C. M., Phillips, C. L., McKinnon, A. C., Lin, P., Duffy, S. L., Mowszowski, L., LaMonica, H. M., Grunstein, R. R., Naismith, S. L., & Gordon, C. J. (2021). Altered heart rate variability during sleep in mild cognitive impairment. *Sleep, 44*(4). https://doi.org/10.1093/sleep/zsaa232

Kurths, J., Voss, A., Saparin, P., Witt, A., Kleiner, H. J., & Wessel, N. (1995). Quantitative analysis of heart rate variability. *Chaos (Woodbury, NY), 5*(1), 88–94. https://doi.org/10.1063/1.166090

La Rovere, M. T., Bigger, J. T., Jr., Marcus, F. I., Mortara, A., & Schwartz, P. J. (1998). Baroreflex sensitivity and heart-rate variability in prediction of total cardiac mortality after myocardial infarction. ATRAMI (Autonomic Tone and Reflexes After Myocardial Infarction) investigators. *Lancet, 351*(9101), 478–484. https://doi.org/10.1016/s0140-6736(97)11144-8

La Rovere, M. T., Pinna, G. D., Hohnloser, S. H., Marcus, F. I., Mortara, A., Nohara, R., Bigger, J. T., Jr., Camm, A. J., & Schwartz, P. J. (2001). Baroreflex sensitivity and heart rate variability in the identification of patients at risk for life-threatening arrhythmias: Implications for clinical trials. *Circulation, 103*(16), 2072–2077. https://doi.org/10.1161/01.cir.103.16.2072

Le, T. Q., & Bukkapatnam, S. T. S. (2017). Correction: Nonlinear dynamics forecasting of obstructive sleep apnea onsets. *PLoS One, 12*(8), e0183422. https://doi.org/10.1371/journal.pone.0183422

Leung, R. S. (2009). Sleep-disordered breathing: Autonomic mechanisms and arrhythmias. *Progress in Cardiovascular Diseases, 51*(4), 324–338. https://doi.org/10.1016/j.pcad.2008.06.002

Li, X., Shaffer, M. L., Rodriguez-Colon, S., He, F., Wolbrette, D. L., Alagona, P., Jr., Wu, C., & Liao, D. (2011). The circadian pattern of cardiac autonomic modulation in a middle-aged population. *Clinical Autonomic Research: Official Journal of the Clinical Autonomic Research Society, 21*(3), 143–150. https://doi.org/10.1007/s10286-010-0112-4

Li, K., Pan, W., Li, Y., Jiang, Q., & Liu, G. (2018). A method to detect sleep apnea based on deep neural network and hidden Markov model using single-lead ECG signal. *Neurocomputing, 294*, 94–101. https://doi.org/10.1016/j.neucom.2018.03.011

Li, K., Rüdiger, H., & Ziemssen, T. (2019a). Spectral analysis of heart rate variability: Time window matters. *Frontiers in Neurology, 10*, 545.

Li, Y., Pan, W., Li, K., Jiang, Q., & Liu, G. (2019b). Sliding trend fuzzy approximate entropy as a novel descriptor of heart rate variability in obstructive sleep apnea. *IEEE Journal of Biomedical and Health Informatics, 23*(1), 175–183. https://doi.org/10.1109/jbhi.2018.2790968

Liang D, Wu S, Tang L, Feng K, Liu G (2021a) Short-term HRV analysis using nonparametric sample entropy for obstructive sleep apnea. 23 (3). doi:https://doi.org/10.3390/e23030267

Liang, X., Xiong, J., Cao, Z., Wang, X., Li, J., & Liu, C. (2021b). Decreased sample entropy during sleep-to-wake transition in sleep apnea patients. *Physiological Measurement, 42*(4). https://doi.org/10.1088/1361-6579/abf1b2

Liao, D., Barnes, R. W., Chambless, L. E., Simpson, R. J., Jr., Sorlie, P., & Heiss, G. (1995). Age, race, and sex differences in autonomic cardiac function measured by spectral analysis of heart rate variability--the ARIC study. Atherosclerosis Risk in Communities. *The American Journal of Cardiology, 76*(12), 906–912. https://doi.org/10.1016/s0002-9149(99)80260-4

Liao, D., Cai, J., Rosamond, W. D., Barnes, R. W., Hutchinson, R. G., Whitsel, E. A., Rautaharju, P., & Heiss, G. (1997). Cardiac autonomic function and incident coronary heart disease: A population-based case-cohort study. The ARIC study. Atherosclerosis Risk in Communities Study. *American Journal of Epidemiology, 145*(8), 696–706. https://doi.org/10.1093/aje/145.8.696

Malpas, S. C., Whiteside, E. A., & Maling, T. J. (1991). Heart rate variability and cardiac autonomic function in men with chronic alcohol dependence. *British Heart Journal, 65*(2), 84–88. https://doi.org/10.1136/hrt.65.2.84

Martín-González, S., Navarro-Mesa, J. L., Juliá-Serdá, G., Ramírez-Ávila, G. M., & Ravelo-García, A. G. (2018). Improving the understanding of sleep apnea characterization using Recurrence Quantification Analysis by defining overall acceptable values for the dimensionality of the system, the delay, and the distance threshold. *PLoS One, 13*(4), e0194462. https://doi.org/10.1371/journal.pone.0194462

Martín-Montero A, Gutiérrez-Tobal GC, Gozal D, Barroso-García V, Álvarez D, Del Campo F, Kheirandish-Gozal L, Hornero R (2021) Bispectral analysis of heart rate variability to characterize and help diagnose pediatric sleep apnea. 23 (8). doi:https://doi.org/10.3390/e23081016

McNicholas, W. T., & Bonsigore, M. R. (2007). Sleep apnoea as an independent risk factor for cardiovascular disease: Current evidence, basic mechanisms and research priorities. *The European Respiratory Journal, 29*(1), 156–178. https://doi.org/10.1183/09031936.00027406

Mendez, M. O., Corthout, J., Van Huffel, S., Matteucci, M., Penzel, T., Cerutti, S., & Bianchi, A. M. (2010). Automatic screening of obstructive sleep apnea from the ECG based on empirical mode decomposition and wavelet analysis. *Physiological Measurement, 31*(3), 273–289. https://doi.org/10.1088/0967-3334/31/3/001

Migliorini, M., Mendez, M. O., & Bianchi, A. M. (2012). Study of heart rate variability in bipolar disorder: Linear and non-linear parameters during sleep. *Frontiers in Neuroengineering, 4*, 22.

Moody, G., Mark, R., Zoccola, A., & Mantero, S. (1985). Derivation of respiratory signals from multilead ECGs. *Computers in Cardiology, 12*.

Naghsh, S., Ataei, M., Yazdchi, M., & Hashemi, M. (2020). Chaos-based analysis of heart rate variability time series in obstructive sleep apnea subjects. *Journal of Medical Signals and Sensors, 10*(1), 53–59. https://doi.org/10.4103/jmss.JMSS_23_19

Narkiewicz, K., & Somers, V. K. (2001). Cardiovascular variability characteristics in obstructive sleep apnea. *Autonomic Neuroscience: Basic & Clinical, 90*(1–2), 89–94. https://doi.org/10.1016/s1566-0702(01)00272-7

Narkiewicz, K., Montano, N., Cogliati, C., van de Borne, P. J., Dyken, M. E., & Somers, V. K. (1998). Altered cardiovascular variability in obstructive sleep apnea. *Circulation, 98*(11), 1071–1077. https://doi.org/10.1161/01.cir.98.11.1071

Nguyen, H. D., Wilkins, B. A., Cheng, Q., & Benjamin, B. A. (2014). An online sleep apnea detection method based on recurrence quantification analysis. *IEEE Journal of Biomedical and Health Informatics, 18*(4), 1285–1293. https://doi.org/10.1109/jbhi.2013.2292928

Noda, A., Yasuma, F., Okada, T., & Yokota, M. (1998). Circadian rhythm of autonomic activity in patients with obstructive sleep apnea syndrome. *Clinical Cardiology, 21*(4), 271–276. https://doi.org/10.1002/clc.4960210408

Nolan, J., Batin, P. D., Andrews, R., Lindsay, S. J., Brooksby, P., Mullen, M., Baig, W., Flapan, A. D., Cowley, A., Prescott, R. J., Neilson, J. M., & Fox, K. A. (1998). Prospective study of heart rate variability and mortality in chronic heart failure: Results of the United Kingdom heart failure evaluation and assessment of risk trial (UK-heart). *Circulation, 98*(15), 1510–1516. https://doi.org/10.1161/01.cir.98.15.1510

Otzenberger, H., Gronfier, C., Simon, C., Charloux, A., Ehrhart, J., Piquard, F., & Brandenberger, G. (1998). Dynamic heart rate variability: A tool for exploring sympathovagal balance continuously during sleep in men. *The American Journal of Physiology, 275*(3), H946–H950. https://doi.org/10.1152/ajpheart.1998.275.3.H946

Pagani, M., Lombardi, F., Guzzetti, S., Rimoldi, O., Furlan, R., Pizzinelli, P., Sandrone, G., Malfatto, G., Dell'Orto, S., Piccaluga, E., et al. (1986). Power spectral analysis of heart rate and arterial pressure variabilities as a marker of sympatho-vagal interaction in man and conscious dog. *Circulation Research, 59*(2), 178–193. https://doi.org/10.1161/01.res.59.2.178

Pan, J., & Tompkins, W. J. (1985). A real-time QRS detection algorithm. *IEEE Transactions on Bio-medical Engineering, 32*(3), 230–236. https://doi.org/10.1109/tbme.1985.325532

Pan, W.-Y., Su, M.-C., Wu, H.-T., Lin, M.-C., Tsai, I. T., & Sun, C.-K. (2015). Multiscale entropy analysis of heart rate variability for assessing the severity of sleep disordered breathing. *Entropy, 17*, 231–243. https://doi.org/10.3390/e17010231

Pawlowski, M. A., Gazea, M., Wollweber, B., Dresler, M., Holsboer, F., Keck, M. E., Steiger, A., Adamczyk, M., & Mikoteit, T. (2017). Heart rate variability and cordance in rapid eye movement sleep as biomarkers of depression and treatment response. *Journal of Psychiatric Research, 92*, 64–73. https://doi.org/10.1016/j.jpsychires.2017.03.026

Peng, C. K., Buldyrev, S. V., Havlin, S., Simons, M., Stanley, H. E., & Goldberger, A. L. (1994). Mosaic organization of DNA nucleotides. *Physical Review. E, Statistical Physics, Plasmas, Fluids, and Related Interdisciplinary Topics, 49*(2), 1685–1689. https://doi.org/10.1103/physreve.49.1685

Peng, C. K., Havlin, S., Stanley, H. E., & Goldberger, A. L. (1995). Quantification of scaling exponents and crossover phenomena in nonstationary heartbeat time series. *Chaos (Woodbury, NY), 5*(1), 82–87. https://doi.org/10.1063/1.166141

Penzel, T., McNames, J., Murray, A., de Chazal, P., Moody, G., & Raymond, B. (2002). Systematic comparison of different algorithms for apnoea detection based on electrocardiogram recordings. *Medical & Biological Engineering & Computing, 40*(4), 402–407. https://doi.org/10.1007/bf02345072

Penzel, T., Kantelhardt, J. W., Grote, L., Peter, J. H., & Bunde, A. (2003). Comparison of detrended fluctuation analysis and spectral analysis for heart rate variability in sleep and sleep apnea. *IEEE Transactions on Bio-medical Engineering, 50*(10), 1143–1151. https://doi.org/10.1109/tbme.2003.817636

Pincus, S. M. (1991). Approximate entropy as a measure of system complexity. *Proceedings of the National Academy of Sciences of the United States of America, 88*(6), 2297–2301. https://doi.org/10.1073/pnas.88.6.2297

Qin, H., Steenbergen, N., Glos, M., Wessel, N., Kraemer, J. F., Vaquerizo-Villar, F., & Penzel, T. (2021a). The different facets of heart rate variability in obstructive sleep apnea. *Frontiers in Psychiatry, 12*, 642333. https://doi.org/10.3389/fpsyt.2021.642333

Qin, H., Keenan, B. T., Mazzotti, D. R., Vaquerizo-Villar, F., Kraemer, J. F., Wessel, N., Tufik, S., Bittencourt, L., Cistulli, P. A., de Chazal, P., Sutherland, K., Singh, B., Pack, A. I., Chen, N. H., Fietze, I., Gislason, T., Holfinger, S., Magalang, U. J., & Penzel, T. (2021b). Heart rate variability during wakefulness as a marker of obstructive sleep apnea severity. *Sleep, 44*(5). https://doi.org/10.1093/sleep/zsab018

Rahman, M. J., Mahajan, R., Morshed, B. I. Severity classification of obstructive sleep apnea using only heart rate variability measures with an ensemble classifier. In *2018 IEEE EMBS International Conference on Biomedical & Health Informatics (BHI)*, 4–7 March 2018. pp 33–36. https://doi.org/10.1109/BHI.2018.8333363

Rajendra Acharya, U., Paul Joseph, K., Kannathal, N., Lim, C. M., & Suri, J. S. (2006). Heart rate variability: A review. *Medical & Biological Engineering & Computing, 44*(12), 1031–1051. https://doi.org/10.1007/s11517-006-0119-0

Ravelo-García, A. G., Saavedra-Santana, P., Juliá-Serdá, G., Navarro-Mesa, J. L., Navarro-Esteva, J., Álvarez-López, X., Gapelyuk, A., Penzel, T., & Wessel, N. (2014). Symbolic dynamics marker of heart rate variability combined with clinical variables enhance obstructive sleep apnea screening. *Chaos (Woodbury, NY), 24*(2), 024404. https://doi.org/10.1063/1.4869825

Richman, J. S., & Moorman, J. R. (2000). Physiological time-series analysis using approximate entropy and sample entropy. *American Journal of Physiology Heart and Circulatory Physiology, 278*(6), H2039–H2049. https://doi.org/10.1152/ajpheart.2000.278.6.H2039

Rioul, O., & Vetterli, M. (1991). Wavelets and signal processing. *IEEE Signal Processing Magazine, 8*(4), 14–38. https://doi.org/10.1109/79.91217

Roche, F., Gaspoz, J. M., Court-Fortune, I., Minini, P., Pichot, V., Duverney, D., Costes, F., Lacour, J. R., & Barthélémy, J. C. (1999). Screening of obstructive sleep apnea syndrome by heart rate variability analysis. *Circulation, 100*(13), 1411–1415. https://doi.org/10.1161/01.cir.100.13.1411

Rolink, J., Kutz, M., Fonseca, P., Long, X., Misgeld, B., & Leonhardt, S. (2015). Recurrence quantification analysis across sleep stages. *Biomedical Signal Processing and Control, 20*, 107–116. https://doi.org/10.1016/j.bspc.2015.04.006

Sadr, N., & de Chazal, P. (2019). A comparison of three ECG-derived respiration methods for sleep apnoea detection. *Biomedical Physics & Engineering Express, 5*(2), 025027. https://doi.org/10.1088/2057-1976/aafc80

Sankari, A., Ravelo, L. A., Maresh, S., Aljundi, N., Alsabri, B., Fawaz, S., Hamdon, M., Al-kubaisi, G., Hagen, E., Badr, M. S., & Peppard, P. (2019). Longitudinal effect of nocturnal R-R intervals changes on cardiovascular outcome in a community-based cohort. *BMJ Open, 9*(7), e030559. https://doi.org/10.1136/bmjopen-2019-030559

Sateia, M. J. (2014). International classification of sleep disorders-third edition: Highlights and modifications. *Chest, 146*(5), 1387–1394. https://doi.org/10.1378/chest.14-0970

Scholz, U. J., Bianchi, A. M., Cerutti, S., & Kubicki, S. (1997). Vegetative background of sleep: Spectral analysis of the heart rate variability. *Physiology & Behavior, 62*(5), 1037–1043. https://doi.org/10.1016/s0031-9384(97)00234-5

Sequeira, V. C. C., Bandeira, P. M., & Azevedo, J. C. M. (2019). Heart rate variability in adults with obstructive sleep apnea: A systematic review. *Sleep Science (Sao Paulo, Brazil), 12*(3), 214–221. https://doi.org/10.5935/1984-0063.20190082

Shinar, Z., Akselrod, S., Dagan, Y., & Baharav, A. (2006). Autonomic changes during wake-sleep transition: A heart rate variability based approach. *Autonomic Neuroscience : Basic & Clinical, 130*(1–2), 17–27. https://doi.org/10.1016/j.autneu.2006.04.006

Singh, N., Moneghetti, K. J., Christle, J. W., Hadley, D., Froelicher, V., & Plews, D. (2018). Heart rate variability: An old metric with new meaning in the era of using mHealth technologies for health and exercise training guidance. *Part Two: Prognosis and Training. Arrhythmia & Electrophysiology Review, 7*(4), 247–255. https://doi.org/10.15420/aer.2018.30.2

Sloan, R. P., Huang, M. H., McCreath, H., Sidney, S., Liu, K., Dale Williams, O., & Seeman, T. (2008). Cardiac autonomic control and the effects of age, race, and sex: The CARDIA study. *Autonomic Neuroscience : Basic & Clinical, 139*(1–2), 78–85. https://doi.org/10.1016/j.autneu.2008.01.006

Soares-Miranda, L., Sattelmair, J., Chaves, P., Duncan, G. E., Siscovick, D. S., Stein, P. K., & Mozaffarian, D. (2014). Physical activity and heart rate variability in older adults. *Circulation, 129*(21), 2100–2110. https://doi.org/10.1161/CIRCULATIONAHA.113.005361

Somers, V. K., Mark, A. L., & Abboud, F. M. (1991). Interaction of baroreceptor and chemoreceptor reflex control of sympathetic nerve activity in normal humans. *The Journal of Clinical Investigation, 87*(6), 1953–1957. https://doi.org/10.1172/jci115221

Somers, V. K., Dyken, M. E., Clary, M. P., & Abboud, F. M. (1995). Sympathetic neural mechanisms in obstructive sleep apnea. *The Journal of Clinical Investigation, 96*(4), 1897–1904. https://doi.org/10.1172/jci118235

Sondermeijer, H. P., van Marle, A. G., Kamen, P., & Krum, H. (2002). Acute effects of caffeine on heart rate variability. *The American Journal of Cardiology, 90*(8), 906–907. https://doi.org/10.1016/s0002-9149(02)02725-x

Sunkaria, R. (2011). Recent trends in nonlinear methods of HRV analysis: A review. *World Academy of Science, Engineering and Technology, 75*, 566–571.

Taranto Montemurro, L., Floras, J. S., Picton, P., Kasai, T., Alshaer, H., Gabriel, J. M., & Bradley, T. D. (2014). Relationship of heart rate variability to sleepiness in patients with obstructive sleep apnea with and without heart failure. *Journal of Clinical Sleep Medicine, 10*(3), 271–276. https://doi.org/10.5664/jcsm.3526

Heart rate variability: standards of measurement, physiological interpretation and clinical use. Task Force of the European Society of Cardiology and the North American Society of Pacing and Electrophysiology (1996). Circulation 93 (5):1043–1065. doi:https://doi.org/10.1161/01.CIR.93.5.1043

Taylor, J. A., Carr, D. L., Myers, C. W., & Eckberg, D. L. (1998). Mechanisms underlying very-low-frequency RR-interval oscillations in humans. *Circulation, 98*(6), 547–555. https://doi.org/10.1161/01.cir.98.6.547

Thomas, R. J., Mietus, J. E., Peng, C. K., & Goldberger, A. L. (2005). An electrocardiogram-based technique to assess cardiopulmonary coupling during sleep. *Sleep, 28*(9), 1151–1161. https://doi.org/10.1093/sleep/28.9.1151

Thomas, R. J., Mietus, J. E., Peng, C. K., Gilmartin, G., Daly, R. W., Goldberger, A. L., & Gottlieb, D. J. (2007). Differentiating obstructive from central and complex sleep apnea using an automated electrocardiogram-based method. *Sleep, 30*(12), 1756–1769. https://doi.org/10.1093/sleep/30.12.1756

Thomas, R. J., Mietus, J. E., Peng, C. K., Guo, D., Gozal, D., Montgomery-Downs, H., Gottlieb, D. J., Wang, C. Y., & Goldberger, A. L. (2014). Relationship between delta power and the electrocardiogram-derived cardiopulmonary spectrogram: Possible implications for assessing the effectiveness of sleep. *Sleep Medicine, 15*(1), 125–131. https://doi.org/10.1016/j.sleep.2013.10.002

Tobaldini, E., Nobili, L., Strada, S., Casali, K., Braghiroli, A., & Montano, N. (2013). Heart rate variability in normal and pathological sleep. *Frontiers in Physiology, 4*, 294.

Tsuji, H., Venditti, F. J., Jr., Manders, E. S., Evans, J. C., Larson, M. G., Feldman, C. L., & Levy, D. (1994). Reduced heart rate variability and mortality risk in an elderly cohort. The Framingham Heart Study. *Circulation, 90*(2), 878–883. https://doi.org/10.1161/01.cir.90.2.878

Ucak, S., Dissanayake, H. U., Sutherland, K., de Chazal, P., & Cistulli, P. A. (2021). Heart rate variability and obstructive sleep apnea: Current perspectives and novel technologies. *Journal of Sleep Research*, e13274. https://doi.org/10.1111/jsr.13274

Vanoli, E., Adamson, P. B., Ba, L., Pinna, G. D., Lazzara, R., & Orr, W. C. (1995). Heart rate variability during specific sleep stages. A comparison of healthy subjects with patients after myocardial infarction. *Circulation, 91*(7), 1918–1922. https://doi.org/10.1161/01.cir.91.7.1918

Voss, A., Kurths, J., Kleiner, H. J., Witt, A., Wessel, N., Saparin, P., Osterziel, K. J., Schurath, R., & Dietz, R. (1996). The application of methods of non-linear dynamics for the improved and predictive recognition of patients threatened by sudden cardiac death. *Cardiovascular Research, 31*(3), 419–433.

Wu, S., Liang, D., Yang, Q., & Liu, G.-Z. (2021). Regularity of heart rate fluctuations analysis in obstructive sleep apnea patients using information-based similarity. *Biomedical Signal Processing and Control, 65*, 102370. https://doi.org/10.1016/j.bspc.2020.102370

Zemaityte, D., Varoneckas, G., & Sokolov, E. (1984). Heart rhythm control during sleep. *Psychophysiology, 21*(3), 279–289. https://doi.org/10.1111/j.1469-8986.1984.tb02935.x

Zhang, L., Fu, M., Xu, F., Hou, F., & Ma, Y. (2019). Heart rate dynamics in patients with obstructive sleep apnea: Heart rate variability and entropy. *Entropy, 21*, 927. https://doi.org/10.3390/e21100927

Zhang, L., Wu, H., Zhang, X., Wei, X., Hou, F., & Ma, Y. (2020). Sleep heart rate variability assists the automatic prediction of long-term cardiovascular outcomes. *Sleep Medicine, 67*, 217–224. https://doi.org/10.1016/j.sleep.2019.11.1259

Zuanetti, G., Neilson, J. M., Latini, R., Santoro, E., Maggioni, A. P., & Ewing, D. J. (1996). Prognostic significance of heart rate variability in post-myocardial infarction patients in the fibrinolytic era. The GISSI-2 results. Gruppo Italiano per lo Studio della Sopravvivenza nell' Infarto Miocardico. *Circulation, 94*(3), 432–436. https://doi.org/10.1161/01.cir.94.3.432

11 Cardiopulmonary Coupling

Mi Lu, Thomas Penzel, and Robert J. Thomas

Abstract

Cardiopulmonary coupling (CPC) is a technique that generates sleep spectrogram by calculating the cross-spectral power and coherence of heart rate variability and respiratory tidal volume fluctuations. There are several forms of CPC in the sleep spectrogram, which may provide information about normal sleep physiology and pathological sleep states. Since CPC can be calculated from any signal recording containing heart rate and respiration information, such as photoplethysmography (PPG) or blood pressure, it can be widely used in various applications, including wearables and non-contact devices. When derived from PPG, an automatic apnea-hypopnea index can be calculated from CPC-oximetry as PPG can be obtained from oximetry alone. CPC-based sleep profiling reveals the effects of stable and unstable sleep on sleep apnea, insomnia, cardiovascular regulation, and metabolic disorders. Here, we introduce, with examples, the current knowledge and understanding of the CPC technique, especially the physiological basis, analytical methods, and its clinical applications.

Keywords

Autonomic nervous system · Cardiopulmonary coupling · Heart rate variability · Sleep apnea · Sleep spectrogram · Insomnia

11.1 Introduction

Everyone sleeps and sleep is essential for a variety of biological functions. However, it is estimated that nearly 2 billion people worldwide suffer from one or both of the two most common sleep disorders – sleep apnea (Benjafield et al., 2019) and insomnia (Roth et al., 2011). Most people with both diseases remain undiagnosed and untreated, thus resulting in major adverse outcomes on health, performance, and safety (Young et al., 1997). The current diagnostic approaches mainly rely on full-night polysomnography (PSG) or home sleep tests utilizing direct cardiopulmonary recording (e.g., effort,

M. Lu (✉)
Department of Otolaryngology-Head and Neck Surgery, Beijing Anzhen Hospital, Capital Medical University, Beijing, China

Interdisciplinary Center of Sleep Medicine, Charité – Universitätsmedizin Berlin, Berlin, Germany
e-mail: mi.lu@charite.de

T. Penzel
Interdisciplinary Center of Sleep Medicine, Charité – Universitätsmedizin Berlin, Berlin, Germany

R. J. Thomas
Division of Pulmonary and Sleep Medicine, Beth Israel Deaconess Medical Center, Boston, MA, USA

T. Penzel, R. Hornero (eds.), *Advances in the Diagnosis and Treatment of Sleep Apnea*, Advances in Experimental Medicine and Biology 1384, https://doi.org/10.1007/978-3-031-06413-5_11

nasal pressure), which are relatively labor-intensive, time-consuming, expensive, and uncomfortable for patients. Moreover, both a first/lab-night effect of PSG and night-to-night variability of sleep are found during recordings (Agnew Jr. et al., 1966; Mosko et al., 1988). Therefore, to better capture the physiological and pathological dynamics of sleep, it is advisable to assess patients' sleep in a natural sleep environment over multiple nights and on multiple occasions. In addition, poor adherence to continuous positive airway pressure (CPAP) as the first-line treatment for obstructive sleep apnea (OSA) may be related to poor subjective sleep quality in patients (Cistulli et al., 2019). As such, a nimble and effective approach to help clinicians and patients evaluate, diagnose, and track sleep disorders is highly desirable.

The cardiopulmonary coupling (CPC) technique, which only requires a continuous electrocardiogram (ECG) or photoplethysmography (PPG) signal as an input, is becoming more widely used in formal medical and consumer wearable devices (Hilmisson et al., 2020; Thomas et al., 2005). The CPC technique is based on analyzing the synchronization intensity of heart rate variability (HRV) and respiration data gathered during sleep, two bio signals that are both highly modulated by the autonomic nervous system (ANS) (Thomas et al., 2005; Waxenbaum et al., 2022), which is in turn highly modulated by sleep state, type, and depth. We can observe coupling between the heart and respiratory systems when external stimuli affect the ANS during sleep, allowing us to measure sleep and sleep stages.

The conventional approach to sleep has the rapid eye movement (REM) and non-rapid eye movement (NREM) stages, with three grades of NREM sleep. However, there are several other methods of quantifying sleep, including cyclic alternating pattern (CAP, a measure of sleep electroencephalogram [EEG] stability), the Odds Ratio Product (ORP, a measure of continuous sleep depth), and fine movement analysis beyond conventional actigraphy. While type, grade, and depth are useful metrics, sleep also has spontaneously shifted bimodal characteristics, independent of conventional grading, readily evident from respiratory stability or CPC analysis (Wood et al., 2020). High-frequency coupling (HFC) is one of them, and it's linked to stable NREM sleep, while low-frequency coupling (LFC) is associated with unstable NREM sleep (Thomas et al., 2005). A third CPC form, very-low-frequency coupling (VLFC), occurs during both REM sleep and wakefulness and can be distinguished by signal quality and motion artifact analysis (Al Ashry et al., 2021). These distinct CPC patterns logically vary with disease state and treatment. For example, patients with OSA have increased LFC, whereas successful CPAP therapy decreases it (Cho & Kim, 2017). In a recent study, CPC data and oxygen desaturation data were combined to calculate the automatic apnea-hypopnea index (AHI), which has been clinically validated and FDA-approved for the diagnosis and management of OSA in both children and adults (Al Ashry et al., 2021; Hilmisson et al., 2020).

In this chapter, we will introduce the current state of knowledge and understanding of the CPC technique. A greater focus is directed on the physiological basics, standard analytical methods, and the clinical application of the CPC technique in the evaluation, diagnosis, and management of sleep disorders.

11.2 Physiological Basics

The interaction between heart rate and respiratory tidal volumes was first documented in 1733 by Stephen Hales (Hales, 1733). Subsequent studies have shown that the synchronization between HRV and respiration improves gas exchange at the lung level through efficient ventilation/perfusion matching while minimizing the workload on the heart (Ben-Tal et al., 2012; Yasuma & Hayano, 2004). Such cardiopulmonary synchronization is optimal during deep sleep, sedation, and anesthesia (Dick et al., 2014). The degree of cardiopulmonary coupling is modulated by the ANS, and its characteristics vary by the type and depth of sleep. In comparison to the waking state, normal NREM sleep is associated with reduced sympathetic-nerve activity and heart rate (Somers et al., 1993), covarying with increased depth of sleep. High vagal tone, sinus arrhythmia, stable breathing, high relative delta power, blood pressure dipping, and stable arousal threshold are all characteristics of stable NREM sleep, whereas unstable NREM sleep has the opposite characteristics, including low-frequency tidal volume fluctuations, cyclic variation in heart rate, low relative delta power, non-dipping of blood pressure, and variable arousal thresholds. In contrast, during rapid eye movement (REM) sleep, sympathetic-nerve activity increases above that observed during wakefulness, and blood pressure and heart rates are similar to those observed during wakefulness (Somers et al., 1993). According to spectral analysis, high-frequency power components have been associated with parasympathetic activity dominance, while low-frequency power components have been associated with the dominance of sympathetic activity.

11.3 Analytical Methods for Cardiopulmonary Coupling

The CPC technique extracts HRV/pulse rate variability and an ECG/PPG-derived respiration (EDR/PDR) signal from a single-channel ECG or PPG. The cross-power and coherence of these two signals are then calculated using the Fourier transform to generate a sleep spectrogram of cardiopulmonary coupling dynamics (Thomas et al., 2005).

The following are the detailed steps in calculating the cardiopulmonary coupling measure: (1) An automated beat detection algorithm is applied to detect beats, classifying them as normal or ectopic, and to determine amplitude fluctuations in the QRS complex. An EDR was calculated using these amplitude fluctuations. (2) From the RR interval time series, the time series of the normal sinus to normal sinus (N-N) interval and its associated EDR interval are extracted. (3) A sliding window average filter is used to remove outliers resulting from false or missed R-wave detections. This filter has a window of 41 data points, and center points that lie outside 20% of the window mean are rejected. (4) A cubic spline is used to resample the resulting N-N interval sequence and its associated EDR at 2 Hz. (5) The fast Fourier transform is performed to the 3 overlapping 512 sample sub-windows within the 1024-sample coherence window to calculate the cross-spectral power and coherence of these 2 signals across a 1024-sample (8.5 min) frame. (6) After that, the 1024-sample coherence window is advanced by 256 samples (2.1 min), and the calculation is repeated until the entire N-N interval/EDR series is analyzed. For each 1024-sample window, the product of the coherence and cross-spectral power is used to calculate the ratio of coherent cross-power in the low-frequency (0.01–0.1 Hz) band to that in the high-frequency (0.1–0.4 Hz) band. The logarithm of the high- to low-frequency cardiopulmonary coupling ratio [log (HFC/LFC)] is then computed to yield a continuously varying measure of cardiopulmonary coupling. Although the ECG signal was originally utilized as an input, the CPC sleep spectrogram can now be computed using any signal recordings that include an ECG signal or a similar information-content signal, such as PPG. The steps of calculating CPC are depicted in Fig. 11.1.

11.4 Distinct Patterns of Cardiopulmonary Coupling and Its Association with CAP and PSG

High-frequency (0.1–0.4 Hz) coupled pattern appears as the (upper) dark blue peaks on the sleep spectrogram (Fig. 11.2), which represents integrated, stable NREM sleep with the characteristics of stable breathing, high vagal tone, generally a non-CAP on the EEG, high relative delta power, and blood pressure dipping. Stable NREM sleep is equivalent to part of stage 2 and usually all of stage 3 NREM sleep derived from PSG; N3 can be unstable and exhibit LFC in conditions such as epilepsy and NREM parasomnias. There is a link between stable NREM sleep (HFC) and delta waves (deep sleep). We therefore consider this pattern as "effective" NREM sleep. Effective

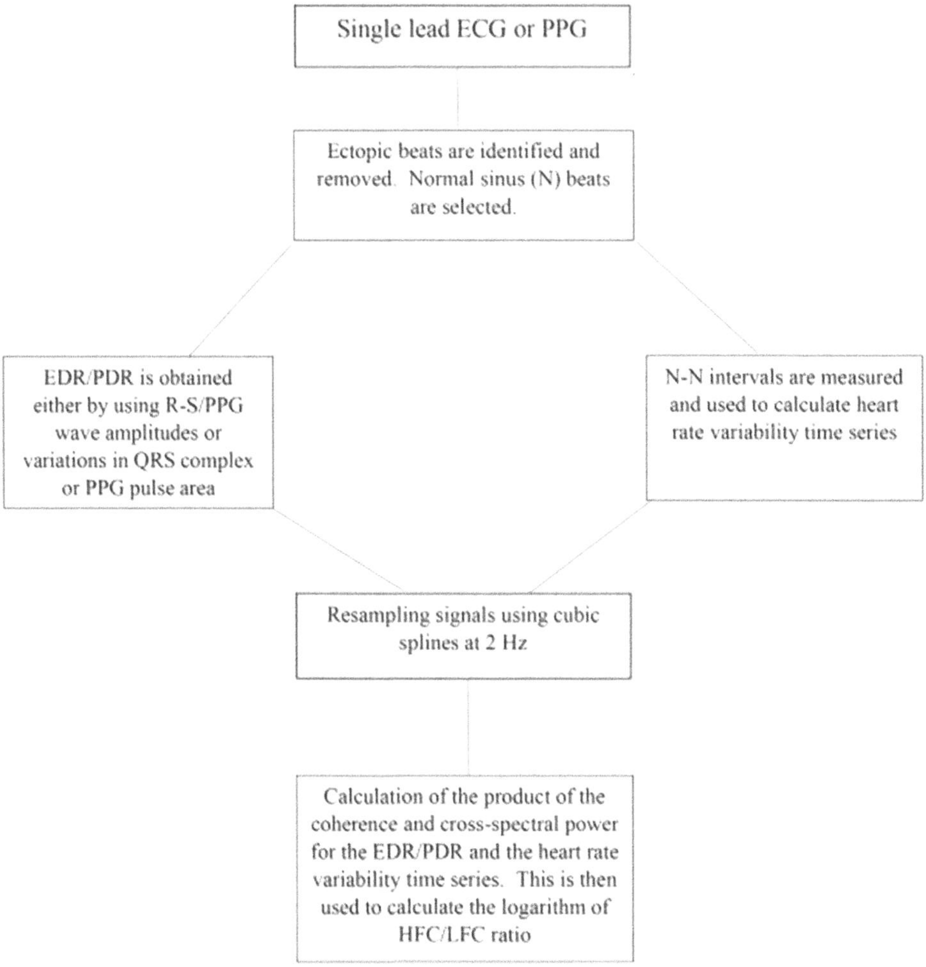

Fig. 11.1 Algorithm outline for CPC analysis. ECG electrocardiogram, PPG photoplethysmogram, EDR ECG-derived respiration, PDR photoplethysmogram-derived respiration, R-S and QRS are ECG waveforms, N-N intervals normal sinus to normal sinus intervals, Hz frequency, HFC high-frequency coupling, LFC low-frequency coupling

sleep enables the desired functions of sleep across multiple dimensions (e.g., metabolic, immune, etc.), allowing for recovery and restorative processes to occur. Low-frequency (0.01–0.1 Hz) coupled patterns appear as light blue peaks on the sleep spectrogram (Fig. 11.2), which represent unstable NREM with the exact opposite characteristics of stable sleep: low-frequency tidal volume fluctuations, cyclic variation in heart rate, CAP, EEG low relative delta power, and non-dipping of blood pressure and variable arousal thresholds. Unstable NREM sleep equates to all of stage 1 and part of stage 2 NREM sleep from PSG, and it is also considered as "ineffective" NREM sleep. Ineffective sleep fails to accomplish the functions that a healthy sleep should. A subset of LFCs called elevated low-frequency coupling (e-LFC) pattern has two further subsets: one with broad-band coupling spectra and the other with narrow-band coupling spectra (e-LFC_{BB} and e-LFC_{NB}). Very-low-frequency (0.004–0.01 Hz) coupling (VLFC), shown as orange peaks on the sleep spectrogram (Fig. 11.2), occurs during both awake and healthy REM sleep; fragmented REM sleep is characteristic of LFC.

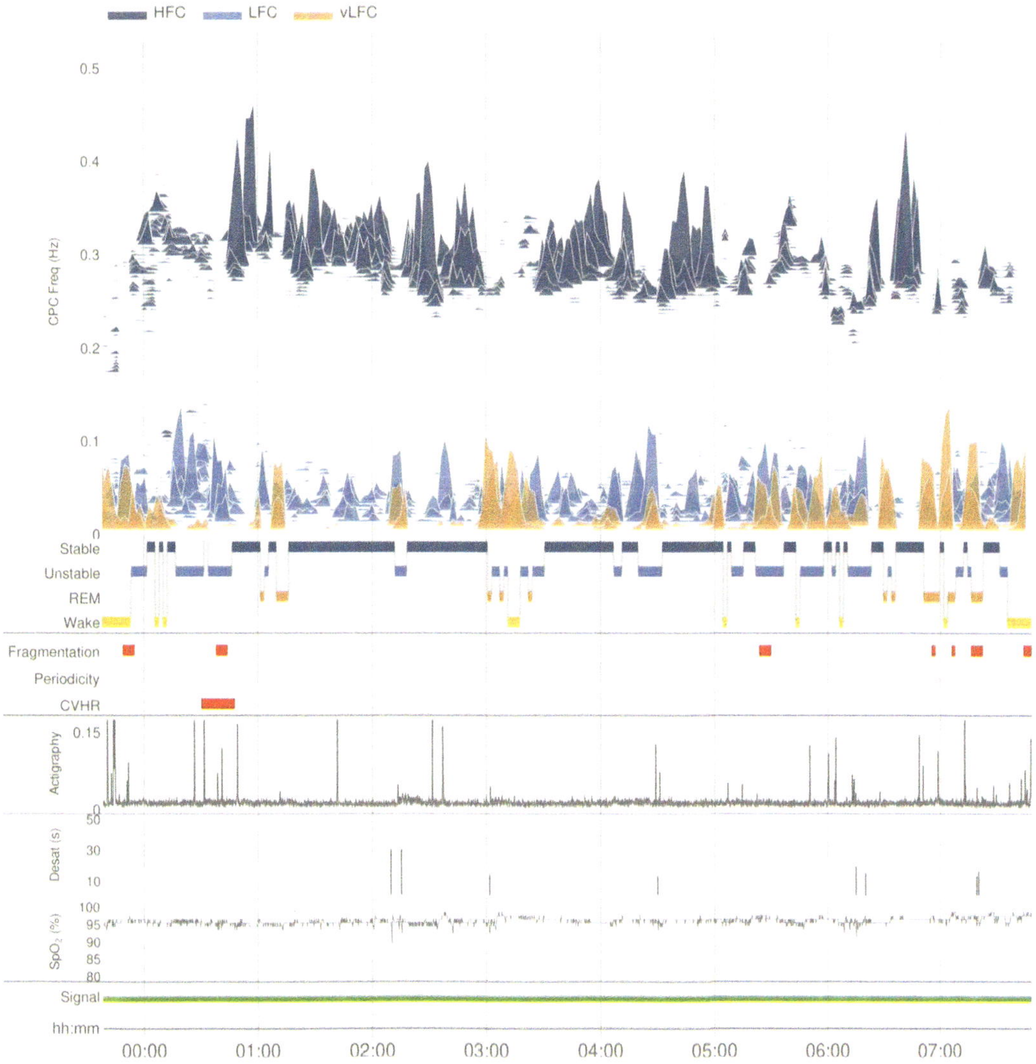

Fig. 11.2 The oximeter-extracted CPC spectrogram. The basic graphical representation of the CPC spectrogram has high-, low-, and very-low-frequency coupling (HFC, LFC, and VLFC, respectively) components

HFC and LFC are mutually incompatible and do not coexist. Some LFC, up to about 20–30% of NREM sleep in adults, is normal, occurring at sleep onset, during brief periods within a given NREM cycle, and just prior to REM sleep. These "lightening" periods may serve a biological disengagement function as sleep processes move from NREM to REM sleep and exhibit other kinetics. Sleep fragmenting disease states "hijack" LFC and increase both duration and biological hostility at the expense of HFC. Similarly, conditions that enhance sleep drive and continuity suppress LFC while amplifying HFC.

The CPC spectrogram showed a strong correlation with CAP scoring, with LFC associated with CAP and HFC with non-CAP (Thomas et al., 2005). The kappa statistic, a measure of interscorer reliability, showed higher agreement between the ECG-based detector and visual scoring of CAP/non-CAP (training set, 74%, and test set, 77.3% agreement, respectively) than between the ECG-based state estimate and standard

NREM stages (training set, 62.7%; test set, 43.9% agreement). The agreement between visual CAP/non-CAP scoring and stage 2/delta sleep (conventional stages 3 + 4) was not significantly better than chance (54%).

Even though CPC and PSG analyze and present biological activity during sleep from different brain structures (ANS regulation vs. cortical brain wave regulation, respectively), they both reflect sleep. As shown in Fig. 11.3, the two methods share important similarities but also exhibit some key differences.

11.5 Sleep Stability Is Independent of Continuous Sleep Depth

The CPC analysis provides a measure of sleep stability and complements conventional polysomnographic analysis. The 30-second epoch-based scoring of sleep heavily down-samples the relevant biology and provides a low-resolution view of the continuous nature of sleep. Stage N2 is especially problematic as this stage can show a wide range of morphologies and oscillatory information content across low amplitude slow waves, spindles, and K-complexes. The ORP is a novel approach to estimate continuous sleep depth in 3-second epochs, utilizing the power content at classic sleep-related frequencies, estimating the probability of arousability (Penner et al., 2019; Younes et al., 2015). A natural question is the correlation of while night sleep depth, especially in NREM sleep, between CPC and ORP measures – there is virtually none. This can be readily understood by consideration of what is being measured – the proportion of stable sleep vs. the overall EEG sleep depth. Stable sleep (HFC) covaries with relative slow-wave power and may be expected to align with low ORP (deeper sleep), but K-complex enriched periods of sleep, for example, can be profoundly unstable yet have increased sleep depth. Thus, in the most extreme instances, such as marked increases in slow-wave sleep or severe whole night sleep fragmentation, the measures may agree somewhat, but not across the entire range of sleep stability and sleep depth. This idea is confirmed by analysis of the Sleep Heart Health Study-I dataset (5781 subjects, age: 63.1 ± 11.2 years, 46.7% male), where all correlations between the Sleep Quality Index (a measure that integrates high- and low-frequency coupling, sleep fragmentation, and total sleep time) and HFC with NREM, REM, or whole night ORP were all statistically non-significant (all correlation coefficients <0.05). Thus, CPC and ORP provide information about non-overlapping dimensions of sleep physiology and pathology.

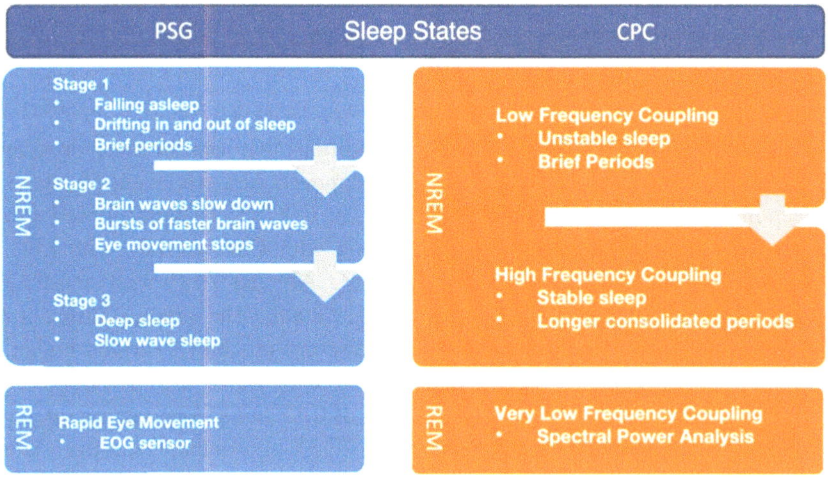

Fig. 11.3 The relationship between the CPC scoring system and conventional sleep scoring system

11.6 Clinical Application of Cardiopulmonary Coupling Technique

11.6.1 Diagnosis of Sleep Apnea

Sleep apnea disrupts rhythmic breathing and increases sympathetic-nerve activity, which results in pathological oscillations in heart rate and breathing. They are represented in the CPC sleep spectrogram as the LF-coupled band spectra. Analysis of the PhysioNet Sleep Apnea Database showed that e-LFC (a subset of LFC) coincided highly with the manually scored apneas and hypopneas. There are two further bands within e-LFC, namely, e-LFC_{BB} and e-LFC_{NB}. Given that other causes of sleep fragmentation may also contribute to the e-LFC spectrum, especially e-LFC_{NB}, the latest methods of AHI calculation have combined oxygen desaturation analysis and CPC analysis to minimize this limitation. Thus, a spectrographic apnea-hypopnea index (sAHI) is defined as (broad-band index + narrow-band index + oxygen desaturation index) per hour of sleep as determined by CPC, which has been approved by the USA FDA (K182618) in 2019 and to be accepted as comparable to manual scoring of AHI from PSG in adults and children.

Numerous studies have validated the diagnostic performance of the CPC technique against PSG in the adult populations (Table 11.1; Liu et al., 2012; Magnusdottir & Hilmisson, 2018; Hilmisson et al., 2019; Lu et al., 2019; Ma et al., 2020; Seo et al., 2021; Al Ashry et al., 2021; Xie et al., 2018; Feng et al., 2017). We further performed a meta-analysis of relevant studies published in the past 10 years, to summarize pooled diagnostic performance. The comprehensive meta-analysis of 6 validation studies (including 1524 patients) that recorded CPC and PSG simultaneously demonstrated that the pooled sensitivity, specificity, positive likelihood ratio, negative likelihood ratio, and diagnostic odds ratio were 94% (95%CI: 92–95%), 63% (95%CI: 56–71%), 2.4 (95%CI: 1.97–2.92), 0.12 (95%CI: 0.04–0.35), and 20.43 (95%CI: 6.49–64.34), respectively, when we used PSG-AHI $\geq$ 5 as the threshold. The summary receiver operating characteristic curve was shown in Fig. 11.4, and the area under the curve was 0.76. In addition, there are two validation studies performed on children (Guo et al., 2011; Hilmisson et al., 2020). The recent large study showed that the novel sAHI combining PPG data with oximetry desaturation data has a significant correlation with manually AHI derived from PSG studies (Pearson correlation = 0.954, $P < 0.0001$) (Hilmisson et al., 2020).

11.6.2 Distinguishing Sleep Apnea Types

Sleep apnea can be caused by several driver endotypes, including high loop gain, a low arousal threshold, an inadequate negative pressure reflex, and increased upper airway collapsibility. At least two types of sleep apnea can be distinguished using spectral profiles of CPC (Thomas et al., 2007). In the sleep spectrogram, the differences between OSA (broad spectral band pattern e-LFC) and CSA or periodic breathing (narrow spectral band pattern e-LFC, high loop gain sleep apnea) are both computationally and visually distinctive and easily quantifiable. As seen in Fig. 11.5a, a broad band of gray peaks suggests that the upper airway obstruction is the primary pathophysiological factor causing the patient's sleep apnea. The presence of a narrow spectral band indicates abnormal chemoreflex regulation of respiration during sleep, which is a hallmark of high loop gain expression (Thomas et al., 2007). It is represented by a narrow red peak in the 3D spectrogram view (Fig. 11.5b). Cheyne-Stokes respiration, a subtype of CSA, shows similar peaks on the CPC sleep spectrogram. As seen in Fig. 11.6, both pathologies can coexist.

11.6.3 Treatment Tracking in Sleep Apnea

Several studies were conducted to evaluate the efficacy of various treatments for OSA, such as CPAP, upper airway surgery, and mandibular

Table 11.1 Studies that used cardiopulmonary coupling to diagnose sleep apnea

Author and year	Sample size	Age, year	Male	BMI, kg/m^2	Main finding
Adults					
Al Ashry et al. (2021)	833	51 ± 13	554	NA	Receiver operating characteristic (ROC) curves demonstrated strong agreement in all OSA categories: 98.5% in mild OSA (95% CI, 97.6–99.3%), 96.4% in moderate OSA (95% CI, 95.3–97.5%), and 98.5% in severe OSA (95% CI, 97.8–99.2%)
Seo et al. (2021)	194	18–72	NA	NA	The spearman correlation coefficient showed that the sAHI was significantly positively correlated with the AHI ($r = 0.973$, $P < 0.05$)
Ma et al. (2020)	205	46.8 ± 12.8	149	27.53 ± 4.28	CPC-REI positively correlated with PSG-AHI ($r = 0.851$, $P < 0.001$). After adjusting for age and gender, CPC-REI and PSG-AHI were still significantly correlated ($r = 0.840$, $P < 0.001$)
Lu et al. (2019)	179	44.9 ± 11.8	152	28.0 ± 4.1	Area under the curve (AUC) for the CPC device in the whole cohort patients was 0.79 (mild), 0.79 (moderate), and 0.86 (severe OSA), respectively (all $P < 0.001$). For patients with cardiovascular disease, AUC was 0.86 (mild), 0.73 (moderate), and 0.83 (severe OSA), respectively (all $P < 0.0001$), and 0.74 (mild), 0.85 (moderate), and 0.91 (severe OSA), respectively, in patients without cardiovascular disease (all $P < 0.0001$)
Hilmisson et al. (2019)	68	45.1 ± 10.9	55	27.6 ± 6.0	CPC identified patients with moderate to severe SA with the sensitivity of 100%, specificity of 81%, and agreement of 93% compared with manual scoring of AHI
Magnusdottir and Hilmisson, (2018)	47	48.6 ± 12.6	14	33.9 ± 9.2	Compared with the manually scored PSG, the combined CPC + CVHR algorithm had a sensitivity of 89%, a specificity of 79%, an agreement of 85%, a PPV of 0.86, an NPV of 0.83, and a Kappa of 0.70
Xie et al. (2018)	44	47.7 ± 13.3	37	25.6 ± 3.5	The corresponding areas under the ROC curves were 0.868, 0.892, 0.915, 0.942, and 0.921, respectively, when PSG-AHI ≥ 5/h, ≥10/h, ≥15/h, ≥20/h, and ≥ 30/h, respectively
Feng et al. (2017)	292	50.26 ± 13.28	212	NA	The correlation between CPC-RDI and PSG-AHI was excellent ($r = 0.801$, $P < 0.01$)
Liu et al. (2012)	69	40.4 ± 10.7	56	28.1 ± 6.5	The AUC was 0.79 when apneas and hypopneas were detected
Children					
Hilmisson et al. (2020)	805	6.81 ± 1.78	378	Z score 1.00 ± 1.27	ROC curve demonstrated strong agreement in all OSA categories (mild, moderate, severe) 91.4% [95% CI: 89.5, 93.4], 96.7% [95% CI: 95.4, 97.9], and 98.6% [95% CI: 97.8, 99.4]; sensitivities 95.4% [95% CI: 93.2, 97.0], 86.5% [95% CI: 80.3, 91.3], and 88.4% [95% CI: 78.4, 94.9]; and specificities 84.4% [95% CI: 79.7, 88.4], 99.2% [95% CI: 98.2, 99.7], and 99.6% [95% CI: 98.8, 99.9], respectively
Guo et al. (2011)	63	6.2 ± 2.5	41	NA	CPC-RDI has a strong positive correlation with the conventional nasal flow RDI (correlation coefficient 0.70)

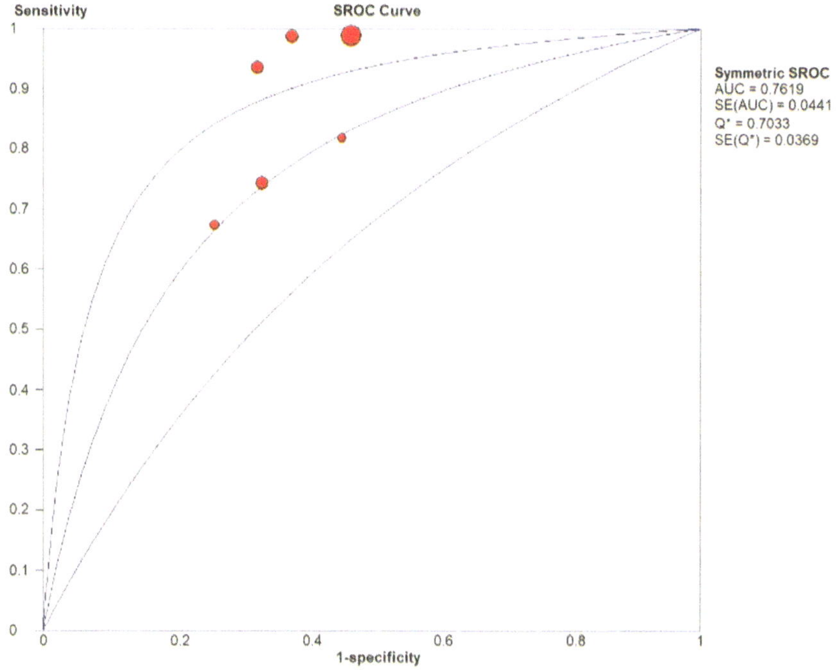

Fig. 11.4 Summary receiver operating characteristic (ROC) curves comparing CPC and PSG studies. ROC for apnea-hypopnea index ≥5 events/h

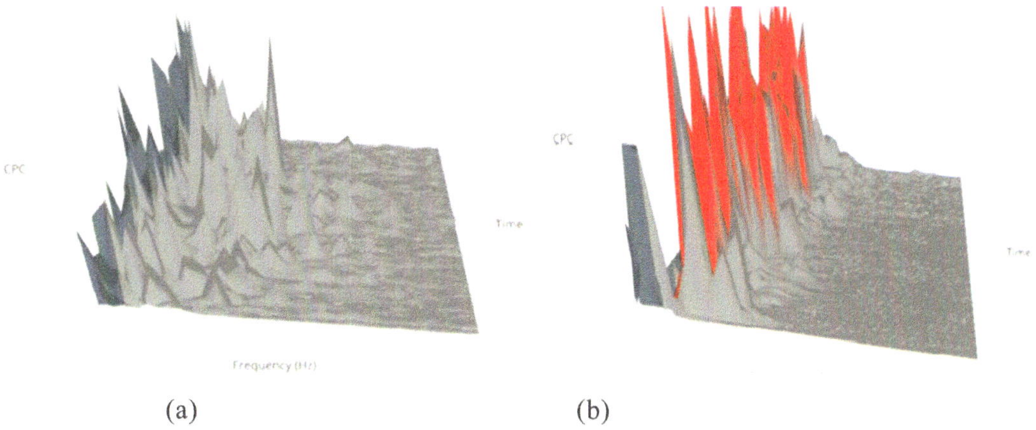

Fig. 11.5 The 3D view spectrogram – (**a**) Obstructive sleep apnea is presented as a "broad" distribution of the peaks colored gray. (**b**) Central sleep apnea is presented as a line of narrow peaks colored red

advancement (Table 11.2). Harrington et al. (2013) revealed that patients with successful CPAP therapy have more HFC, less LFC, and e-LFC_{BB} than those with unsuccessful CPAP therapy. Cho and Kim (2017) looked at how CPC variables changed after CPAP titrations and discovered that HFC increased while LFC and e-LFC decreased. Treatment of OSA with an oral appliance or upper airway surgery produces similar results (Choi et al., 2015; Lee et al., 2016). In addition, Lee et al. (2012) and Chen and He (2019) both found that adenotonsillectomy resulted in a significant change in CPC parameters (increased HFC, decreased

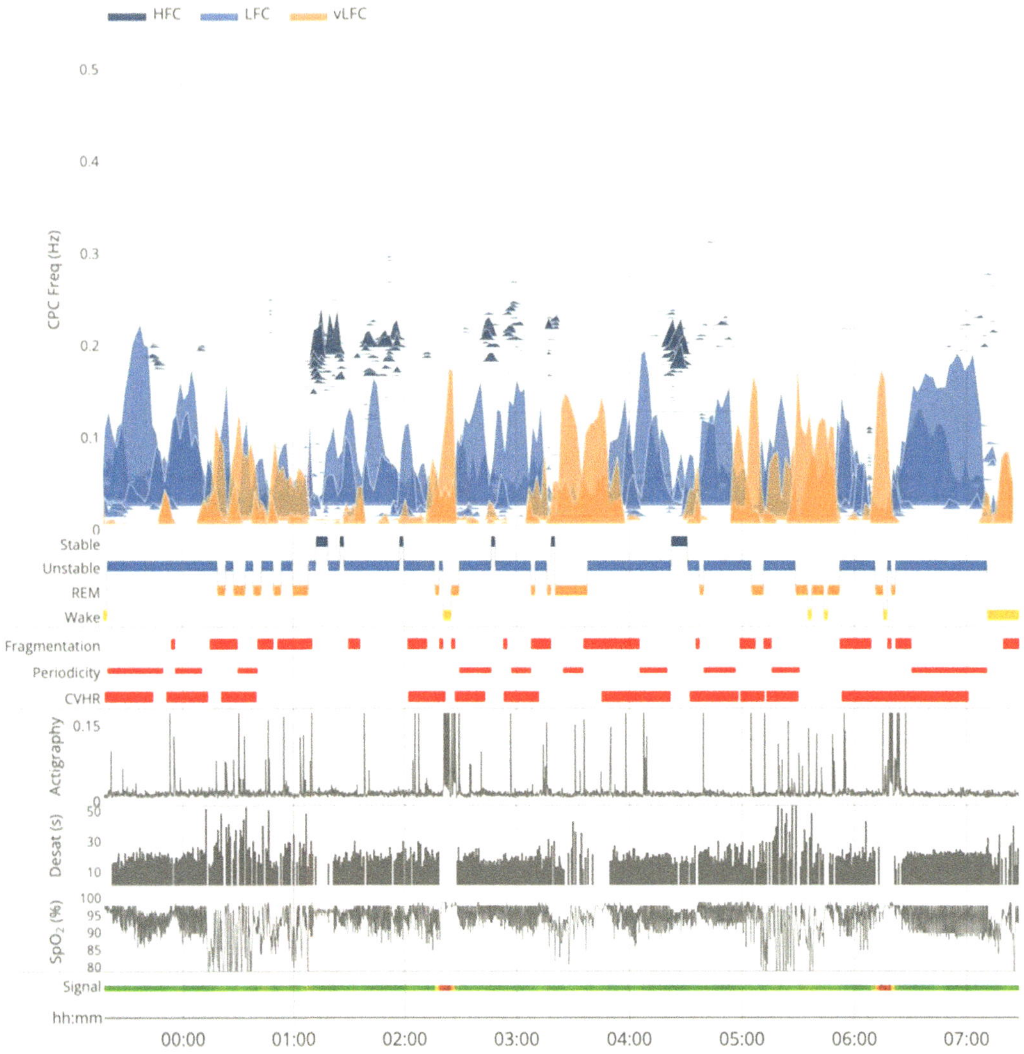

Fig. 11.6 Mixed physiology sleep apnea. A 55-year-old male with classic sleep apnea symptoms, CPC from a ring oximeter. Note (1) poor sleep quality and (2) two patterns of oxygen desaturation: V-shaped in REM sleep consistent with obstructive pathology and band-like oxygen desaturation in NREM sleep associated with detected "periodicity" (narrow-band e-LFC), consistent with additional high loop gain effects

LFC) in pediatric OSA patients. As previously mentioned, in addition to dynamically tracking the change in sleep stability after treatment, the CPC technique can also detect and phenotype residual apnea, predicting PAP failure (Thomas et al., 2007).

In the sleep apnea population, there are several advantages to using the CPC technique through wearable devices, particularly the current device of a ring-form oximeter. These include (1) easy to use, low cost, and comfortable for patients, allowing for repeated testing and ambulatory tracking of sleep apnea; (2) the automatically generated AHI reduces the scoring burdens; (3) detecting expressed high loop gain (central apnea and periodic breathing) may help improve risk stratification and capture therapy effects, such as treatment-emergent CSA; (4) aging does not appear to negatively affect the ability of CPC technology to detect OSA accurately. According to AI Ashry et al. (2021), the patients were divided into three

Table 11.2 Studies that used cardiopulmonary coupling in following treatment effect of sleep apnea

Author and year	Number of subjects	Main finding
Adults		
Lee et al. (2016)	98 OSA patients treated with surgery or with a MAD therapy	The reduction in the apnea-hypopnea index greater than 50% was significantly associated with the reduction in LFC and increment in HFC
Cho and Kim (2017)	115 OSA patients with CPAP	In the CPAP titration group, HFC increased, while LFC and e-LFC decreased linearly as AHI decreased
Choi et al. (2015)	62 OSA patients treated with surgery	Patients with surgical success were found to have a significant increase in HFC and a significant decrease in LFC compared to those without successful surgery
Lee et al. (2014)	52 OSA patients with MAD therapy	LFC decreased, while HFC increased as AHI improved by MAD therapy
Harrington et al. (2013)	24 OSA patients with CPAP	The successful CPAP therapy group had more HFC, less LFC, and e-LFC_{BB} compared to the unsuccessful CPAP therapy group
Ramar et al. (2013)	106 complex sleep apnea patients with ASV	The percentage of e-LFC_{NB} did not relate to the success of ASV treatment
Schramm and Thomas (2012)	Case report of 1 patient with mild OSA with the mandibular advancing appliance, sleep position restriction, oxygen therapy	The HFC/LFC ratio was higher on mandibular advancing appliance nights than oxygen therapy and positional therapy
Children		
Chen and He (2019)	126 children with OSA	There was an improvement in RDI collected from CPC after surgery
Lee et al. (2012)	37 children with OSA	Adenotonsillectomy significantly increased HFC and decreased LFC, which were paralleled by the improvement in the apnea-hypopnea and arousal index

groups based on their age: <45, 45–55, and >55. They discovered that none of these age groups had a significant effect on the accuracy of AHI. As a result, CPC is a desirable method for evaluating sleep apnea in elderly adults because it is not constrained by the dependence of conventionally scored slow-wave sleep which deteriorates with age when measured through EEG from the cortex; (5) it can be applied regardless of autonomic dysfunction. Even with a flat heart rate, the EDR comes through. Of course, it is also worth considering its potential limitation. CPC output is less meaningful in patients with chronic atrial fibrillation, due to complex patterns that cannot be identified and the chaos of the ANS. Therefore, the results should be interpreted cautiously. Figures 11.7, 11.8, and 11.9 show the sleep apnea and sleep quality phenotyping in apnea utility of the CPC technique.

11.7 Cardiopulmonary Coupling Spectrogram in Other Disorders

11.7.1 Insomnia/Mental Health

In the field of insomnia and related adverse mental and psychological diseases, CPC technology has been widely applied (Table 11.3). Primary insomnia patients had lower HFC and higher LFC, VLFC, and e-LFC compared to good sleepers, according to Schramm et al. (2013). A similar finding has been observed by Thomas et al. (2018), and they found that patients with insomnia had a higher e-LFC_{BB} percentage than healthy participants. Zhang et al. (2021) assessed the relationship between cognitive function and sleep stability in insomnia patients and discovered that insomnia patients with cognitive impairment had lower HFC and higher LFC than

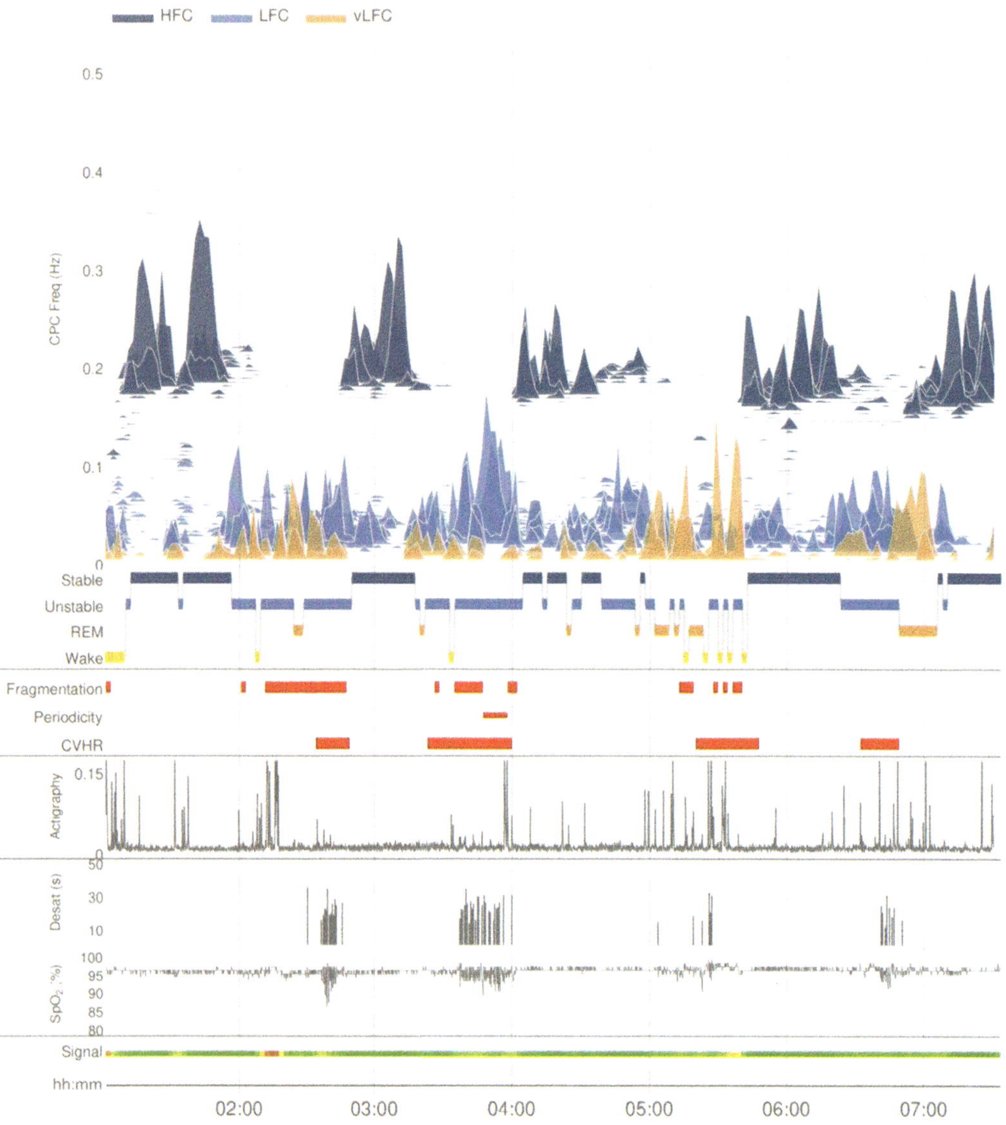

Fig. 11.7 Diagnostic assessment of milder sleep apnea. A 44-year-old male. Note generally good sleep quality but clusters of oxygen desaturation and cyclic variation in heart rate

insomnia patients with normal cognition. However, CPC characteristics did not differ substantially between participants with restless legs syndrome and those with insomnia (Na et al., 2015). Furthermore, Jarrin et al. (2016) evaluated the potential benefits of cognitive-behavioral therapy for insomnia and found that sleep improvements were related to reduced HF following therapy. Because insomnia is linked to a variety of medical and psychiatric conditions (Sivertsen et al., 2014), the CPC technique has been utilized to study and track therapy responses in these patients. In comparison to controls, unmedicated depressive patients exhibited a lower HFC and a higher LFC, according to Yang et al. (2011). Ma et al. (2018) studied the effects of tai chi training on sleep quality in patients with depression. When the patients got tai chi

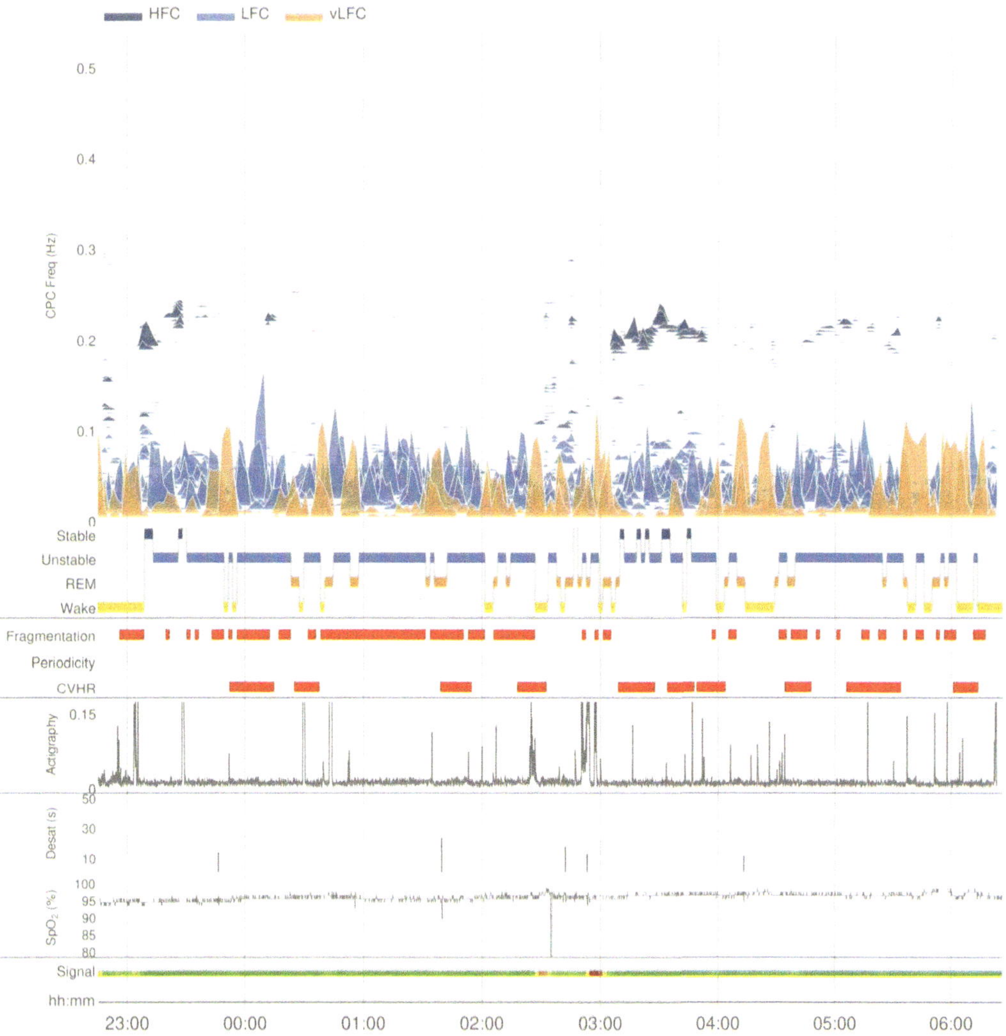

Fig. 11.8 Failure of CPAP to improve sleep quality. The same patient as in Fig. 11.6, after 3 months of CPAP, with a complaint of persistent fatigue despite good use of CPAP and low (less than 5) event index on CPAP. Note the severe loss of HFC, suggesting worse sleep quality, associated with an increase in cyclic variation in heart rate. This could be from non-apnea causes (such as anxiety) or CPAP-induced respiratory instability or sleep fragmentation

training, their CPC analysis revealed an increase in stable sleep percentages and a decrease in unstable sleep percentages. Sun et al. (2019) looked at 41 depressed patients and found that there were significant associations between CPC characteristics at baseline and depression symptom improvement after 2 weeks of antidepressant drug treatment. As sleep apnea syndromes often have comorbid insomnia and mood disorders, CPC spectrograms provide a method to assess sleep quality relatively independent of respiratory abnormality.

11.7.2 Cardio-Cerebral Metabolic Health

Sleep health, as measured by CPC analysis, including sleep duration, sleep quality, and OSA, has been linked to cardio-cerebral meta-

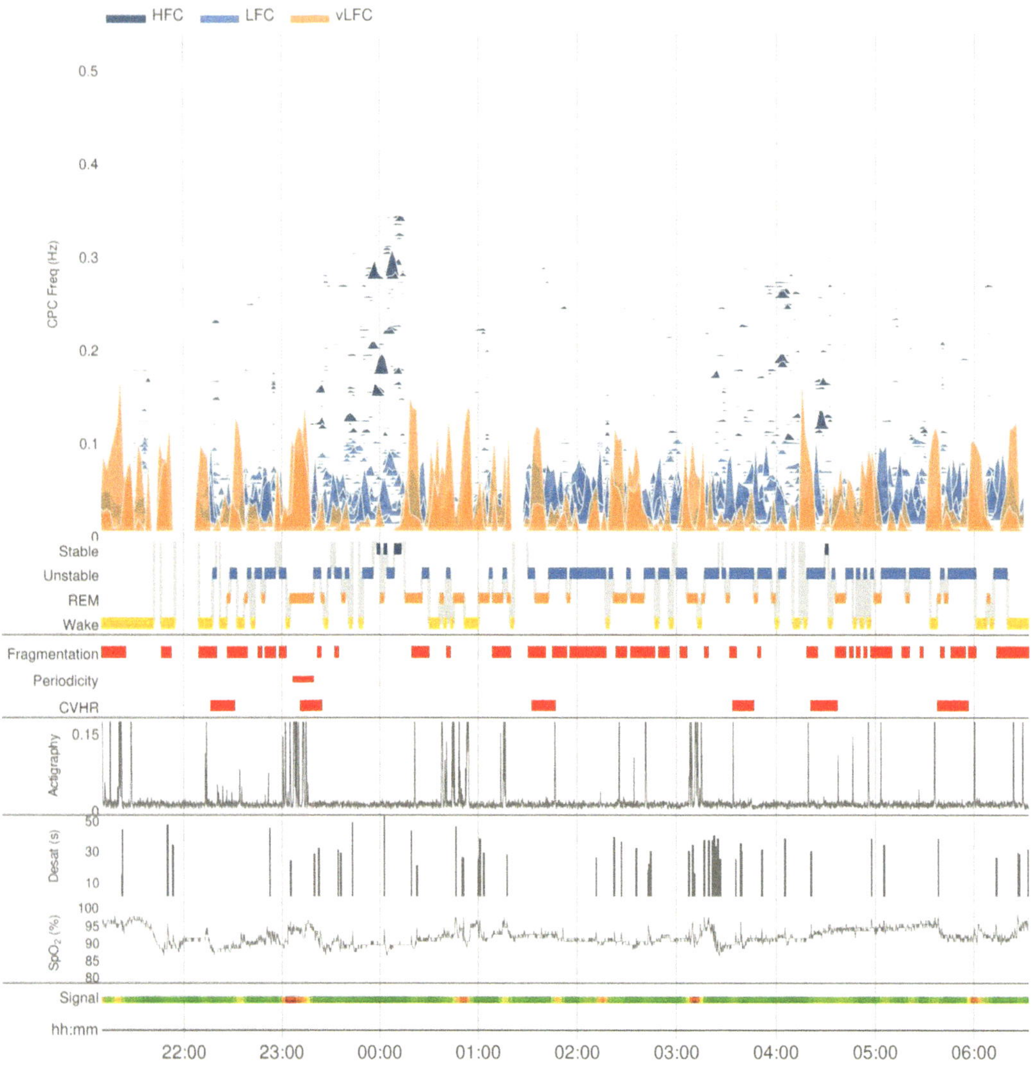

Fig. 11.9 Failure of adaptive servo-ventilation for treatment-emergent central sleep apnea. A 55-year-old male who has been using an ASV for over 10 years, with improvement from CPAP yet residual fatigue. Machine residual AHI is less than 1/h of sleep. Note near absence of stable NREM sleep. Note also unstable oximetry trace. Though residual machine estimated AHI is low, clearly there is ongoing sleep disruption, which can occur from excessive pressure cycling of the ventilator

bolic illnesses in numerous studies (Table 11.4). Thomas and colleagues (2009) found that e-LFC_{NB} is linked to more severe sleep apnea, as well as a higher prevalence of hypertension and stroke. Pogach et al. demonstrated that HFC is an independent driver of the glucose disposal index (Pogach et al., 2012). In a study of 615 patients with acute non-cardioembolic ischemic stroke, Kang et al. (2020) discovered narrow-band coupling could predict severe and protracted functional impairment at 3 months. Magnusdottir et al. (2020) found that CPC-derived sleep quality influenced 24-h mean arterial blood pressure and mean diastolic blood

Table 11.3 Studies that used cardiopulmonary coupling in insomnia/mental health

Author and year	Simple size	Main finding
Zhang et al. (2021)	43 patients with insomnia	Insomnia-cognitive impairment patients had lower HFC and higher LFC compared to the insomnia-normal cognition patients
Sun et al. (2019)	41 patients with depression	Significant correlations were found between CPC variables at baseline and depression symptom improvement after 2 weeks of treatment
Hilmisson et al. (2019)	110 patients with chronic insomnia	The prevalence of moderate-severe SDB (REI > 15) was 25% based on HSAT. Surrogate markers of moderate-severe SDB detected by CPC analysis identified the prevalence of 33%, with a negative predictive value of 96%
Ma et al. (2018)	12 depressed patients	CPC analysis showed decreased stable sleep onset latency, increased stable sleep percentages, and decreased unstable sleep percentages after tai chi training
Thomas et al. (2018)	20 insomnia patients, 10 healthy participants	Patients with insomnia had increased LFC duration and increased e-LFCBB percentage than healthy participants
Schramm et al. (2016)	25 chronically depressed patients	By post-treatment night 6, the Cognitive Behavioral Analysis System of Psychotherapy group had more stable sleep and less wake compared with Treatment as Usual group and less wake than Mindfulness-based Cognitive Therapy group
Jarrin et al. (2016)	65 patients with chronic insomnia	Following cognitive-behavioral therapy, sleep improvements were related to reduced HF in S2 and REM
Park et al. (2015)	200 OSA subjects divided into OSA with insomnia group and OSA without insomnia group	There was no significant difference in CPC parameters between the two groups after adjustment of AHI
Na et al. (2015)	109 subjects with restless legs syndrome and 86 with insomnia	CPC parameters were not significantly different between groups
Sylvia et al. (2014)	8 patients with bipolar disorder	SleepImage M1 device is a feasible means to obtain objective sleep quality and quantity data in individuals with bipolar disorder
Schramm et al. (2014)	19 subjects with depression	Bupropion did not impact CPC variables
Schramm et al. (2013)	50 subjects with primary insomnia and 36 good sleepers	Relative to good sleepers, primary insomnia patients on adaptation night had lower HFC and HFC/LFC ratio and higher LFC, VLFC, and e-LFC. On baseline night, the primary insomnia group had increased LFC, VLFC, and e-LFC and a lower HFC/LFC ratio. Except for HFC, good sleepers had larger CPC variable differences between adaptation and baseline nights compared to the primary insomnia group
Yang et al. (2011)	100 patients with major depressive disorder and 91 healthy controls	Relative to controls, unmedicated depressed patients had a reduction in high-frequency coupling and an increase in low-frequency coupling and very-low-frequency coupling. The medicated depressed group showed a restoration of stable sleep to a level comparable with that of the control group

pressure, as well as blood pressure during wakefulness, in a study of 241 patients with OSA at high cardiovascular risk. They also found that better sleep quality was associated with increased serum adiponectin levels and decreased insulin levels (Magnusdottir et al., 2021). For patients with chronic heart failure, tai chi training is likely to increase HFC and decrease LFC (Yeh et al., 2008). Similarly, in patients with paroxysmal atrial fibrillation, the HFC and VLFC were significantly elevated after radio-frequency catheter ablation, whereas LFC

Table 11.4 Studies that used cardiopulmonary coupling in cardio-brain-metabolic health and diseases

Author and year	Simple size	Main finding
Thomas et al. (2021)	504 patients from Offspring/Omni-1 database	Stable sleep computed using CPC was positively associated with white matter health
Magnusdottir et al. (2021)	241 patients with OSA at high cardiovascular risk	Improvements in CPC-sleep quality were associated with higher serum adiponectin levels and improved measures of glycemic metabolism
Magnusdottir et al. (2020)	241 patients with OSA at high cardiovascular risk	CPC-derived sleep quality impacted 24-h mean arterial blood pressure and mean diastolic blood pressure, as well as blood pressure during wake, in patients participating in the Heart Biomarker Evaluation in Apnea Treatment study
Kim et al. (2020)	225 patients with paroxysmal atrial fibrillation	Six months after radio-frequency catheter ablation, the HFC and VLFC were significantly increased, while LFC was decreased. The recurrence rate of atrial fibrillation was significantly lower in the patient who had unstable sleep before radio-frequency catheter ablation
Kang et al. (2020)	615 patients with acute non-cardioembolic ischemic stroke	Narrow-band coupling was an independent predictor of a higher risk of severe and persistent functional impairment at 3 months
Pogach et al. (2012)	118 nondiabetic subjects with and without SDB	HFC duration was associated with increased and VLFC was associated with reduced disposition index
Thomas et al. (2009)	5247 patients from the SHHS database	(1) Increasing age and male sex are associated with an increase in the prevalence of e-LFCNB. (2) The presence of e-LFCNB is a biomarker of severity of sleep-disordered breathing. (3) Use of diuretics, calcium blockers, and β-blockers was associated with increased e-LFCNB. (4) e-LFCNB was associated with prevalent stroke and hypertension
Yeh et al. (2008)	18 patients with chronic stable heart failure	At 12 weeks, those who participated in tai chi showed a significant increase in HFC and a significant reduction in LFC compared to patients in the control group

decreased (Kim et al., 2020). A recent study of Thomas et al. (2021) found stable sleep computed using CPC was positively associated with white matter health.

11.8 Conclusion

The CPC technique provides an accurate, practical, and low-cost alternative to traditional PSG and home sleep apnea testing for the objective assessment, diagnosis, and tracking of sleep health and disease over time. The technology may be used in both adults and children. It also offers the potential for individualized management of sleep disorders as it allows for repeatable sleep monitoring in the patient's natural sleep environment, as well as automated analysis.

11.9 Clinical Practice Points

- CPC technique generates sleep spectrograms by calculating the cross-spectral power and coherence of HRV and respiratory tidal volume fluctuations.
- The CPC spectrogram shows only a weak correlation with conventional sleep staging, but better follows CAP scoring, with LFC associated with CAP and HFC with non-CAP.
- The CPC sleep spectrogram provides a clear visual view of sleep health during the sleep period and helps healthcare providers manage sleep disorders in their patients, including evaluating sleep quality, diagnosing sleep apnea, and tracking therapy response.
- For the diagnosis of sleep apnea, the spectrographic AHI calculated combining CPC output

and hypoxic events shows strong agreement with AHI calculated manually from PSG.

11.10 Research Points

- The CPC analysis shows a fundamental sleep characteristic – that of bimodal stability, most clearly evident in NREM sleep. This dimension of sleep does not have a known neurobiological explanation, posing a unique research opportunity.
- The presence of a narrow spectral band indicates abnormal chemoreflex regulation of respiration during sleep, which is a hallmark of high loop gain expression.
- The pre- and post-treatment effects of sleep apnea with CPAP or upper airway surgery can be traced by changes in the ratio of HFC to LFC.
- e-LFC_{NB} is associated with higher prevalence of hypertension and stroke.
- HFC is an independent driver of the glucose disposal index.
- Narrow-band coupling was an independent predictor of a higher risk of severe and persistent functional impairment in acute ischemic stroke.
- Better sleep quality was associated with increased serum adiponectin levels and decreased insulin levels.
- Sleep quality and sleep hypoxia were associated with white matter injury.

Acknowledgements The authors gratefully thank MyCardio LLC, dba SleepImage, USA, for providing the figures.

Disclosures and Conflict of Interest Dr. Robert Thomas has the following disclosures:

1. Patent for a device to regulate CO_2 in the positive airway pressure circuit, for the treatment of central/complex apnea.
2. Patent and license for an ECG-based method to phenotype sleep quality and sleep apnea (to MyCardio, LLC, through Beth Israel Deaconess Medical Center).
3. Patent, past consultant – Drive DeVilbiss, CPAP auto-titrating algorithm.
4. GLG Councils and Guidepoint Global – general sleep medicine consulting.

ML and TP have no disclosures.

Author Contributions ML and RT contributed to the study design. ML wrote the draft. ML, RT, and TP revised the manuscript.

References

Agnew, H. W., Jr., Webb, W. B., & Williams, R. L. (1966). The first night effect: An EEG study of sleep. *Psychophysiology, 2*, 263–266.

Al Ashry, H. S., Hilmisson, H., Ni, Y., Thomas, R. J., & Investigators, A. (2021). Automated apnea-hypopnea index from oximetry and spectral analysis of cardiopulmonary coupling. *Annals of the American Thoracic Society, 18*, 876–883.

Benjafield, A. V., Ayas, N. T., Eastwood, P. R., Heinzer, R., Ip, M. S. M., Morrell, M. J., Nunez, C. M., Patel, S. R., Penzel, T., Pepin, J. L., Peppard, P. E., Sinha, S., Tufik, S., Valentine, K., & Malhotra, A. (2019). Estimation of the global prevalence and burden of obstructive sleep apnoea: A literature-based analysis. *The Lancet Respiratory Medicine, 7*, 687–698.

Ben-Tal, A., Shamailov, S. S., & Paton, J. F. (2012). Evaluating the physiological significance of respiratory sinus arrhythmia: Looking beyond ventilation-perfusion efficiency. *The Journal of Physiology, 590*, 1989–2008.

Chen, J., & He, S. (2019). Drug-induced sleep endoscopy-directed adenotonsillectomy in pediatric obstructive sleep apnea with small tonsils. *PLoS One, 14*, e0212317.

Cho, J. H., & Kim, H. J. (2017). The effect of continuous positive airway pressure on cardiopulmonary coupling. *Sleep & Breathing, 21*, 341–345.

Choi, J. H., Thomas, R. J., Suh, S. Y., Park, I. H., Kim, T. H., Lee, S. H., Lee, H. M., Yun, C. H., & Lee, S. H. (2015). Sleep quality change after upper airway surgery in obstructive sleep apnea: Electrocardiogram-based cardiopulmonary coupling analysis. *Laryngoscope, 125*, 1737–1742.

Cistulli, P. A., Armitstead, J., Pepin, J. L., Woehrle, H., Nunez, C. M., Benjafield, A., & Malhotra, A. (2019). Short-term CPAP adherence in obstructive sleep apnea: A big data analysis using real-world data. *Sleep Medicine, 59*, 114–116.

Dick, T. E., Hsieh, Y. H., Dhingra, R. R., Baekey, D. M., Galan, R. F., Wehrwein, E., & Morris, K. F. (2014). Cardiorespiratory coupling: Common rhythms in cardiac, sympathetic, and respiratory activities. *Progress in Brain Research, 209*, 191–205.

Feng, J., Wu, H., Wang, Z., Chen, K., Huang, B., & Zhao, Z. (2017). Correlation analysis of two different techniques for the score of sleep breathing events: Cardiopulmonary coupling vs polysomnography. *Chinese Journal of Neurology, 50*, 606–612.

Guo, D., Peng, C. K., Wu, H. L., Mietus, J. E., Liu, Y., Sun, R. S., & Thomas, R. J. (2011). ECG-derived cardiopulmonary analysis of pediatric sleep-disordered breathing. *Sleep Medicine, 12*, 384–389.

Hales, S. (1733). *Statical essays: Containing haemostaticks, or an account of some hydraulick and hydrostatical experiments on the blood and blood-vessels of animals*. W. Innys, R. Manby, and T. Woodward.

Harrington, J., Schramm, P. J., Davies, C. R., & Lee-Chiong, T. L., Jr. (2013). An electrocardiogram-based analysis evaluating sleep quality in patients with obstructive sleep apnea. *Sleep & Breathing, 17*, 1071–1078.

Hilmisson, H., Lange, N., & Duntley, S. P. (2019). Sleep apnea detection: Accuracy of using automated ECG analysis compared to manually scored polysomnography (apnea hypopnea index). *Sleep & Breathing, 23*, 125–133.

Hilmisson, H., Berman, S., & Magnusdottir, S. (2020). Sleep apnea diagnosis in children using software-generated apnea-hypopnea index (AHI) derived from data recorded with a single photoplethysmogram sensor (PPG): Results from the Childhood Adenotonsillectomy Study (CHAT) based on cardiopulmonary coupling analysis. *Sleep & Breathing, 24*, 1739–1749.

Jarrin, D. C., Chen, I. Y., Ivers, H., Lamy, M., Vallieres, A., & Morin, C. M. (2016). Nocturnal heart rate variability in patients treated with cognitive-behavioral therapy for insomnia. *Health Psychology, 35*, 638–641.

Kang, D. O., Kim, C. K., Park, Y., Jang, W. Y., Kim, W., Choi, J. Y., Roh, S. Y., Choi, C. U., Kim, E. J., Rha, S. W., Park, C. G., Seo, H. S., Oh, K., & Na, J. O. (2020). Impact of sleep-disordered breathing on functional outcomes in ischemic stroke: A cardiopulmonary coupling analysis. *Stroke, 51*, 2188–2196.

Kim, W., Na, J. O., Thomas, R. J., Jang, W. Y., Kang, D. O., Park, Y., Choi, J. Y., Roh, S. Y., Choi, C. U., Kim, J. W., Kim, E. J., Rha, S. W., Park, C. G., Seo, H. S., & Lim, H. E. (2020). Impact of catheter ablation on sleep quality and relationship between sleep stability and recurrence of paroxysmal atrial fibrillation after successful ablation: 24-hour Holter-based cardiopulmonary coupling analysis. *Journal of the American Heart Association, 9*, e017016.

Lee, S. H., Choi, J. H., Park, I. H., Lee, S. H., Kim, T. H., Lee, H. M., Park, H. K., Thomas, R. J., Shin, C., & Yun, C. H. (2012). Measuring sleep quality after adenotonsillectomy in pediatric sleep apnea. *Laryngoscope, 122*, 2115–2121.

Lee, W. H., Ahn, J. C., We, J., Rhee, C. S., Lee, C. H., Yun, P. Y., Yoon, I. Y., & Kim, J. W. (2014). Cardiopulmonary coupling analysis: changes before and after treatment with a mandibular advancement device. *Sleep Breath, 18*, 891–896.

Lee, W. H., Hong, S. N., Kim, H. J., Rhee, C. S., Lee, C. H., Yoon, I. Y., & Kim, J. W. (2016). A comparison of different success definitions in non-continuous positive airway pressure treatment for obstructive sleep Apnea using cardiopulmonary coupling. *Journal of Clinical Sleep Medicine, 12*, 35–41.

Liu, D., Yang, X., Wang, G., Ma, J., Liu, Y., Peng, C. K., Zhang, J., & Fang, J. (2012). HHT based cardiopulmonary coupling analysis for sleep apnea detection. *Sleep Medicine, 13*, 503–509.

Lu, M., Fang, F., Sanderson, J. E., Ma, C., Wang, Q., Zhan, X., Xie, F., Xiao, L., Liu, H., Liu, H., & Wei, Y. (2019). Validation of a portable monitoring device for the diagnosis of obstructive sleep apnea: Electrocardiogram-based cardiopulmonary coupling. *Sleep & Breathing, 23*, 1371–1378.

Ma, Y., Yeung, A., Yang, A. C., Peng, C. K., Clain, A., Alpert, J., Fava, M., & Yeung, A. S. (2018). The effects of Tai Chi on sleep quality in Chinese American patients with major depressive disorder: A pilot study. *Behavioral Sleep Medicine, 16*, 398–411.

Ma, Y., Sun, S., Zhang, M., Guo, D., Liu, A. R., Wei, Y., & Peng, C. K. (2020). Electrocardiogram-based sleep analysis for sleep apnea screening and diagnosis. *Sleep & Breathing, 24*, 231–240.

Magnusdottir, S., & Hilmisson, H. (2018). Ambulatory screening tool for sleep apnea: Analyzing a single-lead electrocardiogram signal (ECG). *Sleep & Breathing, 22*, 421–429.

Magnusdottir, S., Hilmisson, H., & Thomas, R. J. (2020). Cardiopulmonary coupling-derived sleep quality is associated with improvements in blood pressure in patients with obstructive sleep apnea at high-cardiovascular risk. *Journal of Hypertension, 38*, 2287–2294.

Magnusdottir, S., Thomas, R. J., & Hilmisson, H. (2021). Can improvements in sleep quality positively affect serum adiponectin-levels in patients with obstructive sleep apnea? *Sleep Medicine, 84*, 324–333.

Mosko, S. S., Dickel, M. J., & Ashurst, J. (1988). Night-to-night variability in sleep apnea and sleep-related periodic leg movements in the elderly. *Sleep, 11*, 340–348.

Na, G. Y., Choi, S. J., Joo, E. Y., & Hong, S. B. (2015). Objective sleep quality in subjects with restless legs syndrome versus with psychophysiological insomnia: Polysomnography and cardiopulmonary coupling analysis. *Journal of Sleep Medicine, 12*, 13–17.

Park, H. R., Joo, E. Y., & Hong, S. B. (2015). Electrophysiological characteristics of obstructive sleep apnea syndrome with insomnia:

Polysomnography and cardiopulmonary coupling analysis. *Journal of Sleep Medicine, 12*, 53–58.

Penner, C. G., Gerardy, B., Ryan, R., & Williams, M. (2019). The odds ratio product (an objective sleep depth measure): Normal values, repeatability, and change with CPAP in patients with OSA. *Journal of Clinical Sleep Medicine, 15*, 1155–1163.

Pogach, M. S., Punjabi, N. M., Thomas, N., & Thomas, R. J. (2012). Electrocardiogram-based sleep spectrogram measures of sleep stability and glucose disposal in sleep disordered breathing. *Sleep, 35*, 139–148.

Ramar, K., Desrues, B., Ramar, P., & Morgenthaler, T.I. (2013). Analysis of cardiopulmonary coupling to assess adaptive servo-ventilation success in complex sleep apnea management. *Sleep Breath, 17*, 861–866.

Roth, T., Coulouvrat, C., Hajak, G., Lakoma, M. D., Sampson, N. A., Shahly, V., Shillington, A. C., Stephenson, J. J., Walsh, J. K., & Kessler, R. C. (2011). Prevalence and perceived health associated with insomnia based on DSM-IV-TR; international statistical classification of diseases and related health problems, tenth revision; and research diagnostic criteria/international classification of sleep disorders, second edition criteria: Results from the America insomnia survey. *Biological Psychiatry, 69*, 592–600.

Schramm, P. J., Poland, R. E., & Rao, U. (2014). Bupropion response on sleep quality in patients with depression: implications for increased cardiovascular disease risk. *European Neuropsychopharmacology, 24*, 207–214.

Schramm, P. J., & Thomas, R. J. (2012). Assessment of therapeutic options for mild obstructive sleep apnea using cardiopulmonary coupling measures. *Journal of Clinical Sleep Medicine, 8*, 315–320.

Schramm, P. J., Thomas, R., Feige, B., Spiegelhalder, K., & Riemann, D. (2013). Quantitative measurement of sleep quality using cardiopulmonary coupling analysis: A retrospective comparison of individuals with and without primary insomnia. *Sleep & Breathing, 17*, 713–721.

Schramm, P. J., Zobel, I., Monch, K., Schramm, E., & Michalak, J. (2016). Sleep quality changes in chronically depressed patients treated with Mindfulness-based Cognitive Therapy or the Cognitive Behavioral Analysis System of Psychotherapy: A pilot study. *Sleep Medicine, 17*, 57–63.

Seo, M. Y., Yoo, J., Hwang, S. J., & Lee, S. H. (2021). Diagnosis of obstructive sleep apnea in adults using the cardiopulmonary coupling-derived software-generated apnea-hypopnea index. *Clinical and Experimental Otorhinolaryngology, 14*, 424–426.

Sivertsen, B., Lallukka, T., Salo, P., Pallesen, S., Hysing, M., Krokstad, S., & Simon, O. (2014). Insomnia as a risk factor for ill health: Results from the large population-based prospective HUNT Study in Norway. *Journal of Sleep Research, 23*, 124–132.

Somers, V. K., Dyken, M. E., Mark, A. L., & Abboud, F. M. (1993). Sympathetic-nerve activity during sleep in normal subjects. *The New England Journal of Medicine, 328*, 303–307.

Sun, Q. M., Xing, L., Wang, C., & Liang, W. (2019). Cardiopulmonary coupling analysis predicts early treatment response in depressed patients: A pilot study. *Psychiatry Research, 276*, 6–11.

Sylvia, L. G., Salcedo, S., Bianchi, M. T., Urdahl, A., Nierenberg, A. A., & Deckersbach, T. (2014). A novel home sleep monitoring device and brief sleep intervention for bipolar disorder: feasibility, tolerability, and preliminary effectiveness. *Cognitive Therapy Research, 38*, 55–61.

Thomas, R. J., Mietus, J. E., Peng, C. K., & Goldberger, A. L. (2005). An electrocardiogram-based technique to assess cardiopulmonary coupling during sleep. *Sleep, 28*, 1151–1161.

Thomas, R. J., Mietus, J. E., Peng, C. K., Gilmartin, G., Daly, R. W., Goldberger, A. L., & Gottlieb, D. J. (2007). Differentiating obstructive from central and complex sleep apnea using an automated electrocardiogram-based method. *Sleep, 30*, 1756–1769.

Thomas, R. J., Weiss, M. D., Mietus, J. E., Peng, C. K., Goldberger, A. L., & Gottlieb, D. J. (2009). Prevalent hypertension and stroke in the Sleep Heart Health Study: Association with an ECG-derived spectrographic marker of cardiopulmonary coupling. *Sleep, 32*, 897–904.

Thomas, R. J., Wood, C., & Bianchi, M. T. (2018). Cardiopulmonary coupling spectrogram as an ambulatory clinical biomarker of sleep stability and quality in health, sleep apnea, and insomnia. *Sleep, 41*, zsx196.

Thomas, R. J., Kim, H., Maillard, P., DeCarli, C. S., Heckman, E. J., Karjadi, C., Ang, T. F. A., & Au, R. (2021). Digital sleep measures and white matter health in the Framingham Heart Study. *Exploration of Medicine, 2*, 253–267.

Waxenbaum, J. A., Reddy, V., & Varacallo, M. (2022). Anatomy, autonomic nervous system. In *StatPearls*. StatPearls Publishing.

Wood, C., Bianchi, M. T., Yun, C. H., Shin, C., & Thomas, R. J. (2020). Multicomponent analysis of sleep using electrocortical, respiratory, autonomic and hemodynamic signals reveals distinct features of stable and unstable NREM and REM sleep. *Frontiers in Physiology, 11*, 592978. https://doi.org/10.3389/fphys.2020.592978

Xie, M., Lei, F., Guo, D., Ma, Y., Tang, X., & Zhou, J. (2018). Diagnosis of obstructive sleep apnea using cardiopulmonary coupling analysis. *National Medical Journal of China, 98*, 1565–1569.

Yang, A. C., Yang, C. H., Hong, C. J., Tsai, S. J., Kuo, C. H., Peng, C. K., Mietus, J. E., Goldberger, A. L., & Thomas, R. J. (2011). Sleep state instabilities in major depressive disorder: Detection and quantification with electrocardiogram-based cardiopulmonary coupling analysis. *Psychophysiology, 48*, 285–291.

Yasuma, F., & Hayano, J. (2004). Respiratory sinus arrhythmia: Why does the heartbeat synchronize with respiratory rhythm? *Chest, 125*, 683–690.

Yeh, G. Y., Mietus, J. E., Peng, C. K., Phillips, R. S., Davis, R. B., Wayne, P. M., Goldberger, A. L., & Thomas, R. J. (2008). Enhancement of sleep stability with Tai Chi exercise in chronic heart failure: Preliminary findings using an ECG-based spectrogram method. *Sleep Medicine, 9*, 527–536.

Younes, M., Ostrowski, M., Soiferman, M., Younes, H., Younes, M., Raneri, J., & Hanly, P. (2015). Odds ratio product of sleep EEG as a continuous measure of sleep state. *Sleep, 38*, 641–654.

Young, T., Evans, L., Finn, L., & Palta, M. (1997). Estimation of the clinically diagnosed proportion of sleep apnea syndrome in middle-aged men and women. *Sleep, 20*, 705–706.

Zhang, X., Song, B., Liu, Y., Wan, Y., Zhou, K., & Xue, R. (2021). Cognitive deficit is correlated with sleep stability in insomnia: A cardiopulmonary coupling study. *Brain and Behavior: A Cognitive Neuroscience Perspective, 11*, e02068.

Pulse Oximetry: The Working Principle, Signal Formation, and Applications

12

Timo Leppänen, Samu Kainulainen, Henri Korkalainen, Saara Sillanmäki, Antti Kulkas, Juha Töyräs, and Sami Nikkonen

Abstract

Pulse oximeters are routinely used in various medical-grade and consumer-grade applications. They can be used to estimate, for example, blood oxygen saturation, autonomic nervous system activity and cardiac function, blood pressure, sleep quality, and recovery through the recording of photoplethysmography signal. Medical-grade devices often record red and infra-red light-based photoplethysmography signals while smartwatches and other consumer-grade devices usually rely on a green light. At its simplest, a pulse oximeter can consist of one or two photodiodes and a photodetector attached, for example, a fingertip or earlobe. These sensors are used to record light absorption in a medium as a function of time. This time-varying absorption information is used to form a photoplethysmography signal. In this chapter, we discuss the working principles of pulse oximeters and the formation of the photoplethysmography signal. We will further discuss the advantages

Leppänen Timo and Kainulainen Samu contributed equally for this chapter.

The original version of this chapter was previously published non-open access. A Correction to this chapter is available at https://doi.org/10.1007/978-3-031-06413-5_22

T. Leppänen (✉)
Department of Applied Physics, University of Eastern Finland, Kuopio, Finland

Diagnostic Imaging Center, Kuopio University Hospital, Kuopio, Finland

School of Information Technology and Electrical Engineering, The University of Queensland, Brisbane, Australia
e-mail: timo.leppanen@uef.fi

S. Kainulainen (✉) · H. Korkalainen · S. Nikkonen
Department of Applied Physics, University of Eastern Finland, Kuopio, Finland

Diagnostic Imaging Center, Kuopio University Hospital, Kuopio, Finland
e-mail: samu.kainulainen@uef.fi; henri.korkalainen@uef.fi; sami.nikkonen@uef.fi

S. Sillanmäki
Diagnostic Imaging Center, Kuopio University Hospital, Kuopio, Finland
e-mail: saara.sillanmaki@kuh.fi

A. Kulkas
Department of Applied Physics, University of Eastern Finland, Kuopio, Finland

Department of Clinical Neurophysiology, Seinäjoki Central Hospital, Seinäjoki, Finland
e-mail: antti.kulkas@cpshp.fi

J. Töyräs
Department of Applied Physics, University of Eastern Finland, Kuopio, Finland

School of Information Technology and Electrical Engineering, The University of Queensland, Brisbane, Australia

Science Service Center, Kuopio University Hospital, Kuopio, Finland
e-mail: juha.toyras@kuh.fi

T. Penzel, R. Hornero (eds.), *Advances in the Diagnosis and Treatment of Sleep Apnea*, Advances in Experimental Medicine and Biology 1384, https://doi.org/10.1007/978-3-031-06413-5_12

and disadvantages of pulse oximeters, which kind of applications exist in the medical field, and how pulse oximeters are utilized in daily health monitoring.

Keywords

Pulse oximetry · Photoplethysmography · Oxygen saturation · Application

Abbreviations

AC	Alternating current
AHI	Apnea-hypopnea index
C	Concentration
COHb	Carboxyhemoglobin
DC	Direct current
ECG	Electrocardiography
EEG	Electroencephalogram
H_2O	Water
HF-AC	High-frequency alternating current
HRV	Heart rate variability
ICU	Intensive care unit
IR	Infrared
I_{trans}	Transmitted intensity
LED	Light-emitting diode
LF-AC	Low-frequency alternating current
MetHb	Methemoglobin
OHb	Oxygenated hemoglobin
OSA	Obstructive sleep apnea
PPG	Photoplethysmography
PRV	Pulse rate variability
PTT	Pulse transit time
RHb	Deoxygenated hemoglobin
SaO_2	Arterial oxygen saturation
SpO_2	Peripheral blood oxygen saturation

12.1 Working Principle

Pulse oximetry is a method initially developed for the measurement of peripheral blood oxygen saturation (SpO_2). It is an optical technique based on differences in light absorption spectra of oxygenated (OHb) and deoxygenated (RHb) hemoglobin (Nitzan et al., 2014). More precisely, the estimation of the SpO_2 is based on photoplethysmography (PPG, see Sect. 12.2). As a noninvasive method, having a high correlation with invasive arterial oxygen saturation (SaO_2) defined based on arterial blood gas analysis (Nitzan et al., 2014), it has become a valuable technique for measuring oxygen saturation in clinical settings.

In addition to being noninvasive, pulse oximetry has several other advantages. It is a safe, comfortable, and inexpensive method with no need for end-user calibration. As the oximeter is usually placed to the fingertip or earlobe in a medical setting, it can usually be self-applied and does not require a medical specialist. Furthermore, various consumer-grade health technology solutions, such as smartwatches and smartphones, are also capable of estimating SpO_2 with a reasonable correlation to SaO_2. As pulse oximetry is a simple and inexpensive method already integrated into various settings, it is ideal for long-term monitoring of overall well-being, stress, recovery, quality of sleep, and based on the recent research, also for detection of sleep disturbances and disorders.

Pulse oximetry can rely either on the transmission or the reflection of light (Fig. 12.1). Out of these, transmissive pulse oximetry is the most common in medical devices. In transmissive pulse oximetry, light sensors are placed usually on a fingertip or earlobe. The response time of conventional oximeter probes varies and, for example, ear probes respond quicker to a change in blood oxygen saturation than finger probes (Young et al., 1992). Also, the sensors can be attached to a toe in newborns. The sensors emit light usually with two different wavelengths, and the light passes through the skin and reaches a photodetector that measures the changes in absorption of both wavelengths. The reflective mode, on the other hand, can be applied to different parts of the body, not only on the fingertips or the thin portion of the ear, to measure the saturation and PPG signal. Similarly, as in the transmissive pulse oximetry, the sensors emit light with two different wavelengths. However, the main difference is that in reflectance pulse oximetry the photodetector is located next to the light-emitting sensors and detects reflected and backscattered photons of both emitted wave-

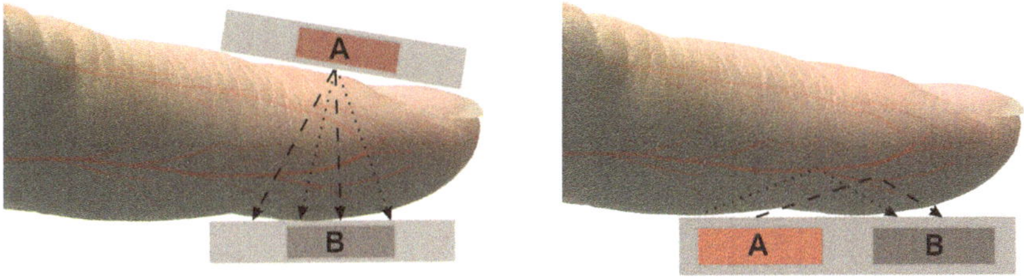

Fig. 12.1 Obtaining the photoplethysmogram and saturation signal with a pulse oximeter. A transmissive mode, on the left-hand side, and a reflective mode, on the right-hand side, are presented. A: a light source (emitting, for example, red and infrared light) and B: a detector

lengths. In some cases, pulse oximeters may utilize more than two different wavelengths of light.

Pulse oximetry is widely used in different clinical domains and is, for example, a basic measurement at intensive care units (ICUs) and always included in polysomnographic evaluations. Pulse oximetry is mainly used to measure SpO_2, but the measured PPG signal contains a vast amount of information and works as a proxy for several physiological functions. For example, it can be used to evaluate indirectly blood pressure (Nachman et al., 2020), depth of anesthesia (Shelley, 2007), pulse rate (Shelley, 2007), heart rate variability (Gil et al., 2010), cardiac arrhythmias (Blanc et al., 1993), respiratory rate (Shelley, 2007), sleep stages (Huttunen et al., 2021), presence of sleep disorders (Lazazzera et al., 2021; Nikkonen et al., 2019), and sleep disorder-related daytime symptoms (Kainulainen et al., 2020a, b). The sampling frequency of a pulse oximeter depends on its intended use and is typically between 1 and 256 Hz. For the determination of SpO_2 in clinical settings, 10–25 Hz sampling frequency is currently recommended (Iber et al., 2007). However, a minimum sampling frequency of 200 Hz is required to get more accurate estimation on, for example, pulse rate and heart rate variability from PPG signal (Béres et al., 2019). Thus, the typical sampling frequency for PPG signal acquisition is 256 Hz.

12.1.1 Green, Red, and Infrared Light

Transmissive pulse oximeters with two light sources are the most commonly used type of oximeters in clinical practice and their working principle is based on the light absorption in the peripheral arterial blood (Fig. 12.2), more specifically, the absorption induced by the OHb and RHb (Damianou, 1995; Mannheimer, 2007). When oxygen binds with the iron ion in the blood, the structure of the heme group in the binding site changes from a non-planar to a planar orientation (Benesch et al., 1975). These molecular structure changes with oxygenation result in differences in light absorption between OHb and RHb (Fig. 12.2).

As illustrated in Fig. 12.2., at 660 nm (red light), the absorption of light is mainly caused by RHb in blood. In contrast, at 530 nm and 900 nm (green and infrared, respectively), the absorption of light is mainly caused by OHb as the molar extinction coefficient of OHb is higher compared to that of RHb. The green, red, and infrared light can penetrate through soft tissue (mostly consisting of water) and the amount of transmitted light can be measured. This information can be used to estimate the level of oxygen in the blood. The measurement is, of course, disturbed by the light scattering within the medium and reflections from the light source-skin surfaces, but these interactions are not significant (Tuchin, 2015).

In transmissive pulse oximeters, red light is more often used than green light. This is because

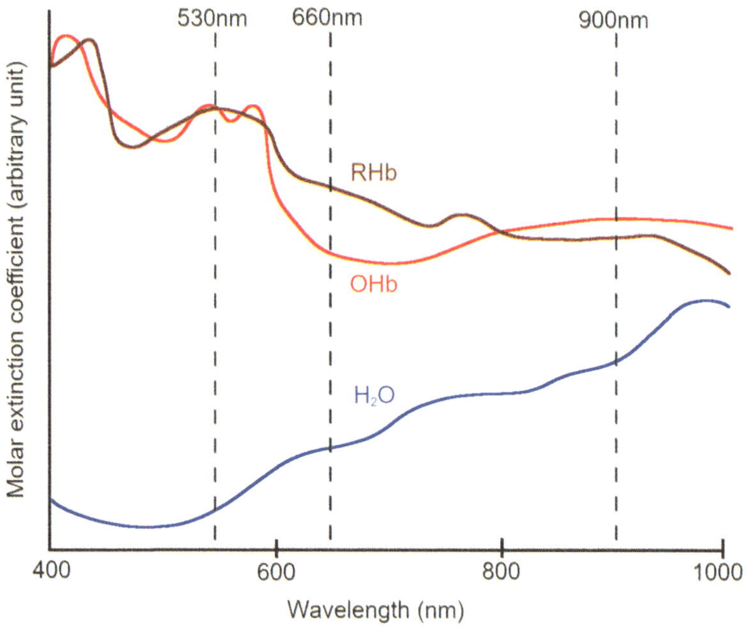

Fig. 12.2 Illustration of the logarithmic molar extinction coefficients as a function of light wavelength. 530 nm, 660 nm, and 900 nm correspond to the green, red, and infrared light wavelengths commonly used in pulse oximeters, respectively (Tuchin, 2015; Kainulainen, 2020). OHb oxygenated hemoglobin, RHb deoxygenated hemoglobin, H_2O water

the use of red light increases the accuracy and precision of the measurements of, for example, heart rate and blood oxygen saturation. Also, the human body poorly absorbs red light allowing it to penetrate much deeper than green light. Therefore, the usefulness of green light in determining muscle saturation or total hemoglobin is limited. Also, red light is not affected as much by dark skin tone or tattoos which can distort the measurements done with green light. However, consumer-grade devices are most often reflective pulse oximeters utilizing green light to measure PPG. This is because the use of green light has several benefits. First, green PPG amplitudes are the strongest across the range of visible light (Verkruysse et al., 2008). Second, the human tissues are good absorbers of green light and, thus, green light coming from external sources does not disturb the measurement affecting the signal quality.

Pulse oximeters utilizing more than two wavelengths (usually four) also exist. They are usually transmissive pulse oximeters and can differentiate between more than two blood components. This is, they can detect other hemoglobin types in addition to RHb and OHb, such as carboxyhemoglobin (COHb) and methemoglobin (MetHb) also called dyshemoglobins. Therefore, multiple wavelength pulse oximeters are often called CO-oximeters (Zaouter & Zavorsky, 2012). Four-wavelength CO-oximeters are typically used as a gold standard reference when evaluating the accuracy of standard oximeters as they are more accurate (Sinex, 1999) or when the standard oximeter is not sufficient, for example, when carbon monoxide poisoning is suspected.

12.2 Photoplethysmogram

With transmissive pulse oximeters, an absorption signal is formed at a photodetector which measures transmitted light passed through a medium. For example, soft tissue, bony structures, and blood absorb light differently and these structures affect the absorption signal characteristics when measuring the absorption signal from a fingertip. Cardiac and respiratory functions cause changes in the blood volume and composition in the arterial blood. Therefore, the concentrations of OHb and RHb, and the optical path length of the light change constantly as a function of time (Damianou, 1995; Mannheimer, 2007). This light absorption signal varying as a function of time is called a photoplethysmogram.

The wavelength of the light affects the PPG signal, i.e., the absorbance. Red light is more sensitive to changes in oxygenation than infrared light as the molar extinction coefficient of RHb is higher and water absorption lower (Fig. 12.2). The absorption of infrared light is similar to water and both oxygenated and deoxygenated hemoglobin. Moreover, the total absorption is lower with infrared light compared to red light. Therefore, the PPG signal measured with infrared wavelength is more stable and is more commonly used than red light (Alian & Shelley, 2014). However, both red and infrared light measurements are needed to estimate SpO_2.

The absorbance signal is formed by three components: the direct current (DC), the low-frequency alternating current (LF-AC), and the high-frequency alternating current (HF-AC) components (Fig. 12.3). The DC component represents the absorption of light in static mediums and has no pulsatile time-varying component. In contrast, the LF-AC and HF-AC vary temporarily. The LF-AC component carries information on changes in blood volume which can be caused by alterations in breathing, thermal regulation, autonomic nervous system activity, and hemoglobin concentration. The HF-AC component mainly represents the pulsatility of the arteries and forms the basis of the PPG signal.

One PPG waveform can be separated into systolic and diastolic phases (Gamrah et al., 2020). The systolic phase starts when the aortic valve opens. During the systolic phase, blood pressure increases as blood flows into the aorta. This can be seen as a decrease in the transmitted light intensity as a function of time. After the systolic maximum pressure is reached (the first maximum peak), the blood pressure starts to decrease and the aortic valve closes. When the aortic valve closes, another increase in the blood pressure can be seen (the second maximum peak) (Hogan et al., 2014). Between these two maximum peaks is the Dicrotic Notch representing the end of the systolic phase and the start of the diastolic phase (Gamrah et al., 2020). During the diastolic phase, blood flows out from the aorta and blood pressure decreases (transmitted light intensity increases as a function of time) until the minimum diastolic pressure is reached. Then, the aortic valve opens again, and a new cycle starts from the beginning.

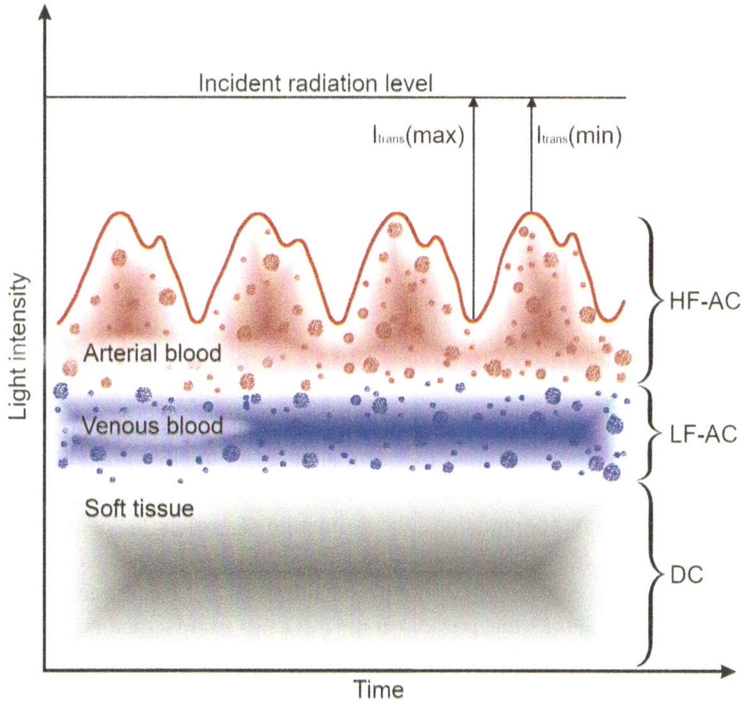

Fig. 12.3 Illustrative example of the formation and waveform of photoplethysmography signal. (Figure modified from Tiihonen (2008), Kainulainen (2020). AC alternating current, DC direct current, LF-AC low-frequency AC component, HF-AC high-frequency AC component, I_{trans}(max) transmitted intensity when the optical path length is the longest, I_{trans}(min) transmitted intensity when the optical path length is shortest

However, it has to be noted that the strengths of both LF-AC and HF-AC components are only a few percentages from the total absorption.

Green light with a wavelength of 530 nm cannot penetrate as deep in tissue as red (660 nm) or infrared light (900 nm) (Huang et al., 2011). The green light can only penetrate through the epidermis and papillary layer while red and infrared lights can reach the dermis, subcutaneous, and deeper layers (Sviridova et al., 2018). When using red or infrared light, the PPG waveform is mainly determined by blood volume changes in the peripheral arteries. However, especially with large vessels, green light cannot penetrate deep enough to reach the arteries, and information on blood volume and flow changes cannot be directly obtained. Therefore, contrary to what has been thought before (Damianou, 1995; Mannheimer, 2007), it has been proposed that the PPG signal is formed based on the pulsatile transmural pressure of the arteries, i.e., the pressure difference between walls of the thinnest capillaries in the papillary dermis (Kamshilin et al., 2015). The distance between nearby capillaries changes as the inter-capillary tissue stretches and compresses caused by the beating of the heart. Thus, changes in the density of the capillaries in the papillary dermis lead to changes in the optical properties of the tissue (Kamshilin et al., 2015; Volkov et al., 2017).

12.2.1 Blood Oxygen Saturation

The blood oxygen saturation is defined as a ratio between concentrations of OHb and RHb:

$$\mathrm{SpO_2} = \frac{C(\mathrm{OHb})}{C(\mathrm{OHb}) + C(\mathrm{RHb})} \times 100\%$$

where C is the concentration. However, invasive measurement is required for accurate definition of the concentrations and, thus, in noninvasive transmissive mode applications, the estimation of SpO_2 is based on the absorption of red and infrared light. PPG signals of red light and infrared light oscillate at the same phase, but the PPG signal of infrared light has a smaller amplitude. During one heart cycle, hemoglobin concentration and body temperature are nearly constant; thus, the only time-varying component is the optical path length of light that is mainly affected by the arterial blood volume changes. This leads to a situation where the AC component (Fig. 12.3), for both PPG signals, can be defined based on the differential absorption during one pulse wave (Damianou, 1995; Mannheimer, 2007). By further dividing the differential absorption of red light with differential absorption of infrared light, the SpO_2 value can be estimated (Damianou, 1995, Mannheimer, 2007). To summarize, both red and infrared PPG signals are required to estimate peripheral blood oxygen saturation noninvasively.

12.2.2 Pulse

Although pulse oximeters are colloquially said to also measure heart rate, this is not strictly true. Instead, pulse oximeters can only record pulse rate (Schäfer & Vagedes, 2013). Generally, the pulse rate and heart rate are highly correlated. Thus, in most everyday applications, heart rate can accurately be estimated by pulse rate. However, minor differences do exist. Since the pressure wave generated by the heart will take a short time to reach the fingertip, there is a small delay between the systole and the time the pulse is detected at the periphery (Smith et al., 1999). This time delay is also called the pulse transit time (PTT), and it can be accurately recorded with simultaneous electrocardiography (ECG) and PPG recording (Smith et al., 1999). Since the pulse through the artery is a mechanical wave, it travels at the speed of sound, making the PTT very short, but still significant. PTT varies between study subjects and is affected by the elasticity of the artery wall and blood pressure among many other factors (Mukkamala et al., 2015). Thus, PTT can also be used for various purposes such as continuous blood pressure estimation (Mukkamala et al., 2015; Smith et al., 1999).

Heart rate variability (HRV) analysis is a common method for evaluating autonomic nervous system function (Pinheiro et al., 2016; Hietakoste et al., 2020). HRV parameters are obtained from

ECG by detecting the fiducial point of R waves and generating the R-R intervals. Since ECG signal is not always available during sleep recordings and there can be artifacts influencing R wave detection, the information on HRV can be also obtained by using the pulse signal. As the detection of the exact time point of the R-peak is important, accurate pulse peak detection is equally essential. However, as the pulse wave is relatively wide (Fig. 12.3) and not as sharp as the R-peak, there is intrinsically more inaccuracy in the pulse wave detection (Schäfer & Vagedes, 2013). However, since there are factors beyond cardiac electrophysiology affecting the PPG signal (Gil et al., 2010), pulse rate variability (PRV) is not directly comparable to HRV. Still, it can be used as an indicator to quantify the activity of the autonomic nervous system (Pinheiro et al., 2016; Mejía-Mejía et al., 2020).

12.3 Error Sources and Limitations

Even though pulse oximetry is a powerful tool, it also has certain limitations (Table 12.1) some of which are discussed in this section. One major limitation of pulse oximetry is that it measures the ratio of oxygenated and deoxygenated hemoglobin rather than ventilation or the amount of oxygen in the tissue (Mcmorrow & Mythen, 2006). While these are normally tightly linked in a healthy subject, some conditions can cause hypoxia in the tissue level even if the measured

Table 12.1 Possible error sources of a pulse oximeter

Error source	Type of error	Reference
Poor peripheral circulation Hypotension Hypothermia/cold periphery Vasoconstriction	Intermittent drop-outs or inability to read SpO_2 or false readings	Hakemi and Bender (2005), Chan and Chan (2013)
Movement artifacts Especially when estimated from a fingertip	Falsely low SpO_2	Mcmorrow and Mythen (2006), Tsien and Fackler (1997), Chan and Chan (2013)
Venous pulsation Probe too tight around the finger Severe tricuspid regurgitation Heart failure	Falsely low SpO_2	Chan and Chan (2013)
Dyshemoglobinemia Carboxyhaemoglobinaemia Methemoglobinemia	False readings	Tremper (1989)
Carbon monoxide poisoning	Falsely normal or elevated SpO_2	Vegfors and Lennmarken (1991), Barker and Tremper (1987)
Lightning Light-emitting diodes, infrared, ultraviolet, fluorescent lamps (the effect can be tested by covering the probe)	False readings	Amar et al. (1989), Schulz and Ham (2019)
Light barriers Nail polish, artificial nails (influence depends on the color) Tattoos Skin discoloration (caused by tobacco, dirt, or paint)	False readings	Coté et al. (1988), Samman et al. (2006), Ralston et al. (1991b)
Dark skin tone	Possible false readings	Ries et al. (1989, Cecil et al. (1988), Ralston et al. (1991b), Bickler et al. (2005)
Severe anemia	Falsely low SpO_2 in hypoxemic patients	Chan and Chan (2013)

SpO_2: peripheral blood oxygen saturation

oxygen saturation is normal. For example, undetected major loss of blood (e.g., internal bleeding) may cause a lack of oxygen in the tissue level, even though the SpO_2 reading is still high. Severe anemia can cause a similar condition. However, it is critical to note that in these cases, the pulse oximeter reading is not false and the blood can be highly oxygenated; there may just not be enough blood volume to keep the tissue fully oxygenated (Mcmorrow & Mythen, 2006).

Pulse oximetry does not require calibration before use for different individuals (Ralston et al., 1991a). The zero calibration is performed automatically due to measuring light absorption both during the diastolic and systolic phases. Gain calibration is not required either as the absolute light intensity at the photoreceptor does not affect the estimated oxygen saturation. Only the ratio of the absorbed light in the two (or more) wavelengths is considered. While the lack of calibration is certainly one of the greatest advantages of pulse oximetry, it is also a major error source. As no calibration is needed, the device simply uses a pre-determined look-up table to estimate the peripheral oxygen saturation based on the ratio of the absorbed light (Ralston et al., 1991a). The values in the look-up table are derived experimentally by simultaneous measurement of true SaO_2 and the light absorbance (Jubran, 1998). An absorption ratio curve can then be formed based on these measurements (Sinex, 1999). Thus, the result of using this standardized look-up table is that if some unaccounted factors are affecting the light absorption ratio other than the fraction of oxygenated hemoglobin, the pulse oximeter can give false readings. In addition, the population in which this curve is produced, i.e., the calibration population, will affect the ultimate accuracy of the pulse oximeter (Sinex, 1999; Mcmorrow & Mythen, 2006). As the experimental data cannot be obtained for arbitrarily low saturation values, the accuracy of pulse oximeters decreases with lower saturation (Sinex, 1999; Jubran & Tobin, 1990). Accuracy rates reported for individual instruments are often in the range of ±2–3% (Balady et al., 2010). Wider confidence limits are not unusual, particularly in the saturation less than 85%, but SpO_2 values of less than 70% are usually considered unreliable (Mcmorrow & Mythen, 2006).

The theoretical foundation behind transmissive pulse oximetry assumes that the blood contains only oxygenated and deoxygenated hemoglobin, which is strictly not true as the hemoglobin can also form other compounds such as carboxyhemoglobin or methemoglobin (Ralston et al., 1991a; Zaouter & Zavorsky, 2012). These compounds, also called dyshemoglobins, affect the absorption and, thus, affect the oximeter reading. Normally, the dyshemoglobins have little effect on the oximeter reading, but in some cases, they can be a major source of error (Tremper, 1989). If the person is suffering from carbon monoxide poisoning, the amount of carboxyhemoglobin in the blood is high. However, standard pulse oximeters cannot reliably distinguish between carboxyhemoglobin and oxygenated hemoglobin (Vegfors & Lennmarken, 1991). True oxygenation of as low as 30% can still give >90% saturation readings when carboxyhemoglobin concentration in the blood is high (Barker & Tremper, 1987). Smoking a cigarette also increases the level of carboxyhemoglobin and can thus temporarily elevate the pulse oximeter reading (DeMeulenaere, 2007). If the concentration of methemoglobin is high, it can dominate the absorption spectrum and erroneously cause a reading of around 85% regardless of the true saturation (Barker et al., 1989). The errors caused by the presence of dyshemoglobin in blood could be eliminated by using multi-wavelength CO-oximeters. However, due to their higher cost, they are commonly only used in specialized applications, such as when carbon monoxide poisoning is suspected or as a reference (Sinex, 1999).

Newborn infants have a different type of hemoglobin in their blood called fetal hemoglobin. It has a different composition and bonds to oxygen more strongly than adult hemoglobin. However, it has a similar absorption spectrum in the wavelength 650-1000 nm (Ralston et al., 1991b). As this is the wavelength range most two-wavelength oximeters operate in, the presence of fetal hemoglobin has little effect on the oximeter reading. However, the fetal hemoglobin will affect the reading of a multi-wavelength CO-oximeter and, thus, its presence should be considered and the oximeter reading corrected (Ralston et al., 1991b).

As the transmissive pulse oximeter assumes that the other tissue and venous blood absorption are constant, limb motion can also cause movement artifacts and is a significant error source (Mcmorrow & Mythen, 2006). Movement artifacts are the most common type of error in the oximeter reading and have been reported to cause as much as half of the false alarms in pulse oximeters in the ICUs (Tsien & Fackler, 1997). In a diagnostic test setting (for example during a cardiac-pulmonary test), the amount of movement artifacts can be reduced by placing a sensor on to forehead skin. Furthermore, excess sweating can affect SpO_2 readings; however, the effect of sweating is easily reduced by wiping the electrode.

Pulse oximetry can also be affected by skin tone and the accuracy of measurements can be decreased in persons with dark skin tone (Ries et al., 1989; Cecil et al., 1988; Ralston et al., 1991b), although contradicting results have also been reported (Bothma et al., 1996; Adler et al., 1998). However, a slight decrease in accuracy due to skin tone is not usually clinically significant (Cecil et al., 1988). Tattoos, artificial nails, and nail polish can also decrease the pulse oximeter accuracy as they can change the light absorption (Jubran, 1999; Coté et al., 1988; Ralston et al., 1991b). However, these issues are usually easily avoided simply by using a non-tattooed skin surface and by removing nail polish or artificial nails.

Low perfusion can also cause errors in pulse oximetry readings (Hakemi & Bender, 2005). Low perfusion could be caused by vasoconstriction, low cardiac output, or even hypothermia. Extremities can be warmed with a mitten (for example during tilt table test) to maintain a reliable transmittance of pulse oximetry and PPG signals. In a case of low perfusion, it might be difficult for the pulse oximeter to reliably detect the HF-AC component of the signal from the noise and, thus, may give erroneous readings. To be considered reliable, the pulse signal from the oximeter should be regular and match the heart rate from the ECG if available. Similarly, if a subject has very low blood pressure, the pulse oximeter readings should be considered unreliable (DeMeulenaere, 2007).

The components used in the pulse oximeters may affect the accuracy of the device, especially if the components are not of the highest quality (Mcmorrow & Mythen, 2006). For example, cheap light-emitting diodes (LEDs) may emit a slightly different frequency of light as designed or the photodetectors may report slightly erroneous intensities. These error sources could be eliminated by only using high-quality components but should still be accounted for especially in consumer-grade devices.

Although there are several limitations in pulse oximetry, the error sources can usually be accounted for or corrected if the error sources are acknowledged. However, a considerable proportion of users, either healthcare professionals or consumers, may not be aware of what exactly a pulse oximeter is recording and that the pulse oximeter is not a direct measure of partial pressure of oxygen in the blood (Stoneham et al., 1994). Therefore, some of the limitations in pulse oximetry may be magnified by these misconceptions as the error sources may not be acknowledged and, thus, false readings may not be noticed.

12.4 Applications

12.4.1 Consumer Use

Consumer pulse oximeters are widely available. Most of these low-cost pulse oximeters have not been rigorously tested and validated against clinical measurement and do not meet standards for medical devices. However, most devices can safely rule out hypoxemia in the vast majority of patients (Harskamp et al., 2021). This is important since the demand for oximeters has increased during the COVID-19 pandemic, and the World Health Organization has recommended home oximetry monitoring for patients with COVID-19 and with risk factors for progression to severe disease (Organization, 2021). Simultaneously, PPG recording has gained popularity in consumer-grade wearable devices. The solutions come with multiple forms and arrangements depending on the manufacturer, but most often the basic principles are the same. Wearables most often utilize reflective mode and green light, or a combination of different light colors. Vulcan et al. (2021) published an extensive review that summarizes the current status of wrist-worn consumer-grade products for sleep tracking. The study reveals that devices perform well when considering the basic PRV estimation, heart rate tracking, and other physical activity, especially if the tracker is equipped with an accelerometer. When tested against PSG measurements, most of the consumer-grade sleep trackers perform relatively poor in epoch-by-epoch sleep staging (Vulcan et al., 2021; Chinoy et al., 2021). Furthermore, it has to be noted that most of the sleep trackers use multiple information sources for the sleep stage estimation, not just the PPG. However, they can measure total sleep time and differentiate wake from sleep with sufficient accuracy surpassing, for example, plain actigraphy. Therefore, it could be speculated that these devices would fit well in long-term evaluations together with sleep diaries.

In the clinical setting, the PPG is most often measured with transmissive mode from the fingertip or earlobe. In contrast, most of the trackers are wrist-worn devices and exploit reflective mode. A variety of studies show that the measurement location affects the measured signal quality and the achieved correspondence with the gold standard measurement. For example, the respiratory rate, heart rate, and RR-interval evaluation are not as accurate from the wrist with reflective mode oximeters as from the fingertip with transmissive mode oximeters (Hartmann et al., 2019; Longmore et al., 2019).

The PPG-based consumer-grade approaches show a lot of potential but most of them lack validation against the gold standard measurements. Other disadvantages are the non-harmonized measurement techniques, measurement locations, and signal processing algorithms (Vulcan et al., 2021). However, it has to be noted that these devices are designed for commercial, not medical use, and, therefore, the performance demands are very different.

12.4.2 Clinical Use

A pulse oximeter is a vital part of standard clinical assessment. It is routinely used in emergency rooms and patient wards because it can quickly give information on ventilation and perfusion deficits. Saturation measurement is also an important tool for diagnostics. For example, pulse oximeter measurements are obtained routinely during cardiopulmonary exercise testing (Balady et al., 2010). A decrease in saturation by >5% is commonly used as an indication of pulmonary limitation to exercise (Balady et al., 2010). Normally, the SpO_2 of a healthy subject is at least 95%, while values <90% are considered low and possibly alarming (Broaddus, 2016; Duncan, 2017). The most common causes for decreased SpO_2 levels are ventilation-perfusion mismatch (for example due to apnea or obstructive pulmonary disease), cardiovascular shunt, abnormal pulmonary diffusion capacity, hypoventilation, or decreased oxygen content of the inhaled air (Bhutta et al., 2021).

The SpO_2 is constantly measured and followed in the ICU. However, critically ill patients' pulse oximeter-based oxygenation estimation has been criticized for its tendency to overestimate the readings (Van de Louw et al., 2001). Also in these patients, periphery vasocontraction is prevalent. For example, low perfusion and sepsis among other conditions can lead to over 90% pulse oximeter readings, even though the true SpO_2 is below 90% (Wilson et al., 2010). For these reasons, ICU perfusion is often screened with more robust methods. Global oxygenation may be also monitored intermittently through blood gas analysis or continuously with specialized catheters (Kipnis & Valle, 2016).

In sleep medicine, especially when assessing obstructive sleep apnea (OSA), a pulse oximeter is a cornerstone of the measurements. When considering the SpO_2, conventional parameters such as the apnea-hypopnea index (AHI), oxygen desaturation index, time spent under 90% saturation, mean SpO_2, and nadir SpO_2 can be estimated. Based on the latest research, the AHI can be evaluated utilizing only a pulse oximeter recording (Nikkonen et al., 2019; Leino et al., 2021; Álvarez et al., 2017). In addition, the frequency domain information in PPG and the desaturation areas can be exploited independently to assess the severity of sleep apnea beyond the AHI (Kainulainen et al., 2020a, b; Kainulainen et al., 2019; Azarbarzin et al., 2019). Thus, it could be speculated that OSA screening in various patient populations and measurement conditions could be conducted based even solely on a pulse oximeter.

However, a pulse oximeter provides a lot more information besides the SpO_2. The PPG signals contain information on the pulse rate, PRV, peripheral vasoconstriction, and perfusion (Jubran, 2015; Lima et al., 2002; Pinheiro et al., 2016; Rauh et al., 2004; Tusman et al., 2019). The recurrent breathing cessations and flow limitations cause the cardiorespiratory system to respond to changes in intrathoracic pressure and deoxygenation. The responses are further seen in the PPG signal as deviations from normal waveform, amplitude drops, and frequency modulations. For simplistic analysis and to gain a better overview of the condition of the patient, normal PRV metrics can be assessed while keeping in mind its limitations (see Sect. 12.2.2).

Moreover, the parasympathetic nervous system controls not just the cardiorespiratory system and various body functions but also sleep stages (Fink et al., 2018). For these reasons, the PPG signal carries information also on the depth of sleep. PPG could be used as a surrogate for electroencephalogram (EEG) in detecting rapid eye movement (REM) sleep, non-REM sleep, and wakefulness (Huttunen et al., 2021; Korkalainen et al., 2020). In addition to sleep stage scoring, PPG could be used for arousal detection (Karmakar et al., 2014). Arousals cause rapid changes in pulse and short periods of vasoconstriction. These phenomena are visible in the PPG signal as frequency increases and amplitude decreases, respectively. Simultaneously, the amplitude drops in PPG are surrogates for subcortical brain activity, even though EEG-based arousal is undetectable (Delessert et al., 2010). In summary, the utility of the pulse oximeter in sleep medicine is evident.

Acknowledgments The funding was received from the European Union's Horizon 2020 research and innovation program under grant agreement no. 965417, NordForsk (NordSleep project 90458) via Business Finland (5133/31/2018), the Academy of Finland (323536), the Research Committee of the Kuopio University Hospital Catchment Area for the State Research Funding (5041767, 5041768, 5041794, 5041797, 5041798, and 5041803), from Seinäjoki Central Hospital, the Competitive State Research Financing of Expert Responsibility Area of Tampere University Hospital (VTR3242, VTR3249, and EVO2089), and Tampere Tuberculosis Foundation.

References

Adler, J. N., Hughes, L. A., Vivilecchia, R., & Camargo, C. A. (1998). Effect of skin pigmentation on pulse oximetry accuracy in the emergency department. *Academic Emergency Medicine, 5*, 965–970.

Alian, A. A., & Shelley, K. H. (2014). Photoplethysmography. *Best Practice & Research. Clinical Anaesthesiology, 28*, 395–406.

Álvarez, D., Alonso-Álvarez, M. L., Gutiérrez-Tobal, G. C., Crespo, A., Kheirandish-Gozal, L., Hornero, R., Gozal, D., Terán-Santos, J., & Del Campo, F. (2017). Automated screening of children with obstructive sleep apnea using nocturnal oximetry: An alternative to respiratory polygraphy in unattended settings. *Journal of Clinical Sleep Medicine, 13*, 693–702.

Amar, D., Neidzwski, J., Wald, A., & Finck, A. D. (1989). Fluorescent light interferes with pulse oximetry. *Journal of Clinical Monitoring, 5*, 135–136.

Azarbarzin, A., Sands, S. A., Stone, K. L., Taranto-Montemurro, L., Messineo, L., Terrill, P. I., Ancoli-Israel, S., Ensrud, K., Purcell, S., White, D. P., Redline, S., & Wellman, A. (2019). The hypoxic burden of sleep apnoea predicts cardiovascular disease-related mortality: The osteoporotic fractures in men study and the sleep heart health study. *European Heart Journal, 40*, 1149–1157.

Balady, G. J., Arena, R., Sietsema, K., Myers, J., Coke, L., Fletcher, G. F., Forman, D., Franklin, B., Guazzi, M., Gulati, M., Keteyian, S. J., Lavie, C. J., Macko, R., Mancini, D., Milani, R. V., American Heart Association Exercise, C. R. R., and Prevention Committee of the Council on Clinical Cardiology, Prevention, C. O. E. A., DISEASE, C. O. P. V. & Research, I. C. O. Q. O. C. A. O. (2010). Clinician's guide to cardiopulmonary exercise testing in adults: A scientific statement from the American Heart Association. *Circulation, 122*, 191–225.

Barker, S. J., & Tremper, K. K. (1987). The effect of carbon monoxide inhalation on pulse oximetry and transcutaneous PO2. *Anesthesiology, 66*, 677–679.

Barker, S. J., Tremper, K. K., & Hyatt, J. (1989). Effects of methemoglobinemia on pulse oximetry and mixed venous oximetry. *Anesthesiology, 70*, 112–117.

Benesch, R., Benesch, R. E., Yung, S., & Edalji, R. (1975). Hemoglobin covalently bridged across the polyphosphate binding site. *Biochemical and Biophysical Research Communications, 63*, 1123–1129.

Béres, S., Holczer, L., & Hejjel, L. (2019). On the minimal adequate sampling frequency of the photoplethysmogram for pulse rate monitoring and heart rate variability analysis in mobile and wearable technology. *Measurement Science Review, 19*, 232–240.

Bhutta, B., Alghoula, F., & Berim, I. (2021). *StatPearls*. StatPearls Publishing.

Bickler, P. E., Feiner, J. R., & Severinghaus, J. W. (2005). Effects of skin pigmentation on pulse oximeter accuracy at low saturation. *Anesthesiology, 102*, 715–719.

Blanc, V. F., Haig, M., Troli, M., & Sauvé, B. (1993). Computerized photo-plethysmography of the finger. *Canadian Journal of Anaesthesia, 40*, 271–278.

Bothma, P. A., Joynt, G. M., Lipman, J., Hon, H., Mathala, B., Scribante, J., & Kromberg, J. (1996). Accuracy of pulse oximetry in pigmented patients. *South African Medical Journal, 86*, 594–596.

Broaddus. (2016). *Murray & Nadel's textbook of respiratory medicine*. Saunders, an imprint of Elsevier Incorporated.

Cecil, W. T., Thorpe, K. J., Fibuch, E. E., & Tuohy, G. F. (1988). A clinical evaluation of the accuracy of the Nellcor N-100 and Ohmeda 3700 pulse oximeters. *Journal of Clinical Monitoring, 4*, 31–36.

Chan, E. D., & Chan, M. M. (2013). Pulse oximetry: Understanding its basic principles facilitates appreciation of its limitations. *Respiratory Medicine, 107*, 789–799.

Chinoy, E. D., Cuellar, J. A., Huwa, K. E., Jameson, J. T., Watson, C. H., Bessman, S. C., Hirsch, D. A., Cooper, A. D., Drummond, S. P. A., & Markwald, R. R. (2021). Performance of seven consumer sleep-tracking devices compared with polysomnography. *Sleep, 44*.

Coté, C. J., Goldstein, E. A., Fuchsman, W. H., & Hoaglin, D. C. (1988). The effect of nail polish on pulse oximetry. *Anesthesia and Analgesia, 67*, 683–686.

Damianou, D. (1995). *The wavelength dependence of the photoplethysmogram and its implication to pulse oximetry*. Ph.D. thesis, University of Nottingham.

Delessert, A., Espa, F., Rossetti, A., Lavigne, G., Tafti, M., & Heinzer, R. (2010). Pulse wave amplitude drops during sleep are reliable surrogate markers of changes in cortical activity. *Sleep, 33*, 1687–1692.

Demeulenaere, S. (2007). Pulse oximetry: Uses and limitations. *The Journal for Nurse Practitioners, 3*, 312–317.

Duncan, D. (2017). *Respiratory care: Assessment and management*. M & K Update Limited.

Fink, A. M., Bronas, U. G., & Calik, M. W. (2018). Autonomic regulation during sleep and wakefulness: A review with implications for defining the pathophysiology of neurological disorders. *Clinical Autonomic Research, 28*, 509–518.

Gamrah, M. A., Xu, J., El Sawy, A., Aguib, H., Yacoub, M., & Parker, K. H. (2020). Mechanics of the dicrotic notch: An acceleration hypothesis. *Proceedings of the Institution of Mechanical Engineers. Part H, 234*, 1253–1259.

Gil, E., Orini, M., Bailón, R., Vergara, J. M., Mainardi, L., & Laguna, P. (2010). Photoplethysmography pulse rate variability as a surrogate measurement of heart rate variability during non-stationary conditions. *Physiological Measurement, 31*, 1271–1290.

Hakemi, A., & Bender, J. A. (2005). Understanding pulse oximetry, advantages, and limitations. *Home Health Care Management & Practice, 17*, 416–418.

Harskamp, R. E., Bekker, L., Himmelreich, J. C. L., De Clercq, L., Karregat, E. P. M., Sleeswijk, M. E., & Lucassen, W. A. M. (2021). Performance of popular pulse oximeters compared with simultaneous arterial oxygen saturation or clinical-grade pulse oximetry: A cross-sectional validation study in intensive care patients. *BMJ Open Respiratory Research, 8*.

Hartmann, V., Liu, H., Chen, F., Hong, W., Hughes, S., & Zheng, D. (2019). Toward accurate extraction of respiratory frequency from the photoplethysmogram: Effect of measurement site. *Frontiers in Physiology, 10*, 732.

Hietakoste, S., Korkalainen, H., Kainulainen, S., Sillanmäki, S., Nikkonen, S., Myllymaa, S., Duce, B., Töyräs, J., & Leppänen, T. (2020). Longer apneas and hypopneas are associated with greater ultra-short-term HRV in obstructive sleep apnea. *Scientific Reports, 10*, 21556.

Hogan, J., O'Dwyer, V., Farah, N., Kennelly, M., Turner, M., & Stuart, B. (2014). Early pregnancy maternal cardiovascular profiling in the prediction of hypertensive disease in pregnancy. *Open Journal of Obstetrics and Gynecology, 4*, 61–65.

Huang, F. H., Yuan, P., Lin, K., Chang, H., & Tsai, C. (2011). Analysis of reflectance Photoplethysmograph sensors. *International Journal of Medical, Health, Biomedical, Bioengineering and Pharmaceutical Engineering, 5*, 622–625.

Huttunen, R., Leppänen, T., Duce, B., Oksenberg, A., Myllymaa, S., Töyrös, J., & Korkalainen, H. (2021). Assessment of obstructive sleep apnea-related sleep fragmentation utilizing deep learning-based sleep staging from photoplethysmography. *Sleep*.

Iber, C., Ancoli-Israel, S., Chesson, A. L., & Quan, S. F. (2007). The AASM manual for the scoring of sleep and associated events: Rules, terminology, and technical specification. In *Version 1*. American Academy of Sleep Medicine.

Jubran, A. (1998). *Pulse oximetry*. McGraw Hill, Inc..

Jubran, A. (1999). Pulse oximetry. *Critical Care, 3*, R11–R17.

Jubran, A. (2015). Pulse oximetry. *Critical Care, 19*, 272.

Jubran, A., & Tobin, M. J. (1990). Reliability of pulse oximetry in titrating supplemental oxygen therapy in ventilator-dependent patients. *Chest, 97*, 1420–1425.

Kainulainen, S. 2020. *Pulse oximetry-derived biomarkers for severity assessment of obstructive sleep apnea: Associating parametric and frequency-domain features of SpO2 and PPG signals with daytime sleepiness and impaired vigilance.* University of Eastern Finland.

Kainulainen, S., Töyräs, J., Oksenberg, A., Korkalainen, H., Sefa, S., Kulkas, A., & Leppänen, T. (2019). Severity of desaturations reflects OSA-related daytime sleepiness better than AHI. *Journal of Clinical Sleep Medicine, 15*, 1135–1142.

Kainulainen, S., Duce, B., Korkalainen, H., Leino, A., Huttunen, R., Kalevo, L., Arnardottir, E. S., Kulkas, A., Myllymaa, S., Töyräs, J., & Leppänen, T. (2020a). Increased nocturnal arterial pulsation frequencies of obstructive sleep apnoea patients is associated with an increased number of lapses in a psychomotor vigilance task. *ERJ Open Res, 6*.

Kainulainen, S., Töyräs, J., Oksenberg, A., Korkalainen, H., Afara, I. O., Leino, A., Kalevo, L., Nikkonen, S., Gadoth, N., Kulkas, A., Myllymaa, S., & Leppänen, T. (2020b). Power spectral densities of nocturnal pulse oximetry signals differ in OSA patients with and without daytime sleepiness. *Sleep Medicine, 73*, 231–237.

Kamshilin, A. A., Nippolainen, E., Sidorov, I. S., Vasilev, P. V., Erofeev, N. P., Podolian, N. P., & Romashko, R. V. (2015). A new look at the essence of the imaging photoplethysmography. *Scientific Reports, 5*, 10494.

Karmakar, C., Khandoker, A., Penzel, T., Schöbel, C., & Palaniswami, M. (2014). Detection of respiratory arousals using photoplethysmography (PPG) signal in sleep apnea patients. *IEEE Journal of Biomedical and Health Informatics, 18*, 1065–1073.

Kipnis, E., & Valle, B. (2016). *Oxford textbook of critical care*. Oxford University Press.

Korkalainen, H., Aakko, J., Duce, B., Kainulainen, S., Leino, A., Nikkonen, S., Afara, I. O., Myllymaa, S., Töyräs, J., & LEPPÄNEN, T. (2020). Deep learning enables sleep staging from photoplethysmogram for patients with suspected sleep apnea. *Sleep*.

Lazazzera, R., Deviaene, M., Varon, C., Buyse, B., Testelmans, D., Laguna, P., Gil, E., & Carrault, G. (2021). Detection and classification of sleep apnea and hypopnea using PPG and SpO. *IEEE Transactions on Biomedical Engineering, 68*, 1496–1506.

Leino, A., Nikkonen, S., Kainulainen, S., Korkalainen, H., Töyräs, J., Myllymaa, S., Leppänen, T., Ylä-Herttuala, S., Westeren-Punnonen, S., Muraja-Murro, A., Jäkälä, P., Mervaala, E., & Myllymaa, K. (2021). Neural network analysis of nocturnal SpO2 signal enables easy screening of sleep apnea in patients with acute cerebrovascular disease. *Sleep Medicine, 79*, 71–78.

Lima, A. P., Beelen, P., & Bakker, J. (2002). Use of a peripheral perfusion index derived from the pulse oximetry signal as a noninvasive indicator of perfusion. *Critical Care Medicine, 30*, 1210–1213.

Longmore, S. K., Lui, G. Y., Naik, G., Breen, P. P., Jalaludin, B., & Gargiulo, G. D. (2019). A comparison of reflective Photoplethysmography for detection of heart rate, blood oxygen saturation, and respiration rate at various anatomical locations. *Sensors (Basel), 19*.

Mannheimer, P. D. (2007). The light-tissue interaction of pulse oximetry. *Anesthesia and Analgesia, 105*, S10–S17.

Mcmorrow, R. C., & Mythen, M. G. (2006). Pulse oximetry. *Current Opinion in Critical Care, 12*, 269–271.

Mejía-Mejía, E., Budidha, K., Abay, T. Y., May, J. M., & Kyriacou, P. A. (2020). Heart rate variability (HRV) and pulse rate variability (PRV) for the assessment of autonomic responses. *Frontiers in Physiology, 11*, 779.

Mukkamala, R., Hahn, J. O., Inan, O. T., Mestha, L. K., Kim, C. S., Töreyin, H., & Kyal, S. (2015). Toward ubiquitous blood pressure monitoring via pulse transit time: Theory and practice. *IEEE Transactions on Biomedical Engineering, 62*, 1879–1901.

Nachman, D., Gepner, Y., Goldstein, N., Kabakov, E., Ishay, A. B., Littman, R., Azmon, Y., Jaffe, E., & Eisenkraft, A. (2020). Comparing blood pressure measurements between a photoplethysmography-based and a standard cuff-based manometry device. *Scientific Reports, 10*, 16116.

Nikkonen, S., Afara, I. O., Leppänen, T., & Töyräs, J. (2019). Artificial neural network analysis of the oxygen saturation signal enables accurate diagnostics of sleep apnea. *Scientific Reports, 9*, 13200.

Nitzan, M., Romem, A., & Koppel, R. (2014). Pulse oximetry: Fundamentals and technology update. *Medical Devices (Auckl), 7*, 231–239.

Organization, W. H. (2021). *COVID-19 clinical management: Living guidance,* WHO/2019-nCoV/clinical/2021.1, World Health Organization 2021.

Pinheiro, N., Couceiro, R., Henriques, J., Muehlsteff, J., Quintal, I., Goncalves, L., & Carvalho, P. (2016). Can PPG be used for HRV analysis? *Annu Int Conf IEEE Eng Med Biol Soc, 2016*, 2945–2949.

Ralston, A. C., Webb, R. K., & Runciman, W. B. (1991a). Potential errors in pulse oximetry. I. Pulse oximeter evaluation. *Anaesthesia, 46*, 202–206.

Ralston, A. C., Webb, R. K., & Runciman, W. B. (1991b). Potential errors in pulse oximetry. III: Effects of interferences, dyes, dyshaemoglobins and other pigments. *Anaesthesia, 46*, 291–295.

Rauh, R., Limley, R., Bauer, R.-D., Radespiel-Troger, M. & Mueck-Weymann, M. (2004). Comparison of heart rate variability and pulse rate variability detected with photoplethysmography. *Proceedings of the SPIE.*

Ries, A. L., Prewitt, L. M., & Johnson, J. J. (1989). Skin color and ear oximetry. *Chest, 96*, 287–290.

Samman, Y. S., Rahimi, J. S., Wali, S. O., Krayem, A. B., & Abdelaziz, M. M. (2006). Effects of henna dye on oxygen saturation reading using pulse oximetry. *Saudi Medical Journal, 27*, 268–269.

Schäfer, A., & Vagedes, J. (2013). How accurate is pulse rate variability as an estimate of heart rate variability? A review on studies comparing photoplethysmographic technology with an electrocardiogram. *International Journal of Cardiology, 166*, 15–29.

Schulz, E. B., & Ham, J. A. (2019). Light-emitting diode surgical light interference with pulse oximetry. *British Journal of Anaesthesia, 123*, e490–e491.

Shelley, K. H. (2007). Photoplethysmography: Beyond the calculation of arterial oxygen saturation and heart rate. *Anesthesia & Analgesia, 105*, S31–S36, tables of contents.

Sinex, J. E. (1999). Pulse oximetry: Principles and limitations. *The American Journal of Emergency Medicine, 17*, 59–67.

Smith, R. P., Argod, J., Pépin, J. L., & Lévy, P. A. (1999). Pulse transit time: An appraisal of potential clinical applications. *Thorax, 54*, 452–457.

Stoneham, M. D., Saville, G. M., & Wilson, I. H. (1994). Knowledge about pulse oximetry among medical and nursing staff. *Lancet, 344*, 1339–1342.

Sviridova, N., Zhao, T., Aihara, K., Nakamura, K., & Nakano, A. (2018). Photoplethysmogram at green light: Where does chaos arise from? *Chaos, Solitons & Fractals, 116*, 157–165.

Tiihonen, P. (2008). *Novel portable devices for recording sleep apnea and evaluation altered consciousness*. Ph.D. thesis, The University of Kuopio.

Tremper, K. K. (1989). Pulse oximetry. *Chest, 95*, 713–715.

Tsien, C. L., & Fackler, J. C. (1997). Poor prognosis for existing monitors in the intensive care unit. *Critical Care Medicine, 25*, 614–619.

Tuchin, V. V. (2015). Tissue optics and photonics: Light-tissue interaction. *Journal of Biomedical Photonics & Engineering, 1*, 98–134.

Tusman, G., Acosta, C. M., Pulletz, S., Böhm, S. H., Scandurra, A., Arca, J. M., Madorno, M., & Sipmann, F. S. (2019). Photoplethysmographic characterization of vascular tone mediated changes in arterial pressure: An observational study. *Journal of Clinical Monitoring and Computing, 33*, 815–824.

Van De Louw, A., Cracco, C., Cerf, C., Harf, A., Duvaldestin, P., Lemaire, F., & Brochard, L. (2001). Accuracy of pulse oximetry in the intensive care unit. *Intensive Care Medicine, 27*, 1606–1613.

Vegfors, M., & Lennmarken, C. (1991). Carboxyhaemoglobinaemia and pulse oximetry. *British Journal of Anaesthesia, 66*, 625–626.

Verkruysse, W., Svaasand, L. O., & Nelson, J. S. (2008). Remote plethysmographic imaging using ambient light. *Optics Express, 16*, 21434–21445.

Volkov, M. V., Margaryants, N. B., Potemkin, A. V., Volynsky, M. A., Gurov, I. P., Mamontov, O. V., & Kamshilin, A. A. (2017). Video capillaroscopy clarifies mechanism of the photoplethysmographic waveform appearance. *Scientific Reports, 7*, 13298.

Vulcan, R. S., André, S., & Bruyneel, M. (2021). *Photoplethysmography in normal and pathological sleep* (p. 21).

Wilson, B. J., Cowan, H. J., Lord, J. A., Zuege, D. J., & Zygun, D. A. (2010). The accuracy of pulse oximetry in emergency department patients with severe sepsis and septic shock: A retrospective cohort study. *BMC Emergency Medicine, 10*, 9.

Young, D., Jewkes, C., Spittal, M., Blogg, C., Weissman, J., & Gradwell, D. (1992). Response time of pulse oximeters assessed using acute decompression. *Anesthesia and Analgesia, 74*, 189–195.

Zaouter, C., & Zavorsky, G. S. (2012). The measurement of carboxyhemoglobin and methemoglobin using a non-invasive pulse CO-oximeter. *Respiratory Physiology & Neurobiology, 182*, 88–92.

13 Oximetry Indices in the Management of Sleep Apnea: From Overnight Minimum Saturation to the Novel Hypoxemia Measures

Daniel Álvarez, Gonzalo C. Gutiérrez-Tobal, Fernando Vaquerizo-Villar, Fernando Moreno, Félix del Campo, and Roberto Hornero

Abstract

Obstructive sleep apnea (OSA) is a multidimensional disease often underdiagnosed due to the complexity and unavailability of its standard diagnostic method: the polysomnography. Among the alternative abbreviated tests searching for a compromise between simplicity and accurateness, oximetry is probably the most popular. The blood oxygen saturation (SpO_2) signal is characterized by a near-constant profile in healthy subjects breathing normally, while marked drops (desaturations) are linked to respiratory events. Parameterization of the desaturations has led to a great number of indices of severity assessment commonly used to assist in OSA diagnosis. In this chapter, the main methodologies used to characterize the overnight oximetry profile are reviewed, from visual inspection and simple statistics to complex measures involving signal processing and pattern recognition techniques. We focus on the individual performance of each approach, but also on the complementarity among the great amount of indices existing in the state of the art, looking for the most relevant oximetric feature subset. Finally, a quick overview of SpO_2-based deep learning applications for OSA management is carried out, where the raw oximetry signal is analyzed without previous parameterization. Our research allows us to conclude that all the methodologies (conventional, time, frequency, nonlinear, and hypoxemia-based) demonstrate high ability to provide relevant oximetric indices, but only a reduced set provide non-redundant complementary information leading to a significant performance increase. Finally, although oximetry is a robust

D. Álvarez (✉) · F. del Campo
Biomedical Engineering Group (GIB), University of Valladolid, Valladolid, Spain

Pneumology Department, Río Hortega University Hospital, Valladolid, Spain

Centro de Investigación Biomédica en Red de Bioingeniería, Biomateriales y Nanomedicina, Instituto de Salud Carlos III, Valladolid, Spain
e-mail: daniel.alvarez.gonzalez@uva.es

G. C. Gutiérrez-Tobal · F. Vaquerizo-Villar
R. Hornero
Biomedical Engineering Group (GIB), University of Valladolid, Valladolid, Spain

Centro de Investigación Biomédica en Red de Bioingeniería, Biomateriales y Nanomedicina, Instituto de Salud Carlos III, Valladolid, Spain

F. Moreno
Pneumology Department, Río Hortega University Hospital, Valladolid, Spain

T. Penzel, R. Hornero (eds.), *Advances in the Diagnosis and Treatment of Sleep Apnea*, Advances in Experimental Medicine and Biology 1384, https://doi.org/10.1007/978-3-031-06413-5_13

tool, greater standardization and prospective validation of the measures derived from complex signal processing techniques are still needed to homogenize interpretation and increase generalizability.

Keywords

Apnea · Blood oxygen saturation · Deep learning · Desaturation · Hypopnea · Hypoxemia · Hypoxic burden · Nonlinear dynamics · Obstructive sleep apnea · Oximetry, oxygen desaturation index · Resaturation · Signal processing · Spectral analysis

13.1 Introduction

The blood oxygen saturation (SpO_2) signal from nocturnal oximetry is one of the most valuable tools in the framework of abbreviated diagnosis of obstructive sleep apnea (OSA). Overnight SpO_2 has shown to gather the relevant changes in the respiratory pattern linked to the presence and severity of OSA while being non-intrusive for patients, portable, and highly available (Del Campo et al., 2018; Terrill, 2020). Parameterization of the overnight oximetry profile focused the effort of many researchers, looking for indices able to characterize the number, duration, and severity of desaturations (Levy et al., 2021). The oxygen desaturation index (ODI) and the cumulative time (CT) below a certain saturation level have been traditionally used due to their simple computation and interpretation (Terrill, 2020). Indeed, many commercial oximeters provide the values of these indices in their summary reports, besides the mean and the minimum saturation values throughout the night (Otero et al., 2012).

Despite its widely known usefulness (Tsai et al., 2013; Dawson et al., 2015; Sharma et al., 2017), traditional measures just based on the number of events or the cumulative duration of the desaturations seem to be insufficient to completely characterize the severity of the disease, particularly in complex patients. In order to overcome this limitation, new measures have been proposed in the last years, aimed at quantifying the amount of hypoxemia involving the depth and duration of the events jointly (Kulkas et al., 2013a). Simultaneously, advanced signal processing techniques have provided clinicians with a number of automated oximetry indices during the past two decades, looking for a better characterization of oximetry dynamics in OSA patients (Zamarrón et al., 1999, 2003; Álvarez et al., 2006, 2010, 2012, 2013; Hornero et al., 2017; Terrill, 2020). Besides avoiding the problem of lack of standardization (as they are computed from well-known mathematical algorithms), these indices have demonstrated significant effectiveness in OSA diagnosis, though are sometimes difficult to interpret in terms of the physiopathology of the disease. Similarly, machine learning has been also applied to build oximetry-based models for decision support in the framework of OSA (Marcos et al., 2009; Álvarez et al., 2013; Uddin et al., 2018; Gutiérrez-Tobal et al., 2019). Additionally, the joint analysis of oximetric features by means of pattern recognition and artificial intelligence has allowed the detection of complementary (non-redundant) indices able to improve the diagnostic capability of oximetry (Álvarez et al., 2012, 2013). Concerning the usefulness of artificial intelligence, the raising of deep learning approaches has recently opened a new way to exhaustively analyze biomedical signals. Particularly, deep neural networks have been found appropriate tools to automatically learn discriminant features from the raw oximetry signal (Vaquerizo-Villar et al., 2021), potentially allowing to provide new oximetric indices if these new models (highly complex) are thoroughly interpreted.

Performance of oximetry-based methods for OSA diagnosis shows significant variability among studies (Uddin et al., 2018; Del Campo et al., 2018). In order to minimize this variability, it would be interesting to clarify some major points: search for the top-performance index; quantify the actual performance increase linked to the use of complex mathematical algorithms compared to the conventional ones (i.e., the complexity-effectiveness balance); or investigate the degree of complementarity among available indices, owing that they are all obtained from the same data source. The main goal of this chapter is to review and analyze the current knowledge concerning the diagnostic information derived from automatic processing of nocturnal oximetry recordings in the context of OSA management: from the conventional desaturation indices to the popular new measures of hypoxemia and the novel deep learning approaches. Accordingly, both individual (univariate) and joint (multivariate) performance of the most common characteristics found in the literature are assessed, which are categorized in the following groups: conventional indices, frequency-domain features, nonlinear measures, morphology-based parameters, and deep learning architectures. Furthermore, when possible, the performance increase concerning the inclusion of a particular set of features (statistical, spectral, nonlinear) is analyzed, in order to gain insight into the complementarity of the available oximetry indices.

13.2 Approaches for Parameterizing Changes in the Dynamics of the Oximetry Signal

A plethora of oximetric features derived from many different manual and automatic methodologies can be found in the literature. In order to facilitate its analysis throughout this chapter, oximetry measures are grouped in the following categories: (i) conventional measures (visual inspection, simple statistics, and the oxygen desaturation index); (ii) frequency-domain features (power spectral density, high-order spectra, wavelet); (iii) nonlinear measures (entropy, complexity, multiscale analysis); (iv) morphology-based parameters (hypoxic burden measures, characteristics of the desaturation curve); and (v) application of deep learning to the raw oximetry signal.

13.2.1 Conventional Approaches to Characterize the Overnight Oximetry Profile: Visual Inspection, Common Statistics, and the Oxygen Desaturation Index

Table 13.1 summarizes the main traditional methodologies applied for easy assessment and interpretation of the nocturnal oximetry profile. Visual inspection of the overnight SpO_2 tracing was predominantly used in the 1990s, in order to identify consecutive drops in the saturation value leading to the common "saw-tooth" pattern linked to the presence of OSA. The specificity of manual analysis is particularly high in severe OSA subjects. Subjectivity and complexity when analyzing long nocturnal profiles are major limitations even for trained sleep experts. Nevertheless, examination of overnight oximetry tracings is still used to perform preliminary OSA screening in especial patient groups, such as children (Brouillette et al., 2000; Velasco-Suarez et al., 2013; Tsai et al., 2013; Van Eyck et al., 2015) and those with concomitant respiratory comorbidities, such as chronic obstructive pulmonary disease (COPD) (Scott et al., 2014). In these studies, sensitivity ranges from 40.6% to 91.6%, while specificity varies between 40.6% and 98.9%.

Simple statistics derived from the data histogram of the nocturnal oximetry signal are also commonly used to characterize averages and trends potentially indicative of pathological states. Overall mean, variance, skewness (a measure of histogram asymmetry), and kurtosis (a measure of data concentration) have been proposed as an easy to obtain measures able to further complement more advanced automated features in numerous studies (Marcos et al., 2010a; Marcos et al., 2012; Álvarez et al., 2010, 2012, 2013, 2020; Gutiérrez-Tobal et al., 2019, 2021a). It is remarkable that at least one of these indexes is systematically included in the final optimum feature subset when a variable selection procedure is implemented,

Table 13.1 Visual inspection approaches and conventional oximetric parameters commonly used to characterize oximetry in the context of OSA diagnosis

Approach	Indices
Visual inspection Brouillette et al. (2000), Nixon et al. (2004), Velasco-Suarez et al. (2013), Tsai et al. (2013), Van Eyck et al. (2015), Scott et al. (2014), Villa et al. (2015)	• Recurrent drops in the SpO_2 profile along the night • Saw-tooth pattern • Clusters of desaturations (mainly in pediatric OSA)
SpO_2 data histogram and simple statistics Lévy et al. (1996), Olson et al. (1999), Magalang et al. (2003), Marcos et al. (2010a, 2012), Álvarez et al. (2010, 2012, 2013, 2017, 2018, 2020), Garde et al. (2014), Crespo et al. (2018), Vaquerizo-Villar et al. (2018c), Gutiérrez-Tobal et al. (2019, 2021a)	• Mean (central tendency), variance (dispersion), skewness (asymmetry), kurtosis (peakedness) • Median (central tendency), quantiles, and interquartile range (dispersion) are less used • Delta index (variability measure)
Intermittent hypoxemia Gyulay et al. (1993), Magalang et al. (2003), Rofail et al. (2010), Chung et al. (2012), Schlotthauer et al. (2014), Dawson et al. (2015), Kirk et al. (2003), Chang et al. (2013), Malbois et al. (2010), Ward et al. (2012), Aaronson et al. (2012), Mazière et al. (2014), Sharma et al. (2017)	• Oxygen desaturation index (ODI) of 2% (children) and 3% or 4% (adults)
Persistent hypoxemia Chaudhary et al. (1998), Golpe et al. (1999), Magalang et al. (2003)	• Overnight minimum saturation • Percentage of cumulative time (CT) spent with a saturation below a threshold: in the range 80–90% in adults and 95% in children

which highlights the importance and complementarity of the information provided by these simple measures. In the pediatric framework, the same behavior has been reported (Álvarez et al., 2017, 2018; Crespo et al., 2018; Vaquerizo-Villar et al., 2018c), showing significant complementarity among these common statistical moments and other techniques, such as spectral, wavelet, and nonlinear measures.

In addition to conventional standard deviation and variance, the delta index was also proposed to estimate the variability of the overnight SpO_2 recording (Lévy et al., 1996). It quantifies variation as the sum of the absolute differences between the saturation values corresponding to the upper and lower limits of each SpO_2 segment (commonly 12 s length with no overlap), normalized by the total number of intervals. Large imbalance in the sensitivity-specificity pair was shown, with high sensitivities ranging 88–98% and notably lower specificities ranging 40–59% (Lévy et al., 1996; Olson et al., 1999; Magalang et al., 2003).

Parameterization of the desaturations by quantifying their number, duration, and depth, either manually or automated, has been traditionally used to characterize oximetry patterns in pathological patients. The number of desaturations from baseline greater than a threshold (usually 3% or 4%) per hour of sleep, i.e., the widely known oxygen desaturation index (ODI) (Gyulay et al., 1993; Magalang et al., 2003), and the overall minimum saturation value and the cumulative time (CT) with a saturation below a cutoff value (usually 90% for adults and 95% for children) relative to the total recording time (Chaudhary et al., 1998; Golpe et al., 1999) have been extensively used and commonly embedded in commercial pulse oximeters. Overall, the ODI has been found to remarkably outperform CT (Magalang et al., 2003). Individually, the ODI has demonstrated to be a high-performance oximetric feature for OSA detection, both in the adult (Rofail et al., 2010; Chung et al., 2012; Schlotthauer et al., 2014; Dawson et al., 2015) and the pediatric context (Kirk et al., 2003; Chang et al., 2013) and also even in the presence of comorbidities (Malbois et al., 2010; Ward et al., 2012; Aaronson et al., 2012; Mazière et al., 2014; Sharma et al., 2017). Reported sensitivities and specificities ranged 70.0–96.3% and 67.3–97.2% for adults, 59.26–70.59% and 60.0–86.0% for children, and 33.3–100% and 32.0–100% in the presence of comorbidities.

In addition, the ODI raised as an essential index in studies using multivariate approaches, being systematically selected to be part of the final optimum models, for both adult (Álvarez et al., 2020; Gutiérrez-Tobal et al., 2019, 2021a) and pediatric (Hornero et al., 2017; Crespo et al., 2017, 2018; Álvarez et al., 2017, 2018; Vaquerizo-Villar et al., 2018a, b, c) OSA automated detection. Similarly, the ODI has been combined with features from other biomedical signals in the context of pediatric sleep apnea diagnosis, showing significant correlation with novel spectral cardiac indices (Martín-Montero et al. 2021a, b) and remarkable complementarity with frequency/scale (power spectrum, bispectrum, and wavelet) and nonlinear (recurrence plots) characteristics from airflow recordings (Gutiérrez-Tobal et al., 2015; Barroso-García et al., 2020, 2021a, b; Jiménez-García et al., 2020). In the latter case, it is important to note that the ODI was selected 100% of times within the optimum feature subset. Interestingly, the studies by Barroso-García et al. (2020, 2021a, b) and Jiménez-García et al. (2020) design and assess their models with and without including the ODI, allowing to quantitatively measure the complementarity of this index in terms of the performance increase. In this regard, Jiménez-García et al. (2020) reported minor accuracy increments (from +0.77% to +1.28%) for the most restrictive cutoff for positive OSA (1 event/h) when the ODI is included in the analysis, while the increase was notably higher for larger cutoffs, particularly when the ODI is combined with airflow-derived measures (+19.23% for 5 events/h and + 11.29% for 10 events/h). Similarly, Barroso-García et al. (2020, 2021a, b) also reported higher performance increase when using cutoffs for detecting moderate-to-severe cases, achieving increments in the accuracy value ranging from +10.4% to +25.0%.

13.2.1.1 An Especial Oximetric Index in Childhood OSA: Clusters of Desaturations

This approach exploits the widely known recurrent behavior of desaturations, which tend to group in different time periods along the sleep time. This characteristic is closely related with the periodicity of desaturations and hence with the analysis of the signal in the frequency domain. However, while spectral analysis has been extensively applied regardless the context (either adult or pediatric), the characterization of the depth, number, and clustering of desaturations has been mostly used as a marker of childhood OSA. The intuition is that the larger the number of clusters, the larger the probability of OSA. However, there is not a clear definition of what a cluster is, and they are mainly detected by visual inspection (Nixon et al., 2004; Velasco-Suarez et al., 2013; Van Eyck et al., 2015; Villa et al., 2015). Recent reviews of the state of the art pointed out that this approach is particularly useful for the detection of moderate-to-severe OSA cases (Van Eyck & Verhulst, 2018).

Brouillette et al. (2000) firstly pointed out the screening ability of clusters of desaturations in children. They reported that the presence of three or more clusters showing falls greater than 4% from baseline and three or more falls in the saturation value below the threshold of 90% was predictive of pediatric OSA, though sensitivity was notably lower than specificity (42.9% vs. 97.8%, respectively). Based on this study, Nixon et al. (2004) defined the McGill oximetry score (MOS), reporting that the number and depth of the clusters could be used to estimate the severity of pediatric OSA, prioritize treatment, and schedule perioperative interventions. Similarly, Velasco-Suarez et al. (2013) reported higher and more balanced sensitivity and specificity values (86.6% vs. 98.9%) using a lower number of clusters (>2) and drops below 90% (>1) for positive OSA in children with adenotonsillar hypertrophy. Recently, Van Eyck et al. (2015) prospectively assessed the methods by Brouillette et al. (2000) and Velasco-Suarez et al. (2013) for childhood OSA detection based on the characterization of clusters of desaturations. Accuracies ranging 68–78% were reported using a conservative diagnostic threshold of 2 events/h for childhood OSA. Looking for a performance increase, Villa et al. (2015) combined the parameterization of clusters with data from the patient's clinical history, reaching 85.8% accuracy in the detection of this condition, while accurateness decreases to 69.4% for the detection of moderate-to-severe cases.

13.2.2 Analysis of Nocturnal Oximetry in the Frequency Domain

In addition to simple statistics and the ODI, one of the first attempts to automatically characterize the SpO_2 signal relied on the use of tools in the frequency domain. Nocturnal desaturations commonly present in the oximetry signal from OSA patients show a relative periodicity. The parameterization of the changes in the power spectrum of the signal linked to this pseudo-periodicity has been found to provide relevant and discriminative features able to discern OSA patients from healthy subjects.

In the framework of frequency analysis, a major decision is to define the spectral band of frequencies that is going to be analyzed. In this regard, standardized spectral bands exist in other biomedical signals, such as heart rate variability (HRV) or electroencephalogram (EEG). The low-frequency (LF: 0.04–0.15 Hz) and the high-frequency (HF: 0.15–0.40 Hz) bands were proposed many years ago to assess the influence of diseases in the cardiac autonomic function using the HRV signal as a surrogate of more intrusive techniques (Stein & Pu, 2012). Similarly, the power spectra in the classical EEG bands delta (0.1–3.5 Hz), theta (4–7.5 Hz), alpha (8–13 Hz), and beta (14–30 Hz) have been widely used to quantitatively measure the impact of diseases in the brain activity (Penttonen & Buzsáki, 2003). On the contrary, no standardized frequency bands are defined concerning the spectral analysis of the oximetry signal.

In the literature, different oximetry-based spectral bands have been proposed to characterize the severity of the disease. Table 13.2 summarizes the main spectral bands of interest used to assess the oximetry signal in the frequency domain. In the context of adult OSA, the frequency band 0.014–0.033 Hz has been predominantly used (Zamarrón et al., 2003; Chen-Liang et al., 2009; Álvarez et al., 2010, 2012, 2013). Shiomi et al. (1996) firstly identified a synchronization between decreased arterial oxygen satu-

Table 13.2 Most common spectral bands of interest of the oximetry signal in the frequency domain

Frequency bands for adults	Frequency bands for children
• 0.008–0.04 Hz (VLF) Shiomi et al. (1996)	• ±0.02 around the peak in 0.005–0.1 Hz Garde et al. (2014)
• 0.014–0.033 Hz (T_{30-70}) Zamarrón et al. (1999)	• 0.01755–0.03433 Hz (for AHI ≥1 event/h) Álvarez et al. (2017)
• 0.017–0.1 Hz (T_{10-60}) Sánchez-Morillo and Gross (2013)	• 0.02136–0.03967 Hz (for AHI ≥3 events/h) Álvarez et al. (2017)
• 0.017–0.05 Hz (T_{20-60}) Sánchez-Morillo and Gross (2013)	• 0.01755–0.03357 Hz (for AHI ≥5 events/h) Álvarez et al. (2017)
• 0.013–0.1 Hz (T_{10-75}) Sánchez-Morillo and Gross (2013)	• 0.02136–0.03357 Hz Álvarez et al. (2017)
• 0.013–0.05 Hz (T_{20-75}) Sánchez-Morillo and Gross (2013)	• 0.018–0.050 Hz Vaquerizo-Villar et al. (2018a, c)
• <0.2 Hz (artifact removal) Schlotthauer et al. (2014)	• 0.020–0.044 Hz Hornero et al. (2017)
	• 0.021–0.040 Hz Crespo et al. (2018)

ration and a power increase in the very-low-frequency components of the heart rate variability signal (VLF: 0.008–0.04 Hz). In this study, the authors laid the foundations of subsequent analysis of oximetry in the frequency domain, relating the upper and lower limits of the VLF band to the maximum and minimum durations of apneas, respectively: 120 s is stated as the maximum cycle length (i.e., 0.008 Hz), while 25 s is considered the minimum duration (i.e., 0.04 Hz), including the recovery (awakening response) after apnea episodes (Shiomi et al., 1996). Then, Zamarrón et al. (1999, 2003) thoroughly analyzed the spectral content of the oximetry signal within this band, reporting a characteristic and highly discriminative power increase in the period 30–70 s, i.e., the widely used 0.014–0.033 Hz band. Other authors used similar approaches, leading to slightly different bands. Sánchez-Morillo and Gross (2013) and Sánchez-Morillo et al. (2014) analyzed the histogram of the duration of the desaturations to determine the most common desaturation periods. They reported that 83.4% of the desaturations last between 10 and 60 s and that 90.5% between 10 and 75 s (Sánchez-Morillo and Gross, 2013). Accordingly, they defined the following periods of interest in order to parameterize the power spectrum of the oximetry signal: 10–60 s (i.e., 0.017–0.1 Hz), 20–60 s (i.e., 0.017–0.05 Hz), 10–75 s (0.013–0.1 Hz), and 20–75 s (0.013–0.05 Hz). Other authors used a more conservative approach when locating the spectral content of oximetry. In this regard, Schlotthauer et al. (2014) considered that desaturations linked to apneas have periods larger than 5 s, leading to relevant frequency components below 0.2 Hz.

In the framework of childhood OSA, there is a larger variability regarding the spectral band of interest of the oximetry signal compared to adults (see Table 13.2). Garde et al. (2014) used a frequency interval of 0.02 Hz centered around the peak amplitude of the power spectrum that they searched from 0.005 to 0.1 Hz. Álvarez et al. (2017) performed a statistical analysis searching for the frequencies leading to the highest discriminant ability between OSA groups. Accordingly, they identified three different bands of interest for pediatric OSA: 0.01755–0.03433 Hz for a cutoff of 1 event/h, 0.02136–0.03967 Hz for a cutoff of 3 events/h, and 0.01755–0.03357 Hz for 5 events/h. Finally, they proposed a single spectral frequency range as the broadest interval showing significant differences regardless the clinical threshold for positive OSA: 0.02136–0.03357 Hz. Similarly, Vaquerizo-Villar et al. (2018a, c) searched for a spectral band of interest able to maximize the differences between different OSA severity groups (AHI < 5 events/h, 5 ≤AHI < 10 events/h, and AHI ≥10 events/h), leading to the interval 0.018–0.050 Hz. Other authors used slight variations of these bands, such as Hornero et al. (2017) (0.020–0.044 Hz) and Crespo et al. (2018) (0.021–0.040 Hz).

Concerning the methodology used to inspect the frequency content of oximetry, different approaches have been assessed. The estimation of the power spectral density (PSD) using the non-parametric Welch method based on the fast Fourier transform has been predominantly used (Zamarrón et al., 1999, 2003; Chen-Liang et al., 2009; Álvarez et al., 2010, 2012, 2013, 2017; Hornero et al., 2017; Crespo et al., 2018). Alternatively, autoregressive methods were used by Sánchez-Morillo and Gross (2013) and Garde et al. (2014) to estimate the PSD. A number of measures have been used to parameterize the power spectrum (see Table 13.3), mainly based on amplitudes and total or relative power in the spectral band of interest. Additionally, common statistics, such as first-to-fourth statistical moments and the median frequency, as well as regularity measures as the Shannon spectral entropy have been also widely used to further characterize the spectral content of the signal. In this regard, peak amplitude, relative power, skewness, and spectral entropy have been found to jointly summarize oximetry dynamics in the frequency domain, for both adults (Álvarez et al., 2010, 2012, 2013, 2020; Sánchez-Morillo & Gross, 2013) and children (Garde et al., 2014; Hornero et al., 2017; Álvarez et al., 2017; Crespo et al., 2018).

On the other hand, novel and complementary approaches have been recently proposed to further assess the recurrent behavior of desatura-

Table 13.3 Measures commonly used to characterize the power spectrum of oximetry in the frequency domain

Method	Indices
Power spectral density (PSD) • Non-parametric fast Fourier transform (FFT)-based methods (Welch, Blackman-Tukey) Zamarrón et al. (1999, 2003), Chen-Liang et al. (2009), Álvarez et al. (2010, 2012, 2013, 2017), Hornero et al. (2017), Crespo et al. (2018) • Autoregressive methods (Yule-Walker) Sánchez-Morillo and Gross (2013), Garde et al. (2014)	• First-to-fourth statistical moments (mean, variance, skewness, kurtosis), median frequency • Shannon spectral entropy (SSE), mobility, Wootters' distance, Euclidean distance (measures of the concentration of the signal power) • Peak and minimum amplitudes, total power, relative power in the band of interest
Bispectrum (high-order spectra) Vaquerizo-Villar et al. (2018a)	• Mean amplitude • Sum of the logarithmic amplitudes of the whole bispectrum, sum of the logarithmic amplitudes in the main diagonal, first-order spectral moment of amplitudes in the main diagonal • Normalized bispectral entropy and normalized bispectral squared entropy • Phase entropy • Mean and variance of the bispectrum invariant
Wavelet transform Vaquerizo-Villar et al. (2018c), Poupard et al. (2012)	• First-to-fourth-order moments of the wavelet coefficients in the 9th detail band (D_9: 0.0244–0.0488 Hz) • Maximum amplitude of wavelet coefficients in D_9 • Energy of the coefficients in D_9 • Wavelet entropy • Ventilatory hypoxemic index

tions and to obtain complementary information to that provided by the PSD. In this regard, Vaquerizo-Villar et al. (2018a) used high-order spectra (HOS) to detect deviations from Gaussianity, linearity, and stationarity of the oximetry signal potentially linked to the apneic events. Particularly, they applied the bispectrum, a representation of the spectral decomposition of the third-order cumulant (skewness) of a signal over the frequency. In this study, two bispectral measures showed complementarity with PSD, the mean amplitude of the bispectrum and the mean of the bispectrum invariant, which account for magnitude differences and for phase coupling between spectral components, respectively. The authors reported a remarkable performance increase (+6.7% three-class accuracy) when including these bispectral features in a model for automated pediatric OSA diagnosis. Similarly, in a subsequent study in the same research line, Vaquerizo-Villar et al. (2018c) applied wavelet analysis in order to further characterize the spectral content of the signal, particularly in the very low frequencies owing that oximetry is characterized by very slow variations. In such low frequencies, traditional methods lack for appropriate spectral resolution, while the wavelet transform performs a multilevel analysis able to provide high frequency resolution at low frequencies and high time resolution at high frequencies. In the study by Vaquerizo-Villar et al. (2018c), the skewness and the energy of the coefficients in the level 9 detail signal (D_9, corresponding to the frequency range 0.0244–0.0488 Hz), as well as the overall wavelet energy, showed complementarity with conventional oximetric indices, including ODI, statistical moments, and features from the PSD.

In the context of adult OSA, the wavelet transform has been also applied to the oximetry signal to obtain a new measure of hypoxemia. Poupard et al. (2012) implemented a wavelet-aggregation procedure to quantify the overall absolute variations (both increases and decreases of amplitude) along the overnight oximetry recording. The ventilatory hypoxemic index (VHI) was defined as the cumulative time with absolute variations >4%, divided by the theoretical apnea cycle period, which was defined as the middle point in the interval 30–70 s identified by Zamarrón et al. (2003), i.e., 50 s. The VHI showed higher correlation with standard AHI than ODI (0.87 vs. 0.81), as well as lower bias (+5.7 vs. +13.5). In the same regard, VHI achieved more balanced sensitivity-specificity pair than the ODI for the common cutoffs for OSA (91–88% vs. 65–100%, AHI ≥5 events/h; 81–98% vs. 58–100%, AHI ≥15 events/h; 67–99% vs. 59–100%, AHI ≥30 events/h).

13.2.3 Methods Derived from Nonlinear Dynamics in the Oximetry Signal

Despite the usefulness shown by conventional oximetric indices, statistics, and frequency-domain methods, they are unable to completely explain all the dynamics of the oximetry signal. In addition to periodicities linked to the recurrent apneic events, there are also nonlinear changes typical of natural systems present in biomedical signals. In this regard, nonlinear methods derived from the chaos theory have demonstrated to provide relevant and complementary information in the automated diagnosis of OSA from oximetry. Table 13.4 shows the methods predominantly used to quantify nonlinear dynamics in the SpO_2 signal.

Approximate (ApEn) and sample (SampEn) entropies, central tendency measure (CTM), and Lempel-Ziv complexity (LZC) have been predominantly used. Individually, the quantification of irregularity in the overnight oximetry recording by means of ApEn shows remarkable performance in the detection of adult OSA (Del Campo et al., 2006; Hornero et al., 2007), reaching balanced sensitivity and specificity, as well as area under the ROC curve >0.90. In the same regard, CTM matches the behavior of ApEn, achieving accuracy values >87% with balanced sensitivity-specificity pair and area under the ROC curve >0.90 (Álvarez et al., 2006, 2007). Finally, LZC has shown slightly lower performance than single CTM or ApEn, although reaching notable accuracy (>82%) and area under the curve (>0.85) (Álvarez et al., 2006).

Concerning multivariate approaches, nonlinear measures have shown valuable complementarity with conventional oximetric indices for automated OSA diagnosis in both adults and children. In the adult context, nonlinear measures are systematically included in the final optimum subset from automated feature selection procedures. Particularly, the width of the Poincaré plot (SD_1) is complemented with different desaturation and resaturation indices as well as with spectral power (Sánchez-Morillo & Gross, 2013), LZC shows remarkable joint relevance with statistical moments in the time domain and the spectral power (Álvarez et al., 2010, 2013), and CTM fits with statistical moments in both the time and frequency domains and the peak spectral amplitude (Álvarez et al., 2012, 2013). Under a multiclass approach, SD_1, SampEn, CTM, and LZC all together combined with the ODI and a histogram-based index to discern between no-OSA and mild OSA individuals, whereas ApEn combined with histogram-based indexes, resaturation measures, and the ODI to classify moderate and severe OSA patients (Sánchez-Morillo et al., 2014). Without an appropriate feature selection stage, complementarity of nonlinear measures is not properly exploited, as shown in the study by Marcos et al. (2009), where the combination of nonlinear and spectral features did not significantly improve the accuracy reached with each individual approach (spectral vs. nonlinear). Alternatively, dimensionality reduction by means of principal component analysis showed a remarkable performance increase (+6.20% accuracy) when combining spectral and nonlinear features (Marcos et al., 2010b). Under a regression approach aimed at estimating the apnea-hypopnea index (AHI),

Table 13.4 Nonlinear methods commonly used to quantify changes in nonlinear dynamics of oximetry

Method	Indices
Irregularity or disorderliness measures by means of entropy del Campo et al. (2006), Hornero et al. (2007), Álvarez et al. (2006, 2010, 2012, 2013, 2017, 2020), Marcos et al. (2009, 2012); Marcos et al. (2010b), Sánchez-Morillo et al. (2014), Garde et al. (2014), Hornero et al. (2017), Crespo et al. (2018)	• Approximate entropyEntropies (ApEn) and cross-approximate entropy (cross-ApEn) • Sample entropy (SampEn) • Kernel entropy (KerEn)
Variability measures from scatter plots Álvarez et al. (2006, 2007, 2010, 2012, 2013, 2017, 2020), Marcos et al. (2009, 2012), Marcos et al. (2010b), Sánchez-Morillo and Gross (2013), Sánchez-Morillo et al. (2014), Garde et al. (2014), Hornero et al. (2017), Crespo et al. (2018)	• Length of the main (SD_1) and secondary (SD_2) axes of the ellipse that encloses the points in Poincaré plots • Central tendency measure (CTM) from second-order difference plots
Complexity measures Álvarez et al. (2006, 2010, 2012, 2013, 2020), Marcos et al. (2009, 2010b, 2012), Sánchez-Morillo et al. (2014), Hornero et al. (2017), Crespo et al. (2018)	• Lempel-Ziv complexity
Multiscale approaches Crespo et al. (2017), Vaquerizo-Villar et al. (2018a, b)	• Multiscale entropy (MSE): Individual entropy values in single scales; entropy value in the scale reaching the maximum margin between groups under study; slope of the MSE curve between a pair of scales; area enclosed under the MSE curve between a pair of scales; area enclosed between the first and the maximum margin scales; time scale where maximum entropy is reached • Detrended fluctuation analysis (DFA): slopes (scaling exponents) of the lines fitting the regions identified in the DFA curve; coordinates of the intersection of the line fitting these regions; value of the fluctuation function in the scale that maximizes its correlation with the AHI
Symbolic dynamics Álvarez et al. (2018)	• Probability of the words (particular sequence of symbols) representative of different states (high and low saturation values) and changes (desaturations and resaturations) of the signal • Forbidden words • Symbolic entropy

SampEn, CTM, and LZC from oximetry showed reliable completeness with statistical, spectral, and conventional oximetric indices (Marcos et al., 2012; Álvarez et al., 2020).

In the framework of pediatric OSA, the usefulness of traditional nonlinear indexes (SampEn, CTM, and LZC) from overnight oximetry has been less investigated, and their relevance seems to be slightly lower than in the adult context. The studies by Álvarez et al. (2017) and Crespo et al. (2018) include these measures in the beginning of a feature selection process, and only SampEn was found non-redundant and finally selected to be part of the optimum model for binary classification of children (non-OSA vs. OSA) using different cutoffs for the disease (1, 3, and 5 events/h). Additionally, Garde et al. (2014) and Hornero et al. (2017) also used nonlinear measures to characterize the nocturnal oximetry profile of children with suspicion of OSA, but no nonlinear index was included in the optimum model (binary classification and regression approaches) due to redundancy.

In addition to these conventional nonlinear measures of irregularity, variability, and complexity, novel nonlinear methods have been recently applied to the oximetry signal in order to obtain as much information as possible from the recording. This is particularly important in the pediatric context, where pattern recognition and machine learning methods face a more challenging task compared to adults. In the study by Crespo et al. (2017), multiscale sample entropy (MSE) was applied to quantify entropy changes in the oximetry signal along larger time scales. Features derived from the MSE curve shown high performance (AUC 0.80) both individually and jointly. It is remarkable the complementarity of MSE variables and conventional oximetric indices (ODI, CT, minimum and average saturation), leading to a significant performance increase (+4.5% accuracy; +6% AUC) when properly combined using a stepwise approach. Similarly, Vaquerizo-Villar et al. (Vaquerizo-Villar et al., 2018b) applied detrended fluctuation analysis (DFA) to analyze changes in the correlation properties of the nocturnal oximetry profile for different ranges of scales. The slope in the first scaling region of the DFA curve showed high relevancy and complementarity with the ODI. Both were combined using a regression neural network aimed at estimating the AHI, reaching high agreement with actual AHI (0.891 intra-class correlation coefficient; 0.412 kappa) and notably outperforming the ODI alone (0.866 intra-class correlation coefficient; 0.355 kappa). Finally, Álvarez et al. (2018) analyzed nonlinearities present in the oximetry recording using a symbolic dynamics approach, which stablishes an alternative framework for investigating complex nonlinear systems. Features from the histogram of symbols reached the highest performance compared to conventional indexes, anthropometric measures, and common statistical moments. Moreover, symbolic dynamics features showed significant complementarity with these variable subsets, leading to a significant performance increase (+4.8% accuracy; +7% AUC) when used together after appropriate feature selection.

13.2.4 Quantifying the Morphology of Desaturation: Influence of the Area and the Velocity of Events

The conventional oxygen desaturation index has demonstrated to provide highly relevant information on the severity of OSA, reaching high performance when used individually as well as being systematically selected within the optimum feature subset under multivariate approaches. Nevertheless, the ODI is just based on counting the number of desaturations, regardless the total depth and length of these events. Hence, there is room for improvement if all these characteristics are put together in the same index. Table 13.5 shows several indices found in the literature aimed at parameterizing the morphology of the desaturation.

First attempts for joint characterization of both the length and depth of the desaturations were made by Chesson et al. (1993, 2001). They proposed the so-called saturation impairment time (SIT), an automated index aimed at quantifying cumulative nocturnal oxygen desaturation as a measure of hypoxemia in the context of respiratory-related breathing disorders. Contrary to traditional indices of hypoxemia just based on the percentage of time spent below a predetermined threshold (CTx%, being x% the cutoff), the SIT index integrates both time (length) and severity (depth) of the desaturations (Chesson et al., 1993). To measure the joint contribution of both characteristics, SIT is computed as the area enclosed under a fixed saturation value (similar to the threshold in CT indices) and the saturation curve. The authors reported good correlation

Table 13.5 Measures used to parameterize the morphology of the desaturation curve

Method	Indices
Severity of desaturations (quantification of the total or partial area of the desaturation) Chesson et al. (1993, 2001), Kulkas et al. (2013a, b, 2017), Muraja-Murro et al. (2014), Leppänen et al. (2017), Kainulainen et al. (2019, 2020), Linz et al. (2018), Khoshkish et al. (2018), Azarbarzin et al. (2019), Kim et al. (2020)	• Saturation impairment time (SIT) • Apnea severity, hypopnea severity, obstruction severity, and desaturation severity • Hypoxia load (HL) • Hypoxic burden
Parameterization of the sections of the desaturation Otero et al. (2012)	• Duration of the desaturation • Average and minimum (nadir) values of the saturation throughout the event • Elapsed time from the beginning of the event until the nadir point and from the nadir to the end of the desaturation • Overall drop in the saturation during the fall part of the event and overall increase in the rise section • Slope of both the fall and the rise parts of the event • Desaturation area

with CT ($r^2 > 0.8$) as well as complementarity with the respiratory disturbance index (RDI), i.e., patients with similar RDI showed variability in their SIT values. Accordingly, they concluded that the SIT index may provide additional and useful information in the characterization of desaturations during sleep.

The standard AHI from nocturnal PSG is commonly criticized due to its low correlation with physiological symptoms and consequences of OSA. In a similar way to the ODI, this problem is attributed to the own definition of the parameter, which is just based on counting the number of apneas and hypopneas throughout the time of sleep regardless their severity. Motivated by the increasing demand for alternatives to the standard AHI due to these limitations, different respiratory disturbance indices have been recently proposed aimed at gathering the severity of each individual event. These indices are commonly known as measures of "hypoxic burden." In 2013, Kulkas et al. (2013a) proposed a set of indices they named severity parameters, aimed at accounting for both the morphology and the duration of desaturations: *apnea severity*, *hypopnea severity*, *obstruction severity*, and *desaturation severity*. They are all based on the quantification of the *desaturation area* for each single event, which is the area enclosed between a saturation level determined by the starting point of the event and the oximetry curve until the minimum saturation value (nadir), i.e., the resaturation part of the event is not considered. The *desaturation severity* index is computed as the cumulative sum of the desaturation area of each single event and normalized by the total analyzed time. *Apnea severity*, *hypopnea severity*, and *obstruction severity* are based on the same definition, but the *desaturation area* is weighted by the duration of each kind of event, and only those events (apnea, obstructive, or hypopnea) followed by a desaturation event within the next 60 s are considered (Kulkas et al., 2013a). The authors reported moderate correlation of the novel severity indices with the standard AHI ($r^2 < 0.7$; $p < 0.001$) and with the ODI ($r^2 < 0.75$; $p < 0.001$) (Kulkas et al., 2013a), as well as remarkable variability for patients within the same AHI/severity range (Kulkas et al. 2013a, b), suggesting that the proposed severity parameters might provide complementary information on the assessment and management of the severity of OSA. In a subsequent study by the same group, Muraja-Murro et al. (2014) used the *obstruction severity* parameter to adjust the AHI. Interestingly, the adjusted AHI correlated better than standard AHI with mortality (both all-cause and cardiovascular) and non-fatal cardiovascular events, leading to significantly higher association (higher risk ratios) for these outcomes in the corrected moderate and severe groups. In addition, using the novel severity indices, Kulkas et al. (2017) and Leppänen et al. (2017) were able to gain insight into the differences among OSA patients concerning gender, while Kainulainen et al. (2019, 2020) found that severity of desaturations had a great impact on the level of daytime sleepiness and vigilance/reaction time in patients with OSA.

More recently, Linz et al. (2018) proposed a new measure for the quantification of the hypoxemic burden during sleep that they termed *hypoxia load*. The *hypoxia load* is defined as the integrated area of the desaturation curve to the theoretical maximal saturation, i.e., 100%. This way, the *hypoxia load* encompasses all the changes in the saturation signal linked to respiratory events (baseline saturation, number and length of desaturations, time below 90%, and minimum saturation value) regardless any threshold. This index is presented in the context of cardiovascular risk assessment in patients with sleep-disordered breathing (SDB). Linz et al. (2018) reported that the *hypoxia load* showed significant moderate correlation ($r^2 = 0.316$; $p < 0.05$) with epicardial fat volume, an established marker of cardiovascular risk, in patients with SDB after acute myocardial infarction. On the contrary, the AHI and conventional measures of hypoxemia did not show significant association. Additionally, Khoshkish et al. (2018) reported significant correlation ($r^2 \approx 0.1$; $p < 0.05$) between hypoxia load and pulse pressure during both the day and the night, while standard AHI did not. These findings led the authors to suggest that the new measures of hypoxic burden could be used to predict blood pressure patterns and help in the management of hypertensive patients.

In 2019, Azarbarzin et al. (2019) defined a similar index of OSA-related hypoxemia, the *hypoxic burden*, which was presented as a potential predictor of cardiovascular disease (CVD)-related mortality. The *hypoxic burden* index aims to characterize just intermittent hypoxemia typical of OSA and not persistent hypoxemia commonly present in other respiratory diseases. Accordingly, it was defined as the area under the oxygen saturation curve only in the desaturations associated with apneas or hypopneas. A subject-specific search window is defined by segmenting, overlapping using a common synchronization point at the end of each event, and finally averaging all the oximetry segments linked to annotated respiratory events of the individual (i.e., apneas and hypopneas both obstructive and central, regardless their association to a desaturation or an arousal). Finally, the total *hypoxic burden* is computed as the cumulative sum of individual areas normalized by the total sleep time. The authors found that the *hypoxic burden* index was a strong predictor of CVD mortality in different populations (Osteoporotic Fractures in Men Sleep Study, hazard ratio 2.73, 95%CI 1.71–4.36; Sleep Heart Health Study, hazard ratio 1.96, 95%CI 1.11–3.43) independent of the AHI/ODI and traditional measures of hypoxemia (CT90, minimum saturation). In a subsequent study, Kim et al. (2020) found a significant association between an increment (1 SD increment in a log-transformed space) in the *hypoxic burden* index and the increase in blood pressure (1.1% increase in systolic blood pressure, 95%CI 0.1–2.1%; 1.9% increase in diastolic blood pressure, 95%CI 1.0–2.8%) in patients not using hypertensive medication.

Concerning the morphology of events, Otero et al. (2012) proposed a set of indices aimed at parameterizing additional features of desaturations that are not usually considered in the diagnosis and characterization of OSA severity. These indices include not only measures of duration and depth but also features related to the velocity of both the fall and rise parts of the desaturation. The following measures were defined: (i) duration of the desaturation; (ii) average and minimum (i.e., nadir point) values of the saturation throughout the whole event; (iii) elapsed time from the beginning of the event until the nadir point is reached as well as from the nadir to the end of the desaturation; (iv) overall drop in the saturation during the fall part of the event and overall increase in the rise section; (v) slope of both the fall and the rise parts of the event; and

(vi) the desaturation area, measured as the area enclosed between the straight line joining the starting and ending points on the event and the saturation curve. In addition, these oximetry-based measures were computed to characterize the whole overnight oximetric recording: (i) mean saturation throughout the recording; (ii) basal saturation; (iii) difference between the basal value and the mean value; (iv) percentage of the sleep time while the patient is in hypoxemia; and (v) area between a straight line set to the basal saturation value and the oximetry curve.

When assessing the relevance of these indices in the context of OSA diagnosis, the authors found that the most common oximetry features (duration of the desaturation, average, and minimum values) were not selected using a battery of automated feature selection processes. The most relevant oximetry indices were the following: (i) percentage of time in hypoxemia; (ii) difference between basal and average values; (iii) area between basal level and the oximetry profile; (iv) saturation increase during the rise part of the desaturation associated with apnea events; and (v) saturation drop during the fall section of the desaturation associated with apnea events. Individually, the performance of these morphology-related desaturation indices in the detection of OSA ranged 81–90.9% accuracy (86.5–95.7% sensitivity; 47.8–76.1% specificity).

13.2.5 Oximetry and Deep Learning Approaches

Deep learning is changing the paradigm of both image and signal processing in the field of medicine. Traditional machine learning methods rely on the so-called feature engineering, where models are fed with features previously derived from the signals based on the knowledge of the problem under study. Thus, this is a human-driven approach, and so it is highly dependent on the skills of the researchers to compose a relevant feature set. On the contrary, deep learning is able to learn hidden complex patterns directly from the raw signal (Faust et al., 2018), avoiding the bias linked to an a priori known limited set of indices. To do that, deep learning techniques use architectures with multiple levels of representation or data abstraction (Goodfellow et al., 2016), commonly different types of neural networks, such as convolutional or recurrent deep neural networks.

In the context of OSA, deep neural networks have been used in the last years for automated decision-making (Mostafa et al., 2019). Table 13.6 summarizes the main goals of deep learning approaches in the framework of OSA involving the oximetry signal. Main tasks focus on abbreviated OSA diagnosis and automated sleep staging. Nikkonen et al. (2019) applied a fully connected deep neural network to estimate the AHI directly from overnight oximetry (10-

Table 13.6 Techniques and approaches involving deep learning and oximetry in the management of OSA

Goals	Methods
Automated diagnosis Classification of segments (normal vs. apneic) and subsequent estimation of the AHI (short segments: 1- to 5-min length epochs, with or without overlapping) Mostafa et al. (2020a, b), Bernardini et al. (2021) Direct regression of the AHI (larger segments: 10- to 20-minute length epochs, with or without overlapping) Nikkonen et al. (2019), Leino et al. (2021), Vaquerizo-Villar et al. (2021)	Dense fully connected neural networks Convolutional neural networks (CNN) Recurrent neural networks (RNN) Long short-term memory (LSTM) Convolutional + dense (CNN+ dense) Convolutional + recurrent (CNN + RNN)
Sleep staging (short segments: 30 s epochs) Casal et al. (2021) 2-class categorization (wake vs. sleep) 3 class (wake vs. NREM (N1/N2/N3) vs. REM) 4-class (wake vs. light sleep (N1/N2) vs. deep sleep (N3) vs. REM) 5-class (wake vs. N1 vs. N2 vs. N3 vs. REM)	Recurrent neural networks (RNN) Long short-term memory (LSTM) Gated recurrent unit (GRU) Convolutional + recurrent (CNN+ RNN)

min epochs with 98% overlap were used), achieving 0.96 intra-class correlation coefficient (ICC) with actual AHI. Using the estimated AHI, 90.9% of patients were classified in the correct OSA severity group. On a subsequent study by this research group, Leino et al. (2021) proposed a convolutional neural network instead, in order to estimate the rate of respiratory events overnight using single-channel oximetry (10-min epochs with 98% overlap). They obtained 0.97 ICC and 88.3% overall four-class accuracy in a test set composed of patients commonly referred to the sleep unit due to suspicion of OSA, while 0.97 ICC and 77.9% four-class accuracy were reached in a test set of patients with acute cardiovascular disease. Similarly, Vaquerizo-Villar et al. (2021) implemented a convolutional neural network aimed at estimating the AHI from oximetry alone (20-minute segments with no overlapping), reaching ICC values ranging 0.58–0.96 in different extensive test datasets. Using the estimated AHI, the overall three-class accuracy varies between 60.2% and 72.8%.

Mostafa et al. (2020a) used shorter epochs (1, 3, and 5 min with 1-minute overlap) to implement an event-based approach (detection of apneas) using different architectures of convolutional neural networks, achieving accuracies ranging 84.8–94.2%. In a subsequent study (Mostafa et al., 2020b), the same group implemented a new convolutional neural network architecture based in the same approach, reporting patient-based accuracies of 95.7% and 100% in different test datasets. In a recent study by Bernardini et al. (2021), oximetry (2.5-minute epochs) is analyzed under a deep learning approach both alone (using a recurrent neural network) and together with ECG (combining convolutional and recurrent neural networks). A long short-term memory (LSTM) neural network (a kind of recurrent deep neural network) reached 67.6% and 63.3% accuracies under per-second and per-patient classification approaches, respectively. Using ECG segments as inputs to a deep learning architecture that combines convolutional and recurrent blocks, 76.9% and 73.3% accuracies were reported for the same performance assessment schemes. Interestingly, when both ECG and oximetry were analyzed jointly (2D input), the network achieved 81.5% and 93.3% accuracies for per-second and per-patient classification, respectively, i.e., +4.6% and +20.0% increase compared to the use of ECG alone.

In automatic sleep staging, only one study has used the raw oximetry signal (Casal et al., 2021). In this work, recurrent neural networks are used to discern wakefulness from sleep (binary classification) using blood oxygen saturation and heart rate from pulse oximetry (30-sec epochs), reporting 90.1% accuracy and 0.74 Cohen's kappa. Similar approaches exist using the photoplethysmogram (PPG) signal from pulse oximetry instead of the blood oxygen saturation or pulse rate times series, achieving promising performance (84.2% accuracy) in a two-class classification problem (Malik et al., 2018), while the accuracy decreases to 80.1%, 68.5%, and 64.1% for three-, four-, and five-stage classification tasks, respectively (Korkalainen et al., 2020).

13.3 Discussion and Conclusions

The great amount of indices derived from the oximetry signal existing in the literature (Del Campo et al., 2018; Terrill, 2020; Levy et al., 2021) is representative of the high relevance of this biomedical recording in the framework of OSA. Recent reviews and meta-analyses summarizing all the research made around oximetry during the last years confirm this intuition. In a systematic review by Uddin et al. (2018), single-channel oximetry raises as an effective biomedical signal to implement binary expert systems for automated OSA detection (OSA positive vs. OSA negative). Similarly, in the recent meta-analysis by Wu et al. (2020), oximetry is found to yield remarkable specificity in the detection of all OSA severity groups. In the context of pediatric OSA, the meta-analysis by Gutiérrez-Tobal et al. (2021b) revealed that top performance methodologies were those involving oximetry, particularly for the detection of moderate and severe cases, showing also less variability among studies. Beyond its diagnostic ability, nocturnal oximetry dynamics have been also found to be associated with clinical and epidemiological outcomes (Suen et al., 2019; Terrill, 2020), increasing its usefulness as essential tool for integrated management of OSA.

Since first attempts to characterize changes in overnight oximetry by means of the minimum saturation value, ODIs, and CTs, different approaches have been applied to obtain as much information as possible from the recording, including characterization of the data histogram using different statistics, analysis in the frequency domain with different methods of power spectral density estimation, nonlinear analysis, parameterization of the parts of the desaturation, quantification of the area enclosed within each desaturation, and, recently, deep learning. All these methodologies have yielded relevant indices in the characterization of OSA severity. Despite being one of the first measures developed for that aim, usually used as benchmark for comparison purposes, the ODI stands out for its great individual accuracy, rarely outperformed by univariate approaches. Moreover, the ODI has been found as an essential predictor under multivariate schemes, being systematically selected within the final optimum feature subsets. In the same regard, statistical, spectral, and nonlinear variables, as well as the novel hypoxic burden measures, have shown major complementarity, leading to a significant performance increase when input features are properly selected via automated variable selection procedures (Álvarez et al., 2010, 2012, 2013). Additionally, oximetry in general, and particularly ODI, has also shown significant complementarity with other cardiorespiratory signals related to OSA, such as pulse rate (Álvarez et al., 2009; Garde et al., 2014) and airflow (Gutiérrez-Tobal et al., 2015; Álvarez et al., 2020; Barroso-García et al., 2021b) in both adults and children.

Besides the nonlinear methods based on traditional entropy and complexity measures, novel nonlinear methods, such as multiscale entropy, detrended fluctuation analysis, and symbolic dynamics, recently demonstrated major efficiency when applied to the oximetry signal. Nevertheless, ApEn, SampEn, CTM, and LZC are predominantly used under multivariate approaches instead of MSE scales, DFA slopes, and symbolic entropy. Although these methods are computationally demanding, it would be important to promote their use to prospectively validate their accuracy and to include them in the available toolboxes for automated signal processing of oximetry.

Concerning the novel measures of hypoxemia named hypoxic burden indices (saturation impairment time, desaturation severity, hypoxia load, and hypoxic burden), they have been found to provide complementary data to the AHI/ODI and conventional hypoxemia measures (CT90, minimum saturation). This suggests that not only the recurrence but also the morphology (depth and duration) of the events have jointly a significant impact on the characteristics of respiratory-related diseases and associated comorbidities. Nevertheless, this "information gain" has not become a significant performance increase regarding automated detection of OSA from oximetry. Thus, further research is encouraged to exploit all the diagnostic capability available in these indices. On the other hand, hypoxic burden measures have demonstrated to be robust predictors of cardiovascular status and mortality due to the intermittent hypoxemia typical of OSA (Muraja-Murro et al., 2014; Khoshkish et al., 2018; Azarbarzin et al., 2019; Kim et al., 2020). These novel hypoxemia measures seem to outperform conventional indices (overnight minimum saturation, CT90, and ODIs), which performed modestly as predictors of cardiovascular events (stroke, heart failure) and related mortality (Kendzerska et al., 2014; Stone et al., 2016; Gellen et al., 2016). Nonetheless, due to the dissimilarities in the computation of these new parameters, thorough research and prospective validation are still needed to fully understand the link between each particular index and patient outcomes.

In regard to the usefulness of artificial intelligence, relevant recent reports highlight its potential to boost sleep medicine (Goldstein et al., 2020; Watson & Fernandez, 2021; Malhotra et al., 2021). Concerning the oximetry signal, a number of automated expert systems have been developed for OSA diagnosis, mostly using feature engineering and traditional machine learning models for both binary and multiclass classification of patients, as well as regression of the AHI. Nevertheless, deep learning approaches recently raised as valuable tool able to boost the diagnostic ability of oximetry, mainly when applied to categorize segments (apneic vs. normal) and predict the AHI using oximetry alone. The next step (individual indices, multivariate analysis, artificial intelligence, deep learning) should be the application of eXplainable Artificial Intelligence (XAI) techniques to thoroughly interpret the particularly complex models derived from deep learning. XAI methods are able to identify which parts of the oximetry signal mainly contribute to the final decision. Thus, XAI might be used to confirm the relevance of sections of the desaturation event highlighted in some studies, whose widespread application is commonly hindered by more popular indices. For example, higher saturation values and resaturations have shown significant relevancy and complementarity (Sánchez-Morillo & Gross, 2013; Sánchez-Morillo et al., 2014; Álvarez et al., 2018), although they have been marginally used in multivariate subsequent studies. Furthermore, XAI approaches have the potential to provide clinicians with new oximetric features with the upmost diagnostic capability hidden until now in the raw oximetry signal.

During the last two decades, the oximetry signal has been found to provide high-performance indices in the framework of OSA management. With the improvement of medical technology in terms of portability, autonomy, and computational capability, and taking into account the simplicity, low cost, and high availability of oximeters, peripheral blood oxygen saturation raises as a key signal in the development of simple as well as accurate diagnostic tests for OSA. Moreover, oximetry could be an essential tool to foster sleep medicine toward the concept of precision and personalized medicine. However, both greater standardization in the definition of available indices and extensive validation of the novel measures derived from the signal processing theory are still needed to increase generalizability of overnight oximetry as an alternative abbreviated test for OSA.FundingThis work was supported in part by "Ministerio de Ciencia e Innovación-Agencia Estatal de Investigación" and the European Regional Development Fund (FEDER) under projects PID2020-115468RB-I00 and PDC2021-12775-I00, in part by "Sociedad Española de Neumología y Cirugía Torácica" (SEPAR) under project 649/2018, in part by "Sociedad Española de Sueño" (SES) under project "Beca de Investigación SES 2019," and in part by CIBER - Consorcio Centro de Investigación Biomédica en Red - (CB19/01/00012), Instituto de Salud Carlos III, Ministerio de Ciencia e Innovación. The work of Daniel Álvarez was supported by a "Ramón y Cajal" grant (RYC2019-028566-I) from the "Ministerio de Ciencia e Innovación-Agencia Estatal de Investigación" co-funded by the European Social Fund.

References

Aaronson, J. A., Van Bezeij, T., Van den Aardweg, J. G., et al. (2012). Diagnostic accuracy of nocturnal oximetry for detection of sleep apnea syndrome in stroke rehabilitation. *Stroke, 43*(9), 2491–2493.

Álvarez, D., Hornero, R., Abásolo, D., del Campo, F., & Zamarrón, C. (2006). Nonlinear characteristics of blood oxygen saturation from nocturnal oximetry for obstructive sleep apnoea detection. *Physiological Measurement, 27*, 399–412.

Álvarez, D., Hornero, R., García, M., del Campo, F., & Zamarrón, C. (2007). Improving diagnostic ability of blood oxygen saturation from overnight pulse oximetry in obstructive sleep apnea detection by means of central tendency measure. *Artificial Intelligence in Medicine, 41*, 13–24.

Álvarez, D., Hornero, R., Abásolo, D., el Campo, F., Zamarrón, C., & López, M. (2009). Nonlinear measure of synchrony between blood oxygen saturation and heart rate from nocturnal pulse oximetry in obstructive sleep apnoea syndrome. *Physiological Measurement, 30*, 967–982.

Álvarez, D., Hornero, R., Marcos, J. V., & del Campo, F. (2010). Multivariate analysis of blood oxygen saturation recordings in obstructive sleep apnea diagnosis. *IEEE Transactions on Biomedical Engineering, 57*, 2816–2824.

Álvarez, D., Hornero, R., Marcos, J. V., & del Campo, F. (2012). Feature selection from nocturnal oximetry using genetic algorithms to assist in obstructive sleep apnoea diagnosis. *Medical Engineering & Physics, 34*, 1049–1057.

Álvarez, D., et al. (2013). Assessment of feature selection and classification approaches to enhance information from overnight oximetry in the context of sleep apnea diagnosis. *International Journal of Neural Systems, 23*, 1350020.

Álvarez, D., Alonso-Álvarez, M. L., Gutiérrez-Tobal, G. C., Crespo, A., Kheirandish-Gozal, L., Hornero, R., Gozal, D., Terán-Santos, J., & del Campo, F. (2017). Automated screening of children with obstructive sleep apnea using nocturnal oximetry: An alternative to respiratory polygraphy in unattended settings. *Journal of Clinical Sleep Medicine, 13*(5), 693–702.

Álvarez, D., Crespo, A., Vaquerizo-Villar, F., Gutiérrez-Tobal, G. C., Cerezo-Hernández, A., Barroso-García, V., Ansermino, J. M., Dumont, G. A., Hornero, R., del Campo, F., & Garde, A. (2018). Symbolic dynamics to enhance diagnostic ability of portable oximetry from the Phone Oximeter in the detection of paediatric sleep apnoea. *Physiological Measurement, 39*(10), 104002.

Álvarez, D., Cerezo-Hernández, A., Crespo, A., Gutiérrez-Tobal, G. C., Vaquerizo-Villar, F., Barroso-García, V., Moreno, F., Arroyo, C. A., Ruiz, T., Hornero, R., & del Campo, F. (2020). A machine learning-based test for adult sleep apnoea screening at home using oximetry and airflow. *Scientific Reports, 10*, 5332.

Azarbarzin, A., Sands, S. A., Stone, K. L., Taranto-Montemurro, L., Messineo, L., Terrill, P. I., Ancoli-Israel, S., Ensrud, K., Purcell, S., White, D. P., Redline, S., & Wellman, A. (2019). The hypoxic burden of sleep apnoea predicts cardiovascular disease-related mortality: The Osteoporotic Fractures in Men Study and the Sleep Heart Health Study. *European Heart Journal, 40*, 1149–1157.

Barroso-García, V., Gutiérrez-Tobal, G. C., Kheirandish-Gozal, L., Álvarez, D., Vaquerizo-Villar, F., Núñez, P., del Campo, F., Gozal, D., & Hornero, R. (2020). Usefulness of recurrence plots from airflow recordings to aid in paediatric sleep apnoea diagnosis. *Computer Methods and Programs in Biomedicine, 183*, 105083.

Barroso-García, V., Gutiérrez-Tobal, G. C., Gozal, D., Vaquerizo-Villar, F., Álvarez, D., del Campo, F., Kheirandish-Gozal, L., & Hornero, R. (2021a). Wavelet analysis of overnight airflow to detect obstructive sleep apnea in children. *Sensors, 21*(4), 1491.

Barroso-García, V., Gutiérrez-Tobal, G. C., Kheirandish-Gozal, L., Vaquerizo-Villar, F., Álvarez, D., del Campo, F., Gozal, D., & Hornero, R. (2021b). Bispectral analysis of overnight airflow to improve the pediatric sleep apnea diagnosis. *Computers in Biology and Medicine, 129*, 104167.

Bernardini, A., Brunello, A., Gigli, G. L., Montanari, A., & Saccomanno, N. (2021). AIOSA: An approach to the automatic identification of obstructive sleep apnea events based on deep learning. *Artificial Intelligence in Medicine, 118*, 102133.

Brouillette, R. T., et al. (2000). Nocturnal pulse oximetry as an abbreviated testing modality for pediatric obstructive sleep apnea. *Pediatrics, 105*, 405–412.

Casal, R., Di Persia, L. E., & Schlotthauer, G. (2021). Classifying sleep–wake stages through recurrent neural networks using pulse oximetry signals. *Biomedical Signal Processing and Control, 63*, 102195.

Chang, L., Wu, J., & Cao, L. (2013). Combination of symptoms and oxygen desaturation index in predicting childhood obstructive sleep apnea. *International Journal of Pediatric Otorhinolaryngology, 77*, 365–371.

Chaudhary, B., Dasti, S., Park, Y., Brown, T., Davis, H., & Akhtar, B. (1998). Hour-to-hour variability of oxygen saturation in sleep apnea. *Chest, 113*, 719–722.

Chen-Liang, L., Chinson, Y., Chen-Wen, Y., Wu-Huei, H., & Liang-Wen, H. (2009). Comparison of the indices of oxyhemoglobin saturation by pulse oximetry in obstructive sleep apnea hypopnea syndrome. *Chest, 135*, 86–93.

Chesson, A. L., Anderson, W. M., Walls, R. C., & Bairnsfather, L. E. (1993). Assessment of hypoxemia in patients with sleep disorders using Saturation Impairment Time (SIT). *The American Review of Respiratory Disease, 148*(6), 1592–1598.

Chesson, A. L., Anderson, W. M., Walls, R. C., & Bairnsfather, L. E. (2001). Comparison of two methods of quantitative assessment of hypoxemia in patients with sleep disorders. *Sleep Medicine, 2*, 37–45.

Chung, F., Liao, P., Elsaid, H., et al. (2012). Oxygen desaturation index from nocturnal oximetry: A sensitive and specific tool to detect sleep disordered breathing in surgical patients. *Anesthesia and Analgesia, 114*(5), 993–1000.

Crespo, A., Álvarez, D., Gutiérrez-Tobal, G. C., Vaquerizo-Villar, F., Barroso-García, V., Alonso-Álvarez, M. L., Terán-Santos, J., Hornero, R., & del Campo, F. (2017). Multiscale entropy analysis of unattended oximetric recordings to assist in the screening of paediatric sleep apnoea at home. *Entropy, 19*, 284.

Crespo, A., Álvarez, D., Kheirandish-Gozal, L., Gutiérrez-Tobal, G. C., Gozal, D., Hornero, R., & del Campo, F. (2018). Assessment of oximetry-based statistical classifiers as simplified screening tools in the management of childhood obstructive sleep apnea. *Sleep & Breathing, 22*(4), 1063–1073.

Dawson, A., Loving, R. T., Gordon, R. M., et al. (2015). Type III home sleep testing versus pulse oximetry: Is the respiratory disturbance index better than the oxygen desaturation index to predict the apnoea-hypopnoea index measured during laboratory polysomnography? *BMJ Open, 5*, e007956.

Del Campo, F., Hornero, R., Zamarrón, C., Abasolo, D. E., & Álvarez, D. (2006). Oxygen saturation regularity analysis in the diagnosis of obstructive sleep apnea. *Artificial Intelligence in Medicine, 37*, 111–118.

Del Campo, F., Crespo, A., Cerezo-Hernández, A., Gutiérrez-Tobal, G. C., Hornero, R., & Álvarez, D. (2018). Oximetry use in obstructive sleep apnea. *Expert Review of Respiratory Medicine, 12*(8), 665–681.

Faust, O., Hagiwara, Y., Hong, T. J., Lih, O. S., & Acharya, U. R. (2018). Deep learning for healthcare applications based on physiological signals: A review. *Computer Methods and Programs in Biomedicine, 161*, 1–13.

Garde, A., Dehkordi, P., Karlen, W., Wensley, D., Ansermino, J. M., & Dumont, G. A. (2014). Development of a screening tool for sleep disordered breathing in children using the Phone Oximeter™. *PLoS One, 9*(11), e112959.

Gellen, B., Canoui-Poitrine, F., Boyer, L., Drouot, X., Le Thuaut, A., Bodez, D., Covali-Noroc, A., D'Ortho, M. P., Guendouz, S., Rappeneau, S., et al. (2016). Apnea-hypopnea and desaturations in heart failure with reduced ejection fraction: Are we aiming at the right target? *International Journal of Cardiology, 203*, 1022–1028.

Goldstein, C. A., Berry, R. B., Kent, D. T., et al. (2020). Artificial intelligence in sleep medicine: An American Academy of Sleep Medicine position statement. *Journal of Clinical Sleep Medicine, 16*(4), 605–607.

Golpe, R., Jimenez, A., Carpizo, R., & Cifrian, J. M. (1999). Utility of home oximetry as a screening test for patients with moderate to severe symptoms of obstructive sleep apnea. *Sleep, 22*, 932–937.

Goodfellow, I., Bengio, Y., & Courville, A. (2016). *Deep learning*. MIT Press.

Gutiérrez-Tobal, G. C., et al. (2015). Diagnosis of pediatric obstructive sleep apnea: Preliminary findings using automatic analysis of airflow and oximetry recordings obtained at patients' home. *Biomedical Signal Processing and Control, 18*, 401–407.

Gutiérrez-Tobal, G. C., Álvarez, D., Crespo, A., del Campo, F., & Hornero, R. (2019). Evaluation of machine-learning approaches to estimate sleep apnea severity from at-home oximetry recordings. *IEEE Journal of Biomedical and Health Informatics, 23*(2), 882–892.

Gutiérrez-Tobal, G. C., Álvarez, D., Vaquerizo-Villar, F., Crespo, A., Kheirandish-Gozal, L., Gozal, D., del Campo, F., & Hornero, R. (2021a). Ensemble-learning regression to estimate sleep apnea severity using at-home oximetry in adults. *Applied Soft Computing, 111*, 107827.

Gutiérrez-Tobal, G. C., Álvarez, D., Kheirandish-Gozal, L., del Campo, F., Gozal, D., & Hornero, R. (2021b). Reliability of machine learning to diagnose pediatric obstructive sleep apnea: Systematic review and meta-analysis. *Pediatric Pulmonology*, In press.

Gyulay, S., Olson, L. G., Hensley, M. J., et al. (1993). A comparison of clinical assessment and home oximetry in the diagnosis of obstructive sleep apnea. *The American Review of Respiratory Disease, 147*, 50–53.

Hornero, R., Álvarez, D., Abásolo, A., del Campo, F., & Zamarrón, C. (2007). Utility of approximate entropy from overnight pulse oximetry data in the diagnosis of obstructive sleep apnea syndrome. *IEEE Transactions on Biomedical Engineering, 54*, 107–113.

Hornero, R., et al. (2017). Nocturnal oximetry-based evaluation of habitually snoring children. *American Journal of Respiratory and Critical Care Medicine, 196*(12), 1591–1598.

Jiménez-García, J., Gutiérrez-Tobal, G. C., García, M., et al. (2020). Assessment of airflow and oximetry signals to detect pediatric sleep apnea-hypopnea syndrome using AdaBoost. *Entropy, 22*(6), e22060670.

Kainulainen, S., Töyräs, J., Oksenberg, A., et al. (2019). Severity of desaturations reflects OSA-related daytime sleepiness better than AHI. *Journal of Clinical Sleep Medicine, 15*(8), 1135–1142.

Kainulainen, S., Duce, B., Korkalainen, H., et al. (2020). Severe desaturations increase psychomotor vigilance task-based median reaction time and number of lapses in obstructive sleep apnoea patients. *The European Respiratory Journal, 55*(4), 1901849.

Kendzerska, T., Gershon, A. S., Hawker, G., Leung, R. S., & Tomlinson, G. (2014). Obstructive sleep apnea and risk of cardiovascular events and all-cause mortality: A decade-long historical cohort study. *PLoS Medicine, 11*, e1001599.

Khoshkish, S., Hohl, M., Linz, B., Arzt, M., Mahfoud, F., Baumert, M., Schöpe, J., Böhm, M., & Linz, D. (2018). The association between different features of sleep-disordered breathing and blood pressure: A cross-sectional study. *Journal of Clinical Hypertension, 20*, 575–581.

Kim, J. S., Azarbarzin, A., Wang, R., Djonlagic, I. E., Punjabi, N. M., Zee, P. C., Koo, B. B., Soliman, E. Z., Younes, M., & Redline, S. (2020). Association of novel measures of sleep disturbances with blood pressure: The Multi-Ethnic Study of Atherosclerosis. *Thorax, 75*, 57–63.

Kirk, V. G., et al. (2003). Comparison of home oximetry monitoring with laboratory polysomnography in children. *Chest, 124*, 1702–1708.

Korkalainen, H., Aakko, J., Duce, B., Kainulainen, S., Leino, A., Nikkonen, S., Afara, I. O., Myllymaa, S., Töyräs, J., & Leppänen, T. (2020). Deep learning enables sleep staging from photoplethysmogram for patients with suspected sleep apnea. *Sleep, 43*(11), 1–10.

Kulkas, A., Tiihonen, P., Eskola, K., Julkunen, P., Mervaala, E., & Töyräs, J. (2013a). Novel parameters for evaluating severity of sleep disordered breathing and for supporting diagnosis of sleep apnea-hypopnea syndrome. *Journal of Medical Engineering & Technology, 37*(2), 135–143.

Kulkas, A., Tiihonen, P., Julkunen, P., et al. (2013b). Novel parameters indicate significant differences in severity of obstructive sleep apnea with patients having similar apnea–hypopnea index. *Medical & Biological Engineering & Computing, 51*(6), 697–708.

Kulkas, A., Duce, B., Leppänen, T., et al. (2017). Gender differences in severity of desaturation events following hypopnea and obstructive apnea events in adults during sleep. *Physiological Measurement, 38*(8), 1490–1150.

Leino, A., Nikkonen, S., Kainulainen, S., Korkalainen, H., Töyräs, J., Myllymaa, S., Leppänen, T., Ylä-Herttuala, S., Westeren-Punnonen, S., Muraja-Murro, A., Jäkälä, P., Mervaala, E., & Myllymaa, K. (2021). Neural network analysis of nocturnal SpO_2 signal enables easy screening of sleep apnea in patients with acute cerebrovascular disease. *Sleep Medicine, 79*, 71e78.

Leppänen, T., Kulkas, A., Duce, B., et al. (2017). Severity of individual obstruction events is gender dependent in sleep apnea. *Sleep & Breathing, 21*(2), 397–404.
Lévy, P., Pépin, J. L., Deschaux-Blanc, C., Paramelle, B., & Brambilla, C. (1996). Accuracy of oximetry for detection of respiratory disturbances in sleep apnea syndrome. *Chest, 109*, 395–399.
Levy, J., Álvarez, D., Rosenberg, A. A., Alexandrovich, A., del Campo, F., & Behar, J. A. (2021). Digital oximetry biomarkers for assessing respiratory function: Standards of measurement, physiological interpretation, and clinical use. *NPJ Digital Medicine, 4*, 1.
Linz, D., Colling, S., Nußstein, W., Debl, K., Hohl, M., Fellner, C., Böhm, M., Maier, L. S., Hamer, O. W., Buchner, S., & Arzt, M. (2018). Nocturnal hypoxemic burden is associated with epicardial fat volume in patients with acute myocardial infarction. *Sleep & Breathing, 22*, 703–711.
Magalang, U. J., Dmochowski, J., Veeramachaneni, S., et al. (2003). Prediction of the apnea-hypopnea index from overnight pulse oximetry. *Chest, 124*, 1694–1701.
Malbois, M., Giusti, V., Suter, M., et al. (2010). Oximetry alone versus portable polygraphy for sleep apnea screening before bariatric surgery. *Obesity Surgery, 20*(3), 326–331.
Malhotra, A., Ayappa, I., Ayas, N., Collop, N., Kirsch, D., Mcardle, N., Mehra, R., Pack, A. I., Punjabi, N., White, D. P., & Gottlieb, D. J. (2021). Metrics of sleep apnea severity: Beyond the apnea-hypopnea index. *Sleep, 44*(7), 1–16.
Malik, J., Lo, Y.-L., & Wu, H.-T. (2018). Sleep-wake classification via quantifying heart rate variability by convolutional neural network. *Physiological Measurement, 39*(8), 085004.
Marcos, J. V., Hornero, R., Álvarez, D., del Campo, F., & Zamarrón, C. (2009). Assessment of four statistical pattern recognition techniques to assist in obstructive sleep apnoea diagnosis. *Medical Engineering & Physics, 31*, 971–978.
Marcos, J. V., Hornero, R., Álvarez, D., Nabney, I. T., del Campo, F., & Zamarrón, C. (2010a). The classification of oximetry signals using Bayesian neural networks to assist in the detection of obstructive sleep apnoea syndrome. *Physiological Measurement, 31*, 375–394.
Marcos, J. V., Hornero, R., Álvarez, D., del Campo, F., & Aboy, M. (2010b). Automated detection of obstructive sleep apnoea syndrome from oxygen saturation recordings using linear discriminant analysis. *Medical & Biological Engineering & Computing, 48*, 895–902.
Marcos, J. V., Hornero, R., Álvarez, D., Aboy, M., & Del Campo, F. (2012). Automated prediction of the apnea-hypopnea index from nocturnal oximetry recordings. *IEEE Transactions on Biomedical Engineering, 59*, 141–149.
Martín-Montero, A., Gutiérrez-Tobal, G. C., Gozal, D., Barroso-García, V., Álvarez, D., del Campo, F., Kheirandish-Gozal, L., & Hornero, R. (2021a). Bispectral analysis of heart rate variability to characterize and help diagnose pediatric sleep apnea. *Entropy, 23*(8), 1016.
Martín-Montero, A., Gutiérrez-Tobal, G. C., Kheirandish-Gozal, L., Jiménez-García, J., Álvarez, D., del Campo, F., Gozal, D., & Hornero, R. (2021b). Heart rate variability spectrum characteristics in children with sleep apnea. *Pediatric Research, 89*(7), 1771–1779.
Mazière, S., Pepin, J. L., Siyanko, N., et al. (2014). Usefulness of oximetry for sleep apnea in frail hospitalized elderly. *JAMDA, 15*, 447–4e9.
Mostafa, S. S., Mendonça, F., Ravelo-García, A. G., & Morgado-Dias, F. (2019). A systematic review of detecting sleep apnea using deep learning. *Sensors, 19*(22), 1–26.
Mostafa, S. S., Mendonça, F., Ravelo-García, A. G., Juliá-Serdá, G., & Morgado-Dias, F. (2020a). Multi-objective hyperparameter optimization of convolutional neural network for obstructive sleep apnea detection. *IEEE Access, 8*, 129586–129599.
Mostafa, S. S., Baptista, D., Ravelo-García, A. G., Juliá-Serdác, G., & Morgado-Dias, F. (2020b). Greedy based convolutional neural network optimization for detecting apnea. *Computer Methods and Programs in Biomedicine, 197*, 105640.
Muraja-Murro, A., Kulkas, A., Hiltunen, M., et al. (2014). Adjustment of apnea–hypopnea index with severity of obstruction events enhances detection of sleep apnea patients with the highest risk of severe health consequences. *Sleep & Breathing, 18*(3), 641–647.
Nikkonen, S., Afara, I. O., Leppänen, T., & Töyräs, J. (2019). Artificial neural network analysis of the oxygen saturation signal enables accurate diagnostics of sleep apnea. *Scientific Reports, 9*, 13200.
Nixon, G. M., Kermack, A. S., Davis, G. M., Manoukian, J. J., Brown, K. A., & Brouillette, R. T. (2004). Planning adenotonsillectomy in children with obstructive sleep apnea: The role of overnight oximetry. *Pediatrics, 113*, e19.
Olson, L. G., Ambrogetti, A., & Gyulay, S. G. (1999). Prediction of sleep-disordered breathing by unattended overnight oximetry. *Journal of Sleep Research, 8*, 51–55.
Otero, A., Félix, P., Presedo, J., & Zamarrón, C. (2012). An evaluation of indexes as support tools in the diagnosis of sleep apnea. *Annals of Biomedical Engineering, 40*(8), 1825–1834.
Penttonen, M., & Buzsáki, G. (2003). Natural logarithmic relationship between brain oscillators. *Thalamus & Related Systems, 2*, 145–152.
Poupard, L., Philippe, C., Goldman, M. D., Sartène, R., & Mathieu, M. (2012). Novel mathematical processing method of nocturnal oximetry for screening patients with suspected sleep apnoea syndrome. *Sleep & Breathing, 16*, 419–425.
Rofail, L. M., Wong, K. K., Unger, G., et al. (2010). Comparison between a single channel nasal airflow device and oximetry for the diagnosis of obstructive sleep apnea. *Sleep, 33*(8), 1106–1114.
Sánchez-Morillo, D., & Gross, N. (2013). Probabilistic neural network approach for the detection of SAHS

from overnight pulse oximetry. *Medical & Biological Engineering & Computing, 51*, 305–315.

Sánchez-Morillo, D., López-Gordo, M. A., & León, A. (2014). Novel multiclass classification for home-based diagnosis of sleep apnea hypopnea syndrome. *Expert Systems with Applications, 41*, 1654–1662.

Schlotthauer, G., et al. (2014). Screening of obstructive sleep apnea with empirical mode decomposition of pulse oximetry. *Medical Engineering & Physics, 36*(8), 1074–1080.

Scott, A. S., Baltzan, M. A., & Wolkove, N. (2014). Examination of pulse oximetry tracings to detect obstructive sleep apnea in patients with advanced chronic obstructive pulmonary disease. *Canadian Respiratory Journal, 21*, 171–175.

Sharma, S., Mukhtar, U., Kelly, C., et al. (2017). Recognition and treatment of sleep disordered breathing in obese hospitalized patients may improve survival. HoSMed database. *The American Journal of Medicine, 130*(10), 1184–1191.

Shiomi, T., Guilleminault, C., Sasanabe, R., Hirota, I., Maekawa, M., & Kobayashi, T. (1996). Augmented very low frequency component of heart rate variability during obstructive sleep apnea. *Sleep, 19*(5), 370–377.

Stein, P. K., & Pu, Y. (2012). Heart rate variability, sleep and sleep disorders. *Sleep Medicine Reviews, 16*, 47–66.

Stone, K. L., Blackwell, T. L., Ancoli-Israel, S., Barrett-Connor, E., Bauer, D. C., Cauley, J. A., Ensrud, K. E., Hoffman, A. R., Mehra, R., Stefanick, M. L., et al. (2016). Sleep disordered breathing and risk of stroke in older community-dwelling men. *Sleep, 39*, 531–540.

Suen, C., Ryan, C. M., Mubashir, T., Ayas, N. T., Abrahamyan, L., Wong, J., Mokhlesi, B., & Chung, F. (2019). Sleep study and oximetry parameters for predicting postoperative complications in patients with OSA. *Chest, 155*(4), 855–867.

Terrill, P. I. (2020). A review of approaches for analysing obstructive sleep apnoea-related patterns in pulse oximetry data. *Respirology, 25*, 475–485.

Tsai, C. M., Kang, C. H., Su, M. C., et al. (2013). Usefulness of desaturation index for the assessment of obstructive sleep apnea syndrome in children. *International Journal of Pediatric Otorhinolaryngology, 77*, 1286–1290.

Uddin, M. B., Chow, C. M., & Su, S. W. (2018). Classification methods to detect sleep apnea in adults based on respiratory and oximetry signals: A systematic review. *Physiological Measurement, 39*, 03TR01.

Van Eyck, A., & Verhulst, S. L. (2018). Improving the diagnosis of obstructive sleep apnea in children with nocturnal oximetry-based evaluations. *Expert Review of Respiratory Medicine, 12*(3), 165–167.

Van Eyck, A., Lambrechts, C., & Vanheeswijck, L. (2015). The role of nocturnal pulse oximetry in the screening for obstructive sleep apnea in obese children and adolescents. *Sleep Medicine, 16*, 1409–1412.

Vaquerizo-Villar, F., Álvarez, D., Kheirandish-Gozal, L., Gutiérrez-Tobal, G. C., Barroso-García, V., Crespo, A., del Campo, F., Gozal, D., & Hornero, R. (2018a). Utility of bispectrum in the screening of pediatric sleep apnea-hypopnea syndrome using oximetry recordings. *Computer Methods and Programs in Biomedicine, 156*, 141–149.

Vaquerizo-Villar, F., Álvarez, D., Kheirandish-Gozal, L., Gutiérrez-Tobal, G. C., Barroso-García, V., Crespo, A., del Campo, F., Gozal, D., & Hornero, R. (2018b). Detrended fluctuation analysis of the oximetry signal to assist in paediatric sleep apnoea-hypopnoea syndrome diagnosis. *Physiological Measurement, 39*(11), 114006.

Vaquerizo-Villar, F., Álvarez, D., Kheirandish-Gozal, L., Gutiérrez-Tobal, G. C., Barroso-García, V., Crespo, A., del Campo, F., Gozal, D., & Hornero, R. (2018c). Wavelet analysis of oximetry recordings to assist in the automated detection of moderate-to-severe pediatric sleep apnea-hypopnea syndrome. *PLoS One, 13*(12), e0208502.

Vaquerizo-Villar, F., Álvarez, D., Kheirandish-Gozal, L., Gutiérrez-Tobal, G. C., Barroso-García, V., Santamaría-Vázquez, E., del Campo, F., Gozal, D., & Hornero, R. (2021). A convolutional neural network architecture to enhance oximetry ability to diagnose pediatric obstructive sleep apnea. *IEEE Journal of Biomedical and Health Informatics, 25*(8), 2906–2916.

Velasco-Suarez, C. T., Figueroa-Turienzo, J. M., Len, F., & Mansilla, E. (2013). Pulse oximetry recording in children with adenotonsillar hypertrophy: Usefulness in the diagnostic of obstructive sleep apnea syndrome. *Archivos Argentinos de Pediatría, 111*, 196e201.

Villa, M. P., Pietropaoli, N., Supino, M. C., et al. (2015). Diagnosis of pediatric obstructive sleep apnea syndrome in settings with limited resources. *JAMA Otolaryngology–Head & Neck Surgery, 141*, 990–996.

Ward, N. R., Cowie, M. R., Rosen, S. D., et al. (2012). Utility of overnight pulse oximetry and heart rate variability analysis to screen for sleep-disordered breathing in chronic heart failure. *Thorax, 67*(11), 100–105.

Watson, N. F., & Fernandez, C. R. (2021). Artificial intelligence and sleep: Advancing sleep medicine. *Sleep Medicine Reviews, 59*, 101512.

Wu, C.-R., Tu, Y.-K., Chuang, L.-P., Gordon, C., Chen, N.-H., Chen, P.-Y., Hasan, F., Kurniasari, M. D., Susanty, S., & Chiu, H.-Y. (2020). Diagnostic meta-analysis of the Pediatric Sleep Questionnaire, OSA-18, and pulse oximetry in detecting pediatric obstructive sleep apnea syndrome. *Sleep Medicine Reviews, 54*, 101355.

Zamarrón, C., Romero, P. V., Rodríguez, J. R., & Gude, F. (1999). Oximetry spectral analysis in the diagnosis of obstructive sleep apnoea. *Clinical Science, 97*, 467–473.

Zamarrón, C., Gude, F., Barcala, J., Rodríguez, J. R., & Romero, P. V. (2003). Utility of oxygen saturation and heart rate spectral analysis obtained from pulse oximetric recordings in the diagnosis of sleep apnea syndrome. *Chest, 123*, 1567–1576.

14 Airflow Analysis in the Context of Sleep Apnea

Verónica Barroso-García, Jorge Jiménez-García, Gonzalo C. Gutiérrez-Tobal, and Roberto Hornero

Abstract

The airflow (AF) is a physiological signal involved in the overnight polysomnography (PSG) that reflects the respiratory activity. This signal is able to show the particularities of sleep apnea and is therefore used to define apneic events. In this regard, a growing number of studies have shown the usefulness of employing the overnight airflow as the only or combined information source for diagnosing sleep apnea in both children and adults. Due to its easy acquisition and interpretation, this biosignal has been widely analyzed by means of different signal processing techniques. In this chapter, we review the main methodological approaches applied to characterize and extract relevant information from this signal. In view of the results, we can conclude that the overnight airflow successfully reflects the particularities caused by the occurrence of apneic and hypopneic events and provides useful information for obtaining relevant biomarkers that characterize this disease.

Keywords

Airflow · Automatic analysis · Sleep apnea · Sleep disorders

14.1 Introduction

Simplification of sleep apnea diagnosis has become a major concern in the field of sleep medicine and the motivation of several investigations in recent years. Currently, the standard method for diagnosing the disease in children and adults remains overnight polysomnography (PSG) (Jon, 2009; Patil et al., 2007). This is an effective medical test, but it has some limitations that should be pointed out. Firstly, a high number of physiological parameters are monitored during PSG, which requires appropriate and expensive

V. Barroso-García (✉) · G. C. Gutiérrez-Tobal
Biomedical Engineering Group, University of Valladolid, Valladolid, Spain

Centro de Investigación Biomédica en Red en Bioingeniería, Biomateriales y Nanomedicina (CIBER-BBN), Valladolid, Spain
e-mail: veronica.barroso@gib.tel.uva.es

J. Jiménez-García
Biomedical Engineering Group, University of Valladolid, Valladolid, Spain

R. Hornero
Biomedical Engineering Group, University of Valladolid, Valladolid, Spain

Centro de Investigación Biomédica en Red en Bioingeniería, Biomateriales y Nanomedicina (CIBER-BBN), Valladolid, Spain

Mathematics Research Institute of the University of Valladolid (IMUVa), Valladolid, Spain

T. Penzel, R. Hornero (eds.), *Advances in the Diagnosis and Treatment of Sleep Apnea*, Advances in Experimental Medicine and Biology 1384, https://doi.org/10.1007/978-3-031-06413-5_14

acquisition equipment (Collop et al., 2007; Ryan et al., 1995). In addition, specialized medical staff should be present during its performance. Then, they should visually assess all acquired recordings, which makes it a complex and intensive task (Collop et al., 2007; Ryan et al., 1995). Another limitation is that patient should spend a night hospitalized in a sleep unit. This involves sleeping in a different environment than usual, which can affect sleep development and characteristics (Bennett & Kinnear, 1999). Moreover, the large number of sensors attached to the patient's body can be very uncomfortable and even distressing, particularly for children (Jon, 2009). It should also be noted that not all hospitals have specialized sleep units, or these are overwhelmed by increasing demand. This fact hinders access to PSG, which causes long waiting lists, thereby leading to diagnosis and treatment delays (Alonso-Álvarez et al., 2015; Ghegan et al., 2006).

Accordingly, great efforts have been made to search and develop simpler alternative methods that help diagnose the sleep apnea. A common approach is to automatically analyze physiological signals with ability to reflect the particularities of the disease (Álvarez et al., 2020; Koley & Dey, 2013c). In this regard, apneic and hypopneic events are defined based on airflow reductions (Berry et al., 2012). When the respiratory cessation is partial (hypopnea), the amount of inspired and expired air is limited. Consequently, airflow (AF) signal experiences a reduction of between 30% and 90% (Berry et al., 2012). When the respiratory cessation is total (apnea), the airflow into the lungs is blocked, causing AF signal to present a $\geq$90% reduction and values ≈ 0 (Barroso-García et al., 2020; Berry et al., 2012). Hence, the analysis of this signal is a natural way of determining the presence and severity of the disease.

Regarding the overnight AF acquisition, the American Academy of Sleep Medicine (AASM) recommends using a thermistor sensor to suitably identify apneas and a nasal pressure sensor to identify hypopneas (Berry et al., 2012). Thermistor sensor is placed in the nostrils and mouth from patient to measure the difference of temperature between inspired (cold air) and expired air (warm air). In the case of nasal pressure sensor, it is also placed in the nostrils from patient to measure the pressure changes that occur during inspiration, when airway pressure is negative respect to atmospheric, and during expiration, when airway pressure becomes positive. Thereby, AF signal acquired by these sensors allows modeling the behavior of respiratory activity and detecting the abnormalities caused by apneic and hypopneic events (Berry et al., 2012). The specifications for routine PSG recordings indicate that AF should be acquired at a minimum sampling rate of 25 Hz, being 100 Hz the recommended rate (Iber et al., 2007).

AF signal can be obtained by means of a portable equipment with built-in thermistor sensor and/or nasal pressure sensor (Collop et al., 2007; Flemons et al., 2003; Tan et al., 2015). Thus, the required equipment is less expensive than for PSG as fewer signals are monitored (Collop et al., 2007). Moreover, the portable equipment can be used at patient's home, without disturbing their usual sleep patterns (Bennett & Kinnear, 1999). This test is also less uncomfortable due to a decreased number of sensors involved. Another advantage is that a single channel is analyzed (AF), resulting in a less complex and less time-consuming task (Ferber et al., 1994). All this would make the diagnostic test more accessible, which would reduce waiting lists and streamline diagnosis. Therefore, the use of AF is a potentially promising way for simplifying sleep apnea diagnosis.

All the above mentioned have led multiple works to be focused on the automatic analysis and characterization of the AF signal, in both pediatric and adult sleep apnea context (Gutiérrez-Tobal et al., 2021; Mendonca et al., 2019). These analyses are conducted from three different perspectives: (i) the evolution of AF signal in the time domain, (ii) its characterization in the frequency domain, and (iii) its study in the time–frequency domain. Thus, the main techniques used to analyze the behavior of AF in the presence of apneic events from these three methodological approaches are reviewed in Sects. 14.2, 14.3, and

14.4 of this chapter. In addition, Sect. 14.5 is devoted to studies that combine the aforementioned approaches. Finally, the discussion and conclusions of our study are presented in Sect. 14.6.

14.2 Analysis in Time Domain

As can be seen in Fig. 14.1, apneic events alter the behavior of AF in the time domain by causing significant reductions in its amplitude. Thus, several studies have focused on automatic detection of these events based on the temporal analysis of AF signal.

One of the first approaches to analyze the information of AF focused on the analysis of the instantaneous respiratory amplitude (IRA) and interval (IRI) signals, directly obtained from AF (Várady et al., 2002). Várady et al. (2002) used the raw AF and respiratory inductive plethysmography (RIP) signals and the IRA and IRI surrogates of them to discriminate between apnea, hypopnea, and normal breathing segments. An artificial neural network (ANN) fed with the IRI and IRA from AF was subsequently trained to perform the detection task.

Cabrero-Canosa et al. (2004) proposed an algorithm based on the identification of respiratory cycles and quantification of AF, in combination with the information provided by other biosignals (Cabrero-Canosa et al., 2004). In their study, symbolic classification was used to determine intervals of normal respiration and different types of airflow reduction (apnea, total reduction; hypopnea, clear reduction). These intervals were subsequently grouped and classified as apneic events or normal respiration.

Other approaches combined the information of AF and thoracic effort signals to detect and classify apneic events as obstructive, mixed, or central (Fontenla-Romero et al., 2005). The apneic segments were identified from AF signal by applying a moving average filter together with an adaptive threshold. Then, these segments were classified according to their origin using the additional information provided by the thoracic effort.

In the study of Pépin et al. (2009), an automatic time-domain analysis of the AF signal from a Holter device with an additional nasal pressure sensor was performed and compared to the medical specialists' annotations. The algorithm relied on the calculation of period, inspiratory surface, and maximum amplitude of breathing cycles. From the information provided by these features, amplitude reductions and cessations lasting at least 10 s were scored, and the apnea–hypopnea index (AHI) was automatically obtained.

The work of Álvarez-Estévez and Moret-Bonillo (2009) proposed the application of fuzzy reasoning methodology to detect apneic events from AF combined with other PSG-derived signals. The method relied on amplitude reductions of AF and oximetric desaturations to build reasoning units, which allow the fuzzy algorithm to determine if these reductions were actual apneic events.

Aydoğan et al. (2016) employed the nasal pressure AF jointly with the thoracic effort and the oxygen saturation (SpO_2) signals to evaluate two automatic scoring algorithms (Aydoğan et al., 2016). These algorithms calculated the

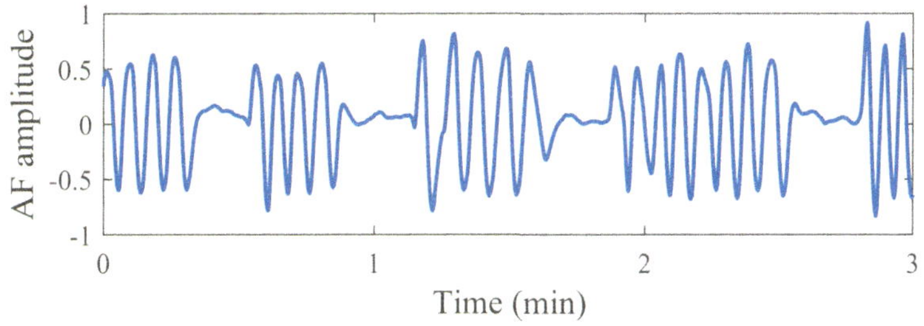

Fig. 14.1 Apneic events presented in overnight airflow (AF) signal. Sleep apnea causes reductions in AF amplitude (≥30%). Consequently, AF signal decreases toward 0 values during the occurrence of these events

mean of the absolute values and power from the signal to derive a rule-based method and an ANN, respectively. It is remarkable that the rule-based method obtained a slightly higher scoring accuracy than the ANN.

In order to predict the pre-apneic and regular breathing events, the work of Ozdemir et al. (2016) extracted 39 statistical and temporal features from the AF signal and its first derivative, since the latter can contribute to reduce false apneic detections (Ozdemir et al., 2016). These features were used to train and evaluate several classifiers, being the support vector machine (SVM) model the one that showed the highest diagnostic performance.

In the case of Kim et al. (2019) and Elmoaqet et al. (2020), they focused their studies on the characterization of the changes caused in the AF signal by the presence of apneic events. They developed an algorithm based on the location of peaks (relative maximum amplitude) and valleys (relative minimum amplitude) of oronasal AF. Then, peak-to-valley amplitudes and peak-to-peak intervals were computed. The apneic event detection relied on the comparison of these metrics in a baseline window and a consecutive detection window. The optimization of this framework was carried out both manually (Kim et al., 2019) and automatically using a Gaussian mixture model (Elmoaqet et al., 2020).

Envelope analysis is also a natural way to characterize the amplitude reductions of AF signal. Hence, several studies have focused on the estimation and analysis of the envelope of AF for detecting apneic events. This is the case of Selvaraj and Narasimhan (2013), who focused on the AF envelope analysis to reflect the changes produced by the apneic events in the respiration. They extracted the amplitude of the envelope and characterized it using three parameters: variability of the respiratory instantaneous amplitudes up to 0.4 Hz, the adaptive trend to quantify the very-low-frequency variations, and the dispersion of the amplitude in a 120-s window. In the study conducted by Diaz et al. (2014), the authors applied the Hilbert transform to derive a respiratory disturbance variable (RDV) from the coefficient of variation of the envelope (Diaz et al., 2014). The RDV was then used as a predictor variable in regression models aimed at estimating the AHI. Other approaches identified the apneas and hypopneas as an amplitude modulation of the normal respiration waveform in AF signal (Ciolek et al., 2015). The apnea detection algorithm relied on the envelope extraction using the following methods: square-law and Hilbert transform. In order to minimize distortions caused by these envelope detectors, standard and recursive median filtering were proposed in substitution of classical linear low-pass filters. The empirical mode decomposition (EMD) was also applied to AF signal to extract and subsequently analyze its envelope. Wang et al. (2019) derived the intrinsic mode functions (IMFs) by means of the EMD algorithm and computed the root-sum-square of the first four (IMFs) (Wang et al., 2019). Then, they obtained the instantaneous respiratory intensity signal and extracted the respiratory fluctuation index. In a recent study conducted by Uddin et al. (2021), a novel method was proposed to detect apneic events based on the analysis of the AF peak excursion (difference between upper and lower envelopes of AF) (Uddin et al., 2021). Thus, an adaptive thresholding was applied to the drops from the maximum peak excursion to determine the presence of apneas and hypopneas. The latter were scored when a drop $\geq$30% in AF was accompanied to a drop $\geq$3% in SpO_2, or a drop >2% during at least 20 s.

Among time-domain characterization techniques applied to AF, non-linear methods have been widely used in the sleep apnea context. The study of Kaimakamis et al. (2016) focused on predicting the AHI from a linear equation of non-linear variables (Kaimakamis et al., 2016). The non-linear variables were derived from the largest Lyapunov exponent (LLE), detrended fluctuation analysis (DFA), and approximate entropy (*ApEn*). Some of these non-linear features showed significant correlation with the AHI. Rathnayake et al. (2010) and Barroso-García et al. (2020) also proposed a methodology based on the non-linear analysis of AF signal. In the first of these studies, the authors segmented the AF signal and extracted several features derived from its corresponding recurrence plots (RPs) to obtain useful apnea-related information (Rathnayake et al., 2010). After, this information was used to compute the respiratory disturbance index (RDI) and discriminate between segments of apneic events and normal breathing in adults. In the case of Barroso-García et al. (2020), the RPs from AF signal were used to characterize the behavior of the pediatric overnight AF in the presence of apneas and hypopneas (Barroso-García et al., 2020). This characterization was carried out by computing up to 9 RP-derived features: 1 from the recurrence density, 5 from the diagonal structures, and 3 from the vertical structures of the RP. The results showed that sleep apnea modifies underlying dynamics and phase space of overnight AF. Particularly, apneic events reduce the variability, stationarity, and complexity of AF, as well as the exponential divergence of its phase space. In addition, this disease also increments the dwell time in the same phase space state, the mean prediction time, and the irregularity of pediatric AF.

14.3 Analysis in Frequency Domain

As can be seen in Fig. 14.2, the recurrent behavior of apneic and hypopneic events modifies the AF spectrum. This has led several studies to focus on the automatic processing of AF signal from a frequency point of view.

Nakano et al. (2007) proposed a method to detect apneas based on the analysis of the power spectrum of 12.8 s AF segments (Nakano et al., 2007). Thus, the AF recordings from 399 subjects were automatically processed to extract the flow power by means of the fast Fourier transform (FFT). Flow power decays in the respiratory band were associated to the presence of apneic events. Once these dips were detected in the overnight AF signal, the RDI was subsequently derived.

The work conducted by Gutiérrez-Tobal et al. (2015) was the first study that applied a spectral analysis to AF signals from pediatric subjects (Gutiérrez-Tobal et al., 2015). Thus, overnight AF was investigated by means of the power spectral density (PSD), and new spectral bands of interest were specifically defined for children: 0.119–0.192 Hz and 0.784–0.890 Hz. These frequency bands were characterized by calculating the maximum and minimum amplitude and first-to-fourth statistical moments. The results indicated that the spectral power in these bands is higher in the presence of sleep apnea, suggesting that the repetitive occurrence of apneic events modifies the spectral components of pediatric AF.

In order to overcome the limitations of classical spectral analysis, such as the assumption of stationarity and linearity, the work of Barroso-García et al. (2021a) proposed the bispectral analysis of AF signal (Barroso-García et al., 2021a). They defined a frequency band adapted to the normal respiratory rate of each pediatric subject. This band was characterized by computing up to 13 features derived from bispectrum: 3 from the amplitude of the bispectral band, 4 from the entropy of distribution, 4 from the bispectral moments of the band, and 2 from the weighted center of the bispectrum. The four types of bispectral features showed complementarity with each other. In addition, the obtained results suggest that the presence of sleep apnea reduces the non-gaussianity and the nonlinear interaction of harmonic components of AF, increments its irregularity, and displaces the activity to lower frequencies that are associated with apnea occurrence.

14.4 Time–Frequency Analysis

A common approach to characterize the presence of sleep apnea in AF is to conduct a time–frequency analysis employing the Hilbert–Huang transform (Fig. 14.3). This method applies an EMD process followed by the Hilbert spectrum computation. In this regard, Salisbury and Sun (2007) obtained the first and second IMFs and computed the Hilbert–Huang spectrum and its histogram in the frequency domain. Afterward, the apnea percentage was derived from the latter (Salisbury & Sun, 2007). Similarly, Caseiro et al. (2010) also employed the Hilbert–Huang transform and extracted features from the spectral histogram: frequency value in the first quarter, ratio between the first and the second halves, and ratio

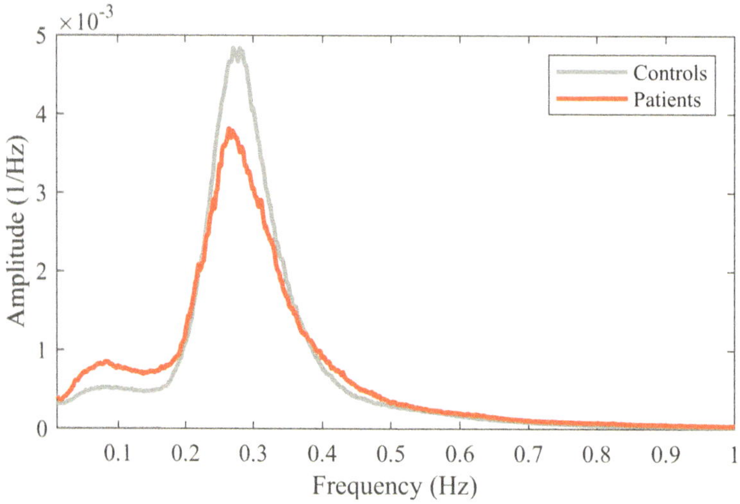

Fig. 14.2 Average spectrum of airflow (AF) signal from 974 subjects with and without sleep apnea (405 patients and 569 controls). The recurrence of apneic events leads to an AF spectrum with less power in the normal breathing band. This spectral power is redistributed in a wide range of frequency components. Particularly, the power increase at certain frequencies would be associated with the repeated occurrence of these event

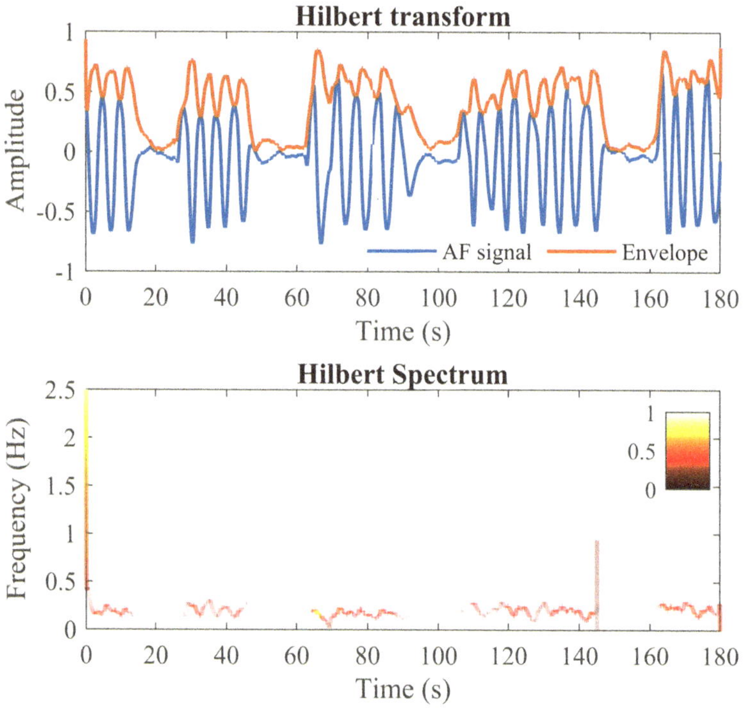

Fig. 14.3 Apneic events in the Hilbert envelope transform and spectrum of AF signal. Sleep apnea causes reductions in AF envelope. Consequently, Hilbert spectrum of AF vanishes during the occurrence of these events

between the maxima found in the first and second halves (Caseiro et al., 2010). Both methods were applied to the AF during an awake period rather than an overnight recording (Caseiro et al., 2010; Salisbury & Sun, 2007).

Another time–frequency mathematical tool is the wavelet transform. Several studies have applied this method to analyze the AF dynamics during sleep apnea. Kermit et al. (2000) applied the Haar wavelet to decompose the AF signal into

16 coefficients, which were directly used to feed a predictive model (Kermit et al., 2000). Further work from the same authors generated templates of apneas and normal respiration using a Haar wavelet decomposition. These templates were used to test the similarity of AF segments with the normal and apneic patterns (Kermit et al., 2002). Based on the continuous wavelet transform, Koley and Dey (2013a) used the cross wavelet analysis for the first time to differentiate between central, obstructive, and mixed apnea (Koley & Dey, 2013a). They assessed three pairwise combinations of signals to be analyzed via cross wavelet transform: AF, thoracic effort, and abdominal effort. The combination of the cross wavelet amplitude and phase coefficients from AF and thoracic effort showed to be more effective differentiating the apnea types. Avci and Akbaş (2015) also employed a wavelet decomposition approach to analyze respiratory signals (Avci & Akbaş, 2015). They selected the Daubechies wavelet family and extracted up to 8 features from the coefficients of 11 detail levels. In the same way, AF recordings from 946 children were analyzed by means of the discrete wavelet transform in Barroso-García et al. (2021b). They used the Haar and Daubechies mother wavelets for the AF analysis and extracted features from the eighth detail level. This detail level approximately corresponds to the normal breathing activity. Consequently, the AF reductions and cessations caused by sleep apnea modified the frequency components and the energy in this frequency band (0.195–0.391 Hz) (Barroso-García et al., 2021b).

McCloskey et al. (2018) employed wavelet spectrograms to generate graphic representations of AF that were analyzed by a convolutional neural network (CNN) for discriminating between normal breathing, apneic, and hypopneic events (McCloskey et al., 2018). The method was also compared with a 1D CNN trained with raw AF data. The 2D CNN trained with wavelet spectrograms outperformed the 1D CNN trained with raw AF data, thus highlighting the usefulness of the wavelet analysis. Similarly, Wu et al. (2021) proposed a method to detect apneic and hypopneic events from the AF signal by computing spectrograms with the short-time Fourier transform (STFT) (Wu et al., 2021). These spectrograms fed a CNN aimed at classifying 15-s segments as normal, hypopnea, and apnea.

14.5 Other Combined Approaches

In addition to the joint time–frequency analysis, other studies have shown the usefulness of combining the time and frequency domain information of the AF signal. In the study of Han et al. (2008), the mean magnitude of the second derivative (MMSD) of AF was analyzed, and an adaptive thresholding method was applied to detect apneic events (Han et al., 2008). The MMSD minimizes the contribution of baseline drifts and offset of the AF signal and, thus, is easier to interpret. The algorithm was designed and tested using the signals from 24 subjects. The normal respiration activity was also analyzed in the frequency domain between 0.2 Hz and 0.4 Hz to establish the reference behavior of the MMSD in normal respiration.

Novel time-domain features were proposed in Bricout et al. (2021), where non-periodic rate, low energy rate, and variance of the dispersion metric were analyzed along with statistical measurements, spectral power from the conventional heart rate variability (HRV) frequency bands (VLF, LF, HF), the ratio LF/HF, and the spectral kurtosis (Bricout et al., 2021). It is important to highlight that diagnostic performance of these features was higher using the AF signal from nasal pressure sensor than that obtained with a novel accelerometry sensor.

Koley and Dey (2013b, c) proposed apneic event detection models based on the analysis of short AF segments (Koley & Dey, 2013b, c). Statistical metrics were computed from the IRA and IRI signals, directly obtained from the airflow. The raw signal was also characterized by means of spectral (total and relative powers in the full spectrum, in the LF, and HF bands, respiratory frequency and its corresponding power, mean, and variance of the spectrum) and nonlinear features (*ApEn*, Lempel–Ziv complexity, LLE-derived features, Higuchi fractal dimension, and correlation dimension).

In the study conducted by Gutiérrez-Tobal et al. (2012, 2013), the authors investigated the diagnostic ability of the AF signal obtained by a thermistor and the respiratory rate variability (RRV) signal derived from AF (Gutiérrez-Tobal et al., 2012, 2013). These two signals were characterized using statistical, spectral, and non-linear features. Two spectral bands of interest were defined from AF and RRV: 0.022–0.059 Hz and 0.095–0.132 Hz, respectively. The features that obtained the highest diagnostic ability were the mean, standard deviation, peak amplitude, and power in the interest band of AF together with the central tendency measurement (*CTM*), skewness, and kurtosis of the full spectrum of RRV (Gutiérrez-Tobal et al., 2012). The combined use of AF and RRV improved the diagnostic performance reached by several classification and regression models to estimate the severity of sleep apnea (Gutiérrez-Tobal et al., 2012, 2013).

The combination of spectral and non-linear features from thermistor-recorded AF and RRV was also conducted in the pediatric sleep apnea context (Barroso-García et al., 2017). In this work, the spectral information was obtained using the first- to third-order spectral entropies (*SE*s) and the non-linear behavior by means of the *CTM*. These measurements characterized the irregularity (through *SE*) and variability (through *CTM*) of AF and RRV. The study showed the complementarity between both methodological approaches and that existing between both respiratory signals. The results suggest that the presence of apneic events reduce the variability and increase the irregularity of AF, while the variability of RRV is increased. The diagnostic ability of pediatric overnight AF was also assessed in combination with the nocturnal SpO_2 signal (Jiménez-García et al., 2020). The authors calculated time-domain statistics, spectral, and non-linear features from both AF and SpO_2, as well as the 3% oxygen desaturation index (*ODI3*). A spectral interest band of AF was obtained: 0.134–0.176 Hz, which is very similar to the low frequency band defined in the study of Gutiérrez-Tobal et al. (2015). A model combining the *CTM* of AF and *ODI3* obtained the highest diagnostic ability, suggesting that the variability of AF provides relevant and complementary information to the *ODI3* to diagnose pediatric sleep apnea.

Finally, other studies jointly employed time and frequency features from a nasal pressure signal. The work of Gutiérrez-Tobal et al. (2016) aimed to distinguish the different severity degrees of sleep apnea in adults. The authors defined a new band of interest characteristic of AF from nasal pressure sensor: 0.025–0.050 Hz, which covers the typical duration of apneic events (20–40 s) and matches the frequency band obtained using thermistor. A total of 12 features were extracted: 9 spectral and 3 non-linear features. The mean, standard deviation, minimum, and maximum from the frequency band, as well as the *CTM*, showed statistically significant differences among severity groups, suggesting that these approaches are useful to establish the severity degree of sleep apnea. Álvarez et al. (2020) combined the nasal pressure-derived AF with SpO_2 to evaluate the diagnostic ability of these two signals jointly (Álvarez et al., 2020). Both signals were characterized using time, spectral, and non-linear features, as well as clinical variables such as conventional oximetric indices and the respiratory disturbance index. The regression algorithm trained with features from both signals largely outperformed the individual diagnostic ability of these signals, suggesting that the information of AF and SpO_2 can be jointly used to diagnose sleep apnea.

14.6 Discussion

In this chapter, we have reviewed a variety of methodological approaches aimed at characterizing and extracting relevant information from the AF signal that can be used to help in the automatic diagnosis of sleep apnea. These approaches have been categorized from the four main perspectives: time domain, frequency domain, time–frequency analysis, and other combined strategies. Each of these perspectives focused on different characteristics that AF manifest in the presence of apneas and hypopneas. We have distinguished these particularities between children and adults throughout this section.

14.6.1 AF Characterization in Adults

The alterations caused by sleep apnea in the AF signal have motivated the development of algorithms for apneic event detection. Most of these algorithms were intended to obtain the localization of apneas and hypopneas by analyzing the changes of amplitude caused by respiratory cessations (Elmoaqet et al., 2020; Kim et al., 2019; Koley & Dey, 2013b). In this regard, algorithms based on the time-domain behavior of AF have been widely investigated. The reductions of amplitude in the AF signal led to the detection of apneic events in most of these algorithms (Uddin et al., 2021). In the same way, the amplitude reductions were characterized by the differences between AF peaks and valleys, as well as the differences between consecutive peaks, which are also reduced in the presence of apneas (Elmoaqet et al., 2020; Kim et al., 2019). Some authors have focused on envelope analysis (Diaz et al., 2014; Uddin et al., 2021). This is an intuitive way to track the amplitude of the AF in the time domain since it is narrowly related with the AASM manual scoring guidelines (Berry et al., 2012). A reduction of the amplitude level of the envelope with respect to the previous baseline described the presence of an apneic event (Ciolek et al., 2015; Selvaraj & Narasimhan, 2013). It is also observed that the apneas and hypopneas increase the long-term correlations of the AF and its irregularity in the time domain. Variability and complexity alterations are other particularities presented by sleep apnea in AF (Gutiérrez-Tobal et al., 2012, 2016). However, there is no consensus to it. While some studies observed a reduction in both variability and complexity, others characterized AF as more variable and more complex as severity increased, even using the same analysis techniques (Gutiérrez-Tobal et al., 2012, 2016).

Regarding frequency-domain approaches, the oscillatory pattern of the respiration and, therefore, the AF signal, have motivated the use of spectral analysis methods. The normal respiration generates activity in a narrow specific frequency band that is altered by the repeated occurrence of apneic events (Nakano et al., 2007). These respiratory bands ranges from 0.2 Hz to 0.4 Hz, which matches the normal breathing periods (every 2.5–5 s). However, the presence of sleep apnea leads to a redistribution of the spectral power, displacing the activity focus of AF to frequencies below the normal respiratory frequency (Gutiérrez-Tobal et al., 2012, 2013). This is also observed in the typical frequency range of the apneic events (around 0.04 Hz), where AF presents a higher spectral power, as well as a more asymmetric and peaked distribution of its frequency components as severity increased. In these cases, the spectral distribution of AF has higher statistical distance to the uniform distribution (Gutiérrez-Tobal et al., 2016).

These two previous approaches can be fused to exploit simultaneously their strengths in order to characterize the particularities of sleep apnea in the AF. Instantaneous variations of the respiratory activity due to apneas and hypopneas lead to changes in the spectrum of AF. The reviewed studies revealed that these changes can be analyzed using time–frequency approaches, since these can characterize spectral alterations that occur in short time intervals (Koley & Dey, 2013a). In this regard, it has been observed that the spectrogram of AF estimated by the STFT presents lower activity in the normal respiratory frequency during the occurrence of apneic events (Wu et al., 2021). This was also observed in the wavelet spectrograms, where the amplitude around the respiratory frequency is significantly reduced during the apneic/hypopneic events (McCloskey et al., 2018). Nevertheless, an exhaustive characterization of the AF signal using the wavelet transform is still lacking in adults.

14.6.2 AF Characterization in Children

The analysis of the AF signal has been much less studied in children than in adults, and some studies suggest that the diagnosis of sleep apnea in children is more challenging due to more restrictive criteria to define apneic events and severity degrees (Rosen et al., 2012). In view of the reviewed studies, apneic events reduce the variability, stationarity, and complexity of AF (Barroso-García et al., 2017, 2020; Jiménez-García et al., 2020). Moreover, when the AF was characterized in the phase space, the exponential divergence was reduced as the severity of sleep apnea increased (Barroso-García et al., 2020). This disease also increments the dwell time in the same phase space state, the mean prediction time, and the irregularity of pediatric AF in the time domain (Barroso-García et al., 2020).

The pediatric overnight AF also experiences variations in the frequency domain (Gutiérrez-Tobal et al., 2015; Jiménez-García et al., 2020). As in the case of adults, the spectral power in specific low frequency bands is higher in the presence of sleep apnea. This suggests that the recurrence of apneic events displaces the spectral power of AF to frequencies below the normal respiratory frequency (Gutiérrez-Tobal et al., 2015). By extension of the classic spectral analysis, the bispectrum also revealed that the severity of the disease localizes more activity around lower frequencies associated with apnea occurrence (Barroso-García et al., 2021a). Moreover, the pediatric AF has a more gaussian behavior as the severity of sleep apnea increases. In addition, the non-linear interaction of harmonic components of AF is reduced in the presence of apneic events, leading to lower phase coupling in the normal breathing band (Barroso-García et al., 2021a).

As far as we know, the combined time–frequency approach was only explored using the wavelet transform (Barroso-García et al., 2021b). In this case, the wavelet coefficients in the detail level related to the normal breathing are reduced as the severity degree of sleep apnea increases. At the same time, the distribution of these wavelet coefficients is more skewed and peaked around lower values. This is in accordance with the reduction of the energy in the frequency band related to normal respiration (Barroso-García et al., 2021b).

Lastly, it was observed that the AF characterization can improve using of a combination of methodological approaches both in adults and children (Álvarez et al., 2020; Barroso-García et al., 2017; Jiménez-García et al., 2020; Koley & Dey, 2013b, c). According to the reviewed studies, the joint use of different analyses is able to provide useful and complementary information to aid in the detection of sleep apnea. This combinational approach has also been applied to the analysis of AF along with other cardiorespiratory signals (Álvarez et al., 2020; Aydoğan et al., 2016; Cabrero-Canosa et al., 2004; Jiménez-García et al., 2020). These studies show that other signals can be complementary to AF and enhance its diagnostic ability.

14.7 Conclusions

In view of the results, we can conclude that the overnight AF successfully reflects the particularities caused by the occurrence of apneic and hypopneic events. The automatic signal processing methods provide useful information to define AF-based biomarkers for characterizing and helping in the diagnosis of this disease.

Regarding future research directions on the AF signal analysis in the sleep apnea context, deep-learning methods have revolutionized the automatic diagnosis of this disease in the last few years. It is true that these techniques are more focused on the classification tasks (apneic events versus normal respiration, or sleep apnea severity degree), rather than for the characterization of AF signal. However, explainable artificial intelligence (XAI) methods are expected to clarify the functional interpretation of deep-learning models, identify novel relevant information from AF signal, and thus improve the diagnosis of sleep apnea in future studies.

Acknowledgments This research has been developed under the grants PID2020-115468RB-I00 and PDC2021-120775-I00 funded by "Ministerio de Ciencia e Innovación/Agencia Estatal de Investigación/10.13039/501100011033/" and ERDF A way of making Europe and by "CIBER en Bioingeniería, Biomateriales y Nanomedicina (CIBER-BBN)" through "Instituto de Salud Carlos III" co-funded with ERDF funds. J. Jiménez-García was in receipt of a PIF grant by the "University of Valladolid."

References

Alonso-Álvarez, M. L., Terán-Santos, J., Ordax Carbajo, E., Cordero-Guevara, J. A., Navazo-Egüia, A. I., Kheirandish-Gozal, L., & Gozal, D. (2015). Reliability of home respiratory polygraphy for the diagnosis of sleep apnea in children. *Chest, 147*(4), 1020–1028. https://doi.org/10.1378/chest.14-1959

Álvarez, D., Cerezo-Hernández, A., Crespo, A., Gutiérrez-Tobal, G. C., Vaquerizo-Villar, F., Barroso-García, V., Moreno, F., Arroyo, C. A., Ruiz, T., Hornero, R., & del Campo, F. (2020). A machine learning-based test for adult sleep apnoea screening at home using oximetry and airflow. *Scientific Reports, 10*(1), 1–12. https://doi.org/10.1038/s41598-020-62223-4

Álvarez-Estévez, D., & Moret-Bonillo, V. (2009). Fuzzy reasoning used to detect apneic events in the sleep apnea-hypopnea syndrome. *Expert Systems with Applications, 36*(4), 7778–7785. https://doi.org/10.1016/j.eswa.2008.11.043

Avci, C., & Akbaş, A. (2015). Sleep apnea classification based on respiration signals by using ensemble methods. *Bio-medical Materials and Engineering, 26*, S1703–S1710. https://doi.org/10.3233/BME-151470

Aydoğan, O., Öter, A., Güney, K., Kıymık, M. K., & Tuncel, D. (2016). Automatic diagnosis of obstructive sleep apnea/hypopnea events using respiratory signals. *Journal of Medical Systems, 40*(12), 1–8. https://doi.org/10.1007/s10916-016-0624-0

Barroso-García, V., Gutiérrez-Tobal, G. C., Kheirandish-Gozal, L., Álvarez, D., Vaquerizo-Villar, F., Crespo, A., del Campo, F., Gozal, D., & Hornero, R. (2017). Irregularity and variability analysis of airflow recordings to facilitate the diagnosis of paediatric sleep apnoea-hypopnoea syndrome. *Entropy, 19*(9), 447. https://doi.org/10.3390/e19090447

Barroso-García, V., Gutiérrez-Tobal, G. C., Gozal, D., Álvarez, D., Vaquerizo-Villar, F., Núñez, P., del Campo, F., Kheirandish-Gozal, L., & Hornero, R. (2020). Usefulness of recurrence plots from airflow recordings to aid in paediatric sleep apnoea diagnosis. *Computer Methods and Programs in Biomedicine, 183*, 105083. https://doi.org/10.1016/j.cmpb.2019.105083

Barroso-García, V., Gutiérrez-Tobal, G. C., Kheirandish-Gozal, L., Vaquerizo-Villar, F., Álvarez, D., del Campo, F., Gozal, D., & Hornero, R. (2021a). Bispectral analysis of overnight airflow to improve the pediatric sleep apnea diagnosis. *Computers in Biology and Medicine, 129*, 104167. https://doi.org/10.1016/j.compbiomed.2020.104167

Barroso-García, V., Gutiérrez-Tobal, G. C., Gozal, D., Vaquerizo-Villar, F., Álvarez, D., del Campo, F., Kheirandish-Gozal, L., & Hornero, R. (2021b). Wavelet analysis of overnight airflow to detect obstructive sleep apnea in children. *Sensors, 21*(4), 1–19. https://doi.org/10.3390/s21041491

Bennett, J. A., & Kinnear, W. J. M. (1999). Sleep on the cheap: The role of overnight oximetry in the diagnosis of sleep apnoea hypopnoea syndrome. *Thorax, 54*(11), 958–959. https://doi.org/10.1136/THX.54.11.958

Berry, R. B., Budhiraja, R., Gottlieb, D. J., Gozal, D., Iber, C., Kapur, V. K., Marcus, C. L., Mehra, R., Parthasarathy, S., Quan, S. F., Redline, S., Strohl, K. P., Ward, S. L. D., & Tangredi, M. M. (2012). Rules for scoring respiratory events in sleep: Update of the 2007 AASM manual for the scoring of sleep and associated events. *Journal of Clinical Sleep Medicine, 08*(05), 597–619. https://doi.org/10.5664/jcsm.2172

Bricout, A., Fontecave-Jallon, J., Pépin, J. L., & Guméry, P. Y. (2021). Accelerometry-derived respiratory index estimating apnea-hypopnea index for sleep apnea screening. *Computer Methods and Programs in Biomedicine, 207*, 106209. https://doi.org/10.1016/j.cmpb.2021.106209

Cabrero-Canosa, M., Hernandez-Pereira, E., & Moret-Bonillo, V. (2004). Intelligent diagnosis of sleep apnea syndrome. *IEEE Engineering in Medicine and Biology Magazine, 23*(2), 72–81. https://doi.org/10.1109/MEMB.2004.1310978

Caseiro, P., Fonseca-Pinto, R., & Andrade, A. (2010). Screening of obstructive sleep apnea using Hilbert-Huang decomposition of oronasal airway pressure recordings. *Medical Engineering and Physics, 32*(6), 561–568. https://doi.org/10.1016/j.medengphy.2010.01.008

Ciolek, M., Niedzwiecki, M., Sieklicki, S., Drozdowski, J., & Siebert, J. (2015). Automated detection of sleep apnea and hypopnea events based on robust airflow envelope tracking in the presence of breathing artifacts. *IEEE Journal of Biomedical and Health Informatics, 19*(2), 418–429. https://doi.org/10.1109/JBHI.2014.2325997

Collop, N. A., Anderson, W. M. D., Boehlecke, B., Claman, D., Goldberg, R., Gottlieb, D. J., Hudgel, D., Sateia, M., & Schwab, R. (2007). Clinical guidelines for the use of unattended portable monitors in the diagnosis of obstructive sleep apnea in adult patients. *Journal of Clinical Sleep Medicine, 3*(7), 737–747. American Academy of Sleep Medicine. https://doi.org/10.5664/jcsm.27032

Diaz, J. A., Arancibia, J. M., Bassi, A., & Vivaldi, E. A. (2014). Envelope analysis of the airflow signal to improve polysomnographic assessment of sleep disordered breathing. *Sleep, 37*(1), 199–208. https://doi.org/10.5665/sleep.3338

Elmoaqet, H., Kim, J., Tilbury, D., Ramachandran, S. K., Ryalat, M., & Chu, C. H. (2020). Gaussian mixture models for detecting sleep apnea events using single oronasal airflow record. *Applied Sciences, 10*(21), 1–15. https://doi.org/10.3390/app10217889

Ferber, R., Millman, R., Coppola, M., Fleetham, J., Murray, C. F., Iber, C., McCall, V., Pressman, M., Sanders, M., Strohl, K., Votteri, B., & Williams, A. (1994). ASDA standards of practice: Portable recording in the assessment of obstructive sleep apnea. *Sleep, 17*(4), 378–392. Associated Professional Sleep Societies, LLC. https://doi.org/10.1093/sleep/17.4.378

Flemons, W. W., Littner, M. R., Rowley, J. A., Gay, P., Anderson, W. M. D., Hudgel, D. W., McEvoy, R. D., & Loube, D. I. (2003). Home diagnosis of sleep apnea: A systematic review of the literature - An evidence review cosponsored by the American Academy of Sleep Medicine, the American College of Chest Physicians, and the American Thoracic Society. *Chest, 124*(4), 1543–1579. https://doi.org/10.1378/chest.124.4.1543

Fontenla-Romero, O., Guijarro-Berdiñas, B., Alonso-Betanzos, A., & Moret-Bonillo, V. (2005). A new method for sleep apnea classification using wavelets and feedforward neural networks. *Artificial Intelligence in Medicine, 34*(1), 65–76. https://doi.org/10.1016/j.artmed.2004.07.014

Ghegan, M. D., Angelos, P. C., Stonebraker, A. C., & Gillespie, M. B. (2006). Laboratory versus portable sleep studies: A meta-analysis. *The Laryngoscope, 116*, 859–864. https://doi.org/10.1097/01.mlg.0000214866.32050.2e

Gutiérrez-Tobal, G. C., Hornero, R., Álvarez, D., Marcos, J. V. V., & Del Campo, F. (2012). Linear and nonlinear analysis of airflow recordings to help in sleep apnoea–hypopnoea syndrome diagnosis. *Physiological Measurement, 33*(7), 1261–1275. https://doi.org/10.1088/0967-3334/33/7/1261

Gutiérrez-Tobal, G. C., Álvarez, D., Marcos, J. V., Del Campo, F., & Hornero, R. (2013). Pattern recognition in airflow recordings to assist in the sleep apnoea-hypopnoea syndrome diagnosis. *Medical and Biological Engineering and Computing, 51*(12), 1367–1380. https://doi.org/10.1007/s11517-013-1109-7

Gutiérrez-Tobal, G. C., Alonso-Álvarez, M. L., Álvarez, D., Del Campo, F., Terán-Santos, J., & Hornero, R. (2015). Diagnosis of pediatric obstructive sleep apnea: Preliminary findings using automatic analysis of airflow and oximetry recordings obtained at patients' home. *Biomedical Signal Processing and Control, 18*, 401–407. https://doi.org/10.1016/j.bspc.2015.02.014

Gutiérrez-Tobal, G. C., Álvarez, D., del Campo, F., & Hornero, R. (2016). Utility of adaboost to detect sleep apnea-hypopnea syndrome from single-channel airflow. *IEEE Transactions on Biomedical Engineering, 63*(3), 636–646. https://doi.org/10.1109/TBME.2015.2467188

Gutiérrez-Tobal, G. C., Álvarez, D., Kheirandish-Gozal, L., del Campo, F., Gozal, D., & Hornero, R. (2021). Reliability of machine learning to diagnose pediatric obstructive sleep apnea: Systematic review and meta-analysis. *Pediatric Pulmonology, February* (pp. 1–13). https://doi.org/10.1002/ppul.25423

Han, J., Shin, H. B., Jeong, D. U., & Park, K. S. (2008). Detection of apneic events from single channel nasal airflow using 2nd derivative method. *Computer Methods and Programs in Biomedicine, 91*(3), 199–207. https://doi.org/10.1016/j.cmpb.2008.04.012

Iber, C., Ancou-Israel, S., Chesson, A. L., & Quan, S. F. (2007). *The AASM manual for the scoring of sleep and associated events rules, terminology and technical specifications*. https://j2vjt3dnbra3ps7ll1clb4q2--wpengine.netdna-ssl.com/wp-content/uploads/2018/04/Summary-of-Updates-in-v2.5-1.pdf

Jiménez-García, J., Gutiérrez-Tobal, G. C., García, M., Kheirandish-Gozal, L., Martín-Montero, A., Álvarez, D., del Campo, F., Gozal, D., & Hornero, R. (2020). Assessment of airflow and oximetry signals to detect pediatric sleep apnea-hypopnea syndrome using adaboost. *Entropy, 22*(6), 670. https://doi.org/10.3390/e22060670

Jon, C. (2009). Polysomnography in children. In R. B. Mitchell & K. D. Pereira (Eds.), *Pediatric otolaryngology for the clinician* (pp. 35–47). Humana Press. https://doi.org/10.1007/978-1-60327-127-1_5

Kaimakamis, E., Tsara, V., Bratsas, C., Sichletidis, L., Karvounis, C., & Maglaveras, N. (2016). Evaluation of a decision support system for obstructive sleep apnea with nonlinear analysis of respiratory signals. *PLoS One, 11*(3), 1–16. https://doi.org/10.1371/journal.pone.0150163

Kermit, M., Eide, Å. J., Lindblad, T., & Waldemark, K. (2000). Treatment of obstructive sleep apnea syndrome by monitoring patients airflow signals. *Pattern Recognition Letters, 21*(3), 277–281. https://doi.org/10.1016/S0167-8655(99)00157-9

Kermit, M., Eide, Å. J., & Lindblad, T. (2002). Early online detection of upper airway obstructions in obstructive sleep apnoea syndrome (OSAS) patients. *Journal of Medical Engineering and Technology, 26*(6), 259–264. https://doi.org/10.1080/03091900210156814

Kim, J., Elmoaqet, H., Tilbury, D. M., Ramachandran, S. K., & Penzel, T. (2019). Time domain characterization for sleep apnea in oronasal airflow signal: A dynamic threshold classification approach. *Physiological Measurement, 40*(5), 054007. https://doi.org/10.1088/1361-6579/aaf4a9

Koley, B. L., & Dey, D. (2013a). Classification of sleep apnea using cross wavelet transform. In *2013 IEEE 1st international conference on condition assessment techniques in electrical systems, IEEE CATCON 2013 - Proceedings* (pp. 275–280). https://doi.org/10.1109/CATCON.2013.6737512

Koley, B. L., & Dey, D. (2013b). Real-time adaptive apnea and hypopnea event detection methodology for portable sleep apnea monitoring devices. *IEEE Transactions on Biomedical Engineering, 60*(12), 3354–3363. https://doi.org/10.1109/TBME.2013.2282337

Koley, B. L., & Dey, D. (2013c). Automatic detection of sleep apnea and hypopnea events from single channel measurement of respiration signal employing ensemble binary SVM classifiers. *Measurement: Journal of the International Measurement Confederation, 46*(7), 2082–2092. https://doi.org/10.1016/j.measurement.2013.03.016

McCloskey, S., Haidar, R., Koprinska, I., & Jeffries, B. (2018). Detecting hypopnea and obstructive apnea events using convolutional neural networks on wavelet spectrograms of nasal airflow. In *Lecture Notes in Computer Science (including subseries Lecture Notes in Artificial Intelligence and Lecture Notes in Bioinformatics): Vol. 10937 LNAI*. Springer International Publishing. https://doi.org/10.1007/978-3-319-93034-3_29

Mendonca, F., Mostafa, S. S., Ravelo-Garcia, A. G., Morgado-Dias, F., & Penzel, T. (2019). A review of obstructive sleep apnea detection approaches. *IEEE Journal of Biomedical and Health Informatics, 23*(2), 825–837. https://doi.org/10.1109/JBHI.2018.2823265

Nakano, H., Tanigawa, T., Furukawa, T., & Nishima, S. (2007). Automatic detection of sleep-disordered breathing from a single-channel airflow record. *European Respiratory Journal, 29*(4), 728–736.

Ozdemir, G., Nasifoglu, H., & Erogul, O. (2016). A time-series approach to predict Obstructive Sleep Apnea (OSA) episodes. In *Proceedings of the 2nd world congress on electrical engineering and computer systems and science* (pp. 1–8). https://doi.org/10.11159/icbes16.117

Patil, S. P., Schneider, H., Schwartz, A. R., & Smith, P. L. (2007). Adult obstructive sleep apnea: Pathophysiology and diagnosis. *Chest, 132*(1), 325–337. https://doi.org/10.1378/CHEST.07-0040

Pépin, J. L., Defaye, P., Vincent, E., Christophle-Boulard, S., Tamisier, R., & Lévy, P. (2009). Sleep apnea diagnosis using an ECG Holter device including a nasal pressure (NP) recording: Validation of visual and automatic analysis of nasal pressure versus full polysomnography. *Sleep Medicine, 10*(6), 651–656. https://doi.org/10.1016/j.sleep.2008.07.002

Rathnayake, S. I., Wood, I. A., Abeyratne, U. R., & Hukins, C. (2010). Nonlinear features for single-channel diagnosis of sleep-disordered breathing diseases. *IEEE Transactions on Biomedical Engineering, 57*(8), 1973–1981.

Rosen, C. L., D'Andrea, L., & Haddad, G. G. (2012). Adult criteria for obstructive sleep apnea do not identify children with serious obstruction. *The American Review of Respiratory Disease, 146*(5), 1231–1234. https://doi.org/10.1164/AJRCCM/146.5_PT_1.1231

Ryan, P. J., Hilton, M. F., Boldy, D. A., Evans, A., Bradbury, S., Sapiano, S., Prowse, K., & Cayton, R. M. (1995). Validation of British Thoracic Society guidelines for the diagnosis of the sleep apnoea/hypopnoea syndrome: Can polysomnography be avoided? *Thorax, 50*(9), 972–975. https://doi.org/10.1136/thx.50.9.972

Salisbury, J. I., & Sun, Y. (2007). Rapid screening test for sleep apnea using a nonlinear and nonstationary signal processing technique. *Medical Engineering and Physics, 29*(3), 336–343. https://doi.org/10.1016/j.medengphy.2006.05.013

Selvaraj, N., & Narasimhan, R. (2013). Detection of sleep apnea on a per-second basis using respiratory signals. In *Proceedings of the annual international conference of the IEEE Engineering in Medicine and Biology Society, EMBS* (pp. 2124–2127). https://doi.org/10.1109/EMBC.2013.6609953

Tan, H. L., Kheirandish-Gozal, L., & Gozal, D. (2015). Pediatric home sleep apnea testing slowly getting there! *Chest, 148*(6), 1382–1395. https://doi.org/10.1378/chest.15-1365

Uddin, M. B., Chow, C. M., Ling, S. H., & Su, S. W. (2021). A novel algorithm for automatic diagnosis of sleep apnea from airflow and oximetry signals. *Physiological Measurement, 42*(1), 015001. https://doi.org/10.1088/1361-6579/abd238

Várady, P., Micsik, T., Benedek, S., & Benyó, Z. (2002). A novel method for the detection of apnea and hypopnea events in respiration signals. *IEEE Transactions on Biomedical Engineering, 49*(9), 936–942. https://doi.org/10.1109/TBME.2002.802009

Wang, F. T., Hsu, M. H., Fang, S. C., Chuang, L. L., & Chan, H. L. (2019). The respiratory fluctuation index: A global metric of nasal airflow or thoracoabdominal wall movement time series to diagnose obstructive sleep apnea. *Biomedical Signal Processing and Control, 49*, 250–262. https://doi.org/10.1016/j.bspc.2018.12.015

Wu, Y., Pang, X., Zhao, G., Yue, H., Lei, W., & Wang, Y. (2021). A novel approach to diagnose sleep apnea using enhanced frequency extraction network. *Computer Methods and Programs in Biomedicine, 206*, 106119. https://doi.org/10.1016/j.cmpb.2021.106119

15 Deep-Learning Model Based on Convolutional Neural Networks to Classify Apnea–Hypopnea Events from the Oximetry Signal

Fernando Vaquerizo-Villar, Daniel Álvarez, Gonzalo C. Gutiérrez-Tobal, C. A. Arroyo-Domingo, F. del Campo, and Roberto Hornero

Abstract

Automated analysis of the blood oxygen saturation (SpO_2) signal from nocturnal oximetry has shown usefulness to simplify the diagnosis of obstructive sleep apnea (OSA), including the detection of respiratory events. However, the few preceding studies using SpO_2 recordings have focused on the automated detection of respiratory events versus normal respiration, without making any distinction between apneas and hypopneas. In this sense, the characteristics of oxygen desaturations differ between obstructive apnea and hypopnea episodes. In this chapter, we use the SpO_2 signal along with a convolutional neural network (CNN)-based deep-learning architecture for the automatic identification of apnea and hypopnea events. A total of 398 SpO_2 signals from adult OSA patients were used for this purpose. A CNN architecture was trained using 30-s epochs from the SpO_2 signal for the automatic classification of three classes: normal respiration, apnea, and hypopnea. Then, the apnea index (AI), the hypopnea index (HI), and the apnea–hypopnea index (AHI) were obtained by aggregating the outputs of the CNN for each subject (AI_{CNN}, HI_{CNN}, and AHI_{CNN}). This model showed a promising diagnostic performance in an independent test set, with 80.3% 3-class accuracy and 0.539 3-class Cohen's kappa for the classification of respiratory events. Furthermore, AI_{CNN}, HI_{CNN}, and AHI_{CNN} showed a high agreement with the values obtained from the standard PSG: 0.8023, 0.6774, and 0.8466 intra-class corre-

F. Vaquerizo-Villar (✉) · G. C. Gutiérrez-Tobal
R. Hornero
Biomedical Engineering Group,
University of Valladolid, Valladolid, Spain

Centro de Investigación Biomédica en Red en Bioingeniería, Biomateriales y Nanomedicina (CIBER-BBN), Instituto de Salud Carlos III, Valladolid, Spain
e-mail: fernando.vaquerizo@gib.tel.uva.es

D. Álvarez
Biomedical Engineering Group, University of Valladolid, Valladolid, Spain

Centro de Investigación Biomédica en Red en Bioingeniería, Biomateriales y Nanomedicina (CIBER-BBN), Instituto de Salud Carlos III, Valladolid, Spain

Pneumology Service, Río Hortega University Hospital, Valladolid, Spain

C. A. Arroyo-Domingo · F. del Campo
Centro de Investigación Biomédica en Red en Bioingeniería, Biomateriales y Nanomedicina (CIBER-BBN), Instituto de Salud Carlos III, Valladolid, Spain

Pneumology Service, Río Hortega University Hospital, Valladolid, Spain

T. Penzel, R. Hornero (eds.), *Advances in the Diagnosis and Treatment of Sleep Apnea*, Advances in Experimental Medicine and Biology 1384, https://doi.org/10.1007/978-3-031-06413-5_15

lation coefficients (ICCs), respectively. This suggests that CNN can be used to analyze SpO_2 recordings for the automated diagnosis of OSA in at-home oximetry tests.

Keywords

Apnea · Apnea index (AI) · Apnea–hypopnea index (AHI) · Blood oxygen saturation (SpO_2) · Convolutional neural networks (CNN) · Deep learning · Hypopnea · Hypopnea index (HI) · Obstructive sleep apnea (OSA) · Oximetry

15.1 Introduction

Obstructive sleep apnea (OSA) has become a major issue in recent years (Senaratna et al., 2017). OSA is marked by recurrent episodes of apneas (complete absences of airflow) and hypopneas (considerable reductions of airflow), which leads to fragmented and restless sleep (Berry et al., 2012). Despite its high prevalence in the adult population (9–38%), OSA is an underdiagnosed condition (Benjafield et al., 2020; Senaratna et al., 2017). This contributes to an increased risk of cardiovascular, metabolic, and psychiatric alterations, such as hypertension, cerebrovascular diseases, diabetes, and depression (Eastwood et al., 2010; Park et al., 2011).

Despite serving as the gold standard for OSA diagnosis, overnight polysomnography (PSG) presents important limitations. PSG is a costly test, highly intrusive for the patients, and technically complex and lacks availability (del Campo et al., 2018; Redline, 2017). In addition, apneas and hypopneas must be manually annotated by trained specialists, which is labor intensive and may lead to errors and inconsistencies in the diagnosis (Shokoueinejad et al., 2017). In order to overcome these PSG limitations, multiple investigations have focused on the use of simplified approaches aimed at the automated detection of OSA from a reduced subset of cardiorespiratory signals. Among these approaches, the automated analysis of the single-channel blood oxygen saturation (SpO_2) signal from nocturnal oximetry has been frequently proposed due to its easy acquisition and interpretation (del Campo et al., 2018). SpO_2 signal provides a continuous measure of the oxygen content in the hemoglobin (McClatchey, 2002), which allows to detect oxygen desaturations induced by OSA-related respiratory events, i.e., apneas and hypopneas (Berry et al., 2012).

Different studies have examined the SpO_2 signal as a simplified alternative to PSG in the automated detection of respiratory events and in the automated diagnosis of OSA (del Campo et al., 2018). A majority of these studies have followed conventional feature-engineering methodologies, which are based on feature extraction and selection stages (del Campo et al., 2018). Nonetheless, these methodologies require a substantial human-based knowledge to identify, a priori, a set of relevant features to extract from the signal under study (Goodfellow et al., 2016), which limits its ability to obtain all the ad-hoc information from the SpO_2 recordings related to respiratory events. This limitation can be overcome by deep-learning methods, which can directly analyze raw data and automatically make decisions based on non-human-driven knowledge (Faust et al., 2018; Goodfellow et al., 2016).

In the last few years, deep-learning algorithms have outperformed conventional approaches in many fields (Goodfellow et al., 2016), such as image recognition, autonomous driving, natural language processing, and time series analysis (Faust et al., 2018; Goodfellow et al., 2016). In the OSA context, recent studies have demonstrated the usefulness of deep-learning approaches to analyze cardiorespiratory signals in the automated detection of apneic events (Mostafa et al., 2019). Particularly, Mostafa et al. (2020a, b) and Vaquerizo-Villar et al. (2019) applied a deep-learning architecture based on convolutional neural networks (CNNs) to the oximetry signal to detect respiratory events in adult and pediatric OSA patients, respectively. However, these studies have only addressed the automated detection of respiratory events versus normal respiration, without making any distinction between apneas and hypopneas (Mostafa, Baptista, et al., 2020a; Mostafa, Mendonca, et al., 2020b; Vaquerizo-

Villar et al., 2019). Conversely, Kulkas et al. (2017) stated that the severity of oxygen desaturations differs between obstructive apnea and hypopnea events.

In the present chapter, a CNN architecture is proposed to automatically identify apnea and hypopnea events. Despite being originally designed for image analysis (Goodfellow et al., 2016), CNNs have become one of the most relevant deep-learning methods for time series classification (Ismail Fawaz et al., 2019) in many fields, including biomedical signal processing (Ebrahimi et al., 2020; Faust et al., 2019; Murat et al., 2020; Roy et al., 2019). Accordingly, we hypothesized that a deep-learning architecture based on CNNs could help to automatically learn the most relevant information from the oximetry signal in the detection and classification of apnea and hypopnea events. Consequently, the main objective of this chapter is to design and evaluate a deep-learning model based on CNNs to automatically classify respiratory events from the SpO_2 signal in OSA patients. In addition, the secondary goal of this research is to assess the usefulness of the CNN model to estimate the apnea–hypopnea index (AHI: the number of apneas and hypopneas per sleep hour), which is the clinical parameter used to establish OSA diagnosis.

15.2 Materials and Methods

15.2.1 Subjects and Signals

This chapter involved a database composed of 398 adult patients diagnosed with OSA (AHI ≥ 5 events per hour). All of them were referred to the sleep laboratory of the Hospital Universitario Río Hortega (Valladolid, Spain), where they underwent overnight PSG. The Ethics and Drugs Research Committee of the hospital approved the protocol (CEIm 47/16).

All subjects were diagnosed by medical specialists following the standards of the American Academy of Sleep Medicine (AASM) (Berry et al., 2012). Accordingly, an episode of apnea was annotated when there was a drop in the amplitude of the oronasal thermal airflow signal higher than 90% during at least 10 seconds (Berry et al., 2012). Similarly, a hypopnea was scored when there was a minimum of 30% reduction in the amplitude of the nasal pressure airflow signal, lasting at least 10 seconds and accompanied by an oxygen desaturation of at least 3% or/and an electroencephalographic arousal (Berry et al., 2012). Subsequently, the apnea index (AI: the number of apneas per hour), hypopnea index (HI: the number of hypopneas per hour), and AHI from each subject were computed as the total number of each type of event divided by the total sleep time.

SpO_2 signals were acquired during PSG at a sampling rate of 16 Hz. In order to reduce the computational requirements, all the SpO_2 recordings were downsampled to a sample rate of 1 Hz. SpO_2 recordings from each subject were then divided into 30-second non-overlapping epochs, being each epoch labelled as normal respiration (N), apnea (A), or hypopnea (H) using the annotations provided by the clinicians. The dataset was divided into three groups: training set (first 199 subjects, 50%), employed to train the CNN architecture; validation set (the following 79 subjects, 20%), used to monitor the convergence of the CNN; and test set (the last 120 subjects, 30%), employed to evaluate the proposed CNN-based methodology. Table 15.1 summarizes polysomnographic and clinical data from the population under study. No statistically significant differences (p-value < 0.05) were found in age, sex, body mass index (BMI), AI, HI, or AHI between the three groups.

15.2.2 Proposed CNN Architecture

Figure 15.1 shows the main components of the proposed CNN architecture. The input section of the CNN consists of the SpO_2 samples for the 30-s epoch (i.e., 30 samples) to be classified, concatenated with the four preceding and the five following epochs, thus having a 10-epoch length (300 samples) 1D input vector. The reason for using preceding and following epochs is twofold: (i) it enhances the identification of oxygen

Table 15.1 Clinical and polysomnographic data of the population under study

	All	Training	Validation	Test
Clinical characteristics in each data subset				
Subjects (*n*)	398	199	79	120
Age (years)	56 [47–65]	56 [47–64]	56 [47–67]	55 [44–65]
Males (*n*)	278 (69.9%)	144 (72.4%)	59 (74.6%)	75 (62.5%)
BMI (kg/m^2)	29.1 [26.1–33.1]	29.4 [26.0–33.5]	28.9 [26.8–31.7]	29.2 [25.6–34.0]
AI (e/h)	8.2 [1.8–24.8]	9.1 [1.6–25.3]	6.1 [2.1–24.1]	7.7 [2.0–24.8]
HI (e/h)	19.4 [10.4–32.5]	19.0 [10.1–29.7]	19.8 [10.9–32.9]	19.9 [10.6–36.5]
AHI (e/h)	35.0 [17.3–59.4]	33.2 [15.9–59.7]	35.6 [18.1–59.2]	36.2 [18.3–61.4]
Number and type of events in each data subset				
Normal (*n*)	250,669 (72.5%)	127,292 (73.2%)	49,932 (72.5%)	73,445 (71.3%)
Apnea (*n*)	40,838 (11.8%)	20,174 (11.6%)	7972 (11.6%)	12,692 (12.3%)
Hypopnea (*n*)	54,165 (15.7%)	26,350 (15.2%)	10,989 (15.9%)	16,826 (16.4%)

Data are presented as median [interquartile range], *n*, or %
BMI: body mass index, *AHI*: apnea index, *HI*: hypopnea index, *AHI*: apnea–hypopnea index, *e/h* events per hour

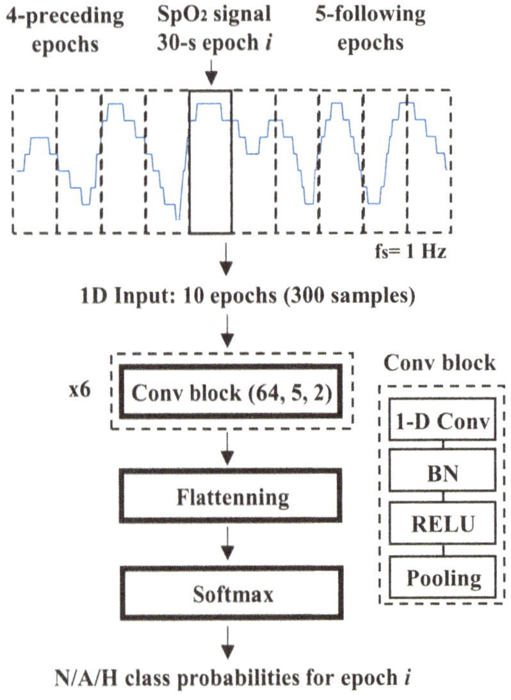

Fig. 15.1 Overview of the proposed CNN architecture. Each convolutional block (conv block) includes a 1D convolution (1D Conv), BN, a RELU activation function, and pooling

desaturations associated to respiratory events, since the onset of oxygen desaturations may occur more than 30 seconds after the start of the respiratory events (Kulkas et al., 2013); and (ii) it allows for a better modeling of the temporal distribution of respiratory events, which are typically grouped in clusters.

The proposed CNN architecture processes this input using six convolutional blocks (conv block), and each one composed of the following:

- 1D convolution (1D conv). This layer extracts feature maps using the 1D convolution operation (Goodfellow et al., 2016):

$$x_i^j[n] = \sum_{k=1}^{k_{size}} w_k^j * a_i[n-k+1] + b_k^j, \quad (15.1)$$

- where x^j_i is the feature map generated by the *j*th convolutional filter ($j = 1, \ldots, 64$) in the *i*th convolutional block ($i = 1, \ldots, 6$); $k_{size} = 5$ is the filter (kernel) size; w^j_k and b^j_k are the filter weights and biases, respectively; and a_i is the input of the *i*th convolutional block. The number of convolutional blocks, the number of filters, and the kernel size were chosen according to the optimum values obtained in Vaquerizo-Villar et al. (2021).
- Batch normalization (BN). BN is applied to normalize the feature maps obtained in the 1D convolution layer (Goodfellow et al., 2016).
- Rectified linear unit (ReLU). ReLU is the standard activation function in CNNs. It is applied to introduce nonlinearity to the normalized feature maps, which provides univer-

sal approximation to any function (Goodfellow et al., 2016):

$$f\left(x_i^j\right) = \max\left(0, x_i^j\right), \tag{15.2}$$

- Pooling. After the ReLU function, a max-pooling layer was applied to the activations with a pooling factor of 2 to reduce dimensionality, while the most relevant features are kept (Goodfellow et al., 2016).

Following the last convolutional block, the 2D feature maps are converted into 1D feature vectors using a flattening operation. Finally, a softmax activation function is used to obtain the output of the CNN architecture, i.e., the probability of belonging to each class (N/A/H) for the input 30-s SpO_2 epoch.

15.2.3 CNN Training Process

The CNN architecture was implemented using the Keras framework with TensorFlow backend. A workstation with a NVIDIA GeForce RTX 2080 GPU running on a Windows 10 environment was used for this purpose. The training data were fed into the CNN using minibatches of size 100 during 200 epochs. The weights of each layer of the network were initialized using He-normal initialization (He et al., 2014). Then, the adaptive moment estimation (Adam) algorithm was used with an initial learning rate of 0.0001 (Kingma & Ba, 2015), and a categorical cross entropy loss function was applied to update the weights and biases at each minibatch. As the whole training data does not fit on the memory of the workstation, training data were fed at each epoch in random order from different patients to the network using 50 reading queues (Sors et al., 2018), which also improves the convergence of the Adam algorithm (Goodfellow et al., 2016; Sors et al., 2018). The validation data was used during the training process to monitor the convergence of the CNN by means of the validation loss. In this respect, the learning rate was reduced by a factor of 2 when the validation loss did not improve for ten consecutive epochs, and early stopping was applied to finish the learning process after 30 epochs of non-improvement in the validation loss, restoring the network weights to those that minimized the validation loss (Goodfellow et al., 2016).

15.2.4 Statistical Analysis

The Kruskal–Wallis test was used to assess statistical differences (p-value < 0.05) between groups. The overall performance of the CNN architecture to automatically classify respiratory events was assessed by means of confusion matrices (3-class), which were used to compute the Cohen's kappa index (kappa) and the 3-class accuracy. The performance for each individual class was measured by means of sensitivity (percentage of epochs belonging to the class rightly classified), specificity (percentage of epochs not belonging to the class rightly classified), positive predictive value (proportion of epochs assigned to the class that are true positives), negative predictive value (proportion of epochs not assigned to the class that are true negatives), and accuracy (proportion of epochs rightly classified). In addition, AI, HI, and AHI were obtained for each subject based on the CNN scoring (AI_{CNN}, HI_{CNN}, and AHI_{CNN}) and compared with those from the standard PSG (AI_{PSG}, HI_{PSG}, and AHI_{PSG}) using Bland–Altman plots and the intra-class correlation coefficient (ICC).

15.3 Results

15.3.1 CNN Model Performance

Figure 15.2 shows the confusion matrix of the CNN model in the test set for the 3-class classification procedure (N/A/H). This model rightly classified 80.3% (82,628/102963) of the 30-s SpO_2 epochs in the test set, with a 3-class kappa value of 0.539. Table 15.2 presents the diagnostic ability for each individual class. Notice that higher performance metrics were obtained for the

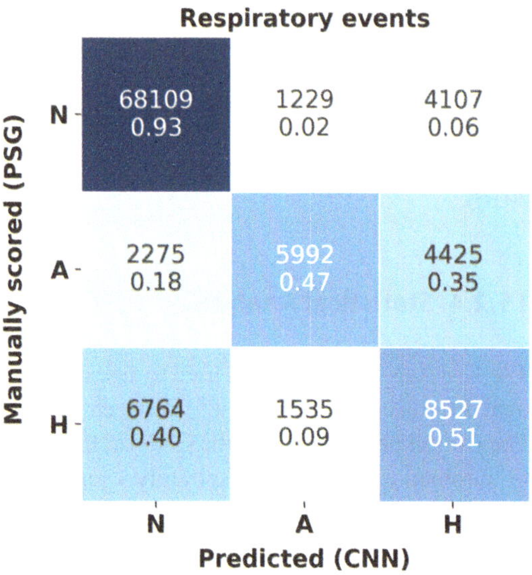

Fig. 15.2 Confusion matrix of the CNN architecture in the test set. This matrix compares the type of respiratory event from standard PSG with the corresponding assignation using the CNN model

Table 15.2 Diagnostic ability of the CNN model in the test set for the detection of normal respiration, apnea, and hypopnea events

Epoch type	Se (%)	Sp (%)	PPV (%)	NPV (%)	Acc (%)
Normal respiration	92.7	69.4	88.3	79.3	86.0
Apnea	47.2	96.9	68.4	92.9	90.8
Hypopnea	50.7	90.1	50.0	90.3	83.7

Se: sensitivity, *Sp:* specificity, *PPV:* positive predictive value, *NPV:* negative predictive value, *Acc:* accuracy

detection of normal respiration than for apnea and hypopnea events.

15.3.2 Estimation of Respiratory Indices

Figure 15.3 shows the Bland–Altman plots comparing AI_{CNN}, HI_{CNN}, and AHI_{CNN} with AI_{PSG}, HI_{PSG}, and AHI_{PSG} in the test set, respectively. ICC is also shown. It can be seen that the respiratory indices predicted by the CNN (AI_{CNN}, HI_{CNN}, and AHI_{CNN}) are underestimating those from standard PSG (AI_{PSG}, HI_{PSG}, and AHI_{PSG}), as reported by their mean difference (bias). HI_{CNN} reached a lower bias (−4.22) than AI_{CNN} (−7.87) and AHI_{CNN} (−12.09), whereas AHI_{CNN} achieved a slightly lower confidence interval (40.82) than AI_{CNN} (45.49) and HI_{CNN} (45.66). In addition, AHI_{CNN} showed a higher agreement with manual scoring (ICC = 0.8466) than AI_{CNN} (ICC = 0.8023) and HI_{CNN} (ICC = 0.6774).

15.4 Discussion

In this chapter, we evaluated the potential usefulness of a CNN architecture to automatically classify respiratory events (apnea, hypopnea, and normal respiration) from the SpO_2 signal in adult OSA patients. To our knowledge, this is the first study applying a deep-learning model to automatically identify apnea and hypopnea events from the oximetry signal.

The proposed CNN-based deep-learning model reached a high performance, with 80.3% 3-class Acc and 0.539 kappa for the classification of respiratory events. According to the guidelines of McHugh (2012), a kappa value between 0.41 and 0.60 indicates that there is a moderate agreement between our CNN architecture and manual PSG-based scoring (McHugh, 2012). Hence, our approach could be potentially used to detect respiratory events in at-home pulse oximetry tests for OSA diagnosis (del Campo et al., 2018).

Looking at the confusion matrix of Fig. 15.2., it can be seen that 93% of normal respiration epochs are rightly classified by the CNN model, which may indicate that oxygen desaturations infrequently occur without being associated to a respiratory event. Furthermore, 47% of apnea and 51% of hypopnea epochs are rightly detected by the CNN architecture, which indicates that the characteristics of SpO_2 desaturations caused by apneas that differ from those related to hypopneas. This agrees with Kulkas et al. (2017), who reported that oxygen desaturations associated to obstructive apneas have significantly larger duration and depth than SpO_2 desaturations related to hypopneas. In this sense, 35% of apnea epochs are misclassified as hypopneas. This can be explained by the fact that oxygen desaturations related to

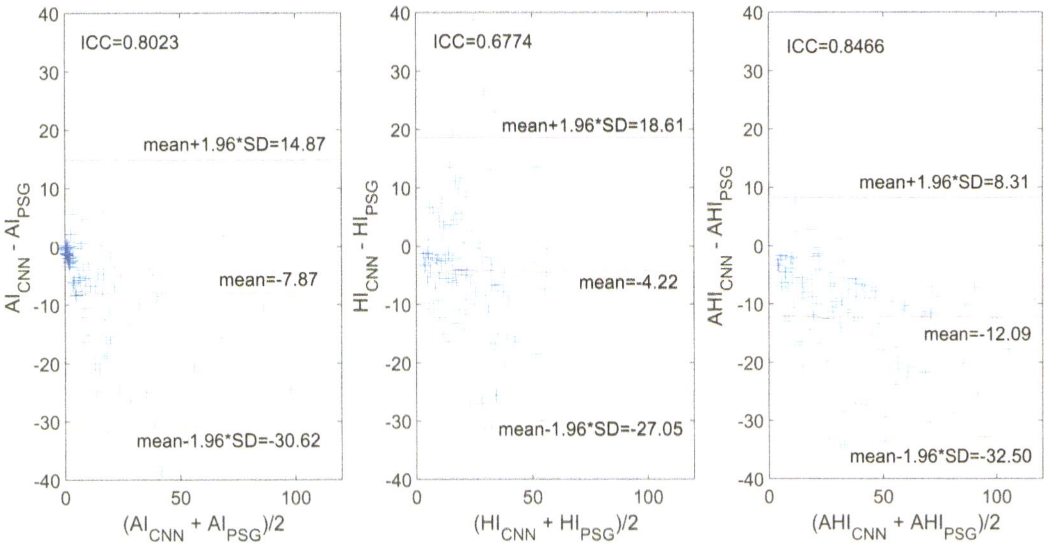

Fig. 15.3 Bland–Altman plots comparing (**a**) AI_{CNN} with AI_{PSG}, (**b**) HI_{CNN} with HI_{PSG}, and (**c**) AHI_{CNN} with AHI_{PSG}

obstructive apnea events of short duration may have similar characteristics to those related to long-duration hypopneas, as the duration and area of SpO_2 desaturations are significantly correlated to the duration of obstructive apnea and hypopnea events (Kulkas et al., 2017). Conversely, 40% of hypopnea epochs are predicted as normal respiration by the CNN. These misclassified hypopneas may be associated to electroencephalographic arousals that do not produce any physiological perturbation in the oximetry signal (Berry et al., 2012).

Regarding the respiratory indices, the CNN model shows a trend to underestimate them, especially AI_{CNN} and AHI_{CNN}. Nonetheless, the CNN model showed promising results, reaching ICCs of 0.8023 (AI_{CNN}), 0.6774 (HI_{CNN}), and 0.8466 (AHI_{CNN}). The higher ICC obtained by AI_{CNN} and AHI_{CNN} can be explained by the fact that their Bland–Altman plots show a linear underestimation trend, whereas HI_{CNN} has outliers in both directions. In this respect, an ICC value in the range 0.50–0.75 indicates a moderate agreement, whereas an ICC value in the range 0.75–0.90 indicates a good reliability (Koo & Li, 2016). Accordingly, our CNN-based deep-learning approach could be used to calculate these respiratory indices in oximetry tests.

Recent studies showed the usefulness of deep-learning techniques to automatically score respiratory events from raw cardiorespiratory signals in OSA patients, outperforming conventional feature-based methodologies (Mostafa et al., 2019). Particularly, some studies faced the automated detection of normal respiration, apneas and hypopneas from airflow; thoracic, abdominal, and chest respiratory signals; and the electrocardiogram (Haidar et al., 2020; McCloskey et al., 2018; Nikkonen et al., 2021; Urtnasan et al., 2018; Van Steenkiste et al., 2020; Yue et al., 2021), reaching a 3-class accuracy (normal, apnea, and hypopnea) in the range 73–91%. In contrast to these studies, our work achieved a 3-class accuracy of 80% using only the SpO_2 signal. In this regard, the oximetry signal has been frequently advocated for OSA screening due to its accessibility, simplicity, and reliability (del Campo et al., 2018).

Vaquerizo-Villar et al. (2019) and Mostafa et al. (2020a, b) have also focused on the automated classification of respiratory events using oximetry-based deep-learning approaches. These studies employed CNNs to differentiate respiratory events from normal respiration episodes using 60-s SpO_2 segments, reaching accuracies in the range 85–95% (Mostafa, Baptista, et al., 2020a; Mostafa, Mendonca, et al., 2020b;

Vaquerizo-Villar et al., 2019). In the present chapter, the 2-class accuracy (normal versus apnea/hypopnea) was included in this range (86%) with a 30-s segment size, which is more appropriate for the detection of clusters of respiratory events that contain more than one respiratory event in a 60-s segment. Furthermore, our CNN-based model addresses for the first time the distinction between apneas and hypopneas from raw oximetry data and the estimation of respiratory indices (AI, HI, and AHI).

Despite the potential usefulness of our proposed approach, some limitations need to be considered. First, the database employed in this work did not contain healthy control subjects (AHI < 5 e/h). The inclusion of these subjects could help to improve the characterization of normal respiration. Another limitation concerns the use of 30-s SpO_2 segments to automatically detect respiratory events, which does not allow to identify the onset and end of apneas and hypopneas. Nonetheless, SpO_2 does not contain this information, as the delay of oxygen desaturations occurring after respiratory events is variable (Kulkas et al., 2013). Similarly, the proposed CNN does not differentiate between obstructive and central respiratory events. However, this would require information about breathing effort (Berry et al., 2012), which is not included in the oximetry signal. In this respect, the acquisition of the photoplethysmography (PPG) signal with the pulse oximetry sensor may contribute to enhance the diagnostic ability of our proposal, as it contains information related to respiratory events (Karmakar et al., 2014; Papini et al., 2020). Furthermore, the use of novel deep-learning techniques (e.g., transformer or generative adversarial networks) may help improve the automatic classification of respiratory events at the cost of higher computational complexity. Finally, the application of eXplainable artificial intelligence techniques could help to further understand the perturbations in the oximetry signal linked with apnea and hypopnea events and the differences between them.

15.5 Conclusions

Our CNN-based deep-learning model exhibited a high performance in the automatic identification of apnea and hypopnea events from the SpO_2 signal. The CNN model also showed a high agreement in the estimation of OSA-related respiratory indices (AI, HI, and AHI). According to our findings, we can conclude that CNN-based SpO_2 approaches could be potentially used to provide an automated diagnosis of OSA in at-home oximetry studies.

Acknowledgments This work was supported by 'Ministerio de Ciencia e Innovación/Agencia Estatal de Investigación/10.13039/501100011033/', ERDF A way of making Europe, and NextGenerationEU/PRTR under projects PID2020-115468RB-I00, PDC2021-120775-I00, and RTC-2017-6516-1, by Sociedad Española de Neumología y Cirugía Torácica (SEPAR) under project 649/2018, by Sociedad Española de Sueño (SES) under project "Beca de Investigación SES 2019," and by "CIBER -Consorcio Centro de Investigación Biomédica en Red-' (CB19/01/00012) through "Instituto de Salud Carlos III". The work of Fernando Vaquerizo-Villar was supported by a "Ayuda para contratos predoctorales para la Formación de Profesorado Universitario (FPU)" grant from the Ministerio de Educación, Cultura y Deporte (FPU16/02938). The work of Daniel Álvarez was supported by a "Ramón y Cajal" grant (RYC2019-028566-I) from the "Ministerio de Ciencia e Innovación–Agencia Estatal de Investigación" co-funded by the European Social Fund.

References

Benjafield, A. V., Eastwood, P. R., Heinzer, R., Morrell, M. J., Federal, U., Paulo, D. S., ... Valentine, K. (2020). Estimation of the global prevalence and burden of obstructive sleep apnoea: A literature-based analysis. *The Lancet Respiratory Medicine, 7*(8), 687–698. https://doi.org/10.1016/S2213-2600(19)30198-5. Estimation

Berry, R. B., Budhiraja, R., Gottlieb, D. J., Gozal, D., Iber, C., Kapur, V. K., ... others. (2012). Rules for scoring respiratory events in sleep: Update of the 2007 AASM manual for the scoring of sleep and associated events: Deliberations of the sleep apnea definitions task force of the American Academy of Sleep Medicine. *Journal of Clinical Sleep Medicine, 8*(5), 597.

del Campo, F., Crespo, A., Cerezo-Hernández, A., Gutiérrez-Tobal, G. C., Hornero, R., & Álvarez, D. (2018). Oximetry use in obstructive sleep apnea.

Expert Review of Respiratory Medicine, 12(8), 665–681. https://doi.org/10.1080/17476348.2018.1495563

Eastwood, P. R., Malhotra, A., Palmer, L. J., Kezirian, E. J., Horner, R. L., Ip, M. S., ... Hillman, D. R. (2010). Obstructive sleep apnoea: From pathogenesis to treatment: Current controversies and future directions. *Respirology, 15*(4), 587–595. https://doi.org/10.1111/j.1440-1843.2009.01699.x

Ebrahimi, Z., Loni, M., Daneshtalab, M., & Gharehbaghi, A. (2020). A review on deep learning methods for ECG arrhythmia classification. *Expert Systems with Applications: X, 7*, 100033. https://doi.org/10.1016/j.eswax.2020.100033

Faust, O., Hagiwara, Y., Hong, T. J., Lih, O. S., & Acharya, U. R. (2018). Deep learning for healthcare applications based on physiological signals: A review. *Computer Methods and Programs in Biomedicine, 161*, 1–13. https://doi.org/10.1016/j.cmpb.2018.04.005

Faust, O., Razaghi, H., Barika, R., Ciaccio, E. J., & Acharya, U. R. (2019). A review of automated sleep stage scoring based on physiological signals for the new millennia. *Computer Methods and Programs in Biomedicine, 176*, 81–91.

Goodfellow, I., Bengio, Y., & Courville, A. (2016). *Deep learning*. MIT Press.

Haidar, R., Koprinska, I., & Jeffries, B. (2020). Sleep apnea event prediction using convolutional neural networks and Markov chains. *Proceedings of the International Joint Conference on Neural Networks*. https://doi.org/10.1109/IJCNN48605.2020.9207345

He, K., Zhang, X., Ren, S., & Sun, J. (2014). Delving deep into rectifiers: Surpassing human-level performance on imagenet classification. In *Proceedings of the IEEE international conference on computer vision* (pp. 1026–1034). https://doi.org/10.1109/ICCV.2015.123

Ismail Fawaz, H., Forestier, G., Weber, J., Idoumghar, L., & Muller, P. A. (2019). Deep learning for time series classification: A review. *Data Mining and Knowledge Discovery, 33*(4), 917–963. https://doi.org/10.1007/s10618-019-00619-1

Karmakar, C., Khandoker, A., Penzel, T., Schobel, C., & Palaniswami, M. (2014). Detection of respiratory arousals using photoplethysmography (PPG) signal in sleep apnea patients. *IEEE Journal of Biomedical and Health Informatics, 18*(3), 1065–1073. https://doi.org/10.1109/JBHI.2013.2282338

Kingma, D. P., & Ba, J. L. (2015). Adam: A method for stochastic optimization. In *3rd International Conference on Learning Representations, ICLR 2015 - Conference Track Proceedings* (pp. 1–15).

Koo, T. K., & Li, M. Y. (2016). A guideline of selecting and reporting intraclass correlation coefficients for reliability research. *Journal of Chiropractic Medicine, 15*(2), 155–163. https://doi.org/10.1016/j.jcm.2016.02.012

Kulkas, A., Tiihonen, P., Julkunen, P., Mervaala, E., & Töyräs, J. (2013). Desaturation delay, parameter for evaluating severity of sleep disordered breathing. *IFMBE Proceedings, 39 IFMBE*, 336–339. https://doi.org/10.1007/978-3-642-29305-4_90

Kulkas, A., Duce, B., Leppänen, T., Hukins, C., & Töyräs, J. (2017). Severity of desaturation events differs between hypopnea and obstructive apnea events and is modulated by their duration in obstructive sleep apnea. *Sleep & Breathing, 21*(4), 829–835. https://doi.org/10.1007/s11325-017-1513-6

McClatchey, K. D. (2002). *Clinical laboratory medicine*. Lippincott Williams & Wilkins.

McCloskey, S., Haidar, R., Koprinska, I., & Jeffries, B. (2018). Detecting hypopnea and obstructive apnea events using convolutional neural networks on wavelet spectrograms of nasal airflow. *Lecture Notes in Computer Science (Including Subseries Lecture Notes in Artificial Intelligence and Lecture Notes in Bioinformatics)*. https://doi.org/10.1007/978-3-319-93034-3_29

McHugh, M. L. (2012). Interrater reliability: The kappa statistic. *Biochemica Medica, 22*(3), 276–282. Retrieved from https://hrcak.srce.hr/89395

Mostafa, S. S., Mendonça, F., Ravelo-García, A. G., & Morgado-Dias, F. (2019). A systematic review of detecting sleep apnea using deep learning. *Sensors (Switzerland), 19*(22), 1–26. https://doi.org/10.3390/s19224934

Mostafa, S. S., Baptista, D., Ravelo-García, A. G., Juliá-Serdá, G., & Morgado-Dias, F. (2020a). Greedy based convolutional neural network optimization for detecting apnea. *Computer Methods and Programs in Biomedicine, 197*, 105640. https://doi.org/10.1016/j.cmpb.2020.105640

Mostafa, S. S., Mendonca, F., Ravelo-Garcia, A. G., Julia-Serda, G., & Morgado-Dias, F. (2020b). Multi-objective hyperparameter optimization of convolutional neural network for obstructive sleep apnea detection. *IEEE Access, 8*, 129586–129599. https://doi.org/10.1109/ACCESS.2020.3009149

Murat, F., Yildirim, O., Talo, M., Baloglu, U. B., Demir, Y., & Acharya, U. R. (2020). Application of deep learning techniques for heartbeats detection using ECG signals-analysis and review. *Computers in Biology and Medicine, 120*(April), 103726. https://doi.org/10.1016/j.compbiomed.2020.103726

Nikkonen, S., Korkalainen, H., Leino, A., Myllymaa, S., Duce, B., Leppanen, T., & Toyras, J. (2021). Automatic respiratory event scoring in obstructive sleep apnea using a long short-term memory neural network. *IEEE Journal of Biomedical and Health Informatics, 25*(8), 2917–2927. https://doi.org/10.1109/JBHI.2021.3064694

Papini, G. B., Fonseca, P., Gilst, M. M. V., Bergmans, J. W. M., Vullings, R., & Overeem, S. (2020). Respiratory activity extracted from wrist-worn reflective photoplethysmography in a sleep-disordered population. *Physiological Measurement, 41*(6). https://doi.org/10.1088/1361-6579/ab9481

Park, J. G., Ramar, K., & Olson, E. J. (2011). Updates on definition, consequences, and management of obstructive sleep apnea concise review for clinicians.

Mayo Clinic Proceedings, 86(6), 549–555. https://doi.org/10.4065/mcp.2010.0810
Redline, S. (2017). Screening for obstructive sleep apnea: Implications for the sleep health of the population. *JAMA, 317*(4), 368–370.
Roy, Y., Banville, H., Albuquerque, I., Gramfort, A., Falk, T. H., & Faubert, J. (2019). Deep learning-based electroencephalography analysis: A systematic review. *Journal of Neural Engineering, 16*(5). https://doi.org/10.1088/1741-2552/ab260c
Senaratna, C. V., Perret, J. L., Lodge, C. J., Lowe, A. J., Campbell, B. E., Matheson, M. C., … Dharmage, S. C. (2017). Prevalence of obstructive sleep apnea in the general population: A systematic review. *Sleep Medicine Reviews, 34*, 70–81. https://doi.org/10.1016/j.smrv.2016.07.002
Shokoueinejad, M., Fernandez, C., Carroll, E., Wang, F., Levin, J., Rusk, S., … others. (2017). Sleep apnea: A review of diagnostic sensors, algorithms, and therapies. *Physiological Measurement, 38*(9), R204.
Sors, A., Bonnet, S., Mirek, S., Vercueil, L., & Payen, J. F. (2018). A convolutional neural network for sleep stage scoring from raw single-channel EEG. *Biomedical Signal Processing and Control, 42*, 107–114. https://doi.org/10.1016/j.bspc.2017.12.001
Urtnasan, E., Park, J. U., & Lee, K. J. (2018). Multiclass classification of obstructive sleep apnea/hypopnea based on a convolutional neural network from a single-lead electrocardiogram. *Physiological Measurement, 39*(6). https://doi.org/10.1088/1361-6579/aac7b7
Van Steenkiste, T., Groenendaal, W., Dreesen, P., Lee, S., Klerkx, S., De Francisco, R., … Dhaene, T. (2020). Portable detection of apnea and hypopnea events using bio-impedance of the chest and deep learning. *IEEE Journal of Biomedical and Health Informatics, 24*(9), 2589–2598. https://doi.org/10.1109/JBHI.2020.2967872
Vaquerizo-Villar, F., Álvarez, D., Kheirandish-gozal, L., Gutiérrez-tobal, G. C., Barroso-garcía, V., Campo, F., … Hornero, R. (2019). Convolutional neural networks to detect pediatric apnea- hypopnea events from oximetry. In *41st annual international conference of the IEEE engineering in medicine and biology society* (pp. 3555–3558).
Vaquerizo-Villar, F., Alvarez, D., Kheirandish-Gozal, L., Gutierrez-Tobal, G. C., Barroso-Garcia, V., Santamaria-Vazquez, E., … Hornero, R. (2021). A convolutional neural network architecture to enhance oximetry ability to diagnose pediatric obstructive sleep apnea. *IEEE Journal of Biomedical and Health Informatics, 25*(8). https://doi.org/10.1109/JBHI.2020.3048901
Yue, H., Lin, Y., Wu, Y., Wang, Y., Li, Y., Guo, X., … Lei, W. (2021). Deep learning for diagnosis and classification of obstructive sleep apnea: A nasal airflow-based multi-resolution residual network. *Nature and Science of Sleep, 13*, 361–373. https://doi.org/10.2147/NSS.S297856

16 Tracheal Sound Analysis

AbdelKebir Sabil and Sandrine Launois

Abstract

Tracheal sound sensors provide multiple respiratory signals that are valuable for studying upper airway characteristics. This chapter reviews the original work and ongoing research on tracheal sound analysis in relation to upper airway obstruction during sleep. Past and current research suggest that being associated with other sleep study recording sensors and advanced signal processing techniques, tracheal sound analysis can extensively contribute to the diagnosis and assessment of sleep-disordered breathing.

Keywords

Tracheal sounds · Sleep-disordered breathing · Apnea and hypopnea detection · Alternative flow sensor · Alternative respiratory effort sensor · Heart rate variability

Supplementary Information The online version contains supplementary material available at [https://doi.org/10.1007/978-3-031-06413-5_16].

A. Sabil (✉)
SOS Oxygène, Paris, France
e-mail: kebir.sabil@cloudsleeplab.com

S. Launois
Centre Médical du Sommeil les Sablons, Paris, France
e-mail: sandrine.launois@protonmail.com

16.1 Introduction

The most common sleep-related respiratory disorder is sleep apnea syndrome, characterized by repeated partial (hypopneas) or complete cessations (apneas) of respiratory flow (Berry et al., 2012). The detection and classification of these respiratory events as obstructive or central are essential for the diagnosis and severity assessment of the disorder and for the choice of treatment. Reliable noninvasive recording of required physiological parameters for the diagnosis of sleep-disordered breathing (SDB) has been made possible through technological advances.

According to international guidelines, the method of choice for detecting flow variations defining respiratory abnormalities during sleep is oronasal airflow or tidal volume reduction (Berry et al., 2012). Pneumotachography (PNT) is considered the gold standard for this measurement but is not used in routine practice in a polysomnography (PSG) or home sleep apnea testing (HSAT). The American Academy of Sleep Medicine (AASM) recommends using an oronasal thermal airflow sensor as the first-choice sensor for detection of apneas and a nasal pressure (NP) transducer as the first-choice sensor for hypopnea detection (Berry et al., 2012). Apneas are defined as a complete airflow cessation or more than 90% amplitude reduction from the reference value for at least 10 second (Berry et al., 2012; Escourrou et al., 2010; Mayer et al., 2017). Hypopneas are sleep-related events where respi-

T. Penzel, R. Hornero (eds.), *Advances in the Diagnosis and Treatment of Sleep Apnea*, Advances in Experimental Medicine and Biology 1384, https://doi.org/10.1007/978-3-031-06413-5_16

ratory flow is reduced by more than 30% with associated oxygen desaturation of more than 3% and/or arousal (Berry et al., 2015).

Thermal airflow sensors use the difference between the temperature of exhaled air and ambient air to estimate airflow and detect mouth breathing. The use of temperature as a surrogate measurement of airflow works well for detecting apnea as it detects both nasal and oral airflow. Nasal cannulas are pressure sensors capable of detecting pressure changes during inspiration and expiration. Most sleep laboratories use signals from both a thermistor and nasal pressure (NP) to ensure adequate oronasal flow measurement. Combined, these sensors improve the identification of apneas that are missed by thermistors or overestimated by NP in the case of mouth breathing, for example. However, placed in the sensitive same area, between the nose and the mouth, these sensors can cause patients much discomfort and, thus, impact their sleep (Goodwin et al., 2001). They are, therefore, often displaced or even removed by patients during sleep studies, potentially leading to lost or unanalyzable signals during part of the recording. The validity of thermistor and NP signals for more than 6 hours of recording is satisfactory for less than 60% for both children (Goodwin et al., 2001) and adults (Redline et al., 1998). This problem is more common in HSAT recording than in a monitored PSG recording at the sleep laboratory.

When nasal pressure and thermistor signals are missing or of poor quality, the AASM recommends the use of signals derived from the respiratory inductance plethysmography (RIP) as a surrogate respiratory flow (Berry et al., 2012). Using two belts placed around the thorax and the abdomen, the RIP method allows semi-quantitative assessment of volume changes through the measurement of thoracic and abdominal movements (Eberhard et al., 2001). Either the RIP-sum- or the RIP-flow-derived signals can be used as a surrogate signal for respiratory flow (Berry et al., 2012).

Recordings of tracheal sounds (TS) correlate well with respiratory flow and provide valuable information about airway structure and respiratory disorders (Beckerman et al., 1982; Cummiskey et al., 1982; Nakano et al., 2004). Tracheal sound analysis is a simple and noninvasive way to study the behavior of the upper airway (UAW). Tracheal sounds have been used for acoustical flow estimation (Gavriely & Cugell, 1996; Que et al., 2002; Yadollahi & Moussavi, 2009) and investigation of the upper airway abnormalities such as airway obstruction (Nakano et al., 2004; Yadollahi et al., 2010) in patients with sleep apnea.

16.2 Tracheal Sounds

Tracheal sounds, recorded at the sternal notch, reflect pressure variations transmitted through the inner surface of the trachea from turbulent airflow in the airways (Penzel & Sabil, 2018). These vibrations are defined by the pressure's magnitude and frequency content and by the physiological characteristics of the tracheal wall and surrounding soft tissue. Thus, microphones recording tracheal sounds on the skin surface detect acoustic sounds reflecting these tracheal wall vibrations. This characteristic has been used not only to detect tracheal breathing sounds (flow and snoring) but also to record suprasternal pressure (SSP), a good surrogate for respiratory effort (Amaddeo et al., 2016; Meslier et al., 2002; Glos et al., 2018; Sabil et al., 2019a). In addition, acoustic characteristics of TS allow the sensors to detect cardiac sounds, and TS signals may also include cardiogenic oscillations (Glos et al., 2018; Freycenon et al., 2021; Priftis et al., 2018).

The specific upper airway characteristics related to obstructive sleep apnea (OSA) influence the sound produced by the UAW superimposed on tracheal breathing sounds. Thus, apnea monitoring using tracheal sounds recording is of great interest to sleep physicians. However, most sleep recording systems that include tracheal sound sensors only use them for the recording and detection of snoring. To our knowledge, the only sleep recording systems currently using tracheal sound sensors for respiratory event detection and characterization are CIDELEC® (Sainte Gemmes sur Loire, France) and Nukute® (Oulu, Finland).

16.3 Tracheal Sound Sensors

Tracheal sound sensors are stethoscope-like transducers with a microphone inserted into a protective housing with a thick cuff creating a deep airtight space between the transducer and the patient's skin (Fig. 16.1). The practicability of the sensor and the good quality of the recorded TS signal are ensured using a cylindrical shape of the sensor's protective chamber. To fit the suprasternal notch curve, the skin contact face of the sensor has to be convex. An air gap of 2–3 mm between the sensor and the contact surface of the protective housing ensures that the microphone does not touch the patient skin during recording. A well-sealed contact surface above the sternal notch insulates against ambient noise, and the sensor should always be secured in place using a double-sided ring tape. Correct positioning of the transducer, about 1 cm right above the sternal notch, is an essential element to obtain a good quality signal. In terms of practicability, the placement of TS sensors on the skin surface just above the tracheal notch is relatively easily. If they are properly secured in place, the sensors are less likely to be displaced or removed by patients compared to other sensors during a recording session, especially for sleep studies. The robustness of TS intensity allows the sensors to achieve an acceptable signal-to-noise ratio even when the airflow is reduced (Yadollahi & Moussavi, 2010). The signal is amplified and band-pass filtered to separate high-pitch frequencies of the breathing sound from low-pitch frequencies of snoring sounds. Frequency cutoffs of the filtered signal depend on the acoustic transducer characteristics. In the MESAM system, for instance, the flow sound was detected at frequencies between 800 and 2000 Hz and snoring between 50 and 800 Hz (Penzel et al., 1990), while the TS sensor in the CIDELEC system filters flow sound between 200 and 2000 Hz and snoring between 20 and 200 Hz (Penzel & Sabil, 2018). However, filtering ranges and threshold level for children may have to be adjusted. Before analysis, the recorded raw signal goes through low-pass antialiasing filter to avoid erroneous sampling. The choice of the sampling rate and the number of the A/D conversion bits vary from one system to another depending on the sensitivity and the frequency range of the microphone. Tracheal sound intensity may vary with sleep stages in the same individual and among individuals of different age, height, and body mass index. These variations may result in reduced signal amplitude that could misclassify regular breathing as hypopneas or apneas. This problem may be solved by adjusting the gain on the acoustic sensor and by filtering out background noise more effectively to optimize the signal-to-noise ratio (Gavriely & Cugell, 1996).

In addition, most TS sensors have linear response characteristics over a wide range of frequencies (Yadollahi & Moussavi, 2005), and if properly processed, they can easily distinguish inspiration from expiration (Beck et al., 2005). For a reliable SDB diagnosis, signal delimitation and

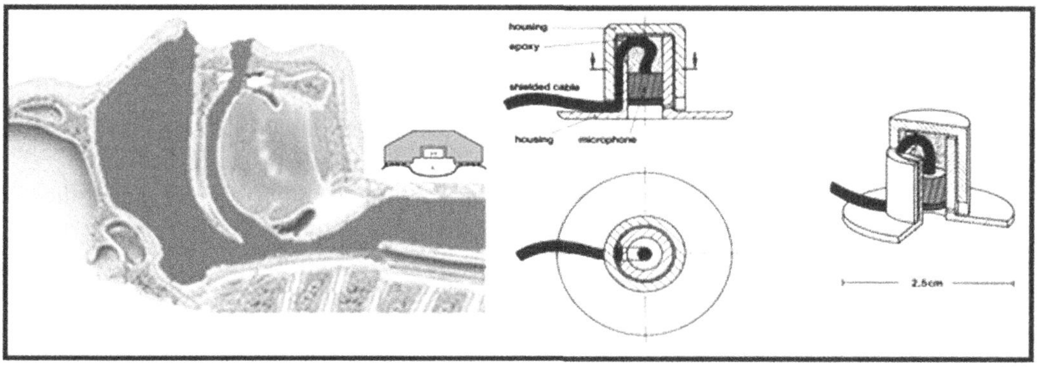

Fig. 16.1 Illustration of the placement of a tracheal sound sensor and a diagram showing a microphone inserted into a protective housing used in the MESAM system (Penzel et al., 1988)

differentiation of inspiratory from expiratory phases are necessary. Different envelope detection techniques, such as Hilbert transform, have been used for the detection of TS respiratory cycle delimitation (Harper et al., 2003; Yadollahi & Moussavi, 2007) even in the presence of complete obstructions, such as during apneas or swallowing (Huq & Moussavi, 2012). Finally, several studies have shown that TS could be used to detect and differentiate complete (apnea) from partial (hypopnea) UAW obstruction during sleep. Thus, the breathing sound signal extracted from TS recording could be used as any other flow signal for the analysis of respiratory events during sleep. While few applications use time domain visual analysis to process tracheal sound signals, most studies use frequency domain automatic analysis.

16.4 Tracheal Sound Technology: A Reliable Recording for SDB Diagnosis

16.4.1 Time Domain Analysis: TS for Classical Manual Detection of Apneas and Hypopneas

Using a TS sensor, an apnea can be easily detected as a cessation of TS and/or the absence of definite respiratory cycles (Fig. 16.2) during respiratory monitoring (Cummiskey et al., 1982; Meslier & Racineux, 1987; Sabil et al., 2019b; Krumpe & Cummiskey, 1980). In the absence of snoring, hypopneas could be easily detected as a drop in the TS intensity (Fig. 16.3). However, for TS, the intensity is not only related to the airflow but also to the UAW characteristics that may change throughout the night for the same patient depending on body positions and sleep stages. For the same airflow, the higher the UAW resistance, the higher the TS intensity during both inspiration and expiration (Chuah & Moussavi, 2001). In the presence of snoring during obstructive hypopneas, for instance (Fig. 16.4), the intensity of the TS signal increases. Thus, compared to apneas, detecting hypopneas directly from TS intensity reduction is challenging, particularly with time domain visual analysis. Many tracheal sound automatic analysis techniques use either time domain or frequency domain TS signal analysis. Time domain analysis tracks tracheal intensity signal changes with the presence of respiratory events throughout the recoding time. Spectral analysis systematically explores frequency contents (Fig. 16.5) of TS signals and examines how much of the signal lies within each given frequency band over a range of frequencies (Alvarez et al., 2008; Solà-Soler et al., 2012; Azarbarzin & Moussavi, 2013).

The recording of TS for the evaluation of sleep-related respiratory disorders has been widely studied. In 1980, Krumpe et al. were the first to show that apneas could be identified by the cessation of laryngeal sounds during continuous monitoring (Krumpe & Cummiskey, 1980). In 1982, Cummiskey et al. used a TS sensor combined with a thermistor, a nasal pressure cannula, and pulse oximeter for the detection of apneas and hypopneas. There was no significant difference in the number of events detected with either TS or the reference sensors (Cummiskey et al., 1982). In 1988, Penzel et al. developed a new technique for real-time snoring analysis using a combination of signals from a laryngeal microphone. Their system could differentiate snoring from physiological breathing sounds and detect obstructive apneic events (Penzel et al., 1988). In 1989, Meslier et al. compared TS and thermistor flow signals to the gold standard PNT signal in healthy patients during sleep. With the PNT as a reference, there was no difference in the number of apneas and their duration recorded using the TS method and the PNT (Meslier & Racineux, 1987). In 1995, a study by Van Surell et al. compared an HSAT system that uses TS for apnea detection and classification with a routine PSG recording. They concluded that the HSAT system with TS sensor can be used to screen patients with severe OSA (Van Surell et al., 1995). Soufflet et al. showed that ST correlated with respiratory flow measured by a PNT and could be used as a noninvasive method to measure airflow (Soufflet et al., 1990). Yadollahi et al. evaluated TS associated with pulse oximetry compared with regular PSG recordings. They classified TS either as the presence of acoustic sound (breathing, snoring, noise)

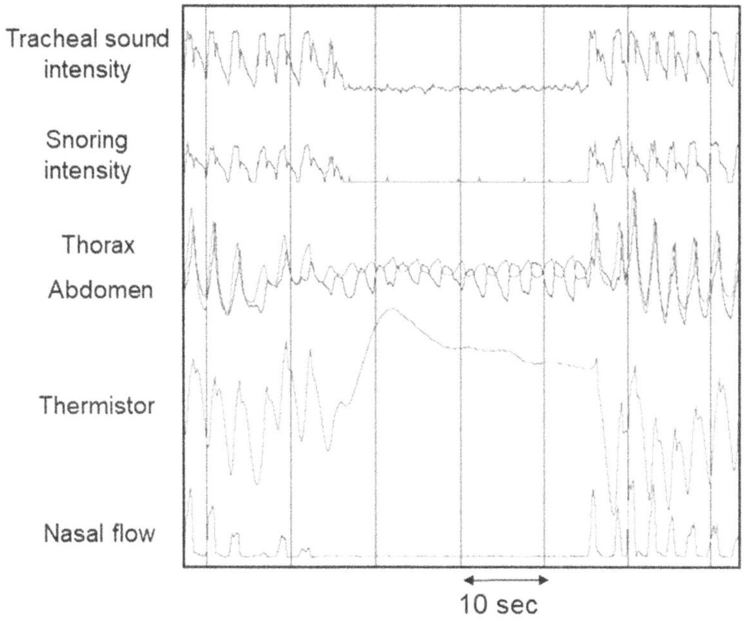

Fig. 16.2 Example of an obstructive apnea detected separately on three different signals (flow sound intensity, nasal pressure, and thermistor). The AASM definition of apnea in terms of signal amplitude decrease and duration was applied to all three signals

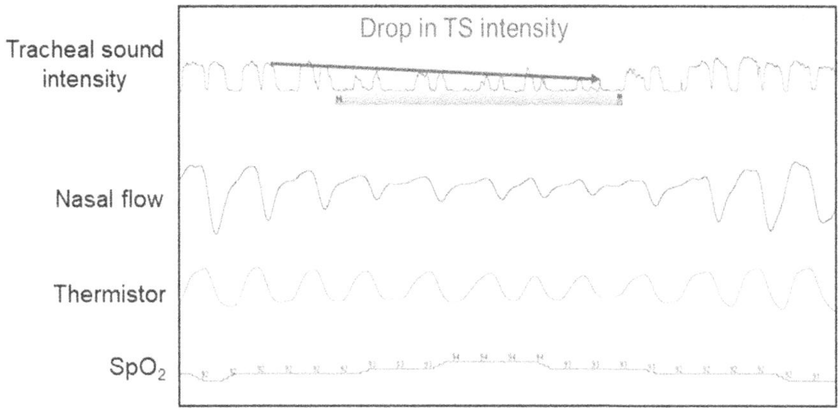

Fig. 16.3 A respiratory event with the criteria for hypopnea without snoring sounds. Persistence of respiratory cycles in the flow sound signal but reduced in amplitude both at inspiration and expiration

or as the absence of sound (silence). Good agreement was found between the events defined by TS and oxygen saturation changes and PSG-recommended scoring rules (Yadollahi et al., 2010).The same team examined the concordance between TS measured directly at the sternal notch and those recorded with the sensor placed at 20–30 cm from the patient and demonstrated a superiority of the suprasternal notch measurement (Yadollahi & Moussavi, 2010).

In a recent study evaluating the CIDELEC® system, apnea detection using a TS sensor, the PneaVox®, during sleep has been compared in OSA patients to apnea detection using thermistor, nasal pressure, and RIP sum signals. This study showed that apneas could be visually identified by the cessation of TS during continuous monitoring. The TS device used in the study provides a sensitive, reliable, technically simple, and easily applied noninvasive means to monitor respiration during sleep. While NP signals tend to overestimate the number of apneas and cannot detect oral breathing, TS can detect apneas seen by a thermistor and/or a RIP sum and additional

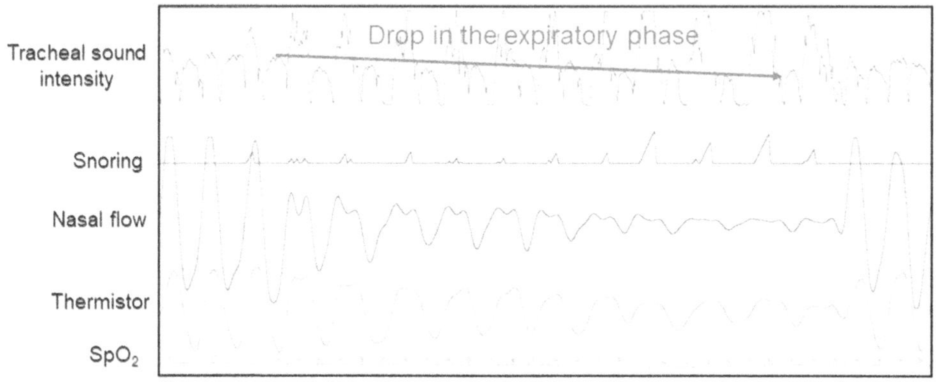

Fig. 16.4 A respiratory event with the criteria for hypopnea with snoring sounds. Persistence of respiratory cycles in the flow sound signal but reduced in amplitude only on expiration

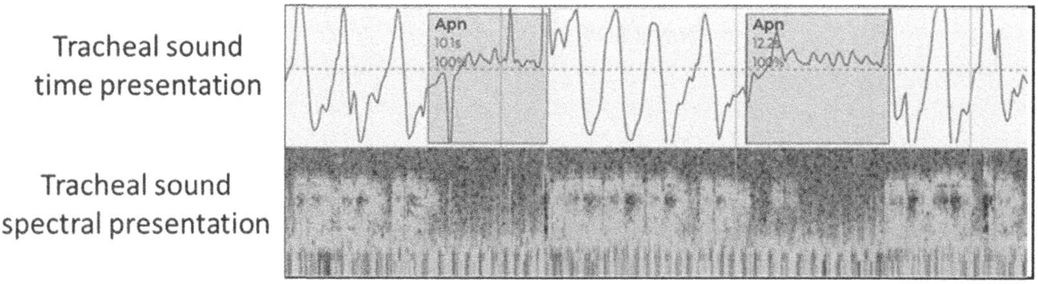

Fig. 16.5 Example of automatic apnea detection using spectral analysis of tracheal sound recorded with the Nukute® HSAT system

events that could be missed by these two sensors. Lastly, combined with NP, TS allow the detection of oral breathing and can reclassify as hypopneas or apneas that would be incorrectly detected by NP alone. TS can therefore be used as a substitute for oral thermistor to reliably detect apneas and associated with NP. TS meets the oronasal flow evaluation required by the AASM for apnea detection and is recommended by French guidelines (Escourrou et al., 2010).

16.4.2 Frequency Domain Analysis: TS Spectral Analysis for Automatic Detection of Apneas and Hypopneas

TS spectral analysis was proposed in many studies. Nakano et al. performed a spectral analysis of TS and recommended the use of TS in HSAT devices, especially for patients with a high probability of OSA (Nakano et al., 2004). Based on spectral analysis, Kulkas et al. proposed a TS analysis algorithm for separation of apneas from normal breathing or snoring sounds (Kulkas et al., 2008). The separation feature was defined as the ratio of the smoothed amplitudes of the two frequency ranges from 0 to 50 Hz and 50 to 600 Hz. The proposed frequency-dependent method was more reliable for apnea analysis than the signal amplitude-dependent time domain analysis (Kulkas et al., 2008). Yadollahi et al. developed an acoustic analysis method for respiratory event detection using a TS coupled with SpO_2 recording (Yadollahi & Moussavi, 2009). Automatically segmenting TS into sound (breathing, snoring, and noise) and silent segments, this method highly correlated (96%) with PSG results and performed well in differentiating simple snorers from OSA patients (Yadollahi & Moussavi, 2009). Other techniques using compressed TS analysis have been used to screen for SDB. Nonlinear filtering of compressed TS provided an effective tool for respiratory events screening during sleep (Kulkas et al., 2008; Rauhala et al., 2008). Sanchez et al. explored the correlation between TS spectra and body length in children and adults. They found that, depending on body length, children had significantly louder sounds and higher frequency components than in adults (Sanchez & Pasterkamp, 1993). This finding was confirmed in another study showing that age, gender, and height are major factors that contribute to the modification of UAW length, wall thickness, and cross-sectional area, which cause narrowing, increased resistance, and UAW collapsibility. These changes have an impact on the TS frequency contents in OSA patients not only when they sleep but also during the day when awake (Yadollahi & Moussavi, 2011). Characteristics of TS spectral features have been used in other studies to separate OSA patients from control subjects during wakefulness (Montazeri & Moussavi, 2010; Montazeri et al., 2012; Elwali & Moussavi, 2017). While clinical validation has yet to confirm the relevance of these novel techniques using TS spectral analysis, findings from these studies extend the potential of TS technology as a possible tool that is reliable, fast, simple, noninvasive, and inexpensive for OSA screening in the awake subject.

Finally, in a recently published study, Mlynczak et al. used a neural network algorithm to analyze TS recorded with a wireless acoustic sensor. Their system was able to differentiate normal breathing sounds from snoring with good discriminatory accuracy (Mlynczak et al., 2017). This system was the first to propose a wireless TS sensor and to use a smartphone application as its interface.

16.5 Respiratory Event Characterization

16.5.1 Respiratory Effort Evaluation: The Gold Standard and Real-Life Practice

To distinguish between obstructive and central sleep apnea, evaluation of inspiratory effort during sleep is required. Esophageal pressure (Pes) is considered as the gold standard for the evaluation of respiratory efforts and classification of events (Berry et al., 2015). However, this method is invasive, is not well tolerated by many patients, and can affect the quality of sleep (Goodwin et al., 2001; Chervin & Aldrich, 1997) and, thus, is not used for routine sleep studies. Indirect respiratory effort measurement techniques have been developed. In the absence of esophageal pressure recording in routine sleep studies, the AASM recommends the use of respiratory inductance plethysmography (RIP) belts as a reference technique for the evaluation of respiratory efforts (Berry et al., 2015). During an apnea, the presence of RIP movements at the respiratory frequency indicates respiratory effort and classifies the apnea as obstructive. In addition, the increased respiratory effort against a collapsed airway may result in an out-of-phase or paradoxical RIP signals. Central apneas are characterized by the absence of RIP movements with, occasionally, cardiogenic oscillations seen on the signals. However, the reliability of the RIP signals depends on accurate placement and stability of the thoracoabdominal belts, which is not always guaranteed, particularly in young children and obese patients.

16.5.2 Suprasternal Pressure: A TS Signal for Respiratory Effort Evaluation

Tracheal sound sensors can be used not only to detect tracheal breathing sounds but also to record suprasternal pressure (SSP), an adequate substitute for assessing respiratory effort (Meslier et al., 2002; Sabil et al., 2019a, 2020; Glos et al., 2018). The SSP is a nonaudible signal with low frequencies resulting from pressure variations induced by respiratory efforts. Respiratory efforts cause variations in pharyngeal pressure leading to pressure changes in the TS sensor chamber. These pressure variations are measured by movements of the skin in contact with the sensor's surface at the sternal notch. Characterization of apneas and hypopneas using the SSP has been compared with the gold standard, Pes, and the recommended alternative, the RIP signal. Provided that the scorer is familiar with the signal, the SSP is a reliable method to characterize apneas and hypopneas in both adults and children (Meslier et al., 2002; Glos et al., 2018; Sabil et al., 2019a). Figure 16.6 shows an example of an apnea that could be misclassified by the RIP signals as central, while it is clearly identified as obstructive on both the Pes and the SSP signals.

16.5.3 Choking Noise Detection: A TS Noise for Apnea Characterization

The placement of the sensor on the suprasternal notch is recommended for a better analysis of snoring sounds. The presence of snoring can help characterize hypopneas as obstructive and avoid misclassifying hypopneas as apneas when the magnitude of airflow reduction is overestimated because of nasal obstruction or poor positioning of the nasal cannula. However, a snoring-like squeaky noise known as "choking sound" (CS) could be heard during apneas. With a different spectrum density than snoring, choking sounds are induced by intense respiratory effort with a slight inspiratory or expiratory transient reopening of the UAW during obstructive apneas and are characterized by the mid-inspiration snoring spectrum that could occur in any hyperventilation following the apnea. In a recent study, we used TS spectral analysis to detect CS (Sabil et al., 2017). Apneas were characterized as mixed or obstructive by the presence of CS at the end of apneas and sometimes during apneic events (Fig. 16.7). During central apneas, these CS are absent.

16.6 Combination of TS with Other Sensors

16.6.1 Tracheal Sounds and RIP Belts for a "Sensor-Face-Free" Sleep Recording

We investigated the use of TS in combination with RIP belt signals for the diagnosis of OSA in adult patients without placing any respiratory sensors on the patient's face (Sabil et al., 2020). Results were compared with those obtained with

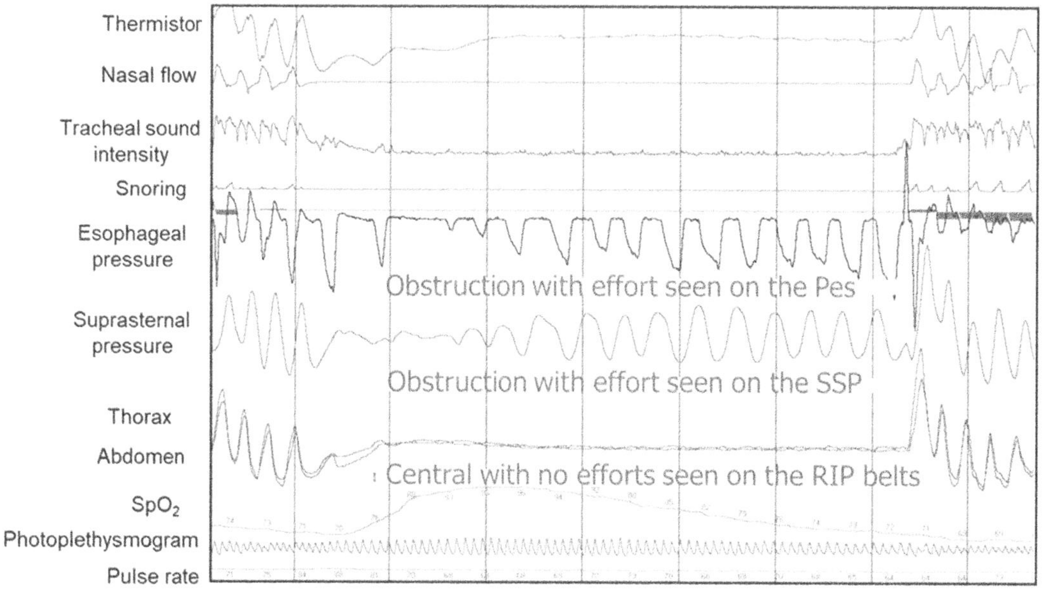

Fig. 16.6 Example of an apnea that could be misclassified by the thoracoabdominal respiratory inductance plethysmograph signals as central, while it is clearly identified as obstructive on both the esophageal pressure and the suprasternal pressure signals

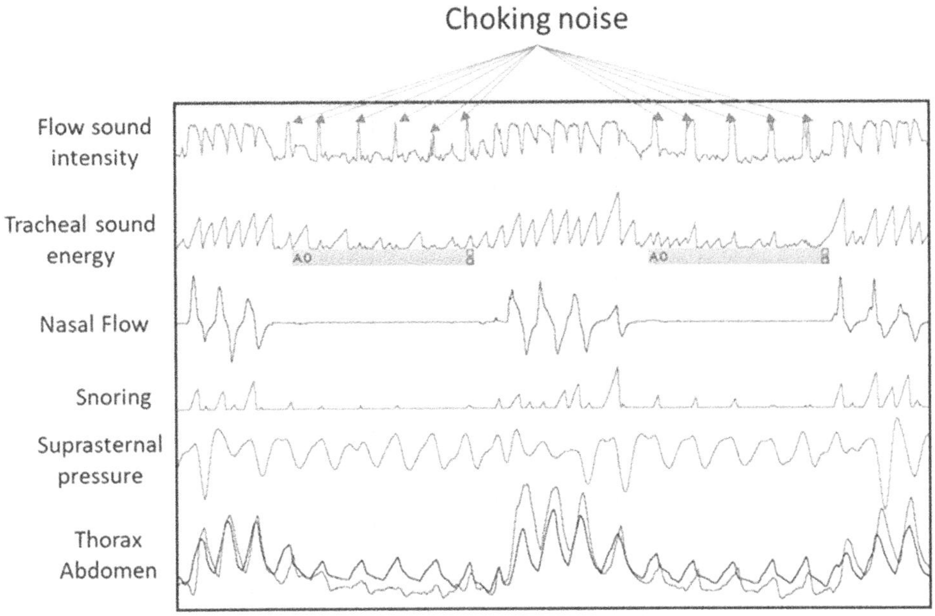

Fig. 16.7 Choking sounds during two consecutive obstructive apneas. Note the inspiratory resumption sound and snoring amplification after the obstructive events

the use of the combination thermistor–nasal cannula–RIP sensors, as recommended by the AASM (Sabil et al., 2020). Assuming that the TS sensor is properly placed just above the sternal notch and that the scorer is familiar with the TS signal analysis, this method provides a correct detection and characterization of respiratory events and, thus, a highly accurate evaluation of OSA (Sabil et al., 2020). Indeed, this "face-free" HSAT recording is practical for patients who do not tolerate a nasal cannula and/or an oronasal thermistor, such as the case in children. The usefulness of a TS sensor as an alternative sensor for the detection and characterization of sleep respiratory events in children has been evaluated (Amaddeo et al., 2020). Amaddeo et al. demonstrated that TS sensors are well tolerated by children for airflow detection compared to standard nasal cannulas and thermistor (Amaddeo et al., 2020). The addition of TS to standard HSAT sensors correctly detected respiratory events even in the absence of the recommended airflow signals, allowing a "face-free" setting that may be of particular interest in children. Moreover, the study demonstrated that TS and SSP signals, in combination with SpO_2 only and without RIP, allow the detection and characterization of respiratory events with a high degree of sensitivity and specificity. Thus, TS and SSP may represent a promising method for screening pediatric patients at risk of SDB (Amaddeo et al., 2020).

16.6.2 Nasal Pressure and TS for the Detection of Oral Breathing

Oral breathing detection is important in sleep studies, and the AASM recommends their detection with oronasal thermistors. Oral breathing is characterized by the absence of nasal pressure signal, while the thermistor signal detects respiratory variations (Fig. 16.8a). However, thermistors can be displaced or removed by patients, particularly children, during sleep recordings. In this situation, the evaluation of respiratory flow by nasal pressure signal alone may result in false detection of apneas. In some patients, oral breathing can also occur exclusively at the inspiratory phase or at the expiratory phase (Fig. 16.8b). As the TS sensor is taped right on the sternal notch, the signal is less likely to be lost during sleep

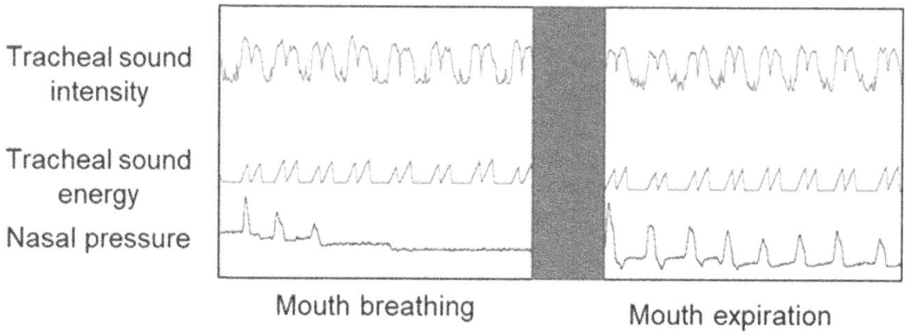

Fig. 16.8 Example of the tracheal sound signal used to confirm exclusive oral breathing or loss of nasal pressure signal. Expiratory mouth breathing could also be seen with the nasal pressure signal dropping to zero, while the sound signal remains present

study recording than thermistor or nasal cannula signals. Thus, in combination with nasal pressure, TS are reliable sensors for oral breathing detection. There is, however, no study to date evaluating the reliability of TS analysis to detect oral breathing.

16.7 Tracheal Sounds Beyond the Usual Respiratory Information

16.7.1 Catathrenia: More Than Just a Regular Snoring

Catathrenia is a particular sleep-related respiratory phenomenon characterized by infrequent groaning sounds during episodes of prolonged expiration that occurs mainly during REM sleep (Guilleminault et al., 2008; Vetrugno et al., 2008). Catathrenia has the characteristics of a complex respiratory motor behavior, and its pathogenesis remains unclear and may be heterogeneous. It consists of a deep inspiration, followed by breath holding and slow release of air through a closed glottis, during which the groaning sound is produced (Guilleminault et al., 2008; Vetrugno et al., 2008; Rodrigues et al., 2021). Catathrenia has been compared to snoring in several studies where the recorded groaning sound was analyzed and characterized as vocal (Guilleminault et al., 2008; Vetrugno et al., 2007; Siddiqui et al., 2008). Iriarte et al. used spectral analysis and oscillogram to compare catathrenia sounds recorded in two patients (one female) with snoring (Iriarte et al., 2011). Unlike snoring, catathrenia had harmonics and a short jitter that establishes it laryngeal origin. None of the previously published studies measured the catathrenia sounds at the tracheal notch, which could probably help better in characterizing it. Figure 16.9 illustrates an example of catathrenia recorded using a TS sensor. Periods of monotonous groaning sound usually lasting much longer than regular snoring are easily seen on the TS signal.

16.7.2 Tracheal Sound Energy Ratio: An Advanced Analysis for Upper Airway Resistance Evaluation

Acoustic intensity increases with friction. When UAW resistance is increased, friction increases and so does the tracheal sound intensity. In a study examining TS acoustic energy in patients with increased UAW resistance, a significant correlation was observed between the acoustic energy ratio Ei/Ee (Ei, inspiratory energy, and Ee, expiratory energy) and UAW resistance, regardless of the presence or absence of snoring. In the absence of UAW resistance, flow sound intensity increases equally throughout inspiratory and expiratory phases, and the energy ratio Ei/Ee remains stable. In the presence of UAW resistance, inspiratory energy increases in comparison to expiratory energy, resulting in an increased energy ratio Ei/Ee (Fig. 16.10). Thus, the variation of the acoustic energy ratio Ei/Ee is a good indicator of UAW resistance evaluation (Racineux, 2006).

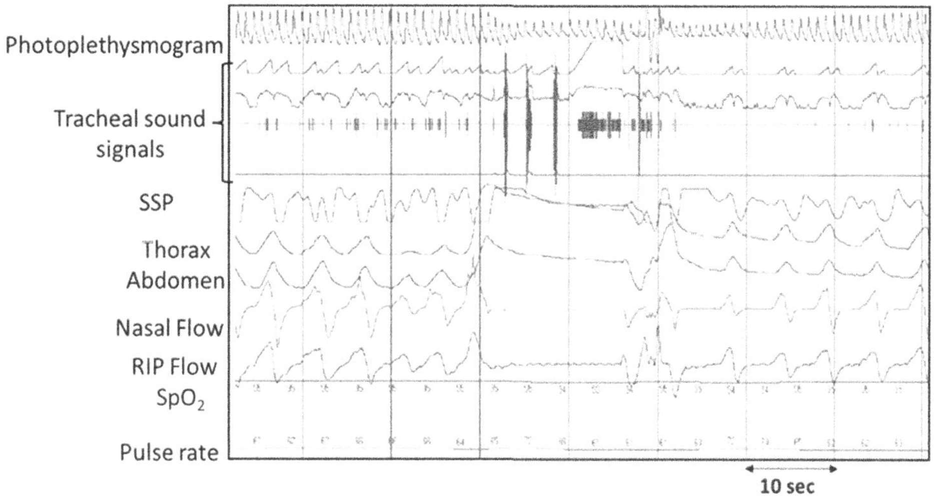

Fig. 16.9 Example of a typical detection of catathrenia using tracheal sounds. Deep inhalation followed by a protracted exhalation. Catathrenia has a respiratory pattern that may mimic central apnea, but during which groaning sounds are produced, usually lasting much longer than regular snoring

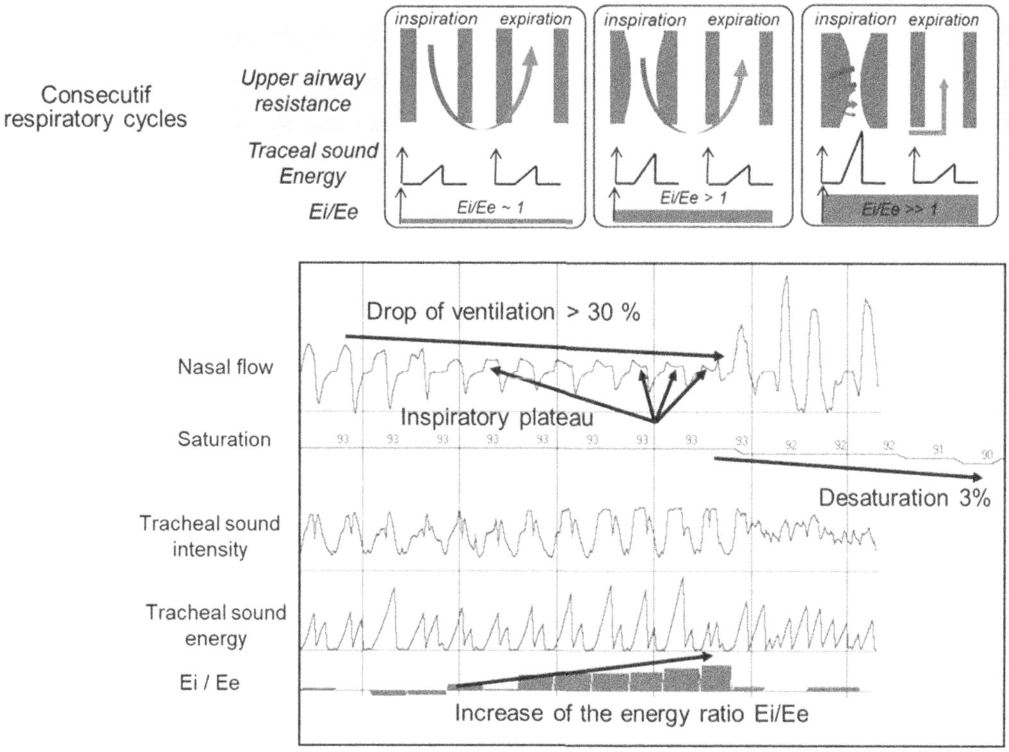

Fig. 16.10 Illustration of tracheal sound energy signal showing how acoustic intensity increases with friction. With increased inspiratory upper airway resistance, the inspiratory energy (Ei) is greater than the expiratory energy (Ee), and the energy ratio Ei/Ee increases

16.7.3 Cardiogenic Oscillations: TS for Heart Rate Variability

Heart rate (HR) variability is an important physiological parameter to be assess during sleep studies, and it is traditionally obtained through electrocardiography (ECG) or photoplethysmography (PPG) recordings included in PSG or HSAT systems. However, phonocardiography (PCG), a recording of sounds and murmurs resulting from heart auscultation (Ismail et al., 2018), is a widely used method of listening to the heart sounds using a microphone sensor placed on the chest of a subject, an auscultation site where the heart sounds are loudest. Each heart cycle consists of two major sounds, S1 and S2, that can be used to determine the heart rate.

TS signals may include cardiogenic oscillations picked up by the sensor (Glos et al., 2018; Freycenon et al., 2021; Gómez et al., 2020). The origin of these oscillations is the same as that of the oscillations observed on the Pes signal or airflow signal, proposed as a marker of central apnea (Ayappa et al., 1999; Morrell et al., 1995). Given that the heart sound frequency is lower than the frequency of the tracheal breath sound, cardiogenic oscillations (Fig. 16.11) could be filtered out, and in turn, the filtered signal could be used to evaluate the heart rate variability. Sharma et al. proposed a novel algorithm for heart rate monitoring using acoustic signals from a TS sensor placed at the suprasternal notch (Sharma et al., 2019). The algorithm constructs the Hilbert energy envelope of the signal by calculating its instantaneous characteristics to segment and classify a cardiac cycle into S1 and S2 sounds using their timing characteristics. Using this method, they achieved an accuracy of 94%, an RMS error of 3.96 bpm, and a correlation coefficient of 0.93 with reference to Somnoscreen® (Randersacker, Germany), a standard commercial FDA-approved heart rate monitor (Sharma et al., 2019). In another study, Freycenon et al. tested algorithms based on optimal filtering to extract the cardiac signal from tracheal sounds (Freycenon et al., 2021). Their method consists of denoising the tracheal signal by optimal filtering followed by a low-pass filter and by estimating the heart rate (HR) using cross-correlation. When compared to a reference ECG signal, these algorithms extracting heart rate from TS signals reached an accuracy of 81–98% and an RMS error from 1.3 to 4.2 bpm depending on the level of snoring (Freycenon et al., 2021). These studies suggest that TS sensors can be used to monitor both breathing and heart rate, making it highly useful for monitoring heart rate variability during sleep studies.

16.7.4 Detection of Obstruction Sites: Could TS Be an Alternative to DISE?

When considering surgical treatment of OSA, it is important to accurately determine the predominant upper airway obstructive site. The relationship between the acoustic property of snoring sound and the site of obstruction has been investigated (Herzog et al., 2015; Peng et al., 2017; Xu et al., 2010). Xu et al. showed that there is a significant difference in sound spectrum between snoring sound immediately following velopharyngeal (above the free margin of the soft palate) and oropharyngeal (below the free margin of the soft palate) obstructive apneas. They suggest that sound spectrum analysis can be used as a method to determine obstruction sites in OSA patients. They demonstrated that an obstruction level above the free margin of the soft palate produces a characteristic frequency and energy in the low frequency domain (Fig. 16.12a), whereas an obstruction level below the free margin of the soft palate generates a characteristic frequency and energy in the high frequency domain (Fig. 16.12b) (Xu et al., 2010). In another study, Herzog et al. proposed an acoustic analysis to classify patterns of obstructions and vibration during drug-induced sleep endoscopy (DISE) and to evalu-

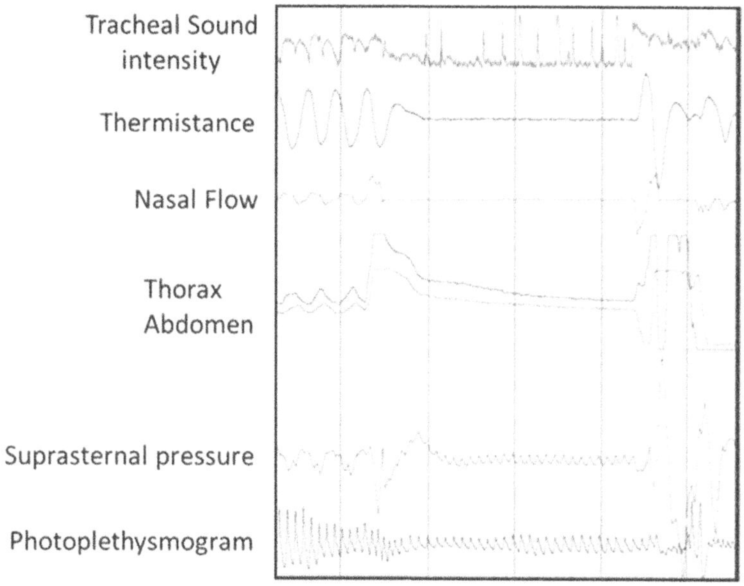

Fig. 16.11 Example of cardiogenic oscillations seen during a central apnea on the suprasternal pressure derived from the tracheal sounds and on the photoplethysmogram signal derived from oximetry

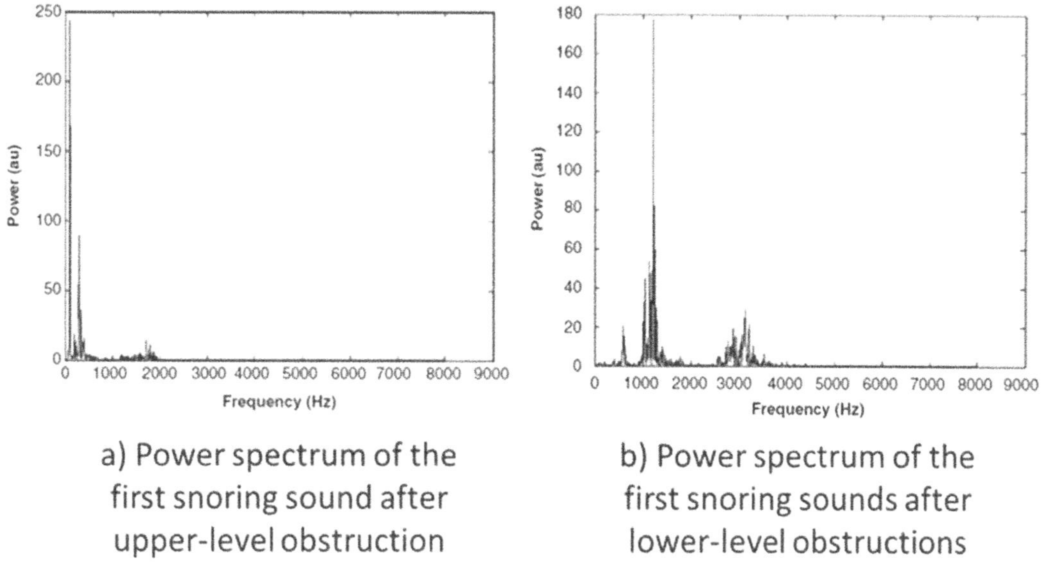

Fig. 16.12 Comparison of power spectrum of the first snoring sound after upper-level (**a**) and lower-level (**b**) obstructions (Herzog et al., 2015)

ate acoustic characteristics between these different patterns of snoring. Obstructive snoring patterns revealed a higher pitch than non-obstructive patterns. Velar snoring showed more turbulence than tonsillar and post-apneic snoring and revealed the lowest center frequency of all patterns. Tonsillar snoring presented the highest sharpness, whereas post-apneic snoring revealed the largest fluctuation strength (Herzog et al., 2015). In a logistic regression analysis comparing snoring sound analysis to DISE results, Lee et al. showed that patients with at least two complete obstruction sites defined by DISE were significantly associated with maximal snoring sound intensity (40–300 Hz) and a higher body mass index. Tonsil obstruction was significantly and inversely correlated with mean snoring sound intensity (301–850 Hz). These findings suggest that snoring sound detection may be helpful in determining obstruction sites.

16.8 Conclusion

The use of TS during sleep has been studied extensively for the last 40 years. The bulk of these studies shows that tracheal sounds recorded with appropriate sensors well placed just above the tracheal notch, rather than simple ambient microphones, can assess variations in breathing, snoring, and respiratory effort and heart rate variations and UAW obstruction sites. All these parameters are essential to diagnostic sleep studies. Compared to other PSG or HSAT respiratory sensors, TS sensors offer simple, noninvasive measurements. Their multiple-derived signals are reliable and may have a significant role in the diagnosis and classification of SDB. The development of new digital acoustic signal processing techniques and the enhancement of tracheal sound sensors over the past decade have led to improved accuracy for SDB diagnosis. Some aspects, such as detection and characterization of respiratory events, are well confirmed in several studies using automatic algorithms based on spectral analysis and time domain visual analysis of TS-derived signals. Thus, TS technology meets the SCOPER criteria (Collop et al., 2011) (sleep, cardiovascular, oximetry, position, effort, and respiratory) for the diagnosis of SDB. However, heart rate variability and detection of UAW obstruction sites based on TS analysis could still be improved, and further studies are needed to establish the validity of these surrogate sensors for this valuable information.

In conclusion, associated with appropriate sensor placement and advanced signal processing techniques, the use of tracheal sound transducers as complementary with routine PSG or HSAT sensors can extensively contribute to the diagnosis and assessment of SDB and, thus, improve treatment strategies. In addition to their relevance in diagnostic sleep studies, standalone tracheal sound technology should also be used as a reliable SDB screening tool.

References

Alvarez, D., et al. (2008). Applying time, frequency and nonlinear features from nocturnal oximetry to OSA diagnosis. In *2008 30th annual international conference of the IEEE engineering in medicine and biology society* (pp. 3872–3875). IEEE. https://doi.org/10.1109/IEMBS.2008.4650055

Amaddeo, A., et al. (2016). Validation of a suprasternal pressure sensor for sleep apnea classification in children. *Journal of Clinical Sleep Medicine, 12*, 1641–1647.

Amaddeo, A., et al. (2020). Tracheal sounds for the scoring of sleep respiratory events in children. *Journal of Clinical Sleep Medicine, 16*, 361–369.

Ayappa, I., Norman, R. G., & Rapoport, D. M. (1999). Cardiogenic oscillations on the airflow signal during continuous positive airway pressure as a marker of central apnea. *Chest, 116*, 660–666.

Azarbarzin, A., & Moussavi, Z. (2013). Snoring sounds variability as a signature of obstructive sleep apnea. *Medical Engineering & Physics, 35*, 479–485.

Beck, R., et al. (2005). Measurements and theory of normal tracheal breath sounds. *Annals of Biomedical Engineering, 33*, 1344–1351.

Beckerman, R. C., Wegmann, M. J., & Waring, W. W. (1982). Tracheal breath sounds for detection of apnea in infants and children. *Critical Care Medicine, 10*, 363–366.

Berry, R. B., et al. (2012). Rules for scoring respiratory events in sleep: Update of the 2007 AASM Manual for the Scoring of Sleep and Associated Events. Deliberations of the Sleep Apnea Definitions Task Force of the American Academy of Sleep Medicine. *Journal of Clinical Sleep Medicine, 8*, 597–619.

Berry, R. B., et al. (2015). AASM scoring manual version 2.2 updates: New chapters for scoring infant sleep staging and home sleep apnea testing. *Journal of Clinical Sleep Medicine, 11*, 1253–1254.

Chervin, R. D., & Aldrich, M. S. (1997). Effects of esophageal pressure monitoring on sleep architecture. *American Journal of Respiratory and Critical Care Medicine, 156*, 881–885.

Chuah, J. S., & Moussavi, Z. (2001). *Automated respiratory phase detection by acoustical means*. https://www.semanticscholar.org/paper/AUTOMATED-RESPIRATORY-PHASE-DETECTION-BY-ACOUSTICAL-Chuah-Moussavi/33557244596d8b1aad8e14c8fd18b32f31b4e43b

Collop, N. A., et al. (2011). Obstructive sleep apnea devices for out-of-center (OOC) testing: Technology evaluation. *Journal of Clinical Sleep Medicine, 07*, 531–548.

Cummiskey, J., Williams, T. C., Krumpe, P. E., & Guilleminault, C. (1982). The detection and quantification of sleep apnea by tracheal sound recordings. *The American Review of Respiratory Disease, 126*, 221–224.

Eberhard, A., Calabrese, P., Baconnier, P., & Benchetrit, G. (2001). Comparison between the respiratory inductance plethysmography signal derivative and the airflow signal. In C.-S. Poon & H. Kazemi (Eds.), *Frontiers in modeling and control of breathing* (Vol. 499, pp. 489–494). Springer US.

Elwali, A., & Moussavi, Z. (2017). Erratum to: Obstructive sleep apnea screening and airway structure characterization during wakefulness using tracheal breathing sounds. *Annals of Biomedical Engineering, 45*, 851–851.

Escourrou, P., et al. (2010). Quelle approche clinique et quelle procédure diagnostique pour le SAHOS? *Revue des Maladies Respiratoires, 27*, S115–S123.

Freycenon, N., Longo, R., & Simon, L. (2021). Estimation of heart rate from tracheal sounds recorded for the sleep apnea syndrome diagnosis. *IEEE Transactions on Biomedical Engineering, 68*, 3039–3047.

Gavriely, N., & Cugell, D. W. (1996). Airflow effects on amplitude and spectral content of normal breath sounds. *Journal of Applied Physiology, 80*, 5–13.

Glos, M., et al. (2018). Characterization of respiratory events in obstructive sleep apnea using suprasternal pressure monitoring. *Journal of Clinical Sleep Medicine, 14*, 359–369.

Gómez, T., et al. (2020). Catathrenia resolved with the lowest CPAP pressure settings. *Pulmonology, 26*, 107–110.

Goodwin, J. L., et al. (2001). Feasibility of using unattended polysomnography in children for research—Report of the Tucson Children's Assessment of Sleep Apnea Study (TuCASA). *Sleep, 24*, 937–944.

Guilleminault, C., Hagen, C. C., & Khaja, A. M. (2008). Catathrenia is not expiratory snoring. *Sleep, 31*, 774–775.

Harper, V. P., Pasterkamp, H., Kiyokawa, H., & Wodicka, G. R. (2003). Modeling and measurement of flow effects on tracheal sounds. *IEEE Transactions on Biomedical Engineering, 50*, 1–10.

Herzog, M., et al. (2015). Evaluation of acoustic characteristics of snoring sounds obtained during drug-induced sleep endoscopy. *Sleep & Breathing, 19*, 1011–1019.

Huq, S., & Moussavi, Z. (2012). Acoustic breath-phase detection using tracheal breath sounds. *Medical & Biological Engineering & Computing, 50*, 297–308.

Iriarte, J., et al. (2011). Sound analysis of catathrenia: A vocal expiratory sound. *Sleep & Breathing, 15*, 229–235.

Ismail, S., Siddiqi, I., & Akram, U. (2018). Localization and classification of heart beats in phonocardiography signals —A comprehensive review. *EURASIP Journal on Advances in Signal Processing, 2018*, 26.

Krumpe, P. E., & Cummiskey, J. M. (1980). Use of laryngeal sound recordings to monitor apnea. *The American Review of Respiratory Disease, 122*, 797–801.

Kulkas, A., et al. (2008). Detection of compressed tracheal sound patterns with large amplitude variation during sleep. *Medical & Biological Engineering & Computing, 46*, 315–321.

Mayer, G., et al. (2017). German S3 Guideline Nonrestorative Sleep/Sleep Disorders, chapter "Sleep-Related Breathing Disorders in Adults," short version: German Sleep Society (Deutsche Gesellschaft für Schlafforschung und Schlafmedizin, DGSM). *Somnologie, 21*, 290–301.

Meslier, N., & Racineux, J. L. (1987). Use of tracheal sound recordings to monitor airflow during sleep. In J. H. Peter, T. Podszus, & P. von Wichert (Eds.), *Sleep related disorders and internal diseases* (pp. 121–124). Springer Berlin Heidelberg. https://doi.org/10.1007/978-3-642-72560-9_11

Meslier, N., et al. (2002). Validation of a suprasternal pressure transducer for apnea classification during sleep. *Sleep, 25*, 753–757.

Mlynczak, M., Migacz, E., Migacz, M., & Kukwa, W. (2017). Detecting breathing and snoring episodes using a wireless tracheal sensor—A feasibility study. *IEEE Journal of Biomedical and Health Informatics, 21*, 1504–1510.

Montazeri, A., & Moussavi, Z. (2010). Acoustical screening for obstructive sleep apnea during wakefulness. In *2010 annual international conference of the IEEE engineering in medicine and biology* (pp. 3662–3665). IEEE. https://doi.org/10.1109/IEMBS.2010.5627653

Montazeri, A., Giannouli, E., & Moussavi, Z. (2012). Assessment of obstructive sleep apnea and its severity during wakefulness. *Annals of Biomedical Engineering, 40*, 916–924.

Morrell, M. J., Badr, M. S., Harms, C. A., & Dempsey, J. A. (1995). The assessment of upper airway patency during apnea using cardiogenic oscillations in the airflow signal. *Sleep, 18*, 651–658.

Nakano, H., Hayashi, M., Ohshima, E., Nishikata, N., & Shinohara, T. (2004). Validation of a new system of tracheal sound analysis for the diagnosis of sleep apnea-hypopnea syndrome. *Sleep, 27*, 951–957.

Peng, H., et al. (2017). Acoustic analysis of snoring sounds originating from different sources determined by drug-induced sleep endoscopy. *Acta Oto-Laryngologica, 137*, 872–876.

Penzel, T., & Sabil, A. (2018). Physics and applications for tracheal sound recordings in sleep disorders. In K. N. Priftis, L. J. Hadjileontiadis, & M. L. Everard (Eds.), *Breath sounds* (pp. 83–104). Springer International Publishing. https://doi.org/10.1007/978-3-319-71824-8_6

Penzel, T., Amend, G., Peter, J. H., Podszus, T., Von Wichert, P., & Zahorka, M. (1988). Objective monitoring of therapeutical success in heavy snorers: A new technique. In C. H. Chouard (Ed.), *Chronic rhonchopathy* (pp. 273–278). John Libbey Eurotext.

Penzel, T., Amend, G., Meinzer, K., Peter, J. H., & von Wichert, P. (1990). MESAM: A heart rate and snoring recorder for detection of obstructive sleep apnea. *Sleep, 13*, 175–182.

Priftis, K. N., Hadjileontiadis, L. J., & Everard, M. L. (2018). *Breath sounds: From basic science to clinical practice*. Springer.

Que, C.-L., Kolmaga, C., Durand, L.-G., Kelly, S. M., & Macklem, P. T. (2002). Phonospirometry for noninvasive measurement of ventilation: Methodology and preliminary results. *Journal of Applied Physiology, 1985*(93), 1515–1526.

Racineux, J. -L. (2006). [Tracheal noise and resistance in the upper airways]. *Revue des Maladies Respiratoires, 23 Spec No 2*, 7S48-47S50.

Rauhala, E., et al. (2008). Compressed tracheal sound analysis in screening of sleep-disordered breathing. *Clinical Neurophysiology, 119*, 2037–2043.

Redline, S., et al. (1998). Methods for obtaining and analyzing unattended polysomnography data for a multicenter study. Sleep Heart Health Research Group. *Sleep, 21*, 759–767.

Rodrigues, D. M., Valério, M. P., & Costa, T. (2021). Catathrenia - a rare but disturbing sleep disorder. *Archivos de bronconeumología, S0300-2896*(21), 00044–00042. https://doi.org/10.1016/j.arbres.2021.01.022

Sabil, A., Baffet, G., Freycenon, C., & Pinguet, J. (2017). Apnoea characterisation using tracheal sounds spectral analysis. In *Obstructive sleep apnoea*. P18. European Respiratory Society. https://doi.org/10.1183/23120541.sleepandbreathing-2017.P18

Sabil, A., et al. (2019a). Apnea and hypopnea characterization using esophageal pressure, respiratory inductance plethysmography, and suprasternal pressure: A comparative study. *Sleep & Breathing, 23*, 1169–1176.

Sabil, A., et al. (2019b). Comparison of apnea detection using oronasal thermal airflow sensor, nasal pressure transducer, respiratory inductance plethysmography and tracheal sound sensor. *Journal of Clinical Sleep Medicine, 15*, 285–292.

Sabil, A., et al. (2020). Diagnosis of sleep apnea without sensors on the patient's face. *Journal of Clinical Sleep Medicine, 16*, 1161–1169.

Sanchez, I., & Pasterkamp, H. (1993). Tracheal sound spectra depend on body height. *The American Review of Respiratory Disease, 148*, 1083–1087.

Sharma, P., Imtiaz, S. A., & Rodriguez-Villegas, E. (2019). An algorithm for heart rate extraction from acoustic recordings at the neck. *IEEE Transactions on Biomedical Engineering, 66*, 246–256.

Siddiqui, F., Walters, A. S., & Chokroverty, S. (2008). Catathrenia: A rare parasomnia which may mimic central sleep apnea on polysomnogram. *Sleep Medicine, 9*, 460–461.

Solà-Soler, J., Fiz, J. A., Morera, J., & Jané, R. (2012). Multiclass classification of subjects with sleep apnoea–hypopnoea syndrome through snoring analysis. *Medical Engineering & Physics, 34*, 1213–1220.

Soufflet, G., et al. (1990). Interaction between tracheal sound and flow rate: A comparison of some different flow evaluations from lung sounds. *IEEE Transactions on Biomedical Engineering, 37*, 384–391.

Van Surell, C., et al. (1995). Evaluation of an ambulatory device, CID 102, in the diagnosis of obstructive sleep apnoea syndrome. *The European Respiratory Journal, 8*, 795–800.

Vetrugno, R., et al. (2007). Catathrenia (nocturnal groaning): An abnormal respiratory pattern during sleep. *European Journal of Neurology, 14*, 1236–1243.

Vetrugno, R., Lugaresi, E., Ferini-Strambi, L., & Montagna, P. (2008). Catathrenia (nocturnal groaning): What is it? *Sleep, 31*, 308–309.

Xu, H., Huang, W., Yu, L., & Chen, L. (2010). Sound spectral analysis of snoring sound and site of obstruction in obstructive sleep apnea syndrome. *Acta Oto-Laryngologica, 130*, 1175–1179.

Yadollahi, A., & Moussavi, Z. (2005). Measuring minimum critical flow for normal breath sounds. *Conference Proceedings: Annual International Conference of the IEEE Engineering in Medicine and Biology Society, 2005*, 2726–2729.

Yadollahi, A., & Moussavi, Z. M. (2007). Acoustical respiratory flow. A review of reliable methods for measuring airflow. *EEE Engineering in Medicine and Biology Magazine, 26*(1), 56–61

Yadollahi, A., & Moussavi, Z. (2009). A. coustic obstructive sleep apnea detection. In *2009 annual international conference of the IEEE engineering in medicine and biology society* (pp. 7110–7113). IEEE. https://doi.org/10.1109/IEMBS.2009.5332870

Yadollahi, A., & Moussavi, Z. (2010). Automatic breath and snore sounds classification from tracheal and ambient sounds recordings. *Medical Engineering & Physics, 32*, 985–990.

Yadollahi, A., & Moussavi, Z. (2011). Detailed analysis of the relationship between tracheal breath sounds and airflow in relation to OSA during wake and sleep. In *2011 annual international conference of the IEEE engineering in medicine and biology society* (pp. 6797–6800). IEEE. https://doi.org/10.1109/IEMBS.2011.6091676

Yadollahi, A., Giannouli, E., & Moussavi, Z. (2010). Sleep apnea monitoring and diagnosis based on pulse oximetry and tracheal sound signals. *Medical & Biological Engineering & Computing, 48*, 1087–1097.

Obstructive Sleep Apnea with COVID-19

17

Ying Huang, DongMing Chen, Ingo Fietze, and Thomas Penzel

Abstract

The novel coronavirus disease-2019 (COVID-19) and the ensuing pandemic have greatly impacted the global healthcare system due to its high infectiousness, associated high mortality, and a complete lack of immunity in the population. Globally, the COVID-19 pandemic has unleashed a health crisis that has not only seriously disrupted people's lives but also affected their normal sleep, along with physical and mental health; this situation is especially exacerbated in people suffering from pre-existing conditions, such as sleep apnea. A recent meta-analysis of 18 studies by Miller et al. (September 2020) showed that obstructive sleep apnea (OSA) is related to higher mortality and morbidity in patients with COVID-19 and is most likely independent of other risk factors. A recent meta-analysis indicated that COVID-19 patients with OSA are more severely affected than those without OSA, thereby providing further evidence that concurrent OSA may elevate the severity of COVID-19 infection, along with the risk of mortality. The COVID-19 pandemic has significantly impacted the diagnosis and therapeutic management of patients with OSA. Thus, it is necessary to identify and develop new diagnostic and therapeutic avenues in the future. In this context, the current study summarizes known associations between COVID-19 and OSA and the regular diagnostic and therapeutic strategies for OSA in the light of COVID-19 pandemic prevention and control.

Keywords

Obstructive sleep apnea · Coronavirus disease-2019 · Diagnosis · Treatment

17.1 Introduction

Coronaviruses are common pathogens in humans associated with intestinal and respiratory diseases (Cui et al., 2019). In most cases, infections due to the following human coronaviruses cause mild symptoms: human coronavirus 229E (HCoV-229E), human coronavirus NL63

Y. Huang (✉) · I. Fietze · T. Penzel
Charité – Universitätsmedizin Berlin, Interdisciplinary Sleep Medicine Center, Berlin, Germany
e-mail: ying.huang@charite.de; ingo.fietze@charite.de; thomas.penzel@charite.de

D. Chen
Charité – Universtitätsmedizin Berlin, Cooperate Member of Freie Universität and Humboldt Universität, Berlin, Germany

Department of Neurosurgery, The First Affiliated Hospital of Anhui University of Traditional Chinese Medicine, Hefei, Anhui Province, China

T. Penzel, R. Hornero (eds.), *Advances in the Diagnosis and Treatment of Sleep Apnea*, Advances in Experimental Medicine and Biology 1384, https://doi.org/10.1007/978-3-031-06413-5_17

(HCoV-NL63), human coronavirus OC43 (HCoV-OC43), and HKU1. However, these coronaviruses can also lead to severe respiratory infections, including pneumonia, as observed previously in patients with the Middle East respiratory syndrome (MERS) and severe acute respiratory syndrome-coronavirus (SARS-CoV) (Cui et al., 2019). In December 2019, a novel respiratory coronavirus, SARS-CoV-2, was identified in Wuhan, China, that triggered a cascade of highly contagious respiratory infections, which spread and transmitted rapidly worldwide and resulted in a global pandemic (Zhu et al., 2020). On March 11, 2020, the global spread of COVID-19 (caused by SARS-CoV-2) forced the World Health Organization (WHO) to declare it as a pandemic (WHO Director, 2020). By November 2021, the worldwide spread of COVID-19 had resulted in more than 256 million positive cases and over 5.13 million deaths in more than 200 countries (World Health Organization, 2021).

The primary routes of transmission of SARS-CoV-2 are close contact and via respiratory droplets. The spike protein S mediates its entry into the host cells. This protein has high affinity, is tightly bound to, and acts as a receptor for the angiotensin-converting enzyme 2 (ACE2), abundantly expressed in the airway epithelial cells. SARS-CoV-2 binds to ACE2 using a host protease (transmembrane protease serine 2, TMPRSS2) to activate the S protein, thereby facilitating entry into the cells (Cevik et al., 2020; Yao et al., 2020; Millet & Whittaker, 2015). After entering the host cells, the viruses use the host machinery to convert RNA into polypeptides, including the RNA-dependent RNA polymerases that the virus consequently uses for its RNA replication processes. Following the structural protein synthesis and particle assembly, new viruses are secreted from the host cells through exocytosis. Host cells can get destroyed during the process, potentially triggering an innate immune response (Liu et al., 2020). Depending on an individual's immune response or the presence of risk factors, the primary clinical manifestations of COVID-19 range from an asymptomatic disease state to acute lung injury, acute respiratory distress syndrome (ARDS), severe pneumonia, multi-organ dysfunction, and eventually death (Yao et al., 2020; Tiwari et al., 2020). A few days post-exposure (average of 4–7 days), people infected with SARS-CoV-2 may develop respiratory symptoms (e.g., runny nose, dyssomnia, cough, and dyspnea), along with fever, fatigue, myalgia, and/or diarrhea (Chhiba et al., 2020). Several known factors are linked to poor prognosis in patients with COVID-19, including advanced age (over 65 years), male gender, obesity, comorbid chronic respiratory disease, and the presence of cardiovascular and metabolic disorders (Chhiba et al., 2020; Altibi et al., 2021; Zhou et al., 2020).

Obstructive sleep apnea (OSA), a disorder where a partial or complete collapse of the upper air tract occurs periodically during sleep, results in interrupted airflow, decreased oxygen saturation during sleep, and nocturnal sleep fragmentation; these are important factors that affect the immune system (Besedovsky et al., 2019; Said et al., 2017). OSA is not only highly prevalent (increases with age) but also tightly linked to gender (male), hypertension, diabetes, obesity, and coronary heart disease (CHD). These diseases are also associated with an unfavorable prognosis for COVID-19 (Gottlieb & Punjabi, 2020; Sweed et al., 2019; Cappuccio & Siani, 2020). Therefore, the correlation between OSA and COVID-19 has attracted great research attention from clinicians and the scientific community. The current review summarizes the known relationship between COVID-19 and OSA, along with diagnostic and therapeutic strategies for OSA in the context of routine mitigation and control of the COVID-19 pandemic.

17.2 Influence of OSA on Incidence, Disease Severity, and Mortality in COVID-19

OSA and COVID-19 share several risk factors and comorbidities in common, including age (increasing with age), gender (male), obesity (high body mass index), hypertension, diabetes, and CHD. In addition, cytokine storm and alveolar macrophage activation are strongly associated with disease exacerbation, resulting even in the death of patients with COVID-19. OSA is a chronic low-level inflammatory disease, and patients with untreated OSA are at a high risk of cardiovascular disease and death (Gottlieb & Punjabi, 2020; Sweed et al., 2019; Cappuccio & Siani, 2020; Mokmeli & Vetrici, 2020). Therefore, it is reasonably hypothesized that OSA may contribute to SARS-CoV-2 infection. Once an infection develops, it may trigger higher incidences of cardiovascular diseases, including heart ischemia, arrhythmias, and hypercoagulability, eventually leading to an unfavorable clinical progression (Tufik et al., 2020).

Due to several changes in their everyday life, loneliness, fear, feelings of helplessness, and high mental burden during the COVID-19 pandemic, many people may have experienced difficulty in sleeping where their sleep quality has been significantly impaired (Pinto et al., 2020). A cross-sectional study consists of 14 countries with a total of 20,598 participants, which is part of the International COVID-19 Sleep Study (ICOSS), has shown that participants with high risk of OSA had higher odds of reporting a COVID-19 diagnosis and were twofold increased odds of hospitalization or ICU treatment (Chung et al., 2021; Partinen et al., 2021). Peker and colleagues have shown that patients with COVID-19 with high-risk OSA had poorer clinical outcomes compared with those with modified low-risk OSA (Peker et al., 2021). A study from France, wherein the researchers analyzed 124 patients with COVID-19, showed that after excluding confounding factors such as age, diabetes, and hypertension, obesity (body mass index [BMI] > 30 kg/m^2) is an independent risk factor for the use of invasive mechanical ventilation (Simonnet et al., 2020). OSA is strongly associated with severe COVID-19-related major comorbidities, including diabetes, hypertension, cardiovascular disease (CVD), and obesity (Jordan et al., 2014; Mashaqi et al., 2021; Gill et al., 2021). Data from two small-sample size-based investigations on patients with severe COVID-19 showed that a quarter of patients suffered from OSA (Bhatraju et al., 2020; Arentz et al., 2020). Strausz et al. using the FinnGen large biorepository (https://www.finngen.fi/en) examined whether OSA independently increased the risk in patients contracting the COVID-19 infection and disease severity after infection after excluding other potential risk factors (including sex, age, BMI, diabetes, hypertension, etc.) (Strausz et al., 2021). The results showed that of the 445 patients positively diagnosed with COVID-19, 38 (8.5%) were previously diagnosed with OSA; of the 91 patients hospitalized with COVID-19, 19 (20.9%) suffered from OSA. Patients with OSA experienced a higher risk of hospitalization owing to the disease severity after contracting the COVID-19 infection as compared to those without OSA; the findings were independent of age, gender, BMI, and other comorbidities that could lead to severe COVID-19. However, their study did not confirm whether OSA increased the risk of COVID-19 infection; however, they reported its correlation with severe COVID-19 infection (odds ratio [OR], 2.37; 95% confidence interval [CI], 1.14–4.95) (Strausz et al., 2021).

Another study reported the risk factors commonly shared between COVID-19 and OSA, including age, CVD, hypertension, pulmonary disease, and diabetes (Pazarlı et al., 2021). A recent study comprising 4756 patients positive for SARS-CoV-2 infection in Iceland showed that among them, 185 had been previously diagnosed with OSA (Rögnvaldsson et al., 2021). Of the 238 COVID-19 patients who were either hospitalized or died, 38 (20.9%) had OSA; after adjusting for the confounding factors of age, gender, and BMI, OSA was found to be related to poor outcome (OR, 2.2; 95% CI, 1.4–3.5). Several studies examined the relationship between OSA and the COVID-19 infection risk, mortality risk, and infection severity. The findings showed that OSA is strongly related to the infection severity of COVID-19 (Tufik et al., 2020; Cade et al., 2020; Maas et al., 2021). In addition, patients with OSA are not at a significantly increased risk of COVID-19 infection as compared to the non-OSA individuals; however, they possess a significantly enhanced risk of developing grave complications and hospitalization following the COVID-19 infection. OSA is an independent risk factor for severe COVID-19 infection. Therefore, OSA can be considered as an independent risk factor for patients with confirmed severe or suspected infection of COVID-19 (Rögnvaldsson et al., 2021).

17.3 Putative Mechanistic Pathways Underlying the Impact of COVID-19 Infection on OSA

Several inflammatory and biochemical mechanisms are associated with the progression of OSA and the related adverse outcomes; among them, some are associated with COVID-19 (Gill et al., 2021). People bearing these risk factors and/or diseases are at an increased risk of developing severe symptoms following COVID-19 infection; the risk of death also increases (Strausz et al., 2021; Miller & Cappuccio, 2021). Among them, obesity is a predisposing risk factor for severe symptoms. A study from France comprising 124 patients with COVID-19 infection reported that after excluding confounding factors such as age, hypertension, and diabetes, obesity (BMI > 30 kg/m^2) is an independent risk factor for the use of IMC (Simonnet et al., 2020). Cytokine storm and alveolar macrophage activation are strongly linked to disease exacerbation and even death in COVID-19 patients. OSA is a chronic low-level inflammatory disease (Gottlieb & Punjabi, 2020; Sweed et al., 2019; Cappuccio & Siani, 2020; Mokmeli & Vetrici, 2020). Fragmented sleep and intermittent blood–qi disturbances (hypercapnia and hypoxemia) caused by OSA can enhance neuronal excitability in the sympathetic nervous system and increase inflammatory marker levels, especially in obese patients, thereby further exacerbating disease severity among the COVID-19 patients (Kohler & Stradling, 2010; Jose & Manuel, 2020).

Melatonin can improve the prognosis of patients with COVID-19 infection, particularly among those diagnosed with OSA. This may be attributed to its attenuating effects on oxidative stress, inflammation, and immune responses (Shneider et al., 2020; El-Missiry et al., 2020). In addition, melatonin also improves sleep quality, thereby facilitating better clinical outcomes among COVID-19-infected patients (Zhang et al., 2020).

Reduced levels of vitamin D are not only associated with the severity of OSA but are also a risk factor for the body's susceptibility to CVD and are related to poor prognoses among COVID-19 patients (Archontogeorgis et al., 2018; Kassi et al., 2013; Derakhshanian et al., 2021). Vitamin D can inhibit the activity of reactive oxygen species, stimulate protective endothelial nitric oxide production, and reduce levels of inflammatory mediators (Kassi et al., 2013). In a previous study comprising 176 children, Kheirandish-Gozal et al. show that OSA is independently related to enhanced levels of high-sensitivity C-reactive protein and lipids and lower levels of vitamin D (Kheirandish-Gozal et al., 2014). High levels of vitamin D also enhance cellular immunity, reduce innate immune-system-induced cytokine storms, and are related to favorable prognoses of COVID-19 patients (Grant et al., 2020). However, a recent study conducted by the UK Biobank refuted this view due to the lack of any such evidence in support (Hastie et al., 2020). Of the 348,598 participants, 449 were positive for COVID-19 infection. Although low vitamin D level was related to COVID-19 infection, this trend disappeared after adjusting for confounding factors. Therefore, these findings refuted the conclusions in favor of the likely association of vitamin D level and the risk of contracting COVID-19 infection.

A meta-analysis by Jin et al. using published reports on the relevance of OSA in the renin–angiotensin–aldosterone system (RAAS) showed that OSA is linked to enhanced levels of angiotensin II and aldosterone, particularly in individuals with hypertension (Jin & Wei, 2016). Therefore, an opinion that OSA may contribute to increased blood pressure through the RAAS system has been proposed (Jin & Wei, 2016). ACE2, a functional SARS-CoV-2 receptor, allows viral access to host cells (Letko et al., 2020). Therefore, patients with OSA have a higher probability of developing CVD following COVID-19 infection.

In the early infection stages, the virus, SARS-CoV-2, often invades the body's immune system. The key to fighting the virus is dependent on the body's immune status. Although sleep disorders alone do not significantly alter the number of immune cells, chronic insomnia changes the relative distribution of immune cell types; dramatic decreases in CD^{3+}, CD^{4+}, and CD^{8+} cell counts have been reported (Irwin, 2015; Mello et al., 2020). Patients with OSA often have disturbed sleep architecture, which negatively affects the immune system and reduces immunity, thereby increasing the risk of viral infections.

Idiopathic pulmonary fibrosis (IPF), a common chronic pulmonary disorder, leads to a decreased saturation of oxygen and pulmonary hypertension. OSA is frequent in patients diagnosed with IPF. Recent guidelines for IPF have identified OSA as crucial comorbidity affecting the survival of patients (Schiza et al., 2015). Therefore, it is advised that the newly diagnosed IPF patients be referred to a sleep center for OSA diagnosis and treatment (Lancaster et al., 2009). Some patients with COVID-19 infection have remanent fibrosis of varying degrees in the lungs, even after recovery (Ye et al., 2020). Therefore, there is a need to screen for OSA in patients with pulmonary fibrosis who have had the COVID-19 infection. However, it is unclear whether patients with OSA who have fibrosis are at an increased risk owing to the effects of COVID-19 infection, and this warrants further investigation.

17.4 OSA Diagnosis During the COVID-19 Pandemic

With the global spread and increased risk of the COVID-19 infection, several sleep centers have been shut down to minimize the virus spread. Thus, initial consultations must now be performed remotely, and home consultations are preferred for diagnosis of OSA, with devices being couriered by post to avoid interpersonal contact with patients. Indeed, the COVID-19 pandemic has significantly impacted sleep services in the United Kingdom and abroad (Morin et al., 2021). Grote et al. in their recent study, analyzed the effects of the COVID-19 pandemic on consultations for individuals with respiratory sleep diseases in 19 European countries; a total of 40 centers were surveyed, and the results showed that patients in 31 centers were unable to attend in-person sessions due to travel restrictions (Grote et al., 2020). In terms of diagnosis, before the pandemic, 92.5% of the sleep laboratories diagnosed OSA using polysomnography (PSG), 87.5% also used portable home-based sleep monitors, and 30% conducted telemedical consultations. During the pandemic, however, the proportion of sleep laboratories that diagnose OSA by PSG decreased significantly to 20%, while the proportion of home-based sleep monitors fell to 32.5%, and that through telemedical consultations was only 27.5%. A significant decrease in the proportion of patients receiving different types of positive airway pressure (PAP) treatment for sleep breathing disorders (SDBs) in sleep centers in the majority of countries is also reported. Additionally, the proportion of physicians and nursing or technical staff practicing sleep medicine fell to 25% and 19%, respectively, relative to the pre-pandemic levels. In areas having a large number of reported infection cases, such as in Wuhan and the surrounding cities in the Hubei Province, China, sleep laboratories have suspended laboratory testing and non-invasive PAP treatment (except in emergencies). In low-risk areas, laboratory sleep monitoring has gradually resumed following a rigorous screening procedure to rule out the possibility of infection. Moreover, other sleep diagnostic and treatment services and follow-up appointments are offered remotely or online (Zhang & Xiao, 2020).

The American Academy of Sleep Medicine (AASM) released its revised recommendations in late April 2020 to assist clinicians practicing sleep medicine during the COVID-19 pandemic (American Academy of Sleep Medicine (AASM), 2021). These recommendations were made under the guidance of the Center for Disease Control and Prevention to control and prevent the risk of COVID-19 infection spread in healthcare settings (Centers for Disease Control and Prevention (CDC), 2021). For necessary testing, the use of disposable or completely disposable devices (components) is recommended. The European Sleep Research Society has released its guidelines on sleep management during the COVID-19 pandemic, particularly, concerning sleep perception (Altena et al., 2020). The British Sleep Society (BSS) has provided a statement on sleep management–related recommendations for patients during the pandemic (British Sleep Society (BSS), 2021). All of these aforementioned guidelines emphasize avoiding any contact between doctors and patients to restrict the spread of the COVID-19 infection. Home sleep testing has received great attention during the COVID-19 pandemic, and sleep studies at different levels are a major part of these guidelines. In this regard, sleep center facilities must be well equipped to serve patients for in-person visits or when sleep centers are reinstated in areas of low risk. It is recommended that long stay in waiting rooms be avoided, hand sanitizers must be placed everywhere, and contact with others should be minimized. Screening of patients for potential symptoms before consultation and on arrival is essential, and temperature testing of patients and staff is recommended. Patients should wear masks, and sleep technicians and doctors should wear personal protective equipment. Sleep studies during the COVID-19 pandemic can be conducted at home or laboratory, based on the state of the sleep disorder and the risk of community transmission.

17.5 Treatment of OSA During the COVID-19 Pandemic

Respiratory droplet is a mode of transmission of the COVID-19 virus, and the size of the exhaled aerosol diameter is influenced by several factors, including fluid properties, exhalation force, velocity, and environmental conditions. Continuous positive airway pressure (CPAP) is the main treatment prescribed for OSA; it is a non-invasive form of ventilation (Labarca et al., 2021; Mutti et al., 2020). CPAP improves the quality of life and health status of patients with OSA, reduces blood pressure to some extent, and is currently the main treatment recommended for patients with moderate-to-severe OSA. However, the efficacy of this treatment is still largely dependent on patient compliance. It is important to ensure that patients with OSA remain on timely and effective CPAP treatment despite the COVID-19 infection. The decision to continue or suspend the CPAP treatment should be made on an individual case basis, considering the comorbidities, the severity of OSA, and the risks associated with suspending treatment (Mutti et al., 2020; Barker et al., 2020; Suen et al., 2020). However, there are advantages and disadvantages in CPAP treatment during the COVID-19 pandemic (Thorpy et al., 2020). Barker et al. concluded that CPAP and other forms of non-invasive ventilation treatments should be discontinued in the community unless in an emergency, as continued use of CPAP in the community may significantly increase the risk of infection in family members or other caregivers (Barker et al., 2020).

In the United Kingdom, the BSS and the OSA alliance (including the British Thoracic Society, the BSS, the Association for Respiratory Technology & Physiology, and the Sleep Apnea Trust Association) have published guidelines for CPAP treatment utility during the pandemic (ARTP, 2021). The guidelines recommend that patients with OSA should continue CPAP treatment as prescribed at home; however, these individuals are advised to take steps to maintain sufficient social distance from the vulnerable family members by moving to other bedrooms or temporarily suspending the use of CPAP. The guidelines encourage patients to continue CPAP if they develop symptoms of respiratory infection. In addition, the current National Health Service guidelines in the United Kingdom recommend that patients at home during the pandemic can continue their prescribed ventilation treatment.

The Association of Francaise Sommeil and Otorhinolaryngology (AFSORL) and the Society of Francaise of Otolaryngology (SFORL) proposed a synopsis of measures to continue the treatment of sleep apnea syndrome during the pandemic (Bastier et al., 2020). During the initial stage of the COVID-19 pandemic, recommendations from international societies underline the risk for CPAP-induced aerosol and droplets, which act as mediums to transmit or spread the virus (Barker et al., 2020; Drummond, 2020). Contamination fears from using PAP may be contributing to deterioration in some patients with OSA (McSharry & Malhotra, 2020). Despite this, among OSA patients, compliance with CPAP therapy increased by 27% (del Campo et al., 2020). A study in France, which consisted of 7485 CPAP users with diagnosis of OSA, has shown that OSA patients significantly increased PAP adherence during the lockdown period, which was compared with data from a month earlier and the same period in the previous year (Attias et al., 2020). The findings suggest that publicity of COVID-19, as a disease affecting the airway, and fear of hospitalization may have motivated patients to comply with treatment. Moreover, spending greater time at home may also lead to an increase in sleep duration and access to CPAP treatment (Bastier et al., 2020).

Alternative treatments for OSA should be considered, such as oral appliances (Ramar et al., 2015). Schwartz et al. have shown that oral appliance therapy, including mandibular advancement devices, without increased risks for transmitting COVID-19, is an effective treatment for OSA and should be considered as first-line treatment during the pandemic (Schwartz et al., 2020). However, the study about oral appliance therapy for OSA is relatively less during the pandemic. In addition, treatment in the stomatological rooms may pose a risk to patients with OSA. In many stomatological practices, aerosols are formed during treatment, and droplets carrying viruses suspended in the air can increase the chances of infection for the staff. Additionally, cleaning the intraoral appliances used to treat OSA (e.g., mandibular advancement devices) can increase the risk of infection spread (Lavigne et al., 2020).

17.6 Outcomes in Patients with OSA and COVID-19 Infection

OSA is a chronic disease linked to COVID-19-related morbidity, mortality, and social outcomes. The COVID-19 pandemic has led to an escalating number of infections and deaths, and the incidence of OSA in COVID-19 patients increases the risk of hospitalization and death. Most patients positive for COVID-19 have mild or no symptoms, but a proportion of the symptomatic patients require admission to intensive care units and have a high mortality rate (Huang et al., 2020). In terms of pathogenesis, OSA shares many similarities with COVID-19, as both are in pro-inflammatory states, which may worsen the consequences of infection with SARS-CoV-2 (Bikov et al., 2017). The novel coronavirus enters the host cells by binding to the ACE2 receptor, the expression of which is enhanced in the adipose tissues due to obesity. Therefore, obese patients with OSA are more susceptible to infection by a novel coronavirus. The cardiac complications related to COVID-19 include acute myocardial infarction, myocarditis, arrhythmias, and heart failure (Bandyopadhyay et al., 2020; Pack & Gislason, 2009); OSA is an identified risk factor of arrhythmias, hypertension, acute heart failure syndromes, and heart failure. Therefore, OSA may increase morbidity and mortality in patients having cardiovascular comorbidities along with COVID-19 infection. OSA is linked to a twofold rise in the risk of developing severe COVID-19 infection (Rögnvaldsson et al., 2021) and is an independent risk factor for severe COVID-19 infection (Strausz et al., 2021).

17.7 Recommendations on the Management of Patients with OSA During the COVID-19 Pandemic

17.7.1 Diagnostic Management

In the early stages of the COVID-19 pandemic, the AASM published guidelines on mitigation strategies in sleep practice, recommending the postponement and rescheduling of in-lab PSG and PAP administration (American Academy of Sleep Medicine, 2021). For essential sleep monitoring, disposable accessories or devices were recommended. It stated that if only reproducible monitoring equipment could be used, disinfect and thoroughly clean the facility per the CDC-recommended disinfection standards and manufacturer's requirements for the facility; staff should wear appropriate personal protective equipment. Moreover, in addition to being rigorously disinfected, reproducible equipment should be taken out of the service for at least 72 h before the next use. The same operation should be implemented in domestic-type monitoring facilities, and courier services were preferred during the pandemic to ensure that patients could receive or return the equipment used without leaving their homes.

To minimize the risk of novel coronavirus infection, many sleep clinics and labs had to stop traditional offline consultations, and the management of OSA in the European sleep medicine sector has been reduced by 80% (Grote et al., 2020). In contrast, Internet-based online teleconsultations have become more active, and national professional associations have encouraged remote diagnosis by testing select patients at home rather than in the laboratory (Bastier et al., 2020; Bikov et al., 2021; Ayas et al., 2020). Sleep telemedicine has been in practice for almost 20 years to provide clinical consultations for patients with sleep disorders. The AASM has defined key criteria related to its application (Singh et al., 2015). At this particular period, the AASM had to also update its recommendations for the diagnosis and treatment of sleep disorders through telemedicine (Shamim-Uzzaman et al., 2021), and the consultation of patients with OSA has been supported well by the sleep telemedicine facility.

17.7.2 Therapeutic Management

The BSS and the UK OSA alliance have jointly published guidelines for the use of CPAP during the COVID-19 pandemic. This recommends that patients with OSA should continue to use CPAP as prescribed at their homes without changing the parameters. However, they may change bedrooms or stop using CPAP briefly if necessary and maintain enough distance from susceptible family members. The UK National Health Service guidelines recommend that the usual mode of ventilation should be continued. The AFSORL and SFORL have also jointly proposed a strategy for the continued treatment of OSA during the COVID-19 pandemic. The Canadian Thoracic Society has also stated to this effect (Ayas et al., 2020).

Chang et al. (2019) evaluated the value of the home sleep apnea test (HSAT) for the diagnosis of concurrent obstructive sleep apnea–hypopnea syndrome (OSAHS) in patients with chronic obstructive pulmonary disease (COPD). HSAT has a higher sensitivity (95%) and a positive predictive value (94%) as compared to PSG. Other home-based monitoring devices, such as handheld pulse oximeter and radar respiratory motion monitor, have been greatly developed during the pandemic, as they were used as the primary screening devices. In addition, electronic questionnaires sent through the Internet can also be used as an aid to assess and better diagnose the patients. Several randomized controlled trials (Rosen et al., 2012; Chai-Coetzer et al., 2013) comparing functional outcomes and changes in compliance in OSAHS patients who received PSG and pressure titration in sleep centers with those treated through HSAT and APAP outside the sleep center show no significant differences between the two groups. In terms of satisfaction with care, Fields et al. (Watson, 2016) conducted a randomized trial for patients with OSAHS and showed no significant differences between telemedicine and face-to-face care; moreover, telemedicine was superior for ease and convenience of communication. No significant differences in clinical decision-making, patient compliance, and sleep functions between the combined physician and community primary care nurse working model and physician-led diagnosis and treatment in a sleep center were observed (Sánchez-Quiroga et al., 2018; Nuria et al., 2017; Ozdere et al., 2015).

In summary, the COVID-19 pandemic exerted a significant effect on the diagnosis and management of SDBs. In the new prevention and control background, there is a need to develop novel models of diagnosis and management while minimizing any possible risk of exposure.

Since its outbreak, the COVID-19 pandemic has resulted in several sleep centers worldwide switching from PSG and PAP treatments to remote online services. Notably, some services have continued to do so until now as a matter of routine, with no reduction in the number of patients served as compared to the traditional model. There has even been an increase in the number of patients due to convenient access and improved service capacity. The role of paramedics in sleep medicine services is particularly important in the context of a great pandemic, resulting in a shortage of medical staff. Furthermore, telemedicine services have been extended from the traditional service audience consisting of the elderly, mobility impaired, and patients from remote and underdeveloped areas to cover the entire population in any part of the world.

The global effects of the COVID-19 pandemic cannot be understated. It has affected the diagnosis and treatment of many diseases, including OSA. Currently, there is limited research on the association between OSA and COVID-19, and more relevant studies are needed in the future. Clinical studies should be further strengthened to verify whether OSA is only related to comorbidities of COVID-19 or is also independently linked to the risk of death.

Clinical Practice Points

1. Patients with OSA experienced a higher risk of hospitalization owing to the disease severity after contracting the COVID-19 infection as compared to those without OSA.
2. The most effective therapeutic option currently available for patients with OSA remains PAP devices.
3. Safety measures should be taken to avoid spread of the SARS-CoV-2 virus to protect both patients and personnel when performing sleep diagnostic and therapeutic procedures in sleep centers.

References

Altena, E., Baglioni, C., Espie, C. A., Ellis, J., Gavriloff, D., Holzinger, B., et al. (2020). Dealing with sleep problems during home confinement due to the COVID-19 outbreak: Practical recommendations from a task force of the European CBT-I Academy. *Journal of Sleep Research, 29*(4), e13052. https://doi.org/10.1111/jsr.13052

Altibi, A. M., Pallavi, B., Liaqat, H., Slota, A. A., Sheth, R., Al Jebbawi, L., et al. (2021). Characteristics and comparative clinical outcomes of prisoner versus non-prisoner populations hospitalized with COVID-19. *Scientific Reports, 11*(1), 6488. https://doi.org/10.1038/s41598-021-85916-w

American Academy of Sleep Medicine. COVID-19 mitigation strategies for sleep clinics and sleep centers – REOPENING [updated January 18, 2021]. https://aasm.org/covid-19-resources/covid-19-mitigation-strategies-sleep-clinics-labs. Accessed 11 Nov 2021.

American Academy of Sleep Medicine (AASM). Summary of CDC recommendations relevant for sleep practices during COVID-19. https://aasm.org/covid-19-resources/covid-19-mitigation-strategies-sleep-clinics-labs/. Accessed 11 Nov 2021.

Archontogeorgis, K., Nena, E., Papanas, N., Zissimopoulos, A., Voulgaris, A., Xanthoudaki, M., et al. (2018). Vitamin D levels in middle-aged patients with obstructive sleep apnoea syndrome. *Current Vascular Pharmacology, 16*(3), 289–297. https://doi.org/10.2174/1570161115666170529085708

Arentz, M., Yim, E., Klaff, L., Lokhandwala, S., Riedo, F. X., Chong, M., et al. (2020). Characteristics and outcomes of 21 critically ill patients with COVID-19 in Washington State. *Journal of the American Medical Association, 323*(16), 1612–1614. https://doi.org/10.1001/jama.2020.4326

ARTP guidance – Respiratory function testing and sleep services during endemic COVID-19. (2021). Available from https://www.artp.org.uk/News/artp-guidance-respiratory-function-testing-and-sleep-services-during-endemic-covid-19. Accessed 11 Nov 2021.

Attias, D., Pepin, J. L., & Pathak, A. (2020). Impact of COVID-19 lockdown on adherence to continuous positive airway pressure by obstructive sleep apnoea patients. *European Respiratory Journal, 56*(1), 2001607. https://doi.org/10.1183/13993003.01607-2020

Ayas, N. T., Fraser, K. L., Giannouli, E., Hanly, P. J., Kendzerska, T., Katz, S. L., et al. (2020). Key highlights from the Canadian Thoracic Society's position statement on optimizing the management of sleep disordered breathing during the coronavirus disease 2019 pandemic. *Chest, 158*(3), 899–900. https://doi.org/10.1016/j.chest.2020.05.572

Bandyopadhyay, D., Akhtar, T., Hajra, A., Gupta, M., Das, A., Chakraborty, S., et al. (2020). COVID-19 pandemic: Cardiovascular complications and future implications. *American Journal of Cardiovascular Drugs, 20*(4), 311–324. https://doi.org/10.1007/s40256-020-00420-2

Barker, J., Oyefeso, O., Koeckerling, D., Mudalige, N. L., & Pan, D. (2020). COVID-19: Community CPAP and NIV should be stopped unless medically necessary to support life. *Thorax, 75*(5), 367. https://doi.org/10.1136/thoraxjnl-2020-214890

Bastier, P. L., Aisenberg, N., Durand, F., Lestang, P., Abedipour, D., Gallet de Santerre, O., et al. (2020). Treatment of sleep apnea by ENT specialists during the COVID-19 pandemic. *European Annals of Otorhinolaryngology, Head and Neck Diseases, 137*(4), 319–321. https://doi.org/10.1016/j.anorl.2020.05.001

Besedovsky, L., Lange, T., & Haack, M. (2019). The sleep-immune crosstalk in health and disease. *Physiological Reviews, 99*(3), 1325–1380. https://doi.org/10.1152/physrev.00010.2018

Bhatraju, P. K., Ghassemieh, B. J., Nichols, M., Kim, R., Jerome, K. R., Nalla, A. K., et al. (2020). Covid-19 in critically ill patients in the Seattle region – Case series. *The New England Journal of Medicine, 382*(21), 2012–2022. https://doi.org/10.1056/NEJMoa2004500

Bikov, A., Losonczy, G., & Kunos, L. (2017). Role of lung volume and airway inflammation in obstructive sleep apnea. *Respiratory Investigation, 55*(6), 326–333. https://doi.org/10.1016/j.resinv.2017.08.009

Bikov, A., Khalil, S., Gibbons, M., Bentley, A., Jones, D., & Bokhari, S. (2021). A fully remote diagnostic and treatment pathway in patients with obstructive sleep apnoea during the COVID-19 pandemic: A single centre experience. *Journal of Clinical Medicine, 10*(19), 4310. https://doi.org/10.3390/jcm10194310

British Sleep Society (BSS). BSS statement on sleep-related advice during the coronavirus (Covid-19) pandemic [Internet]. https://www.sleepsociety.org.uk/covid-19/. Accessed 11 Nov 2021.

Cade, B. E., Dashti, H. S., Hassan, S. M., Redline, S., & Karlson, E. W. (2020). Sleep apnea and COVID-19 mortality and hospitalization. *American Journal of Respiratory and Critical Care Medicine, 202*(10), 1462–1464. https://doi.org/10.1164/rccm.202006-2252LE

Cappuccio, F. P., & Siani, A. (2020). Covid-19 and cardiovascular risk: Susceptibility to infection to SARS-CoV-2, severity and prognosis of Covid-19 and blockade of the renin-angiotensin-aldosterone system. An evidence-based viewpoint. *Nutrition, Metabolism, and Cardiovascular Diseases, 30*(8), 1227–1235. https://doi.org/10.1016/j.numecd.2020.05.013

Centers for Disease Control and Prevention (CDC). Infection control guidance – Summary of changes to the guidance. https://www.cdc.gov/coronavirus/2019-ncov/hcp/infection-control-recommendations.html. Accessed 11 Nov 2021.

Cevik, M., Kuppalli, K., Kindrachuk, J., & Peiris, M. (2020). Virology, transmission, and pathogenesis of SARS-CoV-2. *BMJ, 371*, m3862. https://doi.org/10.1136/bmj.m3862

Chai-Coetzer, C. L., Antic, N. A., Rowland, L. S., Reed, R. L., & Mcevoy, R. D. (2013). Primary care vs specialist sleep center management of obstructive sleep apnea and daytime sleepiness and quality of life: A randomized trial. *JAMA: The Journal of the American Medical Association, 309*(10), 997.

Chang, Y., Xu, L., Han, F., Keenan, B. T., Kneeland-Szanto, E., Zhang, R., et al. (2019). Validation of the Nox-T3 portable monitor for diagnosis of obstructive sleep apnea in patients with chronic obstructive pulmonary disease. *Journal of Clinical Sleep Medicine, 15*(4), 587–596. https://doi.org/10.5664/jcsm.7720

Chhiba, K. D., Patel, G. B., Vu, T. H. T., Chen, M. M., Guo, A., Kudlaty, E., et al. (2020). Prevalence and characterization of asthma in hospitalized and non-hospitalized patients with COVID-19. *The Journal of Allergy and Clinical Immunology, 146*(2), 307–14.e4. https://doi.org/10.1016/j.jaci.2020.06.010

Chung, F., Waseem, R., Pham, C., Penzel, T., Han, F., Bjorvatn, B., et al. (2021). The association between high risk of sleep apnea, comorbidities, and risk of COVID-19: A population-based international harmonized study. *Sleep and Breathing, 25*(2), 849–860. https://doi.org/10.1007/s11325-021-02373-5

Cui, J., Li, F., & Shi, Z. L. (2019). Origin and evolution of pathogenic coronaviruses. *Nature Reviews. Microbiology, 17*(3), 181–192. https://doi.org/10.1038/s41579-018-0118-9

del Campo, F., López, G., Arroyo, C. A., de Frutos, J. F., Crespo, A., Cerezo-Hernández, A., et al. (2020). Study of adherence to continuous positive airway pressure treatment in patients with obstructive sleep apnea syndrome in the confinement during the COVID-19 pandemic. *Archivos de Bronconeumología, 56*(12), 818–819. https://doi.org/10.1016/j.arbr.2020.06.008

Derakhshanian, H., Rastad, H., Ghosh, S., Zeinali, M., Ziaee, M., Khoeini, T., et al. (2021). The predictive power of serum vitamin D for poor outcomes in COVID-19 patients. *Food Science & Nutrition, 9*(11), 6307–6313. https://doi.org/10.1002/fsn3.2591

Drummond, M. (2020). Sleep labs, lung function tests and COVID-19 pandemic – Only emergencies allowed! *Pulmonology, 26*(4), 244–245. https://doi.org/10.1016/j.pulmoe.2020.04.002

El-Missiry, M. A., El-Missiry, Z. M. A., & Othman, A. I. (2020). Melatonin is a potential adjuvant to improve clinical outcomes in individuals with obesity and diabetes with coexistence of Covid-19. *European Journal of Pharmacology, 882*, 173329. https://doi.org/10.1016/j.ejphar.2020.173329

Gill, C. M., Piggott, L. M., & Kent, B. D. (2021). Sleep medicine and coronavirus disease 2019. *Current Opinion in Pulmonary Medicine, 27*(6), 529–534. https://doi.org/10.1097/mcp.0000000000000822

Gottlieb, D. J., & Punjabi, N. M. (2020). Diagnosis and management of obstructive sleep apnea: A review. *Journal of the American Medical Association, 323*(14), 1389–1400. https://doi.org/10.1001/jama.2020.3514

Grant, W. B., Lahore, H., McDonnell, S. L., Baggerly, C. A., French, C. B., Aliano, J. L., et al. (2020). Evidence that vitamin D supplementation could reduce risk of influenza and COVID-19 infections and deaths. *Nutrients, 12*(4), 988. https://doi.org/10.3390/nu12040988

Grote, L., McNicholas, W. T., & Hedner, J. (2020). Sleep apnoea management in Europe during the COVID-19 pandemic: Data from the European Sleep Apnoea Database (ESADA). *The European

Respiratory Journal, 55(6), 2001323. https://doi.org/10.1183/13993003.01323-2020

Hastie, C. E., Mackay, D. F., Ho, F., Celis-Morales, C. A., Katikireddi, S. V., Niedzwiedz, C. L., et al. (2020). Vitamin D concentrations and COVID-19 infection in UK Biobank. *Diabetes and Metabolic Syndrome: Clinical Research and Reviews, 14*(4), 561–565. https://doi.org/10.1016/j.dsx.2020.04.050

Huang, C., Wang, Y., Li, X., Ren, L., Zhao, J., Hu, Y., et al. (2020). Clinical features of patients infected with 2019 novel coronavirus in Wuhan, China. *Lancet, 395*(10223), 497–506. https://doi.org/10.1016/s0140-6736(20)30183-5

Irwin, M. R. (2015). Why sleep is important for health: A psychoneuroimmunology perspective. *Annual Review of Psychology, 66*, 143–172. https://doi.org/10.1146/annurev-psych-010213-115205

Jin, Z. N., & Wei, Y. X. (2016). Meta-analysis of effects of obstructive sleep apnea on the renin-angiotensin-aldosterone system. *Journal of Geriatric Cardiology, 13*(4), 333–343. https://doi.org/10.11909/j.issn.1671-5411.2016.03.020

Jordan, A. S., McSharry, D. G., & Malhotra, A. (2014). Adult obstructive sleep apnoea. *Lancet, 383*(9918), 736–747. https://doi.org/10.1016/s0140-6736(13)60734-5

Jose, R. J., & Manuel, A. (2020). COVID-19 cytokine storm: The interplay between inflammation and coagulation. *The Lancet Respiratory Medicine, 8*(6), e46–ee7. https://doi.org/10.1016/s2213-2600(20)30216-2

Kassi, E., Adamopoulos, C., Basdra, E. K., & Papavassiliou, A. G. (2013). Role of vitamin D in atherosclerosis. *Circulation, 128*(23), 2517–2531. https://doi.org/10.1161/circulationaha.113.002654

Kheirandish-Gozal, L., Peris, E., & Gozal, D. (2014). Vitamin D levels and obstructive sleep apnoea in children. *Sleep Medicine, 15*(4), 459–463. https://doi.org/10.1016/j.sleep.2013.12.009

Kohler, M., & Stradling, J. R. (2010). Mechanisms of vascular damage in obstructive sleep apnea. *Nature Reviews. Cardiology, 7*(12), 677–685. https://doi.org/10.1038/nrcardio.2010.145

Labarca, G., Schmidt, A., Dreyse, J., Jorquera, J., & Barbe, F. (2021). Telemedicine interventions for CPAP adherence in obstructive sleep apnea patients: Systematic review and meta-analysis. *Sleep Medicine Reviews, 60*, 101543. https://doi.org/10.1016/j.smrv.2021.101543

Lancaster, L. H., Mason, W. R., Parnell, J. A., Rice, T. W., Loyd, J. E., Milstone, A. P., et al. (2009). Obstructive sleep apnea is common in idiopathic pulmonary fibrosis. *Chest, 136*(3), 772–778. https://doi.org/10.1378/chest.08-2776

Lavigne, G., Dal Fabbro, C., Herrero Babiloni, A., Huynh, N., Gauthier, L., Arcache, P., et al. (2020). Dental sleep medicine perspectives after COVID-19: Interprofessional adaptation and directions. *Journal of Clinical Sleep Medicine, 16*(8), 1421. https://doi.org/10.5664/jcsm.8546

Letko, M., Marzi, A., & Munster, V. (2020). Functional assessment of cell entry and receptor usage for SARS-CoV-2 and other lineage B betacoronaviruses. *Nature Microbiology, 5*(4), 562–569. https://doi.org/10.1038/s41564-020-0688-y

Liu, P. P., Blet, A., Smyth, D., & Li, H. (2020). The science underlying COVID-19: Implications for the cardiovascular system. *Circulation, 142*(1), 68–78. https://doi.org/10.1161/circulationaha.120.047549

Maas, M. B., Kim, M., Malkani, R. G., Abbott, S. M., & Zee, P. C. (2021). Obstructive sleep apnea and risk of COVID-19 infection, hospitalization and respiratory failure. *Sleep Breath, 25*(2), 1155–1157. https://doi.org/10.1007/s11325-020-02203-0

Mashaqi, S., Lee-Iannotti, J., Rangan, P., Celaya, M. P., Gozal, D., Quan, S. F., et al. (2021). Obstructive sleep apnea and COVID-19 clinical outcomes during hospitalization: A cohort study. *Journal of Clinical Sleep Medicine, 17*(11), 2197–2204. https://doi.org/10.5664/jcsm.9424

McSharry, D., & Malhotra, A. (2020). Potential influences of obstructive sleep apnea and obesity on COVID-19 severity. *Journal of Clinical Sleep Medicine, 16*(9), 1645. https://doi.org/10.5664/jcsm.8538

Mello, M. T., Silva, A., Guerreiro, R. C., da-Silva, F. R., Esteves, A. M., Poyares, D., et al. (2020). Sleep and COVID-19: Considerations about immunity, pathophysiology, and treatment. *Sleep Science, 13*(3), 199–209. https://doi.org/10.5935/1984-0063.20200062

Miller, M. A., & Cappuccio, F. P. (2021). A systematic review of COVID-19 and obstructive sleep apnoea. *Sleep Medicine Reviews, 55*, 101382. https://doi.org/10.1016/j.smrv.2020.101382

Millet, J. K., & Whittaker, G. R. (2015). Host cell proteases: Critical determinants of coronavirus tropism and pathogenesis. *Virus Research, 202*, 120–134. https://doi.org/10.1016/j.virusres.2014.11.021

Mokmeli, S., & Vetrici, M. (2020). Low level laser therapy as a modality to attenuate cytokine storm at multiple levels, enhance recovery, and reduce the use of ventilators in COVID-19. *Canadian Journal of Respiratory Therapy, 56*, 25–31. https://doi.org/10.29390/cjrt-2020-015

Morin, C. M., Bjorvatn, B., Chung, F., Holzinger, B., Partinen, M., Penzel, T., et al. (2021). Insomnia, anxiety, and depression during the COVID-19 pandemic: An international collaborative study. *Sleep Medicine, 87*, 38–45. https://doi.org/10.1016/j.sleep.2021.07.035

Mutti, C., Azzi, N., Soglia, M., Pollara, I., Alessandrini, F., & Parrino, L. (2020). Obstructive sleep apnea, cpap and COVID-19: A brief review. *Acta Biomedica, 91*(4), e2020196. https://doi.org/10.23750/abm.v91i4.10941

Nuria, T., Manuel, S., Nuria, N., Jordi, D. B., Iván, B., Anunciación, C., et al. (2017). Management of obstructive sleep apnoea in a primary care vs sleep unit setting: A randomised controlled trial. *Thorax, 73*, 1152–1160.

Ozdere, E., Sayin, G., & Okutan, Y. (2015). The use of oral appliances in two patients diagnosed with different stages of obstructive sleep apnea. *European*

Journal of Prosthodontics, 3, 80. https://doi.org/10.4103/2347-4610.166186

Pack, A. I., & Gislason, T. (2009). Obstructive sleep apnea and cardiovascular disease: A perspective and future directions. *Progress in Cardiovascular Diseases, 51*(5), 434–451. https://doi.org/10.1016/j.pcad.2009.01.002

Partinen, M., Bjorvatn, B., Holzinger, B., Chung, F., Penzel, T., Espie, C. A., et al. (2021). Sleep and circadian problems during the coronavirus disease 2019 (COVID-19) pandemic: The International COVID-19 Sleep Study (ICOSS). *Journal of Sleep Research, 30*(1), e13206. https://doi.org/10.1111/jsr.13206

Pazarlı, A. C., Ekiz, T., & İlik, F. (2021). Coronavirus disease 2019 and obstructive sleep apnea syndrome. *Sleep & Breathing, 25*(1), 371. https://doi.org/10.1007/s11325-020-02087-0

Peker, Y., Celik, Y., Arbatli, S., Isik, S. R., Balcan, B., Karataş, F., et al. (2021). Effect of high-risk obstructive sleep apnea on clinical outcomes in adults with coronavirus disease 2019: A multicenter, prospective, observational clinical trial. *Annals of the American Thoracic Society, 18*(9), 1548–1559.

Pinto, J., van Zeller, M., Amorim, P., Pimentel, A., Dantas, P., Eusébio, E., et al. (2020). Sleep quality in times of Covid-19 pandemic. *Sleep Medicine, 74*, 81–85.

Ramar, K., Dort, L. C., Katz, S. G., Lettieri, C. J., Harrod, C. G., Thomas, S. M., et al. (2015). Clinical practice guideline for the treatment of obstructive sleep apnea and snoring with oral appliance therapy: An update for 2015. *Journal of Clinical Sleep Medicine, 11*(7), 773–827. https://doi.org/10.5664/jcsm.4858

Rögnvaldsson, K. G., Eyþórsson, E. S., Emilsson, Ö. I., Eysteinsdóttir, B., Pálsson, R., Gottfreðsson, M., et al. (2021). Obstructive sleep apnea is an independent risk factor for severe COVID-19: A population-based study. *Sleep*. https://doi.org/10.1093/sleep/zsab272

Rosen, C. L., Dennis, A., Ruth, B., Nancy, F. S., Conrad, I., Vishesh, K., et al. (2012). A multisite randomized trial of portable sleep studies and positive airway pressure autotitration versus laboratory-based polysomnography for the diagnosis and treatment of obstructive sleep apnea: The HomePAP study. *Sleep, 6*, 757–767.

Said, E. A., Al-Abri, M. A., Al-Saidi, I., Al-Balushi, M. S., Al-Busaidi, J. Z., Al-Reesi, I., et al. (2017). Altered blood cytokines, CD4 T cells, NK and neutrophils in patients with obstructive sleep apnea. *Immunology Letters, 190*, 272–278. https://doi.org/10.1016/j.imlet.2017.08.009

Sánchez-Quiroga, M., Corral, J., Gómez-De-Terreros, F., Carmona-Bernal, C., Asensio-Cruz, M. I., Cabello, M., et al. (2018). Primary care physicians can comprehensively manage sleep apnea patients: A non-inferiority randomized controlled trial. *American Journal of Respiratory and Critical Care Medicine, 198*, 648–656.

Schiza, S., Mermigkis, C., Margaritopoulos, G. A., Daniil, Z., Harari, S., Poletti, V., et al. (2015). Idiopathic pulmonary fibrosis and sleep disorders: No longer strangers in the night. *European Respiratory Review, 24*(136), 327–339. https://doi.org/10.1183/16000617.00009114

Schwartz, D., Addy, N., Levine, M., & Smith, H. (2020). Oral appliance therapy should be prescribed as a first-line therapy for OSA during the COVID-19 pandemic. *Journal of Dental Sleep Medicine, 7*(3), 1.

Shamim-Uzzaman, Q. A., Bae, C. J., Ehsan, Z., Setty, A. R., Devine, M., Dhankikar, S., et al. (2021). The use of telemedicine for the diagnosis and treatment of sleep disorders: An American Academy of Sleep Medicine update. *Journal of Clinical Sleep Medicine, 17*(5), 1103–1107. https://doi.org/10.5664/jcsm.9194

Shneider, A., Kudriavtsev, A., & Vakhrusheva, A. (2020). Can melatonin reduce the severity of COVID-19 pandemic? *International Reviews of Immunology, 39*(4), 153–162. https://doi.org/10.1080/08830185.2020.1756284

Simonnet, A., Chetboun, M., Poissy, J., Raverdy, V., Noulette, J., Duhamel, A., et al. (2020). High prevalence of obesity in severe acute respiratory syndrome coronavirus-2 (SARS-CoV-2) requiring invasive mechanical ventilation. *Obesity (Silver Spring), 28*(7), 1195–1199. https://doi.org/10.1002/oby.22831

Singh, J., Badr, M. S., Diebert, W., Epstein, L., Hwang, D., Karres, V., et al. (2015). American Academy of Sleep Medicine (AASM) position paper for the use of telemedicine for the diagnosis and treatment of sleep disorders. *Journal of Clinical Sleep Medicine, 11*(10), 1187–1198. https://doi.org/10.5664/jcsm.5098

Strausz, S., Kiiskinen, T., Broberg, M., Ruotsalainen, S., Koskela, J., Bachour, A., et al. (2021). Sleep apnoea is a risk factor for severe COVID-19. *BMJ Open Respiratory Research, 8*(1), e000845. https://doi.org/10.1136/bmjresp-2020-000845

Suen, C. M., Hui, D. S. C., Memtsoudis, S. G., & Chung, F. (2020). Obstructive sleep apnea, obesity, and noninvasive ventilation: Considerations during the COVID-19 pandemic. *Anesthesia and Analgesia, 131*(2), 318–322. https://doi.org/10.1213/ane.0000000000004928

Sweed, R. A., Hassan, S., ElWahab, N. H. A., Aref, S. R., & Mahmoud, M. I. (2019). Comorbidities associated with obstructive sleep apnea: A retrospective Egyptian study on 244 patients. *Sleep & Breathing, 23*(4), 1079–1085. https://doi.org/10.1007/s11325-019-01783-w

Thorpy, M., Figuera-Losada, M., Ahmed, I., Monderer, R., Petrisko, M., Martin, C., et al. (2020). Management of sleep apnea in New York City during the COVID-19 pandemic. *Sleep Medicine, 74*, 86–90. https://doi.org/10.1016/j.sleep.2020.07.013

Tiwari, N., Upadhyay, J., Ansari, M. N., & Joshi, R. (2020). Novel β-Coronavirus (SARS-CoV-2): Current and future aspects of pharmacological treatments. *Saudi Pharmaceutical Journal, 28*(10), 1243–1252. https://doi.org/10.1016/j.jsps.2020.08.015

Tufik, S., Gozal, D., Ishikura, I. A., Pires, G. N., & Andersen, M. L. (2020). Does obstructive sleep apnea lead to increased risk of COVID-19 infection and severity? *Journal of Clinical Sleep Medicine, 16*(8), 1425–1426. https://doi.org/10.5664/jcsm.8596

Watson, N. F. (2016). Expanding patient access to quality sleep health care through telemedicine. *Journal of Clinical Sleep Medicine, 12*(2), 155–156.

WHO Director-General's opening remarks at the media briefing on COVID-19 – 11 March 2020. https://www.who.int/director-general/speeches/detail/who-director-general-s-opening-remarks-at-the-media-briefing-on-covid-19%2D%2D-11-march-2020. Accessed 11 Nov 2021.

World Health Organization. https://www.who.int/emergencies/diseases/novel-coronavirus-2019. Accessed 11 Nov 2021.

Yao, Y., Wang, H., & Liu, Z. (2020). Expression of ACE2 in airways: Implication for COVID-19 risk and disease management in patients with chronic inflammatory respiratory diseases. *Clinical and Experimental Allergy, 50*(12), 1313–1324. https://doi.org/10.1111/cea.13746

Ye, Z., Zhang, Y., Wang, Y., Huang, Z., & Song, B. (2020). Chest CT manifestations of new coronavirus disease 2019 (COVID-19): A pictorial review. *European Radiology, 30*(8), 4381–4389. https://doi.org/10.1007/s00330-020-06801-0

Zhang, X. L., & Xiao, Y. (2020). Sleep health service in China during the coronavirus disease outbreak. *Journal of Clinical Sleep Medicine, 16*(7), 1221–1222. https://doi.org/10.5664/jcsm.8472

Zhang, R., Wang, X., Ni, L., Di, X., Ma, B., Niu, S., et al. (2020). COVID-19: Melatonin as a potential adjuvant treatment. *Life Sciences, 250*, 117583. https://doi.org/10.1016/j.lfs.2020.117583

Zhou, F., Yu, T., Du, R., Fan, G., Liu, Y., Liu, Z., et al. (2020). Clinical course and risk factors for mortality of adult inpatients with COVID-19 in Wuhan, China: A retrospective cohort study. *Lancet, 395*(10229), 1054–1062. https://doi.org/10.1016/s0140-6736(20)30566-3

Zhu, N., Zhang, D., Wang, W., Li, X., Yang, B., Song, J., et al. (2020). A novel coronavirus from patients with pneumonia in China, 2019. *The New England Journal of Medicine, 382*(8), 727–733. https://doi.org/10.1056/NEJMoa2001017

Part III

Therapeutic Innovations

APAP, BPAP, CPAP, and New Modes of Positive Airway Pressure Therapy

18

Karin G. Johnson

Abstract

Positive airway pressure (PAP) is the primary treatment of sleep-disordered breathing including obstructive sleep apnea, central sleep apnea, and sleep-related hypoventilation. Just as clinicians use pharmacological mechanism of action and pharmacokinetic data to optimize medication therapy for an individual, understanding how PAP works and choosing the right mode and device are critical to optimizing therapy in an individual patient. The first section of this chapter will describe the technology inside PAP devices that is essential for understanding the algorithms used to control the airflow and pressure. The second section will review how different comfort settings including ramp and expiratory pressure relief and modes of PAP therapy including continuous positive airway pressure (CPAP), autotitrating CPAP, bilevel positive airway pressure, adaptive servoventilation, and volume-assured pressure support control the airflow and pressure. Proprietary algorithms from several different manufacturers are described. This chapter derives its descriptions of algorithms from multiple sources including literature review, manufacture publications and websites, patents, and peer-reviewed device comparisons and from personal communication with manufacturer representatives. Clinical considerations related to the technological aspects of the different algorithms and features will be reviewed.

Keywords

Obstructive sleep apnea · Central sleep apnea · Continuous positive airway pressure · Bilevel positive airway pressure · AutoPAP · Adaptive servoventilation · Volume assured pressure support · Sleep disordered breathing · Respiratory cycle · Adherence · Hypopnea · Flow limitation · Mask · Expiratory pressure relief · Ramp

Abbreviations

AC	Alternating current
AcSV	Anticyclic servoventilation
AHI	Apnea–hypopnea index
APAP	Autotitrating CPAP
ASV	Adaptive servoventilation
AutoBPAP	Autoadjusting BPAP
AVAPS	Average volume-assured pressure support
BPAP	Bilevel positive airway pressure

K. G. Johnson (✉)
Department of Neurology- Baystate Medical Center, University of Massachusetts Medical School-Baystate, Springfield, MA, USA
e-mail: karin.johnson@baystatehealth.org

T. Penzel, R. Hornero (eds.), *Advances in the Diagnosis and Treatment of Sleep Apnea*, Advances in Experimental Medicine and Biology 1384, https://doi.org/10.1007/978-3-031-06413-5_18

COPD	Chronic obstructive pulmonary disease
CPAP	Continuous positive airway pressure
CSR	Cheyne–Stokes respiration
DC	Direct current
dyn	Dynamic
eFl	Epoch with flattening
eMO	Epoch with mild obstruction
EPAP	Expiratory positive airway pressure
EPR	Expiratory pressure relief
eSO	Epoch with severe obstruction
FOT	Forced oscillation technique
HMV	Home mechanical ventilator
IFL	Inspiratory flow limitation
IPAP	Inspiratory positive airway pressure
iVAPS	Intelligent volume-assured pressure support
MEMS	Microelectromechanical
MV	Minute ventilation
NIPPV	Non-invasive positive airway pressure ventilation
NRAH	Non-responsive apnea/hypopnea
OSA	Obstructive sleep apnea
PAP	Positive airway pressure
Pcrit	Critical pressure
PDIFF	Pressure difference
PEEP	Positive end-expiratory pressure
Popt	Optimal pressure
PS	Pressure support
PSmax	Pressure support maximum
PSmin	Pressure support minimum
rAMV	Relative average minute volume
REM	Rapid eye movement
RERA	Respiratory effort–related arousal
RMS	Root mean squared
rRMV	Relative respiratory minute volume
ST	Spontaneous/timed
std.	Standard
SV	Servoventilation
Ti	Inspiratory time
VAPS	Volume-assured pressure support
VB	Variable breathing
WPF	Weighted peak flow

18.1 Introduction

Positive airway pressure (PAP) is the primary treatment of sleep-disordered breathing including obstructive sleep apnea (OSA), central sleep apnea, and sleep-related hypoventilation. This chapter will primarily focus on modes continuous PAP (CPAP), autotitrating CPAP (APAP), bilevel PAP (BPAP), adaptive servoventilation (ASV), and volume-assured pressure support (VAPS) that are available in devices at the Centers for Medicare Services referred to as CPAP or respiratory assist devices, but some of these modalities are also available in non-invasive home mechanical ventilators (HMVs). Device manufacturers have proprietary algorithms so similar modes may work very differently, and limited data about those differences is available. This chapter derives its descriptions of algorithms from multiple sources including literature review, manufacture publications and websites, patents, and peer-reviewed device comparisons. Personal communication with ResMed (ResMed, San Diego, CA), Philips Respironics (Amsterdam, Netherlands) formerly Respironics (Murrysville, PA), DeVilbiss (DeVilbiss Healthcare, Somerset, PA), and Löwenstein (Löwenstein Medical Technologies, Hamburg, Germany) was used to obtain and/or verify all referenced and unreferenced algorithm descriptions presented. There are other manufacturers with APAP devices whose algorithms were unable to be obtained and were not included. This chapter and the included tables were adapted and updated from a previous article by the author (Johnson & Johnson, 2015). As the focus will be primarily on the technology aspects of these modes, we will describe clinical considerations related to the algorithms and features but will not review the overall clinical literature regarding the efficacy of modalities.

18.2 Technology to Control Positive Airway Pressure Devices

Basic PAP equipment involves a device with a flow generator or "blower," a mask that covers either or both the nose and/or the mouth and a tube that connects the device to the mask. Since the original devices, PAP therapy has evolved greatly to enhance the comfort, size, and quietness of the therapy, and current PAP devices are much more complex and may include air filters, sensors (motor speed, airflow rate, pressure, snore transducer), microprocessor-based controller, data storage, multilingual displays, internal modem for data transfer, and humidifiers with heated tubing. Additionally, more technologically advanced masks provide better seal, comfort, and quietness.

18.2.1 Flow Generators

Flow generators or "blowers" are typically made up of a motor and impeller, which are essentially the fan blades. There are two primary functions of the blower: the first is to achieve the desired air pressure to open the airway and the second is to provide enough airflow to adequately ventilate and overcome dead space. The flow rates for some devices can be up to 180 L/min at the mask (Fleming & Grunberg, 2021).

The first generation of commercial CPAP devices utilized vortex blowers designed for whirlpool or spa bath (ResMed Origins, 2021). The size and noise of these motors limited their widespread utility as sleep therapy. Additionally, these used alternating current (AC), which did not have precise control over rotational speed and, thus, the flow and pressures of the PAP therapy. The second-generation blowers were centrifugal designs powered by single-staged 12 V brushed direct current (DC) motors. The voltage waveform, amplitude, and frequency of the DC motor can all be varied to allow for improved control over the pressure, which was essential for more advanced PAP modes and comfort features like ramping. The-third generation devices used by most current PAP systems are DC brushless motors, which allowed for further reduction in size and noise and ability to use electricity at either 120 V/AC/60 Hz or 240 V/AC/50 Hz around the world or with a 12-V battery such as when camping or long-haul trucking. Multistage DC brushless motors have multiple impellers, which allow for a higher flow at lower motor speeds. Lower speeds make the motor quieter and allow the device to achieve desired speeds more rapidly and efficiently (ResMed Origins, 2021). Blowers with different impeller shapes like cones or pyramids are being used to further reduce the size and noise (Kenyon, 2020; Velzy et al., 2018).

An alternative technology that is being designed to create a micro-sized PAP motor is an electrostatic micro air pump using microelectromechanical (MEMS) technology (Marsh, 2021). Micro-bellows are created by applying an electrical charge to a piezoelectric membrane to move the air. Placing thousands of these micro-bellows in sequence amplifies the airflow and, according to a conversation with the inventor, Marsh, allows it to achieve pressure ranges similar to CPAP devices and enough airflow to adequately ventilate. Microsensors and microprocessors controlling the micro pumps can allow for quick adjustments, which could allow for similar functionality as AutoCPAP or bilevel devices without requiring a large device. This technology has yet to be proven to be effective or become commercially available.

18.2.2 Flow Signal Processing

To control the blower speed, data is collected and processed by the device. The airflow is sampled many times per second by the transducer, and the data is then sent to the microprocessor. The first step of flow signal processing is to isolate respiratory flow signal from other artifacts by scaling the data with a low-pass filter to remove large quick deviations in flow (i.e., cough or sneeze artifact) and a high-pass filter to exclude fast frequencies including cardiogenic fluctuations. Feedback limits can be used to determine if expected flow is beyond the expected range of flow variation at a particular motor speed (i.e., break in the tubing) and prevent the device from delivering too much or too little pressure (Colla & Kenyon, 2013).

Next the filtered flow data can be used by the algorithms for detection of flow limitation, snoring, respiratory events, respiratory rate, minute ventilation (MV), and determination of other variables. For many of the algorithms, event detection relies on comparing current flow to recent mean flow. Peak flow can be a poor measure of breath volume, which can lead to over- or underestimation of an apnea or hypopnea, so different techniques are used to more evenly compare the breath size even if the waveform is a different shape (see Fig. 18.1 and Table 18.1).

Philips Respironics' algorithms utilize a weighted peak flow (WPF) method to estimate ventilation. WPF method first determines the inspiratory period and then the inspiratory volume and the points on the inspiratory flow curve that correspond to the 20% and 80% volume. The average flow of all points between the 20% and 80% points is used as the weighted peak flow, a measure of ventilation. The model uses weighted peak flow values over the prior 2 min and determines the average of values between the 80th and 90th percentiles. This baseline is then used to compare to the current weighted peak flow to assess for decreases in amplitude, which would indicate apnea, hypopnea, or other sleep-disordered breathing events (Matthews et al., 2010).

ResMed's algorithms use a scaled low-pass-filtered absolute value of respiratory flow and a root mean squared (RMS) technique of the variance of the flow from the mean to compare one moving time period to another. RMS method determines ventilation from variance of the flow throughout the entire breath by comparing individual flow points to the mean airflow over a defined time period. The mean airflow is the zero point between inspiration and expiration; thus, variance from this mean divided by two equals the amplitude of the inspiratory flow. By taking the square root of the variance squared, outlying values receive less weight (Berthon-Jones, 2010).

Löwenstein's algorithm uses a similar measure of relative respiratory minute volume (rRMV) with an exponentially smoothed inspiratory and expiratory peak flow to compare the minute volumes of different periods. The relative average minute volume (rAMV) is calculated by taking the total inspiratory volume of all inspirations for a 2-min interval and dividing by 2 to get the current average minute volume. The long-term average minute volume is a weighted sum of the long-term average and the current average (Lowenstein sleep therapy products: Clinical benefits whitepaper, 2021).

Table 18.1 Flow sampling and filtering

	Airflow sampling rate (Hz)	Filtered ventilation measure
ResMed AutoSet and Autoset for Her	50	RMS of the variance of moving average scaled low-pass-filtered absolute value of respiratory flow
Philips Respironics APAP	100	WPF of 20–80% of inspiratory volume
DeVilbiss IntelliPAP 2 AutoAdjust	250	Ratio metric relationship of RMS of filtered flow with ranking filter to reduce peak excursions in the 3-min window
Löwenstein Prisma APAP	10	Respiratory flow is filtered by Butterworth low-pass filter with $f_{cutoff} = 0.75$ Hz/t_{insp}/s. rRMV: minute volume of current breath in relation to long-term exponentially smoothed minute volume. AMV: long-term average minute volume is weighted sum of the long-term average and the current average of prior 2 min

RMS root mean squared, *WPF* weighted peak flow, *rRMV* relative respiratory minute volume, *AMV* average minute volume

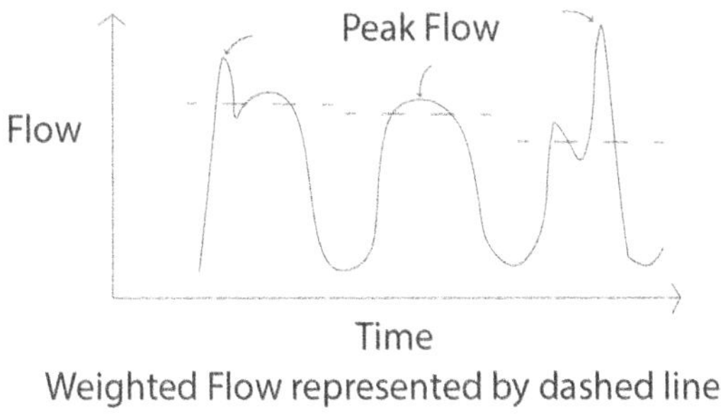

Fig. 18.1 Example of weighted compared to peak flow

DeVilbiss' Intellipap AutoAdjust uses an average scaled peak amplitude, while Intellipap 2 AutoAdjust uses ratio metric relationship of RMS and a ranking filte r applied to moving a 3-min window of flow to adjust for artifacts in peak flow.

18.2.3 Respiratory Cycle Determination

Determination of the inspiratory and expiratory periods of the respiratory cycle is essential not only to provide BPAP but also for expiratory pressure relief (EPR) and determination of inspiratory flow limitation (IFL) in APAP algorithms. Determining respiratory cycle is relatively simple in an invasive mechanical ventilation, because the double-limbed circuit has separate inspiratory and expiratory paths. In non-invasive PAP, inspiratory and expiratory flows are combined, and there is intentional and non-intentional leak that must be accounted for.

A method to determine the start of inspiration and expiration was described in a patent filed in 1989 by Sanders and Zdrojkowski (Sanders & Zdrojkowski, 1992). The average flow rate is compared to the instantaneous flow. An increase in instantaneous flow compared to the average flow rate is the start of inspiration. A decrease in flow compared to the average flow rate is the start of expiration. Because the sensor is removed from the airway, there is an inherent delay in detection. By altering cycle (change to expiratory positive airway pressure [EPAP]) and trigger (change to inspiratory positive airway pressure [IPAP]) sensitivities farther from non-zero value in a flow algorithm can help synchronize the machine with the patient's breathing pattern so that the patient will not encounter resistance to exhalation leading to dyssynchrony. For example, in a typical flow cycle algorithm, the change to EPAP occurs when the flow drops below 25% percent of the peak flow (Sanders & Zdrojkowski, 1995). The trigger to IPAP is typically set above zero flow to sense a significant patient effort so that the pressure change is not triggered by artifacts such as cardiogenic pulsations. Alternatively in a pressure control algorithm, inhalation is assumed when the mask pressure falls below a certain threshold and exhalation when the pressure rises above that threshold (Farrugia & Finn, 2001). In addition to the flow and pressure cycle algorithms, shape cycle (Zdrojkowski & Estes, 2000), volume cycle, and timed cycle algorithms are alternative methods to activate the change to the EPAP.

Some servoventilation (SV) algorithms use more continuous respiratory cycle determination. Table 18.2 shows how ResMed's ASV algorithm uses fuzzy inference rules using flow rate (relative to mean flow), direction, and size to determine the phase of the respiratory cycle (Berthon-Jones, 2014). ResMed's intelligent volume-assured pressure support (iVAPS) algorithm uses trigger and cycle sensitivities to determine respiratory phase. Philips Respironics' servoventilator algorithm determines the length of inspiration and expiration of recent breaths to predict the future breath duration and divides the breaths into short time segments (64 ms) to determine the expected mid-inspiration and other points of the cycle (Hill, 2004).

18.2.4 Pressure Control

Despite the name continuous positive airway pressure, it is important to remember that the motor speed must dynamically change based on respiratory flow and to compensate for leak in order to maintain a steady pressure. Early devices were not able to respond to these changes. In 1936, Poulton described controlling positive end-expiratory pressure (PEEP) through a valve on the mask that controlled not only mask pressure but also provided an exhalation port to prevent rebreathing CO_2 (Poulton, 1936). Sullivan's early devices controlled the pressure by varying the resistance of the airflow port, which could be titrated to a pressure level at the level of the device, but then would not

Table 18.2 ResMed's fuzzy logic for phase determination

Flow	Rate of change	Fuzzy phase
Zero	Increasing	Start inspiration
Small positive	Increasing slowly	Early inspiration
Large positive	Steady	Peak inspiration
Small positive	Decreasing slowly	Late inspiration
Zero	Decreasing fast	Start expiration
Small negative	Decreasing slowly	Early expiration
Large negative	Steady	Peak expiration
Small negative	Increasing slowly	Late expiration
Zero	Steady	Expiratory phase

adjust based on leak or position or respiratory changes to maintain a constant pressure at the level of the patient (Sullivan et al., 1981). In 1984, Berry et al. described a method of controlling pressure level using a column of water (Berry & Block, 1984). In 1987, in both an article and patent, Rapoport described adding a leak port to the mask to prevent rebreathing and moved the PEEP valve to elsewhere within the airflow circuit (Sanders & Zdrojkowski, 1992; Rapoport, 1987, 1991). Allowing the PEEP control to be in the device itself is an important step for allowing the future dynamic (dyn) control of PEEP levels. Brown et al. describe an early device in which PAP valves were swapped out to provide progressively higher PEEP levels until the therapeutic level was found (Brown & Javaheri, 2017).

Axe and colleagues' patent in 1993 first described a CPAP device with a DC motor controlled by microprocessor technology, which allowed for a more dynamic control of pressure needed for ramp and APAP functionality (Axe et al., 1993). Internal sensors (also called transducers) that monitor blower speed, airflow, and pressure provide feedback to the microprocessor controller that then controls blower speed. In order to maintain a stable mask pressure, the microprocessor must adjust the blower speed in response to deviations in pressure that occur from leak or normal swings in air pressure from breathing.

Because the sensors are within the device and not in the mask, in order to provide the desired pressures to the patient's airway, the device must calculate the predicted flow [Flow $(q) = \text{constant}\sqrt{(p1 - p2)}$] at the mask based on the pressure measurements at the transducer within the device (Farrugia & Finn, 2001). The length and diameter of the tubing and mask characteristics affect the constant value pressure and flow, which is why it is important to set the machine with the correct mask and tube type.

With increasing altitude, fan speed needs to be increased to maintain the same pressure (Fromm Jr. et al., 1995). Some older devices had a switch to change from high, medium, or low altitude (Brown & Javaheri, 2017), but most newer devices adjust automatically using an altitude sensor.

For bilevel devices, the inspiratory and expiratory pressures must be able to rapidly change up to 26 cm H_2O in as little as 50–200 ms. This typically involved braking the motor to lower the pressure and accelerating the motor to increase the pressure, which may require a peak load around 60–90 W resulting in the need for a large and expensive power source (Farrugia et al., 2019). Rather than relying solely on braking the motor, devices can allow the motor to freewheel upon detection of expiration, until the pressure falls to a predetermined level and then instructs the motor to control the speed at the second level. Low inertia motors are needed so that the rate of deceleration is fast relative to the typical duration of exhalation—preferably less than 13,600 g/mm^2. Freewheeling with a low inertia motor may result in a drop of about 3–4 cm H_2O in about 1.7 s, while braking can reduce more quickly but uses much more power. The starting pressure will also affect the speed of the pressure drop with a more rapid drop when starting at a higher pressure. Combining both freewheeling and intermittent braking may achieve both the speed and lower power use (Farrugia et al., 2019). Additionally, freewheeling can improve patient comfort over braking with a more natural airflow change.

18.2.5 Leak Compensation

Unlike invasive ventilation through an endotracheal tube, managing excessive air leak is an important factor that must be compensated for to optimize therapy. Leak affects aspects of performance including pressure delivered, cycle and trigger thresholds, and respiratory event determination. Leak is pressure, flow, and mask dependent. The leak can be determined from the flow rate at the end of exhalation (Zdrojkowski, 1994). It can also be calculated by subtracting the estimated flow through the exhaust of the device from the flow through the tubing (Berthon-Jones & Farrugia, 2014).

Air leak falls into two categories: (1) intentional or expected leak, which includes air leak from exhalation ports on the mask and varies by mask type and pressure level and (2) unintentional leak or excessive leak, which typically occurs from mouth opening or around the mask. In general, unintentional leak should be under 24 L/min with nasal masks and under 36 L/min with full face masks, but at higher pressures, intentional leak increases.

Most devices can compensate for excessive leak. Leak compensation works by constantly monitoring flow, looking for deviations from the expected respiratory flow, and compensating by adjusting the motor speed to minimize excessive air leak. Because there are normal variations in the patient's breathing cycle, the expected leak is usually averaged over several breaths. Leak may also be used to influence the control of auto-algorithms. For example, if leak is high, APAP devices may compensate by lowering the pressure, which may seal the mask and reduce unintentional mask leak.

The lips and tongue sometimes act like a one-way valve opening during exhalation when the pharyngeal pressure is at its highest. This is called valve leak or expiratory puffs. Expiratory puffs may falsely imply flow limitation, which can cause APAP to increase pressures unnecessarily. Some algorithms place less reliance on flow limitation when a large leak is present (Armitstead et al., 2011).

Whereas traditional ventilators with double-limbed circuits include a pneumotach that can directly monitor ventilation through an exhalation port, devices with a single airflow channel must estimate the expiratory tidal volume. If there is a leak, the double-limbed circuit tends to overestimate ventilation, which is why single-limbed non-invasive ventilators and newer PAP technology with good leak compensation algorithms may outperform double-limbed ventilators in the setting of non-invasive ventilation (Park & Suh, 2020; Pavone et al., 2020).

18.2.6 Apnea and Hypopnea Determination

Most newer APAP algorithms analyze flow changes to recognize apneas and hypopneas based on relative reductions in flow amplitude. Filtered mean or peak flow, as described previously, is used to compare a short moving time period (e.g., one breath or 2 s) to a moving longer period (e.g., 5 min) to evaluate for apnea or hypopnea (Berthon-Jones, 2010). Apneas and hypopneas are typically defined as a reduction in airflow amplitude below a percentage of recent amplitude for at least 10 s, with varying methods used by different algorithms (see Table 18.2). In some devices, the amplitude drop can be altered to change the sensitivity of the device to recognize events.

There are several studies that have evaluated the reliability of sleep-disordered breathing event detection, specifically the apnea–hypopnea index (AHI), based on the device flow measurements compared to polysomnography (Berry et al., 2012; Ueno et al., 2010; Ikeda et al., 2012; Prasad et al., 2010; Desai et al., 2009; Cilli et al., 2013; Huang et al., 2012; Thomas & Bianchi, 2017; Li et al., 2015). Most found reasonable correlation between the AHIs, with some studies showing device overestimation and others demonstrating underestimations of the AHI (Ueno et al., 2010; Thomas & Bianchi, 2017). Most studies demonstrate a stronger correlation between the PAP and polysomnography-defined AHIs at higher AHI levels.

18.2.7 Differentiating Between Obstructive and Central Apneas

Some manufacturers use algorithms to differentiate central vs obstructive apneas (see Table 18.2). Several algorithms (e.g., ResMed, DeVilbiss IntelliPAP 2 AutoAdjust) use the forced oscillation technique (FOT) to determine airway patency to differentiate central from obstructive apneas (Galetke et al., 2009; Farre et al., 2001; Herkenrath et al., 2020). Figure 18.2 shows how when a potential apnea is detected due to reduced flow, the algorithm provides a single pressure pulse or small oscillation in the flow, e.g., 1 cm, 4–5 Hz, which is only reflected back to the flow sensors if the airway is closed (Berthon-Jones, 2010; Herkenrath et al., 2020; Martin & Oates, 2014; Axe et al., 2000). Low resistance in the absence of

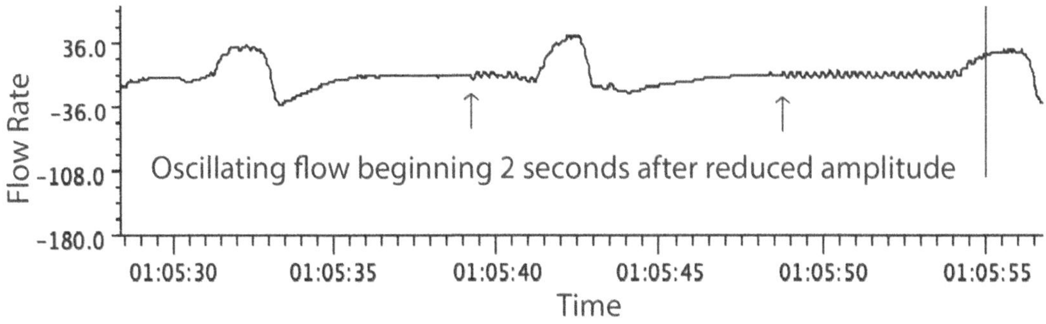

Fig. 18.2 Forced oscillation technique (FOT)

flow (open airway) is interpreted as a central event, and increased resistance in the absence of flow (closed airway) is interpreted as an obstructive event. With a similar concept to FOT, Philips Respironics' algorithm uses pressure pulses. The airway is determined to be clear if the pressure test pulse generates a significant amount of flow; otherwise, the airway is determined to be obstructed. A mixed apnea can be determined if the airway is open for only part of the flow cycle. Another method for differentiating events is to evaluate the signal for cardiogenic pulse artifact in the flow, which is only present if the airway is open, but because it is not as reliable, it is not used by any device to our knowledge.

While most current algorithms differentiate apnea type, most are unable to differentiate obstructive and central hypopnea type. Thus, in the absence of full central apneas, most algorithms will still increase pressure in response to periodic breathing. One algorithm that does identify and respond differently to central hypopneas is DeVilbiss IntelliPAP 2 AutoAdjust algorithm, which defines periodic breathing as a waxing and waning period of greater than 20 s with self-similar repetitive mathematical characteristics. Löwenstein's algorithm uses the presence of snoring or flattening to define hypopneas as obstructive.

18.2.8 Flow Limitation and Snore Determination

Many newer algorithms detect flow limitation via proprietary algorithms using shape and amplitude changes in the pressure transducer signal (see Table 18.3). Snore is typically detected by looking for a vibratory signal within the airflow. Because high leak and high pressure levels can mimic snore signal, the algorithms may be less responsive to snore signal if leak or pressures are high. Some algorithms rate snore severity by the degree of vibratory signal, while others rate by the number of breaths within a time period with snoring. For example, Löwenstein's standard (std) algorithm defines severe snoring as ≥8 breaths and mild snoring as ≥4 breaths with snoring within a 2-min epoch. Their dynamic algorithm rates mild snoring as ≥3 breaths with snoring within a 2-min epoch.

18.2.9 Mask and Humidification and Sound Technology

There are several technological considerations about mask and humidification technology that are relevant to clinical care. One major landmark in the history of technology was the self-sealing "bubble mask" in 1990 (Hansford, 2011). Since then, further improvements in mask seal, limiting nasal bridge irritation, facilitating use with magnets and quick-release hose connections, and reducing noise and intentional leak spraying on bed partners with micro-exhalation ports have improved tolerance.

Mask and humidification choice can affect dead space and potentially CO_2 retention. Oronasal masks have about 205 mL of dead space and nasal masks approximately 120 mL

Table 18.3 Flow limitation and other event determination

Algorithm	Flow limitation detection	Other events	Leak detection
ResMed AutoSet and Autoset for Her	S8: Mid-inspiration flatness S9–S11: Breath-by-breath flow limitation index from breath shape index, RMS flatness index, and ventilation change and breath duty cycle. Ventilation change is the ratio of the current breath ventilation to the recent 3-min ventilation. Breath duty cycle is the ratio of current breath time of inspiration to total breath time of recent 5 min. When a breath is severely flow limited, the flow limitation index will be closer to 1, as opposed to when the breath is "normal" or round, the flow limitation index will be zero (Armitstead et al., 2011)	S9–S11: Unknown apnea—apnea with leak >30 L/min	95% Leak >24 L/min
Philips Respironics APAP	Relative changes in the peak, flatness, roundness, shape (skewness) of inspiratory wave form. Evaluated over short period (4 breaths) and over long period (several minutes) and rates as better, worse, or the same compared to baseline. Roundness is determined by the similarity of the weighted peak flow between 5% and 95% values to a sine wave. Flatness is determined by the absolute value of the variance between 20% and 80% of inspiratory flow from the average of all the values in the same period and dividing by the 80% volume point. Skewness is determined by dividing the average of the highest 5% of flows in the mid-third of the breath by the average of the highest 5% of flows in the first third of the breath (Matthews et al., 2010). Statistical measures are used to help minimize false event detection while allowing the device to be sensitive to even small changes	Variable breathing—standard deviation/adjusted mean flow over a 4 min window above threshold	Leak level exceeds flow limit for a given pressure
DeVilbiss IntelliPAP 2 AutoAdjust	Relative flatness of inspiratory waveform of breath, detecting positive, negative, or zero slope. Score average over 12 s. Scored: none, mild, moderate, severe	Expiratory puff index based on strings of several breaths scores as none, mild, moderate, and severe	95 L/min or expected leak for a given CPAP level
Löwenstein Prisma APAP	Flow limitation index for each breath based on comparison of flow shape with convex hull of inspiration flow shape and amplitude reduction that doesn't meet hypopnea criteria RERA, defined by runs of approximately 5 breaths with flattening followed by an "arousal" based on a sudden high amplitude flow change, with or without an inspiratory snoring that does not otherwise meet criteria for apnea or hypopnea	Unclassified apnea (due to leak, position change, or high pressure), unclassified hypopneas due to leak or high pressure, periodic breathing (marked on data but no pressure change)	High leak >50 L/min. High leak ends if leak is <20 L/min or if <35 L/min for >15 s

RERA respiratory effort–related arousal

(Navalesi et al., 2000). Masks with lower intentional leak can lead to more CO_2 retention, but masks with higher intentional leak can lead to failed triggering.

Mask changes may affect the effectiveness of a given mode. Oronasal mask can be associated with a higher device residual AHI for a given CPAP setting than was determined in the lab for a nasal mask (Ebben et al., 2014). Different cycle and trigger sensitivities may also be needed if mask type is switched.

Nasal masks can promote mouth leak, which can lead to high unidirectional nasal flow that leads to more loss of humidity, increased nasal resistance, nasal/oral dryness, nasal congestion, and more mouth opening, which then can worsen leaks. Heated humidification can reduce nasal resistance and reduce mouth leak (Massie et al., 1999).

Waterless heat and moisture exchangers (HMEs) have been used by Colin Sullivan to humidify the air, and updated versions are now being incorporated into the tubing (e.g., ResMed's AirMini). HMEs wick moisture from exhaled air and then return the moisture to the air on inhalation. The same concept can be used by small humidifier chambers in which a small volume of water is placed in a chamber with a hydrophilic wick. By heating an element at one end of the wick, the amount of water that is vaporized as air is blown over it can be modulated (Harrington et al., 2017). HME humidification can cause an increase in minute ventilation and work of breathing compared to heated humidifier, and higher $PaCO_2$ likely occurs due to increased dead space (Esquinas & Shah, 2012; Jaber et al., 2002). Another technology designed to decrease rainout, save energy, and save water is DeVilbiss PulseDose Humidification, which switches between humidified air during inhalation and dry room air during exhalation.

Over time technology has been used to reduce the sound of PAP devices through blower and mask technology, but also through length of the airflow path within the device and sound dampening materials. A Federal Drug Administration recall in April 2021 led to the recall of almost 2 million Philips Respironics' devices due to the risk of degradation of sound dampening foam that was in the airflow path and risk for irritation from particulates and potential carcinogens from chemical breakdown (Mell, 2021). It is uncertain whether the use of ozone and other PAP cleaners may have exacerbated the breakdown.

18.3 Positive Airway Pressure Modes and Algorithms to Control the Flow of Air

The control of airflow and pressure by microprocessors allows for algorithms that attempt to improve the comfort of air delivery including ramping and EPR and the multiple modes of PAP therapy that allow for a more individualized treatment of different forms of sleep-disordered breathing including CPAP, BPAP, and BPAP with a backup rate and adaptive forms of bilevel PAP with a backup rate including servoventilation (SV) and volume-assured pressure support (VAPS).

18.3.1 Continuous Positive Airway Pressure (CPAP)

Standard CPAP mode today, as previously discussed, is more complex than early algorithms because they involve feedback to maintain pressures at the goal level despite leak and changing respiratory dynamics throughout the night due to medication or alcohol effects, position changes, fluid shifts, and sleep stage changes. Most CPAP devices allow for pressure settings between 4 and 20 cm H_2O. End-expiratory pressure of 4 cm H_2O is the lowest pressure needed to provide enough flow to clear the dead space from the device, tubing, and airway to prevent rebreathing of exhaled air (Ferguson & Gilmartin, 1995). The goal of CPAP is to increase upper airway pressure enough to provide a pneumatic splint to open the airway, which may collapse during inspiration. Typically, the pressure is set to prevent hypopneas, apneas, snoring, flow limitation, and arousals. By providing positive end-expiratory pressure, CPAP may also recruit the alveoli and improve ventilation. Decreases in intrathoracic pressure with CPAP can reduce venous return, decreasing afterload, transmural pressure, and preload and decreasing atrial natriuretic peptide production affecting nocturia. Comfort settings of ramp and EPR are available with many CPAP devices. A recent review by Killick and Marshall summarizes the lack of randomized control trials to show the benefit of any of these modifications over fixed CPAP (Killick & Marshall, 2021). This is consistent with our clinical experience in which patient preference for these settings is variable, but understanding what options are available and how they work can help a provider choose settings to optimize the individual's experience of using PAP therapy.

18.3.2 Autotitrating Continuous Positive Airway Pressure (APAP)

Autotitrating (also known as auto, automated, autoadjusting, or automatic) continuous positive airway pressure (APAP) incorporates the ability of the PAP device to detect and respond to changes in the upper airway flow and/or resistance in real time. APAP can be helpful for patients that may need a higher pressure in rapid eye movement (REM) or supine position but cannot tolerate the higher pressure through the entire night. APAP may also be able to be used diagnostically to determine a fixed pressure setting (Rosen et al., 2012).

Most devices have a therapeutic pressure range between 4 cm H_2O and 20 cm H_2O, giving the clinician the ability to adjust the upper (EPAPmin) and lower (EPAPmax) pressure limits based on the clinical conditions and the patient's response to therapy. APAP gradually increases or decreases the pressure in response to respiratory events or airflow changes between those limits but similarly maintains the same pressure throughout the respiratory cycle unlike BPAP or auto-adjusting BPAP (AutoBPAP; discussed later). When the breathing pattern is stable, the pressure is gradually reduced. Similar to CPAP, EPR is available with some APAP devices.

The responsiveness of the algorithms and the ability to detect (and thus respond to) obstructive and central events vary, which can affect the efficacy (Nilius et al., 2020). Because female and male have different event types including fewer apneas and hypopneas and more flow limitation in females, some algorithms have been developed to be more responsive to these obstructive features (i.e., ResMed AutoSet for Her and Lowestein APAPdyn) (Isetta et al., 2015). Some algorithms (ResMed APAP) have larger and more rapid responses to obstruction than other algorithms (Isetta et al., 2015). Patients with high loop gain or pressure sensitivity may benefit from algorithms that respond more slowly and with small pressure changes (Philips Respironics APAP and Lowestein APAPstd).

Some of the first AutoCPAPs, like Virtuoso LX smart and SOMNOsmart, only detected vibration-based snore changes making them unresponsive to many significant events especially in patients who do not snore (either due to palatal surgery or naturally) (Sullivan & Lynch, 1993; Morgenthaler et al., 2008; Berry et al., 2002; Littner et al., 2002). Most newer APAP algorithms monitor a combination of changes in inspiratory flow patterns, including inspiratory flow limitation, snoring, and reductions or the absence of airflow, using a pneumotachograph, nasal pressure monitors, or alterations in compressor speed as described earlier.

Table 18.4 summarizes the response to events, and Table 18.3 summarizes other pressure changes and comfort settings for several different APAP algorithms including ResMed AutoSet (Berthon-Jones & Farrugia, 2014; Armitstead et al., 2011), ResMed AutoSet for Her (Armitstead et al., 2011), Phillips Respironics AutoCPAP (the same for SystemOne REMstarAuto, Dreamstation, and Dreamstation 2 except RampPlus) (Matthews et al., 2010), and DeVilbiss IntelliPAP 2 AutoAdjust (Johnson & Johnson, 2015; Clinical Overview: DeVilbiss IntelliPAP AutoAdjust, 2014) and Löwenstein Prisma Line APAP (previous Löwenstein/Weinmann devices used different algorithms; Lowenstein sleep therapy products: Clinical benefits whitepaper, 2021).

Philips Respironics' APAP algorithm uses layers of control including ramp, leak, snore, apnea/hypopnea, variable breathing (VB), and flow limitation. After 3–5 min with no obstructive features, it will enter a testing period described in Table 18.5 to search for critical pressure (Pcrit; at which flow characteristics [peak,

Table 18.4 Event determination

Algorithm	A/H flow comparison	Apnea detection	Non-OA detection	Hypopnea detection
ResMed AutoSet and Autoset for Her	Prior 1 min RMS moving average	A 2 s RMS moving average < 25% for 10 s	S8: None S9–S11: 1 cm H_2O 4 Hz FOT throughout apnea with mixed apnea detection	S8: A 12 s RMS scaled average 25–50% for 10 s S9–S11: Above with at least 1 obstructed breath
Philips Respironics AutoCPAP	Average of 80–90th percentile WPFs of prior 4 min moving average	WPF per breath <20% of recent baseline for 10 s, terminating with breath >30%	During the apnea, one or more pressure test pulses. Clear if the pressure test pulse generates a significant amount of flow; otherwise, obstructed	WPF per breath 20–60% of recent baseline for 10 s terminating at 60 s or with breath over 75% of recent WPF
DeVilbiss IntelliPAP 2 AutoAdjust	Three minutes including time before and after event RMS moving average	A 4 s RMS moving average < 10% for 10 s	Modulating micro-oscillation (0.07 cm H_2O at 3½–4½ Hz) throughout apnea	10–40% Default (adjustable to 30–50%) for 10 s
Löwenstein Prisma APAP	No comparison for apneas just reduction to a range of exponential smoothed peak flow. Comparison for hypopneas is a smoothed weighted average minute volume over a variable time period that adapts to different situations	Start of apnea: respiratory flow within −4 L/min to 4 L/min for at least 3.5 s End of apnea: respiratory flow not within −4 L/min ro 4 L/min for at least 2 s within a 4 s period. Min length, 10 s	FOT with 0.3 cm H_2O oscillation at 4 Hz	Hypopnea detected if peak flow remains ≤48%–50% of reference rAMV for >10 s. If flattening/snoring, classified obstructive

A/H apnea/hypopnea, *rAMV* relative average respiratory minute volume, *FOT* forced oscillation technique, *RMS* root mean squared, *WPF* weighted peak flow

flatness, roundness, and skew] worsen) and optimal pressure (Popt; at which there is no further improvement in flow characteristics) within the EPAPmin–EPAPmax limits. The algorithm also uses several mechanisms to avoid overtitration, which include non-responsive apnea/hypopnea (NRAH) logic (Table 18.5), variable breathing (Table 18.6), and leak control (Table 18.6) (Matthews et al., 2010; Puertas et al., 1999; Remmers & Feroah, 2003).

Devilbiss's first IntelliPAP AutoAdjust algorithm allows amplitude and duration cut-points for apneas and hypopneas to be set to change sensitivity. IntelliPAP AutoAdjust algorithm evaluates the previous minute for apneas and hypopneas based on thresholds set from average scaled flow amplitude of the previous 5 min and then either increases pressure by 1 cm H_2O per minute if events of snoring or decreases 0.6 cm H_2O every 6 min if stable breathing. IntelliPAP 2 AutoAdjust similarly adjusts pressures once per minute but also responds to flow limitation based on inspiratory flatness. IntelliPAP 2 prioritizes response first on the presence of periodic breathing, then respiratory events, and then flow limitation. If periodic breathing or central apneas are noted, the algorithm holds or lowers pressures.

Löwenstein's APAP algorithm evaluates 2-min epochs and defines each epoch as epoch with severe obstruction (eSO; which contains apneas, hypopneas, or severe snoring), epoch with mild obstruction (eMO; which contains mild flow limitation or respiratory effort–related arousal [RERA]), epoch with flattening (eFl;

Table 18.5 APAP algorithms for responding to events

Algorithm	OA/H response	Flow limitation response	Vibratory snore response
ResMed AutoSet	Increases pressure based on the current pressure every 10 s of apnea: increment max 3 cm H_2O when pressure is 4 cm H_2O. Increment drops linearly down to 0.5 cm H_2O when pressure is 20 cm H_2O; S8, no increase above 10 cm H_2O	S8: Increment max 0.45 cm H_2O/breath. Lower increment if high leak or as pressure increases further above 10 cm H_2O S9: Uses 3-breath average FL index. Increment typically around 0.6 cm H_2O/breath for severely flow-limited breaths. Lower increment if lower FL index and high leak or as pressure increases further above 15 cm H_2O S10: Increment max 0.6 cm H_2O/breath. Otherwise, the same as S9 (Armitstead et al., 2011)	S8–S9: Increment max 1 cm H_2O/breath. Lower increment if snore is less severe and high leak or as pressure increases further above 10 cm H_2O S10: Increment max 0.6 cm H_2O/breath for a loud snore. Otherwise, the same as S9
ResMed AutoSet for Her	Increases pressure based on the current pressure every 10 s of apnea: Increment max 2.5 cm H_2O when pressure is 4 cm H_2O. Increment drops linearly down to 0.5 cm H_2O when pressure is 20 cm H_2O	Uses single-breath FL index: Increment max 0.5 cm H_2O per breath for severely flow-limited breaths. Lower increment if lower FL index and high leak or as pressure increases further above 10 cm H_2O	Increment max 0.5 cm H_2O per breath. Lower increment if snore is less severe and high leak or as pressure increases further above 10 cm H_2O
Philips Respironics APAP	If 2 apneas or 1 apnea/1 hypopnea or 2 hypopneas: Increases 1 cm H_2O over 15 s and holds for 30 s. NRAH logic activates if pressure is at least 11 cm H_2O. It limits max pressure to 11 or 3 cm H_2O higher than pre-apnea baseline. If more apneas within 8 min, decrease pressure by 2 cm H_2O and then further to 1 cm H_2O over the level that prevents snore and then holds with combined ramp down and hold time of ~15 min. Pressure will continue to increase in response to 2 hypopneas (Matthews et al., 2012)	Pressure increases by 0.5 cm H_2O/min in response to FL. Intermittent upward scans by 1.5 cm H_2O over 3 min to see if improvement in FL and then deceases if no improvement. If pressure not held by snore, A/H, or VB logic, then enters testing protocol that collects 3–5 min data, and then downward search sequence for Pcrit begins ramping down 0.5 cm H_2O/min until Pmin as long as no worsening in FL. If worsening, Pcrit is set, and pressure quickly increases by 1.5 cm H_2O and held for 10 min. Then, Popt search increases pressure by 0.5 cm H_2O/min for at least 2.5 min to test if FL improves, worsens, or stays the same. If improvement, continues 0.5 cm H_2O/min pressure increase; if no improvement, pressure decreases by 1.5 cm H_2O and sets Popt and holds for 5 min. FL or other events end all holds	If 3 snores within 60 s, with no more than 30 s between snores, increase 1 cm H_2O over 15 s and then hold for 1 min with a higher snore threshold at higher pressures
DeVilbiss IntelliPAP 2 AutoAdjust	Increases pressure 1 cm H_2O/min for OA. If event is near end of min, response is delayed until the following minute so centered moving window completes to allow event to be scored. Increases pressure 0.5 cm H_2O/min if hypopnea with 1 other event in a 6 min window or 1/min if hypopnea with >1 events in a 6 min window	Response determined by severity and duration (15 s/min) to evoke a 0.5 cm H_2O/min for moderate–severe FL index. Less response if no OA or hypopnea within the past 8 min. Less response if high leak or high expiratory puffs	If high leak or expiratory, 0.5 cm H_2O/min for moderate–severe snore with no response

(continued)

Table 18.5 (continued)

Algorithm	OA/H response	Flow limitation response	Vibratory snore response
Löwenstein Prisma APAPstd	Pressure increase depends on the starting pressure quartile Lowest quartile: 1.5 cm H_2O Middle quartiles: 1 cm H_2O Highest quartile: 0.5 cm H_2O Increases immediately with event up to 2 times per 2 min epoch	Pressure increase depends on starting pressure quartile Lowest quartile: 0.5 cm H_2O No response for other quartiles Increases at the end of a 2 min epoch	Severe snoring: pressure increases by OA rule at the end of epoch Mild snoring Lowest 2 quartiles: 1 cm H_2O Highest 2 quartiles: 0.5 cm H_2O Increases at the end of a 2 min epoch
Löwenstein Prisma APAPdyn	The same as APAPstd	Pressure increase depends on the starting pressure quartile Lowest 2 quartiles: 0.5 cm H_2O Third quartile: 0.2 cm H_2O Highest quartile: no response Increases at the end of a 2 min epoch	Severe snoring: pressure increases immediately after 6th and 12th breaths with snoring within a 2 min epoch by OA rules Mild snoring: pressure increases at the end of epoch by APAPstd rules

A/H apnea/hypopnea, *APAP* autotitrating positive airway pressure, *dyn* dynamic, *FL* flow limitation, *FOT* forced oscillation technique, *max* maximum, *min* minutes, *NRAH* nonresponse apnea–hypopnea logic, *OA* obstructive apnea, *Pcrit* critical pressure, *Pmax* maximum pressure, *Pmin* minimum pressure, *Popt* optimal pressure, *s* seconds, *std* standard, *VB* variable breathing

defined by average flattening index for all breaths in an epoch >18), or epochs with no events. The pressure change in response to the epoch type is altered by the reactivity of the algorithm that can be set to either standard (APAPstd) and dynamic (APAPdyn). With APAPdyn, there are larger and more rapid pressure changes to flow limitation and snoring than the APAPstd. The response is also affected by which quartile of the EPAPmin–EPAPmax range the current pressure is within.

Löwenstein's algorithm limits pressure increases in response to just milder snoring, RERA, and flow limitation to no more than 4 cm H_2O. The initial range is from the EPAP min (or ramp level) to 4 cm H_2O higher. If there is an eSO or if the ramp level increases, then the limit range increases by the same amount as the pressure increases. If the pressure is decreased, then the limit also decreases. In Löwenstein's APAP algorithm, the degree of pressure reduction also depends on the preceding event type and starting pressure quartile (see Table 18.7).

Another variation of an APAP algorithm, Fischer and Paykel's SensAwake, is designed to sense when a patient has awakenings in the middle of the night using airflow patterns and temporarily lowers the pressure with the aim of improving the ability for the patient to fall back to sleep. A randomized 4-week crossover trial of Fischer and Paykel's SensAwake was found to reduce leak but did not affect adherence or objective sleep quality (Bogan & Wells, 2017).

18.3.3 Clinical Considerations Related to APAP Technology

There are some important considerations to recognize about the use of APAP technology. Differences between algorithms are important to understand for appropriate clinical care as they affect the patient's tolerance of the devices and the clinical efficacy:

Table 18.6 APAP algorithm pressure decreases and comfort features

Algorithm	Other pressure changes	Ramp	Expiratory pressure relief
ResMed AutoSet	S8: gradual decrease to Pmin over 20 min after apnea, over 10 min after FL or snoring event as soon as breathing is stable S9–S11: gradual decrease to Pmin over 40 min after apnea, over 20 min after FL or snoring	S8–S9: 0–45 min ramp S10–11: 0–45 min ramp or AutoRamp starts ramping when sleep onset is inferred	EPR off, 1–3 cm H_2O
ResMed AutoSet for Her	Gradual decrease to Pmin over 40 min after apnea and over 20 min after snoring and 60 min after flow limitation as soon as breathing is stable (Berthon-Jones, 2010; Armitstead et al., 2011)	A 0–45 min ramp or AutoRamp starts ramping when sleep onset is inferred	EPR off, 1–3 cm H_2O
Philips Respironics APAP	If high variable breathing is noted, then if recent (5 min) pressure was stable, then pressure stays the same, and if recent pressure decreases, then pressure increases by 0.5 cm H_2O/min up to 2 cm H_2O, and if recent pressure increases, then pressure decreases by 0.5 cm H_2O/min up to 2 cm H_2O If there is a large leak, it reduces pressure by 1 cm H_2O over 10 s and holds pressure for 2 min	Fixed ramp (0–45 min ramp) or smart ramp (DreamStation): increases faster if obstructive events or FL occur RampPlus (DreamStation 2 APAP advanced): Patient sets starting pressure between 4 and 10 cm H_2O for 15, 30, or 45 min. Full APAP algorithm including Popt and Pcrit searching, after sleep onset. Algorithm attempts to keep pressure close to selected starting pressure within the selected time	C-Flex off, 1–3. Pressure drop varies based on setting airflow and patient effort A-Flex and C-Flex plus have a more gradual increase in pressure with inhalation than C-Flex
DeVilbiss IntelliPAP 2 AutoAdjust	Decides whether to decrease every min based on events in prior 6 min. If no events in a 1 min period, a small decrease of <0.1 cm H_2O/min. If no events in a 6 min period, decrease by 0.1 cm H_2O/min. If central apnea, pressure decreases and no increase for 6 min. If periodic breathing, no increase, and then pressure decreases if persists If high expiratory puffs, no pressure increases at least 1 min	A 0–45 min ramp	Smartflex off, 1–3 cm
Löwenstein Prisma APAP	RERA pressure increase depends on the starting pressure quartile—the same for both std. and dyn algorithms Lower 2 quartiles: 1 cm H_2O Higher 2 quartiles: 0.5 cm H_2O See Table 5 for pressure reduction protocol, which varies based on prior events and starting pressure	SoftSTART Ramp: 0–45 min. Obstructive events cause pressure increase during ramp	SoftPAP: off, 1–1.6 cm H_2O

APAP autotitrating positive airway pressure, *dyn* dynamic, *EPR* expiratory pressure relief, *FL* flow limitation, *Pmin* pressure minimum, *RERA* respiratory effort–related arousal, *std* standard

Table 18.7 Löwenstein Prisma APAP pressure reduction algorithm

	Time (minutes) before pressure reduction based on epoch type and pressure quartile			
Pressure quartile/epoch type	Lowest quartile	Second quartile	Third quartile	Highest quartile
eSO	10	8	8	6
eMO	6	6	4	2
eFl or no event	4	4	2	2
Pressure reduction	−0.25 cm H_2O	−0.25 cm H_2O	−0.5 cm H_2O	−1 cm H_2O

eSO epoch with severe obstruction, *eMO* epoch with mild obstruction, *eFl* epoch with flattening

- Address leak first. Mask leak limits much of the algorithm's ability to recognize flow limitation and snoring and differentiate central from obstructive apneas. A high number of unknown (ResMed) or unclassified (Löwenstein) apneas indicate high leak.
- Don't leave the EPAPmin too low. Obstruction can worsen quickly when a patient enters rapid eye movement (REM) sleep or changes position, and by the time the device is able to adjust to the needed pressure, the patient may have had desaturations or awakenings. This is why most of the protocols testing the equivalence of AutoCPAP to in-lab titration recommend changing the EPAP to the pressure that the device achieves 90–95% of the time (Morgenthaler et al., 2008; Loube et al., 1999). Too low EPAPmin may present with awakenings a couple hours into sleep, residual symptoms, or difficulty tolerating PAP. A large difference between the median pressure and the 95% pressure despite low residual AHI is suggestive. The rate of pressure change to events of different algorithms varies, so this effect may be different with different APAP devices.
- Fixed CPAP or limited APAP ranges should be considered in patients with high loop gain at risk for periodic breathing (i.e., congestive heart failure patients) or in patients with high residual AHI on APAP (Abdenbi et al., 2004; Rigau et al., 2006; Farre et al., 2002; Lofaso et al., 2006). APAP is unable to differentiate between central hypopneas from periodic breathing and obstructive hypopneas, which will lead to overtitration. Less dynamic APAP algorithms can also be considered.
- Low residual AHI does not ensure adequate treatment in patients with non-upper airway obstruction causes of hypoxia and hypoventilation, i.e., congestive heart failure, lung diseases such as chronic obstructive pulmonary disease (COPD), obesity hypoventilation syndrome, and neuromuscular disease. In-lab titration is recommended in these conditions; however, ventilation and respiratory rate data from the devices, overnight oximetry, home $ETCO_2$ or $TCCO_2$ monitoring, arterial blood gas PCO_2, or bicarbonate level may help screen for adequate treatment and guide therapy adjustments.
- APAP may induce sleep fragmentation due to pressure changes (Marrone et al., 2002). This concern has not been substantiated in studies evaluating changes in sleep structure or in clinical trials that have measured subjective sleepiness as a main outcome. Specifically, the frequency of microarousals and sleep fragmentation induced by APAP devices appears to be small, and clinical outcomes related to subjective sleepiness also show no significant differences compared with conventional CPAP therapy (Fuchs et al., 2002; Masa et al., 2004; Ayas et al., 2004; Hukins, 2004; Nussbaumer et al., 2006; Vennelle et al., 2010). However, on an individual basis, if a patient is having a suboptimal response to APAP especially with disrupted sleep, a trial of fixed CPAP should be a consideration.
- Shorter data therapy hours than reported by the patient may be due to high leak causing the device to turn on and off despite being on the patient. Automatic "Smart" On/Off features sense changes in airflow consistent with breathing to turn the device on automatically after the patient puts on the device. If the device senses that mask has been removed (i.e., very high leak for 3 s), then the device

will automatically turn off. This will be evident by frequent starts and stops on adherence data as shown in Fig. 18.3.

18.3.4 Ramp and Starting Pressure Adjustments

Ramp functionality was designed because some patients have difficulty tolerating high pressures preventing them from being able to fall asleep. Basic ramp settings typically start with a pressure of 4 cmH_2O and gradually increase to the goal pressure over a set amount of time, typically 5–45 min. Newer automatic ramp algorithms (e.g., Philips Respironics' SmartRamp and RampPlus, ResMed's AutoRamp, and Löwenstein's SoftSTART) use airflow monitoring to sense breathing patterns consistent with sleep or obstruction such as snoring, flow limitation, or apneas and hypopneas and then start increasing the pressure after the patient falls asleep or events are sensed. The automatic ramp can either then increase the pressure linearly at a set rate or the pressure can be increased more quickly if there are repetitive obstructive events (Ramp Plus Discussion Paper, 2021). ResMed's AutoRamp increases pressure more rapidly if events or flow limitation or 5 consecutive breaths with snoring are noted.

Too low of a starting pressure can cause air hunger and intolerance in some patients. Typically, the starting ramp pressure can only be changed by the provider to a higher level if needed for comfort, but Philips Respironics' RampPlus allows for the patient to select the starting pressure between 4 and 10 cm H_2O and choose a set time (15, 30, 45 min) for the ramp. For both SmartRamp and RampPlus, APAP algorithm is triggered after sleep onset is sensed. RampPlus algorithm attempts to keep patient pressure close to their selected starting pressure within the selected time.

In addition to ramp, gradually changing EPAPmin may help with comfort when PAP therapy is first started and achieve more optimal pressures with ongoing treatment. Philips Respironics' CPAP Check, Opti-Start, and EZ-Start algorithms address this. CPAP Check evaluates patient obstructive respiratory disturbance index (ORDI) every 30 h of therapy use and increments pressure ±1 cm H_2O if needed, to a maximum of ±3 cm H_2O from the set pressure (Respironics P, 2012).

Opti-Start algorithm can be used with APAP. The algorithm evaluates patient respiratory events after 30 h and every 30 h of use thereafter and establishes a new Opti-Start pressure for the next 30 h of use. The pressure is typically close to two-thirds of the patient's previous 90% pressure. Opti-Start begins each new therapy session at the Opti-Start pressure if RampPlus is not enabled. Otherwise, a set RampPlus pressure will override the actual starting Opti-Start pressure. If RampPlus titrates below the Opti-Start pressure, the auto-algorithm will be a little more responsive to obstructive respiratory events.

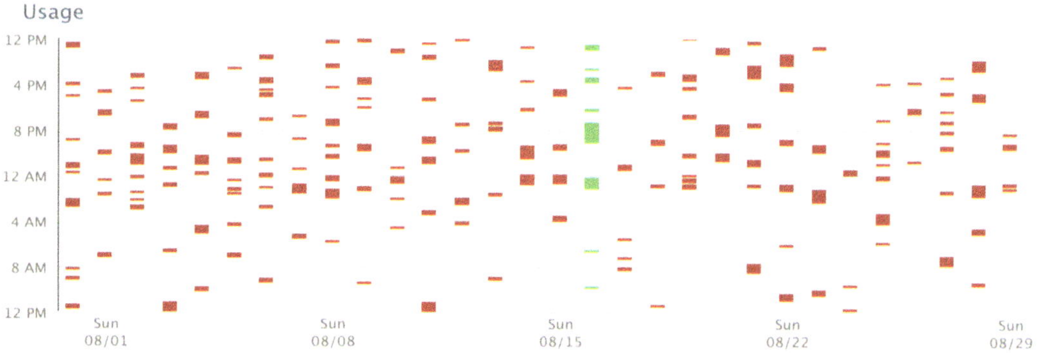

Fig. 18.3 Auto On/Off in the setting of high leak

EZ-Start can be used with CPAP, CPAP-Check, or APAP mode. EZ-Start for fixed CPAP mode reduces the pressure to half of the prescribed setting, but no lower than 5 cm H_2O. In APAP mode, EZ-Start reduces the maximum EPAP to 1 cm H_2O above the minimum EPAP. For either mode, after each day of successful use, defined as a therapy session ≥4 h, the therapy pressure will increase by 1 cm H_2O until the prescription pressure is reached. From that point forward, the therapy device would operate in normal CPAP, CPAP-Check, or auto CPAP mode. If the patient has not reached their prescription pressure after 30 days of EZ-Start, then the therapy pressure will increase by 1 cm H_2O per day until the prescription pressure is reached.

18.3.5 Expiratory Pressure Relief Systems

A common complaint in many patients with OSA using CPAP is the uncomfortable feeling of exhaling against positive pressure. Several PAP manufacturers have developed EPR systems in an attempt to remedy this potential problem. EPR algorithms briefly reduce the PAP pressure, between 1 cm H_2O and 3 cm H_2O, during all (i.e., ResMed) or the start of exhalation (i.e., Philips Respironics) before returning the pressure to the set PAP setting prior to the initiation of inspiration. Only Philips Respironics' EPR algorithm (FLEX) monitors the patient's airflow during exhalation and reduces the expiratory pressure in response to the airflow and patient effort. The amount of pressure relief varies on a breath-by-breath basis, depending on the actual patient's airflow, and is also dictated by the 1–3 setting level. There are currently no randomized trials that demonstrate benefit over fixed PAP therapy (Bakker et al., 2010; Kushida et al., 2011; Dungan et al., 2011; Chihara et al., 2013; Sunderram et al., 2021), but clinically in our experience, some individuals do report better tolerance.

Because the smaller pressure drops with EPR do not require more expensive motors and power sources of bilevel devices, CPAP or APAP with EPR activated may be a cost-effective method to try before considering a switch to BPAP for tolerance. One reason to not use it routinely is that EPR may trigger breathing instability and central events like BPAP can in susceptible patients (Johnson & Johnson, 2005). Additionally, if a patient is titrated to a particular setting in the lab without EPR, it cannot be assumed that the same setting with EPR will still be efficacious (Zhu et al., 2016).

18.3.6 Bilevel PAP (BPAP)

BPAP utilizes the previously described respiratory cycle determination to trigger a higher pressure during inspiration (IPAP) and cycle to a lower pressure during expiration (EPAP). The difference between the EPAP and IPAP is termed pressure support (PS). Pressures generally range from an EPAP minimum of 4 cm H_2O to IPAP maximum of 25–30 cm H_2O with a PS of at least 4 cm H_2O.

BPAP may help with tolerance in patients who have difficulty exhaling against a higher pressure, but there are, in fact, no objective outcome studies that show that BPAP therapy improves adherence and/or daytime sleepiness when compared with CPAP therapy for patients with uncomplicated OSA (Patil et al., 2019a; b; Aloia et al., 2005). Because upper airway obstruction is typically worst during inspiration, airway patency can be maintained despite lower pressure during expiration (Sanders & Kern, 1990). BPAP especially with higher PS can also increase ventilation and reduce work of breathing. However, the overall ventilation for a given pressure will vary based on the patient inspiratory effort, respiratory compliance, sleep stage and position, and inspiratory time so BPAP does not ensure a specific level of ventilation through the night (Pavone et al., 2020). There can be significant variation between devices in terms of how quickly pressurization levels are met and whether a device has a delay or premature cycle, especially in the setting of leak (Battisti et al., 2005).

BPAP can be triggered by spontaneous, timed, or spontaneous/timed (ST) modes. In spontaneous (S) mode, inspiration is only triggered when the device senses a flow change. Large leak may cause the trigger to fail if the device does not appropriately adjust. Timed (T) mode triggers at a fixed rate and makes no attempt to synchronize with the patient's breathing, which can result in breath stacking and patient discomfort. Spontaneous/timed (ST) mode changes pressures with spontaneous breathing efforts, but if the patient has not triggered a breath by a set respiratory rate, then the algorithm will trigger a breath based on time. Spontaneous (S) mode is primarily used for patients with uncomplicated OSA, but ST mode may be useful in patients with comorbid conditions including COPD, obesity hypoventilation syndrome, central apneas due to respiratory depression, or neuromuscular weakness or spinal cord injury when patients may be unable to trigger a sufficient inspiratory effort (Berry et al., 2010).

18.3.7 Respiratory Control Settings: Rise Time, Trigger and Cycle Sensitivity, and Inspiration Time

Many BPAP devices, especially those with backup rate, allow control over the timing and sensitivity of the changes between pressures to help both tolerance and therapeutic benefit. If there is a mismatch between the patient's respiratory cycle and the pressure changes, there can be patient discomfort. Rise time, cycle and trigger sensitivity, and inspiration time settings can be adjusted to enhance effectiveness and patient comfort, especially for patients with COPD, obesity hypoventilation, or neuromuscular disorders (Zdrojkowski & Estes, 2000; see Table 18.8).

In general, BPAP provides a square wave of PS, but manual or automatic adjustments can give more of a smooth pressure change, which may help with comfort. Rise time adjustments from 100 ms to 600 ms change the rate the pressure change as shown in Fig. 18.4. Instead of square wave with rise time control, some devices have alternative waveforms that can be exponential, ramp, or sinusoidal to allow for smoother transitions. One such example is ResMed's shark-fin-shaped "Easy-Breathe" waveform (Douglas et al., 2011). DeVilbiss IntelliPAP allows for multiple waveform shapers through their Flow Rounding® option. The shape of the waveform may be affected by the compliance and resistance of the patient's respiratory system and the breathing effort, as well as mechanical constraints of blower momentum and propagation delays.

Trigger sensitivity refers to the degree of inspiratory flow change needed to change the pressure cycle from EPAP to IPAP. At high sensitivity, small changes in inspiratory flow will trigger the change to IPAP. If the trigger is too sensitive (too high) as shown in Fig. 18.5, then the device may force a breath in response to artifact like leak or abnormally strong heartbeat due to cardiogenic oscillations in the airflow signal. Mask leak may affect the ability to adequately trigger as well. Low PS can lead to early expiration leading to delayed triggering. Cycle sensitivity sets the level of inspiratory flow below, which the device changes from IPAP to EPAP. Inspiratory leaks can delay cycling and reduce inspiratory sensitivity. High PS can delay cycling.

Inspiration time typically ranges from 0.3 s to 2 s, often with default of 1.5 s. The higher the baseline respiratory rate, the shorter the inspiratory time is recommended. Some devices allow for minimum and maximum inspiratory time settings, while others allow for a single setting.

18.3.8 Clinical Considerations Related to BPAP Technology

- Adequate EPAP is needed to maintain airway patency, recruit the lung by preventing bibasilar atelectasis, and in COPD prevent dynamic hyperinflation by overcoming auto or intrinsic PEEP, but a higher EPAP can also limit ventilation by reducing PS. For a given IPAP, a higher EPAP will have a lower PS and generally results in lower ventilation so both lower EPAP or increasing IPAP may lead to improved ventilation (Kinnear et al., 2017). The exception would be if the higher EPAP improved expiratory flow by offsetting auto

PEEP in severe COPD, then a higher EPAP with lower PS may improve ventilation.

- IPAP needed for BPAP settings to prevent obstruction during early inspiration is often higher than those needed by CPAP due to delays in detecting the onset of inspiration and delivery of the pressure adjustment.
- BPAP may worsen central apneas due to Cheyne–Stokes respiration (CSR) by increasing breath size of spontaneous breaths and BPAP ST by forcing a triggered breath during the apneic portion, which is when the PCO_2 level is already at its lowest (Johnson & Johnson, 2005). By further decreasing the PCO_2 during a central apnea, respiratory drive is reduced further and the duration of the apnea will often lengthen, although the oxygenation may improve with the deeper or forced breath. Sometimes the improved oxygenation and PS will help to eventually stabilize the patient's breathing and associated fluctuations in the pulse and electroencephalogram (Willson et al., 2001), but in our experience, many patients with CSR often find BPAP ST intolerable or still have a suboptimal clinical response. SV or VAPS (described later) may be better for these patients.
- BPAP's squared waveform can mimic flow limitation during sleep studies leading to inappropriate interpretation and response by technologist. This is seen most often during device-triggered events or those with reduced patient effort such as at the nadir of central hypopneas as seen in Fig. 18.6.
- Short rise time in COPD patients can allow the lungs to fill more quickly and give enough time to exhale. Long rise time in patients with neuromuscular weakness can improve comfort and help ensure adequate tidal volume and gas exchange (see Fig. 18.4).
- Low trigger sensitivity is recommended if the patient complains that the breath occurs before exhalation is complete or before the patient starts inhalation. High setting is recommended for patients with weak respiratory effort such as in neuromuscular disorders (see Fig. 18.5).
- High cycle (less reduction in flow to induce cycling) can help patients with severe COPD by enhancing exhalation. Loss of alveolar attachments reduces elastic recoil needed to passively force air out of the lung at the end of expiration. This leads to a slower drop in flow rate (longer time to reach 25% reduction of normal setting) especially when high IPAP pressures are pushing air into the lungs. High cycle setting can also help if the patient complains that the breaths are too long. Low cycle is recommended when longer inspiratory time is needed for neuromuscular diseases, weak respiratory effort, or stiff lungs or if the patient complains that the pressure seems to switch from IPAP to EPAP too quickly (see Fig. 18.5).
- Short inspiratory time (Ti) settings can help exhalation in COPD patients (Zdrojkowski & Estes, 2000). Longer Ti can be useful for obesity hypoventilation to maintain IPAP for a longer period to increase tidal volume.
- Don't assume because a patient has a particular comorbidity that they require the suggested respiratory control settings for that condition. Many patients have overlapping comorbidities, for example, a morbidly obese patient with severe COPD, may benefit most from normal settings, because their obesity is enough to end inspiration early. During a titration study, normal respiratory control settings can be used first and then adjusted as needed

Table 18.8 Suggested respiratory control settings

Setting	Normal	COPD	OHS	Restrictive/neuromuscular
Ti	0.3–2.0 s	0.3–1.0 s	0.8–2.0 s	0.8–1.5 s
Rise time	300 ms	150 ms	150–300 ms	300 ms
Trigger	Medium	Medium	Medium–high	High
Cycle	Medium	High	Low–medium	Low

COPD chronic obstructive pulmonary disease, *OHS* obesity hypoventilation syndrome, *s* second, *Ti* inspiratory time (adapted from ResMed Sleep Lab Titration Guide)

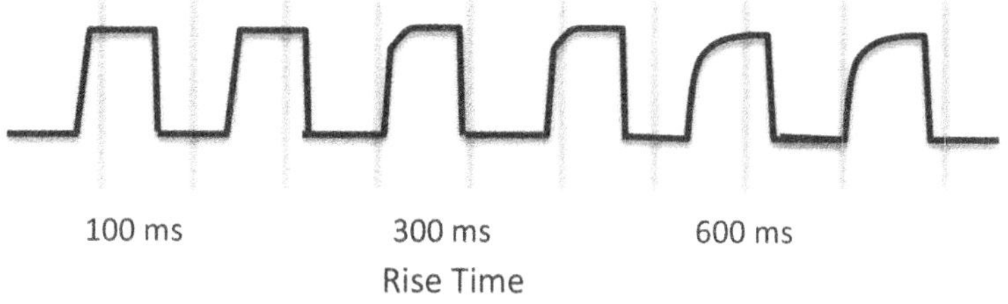

Fig. 18.4 Rise time

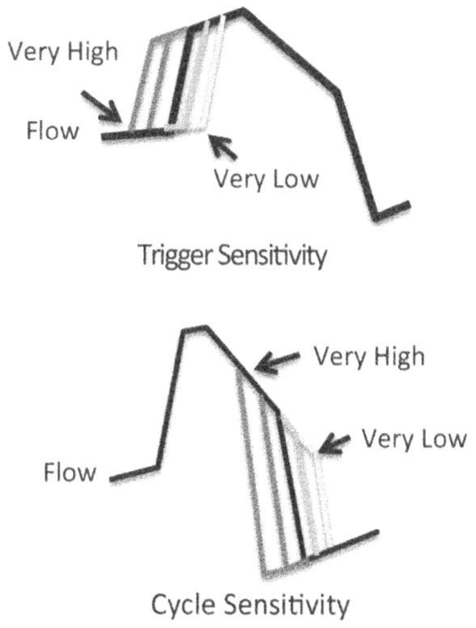

Fig. 18.5 Trigger and cycle sensitivities

for tolerance and dyssynchrony or to improve ventilation.

- Patients with spontaneous breathing only (i.e., obesity hypoventilation) may benefit from extending the Ti to help increase breath size with the ResMed BPAP ST algorithm, but not the Philips Respironics' BPAP ST algorithm, which only affects timed breaths (see Fig. 18.7). Philips Respironics spontaneous or ST Pressure Control (PC) mode, available only in their non-invasive ventilators, allows for respiratory control settings to affect spontaneous breaths. Like BPAP, PC does not ensure a ventilatory level over the course of a night as pressure, not volume, is targeted (Pavone et al., 2020).
- There are 5 main types of dyssynchrony on BPAP described in Table 18.9: ineffective inspiratory efforts, double triggering, auto triggering, short-ventilatory cycling, and long-ventilatory cycling (Al Otair & BaHammam, 2020). A sixth less common type is reverse triggering when insufflation triggers diaphragm muscle contractions by activating patient's respiratory drive (Pavone et al., 2020).
- PAP therapy dilutes the fraction of inspired oxygen (FiO_2) because most of the exiting air is room air entering through the device that is combining with a smaller amount of bled in oxygen. Higher pressures and higher PS lead to more dilution so higher liters of oxygen are typically required to maintain the same degree of oxygen saturation than when oxygen is used without PAP or with lower pressures.

18.3.9 BPAP Expiratory Pressure Relief

EPR is available with some bilevel PAP systems. The Philips BiFlex® device differs from conventional bilevel systems in two major respects. First, the inspiratory pressure is reduced slightly near the end of inspiration, and the expiratory pressure is slightly reduced near the beginning of expiration. Second, the amount of pressure relief change of the EPAP during expiration is proportional to patient effort.

Löwenstein's Trilevel setting has three pressure levels, IPAP (during inspiration), EPAP (during early expiration), and EEPAP (during end of expiration). When the patient expires, pressure is initially reduced to EPAP, but toward the end of expiration, the pressure is increased to help splint open the upper airway and prevent obstruction. In Auto-Trilevel, if the PS difference is at least 6 cm H_2O, the EEPAP min will increase if airway resistance is noted and will decrease if no obstruction.

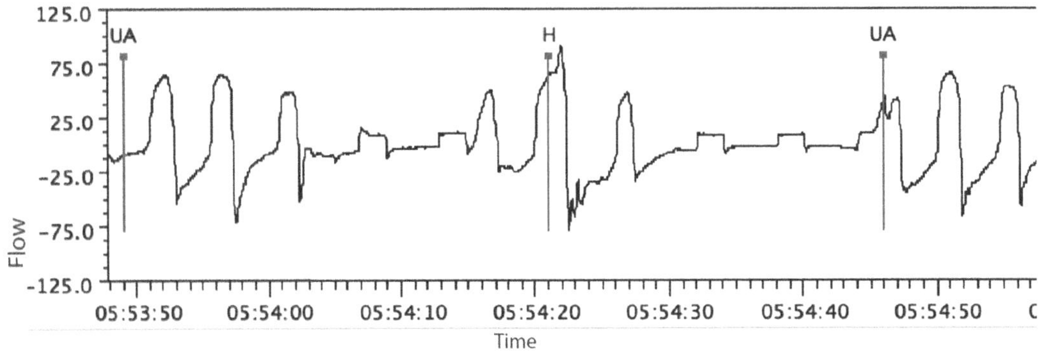

Fig. 18.6 Square pressure waveform with timed BPAP breaths

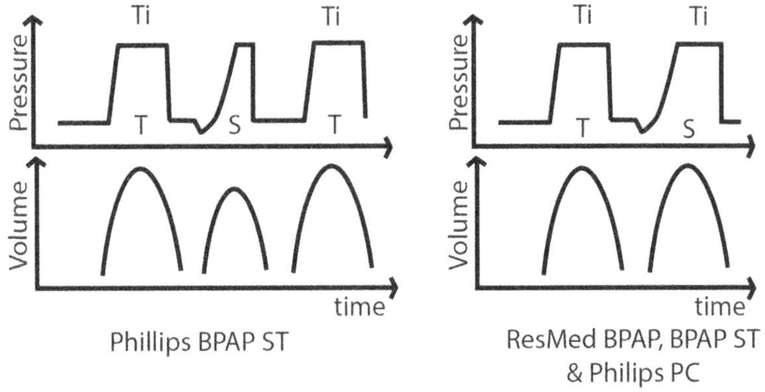

Fig. 18.7 Respiratory control setting activity by mode

18.3.9.1 AutoBPAP

Like AutoCPAP, not all AutoBPAP devices work in the same way. Some devices only allow a fixed pressure support, others only set a PS max, and others allow for both pressure support minimum (PSmin) and pressure support maximum (PSmax). Thus, AutoBPAP may be more likely to provide inadequate ventilator support if PSmin cannot be set. AutoBPAP devices generally do not have an ST option so are not recommended for central apneas. AutoBPAP only adjusts to obstructive events like APAP, and changes in pressures are not triggered by tidal volume or ventilation.

ResMed's AutoBPAP and DeVilbiss' AutoBPAP use a fixed set PS. Philips Respironics' Series 50 AutoBPAP fixes minimum PS (PSmin) at 2 cm H_2O and allows setting maximal PS (PSmax), while Series 60 AutoBPAP allows setting both PSmin and PSmax. Within the limits of PSmin and PSmax, Philips Respironics' newer AutoBPAP changes EPAP in response to apneas (2 apneas or 1 apnea and 1 hypopnea) and snoring and IPAP in response to hypopneas (2 hypopneas) and flow limitation, with algorithms similar to APAP (Matthews et al., 2012). Löwenstein's AutoBPAP (auto-PDIFF) has both autoS and autoST modes. Like APAPstd, EPAP increases obstructive events and flow limitation, but for every 0.5 cm H_2O EPAP increases, IPAP increases by 1 cm H_2O until the PDIFFmax (PSmax) is reached, and then they increase in parallel. For pressure decreases, as long as PS > PSmin, EPAP is reduced by the APAP algorithm, and IPAP is reduced by 1.5X the EPAP reduction. Once at PS min, IPAP and EPAP reduce in parallel.

Table 18.9 BPAP dyssynchrony

Dyssynchrony	Description	Causes	Fixes
Ineffective inspiratory efforts	Patient's inspiratory effort fails to trigger change to IPAP	Low respiratory drive, weak inspiratory muscles, dynamic hyperinflation	Lower trigger, higher EPAP in COPD with intrinsic PEEP to reduce inspiratory threshold load and reduce dynamic hyperinflation
Double-triggering	Two consecutive breaths occur in an interval of less than ½ of inspiratory time	Insufficient level of pressure support when demand is high; effort lasts longer than Ti resulting in retriggering	Increase PS, higher trigger sensitivity
Auto-triggering	Trigger to IPAP in the absence of patient effort	Leak can cause pressure drop interpreted as patient effort—less of a problem with NIV than vents, cardiac contractions, and water in tubing	Reduce trigger sensitivity, fix leak, excess water condensation
Short ventilatory cycling	Premature cycling to EPAP while patient still in inspiratory, Ti <50% of mean inspiratory time	Restrictive diseases have short respiratory cycle (rapid shallow breathing) and an increased inspiratory time leads to premature cycling	Longer Ti min, lower cycle sensitivity
Long ventilatory cycling	Patient is ready to exhale, but machine fails to cycle to EPAP; Ti >2X mean inspiratory time	Due to prolonged expiration, COPD has short inspiratory time and delayed cycling	Shorter Ti, higher cycle sensitivity, shorter rise time to meet demand for higher inspiratory flow

COPD chronic obstructive pulmonary disease, *EPAP* expiratory positive airway pressure, *IPAP* inspiratory positive airway pressure, *NIV* non-invasive ventilation, *PEEP* positive end-expiratory pressure, *PS* pressure support, *Ti* inspiratory time

18.3.10 Adaptive or Anticyclic Servoventilation (SV)

Because higher PAP pressures and high PS can induce periodic breathing and Cheyne–Stokes respiration (CSR), adaptive or anticyclic servoventilation (AcSV) algorithms have been developed to try to even out the breathing over several breaths. These include ResMed's (equivalent to Teijin) ASV and AutoSV, and Philips Respironics' BiPAP AutoSV and BiPAP AutoSV Advanced. Another device, Prisma CR AcSV is available outside of the United States from Löwenstein (Weinmann's SOMNOvent CR [acquired by Löwenstein] uses a less dynamic algorithm and different blower) (see Table 18.10).

ResMed's standard ASV algorithm uses a set fixed EPAP (samples flow 50 times per second) and alters IPAP throughout inspiration to achieve a target minute ventilation of 90% of recent average ventilation. EPAP range is from 4 to 20 cm H_2O. The pressure support range can be 0–20 cm H_2O, but default is usually a PSmin of 3 cm H_2O and PSmax of 15 cm H_2O, but the maximum IPAP is 25 cm H_2O so higher EPAPs may force a reduction of PSmax.

Philips Respironics' BPAP AutoSV Advanced is set with EPAP minimum and maximum, PSmin and PSmax, max pressure, and auto or fixed rate. The level of PS is targeted based on instantaneous average inspiratory peak flow, which is the sum of the inspiratory flows during a time divided by the number of samples during a time in order to adjust for spurious values. Compared to multiple small adjustments in the pressure support throughout the breath cycle with ResMed ASV, Phillips Respironics' BPAP AutoSV Advanced's inspiratory PS is changed at the start of the breath with a mid-breath adjustment if below the target.

The EPAP adjustment algorithm of AutoSV Advanced is similar to Philips Respironics' APAP algorithm but uses a triggered breath rather than a pressure pulse to differentiate obstructive from open airway events (Javaheri et al., 2014).

BPAP AutoSV AutoEPAP has not been shown to be superior over fixed EPAP (Pepin et al., 2018; Javaheri et al., 2011). The older BPAP AutoSV did not automatically titrate the EPAP, and the algorithm for the automatic backup rate was not proportional to the baseline breathing rate, but it would give a breath if no spontaneous breath occurred within 8 s of end of expected breath length or within 4 s if there was recent triggered breath (Javaheri et al., 2011). Philips Respironics' BPAP AutoSV Advanced has been found to be more effective than the older BPAP Auto SV (Javaheri et al., 2011).

Löwenstein's Prisma CR anticyclic servoventilation (AcSV) uses the total inspiratory volume for each 2 min and is divided by 2 to determine a relative average minute volume (rAMV). A long-term rAMV is calculated by giving 50% weight for the most recent 2 min and 10% weight for each of the five 2-min epochs in the preceding 10 min. The increased weight on recent minute volume is designed to not get trapped in expired reference values, for example, after a position change or sleep stage transition. There are two factors that amplify or dampen AcSV's PS adjustments: (1) The strength of the preceding hyperventilation amplifies the PS changes, whereas a period with non-periodic variations, for example, during wake or REM, leads to smaller PS changes. (2) Hyperventilation without obstructive features, for example, after an arousal or periodic breathing, leads to a firm anticyclic reaction to prevent central apneas and the occurrence of periodic breathing.

Because SV targets a reduced ventilation state, there is a risk of hypoventilation especially in REM or supine position. Jaffuel et al. tested Philips Respironics' AutoSV and ResMed's ASV with fixed EPAP devices and confirmed hypoventilation risk despite underlying patient sleep apnea type. ResMed's algorithms tended to have lower median minute ventilation and tidal volume compared to Philips Respironics' algorithms (Jaffuel et al., 2020). Phillips Respironics' BPAP AutoSV Advanced has a safety feature that ensures at least 5 L min ventilation, whereby mandatory PS is provided to maintain an average minute ventilation of at least 5 l.

18.3.11 Clinical Considerations with SV

- Appropriately set SV will stabilize loop gain and the periodic breathing. If the IPAP repetitively changes from high to low IPAP for long periods to maintain the target ventilation, it indicates that the underlying periodic pattern has not subsided and often the patient will either not tolerate the device or there will be a suboptimal clinical response (Chokroverty & Thomas, 2014). This pattern can be identified on pressure tracings of detailed adherence data (see Fig. 18.8).
- ResMed's ASV algorithm and Löwenstein's SoftSTART do not fluctuate the IPAP during the ramp period and just provide gradually increasing BPAP at the PS min. If a patient had sleep onset periodic breathing, having the ramp on may allow for persistent events that can contribute to intolerance or high residual AHI as shown in Fig. 18.9.
- SV targets a hypoventilatory state, which may be a factor contributing to the increased mortality with ASV treatment in patients with heart failure and central sleep apnea in the SERVE-HF trial (Jaffuel et al., 2020; Cowie et al., 2016). Higher PSmin can be used to limit hypoventilation but may reduce the functionality of the SV algorithm so other modalities such as BPAP or VAPS should be considered.

18.3.12 Volume-Assured Pressure Support

VAPS is a variable bilevel PAP that adjusts the PS in order to maintain a target average tidal volume or ventilation over several breaths, which allows more control of the ventilation and lowering CO_2 than BPAP or SV (Turk et al., 2018). This is useful for patients with combined periodic breathing and hypoventilation or patients with REM-related hypoventilation related to conditions like COPD, neuromuscular disorders, or obesity hypoventilation, who may need different PS levels during different sleep stages or posi-

Table 18.10 SV algorithms

	ResMed ASV (Berthon-Jones, 2014) and AutoASV (Javaheri et al., 2014)	Philips Respironics BPAP AutoSV Advanced	Löwenstein Prisma CR AcSV
Target	Ninety percent of the average recent scaled low-pass-filtered absolute value of respiratory flow (an indication of ventilation) weighted toward the last 3 min	Over the last 4 min, 90–95% of mean inspiratory peak flow; 60th percentile of mean inspiratory peak flow if obstruction or CSR based on CSR index measuring fluctuations in breath size with at least 2–3 cycles with a period of 40–90 s (Kane et al., 2014)	100% rAMV: long-term exponentially smoothed minute volume over the last 12 min weighted toward the previous 2 min
Cycle determination	Fuzzy logic based on direction and rate of flow throughout the cycle	Average cycle length and midpoint based on previous breaths	Cycle phase by adaptive threshold method based mainly on flow
EPAP adjustments	Fixed EPAP or AutoEPAP adjusts like AutoSet algorithm with apnea defined as MV < 25% baseline for 10 s and hypopneas as MV < 50% of baseline for 10 s	Fixed EPAP or AutoEPAP utilizing APAP algorithm with events defined as apnea (<20% flow), hypopnea (20–60% flow), and snoring	EPAP fixed or AutoEPAP utilizing APAP algorithm
PS adjustments	PS adjusted throughout inspiration to achieve a goal ventilation with a smooth "shark fin" wave. The change in pressure is calculated by multiplying a gain of 0.3 cm H_2O/L/min/s by the difference between the target minute ventilation and the actual minute ventilation. During apneic periods, PS gradually increases with subsequent breaths. Limited by PSmin and PSmax	PS is changed by gain (moving average of pressure needed in the last 30 breaths) times the difference between the target peak inspiratory flow and the current breath's peak inspiratory flow. The PS is increased in the middle of the breath if the actual flow is less than the target flow in the 100 ms prior to the expected halfway point of the inspiration of the current breath. If the actual flow is larger than the target flow, the PS will be decreased for the following breath (Kane et al., 2014) Limited by PSmin and PSmax	Actual volume during inspiration is compared with a third-order polynomial based on target volume and expected inspiration every 10 ms. PS for current breath is continuously recalculated during inspiration ramp and adjusted until target volume is reached. PS limited by PSmax
Backup rate	Auto rate starting at recent spontaneous rate adapting during apnea over several breaths to target rate of 15 bpm	Fixed rate or auto rate delivered relative to spontaneous breaths after ~1/3 of breath period Tbr has elapsed from the end of previous breath. Tbr—average time of 12 recent spontaneous breaths. In the absence of spontaneous breaths, 10 bpm	Fixed between 5 and 30 bpm or auto with 20% below average patient rate. Auto accelerates if rRMV is rather small or slows down in case of hyperventilation. Target rate is 15 bpm
Waveform options	Shark-fin-shaped "Easy-Breathe" waveform	Square wave with rise time and pressure relief adjustments	Square wave with optional Bi SoftPAP (EPR) and Trilevel, Ti/T setting

(continued)

Table 18.10 (continued)

	ResMed ASV (Berthon-Jones, 2014) and AutoASV (Javaheri et al., 2014)	Philips Respironics BPAP AutoSV Advanced	Löwenstein Prisma CR AcSV
Other	Ramp	Ramp, BiFlex expiratory relief proportional to exhalation flow rate. Ti and rise time. Safety backup: 5 L MV	SoftSTART ramp

SV servoventilation, *APAP* autotitrating positive airway pressure, *CSR* Cheyne–Stokes respiration, *EPAP* expiratory positive airway pressure, *EPAPmax* maximum EPAP, *EPAPmin* minimum EPAP, *min* minute, *MV* minute ventilation, *PS* pressure support, *PSmax* PS maximum, *PSmin* PS minimum, *rAMV* relative average minute volume, *rRMV* relative respiratory minute volume, *SBD* sleep-disordered breathing, *s* second, *Tbr* breath time

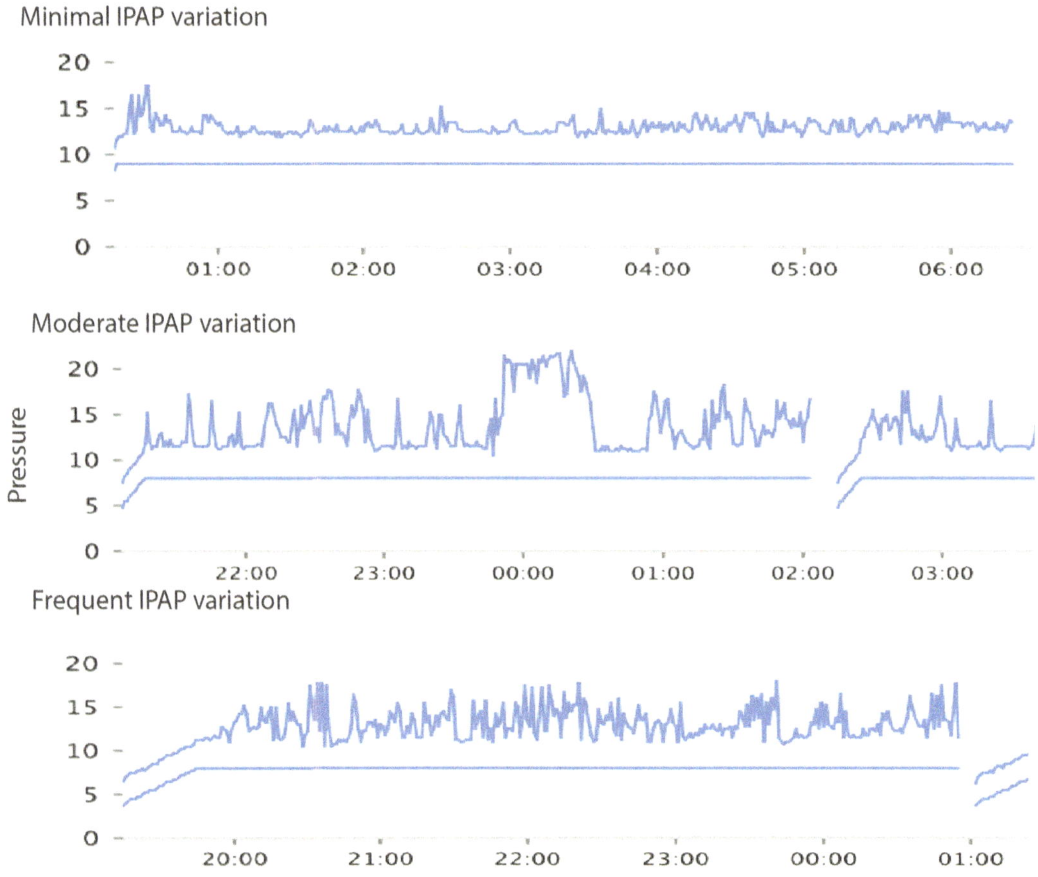

Fig. 18.8 ASV pressure tracings with differing IPAP variability

tions. Because many patients have much worse hypoventilation in REM, BPAP with a fixed PS may provide too much pressure in NREM, which may lead to intolerance or complex sleep apnea and may not provide enough pressure support in REM to control PCO_2 levels and obstruction. Lower PS during wake may increase comfort and aid sleep onset, reduce risk of barotrauma, and lower pressures for more of the time, which can improve adherence and better control oxygenation and hypercarbia (Oscroft et al., 2010; Kelly et al., 2014).

Table 18.11 summarizes ResMed's intelligent volume-assured pressure support (iVAPS)

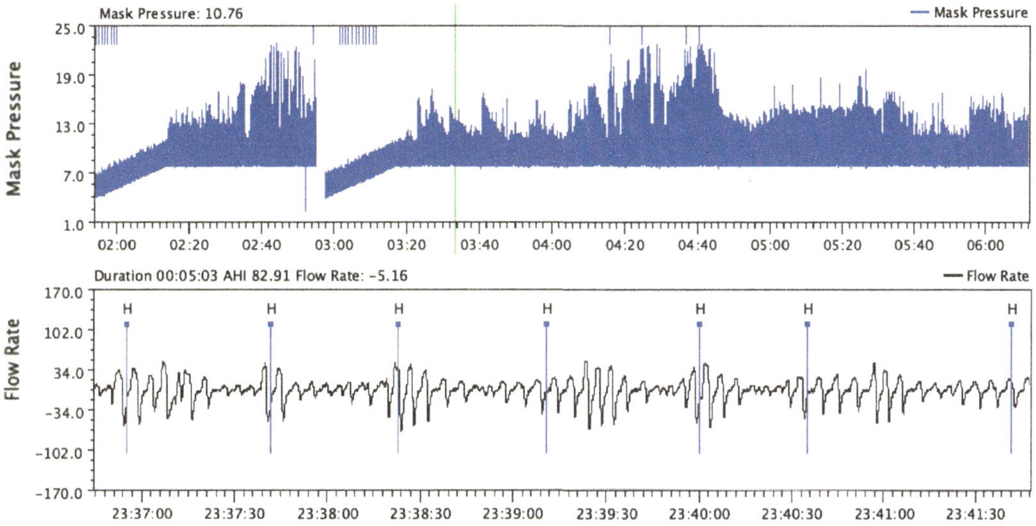

Fig. 18.9 Sleep onset periodic breathing during ramp period with ASV

(Berthon-Jones, 2003; Berthon-Jones et al., 2013; Bassin, 2011) and Philips Respironics' average volume-assured pressure support–AutoEPAP (AVAPS/AVAPS-AE) (Hill et al., 2006) and Löwenstein's Bilevel S/T and autoS/T with set target volume algorithms. Maximum IPAP with VAPS is typically either 25 or 30 mm H_2O often with ventilator algorithms allowing for the higher pressures.

Typically, the target tidal volume or ventilation is set based on 6–10 mL/kg ideal body weight. To have the most precise control over ventilation, alveolar ventilation must be stable. With a target tidal volume (e.g., AVAPS), if there is a large variance in respiratory rate, there can be fluctuations in the alveolar ventilation and thus PCO_2. When minute ventilation is targeted (i.e., ASV), a higher respiratory rate leads to a lower tidal volume and greater dead space ventilation than at a lower respiratory rate, so there may be less alveolar ventilation and higher PCO_2 despite the increase in respiratory rate. By targeting estimated alveolar ventilation (TVa [minute ventilation – estimated dead space ventilation]; i.e., iVAPS), variations in respiratory rate should not affect alveolar ventilation or PCO_2 as long as the estimated dead space equals physiologic dead space. iVAPS estimates the anatomic dead space using height (Hart et al., 1963).

iVAPS' ventilation and rate targets can be configured using the Learn Target® feature. While the patient is awake and comfortably breathing at rest, the clinician initiates the Learn Target session, which typically lasts between 15 min and 20 min, and monitors the breathing while on a base EPAP pressure of 6 cm H_2O. It uses the average respiratory rate, and since metabolic rate decreases in sleep, it uses 90% of the average estimated alveolar ventilation of the last 5 min of the session to propose a target patient rate and target alveolar ventilation, which can then be set by the clinician (Berthon-Jones et al., 2013).

Improvements in ventilation and comfort have been shown in comparative trials of VAPS to both SV and BPAP ST. Compared to BPAP ST, in COPD patients, iVAPS allowed for higher ventilation pressures without affecting sleep quality or inducing ventilation associated events in COPD (Nilius et al., 2017), and AVAPS resulted in a more rapid improvement in Glascow coma scale and greater CO_2 improvement in patients with acute hypercapneic encephalopathy (Briones Claudett et al., 2013) and better quality of life and a 6-min walking distance and $PaCO_2$ reduction at 6 months (Magdy & Metwally, 2020). iVAPS also resulted in more restful sleep that high pressure support BPAP ST and resulted in a larger PCO_2 (Ekkernkamp et al., 2014). In a bench testing

Table 18.11 VAPS algorithms

	ResMed iVAPS (Berthon-Jones, 2003)	Philips Respironics AVAPS and AVAPS-AE (Hill et al., 2006)	Löwenstein Prisma Bilevel S/T and autoS/T with set target volume
Ventilation calculation	Scaled low-pass-filtered absolute value of respiratory flow	From total flow minus leak flow over breathing cycle	Volume calculated from respiratory flow during inspiration
Target	Average set target alveolar ventilation (TVa), range 1–30 L/min. TVa = minute ventilation – anatomical dead space estimated using height so target volume varies with rate	Average set tidal volume over several breaths, range 200–1500 mL	Set tidal volume; range 300–2000 m
EPAP adjustments	Fixed set EPAP AutoEPAP (available in Astral ventilator)	AVAPS, fixed EPAP AVAPS-AE—set EPAPmin and EPAPmax with APAP-like algorithm	BiLevel S/T- fixed EPAP Auto S/T: EPAPmin and max adjusted with APAP algorithm
PS adjustments	Estimates expected ventilation using respiratory cycle position determined by flow cycle and trigger algorithm and current pressure. If expected flow differs from target flow, PS adjustments are made throughout inspiration (every 8th/50th second) to achieve goal ventilation with a smooth transition Limited by PSmin and PSmax	Determines average PS provided over prior 2 min to achieve volume. If target volume differs from average recent ventilation, PS for the next breath is changed at a rate of 1 cm H_2O/min if recent stable breathing and 0.5 cm H_2O/min if unstable breathing (AE model allows a maximum rate of pressure change from 1–5 cm H_2O/min). Limited by PSmin and PSmax	If the average ventilation during the last breaths does not match set target volume, PS is adjusted. As long as the difference between target volume and patient volume is large, a fixed pressure increment is applied. If the error is small, pressure increments/decrements are proportional to error Speed of PS adjustment can be set to slow, medium, or fast. Algorithm considers # breaths for average ventilation and has different rates of pressure change Slow: 8 breaths/0.5 cm H_2O Medium: 5 breaths/1 cm H_2O Fast: 1 breath/1.5 cm H_2O Limited by PSmin and PSmax
Inspiratory waveform	Parabolic pressure profile with Ti, trigger, and cycle controls	Square wave with Ti and rise time settings	Square wave with optional Bi SoftPAP (EPR) and Trilevel, Ti setting, inspiratory and expiratory ramp steepness
Backup rate	Intelligent backup at two-thirds of set rate during spontaneous breathing and set rate during periods of timed breathing (Bassin, 2011)	Fixed or Autoset, which is 2 breaths per min below average rate of recent 50 spontaneous breaths. Also provides sign extension	Off, fixed (S/T) or auto (autoS/T only) with 20% below average patient rate. Auto accelerates if rRMV is rather small and slows down in case of hyperventilation. Target rate is 15 bpm
Others	Learn target setting	–	–

APAP autotitrating positive airway pressure; *AVAPS* average volume-assured pressure support, *bpm* breaths per minute; *EPAP* expiratory positive airway pressure, *EPAPmax* maximum EPAP, *EPAPmin* minimum EPAP, *EPR* expiratory pressure relief, *IPAP* inspiratory pressure, *iVAPS* intelligent volume-assured pressure support, *min* minute, *PS* pressure support, *PSmax* PS maximum, *PSmin* PS minimum, *SBD* sleep-disordered breathing, *s* second, *S/T* spontaneous/timed, *Ti* inspiratory time, *TV* tidal volume, *TVa* target alveolar ventilation

experiment, Lofaso et al. (2020) concluded that ResMed's ASV, Löwenstein's Prisma CR AcSV, and Philips Respironics' AutoSV algorithms responded better to hypopnea/hypoventilation events than AVAPS and iVAPS, but they only tested short hypopneic hypoventilatory durations from 48 s to 144 s more similar to periodic breathing rather than longer-lasting hypoventilation periods as seen in REM (Lofaso et al., 2020). It is not surprising that in the setting of single shorter hypopneic periods, the quicker responding SV algorithms will increase ventilation more during the hypopnea than VAPS, which targets maintaining ventilation over a longer period. However, they did not study prolonged periods of hypoventilation during which the SV devices would start lowering their targeted ventilation as in Jaffuel et al.'s study of SV devices described earlier (Jaffuel et al., 2020).

Comparisons of AutoEPAP to fixed EPAP iVAPS have also been done. In a study of COPD and neuromuscular patients, Auto EPAP iVAPS found no significant difference in respiratory measures including mean EPAP, residual AHI, sleep quality, patient comfort, or preference (McArdle et al., 2017); although in post hoc analysis, patients with COPD were found to have lower $TCCO_2$ on AutoEPAP than fixed EPAP. Mean PS was also higher with AutoEPAP than fixed EPAP. It is unclear whether or not the presence of CHF and risk of central sleep apnea would lead to different results than the studied population. A second study by Orr et al. in 38 patients with more diverse etiologies of hypoventilation found that a single night of AutoEPAP was noninferior to fixed EPAP iVAPS, finding a lower mean 4% oxygen desaturation index with AutoEPAP and similar effects on ventilation and mean shallow breathing index (Orr et al., 2019).

18.3.13 Clinical Considerations with VAPS

- iVAPS TVa estimation based on height will underestimate conditions with increased physiological dead space (i.e., emphysema). Higher TVa may be needed to achieve a desired actual alveolar ventilation. Alternatively, the "height" can be entered artificially high for emphysema patients, so calculated dead space will be closer to physiologic dead space.

18.4 Research Agenda

Future research and product development should be primarily directed at improving adherence, which is the primary limiting factor of PAP therapy. Adherence has also been a major factor limiting randomized trials trying to demonstrate long-term effects of PAP therapy on cardiovascular outcomes. This research should take phenotypic heterogeneity in consideration since understanding which patient will benefit most from a given feature will help personalize care. Further improvements in device size, device and mask noise, mask and air pressure comfort, and the ability to clean the device are still needed. Further comparative trials especially of advanced settings like VAPS are needed to confirm their utility in treating patients with different comorbidities. Reducing the costs of therapy and creating devices with multiple modes would assist with the large regulatory burdens of caring for patients with sleep-disordered breathing. Integration of artificial intelligence and algorithms to incorporate patient feedback to help optimize comfort and data from both the machine and other sources to optimize therapy should be evaluated.

18.5 Conclusion

There are a wide range of PAP devices and different algorithms used to provide PAP therapy to treat sleep-disordered breathing. Algorithms within a particular category may differ in their clinical efficacy and comfort so switching to a new device type may lead to a different response. Understanding how PAP devices function can help the clinician select the best PAP device and appropriately titrate, troubleshoot, and optimize settings for a particular patient. Comfort algorithms, such as ramp and EPR and respiratory control settings like cycle and trigger sensitivities, may help some patients but should not be used as defaults as they may reduce efficacy.

Acknowledgments I want to especially thank Anne Gruenewald and Dr. Jochen Furthmüller at Löwenstein Medical Technology, Brian Palmer at DeVilbiss Healthcare, Dinesh Ramanan at ResMed, and Dr. Teofilo Lee-Chiong, Adrienne Uzonyi, Michael Kane, and Greg Matthews at Philips Respironics. Personal communications and information provided by these individuals were used to inform and verify the descriptions of the specific algorithms.

Declaration Statement The authors have no disclosures or conflict of interest.

References

Abdenbi, F., Chambille, B., & Escourrou, P. (2004). Bench testing of auto-adjusting positive airway pressure devices. *The European Respiratory Journal, 24*, 649–658.

Al Otair, H. A., & BaHammam, A. S. (2020). Ventilator- and interface-related factors influencing patient-ventilator asynchrony during noninvasive ventilation. *Annals of Thoracic Medicine, 15*, 1–8.

Aloia, M. S., Arnedt, J. T., Stepnowsky, C., Hecht, J., & Borrelli, B. (2005). Predicting treatment adherence in obstructive sleep apnea using principles of behavior change. *Journal of Clinical Sleep Medicine, 1*, 346–353.

Armitstead, J. P., Bateman, P. E., & Bassin, D. J., inventors; ResMed Ltd (Bella Vista, AU), assignee. (2011, August 25). *Automated control for detection of flow limitation*. US patent 20110203588.

Axe, J. R., Behbehani, K., Burk, J. R., Loucas, E. A., & Yen, F.-C., inventors; Board of Regents, The University of Texas System, (Austin, TX), assignee. (1993, April 20). *Method and apparatus for controlling sleep disorder breathing*. US patent 5,203,343.

Axe, J. R., Bebehani, K., Burk, J. R., Lucas, E. A., & Yen, F., inventors; Respironics, Inc. (Fort Worth, TX), assignee. (2000, July 11). *Method and apparatus for controlling sleep disorder breathing*. US Patent 6085747.

Ayas, N. T., Patel, S. R., Malhotra, A., et al. (2004). Auto-titrating versus standard continuous positive airway pressure for the treatment of obstructive sleep apnea: Results of a meta-analysis. *Sleep, 27*, 249–253.

Bakker, J., Campbell, A., & Neill, A. (2010). Randomized controlled trial comparing flexible and continuous positive airway pressure delivery: Effects on compliance, objective and subjective sleepiness and vigilance. *Sleep, 33*, 523–529.

Bassin, D. J. C., inventor; Resmed Ltd (Bella Vista, AU), assignee. (2011, November 11). *Methods and apparatus for varying the back-up rate for a ventilator*. US patent 8051852.

Battisti, A., Tassaux, D., Janssens, J. P., Michotte, J. B., Jaber, S., & Jolliet, P. (2005). Performance characteristics of 10 home mechanical ventilators in pressure-support mode: A comparative bench study. *Chest, 127*, 1784–1792.

Berry, R. B., & Block, A. J. (1984). Positive nasal airway pressure eliminates snoring as well as obstructive sleep apnea. *Chest, 85*, 15–20.

Berry, R. B., Parish, J. M., & Hartse, K. M. (2002). The use of auto-titrating continuous positive airway pressure for treatment of adult obstructive sleep apnea. An American Academy of Sleep Medicine review. *Sleep, 25*, 148–173.

Berry, R. B., Chediak, A., Brown, L. K., et al. (2010). Best clinical practices for the sleep center adjustment of noninvasive positive pressure ventilation (NPPV) in stable chronic alveolar hypoventilation syndromes. *Journal of Clinical Sleep Medicine, 6*, 491–509.

Berry, R. B., Kushida, C. A., Kryger, M. H., Soto-Calderon, H., Staley, B., & Kuna, S. T. (2012). Respiratory event detection by a positive airway pressure device. *Sleep, 35*, 361–367.

Berthon-Jones, M. L., inventor; Resmed Ltd, (Bella Vista, AU), assignee. (2003, March 18). *Assisted ventilation to match patient respiratory need*. US Patent 6532957.

Berthon-Jones, M. L., inventor; ResMed Limited (New South Wales, AU), assignee. (2010, June 8). *Determination of patency of the airway*. US Patent 7730886.

Berthon-Jones, M. L., inventor; ResMed Ltd., assignee. (2014, October 14). *Ventilatory assistance for treatment of cardiac failure and Cheyne-Stokes breathing*. US Patent 8857430.

Berthon-Jones, M. L., & Farrugia, S. P., inventors; ResMed Ltd, assignee. (2014, April 1). *Administration of CPAP treatment pressure in presence of apnea*. US patent 8684000.

Berthon-Jones, M. L., Bateman, P., Bassin, D., & Malouf, G., inventor; ResMed Ltd (Bella Vista, AU), assignee. (2013, October 1). *Determining suitable ventilator settings for patients with alveolar hypoventilation during sleep*. US patent 8544467.

Bogan, R. K., & Wells, C. (2017). A randomized crossover trial of a pressure relief technology (SensAwake) in continuous positive airway pressure to treat obstructive sleep apnea. *Sleep Disorders, 2017*, 3978073.

Briones Claudett, K. H., Briones Claudett, M., Chung Sang Wong, M., et al. (2013). Noninvasive mechanical ventilation with average volume assured pressure support (AVAPS) in patients with chronic obstructive pulmonary disease and hypercapnic encephalopathy. *BMC Pulmonary Medicine, 13*, 12.

Brown, L. K., & Javaheri, S. (2017). Positive airway pressure Device technology past and present: What's in the "Black Box"? *Sleep Medicine Clinics, 12*, 501–515.

Chihara, Y., Tsuboi, T., Hitomi, T., et al. (2013). Flexible positive airway pressure improves treatment adherence compared with auto-adjusting PAP. *Sleep, 36*, 229–236.

Chokroverty, S., & Thomas, R. (2014). *Atlas of sleep medicine* (2nd ed.). Elsevier Saunders.

Cilli, A., Uzun, R., & Bilge, U. (2013). The accuracy of autotitrating CPAP-determined residual apnea-hypopnea index. *Sleep & Breathing, 17*, 189–193.

Clinical Overview: DeVilbiss IntelliPAP AutoAdjust. (2014). http://www.devilbisshealthcare.com/files/LT-2089_RevC_FINAL_050814_Web.pdf. Accessed 24 Jan 2022.

Colla, G., & Kenyon, B. J., inventors; ResMed Limited (Bella Vista, AU), assignee. (2013, July 16). *Fault diagnosis in CPAP and NIPPV devices*. United States Patent 8485182.

Cowie, M. R., Wegscheider, K., & Teschler, H. (2016). Adaptive servo-ventilation for central sleep apnea in heart failure. *The New England Journal of Medicine, 374*, 690–691.

Desai, H., Patel, A., Patel, P., Grant, B. J., & Mador, M. J. (2009). Accuracy of autotitrating CPAP to estimate the residual Apnea-Hypopnea Index in patients with obstructive sleep apnea on treatment with autotitrating CPAP. *Sleep & Breathing, 13*, 383–390.

Douglas, R. N., Ujhazy, A. J., Richards, G., Buckley, M. D., & Schindhelm, K. H., inventor; Resmed Ltd., assignee. (2011, September 6). *Mechanical ventilation in the presence of sleep disordered breathing*. United States Patent US 8011365.

Dungan, G. C., Marshall, N. S., Hoyos, C. M., Yee, B. J., & Grunstein, R. R. (2011). A randomized crossover trial of the effect of a novel method of pressure control (SensAwake) in automatic continuous positive airway pressure therapy to treat sleep disordered breathing. *Journal of Clinical Sleep Medicine, 7*, 261–267.

Ebben, M. R., Narizhnaya, M., Segal, A. Z., Barone, D., & Krieger, A. C. (2014). A randomised controlled trial on the effect of mask choice on residual respiratory events with continuous positive airway pressure treatment. *Sleep Medicine, 15*, 619–624.

Ekkernkamp, E., Storre, J. H., Windisch, W., & Dreher, M. (2014). Impact of intelligent volume-assured pressure support on sleep quality in stable hypercapnic chronic obstructive pulmonary disease patients: A randomized, crossover study. *Respiration, 88*, 270–276.

Esquinas, A. M., & Shah, P. S. (2012). Heated Moisture Exchanger (HME) and dead space ventilation. Is Isocapnic conditions unachievable in children? *Korean Journal of Anesthesiology, 63*, 280–281.

Farre, R., Rigau, J., Montserrat, J. M., Ballester, E., & Navajas, D. (2001). Evaluation of a simplified oscillation technique for assessing airway obstruction in sleep apnoea. *The European Respiratory Journal, 17*, 456–461.

Farre, R., Montserrat, J. M., Rigau, J., Trepat, X., Pinto, P., & Navajas, D. (2002). Response of automatic continuous positive airway pressure devices to different sleep breathing patterns: A bench study. *American Journal of Respiratory and Critical Care Medicine, 166*, 469–473.

Farrugia, S. P., & Finn, S. D., inventors; ResMed Limited (North Ryde, AU), assignee. (2001, December 25). *Flow estimation and compensation of flow-induced pressure swings in CPAP treatment and assisted respiration*. United States Patent 6332463.

Farrugia, S. P., Alder, M., Price, A. M., & Richard, R., inventors; ResMed Limited (Bella Vista, AU), assignee. (2019, May 21). *Method and apparatus for improving the comfort of CPAP*. US patent 10,293,127.

Ferguson, G. T., & Gilmartin, M. (1995). CO2 rebreathing during BiPAP ventilatory assistance. *American Journal of Respiratory and Critical Care Medicine, 151*, 1126–1135.

Fleming, D. J., & Grunberg, M., inventors; ResMed Motor Technologies, Inc., (Chatsworth CA), assignee. (2021, July 6). *Control of pressure for breathing comfort*. US patent 11,052,206 B2.

Fromm, R. E., Jr., Varon, J., Lechin, A. E., & Hirshkowitz, M. (1995). CPAP machine performance and altitude. *Chest, 108*, 1577–1580.

Fuchs, F. S., Wiest, G. H., Frank, M., et al. (2002). Auto-CPAP therapy for obstructive sleep apnea: Induction of microarousals by automatic variations of CPAP pressure? *Sleep, 25*, 514–518.

Galetke, W., Randerath, W. J., Stieglitz, S., et al. (2009). Comparison of manual titration and automatic titration based on forced oscillation technique, flow and snoring in obstructive sleep apnea. *Sleep Medicine, 10*, 337–343.

Hansford, A. (2011). Thirty years of CPAP: A brief history of OSA. *ResMedica Clinical Newsletter, 14*. https://www.yumpu.com/en/document/view/35208252/thirty-years-of-cpap-pdf-15mb-resmed. Assessed 1 Sept 2021.

Harrington, M. R., Doudkine, D., Foote, R. M. L., Huby, R. J., Tang, Z. R., Wu, H., Yang, Q., & Bath, A. R., inventors; Resmed Limited (Bella Vista, AU), assignee. (2017, January 5). *A Humidifier for a respiratory therapy device*. US Patent 2017/0000968 A1.

Hart, M. C., Orzalesi, M. M., & Cook, C. D. (1963). Relation between anatomic respiratory dead space and body size and lung volume. *Journal of Applied Physiology, 18*(3), 519–522.

Herkenrath, S. D., Treml, M., Anduleit, N., et al. (2020). Extended evaluation of the efficacy of a proactive forced oscillation technique-based auto-CPAP algorithm. *Sleep & Breathing, 24*, 825–833.

Hill, P. D. M., inventor; Respironics, Inc. (Murrysville, PA), assignee. (2004). *Method and apparatus for providing variable positive airway pressure*. United States.

Hill, P. D., Kissel, M. H., Frank, J., Kane, M. T., Bann, S. L., & Duff, W. K., inventor; RIC Investments, LLC, assignee. (2006, March 14). *Average volume ventilation*. US patent 7011091.

Huang, H. C., Hillman, D. R., & McArdle, N. (2012). Control of OSA during automatic positive airway pressure titration in a clinical case series: Predictors and accuracy of device download data. *Sleep, 35*, 1277–183A.

Hukins, C. (2004). Comparative study of autotitrating and fixed-pressure CPAP in the home: A randomized, single-blind crossover trial. *Sleep, 27*, 1512–1517.

Ikeda, Y., Kasai, T., Kawana, F., et al. (2012). Comparison between the apnea-hypopnea indices determined by the REMstar Auto M series and those determined by standard in-laboratory polysomnography in patients with obstructive sleep apnea. *Internal Medicine, 51*, 2877–2885.

Isetta, V., Navajas, D., Montserrat, J. M., & Farre, R. (2015). Comparative assessment of several automatic CPAP devices' responses: A bench test study. *ERJ Open Research, 1*, 00031.

Jaber, S., Chanques, G., Matecki, S., et al. (2002). Comparison of the effects of heat and moisture exchangers and heated humidifiers on ventilation and gas exchange during non-invasive ventilation. *Intensive Care Medicine, 28*, 1590–1594.

Jaffuel, D., Rabec, C., Philippe, C., et al. (2020). Patterns of adaptive servo-ventilation settings in a real-life multicenter study: Pay attention to volume!: Adaptive servo-ventilation settings in real-life conditions. *Respiratory Research, 21*, 243.

Javaheri, S., Goetting, M. G., Khayat, R., Wylie, P. E., Goodwin, J. L., & Parthasarathy, S. (2011). The performance of two automatic servo-ventilation devices in the treatment of central sleep apnea. *Sleep, 34*, 1693–1698.

Javaheri, S., Brown, L. K., & Randerath, W. J. (2014). Positive airway pressure therapy with adaptive servoventilation: Part 1: Operational algorithms. *Chest, 146*, 514–523.

Johnson, K. G., & Johnson, D. C. (2005). Bilevel positive airway pressure worsens central apneas during sleep. *Chest, 128*, 2141–2150.

Johnson, K. G., & Johnson, D. C. (2015). Treatment of sleep-disordered breathing with positive airway pressure devices: Technology update. *Medical Devices (Auckl), 8*, 425–437.

Kane, M. T., Bann, S. L., Siirola, R. E., Duff, W. K., & Baloa, L. A., inventor; RIC Investments, LLC (Wilmington, DE, US), assignee. (2014, April 15). *Method and apparatus for treating Cheyne-Stokes respiration*. US Patent 8695595.

Kelly, J. L., Jaye, J., Pickersgill, R. E., Chatwin, M., Morrell, M. J., & Simonds, A. K. (2014). Randomized trial of 'intelligent' autotitrating ventilation versus standard pressure support non-invasive ventilation: Impact on adherence and physiological outcomes. *Respirology, 19*, 596–603.

Kenyon, B. J., inventor; ResMed Pty Ltd. (Bella Vista, AU), assignee. (2020, November 24). *Impeller with inclined and reverse inclined blades*. US patent 10, 844,876 B2.

Killick, R., & Marshall, N. S. (2021). The impact of device modifications and pressure delivery on adherence. *Sleep Medicine Clinics, 16*, 75–84.

Kinnear, W., Watson, L., Smith, P., et al. (2017). Effect of expiratory positive airway pressure on tidal volume during non-invasive ventilation. *Chronic Respiratory Disease, 14*, 105–109.

Kushida, C. A., Berry, R. B., Blau, A., et al. (2011). Positive airway pressure initiation: A randomized controlled trial to assess the impact of therapy mode and titration process on efficacy, adherence, and outcomes. *Sleep, 34*, 1083–1092.

Li, Q. Y., Berry, R. B., Goetting, M. G., et al. (2015). Detection of upper airway status and respiratory events by a current generation positive airway pressure device. *Sleep, 38*, 597–605.

Littner, M., Hirshkowitz, M., Davila, D., et al. (2002). Practice parameters for the use of auto-titrating continuous positive airway pressure devices for titrating pressures and treating adult patients with obstructive sleep apnea syndrome. An American Academy of Sleep Medicine report. *Sleep, 25*, 143–147.

Lofaso, F., Desmarais, G., Leroux, K., et al. (2006). Bench evaluation of flow limitation detection by automated continuous positive airway pressure devices. *Chest, 130*, 343–349.

Lofaso, F., Leroux, K., Boussaid, G., Prigent, H., & Louis, B. (2020). Response of home-use adaptive pressure modes to simulated transient hypoventilation. *Respiratory Care, 65*, 1258–1267.

Loube, D. I., Gay, P. C., Strohl, K. P., Pack, A. I., White, D. P., & Collop, N. A. (1999). Indications for positive airway pressure treatment of adult obstructive sleep apnea patients: A consensus statement. *Chest, 115*, 863–866.

Lowenstein sleep therapy products: Clinical benefits whitepaper. Sante, ed. https://online.flippingbook.com/view/825139686/. Assessed 28 Aug 2021.

Magdy, D. M., & Metwally, A. (2020). Effect of average volume-assured pressure support treatment on health-related quality of life in COPD patients with chronic hypercapnic respiratory failure: A randomized trial. *Respiratory Research, 21*, 64.

Marrone, O., Insalaco, G., Bonsignore, M. R., Romano, S., Salvaggio, A., & Bonsignore, G. (2002). Sleep structure correlates of continuous positive airway pressure variations during application of an autotitrating continuous positive airway pressure machine in patients with obstructive sleep apnea syndrome. *Chest, 121*, 759–767.

Marsh, S. A., inventor; Encite LLC, (Burlington, MA), assignee. (2021, August 17). *Micro pump systems and processing techniques*. US patent 11,092,150 B2.

Martin, D. C., & Oates, J. D., inventor; ResMed Ltd (Bella Vista, AU), assignee. (2014, November 13). *Systems, methods, and/or apparatuses for non-invasive monitoring of respiratory parameters in sleep disordered breathing*. US Patent 8646447.

Masa, J. F., Jimenez, A., Duran, J., et al. (2004). Alternative methods of titrating continuous positive airway pressure: A large multicenter study. *American Journal of Respiratory and Critical Care Medicine, 170*, 1218–1224.

Massie, C. A., Hart, R. W., Peralez, K., & Richards, G. N. (1999). Effects of humidification on nasal symptoms and compliance in sleep apnea patients using continuous positive airway pressure. *Chest, 116*, 403–408.

Matthews, G. P., Kane, M. T., Duff, W. K., Siirola, R., Martin, D., Ressler, H., & Shankar, U., inventors; RIC Investments, LLC (Wilmington, DE, US), assignee. (2010, November 9). *Auto-titration pressure support system and method of using same*. US Patent 7827988.

Matthews, G., Duff, W. K., Martin, D., Shankar, U. S., & Ressler, H. (2012, March 20). *Auto-titration bi-level pressure support system and method of using the same*. United States Patent US 8136521.

McArdle, N., Rea, C., King, S., et al. (2017). Treating chronic hypoventilation with automatic adjustable versus fixed EPAP intelligent volume-assured positive airway pressure support (iVAPS): A Randomized Controlled Trial. *Sleep, 40*(10), zsx136. https://doi.org/10.1093/sleep/zsx136

Mell, R. (2021). Medical device recall notification Philips respironics CPAP and Bi-level PAP devices.

Morgenthaler, T. I., Aurora, R. N., Brown, T., et al. (2008). Practice parameters for the use of autotitrating continuous positive airway pressure devices for titrating pressures and treating adult patients with obstructive sleep apnea syndrome: An update for 2007. An American Academy of Sleep Medicine report. *Sleep, 31*, 141–147.

Navalesi, P., Fanfulla, F., Frigerio, P., Gregoretti, C., & Nava, S. (2000). Physiologic evaluation of noninvasive mechanical ventilation delivered with three types of masks in patients with chronic hypercapnic respiratory failure. *Critical Care Medicine, 28*, 1785–1790.

Nilius, G., Katamadze, N., Domanski, U., Schroeder, M., & Franke, K. J. (2017). Non-invasive ventilation with intelligent volume-assured pressure support versus pressure-controlled ventilation: Effects on the respiratory event rate and sleep quality in COPD with chronic hypercapnia. *International Journal of Chronic Obstructive Pulmonary Disease, 12*, 1039–1045.

Nilius, G., Schroeder, M., Domanski, U., & Richter, M. (2020). A comparison of sleep disordered breathing events detected by CPAP devices with in-lab polysomnography. *European Respiratory Journal, 56*, 2491.

Nussbaumer, Y., Bloch, K. E., Genser, T., & Thurnheer, R. (2006). Equivalence of autoadjusted and constant continuous positive airway pressure in home treatment of sleep apnea. *Chest, 129*, 638–643.

Orr, J. E., Coleman, J., Criner, G. J., et al. (2019). Automatic EPAP intelligent volume-assured pressure support is effective in patients with chronic respiratory failure: A randomized trial. *Respirology, 24*, 1204–1211.

Oscroft, N. S., Ali, M., Gulati, A., et al. (2010). A randomised crossover trial comparing volume assured and pressure preset noninvasive ventilation in stable hypercapnic COPD. *COPD: Journal of Chronic Obstructive Pulmonary Disease, 7*, 398–403.

Park, S., & Suh, E. S. (2020). Home mechanical ventilation: Back to basics. *Acute and Critical Care, 35*, 131–141.

Patil, S. P., Ayappa, I. A., Caples, S. M., Kimoff, R. J., Patel, S. R., & Harrod, C. G. (2019a). Treatment of adult obstructive sleep apnea with positive airway pressure: An American Academy of Sleep Medicine systematic review, meta-analysis, and GRADE assessment. *Journal of Clinical Sleep Medicine, 15*, 301–334.

Patil, S. P., Ayappa, I. A., Caples, S. M., Kimoff, R. J., Patel, S. R., & Harrod, C. G. (2019b). Treatment of adult obstructive sleep apnea with positive airway pressure: An American Academy of Sleep Medicine clinical practice guideline. *Journal of Clinical Sleep Medicine, 15*, 335–343.

Pavone, M., Verrillo, E., Onofri, A., Caggiano, S., & Cutrera, R. (2020). Ventilators and ventilatory modalities. *Frontiers in Pediatrics, 8*, 500.

Pepin, J. L., Woehrle, H., Liu, D., et al. (2018). Adherence to positive airway therapy after switching from CPAP to ASV: A big data analysis. *Journal of Clinical Sleep Medicine, 14*, 57–63.

Poulton, E. P. (1936). Left-sided heart failure with pulmonary oedema: It's treatment with the pulmonary plus pressure machine. *The Lancet, 228*, 981–983.

Prasad, B., Carley, D. W., & Herdegen, J. J. (2010). Continuous positive airway pressure device-based automated detection of obstructive sleep apnea compared to standard laboratory polysomnography. *Sleep & Breathing, 14*, 101–107.

Puertas, F. J., Ondze, B., Carlander, B., & Billiard, M. (1999). Arousal of respiratory origin and upper airway resistance syndrome: Pathophysiological and diagnostic aspects. *Revista de Neurologia, 28*, 565–572.

Ramp Plus Discussion Paper. (2021). Philips. https://www.usa.philips.com/healthcare/product/HCEUX520H15C/dreamstation-2-auto-cpap-advanced-auto-cpap#features. Accessed 22 Aug 2021.

Rapoport, D. M. (1987). Techniques for administering nasal CPAP. *Respiratory Management, 17*, 17–19.

Rapoport, D. M., inventor; New York University, (New York, NY), assignee. (1991, November 19). *Method and apparatus for the treatment of obstructive sleep apnea*. US patent 5,065,756.

Remmers, J. E., & Feroah, T. R., inventor; University Technologies International, Inc. (Calgary, CA), assignee. (2003, April 22). *Auto CPAP system profile information*. US Patent 6550478.

ResMed Origins. https://document.resmed.com/en-us/documents/articles/resmed-origins.pdf. Assessed 31 Aug 2021.

Respironics P. (2012). System One Spec Sheet/REMstarPro and Auto. documents.philips.com/assets/20170523/c481a9700b-75419788b9a77c0158365b.pdf. Assessed 1 Sept 2021.

Rigau, J., Montserrat, J. M., Wohrle, H., et al. (2006). Bench model to simulate upper airway obstruction for analyzing automatic continuous positive airway pressure devices. *Chest, 130*, 350–361.

Rosen, C. L., Auckley, D., Benca, R., et al. (2012). A multisite randomized trial of portable sleep studies and positive airway pressure autotitration versus laboratory-based polysomnography for the diagnosis and treatment of obstructive sleep apnea: The HomePAP study. *Sleep, 35*, 757–767.

Sanders, M. H., & Kern, N. (1990). Obstructive sleep apnea treated by independently adjusted inspiratory and expiratory positive airway pressures via nasal mask. Physiologic and clinical implications. *Chest, 98*, 317–324.

Sanders, M. H., & Zdrojkowski, R. J., inventors; Respironics, Inc, Murrysville, PA, assignee. (1992, September 22). *Method and apparatus for maintaining airway patency to treat sleep apnea and other disorders.* US patent 5148802.

Sanders, M. H., & Zdrojkowski, R. J., inventors; Respironics Inc. (Murrysville, PA), assignee. (1995, July 18). *Breathing gas delivery method and apparatus.* United States 5433193.

Sullivan, C. E., & Lynch, C., inventor; Sullivan, C. E., Lynch, C. (AU), assignee. (1993, April 6). *Device for monitoring breathing during sleep and control of CPAP treatment that is patient controlled.* 5199424. United States patent US.

Sullivan, C. E., Issa, F. G., Berthon-Jones, M., & Eves, L. (1981). Reversal of obstructive sleep apnoea by continuous positive airway pressure applied through the nares. *Lancet, 1*, 862–865.

Sunderram, J., Ayappa, I., Lu, S. E., et al. (2021). PAP adherence and nasal resistance. a randomized controlled trial of CPAPflex versus CPAP in world trade center responders. *Annals of the American Thoracic Society, 18*, 668–677.

Thomas, R. J., & Bianchi, M. T. (2017). Urgent need to improve PAP management: The devil is in two (fixable) details. *Journal of Clinical Sleep Medicine, 13*, 657–664.

Turk, M., Aydogdu, M., & Gursel, G. (2018). Effects of modes, obesity, and body position on non-invasive positive pressure ventilation success in the intensive care unit: A randomized controlled study. *Turkish Thoracic Journal, 19*, 28–35.

Ueno, K., Kasai, T., Brewer, G., et al. (2010). Evaluation of the apnea-hypopnea index determined by the S8 auto-CPAP, a continuous positive airway pressure device, in patients with obstructive sleep apnea-hypopnea syndrome. *Journal of Clinical Sleep Medicine, 6*, 146–151.

Velzy, A. F., Mebasser, S. A., Vinokur, R., et al., inventors; ResMed Motor Technologies Inc, (Chatsworth, CA), assignee. (2018, October 9). *Flow generator.* US patent 10,092,716 B2.

Vennelle, M., White, S., Riha, R. L., Mackay, T. W., Engleman, H. M., & Douglas, N. J. (2010). Randomized controlled trial of variable-pressure versus fixed-pressure continuous positive airway pressure (CPAP) treatment for patients with obstructive sleep apnea/hypopnea syndrome (OSAHS). *Sleep, 33*, 267–271.

Willson, G. N., Wilcox, I., Piper, A. J., et al. (2001). Noninvasive pressure preset ventilation for the treatment of Cheyne-Stokes respiration during sleep. *The European Respiratory Journal, 17*, 1250–1257.

Zdrojkowski, R. J., inventor; Respironics Inc. assignee. (1994, May 24). *Leak compensation method and apparatus for a breathing system.* US patent 5313937.

Zdrojkowski, R. J., & Estes, M., inventors; Respironics, Inc. (Pittsburgh, PA), assignee. (2000, February 29). *Breathing gas delivery method and apparatus.* US patent 6029664.

Zhu, K., Aouf, S., Roisman, G., & Escourrou, P. (2016). Pressure-relief features of fixed and autotitrating continuous positive airway pressure may impair their efficacy: Evaluation with a respiratory bench model. *Journal of Clinical Sleep Medicine, 12*, 385–392.

19 Adherence Monitoring Using Telemonitoring Techniques

Sarah Dietz-Terjung, Martina Große-Suntrup, and Christoph Schöbel

Abstract

Telemonitoring is a frequently used tool in the long-term management of many chronic diseases, such as chronic obstructive pulmonary disease or chronic heart failure. The use of new sensors and telemedical tools will shape medical practice in the future, particularly in sleep medicine. During the last decades, the number of people with sleep disordered breathing has been increasing.

Telemedicine (TM) approaches could be used in various ways in sleep medicine: telediagnostics, teleconsultation, teletherapy, and telemonitoring of patients being treated with positive pressure devices.

This chapter aims to summarize the recent scientific progresses of these techniques as well as their potential clinical applications and tries to give consideration to the remaining problems with TM applications.

Keywords

Telediagnostics · Teleconsultation · Teletherapy · Telemonitoring · Telemedicine · Obstructive sleep apnea · Positive airway pressure · Therapy management

S. Dietz-Terjung · M. Große-Suntrup · C. Schöbel (✉)
University Medicine Essen, Ruhrlandklinik - West German Lung Center, Center for Sleep- and Telemedicine, Essen, Germany
e-mail: christoph.schoebel@rlk.uk-essen.de

19.1 Background

Obstructive sleep apnea (OSA) is the most common organic sleep disorder. Due to a collapse of the upper airways during sleep, breathing disturbances occur with resulting short, sleep-disrupting awakenings and consecutive activations of the stress nervous system. The resulting increases in blood pressure and heart rate can stress the cardiovascular system, increasing the risk of developing cardiac disease. OSA patients often complain about snoring, nonrestorative sleep, or daytime sleepiness, which is also associated with an increased risk for road traffic accidents. Furthermore, patients with untreated OSA display a greater burden on the healthcare system (Peppard et al., 2013). The number of breathing episodes per hour of sleep apnea-hypopnea index(AHI) defines OSA severity.

Epidemiological studies based on the AHI show a high prevalence of sleep apnea in terms of a common disease. For instance, the number of affected persons in Germany is currently estimated at 26 million (Benjafield et al., 2019). Due to limited diagnostic capacities, it is assumed that there is a high number of unreported cases.

However, the indication for a therapy should not solely be based on the AHI. This is especially

T. Penzel, R. Hornero (eds.), *Advances in the Diagnosis and Treatment of Sleep Apnea*, Advances in Experimental Medicine and Biology 1384, https://doi.org/10.1007/978-3-031-06413-5_19

true for patients with low-grade OSA, in whom therapy is only indicated in the presence of symptoms or a cardiovascular comorbidity. Nocturnal positive airway pressure therapy is an effective therapeutic method. Positive airway pressure (PAP) is applied via a nasal or a full-face mask to splint the upper airways. This can be done by applying a continuous (CPAP) or an automatically adjusted positive airway pressure (APAP) during sleep. Approximately 1 million patients in Germany are currently using this therapy. This can reduce symptoms and significantly improve quality of life (Woehrle et al., 2017). However, regular use of the therapy is a prerequisite for this.

Studies reveal that tolerance problems, low utilization times, and therapy termination rates of up to 20% in the first year of therapy are frequently observed (Woehrle et al., 2017).

The economic pressure with uncertain counterfinancing as well as the corona pandemic has led to a reduction of inpatient sleep laboratory capacities with long waiting times for a diagnostic study as polysomnography (PSG) remains the reference method for the diagnosis of sleep disordered breathing (SDB) (Kapur er al., 2017). Patients spend one night in a sleep laboratory to enable an attended, in-lab PSG to be performed. As the number of obese people continues to rise, OSA is becoming increasingly prevalent (Peppard et al., 2013), and, as a result, the waiting time to obtain a PSG can be very long. Outpatient sleep centers are often unable to compensate this great amount of patients. According to an analysis of the German Sleep Society DGSM among the accredited German sleep laboratories from March 2020, average waiting times of 4.2 months to a maximum of 25 months could be reported (internal data, not published). Therefore, many potential patients remain undiagnosed without any therapy.

In Germany, after PAP therapy initiation in the sleep laboratory, an outpatient therapy control by means of a portable monitoring by a medical specialist is performed after 6 months (Jacobsen et al., 2017). However, problems affecting long-term treatment compliance often become apparent during the first weeks of therapy (Patil et al., 2019). Therefore, the German medical guideline also recommends earlier monitoring if there are problems with therapy.

19.2 Recent Advances

Modern PAP devices can already send daily therapy data (time of use, mask fit, remaining respiratory disorders) to a cloud of the medical device provider via an integrated SIM card (subscriber identity module). Thus, telemonitoring is a collective term for a diagnostic or therapeutic monitoring using mobile technical tools. Telemedicine extends the term telemonitoring and is defined as "the use of electronic information and communication technologies to provide and support healthcare when distance separates patient and healthcare unit or professional" (https://www.bundesgesundheitsministerium.de/service/begriffe-von-a-z/t/telemedizin.html) by means of all kinds of telecommunication tools, including smartphones, wireless devices, and video and telephone consultations.

Studies show that telemonitoring-assisted follow-up incorporating telemedically transmitted PAP therapy data reduced therapy termination rates and increased PAP usage time. Significant effects on quality-of-life outcomes have also been observed. Also an improved resolution of symptoms such as daytime sleepiness has been demonstrated. Telemonitoring is already technically available from various providers but is hardly used by the physicians in charge due to the interface problems, the lack of remuneration, and the lack of standards of care (Fox et al., 2012; Walker et al., 2018).

Development of medicine has also made great strides toward personalized medicine including the application of telemedical approaches. Various studies have shown the benefits of telemedical applications in terms of patient-centered outcome (e.g., through increased compliance with therapy and more self-determination by patients), reduction of costs, and increasing access to medical care for patients with chronic diseases especially in underserved or rural areas and in low-income countries (Flodgren et al., 2015; Rada, 2015).

During the last decade, the use of telemedical technology has dramatically increased in developed countries. This is also reflected in the number of publications: In 2005, there were still 1090 publications on the subject of TM, whereas in 2015, there were 2307 (Bruyneel, 2016).

Telemedicine requires a strong bidirectional interaction between patients and healthcare providers (Flodgren et al., 2015). Unfortunately, the kind of data transmitted by the patient, the frequency of data transfer, and the frequency of interactions between patient and healthcare providers vary largely across studies (Flodgren et al., 2015). Patient data transmission methods as well as feedback ways include telephone contacts, the Internet, video, and smartphone-based data transfer. In sleep medicine, telemedical solutions for patient management are applicable ranging from diagnosis to the monitoring of treatment. The potential benefits of telemedicine include improved access to healthcare, reduced waiting time for appointments, and increased adherence to chronic illness treatment plans.

A study of Gagnadoux et al. (2002) included 99 patients in a prospective randomized crossover trial. Each patient underwent one home PSG and one telemonitored, PSG (TM-PSG) on two consecutive nights. The TM-PSG was recorded in the medical unit of two peripheral hospitals, with remote control from the central sleep lab. The sleep technicians checked the quality of the recordings every 30 min and instructed the nursing staff at the two hospitals to replace any insufficient electrodes. Failure rate was 11% for TM-PSG vs. 23% for home PSG. Thirteen TM-PSGs required technical intervention to replace lost sensors, but in four cases, the nurse in the medical unit was not able to correct the problem. Without telemonitoring, the failure rate would have been 19% but was reduced to 11% with remote supervision. A cost analysis was also performed and concluded that although telemedicine was more effective (half the number of failures), it was also very expensive ($244 vs. $153 for home PSG; Pelletier-Fleury et al., 2001).

Another study (Pelletier-Fleury et al., 2001) investigated telemonitored polygraphy (TM-PG) in patients with suspected OSAS. TM-PG was performed on 40 patients, in a "Virtual Sleep Unit," in another hospital some 80 km from the central sleep lab. The sleep lab nurses performed real-time continuous TM-PG check with additional continuous video monitoring. No PG failure was observed, but data transmission failed for 2.5% of the recordings. The cost analysis showed also that telemedicine is associated with additional costs: TM-PG costs $277 compared to $145 for a PSG.

Additionally, recent studies demonstrate the positive effect of telemonitoring in patients undergoing PAP therapy.

Chumpangern et al. (2021) performed a prospective randomized controlled trial on 60 Asian adults (70% male) with moderate-to-severe OSA. Two groups (telemonitoring vs. no telemonitoring) with 30 patients each were analyzed. The telemonitoring system functioned by transferring CPAP usage data via cellular network. When there were any triggers occurring 2 nights consecutively (usage hours <4 h per night; leakage >27 L/min or AHI > 5 events/h), the investigator contacted the patients. As the primary outcome of this study, the 4-week CPAP usage hours per night was chosen. As a result, they found that the percentage of good adherence was significantly higher in the telemonitoring group (64.2% vs. 34.4%; $P = 0.024$). Median leakage per night was also significantly lower in the telemonitoring group and a significant sleep quality improvement was observed.

Isetta et al. (2014) tested the feasibility of teleconsultation. Two different schemes were studied to assess whether teleconsultation could replace, first, continuous positive airway pressure (CPAP) follow-up consultation (n = 50) and, second, CPAP training consultation (n = 40 patients, two groups: face-to-face vs. teleconsultation). For CPAP follow-up, 95% of the patients were satisfied with the teleconsultation, and 66% declared that teleconsultation could replace up to 100% of the CPAP therapy follow-up visits. Younger patients (<65 years) were more inclined to recommend teleconsultation to others. For CPAP training, patients trained via videoconference demonstrated the same knowledge about OSAS and CPAP therapy as the face-to-face group (94%

of correct answers vs. 92%). Video-trained patients also showed similar performances on mask placement and mask leak avoidance.

Woehrle et al. (2017) analyzed data from a large German homecare provider with the aim to investigate the effect of a proactive patient management program supported by remote access to PAP therapy data on therapy termination compared with standard care. They found that the long-term rate of therapy termination was significantly lower by using a telemedicine-based proactive management strategy compared with standard care independently of sex, age, type of device, or type of insurance. Therefore, they conclude that the implementation of telemedicine-based strategies has the potential to improve adherence and patient outcomes and may allow a more efficient allocation of scarce healthcare resources.

In a study of Dellaca et al. (2011), 20 severe OSAS patients who were using CPAP for the first time were estimated. Here, CPAP was coupled with a telemetric unit in order to allow remote control of CPAP parameters (flow, pressure, leaks) and CPAP pressure adaptation. After 7 days patients underwent full in-lab PSG with another CPAP titration. Pressure level was similar in both settings: 9.15 at home vs. 9.2 cm H2O. Real-time remote CPAP titration is feasible and offers pressure-setting outcomes similar to in-hospital-attended CPAP titration.

Hwang et al. (2018) performed a four-arm, randomized, factorial design clinical trial which enrolled 1455 patients, mainly woman (51.0%) referred to sleep laboratory due to suspected OSA. Nine hundred and fifty-six patients underwent home sleep apnea testing, and 556 were prescribed CPAP. A total of two telemedicine interventions were implemented: first, a web-based OSA education and, second, a CPAP telemonitoring with automated patient feedback. Patients were randomized to (1) usual care, (2) OSA education, (3) CPAP telemonitoring, or (4) OSA education and CPAP telemonitoring. They found that the use of CPAP telemonitoring with an automated feedback messaging system improved 90-day adherence in patients with OSA. Additionally, telemedicine-based education did not significantly improve CPAP adherence in the presented study but did increase clinic attendance for OSA evaluation.

Patil et al. (2019) studied in their systematic review the effect of telemonitoring-guided PAP interventions. They identified a total of five randomized controlled trials (RCT) that evaluated the use of remote monitoring as an adjunct to PAP therapy in order to optimize treatment effects for OSA patients. They defined adherence to PAP therapy, sleepiness, QoL, and PAP-associated side effects as outcomes. Outcomes were analyzed at 2–3 months after PAP initiation. Triggers for intervention and the interventions themselves varied greatly across the analyzed studies resulting in a great variety of results. Reasons for low usage were, for example, high mask leakages, too high delivered pressures, or a high residual AHI.

The authors highlighted that also interventions triggered by PAP data varied substantially in the analyzed studies: from text messages to telephone calls or in-person visits with sleep staff or even a sleep physician.

Patil et al. also performed meta-analyses to assess the efficacy of telemedical-guided interventions in order to increase PAP adherence. Here, a clinically significant improvement in PAP adherence with the use of telemonitoring could been shown.

Additionally, the efficacy of an intervention guided by remote monitoring of PAP therapy to improve PAP adherence was evaluated using a meta-analysis of five RCTs that reported on hours per night of PAP usage. All analyzed studies used data from PAP devices to guide the intervention. Four of the five studies included newly diagnosed OSA patients with minimal comorbidity, but follow-up was short ranging from 1 to 3 months. The performed meta-analysis demonstrated a clinically significant increase in PAP usage of 1.0 h/night (95% CI: 0.5–1.4 h/night).

19.3 Discussion

There are two potential explanations for this increase in adherence with telemonitoring.

Firstly, telemedicine allows an immediate, low-threshold access to real-time assistance from a clinical provider to address PAP-related issues for patients in comparison to a long waiting time for an appointment to see a clinician face to face.

Secondly, daily monitoring motivates and engages patients and increases their health literacy leading to an improved sense of accountability for their own healthcare.

However, these mechanisms are still not fully evaluated. Overall, the analyses demonstrated a clinically significant improvement in adherence in adults with OSA using telemonitoring compared to usual care. Additionally, the quality of evidence for PAP adherence was high.

The performed meta-analysis did not demonstrate a clinically significant reduction in the Epworth Sleepiness Scale (ESS) with telemonitoring-guided intervention as compared to no such interventions. However, the quality of evidence for self-reported sleepiness was moderate due to imprecision.

Patil et al. (2019) also identified two RCTs that assessed the impact of a telemonitoring-guided PAP adherence intervention on PAP-induced side effects, although data were not reported in a sufficiently standardized format to allow for a meta-analysis.

Identified side effects included CPAP discomfort, difficulty exhaling, mask leakages, aerophagia, allergic reactions to device components, headache, facial pain or bruises, mouth dryness, or nasal congestion independent from the frequency of PAP-related side effects, with the exception of one study which suggested that telemonitoring was associated with fewer complaints of a dry mouth. The quality of evidence was low due to imprecision.

They also found no statistically significant change in the quality of Life (QOL).

The use of telemonitoring has advantages and disadvantages according to Patil et al. The benefits include improvements in PAP adherence to improve control of OSA symptoms and reduce the need for office visits, more patient confidence, and reduced healthcare costs.

A potential harm of a telemedical-guided intervention is the potential loss of privacy. Furthermore, a constant contact with physician and provider could be stressful for some patients.

Onofri et al. (2021) presented in their work the use of telemedicine in children with complex conditions on home ventilation (invasive and noninvasive) during the COVID-19 pandemic. During this intervention, they were able to detect a total of 12 healthcare problems during scheduled telemedicine consulting. Only one problem was not solved by remote intervention. In conclusion, the use of telemedicine in CMC-ventilated patients resulted in a feasible tool to avoid in-person visits during the pandemic without deterioration of clinical care.

Kulkarny et al. (2021) analyzed in their review current pulmonary telemonitoring technologies. They conclude that despite increasing evidence, remote patient monitoring and telehealth could improve patient-provider communication in remote residential populations as well as in rural healthcare settings in the guise of equitable patient-specific healthcare. However, they point out that a successful implementation of a telehealth paradigm requires reliable, accountable, secure, and accurate real-time remote monitoring devices as well as accurate artificial intelligence for support of clinical decision-making.

Beside technical solutions for transmitting and processing therapy data, it is of great importance to define replicable standardized intervention descriptions as a basis for a scalable telemedical management, e.g., in terms of generating further evidence for a positive healthcare effect and for assessing resources in order to implement remuneration for TM (Schöbel et al., 2020).

19.4 Conclusion

Telemedicine is a logical and important step for sleep medicine.

Various methods are available to estimate patients' adherence and compliance with PAP therapy at home. Objective data can be obtained by downloads from the memory card of the PAP device or by direct interrogation of PAP devices to cloud-based systems. Telemedical applications can be used to monitor patients on a large scale and encouraging patients' empowerment. There are various mechanisms for supporting virtual consulting and remote monitoring, although it remains unclear if they can be conducted in an economic way. Inclusion of sham telemedicine control arms might determine whether increased adherence might have been caused by the perception of monitoring by the patients or by the earlier clinical interventions. Additionally, telemedical management of PAP treatment could generate more evidence for a positive dose-response relationship of PAP adherence and patient-reported-outcome measures. Furthermore, it is of great importance to be aware of how to use this technology safely according to GDPR or any other current legal confines in the respective healthcare system. In addition, standardized and replicable intervention descriptions for telemedicine management will provide a more seamless communication flow, to the benefit of medical providers, the healthcare system, and ultimately for patients.

Therefore, a structured framework, e.g., the MAST (model for assessment of telemedicine) is necessary to facilitate the choice of the most efficient and cost-effective device for the treating physician. Yet, inclusion of lifestyle devices as well as new sensors and technologies needs to be validated according to the Medical Device Regulations in order to use them in a medical context.

References

Benjafield, A. V., Pepin, J.-L. D., Valentine, K., et al. (2019). Compliance after switching from CPAP to bilevel for patients with non-compliant OSA: Big data analysis. *BMJ Open Respiratory Research, 6*, e000380. https://doi.org/10.1136/bmjresp-2018-000380

Bruyneel, M. (2016). Technical developments and clinical use of telemedicine in sleep medicine. *Journal of Clinical Medicine, 5*(12), 116. https://doi.org/10.3390/jcm5120116. Published 2016 Dec 13.

Chumpangern, W., Muntham, D., & Chirakalwasan, N. (2021). Efficacy of a telemonitoring system in continuous positive airway pressure therapy in Asian obstructive sleep apnea. *Journal of Clinical Sleep Medicine, 17*(1), 23–29. https://doi.org/10.5664/jcsm.8772. PMID: 32895116; PMCID: PMC7849645.

Dellacà, R., Montserrat, J. M., Govoni, L., Pedotti, A., Navajas, D., & Farré, R. (2011). Telemetric CPAP titration at home in patients with sleep apnea-hypopnea syndrome. *Sleep Medicine, 12*(2), 153–157. https://doi.org/10.1016/j.sleep.2010.07.014. Epub 2011 Jan 17. PMID: 21247800.

Flodgren, G., Rachas, A., Farmer, A. J., Inzitari, M., & Shepperd, S. (2015). Interactive telemedicine: Effects on professional practice and health care outcomes. *Cochrane Database of Systematic Reviews, 9*, CD002098.

Fox, N., et al. (2012). The impact of a telemedicine monitoring system on positive airway pressure adherence in patients with OSA: A randomized controlled trial. *Sleep, 35*(4), 477–481.

Franek, J. (2012). Home telehealth for patients with chronic obstructive pulmonary disease (COPD): An evidence-based analysis. *Ontario Health Technology Assessment Series, 12*(11), 1–58.

Gagnadoux, F., Pelletier-Fleury N., Philippe C., Rakotonanahary, D., & Fleury, B. (2002). Home unattended vs hospital telemonitored polysomnography in suspected obstructive sleep apnea syndrome: A randomized crossover trial. *Chest, 121*(3), 753–758, ISSN 0012-3692. https://doi.org/10.1378/chest.121.3.753

https://www.bundesgesundheitsministerium.de/service/begriffe-von-a-z/t/telemedizin.html. Last accessed 20 Mar 2022.

Hwang, D., Chang, J. W., Benjafield, A. V., Crocker, M. E., Kelly, C., Becker, K. A., Kim, J. B., Woodrum, R. R., Liang, J., & Derose, S. F. (2018). Effect of telemedicine education and telemonitoring on continuous positive airway pressure adherence. The Tele-OSA randomized trial. *American Journal of Respiratory and Critical Care Medicine, 197*(1), 117–126. https://doi.org/10.1164/rccm.201703-0582OC. PMID: 28858567.

Institute of Medicine. (1996). *Telemedicine: A guide to assessing telecommunication in health care*. National Academy Press.

Isetta, V., León, C., Torres, M., Embid, C., Roca, J., Navajas, D., Farré, R., & Montserrat, J. M. (2014). Telemedicine-based approach for obstructive sleep apnea management: Building evidence. *Interactive Journal of Medical Research, 3*(1), e6. https://doi.org/10.2196/ijmr.3060. PMID: 24554392; PMCID: PMC3961625.

Jacobsen, A. R., et al. (2017). Determinants for adherence to continuous positive airway pressure therapy in obstructive sleep apnea. *PLoS One, 12*(12), e0189614.

Kapur, V. K., Auckley, D. H., Chowdhuri, S., Kuhlmann, D. C., Mehra, R., Ramar, K., Harrod, C. G. (2017). Clinical Practice Guideline for Diagnostic Testing for Adult Obstructive Sleep Apnea: An American Academy of Sleep Medicine Clinical Practice Guideline. *Journal of Clinical Sleep Medicine, 13*(3):479–504. https://doi.org/10.5664/jcsm.6506. PMID: 28162150; PMCID: PMC5337595.

Kidholm, K., Bowes, A., Ekeland, A. G., Flottorp, S. A., Jensen, L. K., Rasmussen, J., Pedersen, C. A., &

Bech, M. (2012). A model for assessment of telemedicine applications: MAST. *International Journal of Technology Assessment in Health Care, 28*(1), 44–51.

Kulkarni, K., Sevakula, R. K., Kassab, M. B., Nichols, J., Roberts, J. D., Jr., Isselbacher, E. M., & Armoundas, A. A. (2021). Ambulatory monitoring promises equitable personalized healthcare delivery in underrepresented patients. *European Heart Journal – Digital Health, 2*(3), 494–510. https://doi.org/10.1093/ehjdh/ztab047. PMID: 34604759; PMCID: PMC8482046.

McCurdy, B. R. (2012). Noninvasive positive pressure ventilation for acute respiratory failure patients with chronic obstructive pulmonary disease (COPD): An evidence-based analysis. *Ontario Health Technology Assessment Series, 12*(8), 1–102.

Onofri, A., Pavone, M., De Santis, S., Verrillo, E., Caggiano, S., Ullmann, N., & Cutrera, R. (2021). Telemedicine in children with medical complexity on home ventilation during the COVID-19 pandemic. *Pediatric Pulmonology, 56*(6), 1395–1400. https://doi.org/10.1002/ppul.25289. Epub 2021 Feb 1. PMID: 33524228; PMCID: PMC8013575.

Patil, S. P., Ayappa, I. A., Caples, S. M., Kimoff, R. J., Patel, S. R., & Harrod, C. G. (2019). Treatment of adult obstructive sleep apnea with positive airway pressure: An American Academy of Sleep Medicine systematic review, meta-analysis, and GRADE assessment. *Journal of Clinical Sleep Medicine, 15*(2), 301–334.

Pelletier-Fleury, N., Gagnadoux, F., Philippe, C., Rakotonanahary, D., Lanoé, J., & Fleury, B. (2001). A COST-minimization study of telemedicine: The case of telemonitored polysomnography to diagnose obstructive sleep apnea syndrome. *International Journal of Technology Assessment in Health Care, 17*(4), 604–611. https://doi.org/10.1017/S0266462301107154

Peppard, P. E., Young, T., Barnet, J. H., Palta, M., Hagen, E. W., & Hla, K. M. (2013). Increased prevalence of sleep disordered breathing in adults. *American Journal of Epidemiology, 177*(9), 1006–1014.

Rada, G. (2015). Telemedicine: Are we advancing the science? *Cochrane Database of Systematic Reviews, 9*, ED000105.

Schöbel, C., Werther, S., Teschler, H., & Taube, C. (2020). Telemedicine in respiratory sleep medicine: COVID-19 pandemic unmasks the need for a process-oriented, replicable approach for implementation in clinical routine. *Journal of Thoracic Disease, 12*, S261. https://doi.org/10.21037/jtd-cus-2020-011

Skov Schacksen, C., Dyrvig, A. K., Henneberg, N. C., Dam Gade, J., Spindler, H., Refsgaard, J., Hollingdal, M., Dittman, L., Dremstrup, K., & Dinesen, B. (2021). Patient-reported outcomes from patients with heart failure participating in the future patient telerehabilitation program: Data from the intervention arm of a randomized controlled trial. *JMIR Cardio, 5*(2), e26544. https://doi.org/10.2196/26544. PMID: 34255642; PMCID: PMC8285740.

Smith, S. M., Holland, A. E., & McDonald, C. F. (2019). Beyond forest plots: Clinical gestalt and its influence on COPD telemonitoring studies and outcomes review. *BMJ Open, 9*, e030779. https://doi.org/10.1136/bmjopen-2019-030779

Verbraecken, J. (2016). Telemedicine applications in sleep disordered breathing: Thinking out of the box. *Sleep Medicine Clinics, 11*(4), 445–459. https://doi.org/10.1016/j.jsmc.2016.08.007. Epub 2016 Sep 23. PMID: 28118869.

Walker, P. P., Pompilio, P. P., Zanaboni, P., Bergmo, T. S., Prikk, K., Malinovschi, A., Montserrat, J. M., Middlemass, J., Šonc, S., Munaro, G., Marušič, D., Sepper, R., Rosso, R., Siriwardena, A. N., Janson, C., Farré, R., Calverley, P. M. A., & Dellaca', R. L. (2018). Telemonitoring in chronic obstructive pulmonary disease (CHROMED). A randomized clinical trial. *American Journal of Respiratory and Critical Care Medicine, 198*(5), 620–628. https://doi.org/10.1164/rccm.201712-2404OC. PMID: 29557669.

Woehrle, H., Ficker, J. H., Graml, A., Fietze, I., Young, P., Teschler, H., & Arzt, M. (2017). Telemedicine-based proactive patient management during positive airway pressure therapy: Impact on therapy termination rate. *Somnologie (Berlin), 21*(2), 121–127. https://doi.org/10.1007/s11818-016-0098-9. Epub 2017 Jan 23. PMID: 28706464; PMCID: PMC5486580.

20 Innovations in the Treatment of Pediatric Obstructive Sleep Apnea

Allan Damian and David Gozal

Abstract

Obstructive sleep apnea affects a large proportion of otherwise healthy children in the context of interactions between craniofacial elements, adenotonsillar hypertrophy and other anatomical factors, and neuromuscular reflexes of the upper airway. In light of the adverse consequences of sleep apnea, it is important not only to proceed with early diagnosis but also to implement adequate treatment that is guided by the pathophysiological determinants of the disease in each child. Here, we will describe the current standard of care approaches to the treatment of pediatric obstructive sleep apnea, and will also explore novel management strategies that should enable more personalized therapy in the near future.

A. Damian
Department of Neurology, University of Missouri School of Medicine, Columbia, MO, USA

Comprehensive Sleep Medicine Program, University of Missouri School of Medicine, Columbia, MO, USA

D. Gozal (✉)
Comprehensive Sleep Medicine Program, University of Missouri School of Medicine, Columbia, MO, USA

Department of Child Health and the Child Health Research Institute, University of Missouri School of Medicine, Columbia, MO, USA
e-mail: gozald@health.missouri.edu

Keywords

Adenotonsillar hypertrophy · Obesity · Tonsillectomy · Adenoidectomy · Orthodontics · Intra-oral appliances · Anti-inflammatory · Nasal corticosteroids · Leukotriene receptor modifiers · Myofunctional therapy · DISE · CPAP · Bilevel PAP · High flow nasal cannula · Mandibular advancement · Diet · Weigh management

20.1 Importance of Sleep

The American Academy of Sleep Medicine (AASM) and the National Sleep Foundation set forth recommendations for each age group on the total amount of sleep needed to promote optimal health in children (Paruthi et al., 2016; Hirshkowitz et al., 2015). These technical reports summarizing the extant literature and reflecting the consensus opinion of panels of experts indicated that sleeping less than the number of hours recommended could lead to behavior and academic problems, as well as various health issues, which prompted the endorsement of such recommendations by the American Academy of Pediatrics (AAP) (Hirshkowitz et al., 2015; Jenco, 2016). However, it is not only the duration of sleep that is important for health, but the quality and regularity as well. Indeed, reducing dis-

T. Penzel, R. Hornero (eds.), *Advances in the Diagnosis and Treatment of Sleep Apnea*, Advances in Experimental Medicine and Biology 1384, https://doi.org/10.1007/978-3-031-06413-5_20

ruption of sleep (i.e., sleep fragmentation) (Fatima et al., 2016; Ohayon et al., 2017; Phillips et al., 2020), as well as promoting regular bedtimes and sleep schedules (Spruyt et al., 2011), are critical to derive the best outcome from sleep at all ages, and particularly in growing children. Whereas sleep restriction, insomnia and delayed sleep-wake phase disorder are among the more common causes of short sleep duration in adolescents (Phillips et al., 2020), obstructive sleep apnea (OSA), with its concomitant microarousals is one of the most common causes of sleep fragmentation across the lifespan (Gipson et al., 2019; Zhang et al., 2017). Adult studies have shown that even mild OSA and even snoring with upper airway resistance syndrome can lead to poor sleep quality and can also be associated with comorbidities such as hypertension, obesity, diabetes and mood symptoms (Young et al., 2002; Pinto et al., 2016). Thus, maintaining a regular sleep schedule with the homeostatic need of sleep duration being preserved, and suffering from minimal disruption are all the more of paramount importance in children (Paruthi et al., 2016; Ohayon et al., 2017; Phillips et al., 2020; Gipson et al., 2019).

20.2 Diagnosis of OSA

The gold standard for the diagnosis of OSA remains as an in-laboratory diagnostic polysomnogram (PSG) (Gipson et al., 2019; Marcus et al., 2012; Dehlink & Tan, 2016). This is most effective when clinically correlated with a positive patient history and physical examination (Gipson et al., 2019; Marcus et al., 2012). Alternative approaches to the diagnosis of OSA in children are emerging. For a more detailed coverage of this topic, the reader is referred to a previous chapter by the authors in this book.

20.3 Overview of Treatment

Adenotonsillectomy (AT) remains the most common and first line management procedure for pediatric OSA (Marcus et al., 2012; Kaditis et al., 2016). However, there is currently no consensus algorithm for the management of pediatric OSA, or even a consensus regarding which cut-off values in the polysomnogram designate the need for certain interventions (Kaditis et al., 2012; Dehlink & Tan, 2016). An algorithm attempting to circumvent such issues was first proposed by Kaditis et al in 2011, and is presented in Fig. 20.1 below.

20.3.1 Weight Management for Obesity

Obesity and OSA are tightly interrelated. Obesity is a major risk factor for sleep-disordered breathing in children (Kanney et al., 2020; Redline et al., 1999). There is increasing frequency and severity of OSA when obesity is present, and obese children are much more likely to suffer from persistent OSA after customary treatment, which usually consists of surgical removal of adenoids and tonsils (Amaddeo et al., 2017; Barlow and Expert Committee, 2007). In a previous study, we showed that obese children with OSA ate more fast food, ate less fruits and vegetables, and were significantly less likely to engage frequently in organized sports (Alonso-Álvarez et al., 2015; Spruyt et al., 2010).

The American Academy of Pediatrics (AAP) outlines a four-stage approach to the diagnosis and management of childhood overweight and obesity (Barlow and Expert Committee, 2007). The first three stages comprise of lifestyle and behavior modifications, whereas medications and/ or bariatric surgery are considered during the fourth stage (Barlow and Expert Committee, 2007). Lifestyle changes employed in the treatment of obesity comprise caloric restriction (i.e., hypocaloric diet and other nutritional modifications aimed at minimizing for example the glycemic index) and increasing caloric expenditure (i.e., exercise or non-exercise thermogenesis) (Cuda et al., 2016). Roche et al explored the effect of this first-line intervention for obesity, and showed that a combination of aerobic exercise and a balanced diet led to weight loss, but did not affect the AHI (Roche et al., 2018). They

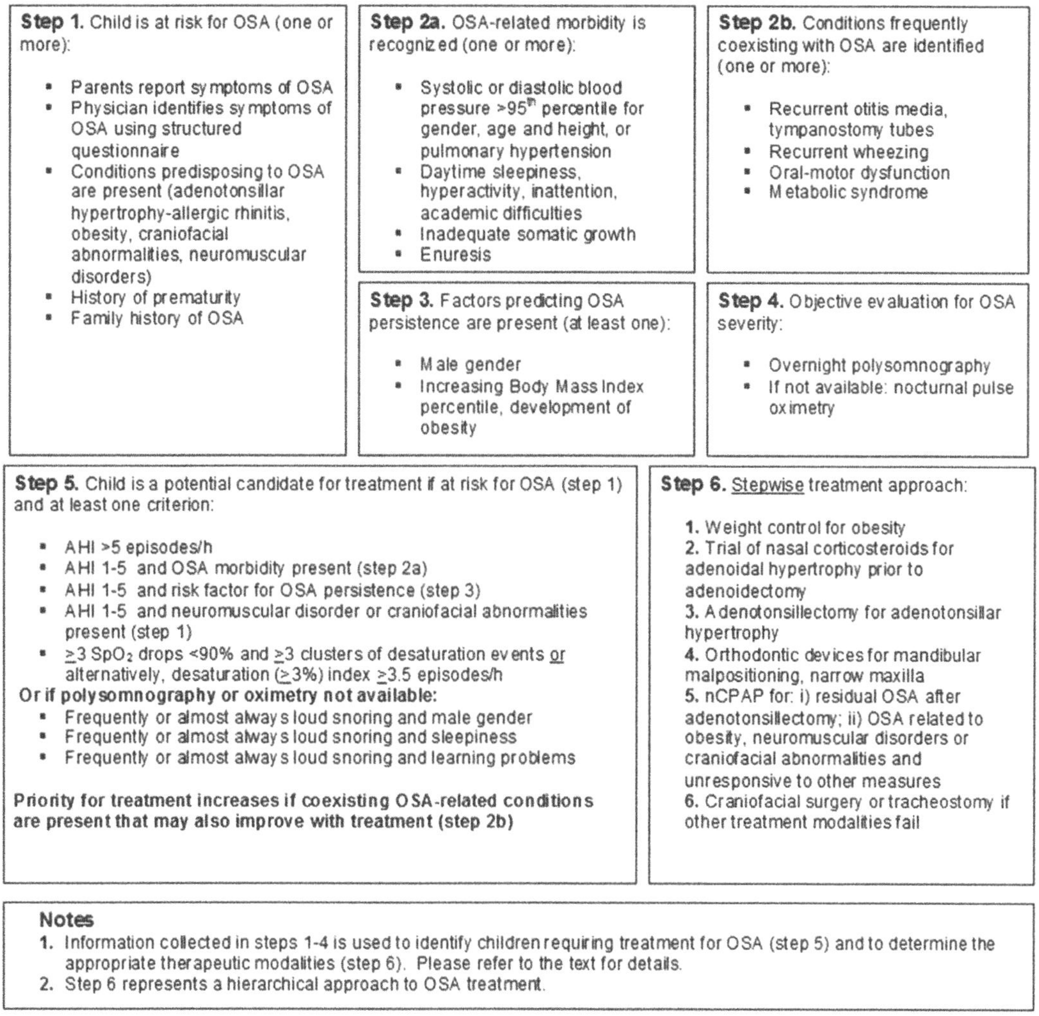

Fig. 20.1 Algorithm for the diagnosis and treatment of pediatric OSA, as proposed by Kaditis et al. (2012)

concluded that their subjects had severe obesity and despite the significant weight loss, remained in the same Body Mass Index (BMI) category, and that greater weight loss was required to see an improvement in AHI (Roche et al., 2018). Conversely, in another study they showed that a comprehensive obesity management program yielded improvements in metabolic dysfunction independently of the changes in AHI (Roche et al., 2020). Verhulst and colleagues found that weight loss was successful in treating OSA in 31 obese teenagers. In addition, a positive association between the severity of OSA at the start of the treatment and the amount of weight loss achieved was detected (Verhulst et al., 2009). Metformin, phentermine, liraglutide, topiramate and orlistat are the medications currently used in pediatrics for weight loss (Cuda et al., 2016; Page & Freemark, 2020; Pharmareviews.in, 2021). Only orlistat is FDA-approved for this indication (Cuda et al., 2016). They have all been studied for weight loss for various pediatric age groups, but to the best of our knowledge none of these medications has specifically been examined as far as its effect on AHI after weight loss. Jaramillo et al conducted a retrospective chart review of patients who underwent laparoscopic sleeve gastrectomy and found that most had resolution of their comorbidities after weight loss from surgery, and this includes 16 of 18 patients not needing their CPAP anymore (Jaramillo et al., 2017). Thus, when OSA is present in the context of obesity, it is important to emphasize that all efforts should be made to improve dietary habits and daily physical activity to not only reduce the likelihood of residual OSA, but also to improve the cardiometabolic dysfunction that is frequently present in these children (Koren et al., 2016a, b).

20.3.2 Anti-Inflammatory Therapy

Systemic and local (topical) anti-inflammatory therapy is now commonly used in the management of pediatric OSA and includes intranasal steroids and montelukast.

It is recognized that in OSA, there is evidence of accelerated proliferation of inflammatory cells in the airways, such as eosinophils, lymphocytes and mast cells (Brockmann & Bertran Salinas, 2021). Corticosteroids will inhibit chemical messengers like cytokines, chemokines, and adhesion molecules ultimately reducing the recruitment of inflammatory cells to tissues such as the adenoids or tonsils (Brockmann & Bertran Salinas, 2021). Corticosteroids will also operate via inhibition of proliferation, hence potentially reducing the hyperplasia and hypertrophy of these tissues that play such a fundamental role in enhancing the risk of OSA (Brockmann & Bertran Salinas, 2021).

Interestingly, only a few randomized controlled studies have been carried out to date. Generally speaking, the extant literature clearly supports the efficacy of the anti-inflammatory approaches in the treatment of OSA in children (Gozal et al., 2021). The initial study by Brouillette et al studied the effects of intranasal fluticasone on the AHI in a triple-blind randomized trial, and found that the AHI in the study group decreased from around 10.7 events per hour to 5.8 events per hour after 6 weeks of use (Brouillette et al., 2001). Since then, most studies have particularly explored the impact of this treatment in mild cases, based on the ethical equipoise that delaying surgical treatment in more severe OSA cases may not be appropriate (Brockmann & Bertran Salinas, 2021). In a recent study, Kajiyama and colleagues confirmed the validity of such approach, and showed that although adenotonsillectomy could be presented in a substantial proportion of mild and moderate OSA cases by medical treatment, severely affected children were significantly less likely to benefit from such approach (Kajiyama et al., 2021). Notwithstanding, there appears to be some advantage in efficacy when combination treatment with intranasal corticosteroids and montelukast is implemented rather than intranasal steroids alone (Liming et al., 2019; Kuhle et al., 2020). Overall, the empirical experience of our group over several years seems to indicate a substantial benefit from anti-inflammatory approaches in mild to moderate OSA, with up to >80% of children not requiring surgical intervention (Kheirandish-Gozal et al., 2014). We should also indicate that the younger ages (<6–7 years) seem to be more responsive to treatment than older children, likely because in older children the lymphadenoid tissues in the upper airway are more likely to be more rigid and less malleable due to increased fibroelastic structural content.

Montelukast is a leukotriene receptor antagonist and acts via cysteinyl leukotriene receptor-1 in tonsils and adenoids (Brockmann & Bertran Salinas, 2021). Several studies have shown that use of montelukast can be used in the short-term treatment of mild to moderate OSA (Brockmann & Bertran Salinas, 2021; Liming et al., 2019; Kuhle et al., 2020; Ji et al., 2021; Kheirandish-Gozal et al., 2016; Bao et al., 2021). Kheirandish-Gozal et al showed in a double-blind randomized trial that use of montelukast for 16 weeks improved AHI from 9.2 events per hour to 4.2 events per hour (Kheirandish-Gozal et al., 2016). In vitro studies have confirmed decreased proliferation in tonsillar and adenoidal tissue after exposure to montelukast (Dayyat et al., 2009). Of note similar results were reported by Goldbart and colleagues in a preceding RCT (Goldbart et al., 2012). Several issues remain unaddressed however and will need to be delineated to allow for a more uniform implementation of these approaches. These issues will need to include whether specific ages are more likely to respond favorably, what anatomical attributes need to be considered (e.g., extent of tonsillar or adenoidal hypertrophy, Mallampati score, degree of retrognathia, etc…), the duration and dosage of treatment, when to use this approach either as single vs. combined therapy, etc…

20.3.3 Orthodontic Management

Certain craniofacial features have been frequently linked to pediatric OSA. Syndromic children usually present with craniofacial malformations, which could include midface hypoplasia, mandibular hypoplasia, macroglossia, or a narrowed oropharynx (Fernandes Fagundes et al., 2021). These children should be screened regularly by both their primary care physicians, and oral health providers (Fernandes Fagundes et al., 2021). Validated questionnaires can be used for screening, albeit being fraught with limited reliability, and clinical imaging and/ or referrals to specialists such as ENT are periodically needed (Fernandes Fagundes et al., 2021). Orthodontic treatment is currently deemed beneficial in the management of OSA in children in the presence of certain craniofacial features or malformations associated with OSA problems (Fernandes Fagundes et al., 2021). However, more expansive training of orthodontic specialists in sleep medicine and development of more standardized interventions along with the appropriate underlying clinical trial evidence may propel orthodontic options to the high priority standing of treatment choices. Presented below are two orthodontic procedures currently employed in the management of some children suffering from OSA.

20.3.3.1 Rapid Maxillary Expansion

Rapid Maxillary Expansion (RME) is used for correction of dental crowding in pre-pubertal children with the presence of maxillary skeletal constriction (Fernandes Fagundes et al., 2021; Júnior et al., 2018). Overall, the purpose of the procedure is to increase the volumetric space within the oral cavity while facilitating improved positioning of the tongue, the latter generally being displaced backwards and reducing further the pharyngeal introitus (Júnior et al., 2018; McNamara Jr et al., 2015). Hence, it may be an adjunct treatment in children who have residual OSA (Fernandes Fagundes et al., 2021). RME is also performed in selected pediatric cases when nasal obstruction symptoms are predominant (Júnior et al., 2018). It then leads to reduced nasal airway resistance and improved respiratory pattern (Júnior et al., 2018; McNamara Jr et al., 2015; White et al., 1989; Kiliç & Oktay, 2008). In recent years the selection of rapid vs. intermediate vs. slow maxillary expansion using the customary expanders has prompted renewed interest in the advantages and disadvantages of each approach (Luiz Ulema Ribeiro et al., 2020; Adobes Martin et al., 2020; Hoxha et al., 2018). Villa et al showed that after 12 months of RME, a 20% reduction in AHI was anticipated, and that RME appears to be a safe and efficacious intervention in selected cases with more complex forms of OSA in which adenotonsillectomy alone is insufficient to normalize the breathing patterns during sleep (Villa et al., 2015).

20.3.3.2 Mandibular Advancement

Mandibular advancement devices (MAD) involve changing the mandible to a more forward position in relation to the maxilla. This increases the sagittal dimension of the oropharyngeal area and reduces collapsibility of the airway there (Fernandes Fagundes et al., 2021). This treatment modality is already regularly employed in the treatment of mild-to-moderate OSA in adults, especially those with a history of non-compliance or non-tolerance of Positive Airway Pressure (PAP) Therapy (Ramar et al., 2015). A systematic review and meta-analysis by Noller et al found a dramatic improvement in AHI (89% decrease) in patients with mandibular insufficiency when treated with either mandibular advancement or mandibular distraction (Noller et al., 2018). Another systematic review and meta-analysis, this time by Schwartz et al, found that CPAP is still more efficient in reducing AHI, but has less compliance (Schwartz et al., 2018). There was no difference in quality of life with MAD (Schwartz et al., 2018).

20.3.4 Surgical Treatment of Pediatric OSA

As mentioned above, adenotonsillectomy (T&A) is considered the first-line treatment for pediatric OSA by the AAP, AASM and the American Academy of Otolaryngology-Head and Neck Surgery (AAO-HNS) (Marcus et al., 2012; Aurora et al., 2011; Mitchell et al., 2019). The gold standard technique involves removal of the tonsils along with the capsule (i.e., tonsillectomy) (Sarber & Ishman, 2021). Another approach, tonsillotomy, refers to the removal of the tonsillar tissues, while leaving the tonsillar capsule in place (Sarber & Ishman, 2021). This has been shown to be associated with less bleeding and pain (Konstantinopoulou et al., 2015; Borgström et al., 2019) but is associated with a higher rate of tonsillar re-proliferation and recurrence of OSA in a substantial proportion of the children (Sarber & Ishman, 2021). T&A is associated with improved AHI, behavior and quality of life (Sarber & Ishman, 2021; Marcus et al., 2013; El-Kersh et al., 2017; Fehrm et al., 2020; Song et al., 2016). Obesity, age > 7 years, black race, OSA severity before surgery, and presence of genetic and metabolic syndromes (especially those associated with craniofacial and neuromuscular disorders) negatively affect the outcomes of T&A and are inherently fraught with a high risk of residual OSA that may require additional interventions (Konstantinopoulou et al., 2015; Sarber & Ishman, 2021; Connolly et al., 2020; Della Vecchia et al., 2020).

20.3.4.1 Drug Induced Sleep Endoscopy (DISE)

DISE is a technique to evaluate the upper airways while the patient is deeply sedated. The sedation state using specific drugs is considered as a viable simulation of natural sleep. DISE is presently used to aid in surgical decision-making for children with persistent OSA, and sometimes used in decision making for those children at risk for persistent OSA (Sarber & Ishman, 2021; Baldassari et al., 2021). Dynamic evaluation of the upper airway during sedation allows for identification of additional sites of obstruction that could be potentially addressed surgically (Sarber & Ishman, 2021; Baldassari et al., 2021). DISE is required for patients being considered for a hypoglossal nerve stimulator to evaluate the degree and pattern of velopharyngeal collapse (Sarber & Ishman, 2021). Of note, nerve stimulation devices are not currently approved by the FDA for use in children, but have gained popularity among adults who are refractory to CPAP.

20.3.4.2 Nasal and Nasopharyngeal Surgery

Nasal surgery has been shown to improve AHI, and reduce CPAP pressures (Hoxha et al., 2018). Septoplasty has historically been avoided in children due to concerns about its effects on nasal growth (Cingi et al., 2016). More recently, limited septoplasty has been shown to be safe especially for children older than 6 years of age (Cingi et al., 2016). Turbinoplasty is a commonly performed procedure to improve nasal breathing in children with turbinate hypertrophy (usually associated with chronic allergic rhinitis) and signs of nasal airway obstruction (Sarber & Ishman, 2021; Wright et al., 2020).

20.3.4.3 Oropharyngeal Surgery

Uvulopalatopharyngoplasty (UPPP) involves removal of excessive tissue of the lower soft palate and uvula (Sarber & Ishman, 2021). Because of complications such as velopharyngeal insufficiency, voice changes, globus, and airway stenosis, the traditional technique has undergone many modifications over the years (Sarber & Ishman, 2021). UPPP has been shown to improve OSA in children who are neurologically impaired (Kosko & Derkay, 1995; Kerschner et al., 2002), and in those with severe obesity (Com et al., 2015).

20.3.4.4 Tongue Surgery

Tongue base reduction (i.e., base-of-tongue surgery, BOT) can be achieved through lingual tonsillectomy (Sarber & Ishman, 2021). Other surgeries in this category include tongue suspension and true reduction (Camacho et al., 2017). All three were evaluated by Camacho and colleagues in a systematic review and meta-analysis. BOT was shown to reduce AHI in both syndromic and non-syndromic children. However, the majority of reports involve very small series or even isolated case reports.

20.3.4.5 Tracheotomy

Tracheotomy allows for complete bypass of the obstructed upper airway structures for the treatment of OSA. It is considered as a salvage treatment option for children with severe OSA after other options have failed (Sarber & Ishman, 2021). Although used in the past, every effort should be made to avoid this intervention.

20.3.5 Positive Airway Pressure (PAP) Therapy

PAP preserves airway patency during sleep by stenting the collapsible segments of the airway. Thus, PAP reduces the inspiratory work of breathing in the setting of increased airway resistance. Though PAP therapy is the first-line treatment for adult OSA (Patil et al., 2019), it is often reserved for cases of persistent OSA (i.e., refractory or residual OSA) after T&A, or in those children with otherwise little or no evidence of adenotonsillar hypertrophy (Dehlink & Tan, 2016; Gozal et al., 2020). Usually, continuous pressures are delivered throughout the complete respiratory cycle (i.e., inspiratory and expiratory phases; CPAP). However, when a very high positive end-expiratory pressure is required, or if the patient has a neuromuscular condition or obesity hypoventilation syndrome, bi-level PAP (BIPAP) may be implemented, whereby the expiratory pressure administered is inferior to the inspiratory pressure and permits more comfortable respiratory efforts while maintaining airway patency throughout the respiratory cycle. Though PAP therapy is very effective, adherence is usually a major problem (Gozal et al., 2020; Hawkins et al., 2016; Bhattacharjee et al., 2020). Behavioral interventions such as desensitization (Gozal et al., 2020) or family member modeling (Puri et al., 2016) have been developed and shown to increase compliance or adherence to treatment.

20.3.6 Myofunctional Approaches

Orofacial myofunctional therapy helps to re-establish correct habits and functioning of the orofacial muscles to avoid residual OSA after surgical and orthodontic treatment (Villa & Evangelisti, 2021). The treatment must be as early as possible for protecting airway health and sleep quality (Villa & Evangelisti, 2021).

Children with OSA traditionally present with alterations in posture and mobility of the orofacial musculature, and oropharyngeal muscle hypotonia is also implicated in the pathogenesis of OSA (Villa & Evangelisti, 2021). Exercises may improve function and reduce such impairments (Gozal et al., 2020). These exercises aim to correct functions such as swallowing, breathing, speech, and chewing. Exercises can be categorized into: (1) nasal breathing rehabilitation; (2) labial seal and lip tone exercises; and (3) tongue posture exercises (Fehrm et al., 2020).

Overall, myofunctional exercises lead to decreased mouth-breathing, re-establish nasal breathing and improve orofacial muscle performance (Villa & Evangelisti, 2021). However, their efficacy has not been critically evaluated even if preliminary results are encouraging in children (Camacho et al., 2015). In addition, adherence to these exercises has been problematic.

In summary, a large array of different approaches has emerged over the last several decades aimed at resolving OSA in the pediatric age group. It is likely that as multi-disciplinary approaches become the standard, rather than the exception, we will witness rapid evolution of integrative personalized therapies that are tailored to the specific patient, finally replacing the one treatment fits all approach which has dominated to the present day.

Acknowledgements This work was partially supported by the Leda Sears Foundation. D.G. is supported by NIH grants AG061824 and HL140548, and by Tier 2 and TRIUMPH grants from the University of Missouri.

References

Adobes Martin, M., Lipani, E., Alvarado Lorenzo, A., Bernes Martinez, L., Aiuto, R., Dioguardi, M., Re, D., Paglia, L., & Garcovich, D. (2020). The effect of maxillary protraction, with or without rapid palatal expansion, on airway dimensions: A systematic review and meta-analysis. *European Journal of Paediatric Dentistry, 21*(4), 262–270. https://doi.org/10.23804/ejpd.2020.21.04.2. PMID: 33337900.

Alonso-Álvarez, M. L., Terán-Santos, J., Navazo-Egüia, A. I., Martinez, M. G., Jurado-Luque, M. J., Corral-Peñafiel, J., Duran-Cantolla, J., Cordero-Guevara, J. A., Kheirandish-Gozal, L., & Gozal, D.; Spanish Sleep Network. (2015). Treatment outcomes of obstructive sleep apnoea in obese community-dwelling children: The NANOS study. *The European Respiratory Journal, 46*(3), 717–727. https://doi.org/10.1183/09031936.00013815. Epub 2015 Jun 11. PMID:26065566.

Amaddeo, A., de Sanctis, L., Olmo Arroyo, J., Giordanella, J. P., Monteyrol, P. J., & Fauroux, B. (2017). Obésité et SAOS de l'enfant [Obesity and obstructive sleep apnea in children]. *Archives de Pédiatrie, 24*(Suppl 1), S34–S38. French. https://doi.org/10.1016/j.arcped.2016.09.003. Epub 2016 Oct 18. PMID: 27769628.

Aurora, R. N., Zak, R. S., Karippot, A., Lamm, C. I., Morgenthaler, T. I., Auerbach, S. H., Bista, S. R., Casey, K. R., Chowdhuri, S., Kristo, D. A., & Ramar, K.; American Academy of Sleep Medicine. (2011). Practice parameters for the respiratory indications for polysomnography in children. *Sleep, 134*(3), 379–388. https://doi.org/10.1093/sleep/34.3.379. PMID: 21359087; PMCID: PMC3041715.

Baldassari, C. M., Lam, D. J., Ishman, S. L., Chernobilsky, B., Friedman, N. R., Giordano, T., Lawlor, C., Mitchell, R. B., Nardone, H., Ruda, J., Zalzal, H., Deneal, A., Dhepyasuwan, N., & Rosenfeld, R. M. (2021). Expert consensus statement: Pediatric drug-induced sleep endoscopy. *Otolaryngology and Head and Neck Surgery, 165*(4), 578–591. https://doi.org/10.1177/0194599820985000. Epub 2021 Jan 5. PMID: 33400611.

Bao, J. L., Xinyuan-Gao, Y. B. H., Zhang, K., & Liu, L. (2021). Efficacy and safety of montelukast for pediatric obstructive sleep apnea syndrome: A protocol for systematic review and meta-analysis. *Medicine (Baltimore), 100*(3), e23958. https://doi.org/10.1097/MD.0000000000023958. PMID: 33545976; PMCID: PMC7837961.

Barlow SE; Expert Committee. (2007). Expert committee recommendations regarding the prevention, assessment, and treatment of child and adolescent overweight and obesity: Summary report. *Pediatrics, 120*(Suppl 4), S164–S192. https://doi.org/10.1542/peds.2007-2329C. PMID: 18055651.

Bhattacharjee, R., Benjafield, A. V., Armitstead, J., Cistulli, P. A., Nunez, C. M., Pepin, J. D., Woehrle, H., Yan, Y., & Malhotra, A.; medXcloud group. (2020). Adherence in children using positive airway pressure therapy: A big-data analysis. *The Lancet Digital Health, 2*(2), e94–e101. https://doi.org/10.1016/S2589-7500(19)30214-6. Epub 2019 Dec 23. Erratum in: *The Lancet Digital Health* 2020, *2*(9), e455. PMID: 33334566; PMCID: PMC7847310.

Borgström, A., Nerfeldt, P., & Friberg, D. (2019). Postoperative pain and bleeding after adenotonsillectomy versus adenotonsillotomy in pediatric obstructive sleep apnea: An RCT. *European Archives of Oto-Rhino-Laryngology, 276*(11), 3231–3238. https://doi.org/10.1007/s00405-019-05571-w. Epub 2019 Aug 3. PMID: 31377901; PMCID: PMC6811389.

Brockmann, P., & Bertran Salinas, K. (2021). Obstructive sleep apnea: Treatment - Anti-inflammatory therapy. In *Pediatric sleep medicine* (pp. 477–481). Springer.

Brouillette, R. T., Manoukian, J. J., Ducharme, F. M., Oudjhane, K., Earle, L. G., Ladan, S., & Morielli, A. (2001). Efficacy of fluticasone nasal spray for pediatric obstructive sleep apnea. *The Journal of Pediatrics, 138*(6), 838–844. https://doi.org/10.1067/mpd.2001.114474. PMID: 11391326.

Camacho, M., Certal, V., Abdullatif, J., Zaghi, S., Ruoff, C. M., Capasso, R., & Kushida, C. A. (2015). Myofunctional therapy to treat obstructive sleep apnea: A systematic review and meta-analysis. *Sleep, 38*(5), 669–675. https://doi.org/10.5665/sleep.4652. PMID: 25348130; PMCID: PMC4402674.

Camacho, M., Noller, M. W., Zaghi, S., Reckley, L. K., Fernandez-Salvador, C., Ho, E., Dunn, B., & Chan, D. (2017). Tongue surgeries for pediatric obstructive sleep apnea: A systematic review and meta-analysis. *European Archives of Oto-Rhino-Laryngology, 274*(8), 2981–2990. https://doi.org/10.1007/s00405-017-4545-4. Epub 2017 Apr 4. PMID: 28378061.

Cingi, C., Muluk, N. B., Ulusoy, S., Lopatin, A., Şahin, E., Passali, D., Bellussi, L., Atilla, H., Hanci, D., Altıntoprak, N., Rusetski, Y., Sarafoleanu, C., Mladina, R., Kalogjera, L., & Manea, C. (2016). Septoplasty in children. *American Journal of Rhinology & Allergy, 30*(2), e42–e47. https://doi.org/10.2500/ajra.2016.30.4289. PMID: 26980385.

Com, G., Carroll, J. L., Tang, X., Melguizo, M. S., Bower, C., & Jambhekar, S. (2015). Characteristics and surgical and clinical outcomes of severely obese children with obstructive sleep apnea. *Journal of Clinical Sleep Medicine, 11*(4), 467–474. https://doi.org/10.5664/jcsm.4608. PMID: 25665695; PMCID: PMC4365461.

Connolly, H. V., Tomaselli, L. T., & McKenna Benoit, M. K. (2020). Adenotonsillectomy for pediatric obstructive sleep apnea: How to predict those at risk for postoperative complications. *Journal of Clinical Sleep Medicine, 16*(1), 3–4. https://doi.org/10.5664/jcsm.8150. Epub 2019 Dec 5. PMID: 31957649; PMCID: PMC7053012.

Cuda, S. E., Censani, M., Joseph, M., Green, R., & Scinta, W. (2016–17). Pediatric obesity algorithm. Available online at www.Pediatricobesityalgorithm.org

Dayyat, E., Serpero, L. D., Kheirandish-Gozal, L., Goldman, J. L., Snow, A., Bhattacharjee, R., & Gozal, D. (2009). Leukotriene pathways and in vitro adenotonsillar cell proliferation in children with obstructive sleep apnea. *Chest, 135*(5), 1142–1149. https://doi.org/10.1378/chest.08-2102. Epub 2008 Dec 31. PMID: 19118273; PMCID: PMC2716539.

Dehlink, E., & Tan, H. L. (2016). Update on paediatric obstructive sleep apnoea. *Journal of Thoracic Disease, 8*(2), 224–235. https://doi.org/10.3978/j.issn.2072-1439.2015.12.04. PMID: 26904263; PMCID: PMC4739955.

Della Vecchia, L., Passali, F. M., & Coden, E. (2020). Complications of adenotonsillectomy in pediatric age. *Acta Biomedica, 91*(1-S), 48–53. https://doi.org/10.23750/abm.v91i1-S.9256. PMID: 32073561; PMCID: PMC7947737.

El-Kersh, K., Cavallazzi, R., & Senthilvel, E. (2017). Outcomes of adenotonsillectomy in severe pediatric obstructive sleep apnea. *Ear, Nose, & Throat Journal, 96*(12), E6–E9. https://doi.org/10.1177/014556131709601202. PMID: 29236274.

Fatima, Y., Doi, S. A., & Mamun, A. A. (2016). Sleep quality and obesity in young subjects: A meta-analysis. *Obesity Reviews, 17*(11), 1154–1166. https://doi.org/10.1111/obr.12444. Epub 2016 Jul 15. PMID: 27417913.

Fehrm, J., Nerfeldt, P., Browaldh, N., & Friberg, D. (2020). Effectiveness of adenotonsillectomy vs watchful waiting in young children with mild to moderate obstructive sleep apnea: A randomized clinical trial. *JAMA Otolaryngology. Head & Neck Surgery, 146*(7), 647–654. https://doi.org/10.1001/jamaoto.2020.0869. Erratum in: *JAMA Otolaryngology. Head & Neck Surgery* 2020, *146*(12), 1181. PMID: 32463430; PMCID: PMC7256861.

Fernandes Fagundes, N. C., Almeida, F., & Flores-Mir, C. (2021). Pediatric obstructive sleep apnea: Orthodontic management. In *Pediatric sleep medicine* (pp. 483–492). Springer.

Gipson, K., Lu, M., & Kinane, T. B. (2019). Sleep-disordered breathing in children. *Pediatrics in Review, 40*(1):3–13. https://doi.org/10.1542/pir.2018-0142. Erratum in: *Pediatrics in Review* 2019, *40*(5), 261. PMID: 30600274; PMCID: PMC6557418.

Goldbart, A. D., Greenberg-Dotan, S., & Tal, A. (2012). Montelukast for children with obstructive sleep apnea: A double-blind, placebo-controlled study. *Pediatrics, 130*(3), e575–e580. https://doi.org/10.1542/peds.2012-0310. Epub 2012 Aug 6. PMID:22869829.

Gozal, D., Tan, H. L., & Kheirandish-Gozal, L. (2020). Treatment of obstructive sleep apnea in children: Handling the unknown with precision. *Journal of Clinical Medicine, 9*(3), 888. https://doi.org/10.3390/jcm9030888. PMID: 32213932; PMCID: PMC7141493.

Gozal, D., Ismail, M., & Brockmann, P. E. (2021). Alternatives to surgery in children with mild OSA. *The World Journal of Otorhinolaryngology – Head & Neck Surgery, 7*(3), 228–235. https://doi.org/10.1016/j.wjorl.2021.03.005. PMID: 34430830; PMCID: PMC8356096.

Hawkins, S. M., Jensen, E. L., Simon, S. L., & Friedman, N. R. (2016). Correlates of pediatric CPAP adherence. *Journal of Clinical Sleep Medicine, 12*(6), 879–884. https://doi.org/10.5664/jcsm.5892. PMID: 27092702; PMCID: PMC4877321.

Hirshkowitz, M., Whiton, K., Albert, S. M., Alessi, C., Bruni, O., DonCarlos, L., Hazen, N., Herman, J., Adams Hillard, P. J., Katz, E. S., Kheirandish-Gozal, L., Neubauer, D. N., O'Donnell, A. E., Ohayon, M., Peever, J., Rawding, R., Sachdeva, R. C., Setters, B., Vitiello, M. V., & Ware, J. C. (2015). National Sleep Foundation's updated sleep duration recommendations: Final report. *Sleep Health, 1*(4), 233–243. https://doi.org/10.1016/j.sleh.2015.10.004. Epub 2015 Oct 31. PMID: 29073398.

Hoxha, S., Kaya-Sezginer, E., Bakar-Ates, F., Köktürk, O., & Toygar-Memikoğlu, U. (2018). Effect of semi-rapid maxillary expansion in children with obstructive sleep apnea syndrome: 5-month follow-up study. *Sleep & Breathing, 22*(4), 1053–1061. https://doi.org/10.1007/s11325-018-1636-4. Epub 2018 Feb 17. PMID: 29453640.

Jaramillo, J. D., Snyder, E., Farrales, S., Stevens, M., Wall, J. K., Chao, S., Morton, J., Pratt, J. S., Hammer, L., Shepard, W. E., & Bruzoni, M. (2017). A multidisciplinary approach to laparoscopic sleeve gastrectomy among multiethnic adolescents in the United States. *Journal of Pediatric Surgery, 52*(10), 1606–1609. https://doi.org/10.1016/j.jpedsurg.2017.06.021. Epub 2017 Jun 28. PMID: 28697852.

Jenco, M. (2016). AAP endorses new recommendations on sleep times [Internet]. *AAP News* [cited 2021 Nov.26]. https://publications.aap.org/aapnews/news/6630?autologincheck=redirected

Ji, T., Lu, T., Qiu, Y., Li, X., Liu, Y., Tai, J., Guo, Y., Zhang, J., Wang, S., Zhao, J., Ni, X., & Xu, Z. (2021). The efficacy and safety of montelukast in children with obstructive sleep apnea: A systematic review and meta-analysis. *Sleep Medicine, 78*, 193–201. https://doi.org/10.1016/j.sleep.2020.11.009. Epub 2020 Nov 10. PMID: 33465554.

Júnior, A. J. M., Crespo, A. N., & Pauna, H. F. (2018). Rapid maxillary expansion in pediatric patients with obstructive sleep apnea: Current and future perspectives. *Sleep Medicine, 51*, 7–8. https://doi.org/10.1016/j.sleep.2018.06.002. Epub 2018 Jun 27. PMID: 30077017.

Kaditis, A., Kheirandish-Gozal, L., & Gozal, D. (2012). Algorithm for the diagnosis and treatment of pediatric OSA: A proposal of two pediatric sleep centers. *Sleep Medicine, 13*(3), 217–227. https://doi.org/10.1016/j.sleep.2011.09.009. Epub 2012 Feb 1. PMID: 22300748.

Kaditis, A. G., Alonso Alvarez, M. L., Boudewyns, A., Alexopoulos, E. I., Ersu, R., Joosten, K., Larramona, H., Miano, S., Narang, I., Trang, H., Tsaoussoglou, M., Vandenbussche, N., Villa, M. P., Van Waardenburg,

D., Weber, S., & Verhulst, S. (2016). Obstructive sleep disordered breathing in 2- to 18-year-old children: Diagnosis and management. *The European Respiratory Journal, 47*(1), 69–94. https://doi.org/10.1183/13993003.00385-2015. Epub 2015 Nov 5. PMID: 26541535.

Kajiyama, T., Komori, M., Hiyama, M., Kobayashi, T., & Hyodo, M. (2021). Changes during medical treatments before adenotonsillectomy in children with obstructive sleep apnea. *Auris Nasus Larynx,* S0385-8146(21)00263-7. https://doi.org/10.1016/j.anl.2021.11.001. Epub ahead of print. PMID: 34840035.

Kanney, M. L., Harford, K. L., Raol, N., & Leu, R. M. (2020). Obstructive sleep apnea in pediatric obesity and the effects of sleeve gastrectomy. *Seminars in Pediatric Surgery, 29*(1), 150887. https://doi.org/10.1016/j.sempedsurg.2020.150887. Epub 2020 Feb 21. PMID: 32238281.

Kerschner, J. E., Lynch, J. B., Kleiner, H., Flanary, V. A., & Rice, T. B. (2002). Uvulopalatopharyngoplasty with tonsillectomy and adenoidectomy as a treatment for obstructive sleep apnea in neurologically impaired children. *International Journal of Pediatric Otorhinolaryngology, 62*(3), 229–235. https://doi.org/10.1016/s0165-5876(01)00623-1. PMID: 11852126.

Kheirandish-Gozal, L., Bhattacharjee, R., Bandla, H. P. R., & Gozal, D. (2014). Antiinflammatory therapy outcomes for mild OSA in children. *Chest, 146*(1), 88–95. https://doi.org/10.1378/chest.13-2288. PMID: 24504096; PMCID: PMC4077412.

Kheirandish-Gozal, L., Bandla, H. P., & Gozal, D. (2016). Montelukast for children with obstructive sleep apnea: Results of a double-blind, randomized, placebo-controlled trial. *Annals of the American Thoracic Society, 13*(10), 1736–1741. https://doi.org/10.1513/AnnalsATS.201606-432OC. PMID: 27439031; PMCID: PMC6944383.

Kiliç, N., & Oktay, H. (2008). Effects of rapid maxillary expansion on nasal breathing and some naso-respiratory and breathing problems in growing children: A literature review. *International Journal of Pediatric Otorhinolaryngology, 72*(11), 1595–1601. https://doi.org/10.1016/j.ijporl.2008.07.014. Epub 2008 Sep 11. PMID: 18789544.

Konstantinopoulou, S., Gallagher, P., Elden, L., Garetz, S. L., Mitchell, R. B., Redline, S., Rosen, C. L., Katz, E. S., Chervin, R. D., Amin, R., Arens, R., Paruthi, S., & Marcus, C. L. (2015). Complications of adenotonsillectomy for obstructive sleep apnea in school-aged children. *International Journal of Pediatric Otorhinolaryngology, 79*(2), 240–245. https://doi.org/10.1016/j.ijporl.2014.12.018. Epub 2014 Dec 22. PMID: 25575425; PMCID: PMC4319650.

Koren, D., Gozal, D., Philby, M. F., Bhattacharjee, R., & Kheirandish-Gozal, L. (2016a). Impact of obstructive sleep apnoea on insulin resistance in nonobese and obese children. *The European Respiratory Journal, 47*(4), 1152–1161. https://doi.org/10.1183/13993003.01430-2015. Epub 2016 Feb 4. PMID: 26846822.

Koren, D., Gozal, D., Bhattacharjee, R., Philby, M. F., & Kheirandish-Gozal, L. (2016b). Impact of adenotonsillectomy on insulin resistance and lipoprotein profile in nonobese and obese children. *Chest, 149*(4), 999–1010. https://doi.org/10.1378/chest.15-1543. Epub 2016 Jan 12. PMID: 26447773; PMCID: PMC4944784.

Kosko, J. R., & Derkay, C. S. (1995). Uvulopalatopharyngoplasty: Treatment of obstructive sleep apnea in neurologically impaired pediatric patients. *International Journal of Pediatric Otorhinolaryngology, 32*(3), 241–246. https://doi.org/10.1016/0165-5876(95)01178-e. PMID: 7665271.

Kuhle, S., Hoffmann, D. U., Mitra, S., & Urschitz, M. S. (2020). Anti-inflammatory medications for obstructive sleep apnoea in children. *Cochrane Database of Systematic Reviews, 1*(1), CD007074. https://doi.org/10.1002/14651858.CD007074.pub3. PMID: 31978261; PMCID: PMC6984442.

Liming, B. J., Ryan, M., Mack, D., Ahmad, I., & Camacho, M. (2019). Montelukast and nasal corticosteroids to treat pediatric obstructive sleep apnea: A systematic review and meta-analysis. *Otolaryngology and Head and Neck Surgery, 160*(4), 594–602. https://doi.org/10.1177/0194599818815683. Epub 2018 Dec 4. PMID: 30513051.

Luiz Ulema Ribeiro, G., Jacob, H. B., Brunetto, M., da Silva Pereira, J., Motohiro Tanaka, O., & Buschang, P. H. (2020). A preliminary 3-D comparison of rapid and slow maxillary expansion in children: A randomized clinical trial. *International Journal of Paediatric Dentistry, 30*(3), 349–359. https://doi.org/10.1111/ipd.12597. Epub 2020 Jan 16. PMID: 31755620.

Marcus, C. L., Brooks, L. J., Davidson Ward, S., Draper, K. A., Gozal, D., Halbower, A. C., Jones, J., Lehmann, C., Schechter, M. S., Sheldon, S., Shiffman, R. N., & Spruyt, K. (2012). Diagnosis and management of childhood obstructive sleep apnea syndrome. *Pediatrics, 130*(3), e714–e755. https://doi.org/10.1542/peds.2012-1672

Marcus, C. L., Moore, R. H., Rosen, C. L., Giordani, B., Garetz, S. L., Taylor, H. G., Mitchell, R. B., Amin, R., Katz, E. S., Arens, R., Paruthi, S., Muzumdar, H., Gozal, D., Thomas, N. H., Ware, J., Beebe, D., Snyder, K., Elden, L., Sprecher, R. C., Willging, P., Jones, D., Bent, J. P., Hoban, T., Chervin, R. D., Ellenberg, S. S., & Redline, S.; Childhood Adenotonsillectomy Trial (CHAT). (2013). A randomized trial of adenotonsillectomy for childhood sleep apnea. *The New England Journal of Medicine, 368*(25), 2366–2376. https://doi.org/10.1056/NEJMoa1215881. Epub 2013 May 21. PMID: 23692173; PMCID: PMC3756808.

McNamara, J. A., Jr., Lione, R., Franchi, L., Angelieri, F., Cevidanes, L. H., Darendeliler, M. A., & Cozza, P. (2015). The role of rapid maxillary expansion in the promotion of oral and general health. *Progress in Orthodontics, 16*, 33. https://doi.org/10.1186/s40510-

015-0105-x. Epub 2015 Oct 7. PMID: 26446931; PMCID: PMC4596248.
Mitchell, R. B., Archer, S. M., Ishman, S. L., Rosenfeld, R. M., Coles, S., Finestone, S. A., Friedman, N. R., Giordano, T., Hildrew, D. M., Kim, T. W., Lloyd, R. M., Parikh, S. R., Shulman, S. T., Walner, D. L., Walsh, S. A., & Nnacheta, L. C. (2019). Clinical practice guideline: Tonsillectomy in children (Update). *Otolaryngology and Head and Neck Surgery, 160*(1_suppl), S1–S42. PMID: 30798778.
Noller, M. W., Guilleminault, C., Gouveia, C. J., Mack, D., Neighbors, C. L., Zaghi, S., & Camacho, M. (2018). Mandibular advancement for pediatric obstructive sleep apnea: A systematic review and meta-analysis. *Journal of Cranio-Maxillo-Facial Surgery, 46*(8), 1296–1302. https://doi.org/10.1016/j.jcms.2018.04.027. Epub 2018 May 4. PMID: 29898824.
Ohayon, M., Wickwire, E. M., Hirshkowitz, M., Albert, S. M., Avidan, A., Daly, F. J., Dauvilliers, Y., Ferri, R., Fung, C., Gozal, D., Hazen, N., Krystal, A., Lichstein, K., Mallampalli, M., Plazzi, G., Rawding, R., Scheer, F. A., Somers, V., & Vitiello, M. V. (2017). National Sleep Foundation's sleep quality recommendations: First report. *Sleep Health, 3*(1), 6–19. https://doi.org/10.1016/j.sleh.2016.11.006. Epub 2016 Dec 23. PMID: 28346153.
Page, L. C., & Freemark, M. (2020). Role of GLP-1 receptor agonists in pediatric obesity: Benefits, risks, and approaches to patient selection. *Current Obesity Reports, 9*(4), 391–401. https://doi.org/10.1007/s13679-020-00409-7. Epub 2020 Oct 21. PMID: 33085056.
Paruthi, S., Brooks, L. J., D'Ambrosio, C., Hall, W. A., Kotagal, S., Lloyd, R. M., Malow, B. A., Maski, K., Nichols, C., Quan, S. F., Rosen, C. L., Troester, M. M., & Wise, M. S. (2016). Consensus statement of the American Academy of Sleep Medicine on the recommended amount of sleep for healthy children: Methodology and discussion. *Journal of Clinical Sleep Medicine, 12*(11), 1549–1561. https://doi.org/10.5664/jcsm.6288. PMID: 27707447; PMCID: PMC5078711.
Patil, S. P., Ayappa, I. A., Caples, S. M., Kimoff, R. J., Patel, S. R., & Harrod, C. G. (2019). Treatment of adult obstructive sleep apnea with positive airway pressure: An American Academy of Sleep Medicine clinical practice guideline. *Journal of Clinical Sleep Medicine, 15*(2), 335–343. https://doi.org/10.5664/jcsm.7640. PMID: 30736887; PMCID: PMC6374094.
Pharmareviews.in. (2021). *FDA approves new treatment for pediatric patients with type 2 diabetes – Pharma Reviews*. [online] Available at http://www.pharmareviews.in/fda-approves-new-treatment-for-pediatric-patients-with-type-2-diabetes/. Accessed 27 Nov 2021.
Phillips, S. R., Johnson, A. H., Shirey, M. R., & Rice, M. (2020). Sleep quality in school-aged children: A concept analysis. *Journal of Pediatric Nursing, 52*, 54–63. https://doi.org/10.1016/j.pedn.2020.02.043. Epub 2020 Mar 14. PMID: 32179378; PMCID: PMC7285623.
Pinto, J. A., Ribeiro, D. K., Cavallini, A. F., Duarte, C., & Freitas, G. S. (2016). Comorbidities associated with obstructive sleep apnea: A retrospective study. *International Archives of Otorhinolaryngology, 20*(2), 145–150. https://doi.org/10.1055/s-0036-1579546. Epub 2016 Mar 10. PMID: 27096019; PMCID: PMC4835326.
Puri, P., Ross, K. R., Mehra, R., Spilsbury, J. C., Li, H., Levers-Landis, C. E., & Rosen, C. L. (2016). Pediatric positive airway pressure adherence in obstructive sleep apnea enhanced by family member positive airway pressure usage. *Journal of Clinical Sleep Medicine, 12*(7), 959–963. https://doi.org/10.5664/jcsm.5924. PMID: 27092699; PMCID: PMC4918996.
Ramar, K., Dort, L. C., Katz, S. G., Lettieri, C. J., Harrod, C. G., Thomas, S. M., & Chervin, R. D. (2015). Clinical practice guideline for the treatment of obstructive sleep apnea and snoring with oral appliance therapy: An update for 2015. *Journal of Clinical Sleep Medicine, 11*(7), 773–827. https://doi.org/10.5664/jcsm.4858. PMID: 26094920; PMCID: PMC4481062.
Redline, S., Tishler, P. V., Schluchter, M., Aylor, J., Clark, K., & Graham, G. (1999). Risk factors for sleep-disordered breathing in children. Associations with obesity, race, and respiratory problems. *American Journal of Respiratory and Critical Care Medicine, 159*(5 Pt 1), 1527–1532. https://doi.org/10.1164/ajrccm.159.5.9809079. PMID: 10228121.
Roche, J., Gillet, V., Perret, F., & Mougin, F. (2018). Obstructive sleep apnea and sleep architecture in adolescents with severe obesity: Effects of a 9-month lifestyle modification program based on regular exercise and a balanced diet. *Journal of Clinical Sleep Medicine, 14*(6), 967–976. https://doi.org/10.5664/jcsm.7162. PMID: 29852904; PMCID: PMC5991961.
Roche, J., Corgosinho, F. C., Isacco, L., Scheuermaier, K., Pereira, B., Gillet, V., Moreira, G. A., Pradella-Hallinan, M., Tufik, S., de Mello, M. T., Mougin, F., Dâmaso, A. R., & Thivel, D. (2020). A multidisciplinary weight loss intervention in obese adolescents with and without sleep-disordered breathing improves cardiometabolic health, whether SDB was normalized or not. *Sleep Medicine, 75*, 225–235. https://doi.org/10.1016/j.sleep.2020.06.030. Epub 2020 Jun 30. PMID: 32861060.
Sarber, K., & Ishman, S. (2021). Surgical treatment of pediatric obstructive sleep apnea. In *Pediatric sleep medicine* (pp. 465–475). Springer.
Schwartz, M., Acosta, L., Hung, Y. L., Padilla, M., & Enciso, R. (2018). Effects of CPAP and mandibular advancement device treatment in obstructive sleep apnea patients: A systematic review and meta-analysis. *Sleep & Breathing, 22*(3), 555–568. https://doi.org/10.1007/s11325-017-1590-6. Epub 2017 Nov 11. PMID: 29129030.

Song, S. A., Tolisano, A. M., Cable, B. B., & Camacho, M. (2016). Neurocognitive outcomes after pediatric adenotonsillectomy for obstructive sleep apnea: A systematic review and meta-analysis. *International Journal of Pediatric Otorhinolaryngology, 83*, 205–210. https://doi.org/10.1016/j.ijporl.2016.02.011. Epub 2016 Feb 20. PMID: 26968078.

Spruyt, K., Sans Capdevila, O., Serpero, L. D., Kheirandish-Gozal, L., & Gozal, D. (2010). Dietary and physical activity patterns in children with obstructive sleep apnea. *The Journal of Pediatrics, 156*(5), 724–730. https://doi.org/10.1016/j.jpeds.2009.11.010. Epub 2010 Feb 6. PMID: 20138306.

Spruyt, K., Molfese, D. L., & Gozal, D. (2011). Sleep duration, sleep regularity, body weight, and metabolic homeostasis in school-aged children. *Pediatrics, 127*(2), e345–e352. https://doi.org/10.1542/peds.2010-0497. Epub 2011 Jan 24. PMID:21262888; PMCID: PMC3025425.

Verhulst, S. L., Franckx, H., Van Gaal, L., De Backer, W., & Desager, K. (2009). The effect of weight loss on sleep-disordered breathing in obese teenagers. *Obesity (Silver Spring), 17*(6), 1178–1183. https://doi.org/10.1038/oby.2008.673. Epub 2009 Mar 5. PMID: 19265797.

Villa, M. P., & Evangelisti, M. (2021). Myofunctional approaches to pediatric sleep medicine. In *Pediatric sleep medicine* (pp. 493–500). Springer.

Villa, M. P., Rizzoli, A., Rabasco, J., Vitelli, O., Pietropaoli, N., Cecili, M., Marino, A., & Malagola, C. (2015). Rapid maxillary expansion outcomes in treatment of obstructive sleep apnea in children. *Sleep Medicine, 16*(6), 709–716. https://doi.org/10.1016/j.sleep.2014.11.019. Epub 2015 Mar 16. PMID: 25934539.

White, B. C., Woodside, D. G., & Cole, P. (1989). The effect of rapid maxillary expansion on nasal airway resistance. *Journal of Otolaryngology [Internet], 18*(4), 137–143.

Wright, L., Grunzweig, K. A., & Totonchi, A. (2020). Nasal obstruction and rhinoplasty: A focused literature review. *Aesthetic Plastic Surgery, 44*(5), 1658–1669. https://doi.org/10.1007/s00266-020-01710-5. Epub 2020 Apr 23. PMID: 32328743.

Young, T., Peppard, P. E., & Gottlieb, D. J. (2002). Epidemiology of obstructive sleep apnea: A population health perspective. *American Journal of Respiratory and Critical Care Medicine, 165*(9), 1217–1239. https://doi.org/10.1164/rccm.2109080. PMID: 11991871.

Zhang, C., Berger, M., Reilly, A., Malhotra, A., & Kales, S. N. (2017). Obstructive sleep apnea in the workplace. In *Principles and practice of sleep medicine* (6th ed., pp. 750–756). Elsevier.

21 Hypoglossal Nerve Stimulation Therapy

Philipp Arens, Toni Hänsel, and Yan Wang

Abstract

Hypoglossal nerve stimulation (HNS) has been shown to be a safe alternative in the treatment of moderate-to-severe obstructive sleep apnea (OSA). A recent meta-analysis of 12 studies by Costantino et al. indicated the surgical success rates at 55–75%, a reduction of the apnea hypopnea index (AHI) of 18 events/h, and a reduction of the Epworth Sleepiness Scale (ESS) of 2.9–5.3. After animal studies in the 1970s, the first trial on humans to decrease upper airway resistance by transcutaneous electrical stimulation of the genioglossus was reported in 1989. A separate stimulation of protruding and retracting muscles was realized in 1995 by fine-wire electrodes that were placed into the tongue transoral. Over the next years, several companies developed implantable devices for hypoglossal stimulation in OSA. Initially, devices were developed that used unilateral stimulation of the hypoglossal nerve. In 2014, a device for unilateral respiratory frequency-controlled hypoglossal stimulation finally received FDA approval after a successful phase III trial. In recent years, a device for bilateral breath rate-independent stimulation of the hypoglossal nerve has been added to these approaches as a new development. Accordingly, hypoglossal nerve stimulation, on the one hand, is now an established tool for patients with OSA when standard treatments are not satisfactory. Beyond that, hypoglossal stimulation is undergoing a continuous and impressive development like hardly any other field of surgical therapy for OSA.

Keywords

Neurostimulation · Hypoglossal nerve stimulation therapy · Obstructive sleep apnea

21.1 Introduction

Standard care for obstructive sleep apnea syndrome is positive airway pressure therapy (PAP therapy). However, a pertinent portion of patients does not benefit of this therapy due to various reasons (Rotenberg et al., 2016). Hypoglossal nerve stimulation (HNS) has been shown to be a safe

P. Arens (✉) · T. Hänsel
Department of Otorhinolaryngology, Charité – Universitätsmedizin Berlin, Freie Universität Berlin, Humboldt-Universität zu Berlin, and Berlin Institute of Health, Berlin, Germany
e-mail: philipp.arens@charite.de; toni.haensel@charite.de

Y. Wang
Charité – Universitätsmedizin Berlin, Freie Universität Berlin, Humboldt-Universität zu Berlin, and Berlin Institute of Health, Interdisciplinary Sleep Medicine Center, Berlin, Germany
e-mail: yan.wang@charite.de

T. Penzel, R. Hornero (eds.), *Advances in the Diagnosis and Treatment of Sleep Apnea*, Advances in Experimental Medicine and Biology 1384, https://doi.org/10.1007/978-3-031-06413-5_21

alternative in the treatment of moderate-to-severe obstructive sleep apnea (OSA) (Costantino et al., 2020). During OSA, the collapse of the upper airway is prevented by a balance between phasic muscle activation and the subatmospheric pharyngeal intraluminal pressure. Especially the genioglossus muscle – innervated by the hypoglossal nerve – is responsible for the oropharyngeal patency (Kuna & Sant'Ambrogio, 1991). After animal studies in the 1970s, the first trial on six humans to decrease upper airway resistance by transcutaneous electrical stimulation of the genioglossus was reported in 1989 (Miki et al., 1989). A separate stimulation of protruding and retracting muscles was realized in 1995 by fine-wire electrodes that were placed into the tongue transorally (Schwartz et al., 1996). Over the next 15 years, several companies developed implantable devices for hypoglossal nerve stimulation in OSA. Initially, devices were developed that used unilateral stimulation of the hypoglossal nerve. In 2014, a device for unilateral respiratory frequency-controlled hypoglossal stimulation finally received FDA approval after a successful phase III trial (Strollo et al., 2014). In recent years, a device for bilateral breath rate-independent stimulation of the hypoglossal nerve has been added to these approaches as a new development (Eastwood et al., 2020). A recent meta-analysis of 12 studies indicated the surgical success rates at 55–75%, a reduction of the apnea hypopnea index (AHI) of 18 events/h, and a reduction of the Epworth Sleepiness Scale (ESS) of 2.9–5.3 (Costantino et al., 2020). Accordingly, hypoglossal nerve stimulation, on the one hand, is now an established tool for OSA patients when standard treatments are not satisfactory. Beyond that, hypoglossal stimulation is undergoing a continuous and impressive development like hardly any other field of surgical therapy for OSA.

21.2 Hypoglossal Nerve Stimulation Techniques

The implantable hypoglossal nerve stimulation devices that have been developed use different technical approaches. From a clinical point of view, a distinction must be made between unilateral and bilateral stimulation as well as stimulation at the proximal hypoglossal nerve and selective stimulation of distal nerve branches. In addition, there are and were devices that stimulate respiratory frequency-controlled and stimulate respiratory frequency-independent (see also Fig. 21.1).

21.2.1 Unilateral Hypoglossal Nerve Stimulation Therapy with Respiratory Sensing

Function of the system (Inspire IV System, Inspire Medical Systems, Inc.) is based on respiratory rate-controlled phasic stimulation of the distal hypoglossal nerve that starts at the end of expiration and lasts the entire phase of inspiration. A respiratory sensing lead in the intercostal space detects mechanical breathing effort to synchronize the stimulation. During surgery, the branching of the distal hypoglossal nerve is being visualized to distinguish between medial and lateral division. Lateral branches innervate musculus hyoglossus and musculus styloglossus – both retracting muscles; therefore, stimulation must be avoided, whereas medial branches activate intrinsic tongue muscles and the musculus genioglossus, which lets the tongue protrude. Due to palatoglossus coupling, oro- and nasopharynx also dilate (Heiser et al., 2017a). The system advanced from the first implant developments in 1996 (Smith et al., 1996) and was able to prove efficacy in the phase III trial for the stimulation therapy for apnea reduction (STAR) at 12 months of treatment (Strollo et al., 2014). Former studies showed that an AHI with >50/h, BMI >32, and a complete concentric collapse at velum level predicted a poor response rate (van de Heyning et al., 2012). New inclusion criteria led to a reduction of AHI by >50% in 124 patients. Since there was no control group, 46 responders to the therapy were later appointed either to continuation of stimulation or to a withdrawal group. Hence, therapy-related results were inferred with an AHI reduction by 68% (p value <0.001) and notable improvements in disease-specific quality of life measures (Strollo et al., 2014). Several publications reported follow-up data for up to

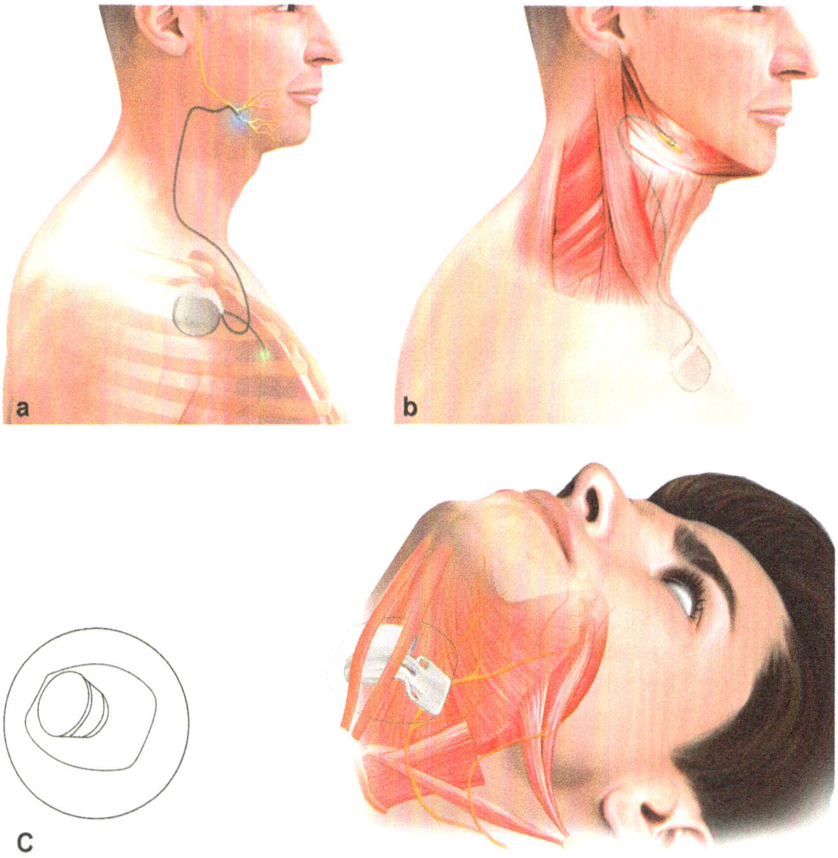

Fig. 21.1 Schematic illustration of different stimulator concepts: (**a**) unilateral distal HNS: breathing cycle controlled (simplified according to Strollo et al. (2015) and Kent et al. (2020)); (**b**) unilateral proximal HNS: non-breathing cycle controlled (simplified according to Friedman et al. (2016)); (**c**) bilateral distal HNS: non-breathing cycle controlled (simplified according to Eastwood et al. (2020))

5 years. Surgical success was defined by a reduction in AHI >50% of baseline (with <20 events/h) and reached by 75% of participants with a response rate to treatment with 63% at 5 years (Strollo et al., 2015; Woodson et al., 2016, 2018). A German post-market study showed an AHI reduction of 28.6 to 8.3 events/h in 60 patients (Heiser et al., 2017b). Until February 2019, the multicenter observational ADHERE Registry included 1017 patients. At the 12-month follow-up, the AHI was reduced from 32.8 to 9.5 in 382 patients. The multivariate analysis identified female sex and lower baseline body mass index as important predictors of therapy response (Boon et al., 2018; Thaler et al., 2019). To achieve blinding integrity, sham stimulation was used in a randomized controlled crossover trial with 89 patients to assess the efficacy of active stimulation and placebo effects. After 1 week of therapy, a difference in AHI of 47.2% (95% CI: 24.4–64.9, $p < 0.001$) and difference in ESS of 4.6 points (95% CI: 3.1–6.1) were reported (Heiser et al., 2021).

Another device with respiratory sensing and a concept of unilateral distal hypoglossal nerve stimulation was developed by Apnex Medical Inc. Despite promising results during the phase II study (Eastwood et al., 2011; Kezirian et al., 2014), the company ceased business activities in 2013 before completion of a phase III study.

21.2.2 Unilateral Hypoglossal Nerve Stimulation Therapy Without Respiratory Sensing

The targeted hypoglossal neurostimulation system (THN) works without respiratory sensing and uses a cyclical unilateral stimulation pattern of the proximal hypoglossal nerve to avoid muscle fatigue. It was developed by ImThera Medical. A small study with 14 participants revealed a reduction in AHI from 45.2 events/h at baseline to 21.0/h at 12 months of follow-up (Mwenge et al., 2013). A subsequent multicenter study similarly underlined safety and effectiveness for 43 participants (Friedman et al., 2016). LivaNova has purchased ImThera with its aura6000™ system in 2018.

21.2.3 Bilateral Hypoglossal Nerve Stimulation Therapy Without Respiratory Sensing

The Nyxoah Genio™ system works with phasic bilateral nerve stimulation without respiratory sensing. The surgical approach is different. Only one incision is needed, while the idea is to stimulate only the most distal part of both hypoglossal nerves to ensure symmetric protrusion of the tongue. Batteries and activation unit are disposable and to be placed under the chin during the night using an adhesive bandage. Published results of the BLAST OSA study showed a reduction in AHI from 23.7/h to 12.9/h in 22 participants at 6 months of follow-up (Eastwood et al., 2020).

21.2.4 Noninvasive Electrical Stimulation

Opposed to implantable devices, there is a notable variety of alternative treatment options with electrical stimulation that lack so far consistent evidence and are difficult to compare. Further research is needed to evaluate this promising field (Rodríguez Hermosa et al., 2021).

21.3 Study Situation on Hypoglossal Nerve Stimulation

21.3.1 Effects Following HNS Therapy

Previous studies demonstrated that hypoglossal nerve stimulation is a safe and effective treatment option for selected adult obstructive sleep apnea patients; long-term follow-up studies have demonstrated a sustained effect on outcomes after 3 and 5 years of HNS therapy (Woodson et al., 2016, 2018). A meta-analysis of 12 studies indicated the surgical success rates at 55–75%, a reduction of the apnea hypopnea index (AHI) of 18 events/h, and a reduction of the Epworth Sleepiness Scale (ESS) of 2.9–5.3 (Costantino et al., 2020). Regarding bilateral distal non-respiratory rate-controlled HNS therapy (BHNS), data are currently available only over a relatively short period of time. The effect was proved to be comparable with HNS and could reduce OSA severity and improve quality of life in the targeted population without device-related complications (Lewis et al., 2019; Eastwood et al., 2020). However, the potential safety and long-term efficacy of the BHNS should be studied further. Table 21.1 provides an overview of the key studies and the longest published follow-up time in each of them, subdivided according to stimulation techniques.

For children and adolescents with Down syndrome (DS) and OSA, short-term results of an early efficacy and safety trial were available, primarily reflecting improvements in AHI, OSA-18 QOL survey scores with good tolerance of HNS (Yu et al., 2021). Adherence to HNS seems better for older adults, but one must expect higher rates of insomnia symptoms, physical dysfunction, cognitive deficits, and depressive symptoms in this population (Dzierzewski et al., 2021). In 2021, Seda et al. found that OSA treatment, especially CPAP, appears to mitigate and slow the rate of cognitive decline and may reduce the risk of dementia (Seda et al., 1910). The potential effect of HNS on improving cognitive function remains unclear but possible, and further research is

Table 21.1 Key studies: longest published follow-up, divided by stimulation technique

	Unilateral stimulation				Bilateral stimulation		Unilateral stimulation					
	Breathing cycle independent stimulation						Breathing cycle controlled stimulation					
	Proximal HNS				Distal HNS							
Study	Mwenge et al. (2013) ($n = 13$)		Friedman et al. (2016) ($n = 43$)		Eastwood et al. (2020) ($n = 27$)		Woodson et al. (2018) ($n = 126$)		Steffen et al. (2018b) ($n = 60$)		Thaler et al. (2019) ($n = 382$)	
	Baseline	12 m	Baseline	6 m	Baseline	6 m	Baseline	5y	Baseline	12 m	Baseline	12 m
AHI	45.2	21	34.9	25.4	23.7	12.9	29.3	6.2	28.6	9.5	32.8	9.5
ODI	29.2	15.3	32.4	23.6	19.1	9.8	25.4	4.6	27	9.8	–	–

required. In addition, the successful application of HNS therapy for the treatment of specific OSA populations with unusual diseases have been reported, including a patient with resolution of restless legs syndrome (Myc et al., 2018), veterans with comorbid insomnia and post-traumatic stress disorder (PTSD) (Patil et al., 2020), and a patient with prior radiation for oropharynx cancer complicated by osteoradionecrosis of the mandible (Zheng et al., 2017). Lewis et al. presented the first case of BHNS therapy successfully controlling OSA symptoms in a patient with complete concentric collapse (CCC) at DISE in 2021 (Lewis et al., 2021). Regarding a study conducted by Bowen et al. (2018), voice and swallowing function were assessed with the use of Voice Handicap Index-10 (VHI-10) and Eating Assessment Tool-10 (EAT-10) questionnaires in a small group of 14 OSA patients following HNS. No sustained changes over 5 months were reported (Bowen et al., 2018).

To sum up, large multicenter prospective clinical trials have demonstrated stable and long-term efficacy of HNS, showing improvement in AHI and quality-of-life measures. Specifically, HNS could significantly decrease AHI and daytime sleepiness (measured with ESS) and improve oxygen desaturation index (ODI) and sleep-related quality of life (e.g., FOSQ (Functional Outcomes of Sleep Questionnaire), SAQLI (Sleep Apnea Quality of Life), and PSQI (Pittsburgh Sleep Quality Index)). Because the short-term benefits of HNS therapy for specific populations are primarily derived from case reports or studies with small samples, these results still need to be confirmed in long-term studies.

21.3.2 Sleep Architecture Changes

The 3-year STAR trial data published by Woodson et al. demonstrated variable improvement in each stage of sleep at 12-, 18-, and 36-month intervals (Woodson et al., 2016). In 2017, Hofauer et al. had similar results, but more detailed conclusions as they saw a reduction of the time spend in N1-sleep stage, the number of arousals, the arousal index and the rebound of REM sleep. In contrast, time in bed (TIB), total sleep time (TST), N2- and N3-sleep period did not change during the observation period (Hofauer et al., 2017). Equally, Bohorquez et al. demonstrated a significant N2 and N3 time improvement among patients who responded successfully to HNS therapy; in contrast, the N1 sleep percentage, arousal index, and wake after sleep onset (WASO) decreased, no significant changes were observed in REM sleep and sleep latency (Bohorquez et al., 2019). This publication indicated that the possible explanation for the difference is that nonresponders were excluded from this analysis. In contrast, the STAR study and Hofauer et al. included nonresponders.

In the future, studies to better understand the longitudinal effects of HNS intervention on sleep architecture are necessary. These results may help demonstrate additional benefits offered by HNS therapy, determine factors preoperatively to predict the success of HNS, and explain HNS-related side effects.

21.3.3 HNS with Down Syndrome

Compared with the prevalence of OSA in the general population, it is highest in pediatric patients with Down syndrome, where the prevalence is 55% to 80%. Anatomical factors, including generalized hypotonia, macroglossia, midface hypoplasia, small tracheal caliber, and lingual tonsil hypertrophy, may contribute to this high morbidity rate in the OSA population. Lacking effective treatment will lead to cardiopulmonary complications, adverse behavior, and reduced quality of life. Furthermore, OSA in children with down syndrome is associated with lower mean verbal IQ scores and lower cognitive flexibility (Simpson et al., 2018). As a result, effective treatment and intervention at an early age are necessary.

However, a lot of OSA patients with DS fail to tolerate CPAP as the standard therapy. To solve the problem, an experimental demonstration of the effect of HNS in this population was first carried out by Diercks et al. A 14-year-old boy with

DS and a longstanding tracheotomy due to severe OSA received HNS therapy, resulting in significant improvement in the patient's AHI, and 5 months after implantation, the patient's tracheotomy was successfully closed (Diercks et al., 2017). In 2019, Caloway et al. performed implantation of the Inspire HNS for 20 nonobese children and adolescents (aged 10–21 years) with DS and severe OSA. A significant reduction in AHI and improvement of OSA-related QOL(OSA-18 score) indicate that HNS is a safe and effective therapy for children with DS and severe refractory OSA (Caloway et al., 2019).

Currently, there is an ongoing clinical trial (NCT02344108, "A Pilot Study to Evaluate the Hypoglossal Nerve Stimulator in Adolescents with Down Syndrome and Obstructive Sleep Apnea") that is conducted by Stenerson et al. to assess the long-term safety and efficacy of HNS among 42 children and young adults with DS and persistent OSA (Stenerson et al., 2021). They expect to prove HNS continues to control OSA in children with DS as they mature effectively because the residual OSA often persists into adulthood.

In 2019, the effect of HNS for adult patients with DS to treat severe OSA was tested and reported (Li et al., 2018). This indicate HNS is very promising for both children and adult OSA patient with down syndrome, in whom PAP adherence is extremely challenging.

21.3.4 HNS and Cardiovascular Disease

Previous studies demonstrate a strong correlation between OSA and deteriorating cardiovascular health. Moderate-to-severe OSA is a risk factor for cardiovascular diseases (CVD), such as coronary artery disease, congestive heart failure, atrial fibrillation, and hypertension. The mechanistic pathway of OSA on poor cardiovascular outcomes is driven by an increased sympathetic tone and oxidative stress due to repetitive airway obstruction and cyclical hypoxia (Mashaqi et al., 2021). Rokkas et al. reported HNS impact on a patient with OSA and heart failure in 2021, indicating that HNS therapy may become an essential adjunct in the comprehensive multidisciplinary treatment of heart failure in patients with OSA (Rokkas et al., 2021). However, the literature data evaluating the cardiovascular outcomes following HNS use are scarce. Further clinical studies to investigate the impact of UAS on HRV, 24-h ambulatory blood pressure, sympathetic activity, and vascular function are required, which could help the better understanding of related pathophysiology and heart disease prognosis following HNS.

21.3.5 HNS and Heart Rate Variability

In 1996, Tsuji et al. studied the association of reduced heart rate variability (HRV) with risk for new cardiac events in a large community-based population and found that a decreased HRV has been associated with an increased risk for incident coronary heart disease, cardiovascular mortality, and all-cause mortality across populations (Tsuji et al., 1996). HRV represents a dynamic measure of autonomic dysfunction, modulated by the interaction at the sinoatrial node between the sympathetic and parasympathetic nervous systems neurotransmitters. Dedhia et al. described the changes of HRV following HNS from the STAR trial and demonstrated that successful HNS therapy significantly improves HRV during sleep after 1 year. The analysis included a standard deviation of the R–R interval (SDNN), the standard measure to domain HRV, low-frequency power of the R–R interval, and high-frequency power of the R–R. In contrast, no significant changes in SDNN were seen during sleep observed in the therapy withdrawal group (Dedhia et al., 2019).

21.3.6 HNS and Hypertension

Several longitudinal studies have shown that OSA is associated with prevalent and incident hypertension (O'connor et al., 2009; Mokhlesi et al., 2014; Marin et al., 2012). Woodson et al.

published a study in 2014 in which they evaluated changes in systolic and diastolic blood pressure readings at 12- and 18-month post-implant, and a significant decrease compared with blood pressure at baseline was reported (Woodson et al., 2014). However, evidence and studies focusing on the relationship between HNS therapy and hypertension are limited. Since 2018, Dedhia et al. are conducting a randomized crossover trial using sham and active HNS therapy to examine the effects of HNS on measures of the sympathetic nervous system (including 24-h ambulatory blood pressure, muscle sympathetic nerve activity, and prerejection period (PEP)) and vascular health (i.e., flow-mediated dilation, pulse wave velocity) (Dedhia et al., 2018). The findings of this ongoing trial of cardiovascular endpoints for OSA patients are anticipated.

21.3.7 HNS and Electrical Cardioversion

Cardiac arrhythmias are common in patients with OSA and require electrical cardioversion. Four OSA patients with cardiac arrhythmias received undergoing external electrical cardioversion. Vasconcellos et al. observed a change in device functionality or complete cessation of functionality after electro-cardioversion in 2019 (Vasconcellos et al., 2019). Similarly, in 2021, Yacono et al. reported a case of HNS-associated neurapraxia after electrical cardioversion of atrial fibrillation, indicating the importance of being aware of an implanted device (Yacono & Hyman, 2021).

21.3.8 HNS with Cardiac Implantable Electronic Device

In 2016, Ong et al. were the first to report a 62-year-old man simultaneously implanted with a cardioverter-defibrillator (ICD) and HNS. No untoward device interference between the two implantable devices was reported (Ong et al., 2016). Following this study, Parikh et al. retrospectively analyzed 14 ad hoc patients with CPAP intolerance, moderate-to-severe OSA, and preexisting transvenous cardiac implantable electronic device (CIED) undergoing HNS implantation that were followed up for a year. During the follow-up, bipolar and unipolar HNS stimulation did not impact CIED sensing, with no oversensing episodes noted on the CIEDs (Parikh et al., 2018). These findings indicate that simultaneous HNS with transvenous CIEDs seemed safe and effective without any device interactions.

21.4 Patient Selection

21.4.1 Baseline Clinical Characteristics

Certain variables such as age and gender seem to have associations with better response rates to HNS. Patel et al. found a positive association to younger age but not to gender (Patel et al., 2020b). Thaler et al. found a positive association to female gender in the ADHERE registry cohort (Thaler et al., 2019). Previous studies showed good response rates at BMI $\leq$ 32 kg/m^2. More recent data from study registries and multicenter studies show good efficacy even at higher values up to 35 kg/m^2 (Steffen et al., 2018b; Heiser et al., 2019). Concomitant diseases that influence therapy should be considered when deciding on therapy. For example, insomniac disorder is usually a contraindication to stimulation therapy. Neuromuscular diseases are usually contraindicated. In addition, not every stimulator model has unrestricted MRI capability. All these aspects should be clarified with the patient in advance by the physician in charge.

21.4.2 Drug-Induced Sleep Endoscopy

Drug-induced sleep endoscopy (DISE) simulates the pharyngeal space during sleep. Common drugs are dexmedetomidine, midazolam, or propofol. DISE is required to determine the location and pattern of obstruction in the upper airway. In general, the examination is not well standardized. A variety of classification systems exist (Dijemeni et al., 2017). In addition, the findings obtained are dependent on the technique of the examination, in especially the depth of sedation achieved (Hong et al., 2013). Under controlled conditions, however, a good test-retest reliability could be determined (Kim & Heo, 2020).

Vanderveken et al. demonstrated that complete concentric collapse (CCC) at the velopharynx led to an increased nonresponder rate in patients undergoing unilateral distal hypoglossal nerve stimulator implantation in a small number of patients (Vanderveken et al., 2013). To ensure a therapeutic effect, CCC of the velum is to date a contraindication for the available unilateral respiratory cycle-controlled hypoglossal nerve stimulation device (Strollo et al., 2014). Even though CCC is a relatively frequent DISE finding in patients with higher BMI and higher AHI, the exclusion of all overweight candidates without performing a sleep endoscopy is not justifiable (Steffen et al., 2015). Also, the primary role in procedure selection was based on an early, small study of this technology even though there were potential confounders, which indicates a possibility that some patients with CCC will respond well. For instance, Lewis et al. reported successful BHNS therapy in a patient with CCC (Lewis et al., 2021). Currently, this circumstance is the subject of a clinical trial (BETTER SLEEP study; NCT03763682).

Refer to a large multicenter cohort study to determine the association between findings of preoperative DISE according to VOTE classification system (velum, oropharynx, tongue base, epiglottis) and outcomes of HNS; the finding of primary tongue contribution to airway obstruction was associated with better response to HNS. In contrast, oropharyngeal lateral wall collapse and epiglottis-related obstruction were associated with the lowest response rates (Huyett et al., 2021). This study suggests that the role of DISE in counseling candidates for HNS extends beyond solely excluding complete concentric collapse related to the velum.

In conclusion, we support the predictive value of DISE to assess findings associated with better or worse outcomes and recommend DISE as a patient selection tool for HNS therapy to treat OSA.

21.4.3 Sleep Lab Testing

Patients with OSA are treated by a sleep medicine specialist. Usually at least once a year, an at-home polygraphy (PG) or in-lab polysomnography (PSG) is scheduled to monitor and adjust the treatment. If the standard therapy with PAP fails and the patient is not PAP compliant or PAP adherent, patients are counseled for hypoglossal nerve stimulation and if needed referred to a head and neck surgeon. Prior to implantation, diagnostic baseline polysomnography should be performed to verify that the patient meets the indication criteria for hypoglossal stimulation.

21.4.3.1 PSG

In general, when determining the indication for HNS therapy, it is important to consider within which AHI range the system is approved. Patel et al. demonstrated that also patients with an AHI >65/h responded to HNS therapy but suffered from residual mild-to-moderate OSAS more frequently than the comparison group (Patel et al., 2020b). The proportion of central events should not exceed 25% to qualify for hypoglossal nerve stimulation (Mashaqi et al., 2021).

Away from the general indication criteria, there is some evidence for further polysomnographic predictors. Especially with regard to the concept of OSA phenotyping, there seem to be exciting opportunities for further development in patient selection.

21.4.3.2 OSA Phenotyping

In addition to anatomical causes such as a narrow or collaptic pharynx, it is now understood that nonanatomical phenotypic mechanisms such as high loop gain, a low respiratory arousal threshold, and impairment in pharyngeal dilator muscle control and function during sleep additional are crucial determinants in the development of OSA. The PALM (Pcrit, arousal threshold, loop gain, and muscle responsiveness) scale was proposed to categorize OSA patients according to the degree of upper airway anatomy impairment and nonanatomical phenotypes (Eckert et al., 2013). Eckert et al. pointed out that each phenotype could be a potential therapeutic target for OSA patients and could be used to apply or develop targeted non-CPAP therapies (Eckert, 2018).

An analysis of the STAR study cohort data by de Beeck et al. showed no differences between responders and nonresponders to HNS therapy at the basal polysomnographic measures except for the arousal index. The more profound analysis of the polysomnographic data revealed that mechanistic PALM factors related to nonanatomical deficits influenced the likelihood of treatment success. A higher arousal threshold was a factor in better treatment efficacy. Higher loop gain was associated with lower HNS efficacy. Interestingly, they also showed that patients with higher muscle compensation had better HNS therapy outcomes, which contradicts the assumption that especially patients with a low muscle responsiveness benefit from stimulation therapy (de Beeck et al., 2021). Contrary to the assumption, they found that patients with low pharyngeal collapsibility experienced poorer treatment efficacy. However, the authors point out that the cohort included too few patients with severe collapsibility to draw further conclusions. Lee et al. demonstrated that patients requiring lower PAP pressures (<8 cm H_2O) to treat their OSA had a better outcome with HNS therapy (Lee et al., 2019).

Thus, patient selection according to pathophysiological OSA characteristics based on the PALM concept seems to potentially generate added value, but at this point can neither replace nor question the current selection parameters. However, these findings highlight the potential to use routinely collected sleep study data and clinical data with machine learning-based approaches underpinned by OSA endotype concepts for further research.

21.4.4 Clinical Anatomical/Radiographic Predictors

To date, there are limited data on anatomical and radiologic factors that influence HNS therapy. Lateral cephalometry/lateral neck X-rays, computed tomography (CT), and ultrasound have been used to assess the upper airway in patients with HNS therapy (Goding et al., 2012; Hofauer et al., 2016; Schwab et al., 2018; Korotun et al., 2020; Arens et al., 2021; Lee et al., 2021). Most of these studies examine action by stimulation of the nerve.

Evidence for predictors of success or patient selection was sought by Schwab et al. in a small cohort using CT. Awake CT was used to identify anatomical differences between responders and nonresponders of HNS. It was found that therapy responders had a significant smaller soft palate volume (Schwab et al., 2018). Similar findings were found by Lee et al. studying lateral cephalograms of patients with an implanted HNS (Lee et al., 2021).

Further research and clinical studies seem to be necessary to find objective methods for the anatomical or dynamic assessment of patient morphology in relation to planned or completed implantation of an HNS system. Both, the increase of MRI capability of the current HNS systems and the already started implementation of ultrasound examination techniques will be helpful.

21.5 Surgical Procedure

The surgical implantation procedures have been modified over the course of several studies, and the devices mentioned differ in their implantation procedure, simply because of the different technical and medical approaches. All systems are implanted under general anesthesia. The systems have in common that they initially are inactive after the surgery and are activated after completing the healing process.

21.5.1 Unilateral Hypoglossal Nerve Stimulation Therapy with Respiratory Sensing

The Inspire IV system (Inspire Medical Systems, Inc.) was initially implanted using three incisions. Currently, the procedure is performed with two incisions. The basic steps of implantation are as follows: The first incision is made submandibular. The hypoglossal nerve is located, and its course is followed under the mylohyoid muscle until it is revealed in its terminal branches.

Now, using neurostimulation and EMG monitoring, the medial-distal branches of the hypoglossal nerve are identified. The cuff of the stimulating electrode is placed around the branches (see Fig. 21.2). If possible, the accompanying C1 nerve is included. The stimulator cable is also fixed to the digastricus muscle at a defined point, forming a loop. A second incision is made at the chest. The sensor electrode is inserted and fixed between the Mm. intercostales externi et interni in the second intercostal space. Tunneling is performed between the cervical access and the thoracic access, and the connector of the stimulating electrode is pulled into the thoracic pocket. The sensor and stimulation cable are connected to the pulse generator (IPG). The IPG is fixed on the fascia of the pectoralis major muscle. The system is tested and initially inactivated. Postoperatively, the integrity of the system is verified by chest X-ray (Heiser et al., 2016b; Kent et al., 2020).

21.5.2 Unilateral Hypoglossal Nerve Stimulation Therapy Without Respiratory Sensing

The aura6000™ system (LivaNova PLC, London, UK; formerly ImThera Medical San Diego, CA, USA) applies neurostimulation independently of the respiratory cycle. It is implanted completely and consists of an IPG in the subcutaneous pocket in the upper chest and subcutaneously connected to a six-electrode lead. The cuff is placed around the proximal hypoglossal nerve, close to the middle tendon of the digastric muscle using a submandibular approach (Mwenge et al., 2013).

21.5.3 Bilateral Hypoglossal Nerve Stimulation Therapy Without Respiratory Sensing

Nyxoah provides the Genio system, a partially implantable device. The implantable part is applied through a single submental incision on both Mm. genioglossi in direct proximity to the distal terminal branches of the hypoglossal nerve. Also with this system, the correct position in the area of the correct nerve branches is verified by intraoperative neurostimulation. The external non-implantable part of the system (pulse generator, energy source) is applied submental by means of an adhesive patch at night after successful initial activation and transfers the energy transdermal to the implanted part of the system according to the programming (Lewis et al., 2019; Eastwood et al., 2020).

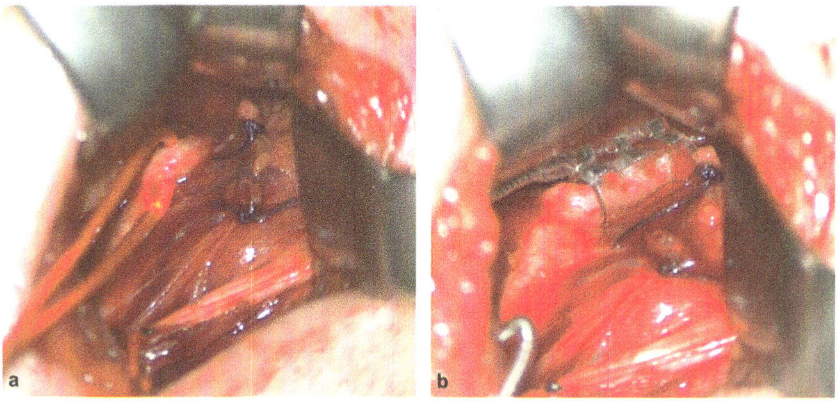

Fig. 21.2 (**a**) Surgical microscopic image of the medial-distal fibers of the right hypoglossal nerve separated from the others with a loop. (**b**) Surgical microscopic image of the cuff of the stimulating electrode, successfully placed around the medial-distal fibers of the right hypoglossal nerve

21.5.4 HNS Therapy Versus Traditional Upper Airway Surgery

In general, it can be said that hypoglossal stimulation has been the most significant innovation in the field of surgical therapies for OSA in the past decade and main innovations are taking place here. In the field of conventional surgical therapies, however, there has also been a significant development. In particular, transoral robotic surgery (TORS) and new pharyngoplasty techniques like expansion sphincter pharyngoplasty and barbed reposition pharyngoplasty represent significant advances (Pang & Woodson, 2007; Vicini & Montevecchi, 2019; Iannella et al., 2022). A recent systematic review by Neruntarat et al. compared HNS with other upper airway procedures (UAP). When comparing HNS therapy with other UAP in patients with moderate-to-severe OSA, they found objective improvement in outcomes of AHI, oxygen saturation nadir, success rate, cure rate, and comparable subjective outcomes of ESS (Neruntarat et al., 2022).

Positive data exist on the combination of HNS therapy and tonsillectomy-uvulopalatopharyngoplasty (TE-UPPP). Steffen et al. demonstrated that additional surgery at the soft palate can improve the outcome in patients with insufficient HNS therapy (Steffen et al., 2019a).

In the future, the inclusion of the concept of OSA phenotyping will play a role in upper airway surgery (de Vito et al., 2021) including also HNS therapy and is possibly leading to more individualized patient-specific targeted therapy.

21.6 Postoperative Management and Care

In order to improve the subjective and objective outcomes following HNS implantation, an effective and long-term postimplantation management and care procedure (therapy titration, adjustments, adherence monitoring, and troubleshooting) need to be addressed and established.

Generally, the HNS device will be activated after 1–2-month postimplantation, and sensory and motor thresholds and limits are determined while awake. Between follow-up visits, the stimulation strength can be adjusted via the remote controller/charger (RCC) for comfort and effectiveness by patients themselves. In-laboratory polysomnography (PSG) evaluation during follow-up guides to confirm or adjust the stimulation parameters. Stimulation parameters and adherence data could be downloaded during RCC communication with the patient's implanted unit and obtained by the surgeons and sleep technologists.

21.6.1 Device Titration and Optimal Stimulus Parameters

As OSA is a long-term chronic disease, the stimulation parameters, including voltage amplitude, impulse voltage, patient control range, and threshold, can be titrated in the clinical or sleep laboratory setting to optimize effectiveness and comfort. Previous studies have pointed that voltage, changed electrode configuration, tongue motion patterns, and respiratory sensing quality affect HNS therapy outcomes (Safiruddin et al., 2015; Steffen et al., 2019b; Meleca & Kominsky, 2020). However, these findings were based on standalone results for either awake endoscope or DISE.

A study conducted by Pawlak et al. compared the effects of various combinations of voltage and electric field both by DISE and awake endoscopy, further proving that electric configuration changes can improve patient airway patency (Pawlak et al., 2021). Meanwhile, Steffen et al. carried out a cohort analysis to reduce the voltage amplitude in three electrode configurations by changing the pulse width and frequency combination. It indicates stimulation parameters of HNS could lower amplitude while still maintaining functional tongue protrusion and a similar patient control range, eventually contributing to the improvement of patient adherence and effect following HNS (Steffen et al., 2021).

Additionally, the understanding of the anatomy and neurophysiology of the tongue and floor of the mouth has improved significantly. Previous data have demonstrated a correlation between the phenotype of tongue motion and therapy response. Bilateral protrusion (BP) or right protrusion (RP) of the tongue phenotype is related to better therapy response, while the response of other tongue motion patterns such as left protrusion (LP) and mixed activation (MA) is lower (Heiser et al., 2016a, 2017a). In 2017, Steffen et al. evaluated the changes in tongue motion during different electrode configuration settings and concluded tongue motion patterns and their shifting may make a difference in therapy outcomes. Patients with shifting tongue movement in response to the change of electrode configuration should attract attention. These findings also suggested reassessing intraoperative cuff placement when tongue movement shifting is observed (Steffen et al., 2018a).

21.6.2 Patients' Adherence and Experience

In contrast with other surgical treatments of OSA, which lead to the permanent modification of the upper airway, the patient's adherence is essential to guarantee the long-term effect following HNS. Refer to the reported data from previous studies; since 2014, the mean usage time has ranged from 4 h to 7.5 h per night, with patient compliance ranging from 29% to 83%.

In 2021, Coca et al. published a paper comparing nonresponder and responder groups following HNS. The responder group adhered significantly better to the recommended duration of therapy (>4 h/night) and had a higher number of hours of nightly use, which indicated the value of usage time and adherence (Coca et al., 2021). Most recently, Hofauer et al. evaluated the therapy adherence and experience of 102 patients from two German implantation centers following HNS implantation, an objective therapy usage of 5.7-h and subjective reports of 6.8-h nights per week with no decrease observed during the follow-up. Accordingly, they concluded that constant use and nightly activation are critical in achieving the long-term success of HNS (Hofauer et al., 2018). A possible explanation for the high adherence and positive attitude toward HNS might be related to outpatient visits, home sleep studies, and telephone-linked communication interventions.

Many pre-surgery factors contribute to adherence to HNS. For example, as discussed above, older patients revealed a better adherence to the stimulation therapy. In addition, Rosenthal et al. underlined an association between insomnia, anxiety, depression, emotional distress, and hypoglossal nerve stimulator adherence, which indicates that screening patients with such circumstances, the GAD-7 and PHQ-9 before implantation may be helpful when evaluating patient adherence to therapy (Rosenthal et al., 2021).

In summary, improving the patient's adherence and experience is essential for long-term effect following HNS, which could be achieved by preimplantation assessment and postoperative monitoring.

21.6.3 Monitoring Methods

Soose et al. analyzed the postimplant care pathway from 5 years of clinical implementation of HNS and addressed that targeted postimplantation care such as patient education, close clinical monitoring, and targeted therapy are essential to successful long-term management (Soose et al., 2021). Nonetheless, due to various reasons, including limited sleep, laboratory's capacity, long waiting list, the schedule of patients, and the impact of the novel coronavirus, follow-up overnight sleep study at sleep labor faces enormous challenges during the monitoring process.

To solve the problem, daytime polysomnography and home sleep testing (HST) have been suggested and recommended alternatively for monitoring and overnight titrations due to their simple accessibility and low cost. In 2021, Bosschieter et al. performed a prospective single-center observational cohort study to evaluate the effect of daytime titration for OSA patients. This study first reported that daytime titrations are a valuable alternative for conventional overnight titrations and suggested implementing daytime titrations as the standard of care (Bosschieter et al., 2021).

At the same time, home sleep testing (HST) has been proved helpful as a second-line titration control method. Huyett et al. was the first to use practical home sleep studies combined with pulse oximetry data and demonstrated a successful application to titrate hypoglossal nerve stimulation therapy (Huyett & Stagnone, 2020). Furthermore, Steffen et al. evaluated the clinical and economic value of HST to monitor and direct therapy optimization of HST after device activation; this study also pointed that effective therapy adjustment using HST depends on good communication between the otorhinolaryngology center performing the implant, the local sleep lab, and the patient (Steffen et al., 2020).

21.7 Complications and Adverse Events

21.7.1 Treatment-Emergent Central Sleep Apnea (TECSA)

Treatment-emergent central sleep apnea (TECSA) is a well-known phenomenon occurring in obstructive sleep apnea (OSA) patients and has been officially recognized in the third edition of the international classification of sleep disorders (ICSD-3) in 2014. The definition of TECSA is as follows: patients with OSA (with a baseline CAI <5 events/h) who demonstrated a central and mixed apnea index (CMAI) of 5 events/h and/or demonstrated Cheyne-Stokes respiration (CSR) becoming prominent or disruptive on PAP treatment, measured during the therapeutic device titration 6–8 weeks after device activation (Berger et al., 2021). But as many central sleep apneas observed with non-PAP therapy with them increasingly used in clinical practice, such as HNS and mandibular advancement device (MAD), the definition of TECSA should refer to the phenomenon of transient and/or persistent CSA after not only PAP treatment but also all kinds of therapy for OSA. Since 2018, there have been three patients reported who developed TECSA following HNS (Chan et al., 2018; Sarber et al., 2019; Hong et al., 2021). Furthermore, a prospective cohort study with 141 patients reported the prevalence of TECSA as 3.3% (n = 5 patients). This study also revealed that demographics, comorbid conditions, and device settings were not associated with an elevated postoperative CMAI. The only factor associated with CMAI ≥5 events/h was an elevated postoperative AHI. Meanwhile, they describe a spontaneous resolution over time of early TECSA (Patel et al., 2020a). However, due to insufficient data from the case reports and small case series, the true prevalence and clinical relevance of TECSA are challenging to assess. The natural course and the underlying pathophysiological mechanisms remain controversial and limited. The possible pathophysiological mechanisms focused on either an unresolved obstruction that may have led to microarousals and overshoot of $PaCO_2$ reduction below the apneic threshold during sleep or a demasking of an OSA comorbidity related elevated loop gain. Nevertheless, this phenomenon seems to spontaneously resolve over time and with continued use or adjustment of stimulation parameters. According to the experience and findings of previous studies, many factors might contribute to the development of CSA following PAP therapy, including stimulation of pulmonary-irritant receptors by pulmonary congestion, increased chemoreceptor sensitivity, reduced cerebrovascular blood flow, and excessive sympathetic nervous system activity (Berger et al., 2021). These aberrations can be seen in congestive heart failure, atrial fibrillation, stroke, and renal failure patients. Meanwhile, the body position and sleep architecture changes were demonstrated as related to the emergency of CSA. Consequently, the possible influence of these factors should be evaluated further to detect the predictors of HNS-related TECSA. In the future, further studies should focus on risk assessment, early detection of TECSA (including comorbidities conditions, stimulation parameters, and other factors), and clinical management, most importantly, patients' specific underlying pathophysiology, which could help avoid and resolve TECSA following HNS.

21.7.2 Cheyne-Stokes Breathing

In 2019, Sarber et al. reported the emergence of Cheyne-Stokes breathing after HNS implantation in a patient with mixed sleep apnea. They presented data from a 60-year-old man with moderate OSA (AHI 22.6 event/h, obstructive AHI (oAHI) of 22.3, and central apnea index (CAI) of 0.3 events/h) and medical history of stage 3 chronic kidney disease, hypertension, hyperlipidemia, type 2 diabetes, bladder and kidney cancer, and depression. Initially, CPAP and auto PAP could not control his central or obstructive events with an AHI of 31.0 events/h (oAHI of 5.5 and CAI of 25.5 events/h). Three months after device activation, an AHI of 83.8 (oAHI of 4.9 and CAI of 78.9 events/h) and Cheyne-Stokes breathing were observed throughout the study, even without HNS activation. Nevertheless, this patient's obstructive events and subjective sleepiness were treated effectively despite the new onset of CSA with Cheyne-Stokes breathing (Sarber et al., 2019). This case report shows that the possible factors contributing to this phenomenon are unclear. Consequently, they will continue close monitoring and consider repeated titration.

21.7.3 Adverse Events

Implantation of HNS has been demonstrated as a safe and effective treatment. Nonetheless, several technical difficulties and complications still exist. Generally, the adverse events can be divided into surgery-related and device-related. Occurring surgery-related complications include infection, hematoma, localized pain, and brief and mild/transient tongue paresis. Frequent device-related adverse events include discomfort due to electrical stimulation, tongue abrasion, mouth dryness, functionality issues with the implanted device, and mechanical pain associated with the device (Bestourous et al., 2020; Bellamkonda et al., 2021). A systematic review of long-term sequelae after HNS reported that tongue discomfort due to repetitive stimulation was the most common. It can be resolved by modification of the stimulation parameters and sometimes also by dental adjustment (Costantino et al., 2020). Three case reports observed some unusual complications, including pleural effusion, iatrogenic pneumothorax (PTX) during the placement of the chest sensor lead (Arteaga et al., 2018), and sensor leads penetrating the pleural space followed by a successful reimplantation (Lou et al., 2021).

21.7.4 Revision Surgery

Hypoglossal nerve stimulators have been implanted increasingly worldwide over the past two decades; however, as most of the devices were implanted only within the last 7 years, long-term data about adverse events is limited; the safety assessment of the revision surgery, including explantation or reimplantation, needs to be considered. A physician must be aware that a patient with an implanted stimulator may need to be revised or explanted in the future for various reasons. For the stimulator models powered by an implanted battery, an IPG change must be performed at least at the end of the battery life. There are reviews of adverse events in hypoglossal stimulator implantation that refer to the Manufacturer and User Facility Device Experience (MAUDE) database from the Food and Drug Administration (FDA) (Bestourous et al., 2020; Bellamkonda et al., 2021). Among the events listed there, 42.3% required surgical revision (Bestourous et al., 2020). However, scientifically published data and field reports with technical assistance for revision or explantation are rare. Arens et al. reported a series of nine explantations with and without single-staged reimplantation in 2020. All foreign material was enclosed in a transparent, solid scar tissue up to 1 mm thick. To free the impulse generator, the leads, and the cuff electrode without trauma, this scar tissue needed to be sharply severed, partly under microscopic view. Due to extensive scar tissue formation, all procedures were technically challenging. The explantation was successful in every case. The complication rate was significantly higher when complete reimplantation was performed or attempted in the same session than explantation alone (Arens et al., 2020).

21.8 Current Developments and Outlook for the Future

Hypoglossal nerve stimulation therapy is currently undergoing rapid forward development. The number of publications in this field has increased exponentially in the last 5–7 years. Existing devices have been further developed, on the one hand, for example, in terms of MRI capability, and, on the other hand, in terms of the surgical procedure. Recently, a two-incision technique has been established for the respiratory frequency-controlled unilateral stimulation system (FA Inspire) (Kent et al., 2020). In addition, a new stimulation technique is available with bilateral medial stimulation of the hypoglossal nerve (Eastwood et al., 2020). The device has been approved as a medical device in Europe and is currently being further investigated in multicenter studies. Besides the improvement of existing stimulation techniques, refining patient selection and optimizing postoperative stimulation parameters, the approach of finding new or additional targets for neurostimulation seems interesting. Kent et al., for example, showed that isolated ansa cervicalis stimulation could increase the maximum inspiratory airflow in the context of a sleep video endoscopy (Kent et al., 2021). In the future, it will be exciting to see to what extent the different approaches – unilateral versus bilateral stimulation, distal versus proximal nerve stimulation, and breathing cycle-controlled versus phasic or breathing cycle-independent stimulation – will have an impact on the results and patient selection.

Additionally, the understanding of the anatomy and neurophysiology of the tongue and floor of the mouth has improved significantly. Based on the work of Mu and Sanders, current research focuses on understanding contralateral tongue muscle activation during unilateral stimulation of the hypoglossal nerve (Mu & Sanders, 2010; Sanders & Mu, 2013). Sturm et al. demonstrated that unilateral stimulation resulted in bilateral activation of the genioglossal muscle in 39% of patients (Sturm et al., 2020). Heiser et al. demonstrated in a small group of patients that bilateral activation of the genioglossal nerve resulted in improved velopalatal opening of the airway. In contrast to the first mentioned work, this study showed that patients with bilateral tongue protusion showed a higher AHI reduction under stimulation therapy (Heiser et al., 2020). Another approach to better understand the function and physiology of the tongue under stimulation is sonography. Some studies have focused on functional imaging under hypoglossal stimulation, while others have attempted to demonstrate effects on morphology (Arens et al., 2019; Hofauer et al., 2019; Korotun et al., 2020). One approach that could help to better understand function and neurophysiology is ultrasound shear wave elastography. Arens et al. have shown that stimulation of the hypoglossal nerve results in a measurable change in shear wave velocity in the tongue muscles as a measure of muscle stiffness (Arens et al., 2021). This provides a quantifiable measure of the effect of stimulation directly on the target muscle, which can be used noninvasively on the implanted awake patient.

Impressively, the introduction of hypoglossal stimulators has led to a major boost in understanding the neuroanatomy of the tongue. It remains an exciting question whether this will influence patient selection and hypoglossal stimulation system selection in the future.

21.9 Conclusion

Over the past decade, more and more hypoglossal nerve stimulators have been implanted worldwide. Initially in the context of studies but then also in standard care. The available data indicate the efficacy and good tolerability of the treatment without long-term complications. To achieve the best subjective and objective results after HNS implantation, a precise procedure for selecting patients before implantation and effective management after implantation are required. This is the only way to ensure patient adherence and determine optimal stimulus parameters. On the other hand, the long-term safety of HNS needs to be further investigated. The physician performing HNS implantation must be aware of the obligations that implantation entails and the potential consequences that must be considered. It remains exciting to see the further development of various stimulation approaches as well as the advancement of existing techniques through an increasingly better understanding of the neurophysiology of the tongue.

References

Arens, P., Uecker, F. C., Fietze, I., Penzel, T., Olze, H., Fischer, T., & Dommerich, S. (2019). Evaluation of the base of tongue in obstructive sleep apnea during electric stimulation therapy of the hypoglossal nerve: Current sonographic methods and initial results. *Somnologie, 23*, 93–97. https://doi.org/10.1007/s11818-019-0193-9

Arens, P., Penzel, T., Fietze, I., Blau, A., Weller, B., Olze, H., & Dommerich, S. (2020). Safety and effectiveness in explantation and re-implantation of hypoglossal nerve stimulation devices. *European Archives of Oto-Rhino-Laryngology, 278*, 477–483. https://doi.org/10.1007/s00405-020-06170-w

Arens, P., Fischer, T., Dommerich, S., Olze, H., & Lerchbaumer, M. H. (2021). Ultrasound shear wave elastography of the tongue during selective hypoglossal nerve stimulation in patients with obstructive sleep apnea syndrome. *Ultrasound in Medicine & Biology, 47*, 2869–2879. https://doi.org/10.1016/j.ultrasmedbio.2021.06.009

Arteaga, A., Pitts, K., & Lewis, A. (2018). Iatrogenic pneumothorax during hypoglossal nerve stimulator implantation. *American Journal of Otolaryngology, 39*, 636–638. https://doi.org/10.1016/J.AMJOTO.2018.06.014

Bellamkonda, N., Shiba, T., & Mendelsohn, A. H. (2021). Adverse events in hypoglossal nerve stimulator implantation: 5-year analysis of the FDA MAUDE database. *Otolaryngology-Head and Neck Surgery, 164*, 443–447. https://doi.org/10.1177/0194599820960069

Berger, M., Solelhac, G., Horvath, C., Heinzer, R., & Brill, A. (2021). Treatment-emergent central sleep apnea associated with non-positive airway pressure therapies in obstructive sleep apnea patients: A systematic review. *Sleep Medicine Reviews, 58*, 101513. https://doi.org/10.1016/J.SMRV.2021.101513

Bestourous, D. E., Pasick, L. J., Benito, D. A., & Zapanta, P. E. (2020). Adverse events associated with the inspire implantable hypoglossal nerve stimulator: A MAUDE database review. *American Journal of Otolaryngology, 41*, 102616. https://doi.org/10.1016/J.AMJOTO.2020.102616

Bohorquez, D., Mahmoud, A. F., Yu, J. L., & Thaler, E. R. (2019). Upper airway stimulation therapy and sleep architecture in patients with obstructive sleep apnea. *The Laryngoscope, 130*, 1085–1089. https://doi.org/10.1002/lary.28057

Boon, M., Huntley, C., Steffen, A., Maurer, J. T., Sommer, J. U., Schwab, R., Thaler, E., Soose, R., Chou, C., Strollo, P., Kezirian, E. J., Chia, S., Withrow, K., Weidenbecher, M., Strohl, K., Doghramji, K., Hofauer, B., Heiser, C., & Investigators on B of the AR. (2018). Upper airway stimulation for obstructive sleep apnea: Results from the ADHERE registry. *Otolaryngology–Head and Neck Surgery, 159*, 379–385. https://doi.org/10.1177/0194599818764896

Bosschieter, P. F. N., Schoustra, E., de Vries, N., Steinbusch, M. J., Kasius, K. M., & Ravesloot, M. J. (2021). Daytime polysomnography to perform titration for upper airway stimulation in patients with obstructive sleep apnea. *Sleep and Breathing, 26*, 707–715. https://doi.org/10.1007/s11325-021-02441-w

Bowen, A. J., Nowacki, A. S., Kominsky, A. H., Trask, D. K., Benninger, M. S., & Bryson, P. C. (2018). Voice and swallowing outcomes following hypoglossal nerve stimulation for obstructive sleep apnea. *American Journal of Otolaryngology, 39*, 122–126. https://doi.org/10.1016/J.AMJOTO.2017.12.011

Caloway, C. L., Diercks, G. R., Keamy, D., de Guzman, V., Soose, R., Raol, N., Shott, S. R., Ishman, S. L., & Hartnick, C. J. (2019). Update on hypoglossal nerve stimulation in children with down syndrome and obstructive sleep apnea. *The Laryngoscope, 130*, E263–E267. https://doi.org/10.1002/lary.28138

Chan, L. C., Chang, M., Aysola, R., Ryden, A., Mendelsohn, A., & Zeidler, M. (2018). 1129 central sleep apnea after hypoglossal nerve stimulation. *Sleep, 41*, A418–A418. https://doi.org/10.1093/SLEEP/ZSY063.1128

Coca, K., Heiser, C., Huntley, C., Boon, M., de Vries, N., Mamidala, M., & Gillespie, M. (2021). Hypoglossal nerve stimulation usage by therapy nonresponders. *Otolaryngology--Head and Neck Surgery*. https://doi.org/10.1177/01945998211036867

Costantino, A., Rinaldi, V., Moffa, A., Luccarelli, V., Bressi, F., Cassano, M., Casale, M., & Baptista, P. (2020). Hypoglossal nerve stimulation long-term clinical outcomes: A systematic review and meta-analysis. *Sleep and Breathing, 24*, 399–411. https://doi.org/10.1007/s11325-019-01923-2

de Beeck, S. O., Wellman, A., Dieltjens, M., Strohl, K. P., Willemen, M., van de Heyning, P. H., Verbraecken, J. A., Vanderveken, O. M., & Sands, S. A. (2021). Endotypic mechanisms of successful hypoglossal nerve stimulation for obstructive sleep apnea. *American Journal of Respiratory and Critical Care Medicine, 203*, 746–755. https://doi.org/10.1164/rccm.202006-2176OC

de Vito, A., Woodson, B. T., Koka, V., Cammaroto, G., Iannella, G., Bosi, M., Pelucchi, S., Filograna-Pignatelli, G. R., el Chater, P., & Vicini, C. (2021). OSA upper airways surgery: A targeted approach. *Medicina, 57*, 690. https://doi.org/10.3390/MEDICINA57070690

Dedhia, R. C., Quyyumi, A. A., Park, J., Shah, A. J., Strollo, P. J., & Bliwise, D. L. (2018). Cardiovascular endpoints for obstructive sleep apnea with twelfth cranial nerve stimulation (CARDIOSA-12): Rationale and methods. *The Laryngoscope, 128*, 2635–2643. https://doi.org/10.1002/lary.27284

Dedhia, R. C., Shah, A. J., Bliwise, D. L., Quyyumi, A. A., Strollo, P. J., Li, Q., da Poian, G., & Clifford, G. D. (2019). Hypoglossal nerve stimulation and heart rate variability: Analysis of STAR trial responders. *Otolaryngology-Head and Neck Surgery, 160*, 165–171. https://doi.org/10.1177/0194599818800284

Diercks, G. R., Wentland, C., Keamy, D., Kinane, T. B., Skotko, B., de Guzman, V., Grealish, E., Dobrowski, J., Soose, R., & Hartnick, C. J. (2017). Hypoglossal nerve stimulation in adolescents with down syndrome and obstructive sleep apnea. *JAMA Otolaryngology–Head & Neck Surgery, 144*, 37–42. https://doi.org/10.1001/jamaoto.2017.1871

Dijemeni, E., D'Amone, G., & Gbati, I. (2017). Drug-induced sedation endoscopy (DISE) classification systems: A systematic review and meta-analysis. *Sleep & Breathing = Schlaf & Atmung, 21*, 983. https://doi.org/10.1007/S11325-017-1521-6

Dzierzewski, J. M., Soto, P., Vahidi, N., & Nord, R. (2021). Clinical characteristics of older adults seeking hypoglossal nerve stimulation for the treatment of obstructive sleep apnea. *Ear Nose & Throat Journal*. https://doi.org/10.1177/01455613211042126

Eastwood, P. R., Barnes, M., Walsh, J. H., Maddison, K. J., Hee, G., Schwartz, A. R., Smith, P. L., Malhotra, A., McEvoy, R. D., Wheatley, J. R., O'Donoghue, F. J., Rochford, P. D., Churchward, T., Campbell, M. C., Palme, C. E., Robinson, S., Goding, G. S., Eckert, D. J., Jordan, A. S., … Hillman, D. R. (2011). Treating obstructive sleep apnea with hypoglossal nerve stimulation. *Sleep, 34*, 1479–1486. https://doi.org/10.5665/sleep.1380

Eastwood, P. R., Barnes, M., MacKay, S. G., Wheatley, J. R., Hillman, D. R., Nguyên, X.-L., Lewis, R., Campbell, M. C., Pételle, B., Walsh, J. H., Jones, A. C., Palme, C. E., Bizon, A., Meslier, N., Bertolus, C., Maddison, K. J., Laccourreye, L., Raux, G., Denoncin, K., … Launois, S. H. (2020). Bilateral hypoglossal nerve stimulation for treatment of adult obstructive sleep apnea. *The European Respiratory Journal*. https://doi.org/10.1183/13993003.01320-2019

Eckert, D. J. (2018). Phenotypic approaches to obstructive sleep apnoea – New pathways for targeted therapy. *Sleep Medicine Reviews, 37*, 45–59. https://doi.org/10.1016/J.SMRV.2016.12.003

Eckert, D. J., White, D. P., Jordan, A. S., Malhotra, A., & Wellman, A. (2013). Defining phenotypic causes of obstructive sleep apnea. Identification of novel therapeutic targets. *American Journal of Respiratory and Critical Care Medicine, 188*, 996–1004. https://doi.org/10.1164/RCCM.201303-0448OC

Friedman, M., Jacobowitz, O., Hwang, M. S., Bergler, W., Fietze, I., Rombaux, P., Mwenge, G. B., Yalamanchali, S., Campana, J., & Maurer, J. T. (2016). Targeted hypoglossal nerve stimulation for the treatment of obstructive sleep apnea: Six-month results. *The Laryngoscope, 126*, 2618–2623. https://doi.org/10.1002/lary.25909

Goding, G. S., Tesfayesus, W., & Kezirian, E. J. (2012). Hypoglossal nerve stimulation and airway changes under fluoroscopy. *Otolaryngology--Head and Neck Surgery, 146*, 1017–1022. https://doi.org/10.1177/0194599812436472

Heiser, C., Maurer, J. T., & Steffen, A. (2016a). Functional outcome of tongue motions with selective hypoglossal nerve stimulation in patients with obstructive sleep apnea. *Sleep and Breathing, 20*, 553–560. https://doi.org/10.1007/s11325-015-1237-4

Heiser, C., Thaler, E., Boon, M., Soose, R. J., & Woodson, B. T. (2016b). Updates of operative techniques for upper airway stimulation. *The Laryngoscope, 126*, S12–S16. https://doi.org/10.1002/lary.26158

Heiser, C., Edenharter, G., Bas, M., Wirth, M., & Hofauer, B. (2017a). Palatoglossus coupling in selective upper airway stimulation. *The Laryngoscope, 127*, E378–E383. https://doi.org/10.1002/lary.26487

Heiser, C., Maurer, J. T., Hofauer, B., Sommer, J. U., Seitz, A., & Steffen, A. (2017b). Outcomes of upper airway stimulation for obstructive sleep apnea in a multicenter German postmarket study. *Otolaryngology-Head and Neck Surgery, 156*, 378–384. https://doi.org/10.1177/0194599816683378

Heiser, C., Steffen, A., Boon, M., Hofauer, B., Doghramji, K., Maurer, J. T., Sommer, J. U., Soose, R., Strollo, P. J., Schwab, R., Thaler, E., Withrow, K., Kominsky, A., Larsen, C., Kezirian, E. J., Hsia, J., Chia, S., Harwick, J., Strohl, K., … ADHERE registry investigators R. (2019). Post-approval upper airway stimulation predictors of treatment effectiveness in the ADHERE registry. *The European Respiratory Journal, 53*, 1801405. https://doi.org/10.1183/13993003.01405-2018

Heiser, C., Vanderveken, O. M., Edenharter, G. M., & Hofauer, B. (2020). Cross motor innervation of the hypoglossal nerve—A pilot study of predictors for successful opening of the soft palate. *Sleep and Breathing*. https://doi.org/10.1007/s11325-020-02112-2

Heiser, C., Steffen, A., Hofauer, B., Mehra, R., Strollo, P. J., Vanderveken, O. M., Maurer, J. T., & Barnes, D. (2021). Clinical medicine effect of upper airway stimulation in patients with obstructive sleep Apnea (EFFECT): A randomized controlled crossover trial. *Journal of Clinical Medicine, 10*, 2880. https://doi.org/10.3390/jcm10132880

Hofauer, B., Strohl, K., Knopf, A., Bas, M., Wirth, M., Stock, K., & Heiser, C. (2016). Sonographic evaluation of tongue motions during upper airway stimulation for obstructive sleep apnea – A pilot study. *Sleep and Breathing*. https://doi.org/10.1007/s11325-016-1383-3

Hofauer, B., Philip, P., Wirth, M., Knopf, A., & Heiser, C. (2017). Effects of upper-airway stimulation on sleep architecture in patients with obstructive sleep apnea. *Sleep and Breathing, 21*, 901–908. https://doi.org/10.1007/s11325-017-1519-0

Hofauer, B., Steffen, A., Knopf, A., Hasselbacher, K., & Heiser, C. (2018). Patient experience with upper airway stimulation in the treatment of obstructive sleep apnea. *Sleep and Breathing*. https://doi.org/10.1007/s11325-018-1689-4

Hofauer, B., Knopf, A., Karl, J., & Heiser, C. (2019). Sonographic evaluation of anatomic landmarks in patients with obstructive sleep apnea. *HNO, 67*, 685–689. https://doi.org/10.1007/s00106-019-0677-8

Hong, S. D., Dhong, H. J., Kim, H. Y., Sohn, J. H., Jung, Y. G., Chung, S. K., Park, J. Y., & Kim, J. K. (2013). Change of obstruction level during drug-induced sleep endoscopy according to sedation depth in obstructive sleep apnea. *The Laryngoscope, 123*, 2896–2899. https://doi.org/10.1002/LARY.24045

Hong, H., Oster, J., Grover, A., & Ismail, K. (2021). A case of Cheyne-Stokes breathing emerging in a patient with atrial fibrillation and an implanted hypoglossal nerve stimulator. *Journal of Clinical Sleep Medicine, 17*, 1731–1735. https://doi.org/10.5664/JCSM.9314

Huyett, P., & Stagnone, R. (2020). Use of overnight pulse oximetry and a type 3 sleep study to titrate hypoglossal nerve stimulation therapy. *Journal of Clinical Sleep Medicine, 16*, 2109–2111. https://doi.org/10.5664/JCSM.8754

Huyett, P., Kent, D. T., D'Agostino, M. A., Green, K. K., Soose, R. J., Kaffenberger, T. M., Woodson, B. T., Huntley, C., Boon, M. S., Heiser, C., Birk, A., Suurna, M. V., Lin, H. S., Waxman, J. A., & Kezirian, E. J. (2021). Drug-induced sleep endoscopy and hypoglossal nerve stimulation outcomes: A Multicenter Cohort Study. *The Laryngoscope, 131*, 1676–1682. https://doi.org/10.1002/LARY.29396

Iannella, G., Lechien, J. R., Perrone, T., Meccariello, G., Cammaroto, G., Cannavicci, A., Burgio, L., Maniaci, A., Cocuzza, S., di Luca, M., Stilo, G., de Vito, A., Magliulo, G., Greco, A., de Vincentiis, M., Ralli, M., Pelucchi, S., Ciorba, A., & Vicini, C. (2022). Barbed reposition pharyngoplasty (BRP) in obstructive sleep apnea treatment: State of the art. *American Journal of Otolaryngology, 43*, 103197. https://doi.org/10.1016/J.AMJOTO.2021.103197

Kent, D. T., Weiner, J. S., Chio, E. G., & Weidenbecher, M. (2020). Hypoglossal nerve stimulator implantation via a 2-incision technique. *Operative Techniques in Otolaryngology-Head and Neck Surgery, 31*, e35–e42. https://doi.org/10.1016/J.OTOT.2020.06.002

Kent, D. T., Zealear, D., & Schwartz, A. R. (2021). Ansa cervicalis stimulation. *Chest, 159*, 1212–1221. https://doi.org/10.1016/j.chest.2020.10.010

Kezirian, E. J., Goding, G. S., Malhotra, A., O'Donoghue, F. J., Zammit, G., Wheatley, J. R., Catcheside, P. G., Smith, P. L., Schwartz, A. R., Walsh, J. H., Maddison, K. J., Claman, D. M., Huntley, T., Park, S. Y., Campbell, M. C., Palme, C. E., Iber, C., Eastwood, P. R., Hillman, D. R., & Barnes, M. (2014). Hypoglossal nerve stimulation improves obstructive sleep apnea: 12-month outcomes. *Journal of Sleep Research, 23*, 77–83. https://doi.org/10.1111/jsr.12079

Kim, J. S., & Heo, S. J. (2020). Test-retest reliability of drug-induced sleep endoscopy using midazolam. *Journal of Clinical Sleep Medicine, 16*, 675–678. https://doi.org/10.5664/JCSM.8314

Korotun, M., Hahn, S., Quintero, L., Rajan, P., Iakovou, A., Mayo, P., & Greenberg, H. (2020). Ultrasound assessment of tongue movement as a predictor of response to hypoglossal nerve stimulation (HGNS). *Sleep, 43*, A247–A248. https://doi.org/10.1093/sleep/zsaa056.645

Kuna, S., & Sant'Ambrogio, G. (1991). Pathophysiology of upper airway closure during sleep. *JAMA, 266*, 1384–1389.

Lee, C., Seay, E., Dedhia, R., Walters, B., & Scalzitti, N. (2019). Therapeutic positive airway pressure level predicts response to hypoglossal nerve stimulation for obstructive sleep apnea. *Journal of Clinical Sleep Medicine, 15*, 1165–1172. https://doi.org/10.5664/jcsm.7814

Lee, C. H., Seay, E. G., Reese, J. W., Wu, X., Schwab, R. J., Keenan, B., & Dedhia, R. C. (2021). Clinical radiographic predictors of response to hypoglossal nerve stimulation for obstructive sleep apnea. *Otolaryngology-Head and Neck Surgery, 164*, 1122–1127. https://doi.org/10.1177/0194599820963141

Lewis, R., Pételle, B., Campbell, M. C., Mackay, S., Palme, C., Raux, G., Sommer, J. U., & Maurer, J. T. (2019). Implantation of the nyxoah bilateral hypoglossal nerve stimulator for obstructive sleep apnea. *Laryngoscope Investigative Otolaryngology, 4*, 703–707. https://doi.org/10.1002/lio2.312

Lewis, R., Le, J., Czank, C., & Raux, G. (2021). Control of OSA in a patient with CCC of soft palate using bilateral hypoglossal nerve stimulation. *Clinical Case Reports, 9*, 2222. https://doi.org/10.1002/CCR3.3990

Li, C., Boon, M., Ishman, S. L., & Suurna, M. V. (2018). Hypoglossal nerve stimulation in three adults with down syndrome and severe obstructive sleep apnea. *The Laryngoscope*. https://doi.org/10.1002/lary.27723

Lou, B., Hahn, S., Korotun, M., Quintero, L., Shikowitz, M., & Greenberg, H. (2021). Space invader: Pleural penetration of a hypoglossal nerve stimulator sensor lead. *Journal of Clinical Sleep Medicine*. https://doi.org/10.5664/JCSM.9492

Marin, J. M., Agusti, A., Villar, I., Forner, M., Nieto, D., Carrizo, S. J., Barbé, F., Vicente, E., Wei, Y., Nieto, F. J., & Jelic, S. (2012). Association between treated and untreated obstructive sleep apnea and risk of hypertension. *JAMA, 307*(20), 2169–2176. https://doi.org/10.1001/jama.2012.3418

Mashaqi, S., Patel, S. I., Combs, D., Estep, L., Helmick, S., Machamer, J., Parthasarathy, S., Iannella, G., & Vicini, C. (2021). The hypoglossal nerve stimulation as a novel therapy for treating obstructive sleep apnea-A literature review. *Public Health, 18*, 1642. https://doi.org/10.3390/ijerph18041642

Meleca, J., & Kominsky, A. (2020). Reconfiguration of upper airway stimulation devices utilizing awake endoscopy. *The Laryngoscope, 130*, 2494–2498. https://doi.org/10.1002/LARY.28569

Miki, H., Hida, W., Chonan, T., Kikuchi, Y., & Takishima, T. (1989). Effects of submental electrical stimulation during sleep on upper airway patency in patients with obstructive sleep apnea. *The American Review of Respiratory Disease, 140*, 1285–1289. https://doi.org/10.1164/AJRCCM/140.5.1285

Mokhlesi, B., Finn, L. A., Hagen, E. W., Young, T., Hla, K. M., van Cauter, E., & Peppard, P. E. (2014). Obstructive sleep apnea during REM sleep and hypertension results of the Wisconsin Sleep Cohort. *American Journal of Respiratory and Critical Care Medicine, 190*, 1158–1167. https://doi.org/10.1164/rccm.201406-1136OC

Mu, L., & Sanders, I. (2010). Human tongue neuroanatomy: Nerve supply and motor endplates. *Clinical Anatomy, 23*, 777–791. https://doi.org/10.1002/ca.21011

Mwenge, G. B., Rombaux, P., Dury, M., Lengelé, B., & Rodenstein, D. (2013). Targeted hypoglossal neurostimulation for obstructive sleep apnoea: A 1-year pilot study. *The European Respiratory Journal, 41*, 360–367. https://doi.org/10.1183/09031936.00042412

Myc, L. A., Churnin, I. T., Jameson, M. J., & Davis, E. M. (2018). Treatment of comorbid obstructive sleep apnea by upper airway stimulation results in resolution of debilitating symptoms of restless legs syndrome. *Journal of Clinical Sleep Medicine, 14*, 1797–1800. https://doi.org/10.5664/jcsm.7400

Neruntarat, C., Wanichakorntrakul, P., Khuancharee, K., Saengthong, P., & Tangngekkee, M. (2022). Upper airway stimulation vs other upper airway surgical procedures for OSA: A meta-analysis. *Sleep and Breathing, 26*, 407–418. https://doi.org/10.1007/S11325-021-02402-3/FIGURES/4

O'connor, G. T., Caffo, B., Newman, A. B., Quan, S. F., Rapoport, D. M., Redline, S., Resnick, H. E., Samet, J., & Shahar, E. (2009). Prospective study of sleep-disordered breathing and hypertension: The Sleep Heart Health Study. *American Journal of Respiratory and Critical Care Medicine, 179*, 1159–1164. https://doi.org/10.1164/rccm.200712-1809OC

Ong, A. A., O'Brien, T. X., Nguyen, S. A., & Gillespie, M. B. (2016). Implantation of a defibrillator in a patient with an upper airway stimulation device. *The Laryngoscope, 126*, E86–E89. https://doi.org/10.1002/LARY.25683

Pang, K. P., & Woodson, B. T. (2007). Expansion sphincter pharyngoplasty: A new technique for the treatment of obstructive sleep apnea. *Otolaryngology--Head and Neck Surgery, 137*, 110–114. https://doi.org/10.1016/J.OTOHNS.2007.03.014

Parikh, V., Thaler, E., Kato, M., Gillespie, M., Nguyen, S., Withrow, K., Calhoun, D., Soose, R., Stevens, D., Stevens, S., Larsen, C., Reddy, M., & Lakkireddy, D. (2018). Early feasibility of hypoglossal nerve upper airway stimulator in patients with cardiac implantable electronic devices and continuous positive airway pressure-intolerant severe obstructive sleep apnea. *Heart Rhythm, 15*, 1165–1170. https://doi.org/10.1016/J.HRTHM.2018.04.016

Patel, J., Daniels, K., Bogdan, L., Huntley, C., & Boon, M. (2020a). Elevated central and mixed apnea index after upper airway stimulation. *Otolaryngology--Head and Neck Surgery, 162*, 767–772. https://doi.org/10.1177/0194599820912740

Patel, J., Daniels, K., Bogdan, L., Huntley, C., & Boon, M. (2020b). Effect of gender, age, and profound disease on upper airway stimulation outcomes. *The Annals of Otology, Rhinology, and Laryngology, 129*, 772–780. https://doi.org/10.1177/0003489420911656

Patil, R. D., Sarber, K. M., Epperson, M. V., Tabangin, M., Altaye, M., Mesa, F., & Ishman, S. L. (2020). Hypoglossal nerve stimulation: Outcomes in veterans with obstructive sleep apnea and common comorbid post-traumatic stress disorder. *The Laryngoscope*. https://doi.org/10.1002/lary.29292

Pawlak, D., Bohorquez, D., König, I., Steffen, A., & Thaler, E. (2021). Effect of electrode configuration and impulse strength on airway patency in neurostimulation for obstructive sleep apnea. *The Laryngoscope, 131*, 2148–2153. https://doi.org/10.1002/LARY.29530

Rodríguez Hermosa, J., Calle, M., Guerassimova, I., Fernández, B., Montero, V., & Álvarez-Sala, J. (2021). Noninvasive electrical stimulation of oropharyngeal muscles in obstructive sleep apnea. *Expert Review of Respiratory Medicine*. https://doi.org/10.1080/17476348.2021.1935244

Rokkas, S., Huntley, C., & Boon, M. (2021). Hypoglossal nerve stimulation impact on a patient with obstructive sleep apnea and heart failure. *Journal of Clinical Sleep Medicine, 17*, 1125–1127. https://doi.org/10.5664/JCSM.9090

Rosenthal, M. E., Lyons, M. M., Schweller, J., Yildiz, V. O., Chio, E. G., & Khan, M. S. (2021). Association between anxiety, depression, and emotional distress and hypoglossal nerve stimulator adherence. *Sleep and Breathing*. https://doi.org/10.1007/s11325-021-02366-4

Rotenberg, B. W., Murariu, D., & Pang, K. P. (2016). Trends in CPAP adherence over twenty years of data collection: A flattened curve. *Journal of Otolaryngology-Head & Neck Surgery, 45*, 1–9. https://doi.org/10.1186/s40463-016-0156-0

Safiruddin, F., Vanderveken, O. M., de Vries, N., Maurer, J. T., Lee, K., Ni, Q., & Strohl, K. P. (2015). Effect of upper-airway stimulation for obstructive sleep apnoea on airway dimensions. *The European Respiratory Journal, 45*, 129–138. https://doi.org/10.1183/09031936.00059414

Sanders, I., & Mu, L. (2013). A three-dimensional atlas of human tongue muscles. *Anatomical Record, 296*, 1102–1114. https://doi.org/10.1002/ar.22711

Sarber, K., Ishman, S., & Patil, R. (2019). Emergence of Cheyne-Stokes breathing after hypoglossal nerve stimulator implant in a patient with mixed sleep apnea. *JAMA Otolaryngology--Head & Neck Surgery, 145*, 389–390. https://doi.org/10.1001/JAMAOTO.2018.4077

Schwab, R. J., Wang, S. H., Verbraecken, J., Vanderveken, O. M., van de Heyning, P., Vos, W. G., DeBacker, J. W., Keenan, B. T., Ni, Q., & DeBacker, W. (2018). Anatomic predictors of response and mechanism of action of upper airway stimulation therapy in patients with obstructive sleep apnea. *Sleep, 41*, zsy021. https://doi.org/10.1093/SLEEP/ZSY021

Schwartz, A. R., Eisele, D. W., Hari, A., Testerman, R., Erickson, D., & Smith, P. L. (1996). Electrical stimulation of the lingual musculature in obstructive sleep apnea. *Journal of Applied Physiology, 81*, 643–652.

Seda, G., Matwiyoff, G., & Parrish, J. S. (1910). Effects of obstructive sleep apnea and CPAP on cognitive function. *Sleep*. https://doi.org/10.1007/s11910-021-01123-0/Published

Simpson, R., Oyekan, A. A., Ehsan, Z., & Ingram, D. G. (2018). Nature and science of sleep dovepress obstructive sleep apnea in patients with Down syndrome: Current perspectives. *Nature and Science of Sleep, 10*, 287. https://doi.org/10.2147/NSS.S154723

Smith, P. L., Eisele, D. W., Podszus, T., Penzel, T., Grote, L., Peter, J. H., & Schwartz, A. R. (1996). Electrical stimulation of upper airway musculature. *Sleep, 19*, S284–S287.

Soose, R. J., Faber, K., Greenberg, H., Boon, M., Woodson, T., Strollo, P., & Ryan Soose, C. J. (2021). Post-implant care pathway: Lessons learned and recommendations after 5 years of clinical implementation of hypoglossal nerve stimulation therapy. *Sleep, 44*, S4–S10. https://doi.org/10.1093/sleep/zsaa279

Steffen, A., Frenzel, H., Wollenberg, B., & König, I. R. (2015). Patient selection for upper airway stimulation: Is concentric collapse in sleep endoscopy predictable? *Sleep and Breathing, 19*, 1373–1376. https://doi.org/10.1007/s11325-015-1277-9

Steffen, A., Kilic, A., König, I. R., Suurna, M. V., Hofauer, B., & Heiser, C. (2018a). Tongue motion variability with changes of upper airway stimulation electrode configuration and effects on treatment outcomes. *The Laryngoscope, 128*, 1970–1976. https://doi.org/10.1002/lary.27064

Steffen, A., Sommer, J. U., Hofauer, B., Maurer, J. T., Hasselbacher, K., & Heiser, C. (2018b). Outcome after one year of upper airway stimulation for obstructive sleep apnea in a multicenter German post-market study. *The Laryngoscope, 128*, 509–515. https://doi.org/10.1002/lary.26688

Steffen, A., Abrams, N., Suurna, M. V., Wollenberg, B., & Hasselbacher, K. (2019a). Upper-airway stimulation before, after, or without uvulopalatopharyngoplasty: A two-year perspective. *The Laryngoscope, 129*, 514–518. https://doi.org/10.1002/lary.27357

Steffen, A., Sommer, J. U., Strohl, K., Hasselbacher, K., Suurna, M. V., Hofauer, B., & Heiser, C. (2019b). Changes in breath cycle sensing affect outcomes in upper airway stimulation in sleep apnea. *Laryngoscope Investigative Otolaryngology, 5*, 326–329. https://doi.org/10.1002/lio2.334

Steffen, A., König, I. R., Baptista, P. M., Abrams, N., Jeschke, S., & Hasselbacher, K. (2020). Home sleep testing to direct upper airway stimulation therapy optimization for sleep apnea. *The Laryngoscope, 131*, E1375–E1379. https://doi.org/10.1002/lary.29043

Steffen, A., Jeschke, S., Soose, R. J., Hasselbacher, K., & König, I. R. (2021). Neuromodulation: Technology at the neural interface impulse configuration in hypoglossal nerve stimulation in obstructive sleep apnea: The effect of modifying pulse width and frequency. https://doi.org/10.1111/ner.13490

Stenerson, M. E., Yu, P. K., Kinane, T. B., Skotko, B. G., & Hartnick, C. J. (2021). Long-term stability of hypoglossal nerve stimulation for the treatment of obstructive sleep apnea in children with Down syndrome. *International Journal of Pediatric Otorhinolaryngology, 149*, 110868. https://doi.org/10.1016/J.IJPORL.2021.110868

Strollo, P. J., Soose, R. J., Maurer, J. T., de Vries, N., Cornelius, J., Froymovich, O., Hanson, R. D., Padhya, T. A., Steward, D. L., Gillespie, M. B., Woodson, B. T., van de Heyning, P. H., Goetting, M. G., Vanderveken, O. M., Feldman, N., Knaack, L., & Strohl, K. P. (2014). Upper-airway stimulation for obstructive sleep apnea. *The New England Journal of Medicine, 370*, 139–149. https://doi.org/10.1056/NEJMoa1308659

Strollo, P. J., Gillespie, M. B., Soose, R. J., Maurer, J. T., de Vries, N., Cornelius, J., Hanson, R. D., Padhya, T. A., Steward, D. L., Woodson, B. T., Verbraecken, J., Vanderveken, O. M., Goetting, M. G., Feldman, N., Chabolle, F., Badr, M. S., Randerath, W., Strohl, K. P., Stimulation Therapy for Apnea Reduction (STAR) Trial Group KP, & Group for the ST for AR (STAR) T. (2015). Upper airway stimulation for obstructive sleep apnea: Durability of the treatment effect at 18 months. *Sleep, 38*, 1593–1598. https://doi.org/10.5665/sleep.5054

Sturm, J. J., Modik, O., Koutsourelakis, I., & Suurna, M. V. (2020). Contralateral tongue muscle activation during hypoglossal nerve stimulation. *Otolaryngology-Head and Neck Surgery, 162*, 985–992. https://doi.org/10.1177/0194599820917147

Thaler, E., Schwab, R., Maurer, J., Soose, R., Larsen, C., Stevens, S., Stevens, D., Boon, M., Huntley, C., Doghramji, K., Waters, T., Kominsky, A., Steffen, A., Kezirian, E., Hofauer, B., Sommer, U., Withrow, K., Strohl, K., & Heiser, C. (2019). Results of the ADHERE upper airway stimulation registry and predictors of therapy efficacy. *The Laryngoscope, 130*, 1333–1338. https://doi.org/10.1002/lary.28286

Tsuji, H., Larson, M., Venditti, F., Manders, E., Evans, J., Feldman, C., & Levy, D. (1996). Impact of reduced heart rate variability on risk for cardiac events. The Framingham Heart Study. *Circulation, 94*, 2850–2855. https://doi.org/10.1161/01.CIR.94.11.2850

van de Heyning, P. H., Badr, M. S., Baskin, J. Z., Cramer Bornemann, M. A., de Backer, W. A., Dotan, Y., Hohenhorst, W., Knaack, L., Lin, H.-S., Maurer, J. T., Netzer, A., Odland, R. M., Oliven, A., Strohl, K. P., Vanderveken, O. M., Verbraecken, J., & Woodson, B. T. (2012). Implanted upper airway stimulation device for obstructive sleep apnea. *The Laryngoscope, 122*, 1626–1633. https://doi.org/10.1002/lary.23301

Vanderveken, O. M., Maurer, J. T., Hohenhorst, W., Hamans, E., Lin, H.-S., Vroegop, A. V., Anders, C., de Vries, N., & van de Heyning, P. H. (2013). Evaluation of drug-induced sleep endoscopy as a patient selection tool for implanted upper airway stimulation for obstructive sleep apnea. *Journal of Clinical Sleep Medicine, 9*, 433–438. https://doi.org/10.5664/jcsm.2658

Vasconcellos, A., Huntley, C., Schell, A., Soose, R., & Boon, M. (2019). Dysfunctional hypoglossal nerve stimulator after electrical cardioversion: A case series. *The Laryngoscope, 129*, 1949–1953. https://doi.org/10.1002/LARY.27488

Vicini, C., & Montevecchi, F. (2019). Transoral robotic surgery for obstructive sleep apnea: Past, present, and future. *Sleep Medicine Clinics, 14*, 67–72. https://doi.org/10.1016/J.JSMC.2018.10.008

Woodson, B. T., Gillespie, M. B., Soose, R. J., Maurer, J. T., de Vries, N., Steward, D. L., Baskin, J. Z., Padhya, T. A., Lin, H., Mickelson, S., Badr, S. M., Strohl, K. P., & Strollo, P. J. (2014). Randomized controlled withdrawal study of upper airway stimulation on OSA: Short- and long-term effect. *Otolaryngology--Head and Neck Surgery, 151*, 880–887. https://doi.org/10.1177/0194599814544445

Woodson, B. T., Soose, R. J., Gillespie, M. B., Strohl, K. P., Maurer, J. T., de Vries, N., Steward, D. L., Baskin, J. Z., Badr, M. S., Lin, H., Padhya, T. A., Mickelson, S., Anderson, W. M., Vanderveken, O. M., Strollo, P. J., & Investigators on B of ST. (2016). Three-year outcomes of cranial nerve stimulation for obstructive sleep apnea. *Otolaryngology-Head and Neck Surgery, 154*, 181–188. https://doi.org/10.1177/0194599815616618

Woodson, B. T., Strohl, K. P., Soose, R. J., Gillespie, M. B., Maurer, J. T., de Vries, N., Padhya, T. A., Badr, M. S., Lin, H., Vanderveken, O. M., Mickelson, S., & Strollo, P. J. (2018). Upper airway stimulation for obstructive sleep apnea: 5-year outcomes. *Otolaryngology–Head and Neck Surgery, 159*, 194–202. https://doi.org/10.1177/0194599818762383

Yacono, C., & Hyman, M. (2021). Hypoglossal nerve stimulator-induced neurapraxia following electrical cardioversion for atrial fibrillation. *JACC Case Reports, 3*, 1128–1131. https://doi.org/10.1016/J.JACCAS.2021.05.014

Yu, P. K., Jayawardena, A. D. L., Stenerson, M., Pulsifer, M. B., Grieco, J. A., Abbeduto, L., Dedhia, R. C., Soose, R. J., Tobey, A., Raol, N., Ishman, S. L., Shott, S. R., Cohen, M. S., Skotko, B. G., Kinane, T. B., Keamy, D. G., & Hartnick, C. J. (2021). Redefining success by focusing on failures after pediatric hypoglossal stimulation in Down syndrome. *The Laryngoscope, 131*, 1663–1669. https://doi.org/10.1002/LARY.29290

Zheng, Z., Hu, S., & Chernobilsky, B. (2017). Hypoglossal nerve upper airway stimulator implantation after radiotherapy for head and neck malignancy. *Otolaryngology--Head and Neck Surgery, 157*, 160–161. https://doi.org/10.1177/0194599817703928

Mandibular Advancement Splint Therapy

22

Anna M. Mohammadieh, Kate Sutherland, Andrew S. L. Chan, and Peter A. Cistulli

Abstract

Mandibular advancement splint (MAS) therapy is the leading alternative to continuous positive airway pressure (CPAP) therapy for the treatment of obstructive sleep apnoea. A MAS is an oral appliance which advances the mandible in relation to the maxilla, thus increasing airway calibre and reducing collapsibility. Although it is less effective than CPAP in reducing the apnoea-hypopnoea index (AHI), it has demonstrated equivalence to CPAP in a number of key neurobehavioural and cardiovascular health outcomes, perhaps due to increased tolerability and patient adherence when compared to CPAP. However, response to MAS is variable, and reliable prediction tools for patients who respond best to MAS therapy have thus far been elusive; this is one of the key clinical barriers to wider uptake of MAS therapy. In addition, the most effective MAS devices are custom-made by a dentist specialising in the treatment of sleep disorders, which may present financial or accessibility barriers for some patients. MAS devices are generally well tolerated but may have side effects including temporomandibular joint (TMJ) dysfunction, hypersalivation, tooth pain and migration as well as occlusal changes. A patient-centred approach to treatment from a multidisciplinary team perspective is recommended. Evidence-based clinical practice points and areas of future research are summarised at the conclusion of the chapter.

Keywords

Mandibular advancement · OSA therapy · MAS titration · MAS customisation · Apnoea-hypopnoea index

22.1 Introduction

A mandibular advancement splint (MAS), also referred to as a mandibular advancement device (MAD) or mandibular repositioning appliance (MRA), is an oral appliance for the treatment of sleep apnoea and snoring. It is recognised as the

A. M. Mohammadieh (✉) · K. Sutherland
P. A. Cistulli
Department of Respiratory & Sleep Medicine, Royal North Shore Hospital, St Leonards, NSW, Australia

Sleep Research Group, Charles Perkins Centre, Faculty of Medicine and Health, The University of Sydney, Sydney, NSW, Australia

Faculty of Medicine and Health, Northern Clinical School, The University of Sydney, Sydney, NSW, Australia
e-mail: amoh2259@uni.sydney.edu.au

A. S. L. Chan
Department of Respiratory & Sleep Medicine, Royal North Shore Hospital, St Leonards, NSW, Australia

Faculty of Medicine and Health, Northern Clinical School, The University of Sydney, Sydney, NSW, Australia

T. Penzel, R. Hornero (eds.), *Advances in the Diagnosis and Treatment of Sleep Apnea*, Advances in Experimental Medicine and Biology 1384, https://doi.org/10.1007/978-3-031-06413-5_22

lead alternative to continuous positive airway pressure (CPAP) therapy for the treatment of obstructive sleep apnoea (OSA). Guidelines of the American Academy and Sleep Medicine (AASM) and American Academy of Dental Sleep Medicine (AADSM) recommend MAS therapy for the treatment of OSA where CPAP is not tolerated, or if there is a patient preference for an alternative device (Ramar et al., 2015). A MAS advances the mandible in relation to the maxilla, increasing the calibre of the upper airway and reducing upper airway collapsibility. Several iterations of the device exist and may differ substantially in terms of design and customisation.

22.2 Mechanism of Action

Imaging studies provide insights into anatomical changes in the upper airway which occur with a MAS device in situ. MAS functions via two principal mechanisms to increase airway size: (1) anterior movement of the tongue and (2) lateral expansion of the airway walls, especially in the velopharynx (Brown et al., 2013; Chan et al., 2010). Although intuitively one might expect a MAS to increase the calibre of the airway in an anterior-posterior dimension, it is the lateral dimension which is increased to the greatest degree. This may be due to the MAS increasing airway wall tension through direct connections between the lateral airway walls and the ramus of the mandible (Brown et al., 2013). A schematic of MAS airway changes and MRI with and without MAS in situ is depicted in Fig. 22.1.

Structural changes with the MAS in situ cause a reduction in airway collapsibility (Chan et al., 2020) and therefore an improvement in the AHI. This has been demonstrated in studies of passive pharyngeal collapsibility with MAS in situ (Ng et al., 2003). Further, when used in conjunction with CPAP therapy, MAS reduces the requirement for CPAP pressure in a dose-response relationship with increasing mandibular advancement. This indicates progressive reduction in airway collapsibility with increasing mandibular protrusion (Bamagoos et al., 2020).

22.3 Efficacy and Adherence: MAS Versus CPAP

MAS therapy is often compared to CPAP therapy as the accepted 'gold standard' for the treatment of OSA. Although CPAP is known to provide superior control of OSA (measured by the AHI) while it is in use (Phillips et al., 2013; Lim et al., 2006; Schwartz et al., 2018), adherence and patient tolerance are generally higher with MAS therapy (Schwartz et al., 2018). Therefore, the reduced efficacy of MAS may be offset by its superior adherence, thus leading to the equivalence in neurobehavioural and cardiovascular outcomes which has been observed in many comparative studies (see below).

Unlike that of CPAP, however, MAS response is variable, with the mean proportion of patients who respond completely to MAS (residual AHI < 5/h) reported between 29 and 71% (Bamagoos et al., 2016). One study suggested that around two thirds of patients have either a complete (37%) or partial (64%) response to MAS, with a partial response defined as a reduction in AHI of

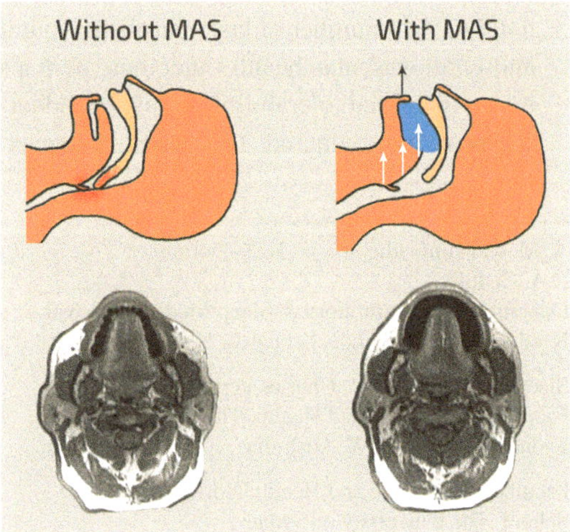

Fig. 22.1 Schematic diagram (*above*) showing upper airway obstruction during sleep in a patient with OSA (without MAS) and mandibular advancement maintaining patency of the upper airway (with MAS). Axial MRI (*below*) shows baseline airway calibre in an awake patient with OSA (without MAS) and mandibular advancement producing an increase in the calibre of the upper airway (with MAS). Note the particular increase in airway calibre in the lateral dimension (*arrows*)

at least 50% (Sutherland et al., 2015). This highlights one of the significant barriers to MAS therapy, which is the selection of patients who are most likely to derive benefit.

For the severe OSA group (AHI ≥30/h), titratable MAS therapy is less effective at reducing the AHI and the oxygen desaturation index (ODI) compared with CPAP; however, this is offset by increased adherence and patient preference in favour of MAS (Trzepizur et al., 2021). Further, MAS is equivalent to CPAP in terms of improvement in sleep architecture, with equivalent increases in slow wave and rapid eye movement (REM) sleep (Trzepizur et al., 2021). A MAS is therefore recognised as a viable alternative to CPAP therapy even in the treatment of severe OSA.

22.4 Patient Selection and Prediction of Response: Endotypes and Phenotypes

MAS therapy is highly efficacious in selected patients, including those with severe OSA. However, reliable prediction tools for treatment success remain elusive. An ideal prediction tool would be readily available from simple anatomical or routine polysomnographic data and would predict MAS treatment response with a high degree of accuracy. Studies looking at predictors of MAS response have investigated endotypic and phenotypic traits.

An endotype refers to a subtype of a condition with a distinct functional or pathobiological mechanism (Edwards et al., 2019). Examples of OSA characteristics which may be used to define endotypes include arousal threshold (defined as the degree of ventilatory drive required to trigger an arousal from sleep), loop gain (defined as instability in ventilatory control in response to a disturbance) and airway collapsibility. Traditionally, these parameters have only been available from physiological studies performed in a highly controlled research setting (Edwards et al., 2016; Bamagoos et al., 2019a); however, more recently endotypic data have been extracted from routine clinical polysomnography. For example, in the case of loop gain, Terril et al. have developed a mathematical algorithm to reliably impute loop gain from the rise in ventilatory drive that follows an obstructive respiratory event (Terrill et al., 2015). The mathematical basis of this method is depicted in Fig. 22.2.

In a group of 93 patients with, on average, moderate OSA, greater MAS efficacy was associated with 5 endotypic traits derived using algorithms applied to clinical polysomnographic data: lower loop gain, higher arousal threshold, lower ventilatory response to arousal, moderate

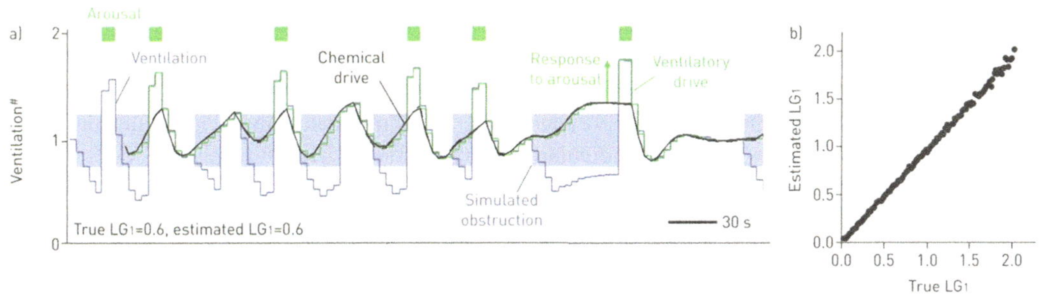

Fig. 22.2 Mathematical basis of the method. (**a**) Schematic of the feedback loop controlling ventilation showing the influence of arousal and airflow obstruction. Ventilatory drive is the sum of chemical drive and the response to arousal (γ). Airflow obstruction provides a disturbance that reduces ventilation from the intended level (i.e., ventilatory drive). In response, chemical drive rises as determined by the chemical control system (loop gain). (**b**) Time course of chemical drive during a step reduction in ventilation (e.g., obstructive hypopnoea). The rise in chemical drive is governed by and the parameters that determine its gain (LG_0), time constant (τ) and delay (δ); these system characteristics are revealed in the time course of ventilation when the airway is reopened. (Reproduced with permission of the © ERS 2022: Terrill et al. (2015))

pharyngeal collapsibility and weaker muscle compensation. The association of lower loop gain and MAS response has been confirmed in another smaller study (Op de Beeck et al., 2021).

Similarly, characteristics which act as direct or surrogate markers for the site of airway collapse have been studied as predictors of response to MAS therapy. For example, the level and specific type of airway collapse observed on drug-induced sleep endoscopy (DISE) have been associated with response to MAS. Tongue base collapse indicates a favourable outcome, whereas complete concentric collapse or complete laterolateral oropharyngeal collapse is seen in those less likely to respond (Op de Beeck et al., 2019), as is complete anteroposterior epiglottic collapse (Zhou et al., 2021). Further, certain 'airflow shapes' derived from routine polysomnography have been used to predict the site of airway collapse and thereby response to MAS. Increased drop in airflow during respiratory events as well as a 'pinched' expiratory flow shape (indicative of palatal prolapse) are associated with the least response to MAS therapy (Vena et al., 2020). Research is ongoing in order to translate these endotypic characteristics into reliable clinical predictive tools.

Phenotype refers to 'observable' anatomical features or consequences of the disease. Small individual studies as well as meta-analyses have identified certain phenotypic characteristics associated with increased response to MAS therapy. These include younger age, female gender, lower body mass index (BMI), shorter neck circumference, milder and supine-dependent OSA (Sutherland et al., 2015; Chen et al., 2020) and the absence of a tendinous pterygomandibular raphe (Brown et al., 2021), the latter of which may allow increased mandibular advancement. Craniofacial characteristics associated with a positive MAS response include a retracted maxilla and mandible, a narrower airway and shorter soft palate (Chen et al., 2020). However, the predictive ability of these individual characteristics is relatively low. Research is ongoing to find more accurate predictors of clinical response to MAS therapy. In clinical practice, many patients undergo a repeat diagnostic sleep study with the MAS device in situ to assess clinical response and guide decisions about ongoing therapy. This strategy is recommended in the guidelines of the AASM/AADSM (Ramar et al., 2015).

MAS relies on dental retention for its efficacy and, therefore, candidates for this therapy should be selected only after careful dental review. Patients are likely to be ineligible if they suffer from significant periodontal disease, insufficient native teeth to ensure device retention, severe TMJ disease or a severe gag reflex (Basyuni et al., 2018).

22.5 Health Outcomes

A number of health outcomes are derived from effective OSA treatment, some of which may have more ascribed importance than others for each individual patient. Health outcomes can be discussed in terms of neurobehavioural outcomes, quality of life and cardiovascular outcomes.

22.6 Neurobehavioural Outcomes

MAS has been shown to improve daytime sleepiness as measured by the Epworth Sleepiness Scale (ESS) when compared to conservative management, though the average effect is modest (Sharples et al., 2016). Nonetheless, a meta-analysis found that the improvement in ESS was not significantly different to that of CPAP (Lim et al., 2006). Not all studies of MAS effects on ESS are consistent, however, with one randomised control trial showing that MAS did not improve the ESS when compared to a placebo device for mild-to-moderate OSA (Marklund et al., 2015).

Small- and medium-sized studies indicate that MAS improves driving simulator performance to the same degree as CPAP (Phillips et al., 2013; Hoekema et al., 2007). One trial compared the performance of patients treated with a titratable, bi-bloc, fully customised MAS device with CPAP therapy on a monotonous driving simulator task. After 8 weeks of therapy, performance in terms of lapses of attention improved to the same degree with both therapies (Hoekema et al., 2007). In a 1-month crossover trial of a titratable bi-bloc MAS vs CPAP, speed deviation and reaction times to divided attention tasks during driving simulation improved to the same extent with both treatments (Phillips et al., 2013).

A meta-analysis found that there was no difference between MAS and CPAP for functional outcomes and neurocognitive tasks (Schwartz et al., 2018). Despite the superiority of CPAP in improving the AHI, it was hypothesised that the similarity in neurocognitive outcomes was due to the increased nightly adherence that was seen with MAS.

Taken together, these studies suggest non-inferiority of MAS to CPAP therapy for neurobehavioural outcomes, though sample sizes have been modest, and many studies were not blinded and/or did not include placebo control groups.

22.7 Quality of Life

MAS improves the quality of life to at least the same degree as CPAP (Phillips et al., 2013; Schwartz et al., 2018). This is true for both the mental component score and the physical component score of the SF-36, a validated quality-of-life questionnaire (Kuhn et al., 2017). Similarly, when considering severe OSA only, a meta-analysis of RCTs comparing MAS to CPAP shows a similar impact on both the SF-36 and FOSQ questionnaires, two validated tools for the assessment of quality of life (Trzepizur et al., 2021).

22.8 Cardiovascular Outcomes

There is a paucity of data examining the effect of MAS therapy on cardiovascular outcomes, though some short-term studies exist. For example, in a subgroup of hypertensive patients, MAS improved blood pressure by 2–4 mmHg and was non-inferior to CPAP (Phillips et al., 2013; Bratton et al., 2015). MAS also improves markers of oxidative stress and cardiac autonomic activity when compared with placebo (Dal-Fabbro et al., 2014). To date there are no large, randomised control trials looking at the effect of MAS therapy on cardiovascular endpoints. In contrast, a few moderately sized randomised control trials have examined the effect of CPAP therapy on cardiovascular endpoints including stroke, myocardial infarction and revascularisation of coronary artery disease, but have failed to demonstrate a benefit (McEvoy et al., 2016; Peker et al., 2016; Sanchez-de-la-Torre et al., 2020). However, these trials have all been characterised by very poor CPAP adherence (less than 4 h per night), which may have contributed to the negative result. In addition, subgroup analysis in groups with higher adherence have shown some significant cardiovascular benefits (Dissanayake et al., 2021). As MAS therapy has increased adherence when compared to CPAP, further research is required to examine the impact of MAS therapy on cardiovascular outcomes.

22.9 Design and Customisation

MAS design falls generally into one of two categories: (i) a single piece (monobloc) design or (ii) an upper and lower component with a coupling mechanism (bi-bloc/duo-bloc design). Bi-bloc designs have the advantage of facilitating some lateral and vertical jaw movement, to varying degrees, and may improve patient comfort and tolerance.

MAS may be prefabricated from thermoplastic material (the 'boil-and-bite' model) or can be formally customised to the patient's own dentition. The 'boil-and-bite' model, in which thermoplastic trays are heated by immersion in hot water and then moulded directly to the patient's mandibular and maxillary dental arches, foregoes the need for specialist dentist impressions and is therefore less costly and more accessible. However, boil-and-bite designs have a higher failure rate due to reduced dental retention and are less effective at lowering the AHI than a fully customised device (Vanderveken et al., 2008; Quinnell et al., 2014). In contrast, customised designs provide superior fit and dental retention and are better tolerated (Johal et al., 2017). For these reasons, fully customised, titratable devices are recommended in MAS guideline statements from the AASM and AADSM (Ramar et al., 2015).

For a fully customised device, dental impressions and measurements of occlusal relationships are taken by a specialist dentist. More recently, dental scan technology has been introduced, allowing for a digitalised model of the teeth and intra-oral tissues (Piskin et al., 2021). Full-arch digital impressions have been shown to be more accurate than traditional impressions (Amin et al., 2017) and also allow a more streamlined digital workflow production process.

Devices also differ significantly in terms of fabrication materials and construction. Materials include hard acrylics, thermal acrylics, laminates, biocompatible polymers and alloys (Chan et al., 2020). Devices may be manufactured by 3D milling or 3D printing. There is little data available to allow comparisons between the various models or brands of fully customised MAS devices in the treatment of OSA (Marklund et al., 2019); hence, data from individual studies should be extrapolated with caution. An example of a customised MAS device is shown in Fig. 22.3.

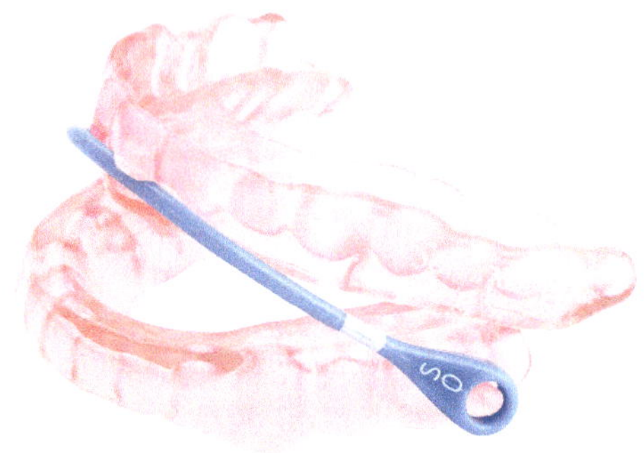

Fig. 22.3 An example of a bi-bloc, fully customised MAS device. This model (Avant device, SomnoMed Australia) uses a series of straps, incrementally shorter in length, to titrate the mandible forward and articulate the upper and lower components. The strap is also designed to limit jaw opening

Fig. 22.4 An example of an in-built adherence recorder (DentiTrac, Braebon Medical) embedded within a MAS device

22.10 Adherence

An additional design feature for some models is an in-built adherence recorder which allows collection and storage of patient adherence data using a thermal-sensor chip. These devices, about the size of a button cell battery, can be embedded into the MAS during the manufacturing process (see Fig. 22.4). When the temperature lies within a certain range (generally 31.5–33 °C to 38.5–39.2 °C), it is inferred that the device is in situ within the oral cavity and therefore in use (Sutherland et al., 2021a). This information can be downloaded and read in a research or clinical setting and may assist with clinical management. An example of downloaded data from an inbuilt compliance recorder is shown in Fig. 22.5. Previous studies of CPAP adherence have shown that subjective patient recollection tends to underestimate objectively recorded adherence data (Kribbs et al., 1993). However, few studies have assessed whether MAS subjectively reported adherence is consistent with objective adherence data. One small study showed that there was no difference between objective and self-reported MAS adherence at 3 months (Vanderveken et al., 2013). At 12 months, close correlation between subjective and objective measures of adherence continued, with an overestimation of 30 min by subjective (self-reported) adherence compared with objective data col-

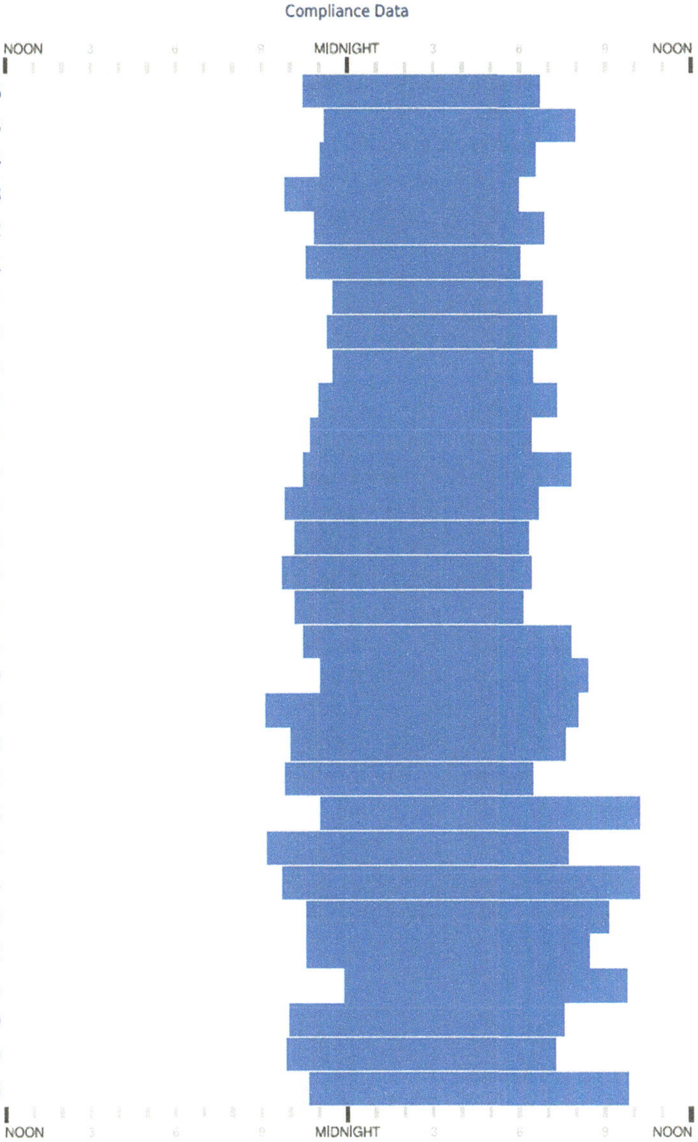

Day	%S	%NS	Hr	±Hr
Nov 06	9	88	8.3	1.3
Nov 05	0	99	8.8	1.8
Nov 04	2	98	7.6	0.6
Nov 03	0	94	8.2	1.2
Nov 02	1	99	8.0	1.0
Nov 01	6	91	7.5	0.5
Oct 31	0	100	7.3	0.3
Oct 30	0	100	8.0	1.0
Oct 29	0	100	7.0	0
Oct 28	0	100	8.3	1.3
Oct 27	0	100	7.8	0.8
Oct 26	1	99	9.3	2.3
Oct 25	0	100	8.8	1.8
Oct 24	0	100	8.2	1.2
Oct 23	0	98	8.8	1.8
Oct 22	0	97	8.0	1.0
Oct 21	0	100	9.3	2.3
Oct 20	0	100	9.3	2.3
Oct 19	2	98	10.9	3.9
Oct 18	0	99	9.6	2.6
Oct 17	0	100	8.7	1.7
Oct 16	1	99	11.2	4.2
Oct 15	0	100	10.5	3.5
Oct 14	0	99	12.5	5.5
Oct 13	0	100	10.6	3.6
Oct 12	1	98	9.9	2.9
Oct 11	0	100	9.9	2.9
Oct 10	0	100	9.6	2.6
Oct 09	0	100	9.4	2.4
Oct 08	0	99	11.2	4.2

Fig. 22.5 An example of downloaded data from an in-built MAS adherence recorder in a highly adherent patient. This particular device (DentiTrac, Braebon Medical) also records sleep position: supine (S) vs non-supine (NS); number of hours used per night (Hr) and the difference between the physician-set 'goal' adherence time, in this case 7 h; and the actual time worn each night (±Hr)

lected by an in-built thermal sensor chip (Dieltjens et al., 2013).

Disadvantages of the in-built adherence monitors include a small increase in bulk added to the device, although this is usually well tolerated, as well as an increased manufacturing cost. In future, in-built monitors may also collect other clinically relevant data including cardiorespiratory parameters. One prototype device captures intra-oral photoplethysmography (PPG) signals and correlates highly to PPG signals obtained from the more traditional and commonly used finger probe (Nabavi et al., 2020). PPG has numerous applications including the acquisition of heart rate, respiratory rate and percent of oxygen-saturated haemoglobin (SpO_2).

As with CPAP, adherence to MAS can be variable, though in general MAS adherence is superior to that of CPAP (Schwartz et al., 2018). Cluster analysis over 60 days of MAS usage identified three adherence subtypes: 48.3% were 'consistent users' who used MAS most of the time (daily usage 7.3 ± 0.8 h), 32.8% were 'inconsistent users' with variable usage (daily usage 4.6 ± 0.8 h), and 19.0% were 'non-users' (daily usage 1.0 ± 0.6 h) (Sutherland et al., 2021b). Patients could be identified into these subtypes within the first 20 days of therapy, suggesting that the early period of MAS therapy is a critical time for close clinical monitoring and support.

22.11 MAS Titration

Titration refers to the incremental advancement of the mandible towards the level of protrusion which provides maximal clinical efficacy. This is an important practice point since increased MAS efficacy has been demonstrated with increasing mandibular protrusion, in a dose-dependent relationship (Kato et al., 2000). Some specialist research centres have used remotely controlled titratable MAS devices in the sleep laboratory, allowing real-time assessment of MAS efficacy at progressive levels of protrusion (Dieltjens et al., 2019; Tsai et al., 2004; Sutherland et al., 2017). A commercially available remotely controlled mandibular protrusion device (MATRx, Zephyr Sleep Technologies Inc., Calgary, Canada) has been used both to predict clinical response to MAS, as well as to determine the optimal level of mandibular protrusion. This device consists of upper and lower dental trays which are fitted to the patient's teeth with an impression material. A small motor which sits just outside the patient's mouth is able to titrate the mandible forward in small increments during sleep, under the control of a remote operator. With the use of this device, one study was able to demonstrate a positive predictive value of 87% for therapeutic success, defined as an AHI <10/h and ≥50% reduction from baseline at the determined effective target protrusive position (Remmers et al., 2013). In addition, a portable version of this device has been used in an at-home setting, consisting of the titratable mandibular protrusion device, nasal cannula to detect flow and pulse oximetry, together with a portable laptop computer for signal processing. Using this system, unattended mandibular titrations were performed at home using real-time feedback from the nasal cannulae and oximetry to guide the titration. Overnight titrations in the patient's home using this device yielded a positive predictive value of 97% and negative predictive value 72% to predict a residual oxygen desaturation index (ODI) ≤10/h (Remmers et al., 2017).

Other researchers have used overnight oximetry devices alone to assess hypoxic burden at various levels of protrusion and titrate accordingly. In the absence of these research tools, for practical clinical purposes, the titration goal may be the maximal level of protrusion that is well tolerated by the patient. Symptoms, such as snoring or daytime somnolence, can also be used as titration guides by the patient at home.

Titration mechanisms will depend upon the individual design of the device and its coupling mechanism. Coupling mechanisms are variable and may include elastic or plastic connectors, metal pin and tube connectors, hook connectors, acrylic extensions or magnets. All titration mechanisms will incrementally advance the mandible forward in relation to the maxilla.

22.12 Side Effects

Patients should be warned of the possibility of both short- and long-term side effects prior to commencement of the MAS therapy. Short-term side effects are common and often temporary. They include hyper-salivation, temporomandibular joint (TMJ) pain and discomfort, a dry mouth and pain or irritation of the intra-oral tissues. Many of these side effects, if not temporary, can be addressed by minor device adjustments.

Long-term side effects include progressive changes in dental occlusion and have been observed for as long as 11 years (Pliska et al., 2014). The prominent changes are a reduction in overbite (−2.3 ± 1.6 mm) and overjet (−1.9 ± 1.9 mm), as well as an expansion of the mandibular arch with reduction in mandibular dental crowding (Pliska et al., 2014). Of note, some of these dental changes may be favourable, depending upon baseline dentition.

MAS devices have a finite lifespan and are commonly reported by providers to last approximately 5 years, though evidence on the lifespan of MAS devices is limited. One study followed 15 patients out to 5 years of MAS therapy and found that the most commonly encountered technical problems requiring review by a dental technician were acrylic breakage at the point of articulation attachment, poor retention and other required adjustments to improve comfort (Martínez-Gomis et al., 2010).

22.13 Patient-Centred Approach

One of the key indications for MAS therapy is patient preference when compared to CPAP (Ramar et al., 2015), and, as with management of any chronic condition, a patient-centred approach is central to treatment success. Patient preferences around treatment must be elucidated after the diagnosis of OSA, incorporating a discussion around the advantages and disadvantages of available suitable therapeutic options. Qualitative analysis has identified MAS convenience and transportability as factors which influence patient choice in favour of MAS therapy, while alterations in bite and concerns about durability are identified as important disadvantages (Almeida et al., 2013). In addition, the cost of MAS therapy will be an insurmountable barrier to some patients, particularly in jurisdictions where CPAP, but not MAS, may attract a government subsidy. Therefore, the choice to proceed with a trial of MAS therapy must be tailored to the individual patient's treatment goals.

When commencing CPAP therapy, patients may undergo a therapeutic trial of a rental or loan device, before making a decision about progression to permanent therapy. With MAS therapy however, this approach is more challenging, since the device is manufactured specifically for the individual patient and normally requires an initial financial investment before the therapy can be trialled. Some patients may choose to undergo an initial trial with a commercially available 'boil-and-bite' MAS, to see whether the sensation of mandibular advancement is tolerable for them and whether they experience any improvement in symptoms. If this approach is undertaken, the patient should be informed that boil-and-bite devices are more likely to cause gum discomfort and may be less adherent and less effective than custom-made devices.

22.14 Multidisciplinary Management

As MAS treatment of OSA lies at the intersection of dentistry and sleep medicine, it is recommended that both of these specialists play an ongoing role in the care of MAS patients (Ramar et al., 2015). Other specialists such as ear nose throat (ENT) surgeons and general practitioners may also play an active role. An initial clinical review will be required to confirm the diagnosis of OSA, to quantify the severity of OSA with a baseline nocturnal sleep study and to confirm patient suitability and preference for MAS therapy. Once the device is fabricated, follow-up is important to ensure ongoing efficacy of the MAS (e.g. using overnight polysomnography with the MAS in situ), to monitor patient satisfaction with the device and to screen for and treat short- and long-term side effects.

22.15 Future Directions

Recognition of OSA as a heterogenous disorder is increasing. As with other areas of medicine, a 'one-size-fits-all' approach to the treatment of OSA is being abandoned, in favour of a more personalised approach, which takes into account specific clinical features and the disease subtype of the individual patient. For example, there is now recognition that some OSA is driven by 'pharyngeal' endotypes (e.g. high airway collapsibility), while other OSA is driven by non-pharyngeal traits including high loop gain and reduced arousal threshold (Bamagoos et al., 2019b). Future research will focus on further differentiating OSA as a whole into clinically relevant subgroups, some of which may be more responsive to MAS therapy than others (Edwards et al., 2019). At the time of writing, there is little personalisation in the choice of MAS therapy for an individual patient in the clinic, other than a failed trial of CPAP or on the basis of patient preference. Future research will focus on the development of reliable endotypic and phenotypic prediction tools for MAS treatment success and translation of these into a clinical setting.

Another advancement in the field of sleep medicine as a whole is the increasing trend towards automated and remote (at-home) diagnosis and management of sleep apnoea (Penzel et al., 2021). This includes the explosion of wearable devices and 'apps' which collect physiological data relating to sleep, including sleep times and staging, snoring, oximetry and ECG data. In future we may anticipate that MAS therapy will also progress in this regard. For example, the technology for remote (at-home) MAS titration studies already exists (Remmers et al., 2017) and may expand to a clinical setting, thus reducing the need for in-laboratory titration studies which are resource intensive. In addition, in-built MAS adherence data recorders described above raise the possibility of remote monitoring and assessments via telemedicine, a technology which is already in common usage with CPAP.

22.16 Clinical Practice Points: Evidence-Based Summary

- There is a variable clinical response to MAS, with approximately one third achieving a complete response and a further one third achieving a partial response (reduction in the AHI of ≥50%).
- Increasing MAS efficacy is observed with increasing mandibular protrusion. In practice, patients may be advised to titrate their MAS to the maximal comfortable limit of protrusion to optimise OSA control.
- Although CPAP is more effective than MAS at lowering the AHI, this may be offset by improved patient tolerance and adherence with MAS therapy, leading to equivalence in key neurobehavioural and cardiovascular outcomes.
- MAS improves key outcomes including daytime somnolence, driving risk and quality of life to the same extent as CPAP. Further research is needed to examine the impact of MAS on cardiovascular endpoints.
- MAS is generally well tolerated. Short-term side effects include hypersalivation, TMJ dysfunction and pain or irritation of the intra-oral tissues. Long-term side effects include tooth migration and dental occlusal changes.

22.17 Areas of Future Research

- Moderately sized CPAP trials have failed to demonstrate a benefit when looking at cardiovascular endpoints; however, these have all been characterised by poor compliance. Since MAS compliance is generally higher than that of CPAP, further research is required to examine the impact of MAS therapy on cardiovascular outcomes.
- A number of phenotypic and endotypic traits have been associated with MAS response, but accurate and reliable clinical predictive tools remain lacking. Further research is required into the development and refinement of reliable prediction tools for MAS treatment success and translation of these into a clinical setting.
- A major barrier to more widespread MAS usage is the expense of the device and the need to have the device custom-made by a specialist dentist, which also increases patient cost and reduces accessibility. Future research should focus on reducing production cost and improving the quality of affordable 'boil-and-

bite' or other preliminary models to allow for effective MAS trial phases.

- Remote monitoring of adherence and other data is already commonplace for CPAP therapy. Future research will focus on the development of MAS in-built adherence recorders including the availability of remote access, as well as the collection of other relevant data to assess nightly efficacy, including haem-oxygen saturation, heart rate, position sensors, snore microphones, respiratory events and even sleep staging.

References

Almeida, F. R., Henrich, N., Marra, C., et al. (2013). Patient preferences and experiences of CPAP and oral appliances for the treatment of obstructive sleep apnea: A qualitative analysis. *Sleep & Breathing, 17*(2), 659–666. https://doi.org/10.1007/s11325-012-0739-6

Amin, S., Weber, H. P., Finkelman, M., El Rafie, K., Kudara, Y., & Papaspyridakos, P. (2017). Digital vs. conventional full-arch implant impressions: A comparative study. *Clinical Oral Implants Research, 28*(11), 1360–1367. https://doi.org/10.1111/clr.12994

Bamagoos, A. A., Sutherland, K., & Cistulli, P. A. (2016). Mandibular advancement splints. *Sleep Medicine Clinics, 11*(3), 343–352. https://doi.org/10.1016/j.jsmc.2016.04.002

Bamagoos, A. A., Cistulli, P. A., Sutherland, K., et al. (2019a). Dose-dependent effects of mandibular advancement on upper airway collapsibility and muscle function in obstructive sleep apnea. *Sleep, 42*(6), zsz049. https://doi.org/10.1093/sleep/zsz049

Bamagoos, A. A., Cistulli, P. A., Sutherland, K., et al. (2019b). Polysomnographic endotyping to select patients with obstructive sleep apnea for oral appliances. *Annals of the American Thoracic Society, 16*(11), 1422–1431. https://doi.org/10.1513/AnnalsATS.201903-190OC

Bamagoos, A. A., Eckert, D. J., Sutherland, K., Ngiam, J., & Cistulli, P. A. (2020). Dose-dependent effects of mandibular advancement on optimal positive airway pressure requirements in obstructive sleep apnoea. *Sleep & Breathing, 24*(3), 961–969. https://doi.org/10.1007/s11325-019-01930-3

Basyuni, S., Barabas, M., & Quinnell, T. (2018). An update on mandibular advancement devices for the treatment of obstructive sleep apnoea hypopnoea syndrome. *Journal of Thoracic Disease, 10*(Suppl 1), S48–S56. https://doi.org/10.21037/jtd.2017.12.18

Bratton, D. J., Gaisl, T., Wons, A. M., & Kohler, M. (2015). CPAP vs mandibular advancement devices and blood pressure in patients with obstructive sleep apnea: A systematic review and meta-analysis. *JAMA, 314*(21), 2280–2293. https://doi.org/10.1001/jama.2015.16303

Brown, E. C., Cheng, S., McKenzie, D. K., Butler, J. E., Gandevia, S. C., & Bilston, L. E. (2013). Tongue and lateral upper airway movement with mandibular advancement. *Sleep, 36*(3), 397–404. https://doi.org/10.5665/sleep.2458

Brown, E. C., Jugé, L., Knapman, F. L., et al. (2021). Mandibular advancement splint response is associated with the pterygomandibular raphe. *Sleep, 44*(4), zsaa222. https://doi.org/10.1093/sleep/zsaa222

Chan, A. S., Lee, R. W., Srinivasan, V. K., Darendeliler, M. A., Grunstein, R. R., & Cistulli, P. A. (2010). Nasopharyngoscopic evaluation of oral appliance therapy for obstructive sleep apnoea. *The European Respiratory Journal, 35*(4), 836–842. https://doi.org/10.1183/09031936.00077409

Chan, A. S. L., Sutherland, K., & Cistulli, P. A. (2020). Mandibular advancement splints for the treatment of obstructive sleep apnea. *Expert Review of Respiratory Medicine, 14*(1), 81–88. https://doi.org/10.1080/17476348.2020.1686978

Chen, H., Eckert, D. J., van der Stelt, P. F., et al. (2020). Phenotypes of responders to mandibular advancement device therapy in obstructive sleep apnea patients: A systematic review and meta-analysis. *Sleep Medicine Reviews, 49*, 101229. https://doi.org/10.1016/j.smrv.2019.101229

Dal-Fabbro, C., Garbuio, S., D'Almeida, V., Cintra, F. D., Tufik, S., & Bittencourt, L. (2014). Mandibular advancement device and CPAP upon cardiovascular parameters in OSA. *Sleep & Breathing, 18*(4), 749–759. https://doi.org/10.1007/s11325-014-0937-5

Dieltjens, M., Braem, M. J., Vroegop, A., et al. (2013). Objectively measured vs self-reported compliance during oral appliance therapy for sleep-disordered breathing. *Chest, 144*(5), 1495–1502. https://doi.org/10.1378/chest.13-0613

Dieltjens, M., Braem, M. J., Op de Beeck, S., et al. (2019). Remotely controlled mandibular positioning of oral appliance therapy during polysomnography and drug-induced sleep endoscopy compared with conventional subjective titration in patients with obstructive sleep apnea: Protocol for a randomized crossover trial. *Trials, 20*(1), 615. https://doi.org/10.1186/s13063-019-3698-4

Dissanayake, H. U., Colpani, J. T., Sutherland, K., et al. (2021). Obstructive sleep apnea therapy for cardiovascular risk reduction-Time for a rethink? *Clinical Cardiology, 44*(12), 1729–1738. https://doi.org/10.1002/clc.23747

Edwards, B. A., Andara, C., Landry, S., et al. (2016). Upper-airway collapsibility and loop gain predict the response to oral appliance therapy in patients with obstructive sleep apnea. *American Journal of Respiratory and Critical Care Medicine, 194*(11), 1413–1422. https://doi.org/10.1164/rccm.201601-0099OC

Edwards, B. A., Redline, S., Sands, S. A., & Owens, R. L. (2019). More than the sum of the respiratory events:

Personalized medicine approaches for obstructive sleep apnea. *American Journal of Respiratory and Critical Care Medicine., 200*(6), 691–703. https://doi.org/10.1164/rccm.201901-0014TR

Hoekema, A., Stegenga, B., Bakker, M., et al. (2007). Simulated driving in obstructive sleep apnoea-hypopnoea; effects of oral appliances and continuous positive airway pressure. *Sleep & Breathing, 11*(3), 129–138. https://doi.org/10.1007/s11325-006-0093-7

Johal, A., Haria, P., Manek, S., Joury, E., & Riha, R. (2017). Ready-made versus custom-made mandibular repositioning devices in sleep apnea: A randomized clinical trial. *Journal of Clinical Sleep Medicine, 13*(2), 175–182. https://doi.org/10.5664/jcsm.6440

Kato, J., Isono, S., Tanaka, A., et al. (2000). Dose-dependent effects of mandibular advancement on pharyngeal mechanics and nocturnal oxygenation in patients with sleep-disordered breathing. *Chest, 117*(4), 1065–1072. https://doi.org/10.1378/chest.117.4.1065

Kribbs, N. B., Pack, A. I., Kline, L. R., et al. (1993). Objective measurement of patterns of nasal CPAP use by patients with obstructive sleep apnea. *The American Review of Respiratory Disease, 147*(4), 887–895. https://doi.org/10.1164/ajrccm/147.4.887

Kuhn, E., Schwarz, E. I., Bratton, D. J., Rossi, V. A., & Kohler, M. (2017). Effects of CPAP and mandibular advancement devices on health-related quality of life in OSA: A systematic review and meta-analysis. *Chest, 151*(4), 786–794. https://doi.org/10.1016/j.chest.2017.01.020

Lim, J., Lasserson, T. J., Fleetham, J., & Wright, J. (2006). Oral appliances for obstructive sleep apnoea. *The Cochrane Database of Systematic Reviews, 2006*(1), Cd004435. https://doi.org/10.1002/14651858.CD004435.pub3

Marklund, M., Carlberg, B., Forsgren, L., Olsson, T., Stenlund, H., & Franklin, K. A. (2015). Oral appliance therapy in patients with daytime sleepiness and snoring or mild to moderate sleep apnea: A randomized clinical trial. *JAMA Internal Medicine, 175*(8), 1278–1285. https://doi.org/10.1001/jamainternmed.2015.2051

Marklund, M., Braem, M. J. A., & Verbraecken, J. (2019). Update on oral appliance therapy. *European Respiratory Review: An Official Journal of the European Respiratory Society, 28*(153). https://doi.org/10.1183/16000617.0083-2019

Martínez-Gomis, J., Willaert, E., Nogues, L., Pascual, M., Somoza, M., & Monasterio, C. (2010). Five years of sleep apnea treatment with a mandibular advancement device. Side effects and technical complications. *The Angle Orthodontist, 80*(1), 30–36. https://doi.org/10.2319/030309-122.1

McEvoy, R. D., Antic, N. A., Heeley, E., et al. (2016). CPAP for prevention of cardiovascular events in obstructive sleep apnea. *New England Journal of Medicine*. https://doi.org/10.1056/NEJMoa1606599

Nabavi, S., Debbarma, S., & Bhadra, S. (2020). A smart mandibular advancement device for intraoral cardiorespiratory monitoring. *Annual International Conference of the IEEE Engineering in Medicine and Biology Society, 2020*, 4079–4084. https://doi.org/10.1109/embc44109.2020.9176520

Ng, A. T., Gotsopoulos, H., Qian, J., & Cistulli, P. A. (2003). Effect of oral appliance therapy on upper airway collapsibility in obstructive sleep apnea. *American Journal of Respiratory and Critical Care Medicine., 168*(2), 238–241. https://doi.org/10.1164/rccm.200211-1275OC

Op de Beeck, S., Dieltjens, M., Verbruggen, A. E., et al. (2019). Phenotypic labelling using drug-induced sleep endoscopy improves patient selection for mandibular advancement device outcome: A prospective study. *Journal of Clinical Sleep Medicine, 15*(8), 1089–1099. https://doi.org/10.5664/jcsm.7796

Op de Beeck, S., Dieltjens, M., Azarbarzin, A., et al. (2021). Mandibular advancement device treatment efficacy is associated with polysomnographic endotypes. *Annals of the American Thoracic Society, 18*(3), 511–518. https://doi.org/10.1513/AnnalsATS.202003-220OC

Peker, Y., Glantz, H., Eulenburg, C., Wegscheider, K., Herlitz, J., & Thunstrom, E. (2016). Effect of positive airway pressure on cardiovascular outcomes in coronary artery disease patients with nonsleepy obstructive sleep apnea. The RICCADSA randomized controlled trial. *American Journal of Respiratory and Critical Care Medicine, 194*(5), 613–620. https://doi.org/10.1164/rccm.201601-0088OC

Penzel, T., Glos, M., & Fietze, I. (2021). New trends and new technologies in sleep medicine: Expanding accessibility. *Sleep Medicine Clinics, 16*(3), 475–483. https://doi.org/10.1016/j.jsmc.2021.05.010

Phillips, C. L., Grunstein, R. R., Darendeliler, M. A., et al. (2013). Health outcomes of continuous positive airway pressure versus oral appliance treatment for obstructive sleep apnea: A randomized controlled trial. *American Journal of Respiratory and Critical Care Medicine., 187*(8), 879–887. https://doi.org/10.1164/rccm.201212-2223OC

Piskin, B., Uyar, A., Yuceer, M., et al. (2021). Fabrication of a mandibular advancement device using a fully digital workflow: A clinical report. *Journal of Prosthodontics, 30*(3), 191–195. https://doi.org/10.1111/jopr.13297

Pliska, B. T., Nam, H., Chen, H., Lowe, A. A., & Almeida, F. R. (2014). Obstructive sleep apnea and mandibular advancement splints: Occlusal effects and progression of changes associated with a decade of treatment. *Journal of Clinical Sleep Medicine, 10*(12), 1285–1291. https://doi.org/10.5664/jcsm.4278

Quinnell, T. G., Bennett, M., Jordan, J., et al. (2014). A crossover randomised controlled trial of oral mandibular advancement devices for obstructive sleep apnoea-hypopnoea (TOMADO). *Thorax, 69*(10), 938–945. https://doi.org/10.1136/thoraxjnl-2014-205464

Ramar, K., Dort, L. C., Katz, S. G., et al. (2015). Clinical practice guideline for the treatment of obstructive sleep apnea and snoring with oral appliance therapy: An update for 2015. *Journal of Clinical Sleep Medicine, 11*(7), 773–827. https://doi.org/10.5664/jcsm.4858

Remmers, J., Charkhandeh, S., Grosse, J., et al. (2013). Remotely controlled mandibular protrusion during sleep predicts therapeutic success with oral appliances in patients with obstructive sleep apnea. *Sleep, 36*(10), 1517–1525, 1525a. https://doi.org/10.5665/sleep.3048

Remmers, J. E., Topor, Z., Grosse, J., et al. (2017). A feedback-controlled mandibular positioner identifies individuals with sleep apnea who will respond to oral appliance therapy. *Journal of Clinical Sleep Medicine, 13*(7), 871–880. https://doi.org/10.5664/jcsm.6656

Sanchez-de-la-Torre, M., Sanchez-de-la-Torre, A., Bertran, S., et al. (2020). Effect of obstructive sleep apnoea and its treatment with continuous positive airway pressure on the prevalence of cardiovascular events in patients with acute coronary syndrome (ISAACC study): A randomised controlled trial. *The Lancet Respiratory Medicine, 8*(4), 359–367. https://doi.org/10.1016/S2213-2600(19)30271-1

Schwartz, M., Acosta, L., Hung, Y. L., Padilla, M., & Enciso, R. (2018). Effects of CPAP and mandibular advancement device treatment in obstructive sleep apnea patients: A systematic review and meta-analysis. *Sleep & Breathing, 22*(3), 555–568. https://doi.org/10.1007/s11325-017-1590-6

Sharples, L. D., Clutterbuck-James, A. L., Glover, M. J., et al. (2016). Meta-analysis of randomised controlled trials of oral mandibular advancement devices and continuous positive airway pressure for obstructive sleep apnoea-hypopnoea. *Sleep Medicine Reviews, 27*, 108–124. https://doi.org/10.1016/j.smrv.2015.05.003

Sutherland, K., Takaya, H., Qian, J., Petocz, P., Ng, A. T., & Cistulli, P. A. (2015). Oral appliance treatment response and polysomnographic phenotypes of obstructive sleep apnea. *Journal of Clinical Sleep Medicine., 11*(8), 861–868. https://doi.org/10.5664/jcsm.4934

Sutherland, K., Ngiam, J., & Cistulli, P. A. (2017). Performance of remotely controlled mandibular protrusion sleep studies for prediction of oral appliance treatment response. *Journal of Clinical Sleep Medicine, 13*(3), 411–417. https://doi.org/10.5664/jcsm.6492

Sutherland, K., Dalci, O., & Cistulli, P. A. (2021a). What do we know about adherence to oral appliances? *Sleep Medicine Clinics, 16*(1), 145–154. https://doi.org/10.1016/j.jsmc.2020.10.004

Sutherland, K., Almeida, F. R., Kim, T., et al. (2021b). Treatment usage patterns of oral appliances for obstructive sleep apnea over the first 60 days: A cluster analysis. *Journal of Clinical Sleep Medicine*. https://doi.org/10.5664/jcsm.9288

Terrill, P. I., Edwards, B. A., Nemati, S., et al. (2015). Quantifying the ventilatory control contribution to sleep apnoea using polysomnography. *The European Respiratory Journal, 45*(2), 408–418. https://doi.org/10.1183/09031936.00062914

Trzepizur, W., Cistulli, P. A., Glos, M., et al. (2021). Health outcomes of continuous positive airway pressure versus mandibular advancement device for the treatment of severe obstructive sleep apnea: An individual participant data meta-analysis. *Sleep, 44*(7), zsab015. https://doi.org/10.1093/sleep/zsab015

Tsai, W. H., Vazquez, J. C., Oshima, T., et al. (2004). Remotely controlled mandibular positioner predicts efficacy of oral appliances in sleep apnea. *American Journal of Respiratory and Critical Care Medicine., 170*(4), 366–370. https://doi.org/10.1164/rccm.200310-1446OC

Vanderveken, O. M., Devolder, A., Marklund, M., et al. (2008). Comparison of a custom-made and a thermoplastic oral appliance for the treatment of mild sleep apnea. *American Journal of Respiratory and Critical Care Medicine, 178*(2), 197–202. https://doi.org/10.1164/rccm.200701-114OC

Vanderveken, O. M., Dieltjens, M., Wouters, K., De Backer, W. A., Van de Heyning, P. H., & Braem, M. J. (2013). Objective measurement of compliance during oral appliance therapy for sleep-disordered breathing. *Thorax, 68*(1), 91–96. https://doi.org/10.1136/thoraxjnl-2012-201900

Vena, D., Azarbarzin, A., Marques, M., et al. (2020). Predicting sleep apnea responses to oral appliance therapy using polysomnographic airflow. *Sleep, 43*(7), zsaa004. https://doi.org/10.1093/sleep/zsaa004

Zhou, N., Ho, J. T. F., de Vries, N., Bosschieter, P. F. N., Ravesloot, M. J. L., & de Lange, J. (2021). Evaluation of drug-induced sleep endoscopy as a tool for selecting patients with obstructive sleep apnea for maxillomandibular advancement. *Journal of Clinical Sleep Medicine*. https://doi.org/10.5664/jcsm.9802

Correction to: Pulse Oximetry: The Working Principle, Signal Formation, and Applications

Timo Leppänen, Samu Kainulainen, Henri Korkalainen,
Saara Sillanmäki, Antti Kulkas, Juha Töyräs, and Sami Nikkonen

Correction to:
Chapter 12 in: T. Penzel, R. Hornero (eds.), *Advances in the Diagnosis and Treatment of Sleep Apnea*, Advances in Experimental Medicine and Biology 1384, https://doi.org/10.1007/978-3-031-06413-5_12

Chapter 12 was previously published non-open access. It has now been changed to open access under a CC BY 4.0 license and the copyright holder updated to 'The Author(s)'. The book has also been updated with this change.

The updated original version of this chapter can be found at
https://doi.org/10.1007/978-3-031-06413-5_12

T. Penzel, R. Hornero (eds.), *Advances in the Diagnosis and Treatment of Sleep Apnea*, Advances in Experimental Medicine and Biology 1384, https://doi.org/10.1007/978-3-031-06413-5_22

Index

T. Penzel, R. Hornero (eds.), *Advances in the Diagnosis and Treatment of Sleep Apnea*, Advances in Experimental Medicine and Biology 1384, https://doi.org/10.1007/978-3-031-06413-5

P

R

S

T

U

V

W

The manufacturer's authorised representative in the EU is Springer Nature Customer Service Centre GmbH, Europaplatz 3, 69115 Heidelberg, Germany. If you have any concerns regarding our products, please contact ProductSafety@springernature.com

Printed and bound by CPI Group (UK) Ltd, Croydon, CR0 4YY
30/09/2026
02225242-0020